Handbook of Phytonutrients in Indigenous Fruits and Vegetables

IFIS

IFIS is a leading international provider of information in the sciences of food and health, and produces FSTA® - Food Science and Technology Abstracts. FSTA is a specialist database of curated content which is indexed using the unique FSTA Thesaurus, the most comprehensive food and beverage thesaurus in the world. As an educational charity, IFIS provides resources and training in information research skills, and provides free access to FSTA in developing countries through Research4Life. IFIS' resources enable researchers, industry practitioners and students in the global food community to find the relevant, reliable research they need. Website: www.ifis.org

IFIS

food and health information

Handbook of Phytonutrients in Indigenous Fruits and Vegetables

Edited by

Dharini Sivakumar

Yasmina Sultanbawa

and

Michael Netzel

CABI

CABI is a trading name of CAB International

CABI
Nosworthy Way
Wallingford
Oxfordshire OX10 8DE
UK

Tel: +44 (0)1491 832111
Fax: +44 (0)1491 833508
E-mail: info@cabi.org
Website: www.cabi.org

WeWork
One Lincoln St
24th Floor
Boston, MA 02111
USA

T: +1 (617)682-9015
E-mail: cabi-nao@cabi.org

A catalogue record for this book is available from the British Library, London, UK.

Library of Congress Cataloging-in-Publication Data

Names: Sivakumar, Dharini, editor. | Netzel, M. E. (Michael E.), editor. | Sultanbawa, Yasmina, editor.
Title: Handbook of phytonutrients in indigenous fruits and vegetables / edited by Dharini Sivakumar, Yasmina Sultanbawa, and Michael Netzel.
Description: Boston, MA : CAB International, 2022. | Includes bibliographical references and index. | Summary: 'Collation of innovative and high-quality research on phytonutrient constituents of traditional fruits and vegetables, including information on the impact of postharvest processing and cooking practices, plus the impact of biological activities of indigenous fruits and vegetables on human health. Key resource to fight 'hidden hunger' and achieve diet diversification'-- Provided by publisher.
Identifiers: LCCN 2021057071 (print) | LCCN 2021057072 (ebook) | ISBN 9781789248043 (hardback) | ISBN 9781789248050 (ebook) | ISBN 9781789248067 (epub)
Subjects: LCSH: Phytonutrients. | Endemic plants.
Classification: LCC QK861 .H345 2022 (print) | LCC QK861 (ebook) | DDC 572/.2--dc23/eng/20211221
LC record available at https://lccn.loc.gov/2021057071
LC ebook record available at https://lccn.loc.gov/2021057072

References to Internet websites (URLs) were accurate at the time of writing.

ISBN-13: 9781789248043 (hardback)
9781789248050 (ePDF)
9781789248067 (ePub)

DOI: 10.1079/9781789248043.0000

Commissioning Editor: Rebecca Stubbs
Editorial Assistant: Kate Hill/Emma McCann
Production Editor: Tim Kapp

Typeset by Exeter Premedia Services Pvt Ltd, Chennai, India
Printed and bound in the UK by Severn, Gloucester

Contents

About the Editors

Professor Dharini Sivakumar was appointed as a Tier 1 SARChI (South African Research Chair Initiative) in the Phytochemical Food Network at the Department of Crop Sciences, The University of Tshwane, Pretoria, South Africa, from 2021 to 2025.

In her research, she seeks to improve consumers' nutritional quality by understanding the changes that occur in functional compounds during postharvest processing, and the bioactive properties of functional compounds present in indigenous fruits and vegetables or in their functional food products or supplements.

A primary focus of her research programme is integrating preharvest factors, postharvest storage, processing, packaging, biological activities, molecular mechanisms and a chemometric approach to select the most effective methods for preparing or processing inexpensive indigenous food to promote diet diversification, sustain food security, eradicate hidden hunger, create employment opportunities and alleviate poverty. She has participated in several community-based cooperative research projects introducing solar dryers for rural areas of South Africa, to reduce postharvest losses of vegetables and to connect famers with the supply chain so they can market their own value-added indigenous food products. Engagement with these communities is a privilege for her, and she is passionate about bringing a positive impact to them. In addition, she serves as an Honorary Associate Professor at The University of Queensland and collaborates with the ARC Training Center for Uniquely Australian Foods, as well as Indigenous enterprise and advisory groups. She has published over 100 peer-reviewed papers and seven book chapters with Springer and Elsevier. She is also an associate editor for the journal *Scientific African*. She acted as one of the topic associate editors for the scientific themes 'Food and Nutrition Security: Underutilized Plant and Animal-Based Foods' and 'Women in Food Chemistry: 2021' for *Frontiers in Nutrition*.

Professor Yasmina Sultanbawa's research is focused within the agribusiness development framework, specifically in the areas of food processing, preservation, food safety and nutrition. Her current research includes the minimization of postharvest losses through value addition and the search for natural preservatives to replace current synthetic chemicals. Her research area also includes the challenge of nutrition security, in particular micronutrient deficiency (hidden hunger), lack of diet diversity and nutritional losses in the food supply chain, which are addressed by her work with underutilized Australian plant species and potential new crops. Her work on Australian native plant foods is focused on incorporation of these plants in mainstream agriculture and diet diversification. Working with Indigenous communities to develop nutritious and sustainable value-added products from native plants for use in the food, feed, cosmetic and health care industries is a key strategy. The creation of employment, economic and social benefits to these remote communities is an anticipated outcome. She considers it a privilege to engage with these communities and is very passionate that her work will have a positive socioeconomic impact. Her extensive experience in leadership and engagement with food industries, Indigenous communities and enterprises has qualified her to become a Centre Director of the ARC Training Centre for Uniquely Australian Foods. Professor Sultanbawa was recognized for her pioneering work in building successful partnerships with commercial outputs by bringing Indigenous communities, industry, funding bodies and academia together, and received the coveted 2017 Business Higher Education Round Table award for Outstanding Community Engagement. She has published 125 research articles, 13 book chapters and two books. For *Frontiers in Nutrition*, she has also served as an associate editor and editor for the research topics 'Food and Nutrition Security: Underutilized Plant and Animal-Based Foods' and 'Women in Food Chemistry: 2021'.

Dr Michael Netzel is a Senior Research Fellow and Registered Nutritionist (RNutr.) at the Centre for Nutrition and Food Sciences, Queensland Alliance for Agriculture and Food Innovation, The University of Queensland, Brisbane, Australia. His research is focused on the assessment of the nutritional quality of plant food – 'From the raw produce to the absorbed and metabolized bioactive compound'. Dr Netzel's expertise is reflected in 115 peer-reviewed journal publications, 11 book chapters, two books, two conference proceedings and more than 300 conference abstracts. He has served on the editorial board of several international peer-reviewed journals in the food and nutrition field such as *Frontiers in Nutrition*, *Foods*, *Nutraceuticals*, *Food Research International* and the *Journal of Food Composition and Analysis*. He is also a guest lecturer at the Technical University of Munich, Germany, as well as an active member of the Nutrition Society of Australia, the International Society for Horticultural Science, the German Nutrition Society and the Australian Association of von Humboldt Fellows. Dr Netzel is a former Feodor Lynen Research Fellow of the Alexander von Humboldt-Foundation, Bonn, Germany.

Contributors

Mohammad Akhtaruzzaman, Institute of Nutrition and Food Science, University of Dhaka, Dhaka-1000, Bangladesh.

Stephen Akinola, Phytochemical Food Network Research Group, Department of Crop Sciences, Tshwane University of Technology, Pretoria West 0001, South Africa.

Saleha Akter, ARC Industrial Transformation Training Centre for Uniquely Australian Foods, Centre for Nutrition and Food Sciences, Queensland Alliance for Agriculture and Food Innovation, The University of Queensland, Elkhorn Building (1024), 80 Meiers Road, Indooroopilly, QLD 4068, Australia.

Mohammad Khairul Alam, Institute of Nutrition and Food Science, University of Dhaka, Dhaka-1000, Bangladesh.

Adeyemi Oladapo Aremu, Indigenous Knowledge Systems Centre, Faculty of Natural and Agricultural Sciences, North-West University, Private Bag X2046, Mmabatho 2790, North West Province, South Africa.

Himansu Baijnath, University of KwaZulu-Natal, School of Life Sciences, Westville Campus, Durban, South Africa.

Manjeshwar Shrinath Baliga, Mangalore Institute of Oncology, Mahavir Circle, Pumpwell, Mangalore, Karnataka, India 575002.

Harish Kumar Bastimal, Department of Pharmacology, Manipal College of Pharmaceutical Sciences, Manipal Academy of Higher Education (MAHE), Karnataka, India Manipal-576104.

Amina Benabdallah, BPE Laboratory, Faculty of Life and Natural Sciences, University of Chadli Bendjedid, BP 73, 36000 El-Tarf, Algeria.

Rajeev Bhat, ERA-Chair for Food (By-) Products Valorization Technologies, Estonian University of Life Sciences, Tartu, Estonia, EU.

Tania Chacón-Ordóñez, CIGRAS, Universidad de Costa Rica, 11501-2060 San Pedro, Costa Rica.

N.N.G. Chiranthika, Department of Food Science and Technology, Faculty of Livestock, Fisheries and Nutrition, Wayamba University of Sri Lanka, Makandura, Gonawila, (NWP), Sri Lanka.

Daniel Cozzolino, Centre for Nutrition and Food Sciences, Queensland Alliance for Agriculture and Food Innovation, The University of Queensland, St. Lucia, Brisbane, QLD 4072, Queensland, Australia.

Maral Seidi Damyeh, ARC Industrial Transformation Training Centre for Uniquely Australian Foods, Centre for Nutrition and Food Sciences, Queensland Alliance for Agriculture and Food Innovation, The University of Queensland, Elkhorn Building (1024), 80 Meiers Road, Indooroopilly, QLD 4068, Australia.

Patricia Esquivel, Escuela de Tecnología de Alimentos, Universidad de Costa Rica, 11501-2060 San Pedro, Costa Rica.

Héctor Estrada-Medina, Departamento de Manejo y Conservación de Recursos Naturales Tropicales, Facultad de Medicina Veterinaria y Zootecnia. Universidad Autónoma de Yucatán. Apartado Postal 4-116 Itzimna Mérida, Yucatán, Mexico.

Olaniyi Amos Fawole, Postharvest Research Laboratory, Department of Botany and Plant Biotechnology, University of Johannesburg, P.O. Box 524, Auckland Park, Johannesburg 2006, South Africa.

Manzur-E-Mohsina Ferdous, Section of Scientific Research, Care to Association, Brobæksgade 12 A, 3550 Slangerup, Denmark.

Miriam Monserrat Ferrer, Departamento de Manejo y Conservación de Recursos Naturales Tropicales, Universidad Autónoma de Yucatán, Mérida, Yucatán, Mexico.

Selina Fyfe, Centre for Nutrition and Food Sciences, Queensland Alliance for Agriculture and Food Innovation, The University of Queensland, Elkhorn Building (1024), 80 Meiers Road, Indooroopilly, QLD 4068, Australia.

Cyrielle Garcia, Qualisud, Univ La Réunion, CIRAD, Univ Montpellier, Montpellier SupAgro, Univ Avignon, 2 rue Wetzell, F-97490 Sainte Clotilde, France; Qualisud, Univ Montpellier, CIRAD, Institut Agro, Avignon Université, Univ de La Réunion, ESIROI, Montpellier, France.

Nadaraj Govender, University of KwaZulu-Natal, School of Education, Edgewood Campus, 121 Marianhill Rd, Pinetown, Durban, 3605, South Africa.

Prasanna P. Gunathilake, Department of Food Science and Technology, Faculty of Livestock, Fisheries and Nutrition, Wayamba University of Sri Lanka, Makandura, Gonawila, (NWP), Sri Lanka.

Ummai Habiba, Section of Scientific Research, Care to Association, Brobæksgade 12 A, 3550 Slangerup, Denmark.

H.A.C.O Hettiarachchi, Department of Food Science and Technology, Faculty of Livestock, Fisheries and Nutrition, Wayamba University of Sri Lanka, Makandura, Gonawila, (NWP), Sri Lanka.

Lee-Hoon Ho, Faculty of Bioresources and Food Industry, Universiti Sultan Zainal Abidin, Malaysia.

Fozia Homa, Bihar Agricultural University, Sabour, Bhagalpur, Bihar 813210, India.

Riadh Ilahy, University of Carthage, National Agricultural Research Institute of Tunisia (INRAT), Laboratory of Horticulture, Tunis, Rue Hédi Karray 2049 Ariana, Tunisia.

Beguvalli Subash Inchara, Sahyadri Science College, Kuvempu University, Shimoga, Karnataka, India 577202.

Sheikh Nazrul Islam, Institute of Nutrition and Food Science, University of Dhaka, Dhaka 1000, Bangladesh.

G. Janarny, Department of Food Science and Technology, Faculty of Livestock, Fisheries and Nutrition, Wayamba University of Sri Lanka, Makandura, Gonawila, (NWP), Sri Lanka.

Víctor M. Jiménez, CIGRAS/IAA, Universidad de Costa Rica, 11501-2060 San Pedro, Costa Rica.

Eugénie Kayitesi, Department of Consumer and Food Sciences, University of Pretoria, Hatfield, Pretoria, South Africa.

Helyes Lajos, Institute of Horticultural Science, Szent István Campus, Hungarian University of Agriculture and Life Sciences, 2100 Gödöllő Páter K. st.1, Hungary

Vimbainashe E. Manhivi, Phytochemical Food Network Research Group, Department of Crop Sciences, Tshwane University of Technology, Pretoria West 0001, South Africa.

Florence. M. Mashitoa, Phytochemical Food Network Research Group, Department of Crop Sciences, Tshwane University of Technology, Pretoria West 0001, South Africa.

Petunia Mashiane, Department of Horticulture, Tshwane University of Technology, Pretoria, P.O. 0001, South Africa.

Bárbara Paes Miglioli da Mata, School of Pharmaceutical Sciences, São Paulo State University – Unesp, Araraquara city, São Paulo state, Brazil.

Patricia Irene Montañez-Escalante, Departamento de Manejo y Conservación de Recursos Naturales Tropicales, Facultad de Medicina Veterinaria y Zootecnia. Universidad Autónoma de Yucatán. Apartado Postal 4-116 Itzimna Mérida, Yucatán, México.

Isabela Jacob Moro, School of Pharmaceutical Sciences, São Paulo State University – Unesp, Araraquara city, São Paulo state, Brazil.

Mack Moyo, Department of Horticulture, Faculty of Applied Sciences, Durban University of Technology, PO Box 1334, Durban 4000, South Africa.

Siphosanele Govender Moyo, Department of Consumer and Food Sciences, University of Pretoria, Hatfield, Pretoria South Africa.

Hosakatte Niranjana Murthy, Department of Botany, Karnatak University, Dharwad 580003, India.

Gabriele Netzel, Centre for Nutrition and Food Sciences, Queensland Alliance for Agriculture and Food Innovation, The University of Queensland, Elkhorn Building (1024), 80 Meiers Road, Indooroopilly, QLD 4068, Australia.

Michael E. Netzel, Centre for Nutrition and Food Sciences, Queensland Alliance for Agriculture and Food Innovation, The University of Queensland, Kessels Rd, Coopers Plains, QLD 4108, Australia.

Olubukola Motunrayo Odeyemi, Department of Horticulture, Federal University of Agriculture, Abeokuta, PMB 2240, Ogun State, Nigeria.

Abiodun Olusola Omotayo, Food Security and Safety Niche Area, Faculty of Natural and Agricultural Sciences, North-West University, Private Bag X2046, Mmabatho 2790, North West Province, South Africa.

Karkala Sreedhara Ranganath Pai, Department of Pharmacology, Manipal College of Pharmaceutical Sciences, Manipal Academy of Higher Education (MAHE), Karnataka, India Manipal-576104.

Karkala Usha Sreedhar Pai, Sahyadri Science College, Kuvempu University, Shimoga, Karnataka, India 577202.

Giriyapura Srikantachar Pavankumar, Sri Lakshmi Group of Institution, Magadi Main Road, Sunkadakatte, Bengaluru, Karnataka, India 560091.

Zoltán Pék, Institute of Horticultural Science, Szent István Campus, Hungarian University of Agriculture and Life Sciences, 2100 Gödöllő Páter K. st.1, Hungary.

Elisa dos Santos Pereira, Embrapa Clima Temperado, Pelotas, Rio Grande do Sul/Brazil.

Adaucto B. Pereira-Netto, Department of Botany, Universidade Federal do Paraná, Curitiba, Brazil.

Anh Dao Thi Phan, ARC Industrial Transformation Training Centre for Uniquely Australian Foods, Queensland Alliance for Agriculture and Food Innovation, The University of Queensland, Elkhorn Building (1024), 80 Meiers Road, Indooroopilly, QLD 4068, Australia.

Charlotte Ralison, LABASAN (Laboratoire de Biochimie Appliquee aux Sciences de l'Alimentation et de Nutrition) Faculté des Sciences, Université d'Antananarivo, Madagascar.

Vonimatra Razafindrazaka, LABASAN (Laboratoire de Biochimie Appliquee aux Sciences de l'Alimentation et de Nutrition) Faculté des Sciences, Université d'Antananarivo, Madagascar.

María del Rocío Ruenes-Morales, Departamento de Manejo y Conservación de Recursos Naturales Tropicales, Facultad de Medicina Veterinaria y Zootecnia. Universidad Autónoma de Yucatán. Apartado Postal 4-116 Itzimna Mérida, Yucatán, México.

Michael Rychlik, Chair of Analytical Food Chemistry, Technical University of Munich, Maximus-von-Imhof-Forum 2, D-85354, Freising, Germany.

Luis Vitor Silva do Sacramento, School of Pharmaceutical Sciences, São Paulo State University – Unesp, Araraquara city, São Paulo state, Brazil.

Mateus Kawata Salgaço, School of Pharmaceutical Sciences, São Paulo State University – Unesp, Araraquara city, São Paulo state, Brazil.

André Gonzaga dos Santos, School of Pharmaceutical Sciences, São Paulo State University – Unesp, Araraquara city, São Paulo state, Brazil.

Horst Joachim Schirra, School of Environment and Science, Griffith University, Nathan, Queensland, Australia.

Mohammed Wasim Siddiqui, Bihar Agricultural University, Sabour, Bhagalpur, Bihar 813210, India.

Faith Seke, Department of Horticulture, Tshwane University of Technology, Pretoria West 0001, South Africa.

Ambreen Naz Shah, Section of Scientific Research, Care to Association, Brobæksgade 12 A, 3550 Slangerup, Denmark.

Md Tariqul Islam Shajib, Section of Scientific Research, Care to Association, Brobæksgade 12 A, 3550 Slangerup, Denmark.

Dharini Sivakumar, Department of Crop Sciences, Phytochemical Food Network Research Group, Tshwane University of Technology, Pretoria, South Africa.

Kátia Sivieri, School of Pharmaceutical Sciences, São Paulo State University – Unesp, Araraquara city, São Paulo state, Brazil.

Heather E. Smyth, ARC Industrial Transformation Training Centre for Uniquely Australian Foods, Queensland Alliance for Agriculture and Food Innovation, The University of Queensland, Elkhorn Building (1024), 80 Meiers Road, Indooroopilly, QLD 4068, Australia.

Sukirtha Srivarathan, ARC Industrial Transformation Training Centre for Uniquely Australian Foods, Queensland Alliance for Agriculture and Food Innovation, The University of Queensland, Elkhorn Building (1024), 80 Meiers Road, Indooroopilly, QLD 4068, Australia and Department of Biosystems Technology, Faculty of Technology, University of Jaffna, Kilinochchi, Sri Lanka.

Yasmina Sultanbawa, ARC Industrial Transformation Training Centre for Uniquely Australian Foods, Queensland Alliance for Agriculture and Food Innovation, The University of Queensland, Elkhorn Building (1024), 80 Meiers Road, Indooroopilly, QLD 4068, Australia.

Thuan-Chew Tan, Food Technology Division, School of Industrial Technology, Universiti Sains Malaysia, Malaysia.

Imen Tlili, University of Carthage, National Agricultural Research Institute of Tunisia (INRAT), Laboratory of Horticulture, Tunis, Rue Hédi Karray 2049 Ariana, Tunisia.

Annegowda Hardur Venkatappa, Sri Adichunchanagiri College of Pharmacy, Adichunchanagiri University, B. G Nagara-571448, Nagamangala, Mandya, Karnataka, India.

Juliana Vinholes, Embrapa Clima Temperado, Pelotas, Rio Grande do Sul/Brazil.

Márcia Vizzotto, Embrapa Clima Temperado, Pelotas, Rio Grande do Sul/Brazil.

Olivia Wright, School of Human Movement and Nutrition Sciences, The University of Queensland, St. Lucia, Brisbane, QLD 4072, Queensland, Australia.

Guggalada Govardhana Yadav, Department of Botany, Karnatak University, Dharwad 580003, India.

Acknowledgements

We would like to thank the Department of Science and Innovation (DSI) and the National Research Foundation (grant number 98352), South Africa, for supporting the SARChI Research Chair programme on the Phytochemical Food Network to Improve Nutritional Quality for Consumers as well as the Australian Research Council (ARC) Industrial Transformation Training Centre (ITTC) for Uniquely Australian Foods (grant Number IC180100045) supported by the Australian Research Council, Australia, for funding the compilation of this book.

We would also like to thank Professor Nurdan Tuna Güneş from the Department of Horticulture at Ankara University for rigorously reviewing and improving the chapters in this book. A special thank you to Dr Saleha Akter, University of Queensland for helping with the formatting of chapters and other editorial assistance.

Preface

The global population is expected to reach 9.7 billion by 2050, which adds to the challenges related to food and nutrition security. This situation has been exacerbated by the Covid-19 pandemic. In 2020, between 720 and 811 million people faced hunger according to the FAO (Food and Agriculture Organization of the United Nations). In addition, natural disasters and war in developing countries have resulted in inadequacy in diversified diets. The effects of inadequate diets on the population include malnutrition, non-communicable diseases (e.g. diabetes) and obesity. 'Hidden hunger', also known as micronutrient deficiencies, leads to various health-related disorders and diseases. Additionally, recent trends indicate that non-communicable diseases such as diabetes, cardiovascular disease and cancer are on the rise due to the consumption of processed food and meat products. Poverty limits access to nutritious food and the availability of balanced diets, which impacts food and nutrition security. Biodiversity plays a crucial role in the provision of nutrient diversity and nutrient adequacy. However, there is limited information available about the phytonutrient composition of cultivars or varieties of the same indigenous fruit or vegetable.

In African rural and urban areas, traditional fruits and vegetables have been identified as a means of diversifying the diet. Indigenous fruits and leafy vegetables are rich in micronutrients, and leafy vegetables from Africa have been studied for their iron and beta-carotene content. To sustain nutrition and food security and to eliminate hidden hunger among rural and urban populations, it is important to address strategies to include these indigenous food crops in nutrition intervention programmes.

In-depth analysis of phytochemicals in indigenous fruits and leafy vegetables, the mechanisms for biological activity and the effects after postharvest processing in relation to health claims, as well as changes in antinutritive compounds, is limited. This restricts the addition of indigenous fruits and vegetables based on health claims to national food databases. According to the literature, these indigenous fruits and leafy vegetables are rich in polyphenols, but their bioaccessibility appears to be more complicated. In addition, the available literature provides evidence that dietary polyphenols play a crucial role in maintaining gut health by promoting the growth of beneficial bacteria (e.g. *Lactobacilli* and *Bifidobacteria*) and inhibiting pathogenic bacteria. Phytochemicals in indigenous fruits and vegetables are not fully understood in terms of postharvest handling, processing, and digestion.

Furthermore, the bioavailability of some functional components is affected by the cellular structure of the food material. It is important to note that although polyphenols have antinutritive effects, they also have many health benefits such as inhibiting non-communicable diseases and reducing obesity. The International Diabetes Federation has predicted type 2 diabetes mellitus to rise to over 642 million or more by 2040. The mode of mechanisms involved in the antidiabetic and antiobesity effects of many indigenous fruits and vegetables is not known. There is, nevertheless, an opportunity for indigenous fruits and vegetables with potential antidiabetic properties to be cultivated for the extraction of bioactive compounds and the development of value-added functional products. Cultivating these crops in concert with rural and subsistence farmers will generate economic, social and environmental benefits. In addition, this approach addresses grassroots innovation, research and development for value addition and entrepreneurship skills. It contributes to the bioeconomy, and to the United Nations Sustainable Development Goals (SDGs) of no poverty, zero hunger and good health and wellbeing.

In this book, *Phytonutrients in Indigenous Fruits and Vegetables*, leading experts in the field have contributed 33 chapters. The aim is to provide understanding of: 1) phytonutrient constituents of traditional fruits and vegetables; 2) the impact of postharvest storage, processing and traditional food preparation methods on phytonutrients and functional compounds; and 3) the biological activity of the functional compounds that will improve productivity and increase utilization of indigenous fruits and vegetables to sustain food security.

We are truly grateful to all the authors for contributing useful and updated information on phytonutrient compounds as well as changes in processing and biological activity. We wish to acknowledge Rebecca Stubbs, commissioning editor at CABI, UK, for initiating the book project and Rachael Russell and Emma McCann of CABI for their dedicated efforts in completing this project on time.

We are convinced that this book will prove to be an invaluable resource for future indigenous food research, both as a textbook for graduate students and teachers and as a reference guide for researchers around the world.

Dharini Sivakumar
Pretoria, South Africa

Yasmina Sultanbawa
Queensland, Australia

Michael Netzel
Queensland, Australia

1 Saltbush (*Atriplex* sp.)

Sukirtha Srivarathan[1,2,3]*, Anh Dao Thi Phan[1,2], Olivia Wright[1,2,4], Daniel Cozzolino[1,2], Yasmina Sultanbawa[1,2] and Michael E. Netzel[1,2]
[1]ARC Industrial Transformation Training Centre for Uniquely Australian Foods, Queensland Alliance for Agriculture and Food Innovation, The University of Queensland, Queensland, Australia; [2]Centre for Nutrition and Food Sciences, Queensland Alliance for Agriculture and Food Innovation, The University of Queensland, Queensland, Australia; [3]Department of Biosystems Technology, Faculty of Technology, University of Jaffna, Kilinochchi, Sri Lanka; [4]School of Human Movement and Nutrition Sciences, The University of Queensland, Queensland, Australia

Abstract
Salinity and shrinkage of water tables are among the major crises in the world which adversely affect food production. It has been reported that about one-fifth of total farming land (45 million ha) is salt-affected worldwide. Currently, immense attention is granted to the domestication of halophytes (salt-tolerant plants) due to their evolved adaptive mechanisms to salinity including the production of secondary compounds. Halophytes are ecologically important plants, of which saltbush (*Atriplex* sp.) has been traditionally used as medicine, food, animal feed and as a 'tool' for rehabilitation of degraded land. This plant species has the potential to be used as a source of functional food/functional food ingredients, but is underutilized due to lack of popularity or limited knowledge. This chapter will give an overview of the traditional use of saltbush, an important edible halophyte, its nutritional composition, bioactive properties and potential food applications.

Introduction

Indigenous plants have gained immense interest as a source of nutrients and bioactive phytochemicals, satisfying both food demand and health needs (Pinela *et al.*, 2017; Sõukand, 2016). One such is saltbush (Fig. 1.1), an indigenous edible halophyte, member of the genus *Atriplex*, of which 400 species are found around the world (Ortíz-Dorda *et al.*, 2005). The highest number of saltbush species was found in South America (70–80 species), while the lowest number was reported in the Mediterranean basin and Europe (40–50 species) (O'Leary and Glenn, 1994; Osmond *et al.*, 1980). *Atriplex* species belong to the subfamily Chenopodioideae of the family Amaranthaceae and are distributed in arid and semi-arid regions with 200 to 400 mm of mean annual rainfall (Ortíz-Dorda *et al.*, 2005). Most species are halophytes and are found in salt marshes, salt deserts, mangroves and sand dunes near coastal areas (Wei *et al.*, 2019). *Atriplex* species are ecologically important plants and have been used for rehabilitation of degraded land (saline/alkaline soils, mine waste, shallow soils and bare saline soil) due to

*Corresponding author: s.srivarathan@uq.edu.au

 Handbook of Phytonutrients in Indigenous Fruits and Vegetables (eds D. Sivakumar *et al.*)
DOI: 10.1079/9781789248067.0001

Fig. 1.1. Saltbush (*Artiplex* sp.).

their ability to tolerate salinity-stressed environments through various biological mechanisms. However, they can also absorb salt from the soil to improve soil structure and increase soil fertility (Le Houérou, 1992; Ortíz-Dorda *et al.*, 2005; Piovan *et al.*, 2014). Though they have been used as human food or animal feed since ancient times, information on their dietary relevance is very limited. Therefore, this chapter reviews the nutritional profiles and dietary relevance of *Atriplex* species and safety issues related to their utilization in the human diet.

Botany

The most common saltbushes are shown in Table 1.1. *Atriplex* species consist of various sizes (1.0–2.5 m) and morphology (Table 1.1). These saltbush species are generally dioecious or monoecious. However, *Atriplex* sp. have specialized salt bladders on their leaves for storing surplus NaCl (Glenn *et al.*, 1998). Most of the *Atriplex sp.* leaves have a green/grey colour (including *A. cinera*, *A. nummularia* and *A. lentiformis*), while leaves of *A. hortensis* can be green, yellow-green, faintly red, dark red and purple. This makes *A. hortensis* an especially attractive addition to salads (Carlsson and Hallqvist, 1981). It is interesting to note that the fruits of *Atriplex* sp. occur in different shapes, from triangular (*A. amnicola*, *A. cinera*) to flat and round (*A. lentiformis*) (Table 1.1).

Traditional Use

Atriplex species have had diverse applications for food and non-food uses from ancient times. For instance, *A. cinera* was popular as a potherb among the Indigenous people of Australia (Maiden, 1889). Furthermore, *Atriplex* sp. have been used in traditional cuisine and as bush food, and are still used raw or cooked in modern recipes. For instance, leaves of *A. hortensis* are used in stews, salads and vegetable spread (Carlsson and Hallqvist, 1981; Bussmann *et al.*, 2017). Some of the *Atriplex* species were consumed not only by humans but also by livestock (*A. nummularia*, *A. halimus*, *A. amnicola* and *A. semibaccata*), with two species identified as nitrogen-rich fodder (*A. halimus* and *A. nummularia*) (Mehdiyeva *et al.*, 2017). It has also been reported that saltbushes have been widely utilized as fodder reserves, particularly in famine periods or periods of forage shortage (in either drought or cold season) and as a supplementary fodder in arid and semi-arid regions (Ortíz-Dorda *et al.*, 2005; Abbeddou *et al.*, 2011). Furthermore, these shrubs have been used as 'emergency' livestock fodder due to their favourable crude protein content (Khan *et al.*, 2000) in addition to their use as a source of fuel wood (Belkheiri and Mulas, 2013). The most important *Atriplex* sp. used for the rehabilitation of degraded lands are *A. nummularia* and *A. canescens*, respectively. These species are also used in ornamental plantations, particularly in seaside resorts (Le Houérou, 1992).

Table 1.1. Botanical information about selected *Atriplex* species.

Scientific name	Common name	Botany	References
Atriplex halimus	Mediterranean saltbush	A perennial native shrub, 1–3 × 3 m, monoecious, inflorescences are in dense spikes, male flowers are at the top of the spike and female flowers at the base	(Ortíz-Dorda *et al.*, 2005; Clauser *et al.*, 2013)
Atriplex amnicola	River saltbush	A perennial shrub, 1 × 1–2.5 × 4 m, dioecious, elongated and spear-shaped leaves (1–3 cm), fruits may be woody (2–6 mm), triangular or spherical	(Barrett-Lennard, 2003)
Atriplex cinera	Grey saltbush	Sprawling, semi-erect, woody, heavily branched, leafy shrub, 1.5 × 4 m and dioecious or subdioecious. Branchlets stout, rooting freely on contact with soil; stems at first ridged and angular, soon becoming terete and woody with age, grey-green leaves (2 cm), hard, fruits (2–6 mm), triangular	(De Lange *et al.*, 1998; Barrett-Lennard, 2003)
Atriplex hortensis	Garden orache or simply orache, red orache, mountain spinach, French spinach or orache	Shrub, 1–2.5 m, large leaves, (14 × 10 cm), flowers in leafless spikes, leaves can be green, yellow-green, faintly red, dark red and purple	(Carlsson, 1975; Carlsson and Hallqvist, 1981; Mehdiyeva *et al.*, 2017)
Atriplex nummularia	Oldman saltbush	Upright shrub, 2 × 1.5 m, irregularly shaped grey-green leaves (2–4 cm), fruits (6 mm), fan shape	(Barrett-Lennard, 2003)
Atriplex semibaccata	Australian saltbush, berry saltbush or creeping saltbush	Sprawling undershrub, xero-hypo halophyte	(Le Houérou, 1992)
Atriplex cana	-	A perennial shrub, multi-branched, 20–50 cm, stems become more or less woody	(Cai *et al.*, 2015; Wei *et al.*, 2019)
Atriplex mollis	-	Shrub, 1 m, pink, erect panicles of bladdery, fruit bracts (1 cm)	(Le Houérou, 1992)
Atriplex undulata	Wavy leaf saltbush	A perennial shrub, 1 × 2 m, wavy or crinkly leaves (0.5–1.5 cm), fruits (1–3 mm), soft and round	(Barrett-Lennard, 2003)
Atriplex lentiformis	Quailbrush	A perennial shrub, 2.5 × 2.5 m, silvery blue-green leaves (2 cm), fruits (2–5 mm), flat and round	(Barrett-Lennard, 2003)

It is interesting to note that *A. hortensis* can be used as part of landscaping gardens and parks since these plants are decorative and beautiful, particularly during the period of fructification (Mehdiyeva *et al.*, 2017). These plants are also considered as annual spinach in North and Middle Asia (Mueller, 1880). According to Shinn (1899), the Australian saltbushes were mentioned for the first time in the Report of the College of Agriculture in 1882. Seeds of *A. vesicaria* and *A. nummularia* were sent from Australia to California for propagation. According to Mueller (1880), *A. nummularia* was mentioned as ‘one of the tallest, more fattening and wholesome of Australian pastoral saltbushes, recommended for artificial rearing' and was found from Queensland through the desert tracts to Victoria and South Australia (Shinn, 1899). In 1894, the nutritional analyses on saltbush were compared with those of other green fodders and the findings suggested that the nutrients were comparable with alfalfa.

Nutritional Composition

A. halimus can be a complement forage with favourable nutritive value during the drought and famine period (Benhammou *et al.*, 2009). Humans have consumed this *Atriplex* sp. since ancient times, but only a few studies have explored its nutritional composition. The leaves of *A. halimus* were found to have 25.1 g/100 g (dry weight: DW) carbohydrate and 16.2 g/100 g (DW) protein (Abbeddou *et al.*, 2011). Interestingly, *Atriplex* species are rich sources of fibre irrespective of their growing locations. The fibre content of *A. halimus* was reported to range from 25.8 to 36.7 g/100 g DW (neutral detergent fibre) and 16.5 to 23.1 g/100 g DW (acid detergent fibre), which is comparable to olive leaves (Abbeddou *et al.*, 2011).

A. nummularia and *A. amnicola* are the most important species utilized for grazing and land rehabilitation in Australia (Kumara Mahipala *et al.*, 2009). As for *A. halimus*, only very little scientific information is available on its nutritional profile and dietary relevance. A recent study with Australian grown Oldman saltbush (*A. nummularia*) showed that saltbush leaves were rich in protein (20.1 g/100 g DW), fibre (41.5 g/100 g DW) and minerals (particularly Ca (1.4 g/100 g DW), Na (4.1 g/100 g DW), Mg (0.9 g/100 g DW) and Fe (11.7 mg/100 g DW); Srivarathan *et al.*, 2020). The reported protein content in *A. nummularia* ranged from 13.4 to 25.2 g/100 g DW and was comparable to that of Australian grown Oldman saltbush (Khalil *et al.*, 1986; Yousif *et al.*, 1994; Ben Salem *et al.*, 2010; Revell *et al.*, 2013; Srivarathan *et al.*, 2020). However, *A. amnicola* and *A. cinera* had a considerably lower protein content (9.0–9.7 g/100 g DW) than *A. nummularia*, whereas their fibre contents were in the same range as found in other *Atriplex* sp. (neutral detergent fibre (25.2–48.8 g/100 g DW); acid detergent fibre (16.1–33.5 g/100 g DW); Warren and Casson, 1993; Kumara Mahipala *et al.*, 2009; Revell *et al.*, 2013).

A. hortensis is considered to be one of the ancient cultivated plants, originated in Europe (Stevens, 1994), and has been used as a potherb in conventional medicine. Although it has been used in the food and pharmaceutical industry (Livadariu, 2013; Bussmann *et al.*, 2017), its nutritional composition and bioactive properties are not well reported (Wright *et al.*, 2002). According to Carlsson and Clarke (1983), leaves of *A. hortensis* are a rich source of protein. The content of crude protein in the leaves was higher than that of spinach (*Spinacia oleracea*) and *A. hortensis* has the potential to be used as a substitute for spinach due to its high yield and prolonged harvest season. Palmitic, oleic, linoleic and α-linolenic acids could be identified as the main fatty acids in the leaves of *A. hortensis* (Carlsson and Clarke, 1983).

The seeds of *A. hortensis* were found to have a relatively high content of carbohydrate (62.0 g/ 100 g DW), followed by protein (28.3 g/100 g DW) and dietary fibre (13.3 g/100 g DW) (Wright *et al.*, 2002). The fibre and protein content of *A. hortensis* seeds was higher than that of sweet and bitter quinoa (Wright *et al.*, 2002). The protein content in the seeds was also higher than that found in the main 'food-cereals' including rice (7.6 g/100 g DW), wheat (14.3 g/100 g DW), maize (10.2 g/100 g DW) and barley (10.8 g/100 g DW) (Duke and Atchley, 1986; Wright *et al.*, 2002). Furthermore, the amino acid profile in both, seeds and leaves, was found to be favourable compared to maize, rice and wheat, having a higher content of lysine, the limiting amino acid in cereals (Wright *et al.*, 2002).

Phytochemicals and Bioactive Properties

The presence of different phytochemicals including saponins, flavonoids (especially glycosides), tannins, terpenoids and alkaloids were reported for *Atriplex* sp. (Table 1.2) (Boutaoui *et al.*, 2018). Halophytes can have relatively high levels of phenolic compounds since they develop

Table 1.2. Bioactive compounds/phytochemicals identified in selected *Atriplex* species.

Plant species	Bioactive compounds/phytochemicals	Bioactive properties	Ethnopharmacological use/ general applications	References
A. halimus	Saponins, betaines, mineral salts, alkaloids, phytoecdysteroids, flavonoids (isorhamnetin-3-*O*-glucopyranoside, rutin-4',7-dimethylether and kaempferol-3,7-dirhaminoside-4'-methyloxide, isoorientin, 4'-methoxy-7-glucoside-5-hydroxyisoflavone, naringenin-4'-*O*-rhamnopyranoside, hesperidin), phenolic acids (*p*-hydroxybenzoic acid, chlorogenic acid, cinnamic acid, ferulic acid and salicylic acid)	Antioxidant, anti-acetylcholinesterase, hypoglycaemic	Leaves are used to treat heart disease, diabetes, rheumatism	(Bayoumi and El-Shaer, 1992; Dinan *et al.*, 1998; Said *et al.*, 2008; Benhammou *et al.*, 2009; Emam, 2011; Clauser *et al.*, 2013)
A. mollis	Rutin, catechin, epicatechin, naringenin, *p*-coumaric acid, gallic acid, chlorogenic acid, *p*-hydroxybenzoic acid, sinapinic acid, syringic acid, ferulic acid and vanillic acid	Anti-inflammatory, antioxidant, hepatoprotective	Phytopharmaceutical use	(Boutaoui *et al.*, 2018)
A. hortensis	Condensed tannins and other phenolic compounds	Antioxidant	Potherb in traditional medicine, used as diuretic and anastasis	(Carlsson and Hallqvist, 1981; Stevens, 1994; Wright *et al.*, 2002; Mehdiyeva *et al.*, 2017)
A.nummularia	Vitamin E, β-carotene	Antioxidant	-	(Pearce *et al.*, 2005; Ben Salem *et al.*, 2010)
A. lentiformis	Quercetin, kaempferol, quercetin and kaemferol glycosides	Antioxidant	-	(Awaad *et al.*, 2012)

a biological antioxidant system as a response to abiotic stress (Ramani *et al.*, 2004; Ksouri *et al.*, 2012; Boutaoui *et al.*, 2018).

A. halimus has been used as a medicinal plant in conventional Arabic medicine for glycaemic control in diabetic patients and to treat heart disease (Table 1.2) (Said *et al.*, 2002). It has also been used as a treatment for rheumatism and to combat parasites in veterinary use (Bayoumi and El-Shaer, 1992). Decoction and/or a remedy bath are the usual administration. *A. hortensis* is used as a diuretic, anastasis and against haemorrhage (Mehdiyeva *et al.*, 2017). It was also reported that it has higher levels of total phenolics and condensed tannins than *S. oleracea* (Carlsson and Hallqvist, 1981). Microwave-assisted extraction of *A. mollis* recovered several phytochemicals including gallic acid, catechin, chlorogenic acid, p-hydroxybenzoic acid, rutin, sinapinic acid, ferulic acid, naringenin and benzoic acid (Boutaoui *et al.*, 2018). No reported bioactivities or therapeutic application of *A. mollis* could be found in the literature. *A. mollis* and *A. nummularia* could be also identified as valuable sources of antioxidants having relatively high levels of vitamin E (139 mg/kg dry matter (DM)) and beta carotene (34.5 mg/kg DM) (Pearce *et al.*, 2005; Ben Salem *et al.*, 2010). According to White and Rewell (2007), *A. nummularia* was used in Australia to treat weaner sheep with vitamin E deficiency. A recent study on *A. lentiformis* showed that the isolated polyphenols, quercetin-6,4′-dimethoxy-3-fructo-rhamnoside and quercetin-4′-methoxy-3-fructo-rhamnoside, exhibited a strong antioxidant activity in rats with no side effects on liver and kidney function (Awaad *et al.*, 2012).

Antinutritive Compounds

According to Watson *et al.* (1987), oxalate presents an important anti-nutrient in *Atriplex* leaves. For instance, *A. hortensis* was reported with an oxalate content of up to 8.7% DM (Carlsson and Hallqvist, 1981). However, the toxicity of saltbush is regarded as low since its high salt content can control the intake of oxalate before toxicity is attained (Ben Salem *et al.*, 2010). As for other plant ingredients, oxalate levels can vary significantly, depending on the maturity, harvest season and specific environmental conditions (Abu-Zanat *et al.*, 2003). It was also reported that oxalate might have lowered the calcium absorption in tilapia fish fed with dehydrated *Atriplex* leaves, resulting in suppressed growth (Yousif *et al.*, 1994). Osmond (1963), Sandberg *et al.* (1967) and Duke and Williams (1977) reported low levels of quaternary alkaloids in *Atriplex* sp. in addition to saponins, tannins and flavonoids. However, it was surprising that no saponins could be found in the seeds of *A. hortensis* (Wright *et al.*, 2002).

Bioaccessibility and Bioavailability

Several *in vitro* studies on the digestibility of halophytes in the context of their 'efficiency' as livestock feed have been conducted (Weston *et al.*, 1970; Abbeddou *et al.*, 2011; Tosto *et al.*, 2015). However, no reports about human studies to assess the actual bioaccessibility and/or bioavailability of the main nutrients and phytochemicals in halophytes could be found.

Products/Formulations/Applications

Based on their nutritional composition, *Atriplex* sp. have the potential to be used as functional foods/functional food ingredients. The seeds of *A. hortensis* could be used to produce saponin-free food products. *A. hortensis* seeds and leaves, with their high protein content and well-balanced amino acid profiles, may have the potential to be used as a novel/alternative protein source (Carlsson and Clarke, 1983; Wright *et al.*, 2002). Since *Atriplex* sp. are generally rich in protein and most of them can be consumed raw, the addition to a meal as a complementary vegetable or incorporation into salads are other possible food 'applications'. Furthermore, due to their salty taste and unique aroma, leaves (fresh and dried) of *Atriplex* sp. can be added to a meal as a potential salt substitute or served together with seafood. Another 'avenue' is the typical functional food and/or phytopharmaceutical applications of *Atriplex* sp., due to their broad range of bioactive phytochemicals such as chlorophylls, carotenoids, betalains and phenolic compounds (Suhaj, 2006; Ksouri *et al.*, 2012).

Conclusions and Future Trends

This chapter has provided an overview of *Atriplex* sp., from their botany to potential (functional) food applications, to showcase their nutritional and bioactive potential to a broader audience. However, the showcased potential needs to be investigated further in human (clinical) studies to substantiate the actual bioavailability and health benefits of the main nutrients, bioactive phytochemicals and derived products. Another focus should be led on the development of different *Atriplex* sp.-based products suitable for the mainstream food market (e.g. salt replacement), but also for specific functional food and/or phytopharmaceutical applications (e.g. supporting diabetes management).

Acknowledgements

The authors acknowledge the Traditional Owners of the lands on which the indigenous plant species were harvested and respect the knowledge and experience the Traditional Owners hold regarding the care, harvest and use of these plants. This research is funded by the Australian Research Council (ARC) Industrial Transformation Training Centre (ITTC) for Uniquely Australian Foods (grant number: IC180100045).

References

Abbeddou, S., Rihawi, S., Hess, H.D., Iñiguez, L., Mayer, A.C. *et al.* (2011) Nutritional composition of lentil straw, vetch hay, olive leaves, and saltbush leaves and their digestibility as measured in fat-tailed sheep. *Small Ruminant Research* 96(2–3), 126–135. DOI: 10.1016/j.smallrumres.2010.11.017.

Abu-Zanat, M.M., Al-Hassanat, F.M., Alawi, M. and Ruyle, G.B. (2003) Oxalate and tannins assessment in *Atriplex halimus* L. and *A. nummularia* L. *Journal of Range Management* 56(4), 370. DOI: 10.2307/4004041.

Awaad, A.S., Maitland, D.J., Donia, A.E.R.M., Alqasoumi, S.I. and Soliman, G.A. (2012) Novel flavonoids with antioxidant activity from a Chenopodiaceous plant. *Pharmaceutical Biology* 50(1), 99–104. DOI: 10.3109/13880209.2011.591806.

Barrett-Lennard, E.G. (2003) Pasture plants for saltland. In: *Saltland Pastures in Australia- A Practical Guide*, 2nd edn. CRC for Plant-based Management of Dryland Salinity, Western Australia.

Bayoumi, M. and El-Shaer, H. (1992) Impact of halophytes on animal health and nutrition. In: Squires, V. and Ayoub, A.T. (eds) *Halophytes as a Resource for Livestock and for Rehabilitation of Degraded Lands*, 1st edn. Springer, Dordrecht. DOI: 10.1007/978-94-011-0818-8.

Belkheiri, O. and Mulas, M. (2013) The effects of salt stress on growth, water relations and ion accumulation in two halophyte *Atriplex* species. *Environmental and Experimental Botany* 86, 17–28. DOI: 10.1016/j.envexpbot.2011.07.001.

Ben Salem, H., Norman, H.C., Nefzaoui, A., Mayberry, D.E., Pearce, K.L. *et al.* (2010) Potential use of oldman saltbush (*Atriplex nummularia* Lindl.) in sheep and goat feeding. *Small Ruminant Research* 91(1), 13–28. DOI: 10.1016/j.smallrumres.2009.10.017.

Benhammou, N., Bekkara, F.A. and Kadifkova Panovska, T. (2009) Antioxidant activity of methanolic extracts and some bioactive compounds of *Atriplex halimus*. *Comptes Rendus Chimie* 12(12), 1259–1266. DOI: 10.1016/j.crci.2009.02.004.

Boutaoui, N., Zaiter, L., Benayache, F., Benayache, S., Cacciagrano, F. *et al.* (2018) *Atriplex mollis* Desf. aerial parts: Extraction procedures, secondary metabolites and color analysis. *Molecules* 23(8), E1962. DOI: 10.3390/molecules23081962.

Bussmann, R.W., Paniagua-Zambrana, N.Y., Sikharulidze, S., Kikvidze, Z., Kikodze, D. *et al.* (2017) Plant and fungal use in Tusheti, Khevsureti, and Pshavi, Sakartvelo (Republic of Georgia), Caucasus. *Acta Societatis Botanicorum Poloniae* 86(2), 1. DOI: 10.5586/asbp.3517.
Cai, D., Yan, C. and Wei, Y. (2015) Seed germination characteristics of the desert subshrub *Atriplex cana* and its ecological significance. *Acta Prataculturae Sinica* 24, 131–138.
Carlsson, R. (1975) *Selection of Centrospermae and Other Species for Production of Leaf Protein Concentrates*. Verlag nicht ermittelbar.
Carlsson, R. and Clarke, E.M.W. (1983) *Atriplex hortensis* L. as a leafy vegetable, and as a leaf protein concentrate plant. *Qualitas Plantarum Plant Foods for Human Nutrition* 33(2–3), 127–133. DOI: 10.1007/BF01091298.
Carlsson, R. and Hallqvist, C.W. (1981) *Atriplex hortensis* L.—revival of a spinach plant. *Acta Agriculturae Scandinavica* 31, 229–234. DOI: 10.1080/00015128109435702.
Clauser, M., Dall'Acqua, S., Loi, M.C. and Innocenti, G. (2013) Phytochemical investigation on *Atriplex halimus* L. from Sardinia. *Natural Product Research* 27(20), 1940–1944. DOI: 10.1080/14786419.2013.793684.
De Lange, P.J., Murray, B.G. and Gardner, R.O. (1998) *Atriplex cinerea*(Chenopodiaceae) in New Zealand. *New Zealand Journal of Botany* 36, 521–529.
Dinan, L., Whiting, P. and Scott, A.J. (1998) Taxonomic distribution of phytoecdysteroids in seeds of members of the Chenopodiaceae. *Biochemical Systematics and Ecology* 26(5), 553–576. DOI: 10.1016/S0305-1978(98)00005-2.
Duke, J.A. and Atchley, A.A. (1986) *Handbook of Proximate Analysis Tables of Higher Plants*. CRC Press.
Duke, J.A. and Williams, M.C. (1977) Phytotoxin tables. *CRC Critical Reviews in Toxicology* 5(3), 189–237. DOI: 10.3109/10408447709082600.
Emam, S.S. (2011) Bioactive constituents of *Atriplex halimus* plant. *Journal of Natural Products* 4, 25–41.
Glenn, E.P., Jed Brown, J. and O'Leary, J.W. (1998) Irrigating crops with seawater. *Scientific American* 279(2), 76–81. DOI: 10.1038/scientificamerican0898-76.
Khalil, J.K., Sawaya, W.N. and Hyder, S.Z. (1986) Nutrient composition of *Atriplex* leaves grown in Saudi Arabia. *Journal of Range Management* 39(2), 104. DOI: 10.2307/3899277.
Khan, M., Ungar, I.A. and Showalter, A.M. (2000) Effects of Salinity on Growth, Water Relations and Ion Accumulation of the Subtropical Perennial Halophyte, *Atriplex griffithii* var. stocksii. *Annals of Botany* 85(2), 225–232. DOI: 10.1006/anbo.1999.1022.
Ksouri, R., Ksouri, W.M., Jallali, I., Debez, A., Magné, C. *et al.* (2012) Medicinal halophytes: potent source of health promoting biomolecules with medical, nutraceutical and food applications. *Critical Reviews in Biotechnology* 32(4), 289–326. DOI: 10.3109/07388551.2011.630647.
Kumara Mahipala, M.B.P., Krebs, G.L., McCafferty, P. and Gunaratne, L.H.P. (2009) Chemical composition, biological effects of tannin and in vitro nutritive value of selected browse species grown in the West Australian Mediterranean environment. *Animal Feed Science and Technology* 153(3–4), 203–215. DOI: 10.1016/j.anifeedsci.2009.06.014.
Le Houérou, H.N. (1992) The role of saltbushes (*Atriplex* spp.) in arid land rehabilitation in the Mediterranean Basin: a review. *Agroforestry Systems* 18(2), 107–148. DOI: 10.1007/BF00115408.
Livadariu, O. (2013) In vitro experimental researches regarding the treatment with phytoregulators of orach (atriplex hortensis L.). *Bulletin of the University of Agricultural Sciences and Veterinary Medicine Cluj-Napoca Animal Science and Biotechnologies* 70, 289–295.
Maiden, J.H. (1889) *The Useful Native Plants of Australia (including Tasmania)*. Turner and Henderson, London. DOI: 10.5962/bhl.title.120959.
Mehdiyeva, N.P., Fayvush, G., Aleksanyan, A., Alizade, V.M., Batsatsashvili, K. *et al.* (2017) *Atriplex hortensis* L. *Atriplex tatarica* L. Amaranthaceae. In: Bussmann, R.W. (ed.) *Ethnobotany of the Caucasus*. Springer International Publishing, Cham, Switzerland. DOI: 10.1007/978-3-319-49412-8.
Mueller, F. von (1880) *Select Extra-tropical Plants Readily Eligible for Industrial Culture or Naturalisation, with Indications of Their Native Countries and Some of their Uses*. Office of the Superintendent of Government, Calcutta, India. DOI: 10.5962/bhl.title.54192.
O'Leary, J. and Glenn, E.P. (1994) Global distribution and potential for halophytes. In: *Halophytes as a Resource for Livestock and for Rehabilitation of Degraded Lands*. Springer, Dordrecht, Netherlands. DOI: 10.1007/978-94-011-0818-8.
Ortíz-Dorda, J., Martínez-Mora, C., Correal, E., Simón, B. and Cenis, J.L. (2005) Genetic structure of *Atriplex halimus* populations in the Mediterranean Basin. *Annals of Botany* 95(5), 827–34. DOI: 10.1093/aob/mci086.

Osmond, B. (1963) Oxalates and ionic equilibria in Australian saltbushes (*Atriplex*). *Nature* 198(4879), 503–504. DOI: 10.1038/198503a0.

Osmond, C.B., Björkman, O. and Anderson, D.J. (1980) *Physiological Processes in Plant Ecology: Toward a Synthesis with Atriplex*. Springer Science & Business Media, Berlin, Heidelberg. DOI: 10.1007/978-3-642-67637-6.

Pearce, K.L., Masters, D.G., Smith, G.M., Jacob, R.H. and Pethick, D.W. (2005) Plasma and tissue α-tocopherol concentrations and meat colour stability in sheep grazing saltbush (*Atriplex* spp.). *Australian Journal of Agricultural Research* 56(7), 663. DOI: 10.1071/AR05001.

Pinela, J., Carvalho, A.M. and Ferreira, I.C.F.R. (2017) Wild edible plants: Nutritional and toxicological characteristics, retrieval strategies and importance for today's society. *Food and Chemical Toxicology* 110, 165–88. DOI: 10.1016/j.fct.2017.10.020.

Piovan, M.J., Zapperi, G.M. and Pratolongo, P.D. (2014) Seed germination of *Atriplex undulata* under saline and alkaline conditions. *Seed Science and Technology* 42(2), 286–292. DOI: 10.15258/sst.2014.42.2.17.

Ramani, B., Zorn, H. and Papenbrock, J. (2004) Quantification and fatty acid profiles of sulfolipids in two halophytes and a glycophyte grown under different salt concentrations. *Zeitschrift Für Naturforschung C* 59(11–12), 835–42. DOI: 10.1515/znc-2004-11-1212.

Revell, D.K., Norman, H.C., Vercoe, P.E., Phillips, N., Toovey, A. *et al.* (2013) Australian perennial shrub species add value to the feed base of grazing livestock in low- to medium-rainfall zones. *Animal Production Science* 53(11), 1221. DOI: 10.1071/AN13238.

Said, O., Fulder, S., Khalil, K., Azaizeh, H., Kassis, E. *et al.* (2008) Maintaining a physiological blood glucose level with "glucolevel", a combination of four anti-diabetes plants used in the traditional Arab herbal medicine. *Evidence-Based Complementary and Alternative Medicine* 5(4), 421–428. DOI: 10.1093/ecam/nem047.

Said, O., Khalil, K., Fulder, S. and Azaizeh, H. (2002) Ethnopharmacological survey of medicinal herbs in Israel, the Golan Heights and the West Bank region. *Journal of Ethnopharmacology* 83(3), 251–265. DOI: 10.1016/s0378-8741(02)00253-2.

Sandberg, F., Michel, K.H., Staf, B. and M, T. (1967) Screening of plants of family chenopodiaceae for alkaloids. *Acta Pharmaceutica Suecica* 4, 51.

Shinn, C.H. (1899) *Australian Salt-bushes: Results of Eighteen Years' Tests, Characteristics, Propagation, and Field Experiments*. University Press.

Sõukand, R. (2016) Perceived reasons for changes in the use of wild food plants in Saaremaa, Estonia. *Appetite* 107, 231–241. DOI: 10.1016/j.appet.2016.08.011.

Srivarathan, S., Netzel, M., Thi Phan, A.D. and Sultanbawa, Y. (2020) Exploring the Nutritional Profile and Bioactive Potential of Australian Grown Saltbush (*Atriplex* sp.). In: *Proceedings of The third International Tropical Agriculture Conference (TROPAG 2019)*, MDPI, Basel, Switzerland, (pp. 83–84). DOI: 10.3390/proceedings2019036083.

Stevens, J. (1994) *Orach-Atriplex hortensis L. Fact sheet HS-637*. Horticultural Sciences Department, Florida Cooperative Extension Service, Institute of Food and Agricultural Sciences, University of Florida, Gainesville, Florida.

Suhaj, M. (2006) Spice antioxidants isolation and their antiradical activity: a review. *Journal of Food Composition and Analysis* 19(6–7), 531–537. DOI: 10.1016/j.jfca.2004.11.005.

Tosto, M.S.L., Araújo, G.G.L., Ribeiro, L.G.P., Henriques, L.T., Menezes, D.R. *et al.* (2015) In vitro rumen fermentation kinetics of diets containing oldman saltbush hay and forage cactus, using a cattle inoculum. *Arquivo Brasileiro de Medicina Veterinária e Zootecnia* 67(1), 149–158. DOI: 10.1590/1678-6937.

Warren, B. and Casson, T. (1993) Saltbush quality and sheep performance. ACIAR proceedings, Australian Centre for International Agricultural Research 71.

Watson, M.C., O'Leary, J.W. and Glenn, E.P. (1987) Evaluation of *Atriplex* lentiformis (Torr.) S. Wats. and *Atriplex nummularia* Lindl. as irrigated forage crops. *Journal of Arid Environments* 13(3), 293–303. DOI: 10.1016/S0140-1963(18)31119-4.

Wei, C., Zhou, S., Li, W., Jiang, C., Yang, W. *et al.* (2019) Chemical composition and allelopathic, phytotoxic and pesticidal activities of *Atriplex cana* Ledeb. (Amaranthaceae) essential oil. Chem Biodivers 16(4), e1800595-n. DOI: 10.1002/cbdv.201800595.

Weston, R., Hogan, J. and Hemsley, J. (1970) Some aspects of the digestion of *Atriplex nummularia* (saltbush) by sheep. In: *Proceedings of the Australian Society of Animal Production*, (pp. 517–521).

White, C.L. and Rewell, L. (2007) Vitamin E and selenium status of sheep during autumn in Western Australia and its relationship to the incidence of apparent white muscle disease. *Australian Journal of Experimental Agriculture* 47(5), 535. DOI: 10.1071/EA05123.

Wright, K.H., Pike, O.A., Fairbanks, D.J. and Huber, C.S. (2002) Composition of *Atriplex* hortensis, Sweet and Bitter Chenopodium quinoa Seeds. *Journal of Food Science* 67(4), 1383–1385. DOI: 10.1111/j.1365-2621.2002.tb10294.x.

Yousif, O.M., Alhadrami, G.A. and Pessarakli, M. (1994) Evaluation of dehydrated alfalfa and salt bush (*Atriplex*) leaves in diets for tilapia (*Oreochromis aureus* L.). *Aquaculture* 126(3–4), 341–347. DOI: 10.1016/0044-8486(94)90050-7.

2 Gumby Gumby (*Pittosporum angustifolium*)

Anh Dao Thi Phan[1,2]*, Dharini Sivakumar[1,3], Daniel Cozzolino[1,2], Michael E. Netzel[1,2] and Yasmina Sultanbawa[1,2]

[1]ARC Industrial Transformation Training Centre for Uniquely Australian Foods, Centre for Nutrition and Food Sciences, Queensland Alliance for Agriculture and Food Innovation, The University of Queensland, Queensland, Australia; [2]Centre for Nutrition and Food Sciences, Queensland Alliance for Agriculture and Food Innovation, The University of Queensland, Australia; [3]Department of Crop Sciences, Phytochemical Food Network Research Group, Tshwane University of Technology, Pretoria, South Africa

Abstract

Belonging to the family Pittosporaceae, *Pittosporum angustifolium* is an Australian indigenous plant that has a long history of usage in traditional Australian indigenous medicine. This plant has been traditionally harvested from the wild and also cultivated recently. Different parts of this plant, including leaves and stem, have been used as either fresh materials or as a natural, dried, powdered accompaniment for supporting general human health and wellbeing as a result of their rich source of phytochemicals. This chapter summarizes important information on the botany, traditional medicinal uses, macro- and micro-nutrients, phytochemicals, antinutritive compounds and health benefits of *P. angustifolium*. In addition, variations in nutritional composition, bioactive compounds, antioxidant capacity and antimicrobial activity against important food poisoning and infective microorganisms, between different plant parts (leaves vs stem) and growing conditions (wild vs cultivation), are also discussed. The information reported in this chapter not only contributes to the literature, but may also be useful for future applications of *P. angustifolium* in the nutraceutical industry.

Introduction

Pittosporum angustifolium is a native tree from Australia, belonging to the genus *Pittosporum* and the family Pittosporaceae, comprising approximately 200 species in nine genera (Cayzer *et al.*, 2000). *Pittosporum* is the greatest diversity genus among the family Pittosporaceae (Negrelle *et al.*, 2018). This plant species is distributed mainly in inland Australia and New Zealand, but also in many other parts of the world. It was introduced to Europe and America for horticultural purposes (Cayzer *et al.*, 2000).

P. angustifolium leaves and fruits have been traditionally harvested from the wild by the Australian indigenous people for thousands of years. Cultivation of this plant has recently been introduced to increase production for commercialization. Although information regarding

*Corresponding author: a.phan1@uq.edu.au

DOI: 10.1079/9781789248067.0002

postharvest handling and practices for the leaves and fruits of this plant is very limited, its traditional medicinal uses are well reported as having been part of indigenous bush medicine across inland Australia for hundreds of years, to enhance general health and wellbeing. Recently, this understudied plant has received attention for therapeutic and pharmaceutical applications in Australia, particularly medicinal application of the leaves.

Plant secondary metabolites such as polyphenols, tannins, saponins and alkaloids have been demonstrated to show bioactivity and health-related benefits. Screening studies and quantitative determination on the phytochemicals of different species of the *Pittosporum* genus have been conducted and it is reported that *P. angustifolium* leaves, seeds and fruit contain triterpenoid saponins, phenolic acids, flavonoids, tannins and essential oils (Errington and Jefferies, 1988; Vesoul and Cock, 2011; Sadgrove and Jones, 2013; Bäcker *et al.*, 2013, Bäcker *et al.*, 2014b, Bäcker *et al.*, 2015; Madikizela and McGaw, 2017; Phan *et al.*, 2020). Among them, triterpenoid saponins, essential oils and non-tannin polyphenols are the main bioactive compounds in the *Pittosporum* genus, whereas tannins and alkaloids are minor. It is unlikely that alkaloids will be present in the leaves and fruit of *P. angustifolium* (Vesoul and Cock, 2011; Sadgrove and Jones, 2013). Recently, information on the nutritional quality, phytochemical profile and bioactivities of *P. angustifolium* leaves and stems, collected either from the wild or cultivated, has been reported (Phan *et al.*, 2020).

This chapter summarizes and discusses information regarding the nutritional quality, phytonutrients and health benefits of *Pittosporum angustifolium*, including the effects of different plant tissues and growing conditions. The chapter concludes with a summary of the latest findings from the literature and recommendations for future research.

Botany

P. angustifolium is a shrub tree with a height of up to 10 m, with secondary and tertiary pendulous branches (Fig. 2.1a). Adult leaves are alternate, linear or narrow-elliptic, falcate and essential flat (50–90 x 6–11 mm). The flowers are individually appearing and unisexual, with a colour ranging from golden yellow to brown-yellow depending on their gender (Fig. 2.1b). The male flower is anthers oblong, whereas the female flower is anthers narrow triangular (Cayzer *et al.*, 2000). Fruit is generally globose capsules (8–14 x 8–14 mm), yellow to orange, with up to six seeds per

(a) (b) (c)

Fig. 2.1. *P. angustifolium*: (a) tree; (b) flowers; and (c) yellow to orange fruit and red-brown seeds inside the fruit.
(a) https://en.wikipedia.org/wiki/Pittosporum_angustifolium (retrieved 20 March 2020).
(b) https://ppnn.org.au/plantlist/pittosporum-angustifolium (retrieved 20 March 2020).
(c) https://www.alphaomegalabs.com/gumby-gumby-pittosporum-augustifolium-100-vegetarian-capsules-x-270-mg-1.html (retrieved 20 March 2020).

loculus (Fig. 2.1c). The seeds are angular-chunky, 4–5 mm long, red-brown, cohering in a sticky mass (Cayzer *et al.*, 2000).

Traditional Medicinal Uses

The genus *Pittosporum* has been reported as a traditional medicinal plant and used for treatment of a wide range of conditions such as inflammatory, spasmodic, malarial and microbial (e.g. chronic bronchitis, leprous infection, rheumatics, bruises, sciatica, chest infection and certain skin diseases) (Madikizela and McGaw, 2017). All parts of the plant, including leaf, bark, root, flower, fruit pulp, seed and even wood have been reported to show potential medicinal applications in many countries such as Australia, China, India and South Africa (Madikizela and McGaw, 2017). Belonging to the genus *Pittosporum*, *P. angustifolium* has been traditionally used either as a complement to mainstream medicines, or to the exclusion of other medicines for treatments of skin infections such as eczema and psoriasis and recent lifestyle diseases such as cardiovascular disease, Crohn's disease and cancer (Sadgrove and Jones, 2016). The Australian Indigenous people have used different botanical tissues of *P. angustifolium* to enhance general health and wellbeing for generations. For instance, the infusion from the seeds and pulps of the fruit has been used to reduce pains and sprains (Cooper, 1956), whereas the intake of leaf infusion can help to treat colds and coughs and to induce lactation (Lassack and Mccarthy, 1983). Decoction from the fruit either taken orally or applied directly has been used to treat skin problems (Latz, 1995; Cayzer *et al.*, 2000). In addition, *P. angustifolium* has been reported to show potential for treatment of rheumatoid arthritis and other inflammatory conditions (Cock *et al.*, 2015).

Nutritional Composition

Proximate

Phan *et al.* (2020) reported nutritional composition of *P. angustifolium* whole branch, combining leaves and stem, collected both from the wild and cultivation in Queensland, Australia. The *P. angustifolium* leaves and stem consist of carbohydrates (26.6–31.3 g/100 g dry weight (DW)) and total dietary fibre (43.3–45.2 g/100 g DW) as predominant nutritional components, whereas protein and fat present at low amounts of 7.2–10.6 and 3.4–4.6 g/100 g DW, respectively (Table 2.1). Furthermore, the leaves and stem can be considered a good source of dietary fibre. A serving of 100 g contains at least 4 to 7 g of fibre, so is considered a good or excellent source, according to Food Standards Australia and New Zealand (FSANZ). A relatively high level of ash (9.6–11.2 g/100 g DW; Table 2.1) was reported for the whole branch sample, indicating the presence of considerable amounts of inorganic elements in the leaves and stem of this plant.

In a comparison between *P. angustifolium* whole branch (leaves and stem together) collected from the wild and one-year-old cultivation, the cultivated sample contained higher amounts of protein and total carbohydrates than the wild sample (Phan *et al.*, 2020) (Table 2.1). In contrast, the leaf and stem sample harvested from the wild contained more fat (4.6 vs 3.4 g/100 g DW) and ash (11.2 vs. 9.6 g/100 g DW). Differences in abiotic conditions and plant feeding nutrients are the main factors causing variations in nutritional profile between the wild harvested and domesticated plants. Another study investigated fatty acid components of *P. angustifolium* and showed that linoleic acid and α-linolenic acid are major free fatty acids in the leaves and twigs of *P. angustifolium* collected in Queensland and New South Wales, with the leaves containing a higher proportion of α-linolenic acid to linoleic acid compared to the twigs (Sadgrove and Jones, 2016). However, the study did not observe the effects of geographical location on the ratio of α-linoleic acid/linoleic acid. Information on nutritional composition of other edible leaves, including spinach, green tea and bay leaves, and standard daily intake has been included in Table 2.1 for comparison.

Minerals and trace elements

Information on minerals and trace elements in different parts of *P. angustifolium* is very limited. Phan *et al.* (2020) reported the levels of five

Table 2.1. Comparison of nutritional composition, minerals and trace elements in leaf and stem powdered samples of *P. angustifolium* collected from the wild and cultivated in Queensland, Australia.

Plant species	*P. angustifolium* leaves and stem[a] Wild harvest	Cultivation	Spinach leaves (*Spinicia oleraceae*)[b]	Green tea leaves[c]	Bay leaves[d]	Nutritional information
Proximate composition (g/100 g DW)						
Protein	7.2	10.6	19.18	18.06	7.62	50 g*
Fat	4.6	3.4	1.04	2.49	8.57	70 g*
Carbohydrates	26.6	31.3	60.34	-	50.83	400 g*
Dietary fibre	45.2	43.3	8.19	15.35	24.4	30 g*
Ash	11.2	9.6	2.99	5.6	3.63	-
Moisture	5.2	1.7	8.26	4.88	4.95	
Minerals and trace elements (mg/kg DW)						
Calcium	27,302	7879	13,039	3748	3770	1.2 g/day AI**
Potassium	15,403	26,180	2338	17,031	5500	4.7 g/day AI**
Magnesium	2547	2590	-	652	1120	0.35 g/day EAR**
Sodium	878	1161	982	76	-	1.3 g/day AI**
Phosphorus	1194	2669	-		1120	0.7 g/day AI**
Iron	116.7	48.3	403.6	205	450	8 mg/day RDA**
Zinc	57	62	133.8	30.6	29	11 mg/day RDA**

[a] Phan *et al.* (2020); [b] Waseem *et al.* (2021); [c] Ahmad *et al.* (2014); [d] AL-Hashimi and Ali (2016); * Daily intake, Australian Government, 2014; ** Otten *et al.* (2006).
(-) not available; RDA, recommended dietary allowance; AI, adequate intake; EAR, estimated average requirement.

minerals in *P. angustifolium* leaves, stem and whole branch including Ca (3,295–27,302 mg/kg DW), K (15,403–26,180 mg/kg DW), Mg (989–3,284 mg/kg DW), Na (686–2,083 mg/kg DW) and P (1,194–2,844 mg/kg DW). In addition, the leaves and stem of this plant contained numerous trace elements such as Cu, B, V, Cr, Mn, Fe, Co, Zn, Se, Sr, Mo, Ag, Sn, Sb, Ba, with all ranging below 100 mg/kg DW, except for Fe having a relatively high level of up to 116.7 mg/kg DW (recommended dietary allowance for Fe (RDA) at 8 mg/day) (Otten *et al.*, 2006; Phan *et al.*, 2020). Compared to spinach, bay and green tea leaves, the content of Fe in *P. angustifolium* leaves is two to four times lower (Table 2.1).

Heavy metals found in the leaves and stem of *P. angustifolium* include Al (7.5–89 mg/kg DW; tolerable weekly intake (TWI) at 1.0 mg/kg body weight (BW)/week), Ni (0.9–1.57 mg/kg DW), As (<0.06 mg/kg DW; TWI at 128 µg/week for a 60 kg BW), Cd (<0.07 mg/kg DW; TWI at 2.5 µg/kg BW/week), Hg (<0.05 mg/kg DW; TWI at 5 µg/kg BW/week) and Pb (0.12 mg/kg DW; TWI at 25 µg/kg BW/week) (Otten *et al.*, 2006; Phan *et al.*, 2020).

There were variations in minerals and trace elements between *P. angustifolium* leaves and stem, with the leaves possessing higher levels of minerals and trace elements compared to the stem. Furthermore, growing conditions contributed to differences in the levels of minerals and chemical elements between the wild harvested leaves and cultivated leaves (Phan *et al.*, 2020).

Vitamins

Vitamins are essential micronutrients that play crucial roles in the metabolism of the most vital pathways. Table 2.2 presents several water-soluble vitamins that could be quantified in *P. angustifolium* leaves and stem. Vitamin C was found in the range of 10.9–137.8 mg/100 g DW (Table 2.2), depending on plant botanical tissues and growing conditions (Phan *et al.*, 2020).

Table 2.2. Water-soluble vitamins reported in wild and cultivated *P. angustifolium* whole branch, leaves and stem collected from different locations in Australia (quantity per 100g DW) (Phan *et al.*, 2020).

	Wild		Cultivation			
	South Australia	Queensland	Queensland			
Vitamins	Whole branch	Whole branch	Whole branch	Leaf	Stem	RDI for adults (*)
C	-	10.9 mg	127.5 mg	137.8 mg	41.1 mg	45 mg/day
Folates (B9)	222.8 µg	327.6 µg	574.9 µg	597 µg	422 µg	400 µg/day
B2	*n/a*	<5 µg	73 µg	*n/a*	*n/a*	1.6 mg/day
B3	*n/a*	840 µg	1300 µg	*n/a*	*n/a*	14–16 mg/day
B5	*n/a*	<5 µg	300 µg	*n/a*	*n/a*	4–6 mg/day
B6	*n/a*	180 µg	59 µg	*n/a*	*n/a*	1.7 mg/day

* Food standards Australian and New Zealand; DW – dry weight; RDI – recommended dietary intake; (-) – not detectable using UHPLC-PDA method (LOD = 0.1 ppm); (n/a) – not available.

The leaves contained higher vitamin C than the stem. The leaves and stem of *P. angustifolium* collected in the wild had lower vitamin C content compared to the domesticated plant, probably due to the lack of standard postharvest handling practices. Vitamin C reported in *P. angustifolium* leaf powder at 137.8 mg/100 g DW (Table 2.2) is twofold higher than green tea leaf powder (60 mg/100 g DW; United States Department of Agriculture USDA (2019)), and approximately three times higher than bay leaf powder (45.33 mg/100 g DW; AL-Hashimi and Ali, 2016). The recommended dietary intake (RDI) of vitamin C for adults is 45 mg/day (FSANZ).

B vitamins are essential in many metabolic and physiological processes and can act as coenzymes for enzyme functions and functioning nervous systems (Khosravi-Largani *et al.*, 2018; Zhang *et al.*, 2018). Vitamin B3 was the highest among determined B vitamins in *P. angustifolium*, followed by folates, B5, B6 and B2 (Table 2.3), whereas B1, B7 and B12 were only found in traces (<5 µg/100g DW; Phan *et al.*, 2020). Higher levels of B vitamins, particularly B2, B3 and B5, were reported for the domesticated *P. angustifolium* leaves and stem compared to the wild sample (Table 2.2).

Like vitamin C, folates were found at moderate levels in the *P. angustifolium* leaves and stem, with the higher concentration being reported in the leaves. Furthermore, different growing conditions and locations induced variations in the total folate content of *P. angustifolium* (Table 2.2), with more folates being reported in the wild leaves. Compared to spinach leaves which contain total folates from 54.1 to 173.2 µg/100 g fresh weight (FW) (Shohag *et al.*, 2011), the folates of *P. angustifolium* leaves and stem reported at 222–597 µg/100 g DW (equivalent to 74–220 µg/100 g FW at a moisture content of 63%; Phan *et al.*, 2020) could be comparable. The RDI of folates for adults is 400 µg/day (FSANZ).

Phytochemicals

There is increasing interest in drug discovery from native medicinal plants, leading to many investigations on extraction, identification and quantification of bioactive compounds in different species of the *Pittosporum* genus, including *P. angustifolium*. Several bioactive compounds have been identified in the crude extracts of *P. angustifolium* such as triterpenoid saponins in leaves and seeds (Errington and Jefferies, 1988; Bäcker *et al.*, 2013, 2014b, 2016), phenolic acids and flavonoids in leaves (Bäcker *et al.*, 2014a; Phan *et al.*, 2020), and tannins and essential oils in leaves (Sadgrove and Jones, 2013).

Polyphenols

Polyphenols are plant secondary metabolites possessing well-known health promoting effects

Table 2.3. Studies determining phenolic compounds in *P. angustifolium* and other species belonging to the Pittosporaceae family.

Parts of plant	Species	Extraction parameters	Analytical methods	Identified phenolic compounds	References
Leaves and stems	*P. angustifolium*	80% MeOH 1 hour sonication room temperature	UHPLC-PDA-MS/MS	Rutin (14.5–246.6 mg/100 g DW) Quercetin-3-glucoside (13.7–86.7 mg/100 g DW) Quercetin-3-*O*-[6”-(3-hydroxy-3-methylglutaroyl)-β-glucoside (42.6–754.2 mg/100 g DW) Dicaffeoylquinic acid (95.7–1,869.5 mg/100 g DW) Caffeic acid (2–2.8 mg/100 g DW) *p*-Coumaric acid (2.6–18.1 mg/100 g DW) Ferulic acid (1.3–3.3 mg/100 g DW) Chlorogenic acid (140.4–1,020.9 mg/100 g DW)	Phan *et al.* (2020)
Leaves	*P. angustifolium*	80% MeOH 10 hour Room temperature	^{1}H NMR, HPLC-DAD, TLC, HRESIMS	Rutin Isoquercitrin 6”-(3-hydroxy-3-methylglutaroyl)-isoquercitrin 3,4-Dicaffeoylquinic acid 4,5-Dicaffeicoyquinic acid	Bäcker *et al.*, 2014a
Whole plant material	*P. angustifolium (former P. phylliraeoides)*	MeOH, water, ethyl acetate, chloroform, hexane 24 hour 4°C	Qualitative screening test based on colour development indicator assays	Indicated the presence of flavonoids without the identity of specific individual compounds	Vesoul and Cock (2011)
Leaves and twigs	*P. angustifolium*	MeOH 24 hour room temperature	HPLC-DAD	Demonstrated the presence of 5 major components (as peaks) and 22 minor compounds (peaks) from UV absorbance at 280 nm. No identity of specific individual compounds	Sadgrove and Jones (2016)
Fruit and leaves	*P. angustifolium*	MeOH Water 30 hour	Qualitative screening test based on colour development indicator assays	Flavonoids and water insoluble phenolic compounds No identity of specific individual compounds	Sadgrove and Jones (2013)

Continued

Table 2.3. Continued

Parts of plant	Species	Extraction parameters	Analytical methods	Identified phenolic compounds	References
Stem bark	*P. dasycaulon*	Petroleum ether Chloroform Acetone MeOH Water Overnight room temperature	Folin-Ciocalteu assay Aluminium chloride assay	Total phenolic content (TPC) = 182 mg GAE/g extract. Total flavonoid content (TFC) = 25.5–33.3 mg QE/g extract	Mani and Dennis (2014)
Seeds	*P. tobira*	80% MeOH 24 hour room temperature	HPLC-DAD Folin-Ciocalteu assay Aluminium chloride assay	Gallic acid = 1.03 mg/g Caffeic acid = 38.6 mg/g Ferulic acid = 20.1 mg/g p-Coumaric acid = 12.9 mg/g Cinnamic acid = 5.1 mg/g TPC = 102.7 mg GAE/g DW TFC = 31.6 mg CAE/g DW	Rjeibi *et al.* (2017)
Leaves	*P. pentandrum*	Aqueous acetone	Folin-Ciocalteu assay	TPC = 40.9 mg chlorogenic acid equivalents/g DW	Shieh *et al.* (2009)
Leaves	*P. mannii*	MeOH	Folin-Ciocalteu assay	TPC = 343.3 mg GAE/g DW	Wandji *et al.* (2016)
Leaves and twigs	*P. viridiflorum*	50% MeOH 1 hour cold temperature	Folin-Ciocalteu assay Aluminium chloride assay	TPC = 26 mg GAE/g DW TFC = 15.6 mg CE/g DW	Amoo *et al.* (2012)
Leaves	*P. viridiflorum*	Acetone 30 min room temperature	Folin-Ciocalteu assay Aluminium chloride assay	TPC = 26.1 mg GAE/g DW TFC = 20.3 mg QE/g DW	Olaokun *et al.* (2017)

Note: MeOH – Methanol; TPC – total phenolic content; GAE – gallic acid equivalents; TFC – total flavonoid content; QE – quercetin equivalents; CAE – catechin equivalents; DW – dry weight; HPLC-DAD – high-performance liquid chromatography with a diode-array detector; NMR – nuclear magnetic resonance; TLC - thin layer chromatography; HRESIMS - high-resolution electrospray ionisation mass spectrometry.

in humans. Recent studies determining phenolic compounds in different botanical parts of *P. angustifolium* and other species belonging to the Pittosporaceae family are summarized in Table 2.3. Most of the investigations mainly focused on quantification of total phenolic content (TPC), total flavonoid content (TFC), and total tannins based on standard spectrophotometric assays.

Recently, more studies on phenolic compounds in *P. angustifolium* using advanced techniques have been conducted. Bäcker *et al.* (2014a) identified main individual phenolic compounds using ^{1}H NMR, HPLC-DAD, thin liquid chromatography and high-resolution electrospray ionization mass spectrometry (HRESIMS) without further quantification. Sadgrove and Jones (2016) indicated five major compounds and 22 minor phenolic constituents (as peak indication acquired at the wavelength 280 nm) from HPLC-PDA analysis; however, no identity and/or quantification of individual phenolic compounds was conducted. Employing a high-resolution Q exactive orbitrap spectrometry, Phan *et al.* (2020) could confirm and quantify eight major phenolic acid and flavonoid compounds (Fig. 2.2) in the leaves and stem of *P. angustifolium* that were wild harvested and cultivated. The authors reported the amount of identified polyphenols, including rutin (14.5–246.6 mg/100 g DW), quercetin-3-glucoside (13.7–86.7 mg/100 g DW), quercetin-3-*O*-[6"-(3-hydroxy-3-methylglutaroyl)-β-glucoside (42.6–754.2 mg/100 g DW), dicaffeoylquinic acid (95.7–1869.5 mg/100 g DW), caffeic acid (2–2.8 mg/100 g DW), *p*-coumaric acid (2.6–18.1 mg/100 g DW), ferulic acid (1.3–3.3 mg/100 g DW) and chlorogenic acid (140.4–1020.9 mg/100 g DW), with variations that could be due to differences in botanical tissues, growing condition and location. Among the phenolics identified, chlorogenic acid, dicaffeoylquinic acid and quercetin-3-*O*-[6'-(3-hydroxy-3-methylglutaroyl)]-β-glucoside were highlighted as predominant compounds (Bäcker *et al.*, 2014a; Phan *et al.*, 2020).

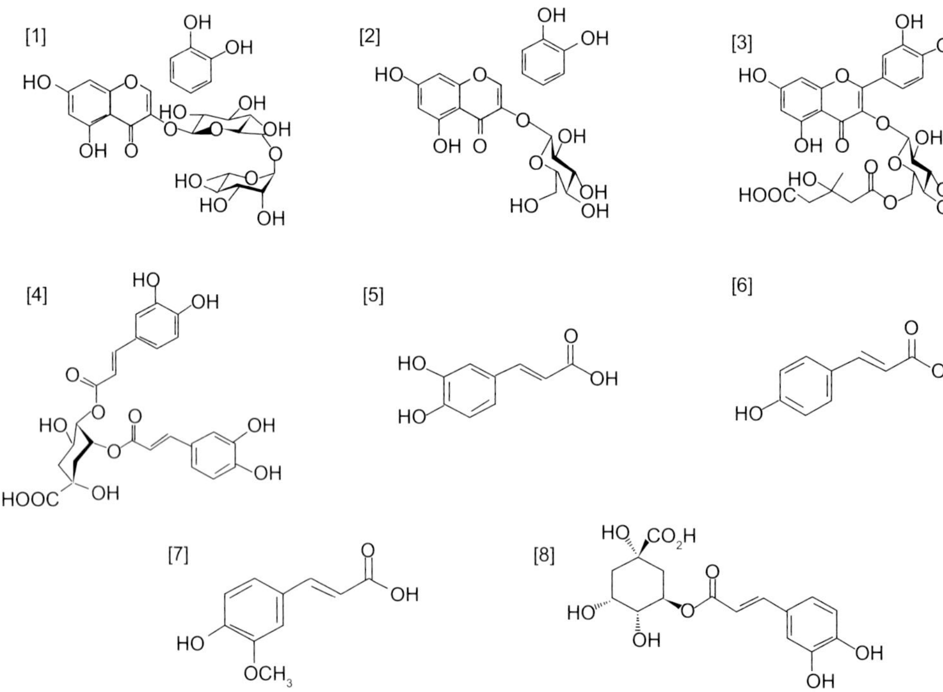

Fig. 2.2. Phenolic compounds in *P. angustifolium* leaves and stem: (1) rutin; (2) quercetin-3-glucoside; (3) quercetin-3-O-[6"-(3-hydroxy-3-methylglutaroyl)]-β-glucoside; (4) dicaffeoylquinic acid; (5) caffeic acid; (6) p-coumaric acid; (7) ferulic acid; (8) chlorogenic acid.

Although health-related benefits of phenolic compounds extracted from *P. angustifolium* have not been investigated intensively, there is scientific evidence of health benefits of chlorogenic acid, dicaffeoylquinic acid and isoquercitrin as important biologically active dietary polyphenols, produced by numerous plants. It has been reported that consumption of chlorogenic acid rich extract from green/decaffeinated coffee bean has been shown to be beneficial for a wide range of chronic diseases of cardiovascular, metabolic, cancerogenic, neurological and other biological functions both in animal models and human clinical studies (Tajik *et al.*, 2017). Dicaffeoylquinic acids have been reported to show potential benefits for skin health in recent published studies. For instance, polyphenol-rich extracts from artichoke comprising 87% of chlorogenic, 3,5-*O*-dicaffeoylquinic, and 1,5-*O*-dicaffeoylquinic acids induced potential health effects on molecular markers that are responsible for anti-inflammation, protection of lymphatic vessels from oxidative damage and upregulation of gene expression involved in anti-ageing mechanisms (D'Antuono *et al.*, 2018). In addition, an enhancement by 19.7% of roughness and 11.5% of elasticity of the skin was observed in a placebo-controlled clinical study involving 20 female volunteers (aged between 35 and 55), who applied a cosmetic formulation containing 0.002% crude artichoke extract for 28 days (D'Antuono *et al.*, 2018). In addition, 4,5-*O*-dicaffeoylquinic acid was proposed to reduce melanin synthesis and tyrosinase activity in a dose-dependent manner in the melanocyte both *in vitro* and in *vivo* (using a zebrafish model) (Tabassum *et al.*, 2016). These authors also suggested that this phenolic compound can be used as a bioactive component in pharmaceutical formulations to treat hyperpigmentation disorders. Isoquercitrin is a glycosidic form of the natural flavonol quercetin belonging to flavonoid subclass. This compound is commonly found in many medicinal herbs, fruit, vegetables and plant-derived food products and has been reported to show a number of chemo-protective effects against oxidative stress, cancer, cardiovascular disorders, diabetes and allergic reactions both *in vitro* and *in vivo* (Valentová *et al.*, 2014).

Carotenoids

Carotenoids are C40 tetraterpenoid compounds and can be found in many parts of plants, including the fruit, leaves, roots and flowers, either in free or esterified form with fatty acids. Carotenoids are yellow-orange-red plant pigments and can be found at relatively high amounts in green leafy vegetables and algae. Some carotenoids possess pro-vitamin A activity and therefore consumption of carotenoid-rich food is beneficial for humans. Like vitamins, carotenoids cannot be synthesized by humans, hence they are supplied by the diet.

There is strong evidence reported in the literature regarding the health benefits of carotenoids due to their strong antioxidant properties. However, different carotenoid constituents exert different health benefits, depending on their molecular structures. For instance, β-carotene has pro-vitamin A function, whereas lutein and zeaxanthin constitute macular pigment in the eye (Saini *et al.*, 2015). Apart from these health-promoting functions, carotenoids could improve cognitive function, cardiovascular health and reduce risks of some types of cancers (e.g. prostate and liver cancers) (Saini *et al.*, 2015).

There is very limited information on carotenoids in *P. angustifolium*. Phan *et al.* (2020) examined and quantitatively determined individual carotenoid compounds both in the wild and cultivated *P. angustifolium* leaves and stem. The authors reported lutein level at 1.2–14.3 mg/100 g DW and β-carotene at 0.7–6.3 mg/100 g DW as the main carotenoid constituents in *P. angustifolium* leaves, followed by zeaxanthin at a lower amount between 0.3 and 1.9 mg/100 g DW. Compared to other green leaves like cowpeas, which have lutein at 74.5–124 mg/100 g DW and β-carotene at 43–92.6 mg/100 g DW (Moloto *et al.*, 2020), carotenoids in *P. angustifolium* leaves are much lower. In addition, there are more carotenoids synthesized and stored in the leaves compared to the stem. Differences in the carotenoid concentrations between wild and cultivated varieties as well as different growing locations were also reported (Phan *et al.*, 2020). Another study on carotenoids of other Pittosporum species conducted by Madikizela and McGaw (2017) indicated the presence of carotenoids in *P. tobira* without any information on identification of individual carotenoid compounds.

Essential oils

P. angustifolium fruits and leaves also contain essential oils which have potential therapeutic properties. A total of 51 different volatile compounds have been identified in the leaf extract of *P. angustifolium*, in which esters and sesquiterpenols were found as main ones (Sadgrove and Jones, 2013). In addition, the chemical composition of *P. angustifolium* essential oils varied with geographical location and botanical tissue (e.g. fruit and leaf). The essential oils extracted from leaves showed greater quantities of esters and sesquiterpenols than those from the fruit (Sadgrove and Jones, 2013). The authors also suggested that the essential oil constituents such as limonene, sabinene, terpinenes, α-pinene and bicyclogermacrene extracted from the fruit and leaves of *P. angustifolium* exhibit potential antimicrobial activity *in vitro*, which may be relevant to the traditional medicinal uses of *P. angustifolium* by the Australian Indigenous people for treatment of colds, eczema and promotion of lactation in humans and animals (as lactagogue).

Antinutritive Compounds

In addition to macro- and micronutrients, *P. angustifolium* also contains antinutritional compounds including saponins and tannins, which might cause a reduction in nutrient bioavailability. Tannins are well known to precipitate proteins by forming reversible and irreversible tannin-protein complexes which can affect protein digestibility (Raes *et al.*, 2014; Samtiya *et al.*, 2020). Similarly to tannins, saponins also showed an inhibitory effect against the activities of several digestive enzymes, potentially inducing indigestion-related health disorders (Samtiya *et al.*, 2020). In addition, saponins were reported to cause a decrease in absorption of fat-soluble vitamins (Jenkins and Atwal, 1994).

Tannins

Tannins are phenolic compounds that can be found in *P. angustifolium*. Literature reporting tannins in *P. angustifolium* and other species of the *Pittosporum* genus is limited. The presence of tannins was qualitatively confirmed in the leaves of *P. phylliraeoides* (a former name of *P. angustifolium*) by Vesoul and Cock (2011). Tannins in the water extract of *P. angustifolium* leaves and fruit were qualitatively indicated using a ferric chloride assay (Sadgrove and Jones, 2013). Employing the Folin-Denis method (a spectrophotometric-based assay), total tannin content of the water and methanolic extracts of *P. dasycaulon* (belonging to Pittosporaceae) was determined at 215 and 100 mg of tannic acid equivalents per g extract, respectively (Mani and Dennis, 2014). In contrast, Amoo *et al.* (2012) reported the absence of condensed tannins in both the leaves and twigs of *P. viridiflorum* collected from KwaZulu-Natal, South Africa. The presence or absence of tannins is reported differently in the literature, mostly due to cultivar effects. Recently, (condensed) tannin levels in *P. angustifolium* whole branch, leaves and stem collected in Queensland and South Australia were reported in the range of 15.9–52 mg catechin equivalents/100 g DW, using a vanillin-HCl assay, but no further identification and quantification of individual tannin compounds was conducted (Phan *et al.*, 2020). The tannin content of *P. angistifolium* leaves could be comparable with other common leafy vegetables like Chinese cabbage (57.6 mg/100 g DW) and nightshade leaves (55.6 mg/100 g DW) (Managa *et al.*, 2020).

Saponins

There is a growing interest in the saponins of *P. angustifolium*. Several studies have been conducted to characterize individual saponin compounds in this plant. However, information on identification and structural elucidation of saponins in *P. angustifolium* is still limited due to the challenges in extracting and isolating the corresponding compounds, and the lack of available commercial standards. Bäcker *et al.* (2013) conducted ^{1}H NMR and high-resolution electrospray ionization mass spectrometry to structurally identify a total of 10 triterpenoid saponin compounds in the leaf extract of *P. angustifolium*. In a follow-up study by the same authors, two new taraxastane-type triterpenoid saponins, namely pittangretosides L and C_1, could be identified in the leaf extract of *P. angustifolium* (Bäcker *et al.*, 2015). Interestingly, these two compounds did not show any cytotoxic activity *in vitro* (up to a concentration of 130 μM) (Bäcker *et al.*, 2015). In another study, the following seven triterpenoid sapogenins were identified in the leaves of *P. phillyraeoides* using ^{1}H NMR spectroscopy: 27-desoxyphillyrigenin (3*β*-hydroxytaraxastan-28,20*β*-olide); 23-hydroxyphillyrigenin (3*β*,23,27-trihydroxytaraxastan-28,20*β*-olide); dihydropriverogenin A; 16-desoxybarringtogenol C; barringtogenol; phillyrigenin; and R_1-barrigenol (Errington and Jefferies, 1988).

Phan *et al.* (2020) assessed the total saponin levels in different parts of *P. angustifolium* by employing a vanillin-H_2SO_4 assay, showing that *P. angustifolium* leaves (3250–3871 mg oleanolic equivalents [OE]/100 g DW) and stem (1590 mg OE/100 g DW) are saponin-rich sources. The reported saponin content in *P. angustifolium* leaf is approximately ten times higher than other common saponin-rich foods such as legume seeds, including chickpeas (230 mg/100 g DW), kidney beans (410 mg/100 g DW) and soybeans (650 mg/100 g DW) (Savage, 2003). Studies on isolation and structural elucidation of saponin compounds in other *Pittosporum* species belonging to the Pittosporaceae family have been reported. For instance, triterpenoid saponins and triterpenoid estersaponins present in the leaves of *P. viridiflorum* (David *et al.*, 2002; Seo *et al.*, 2002) and *P. verticillatum* (Manase *et al.*, 2013), and both in the leaves and fruit of *P. tobira* (D'Acquarica *et al.*, 2002; El Dib *et al.*, 2015).

Bioactive Properties and Potential Health Benefits

Antimicrobial activity

Table 2.4 summarizes *in vitro* studies that demonstrated the antimicrobial activity of *P. angustifolium* and other species belonging to Pittosporaceae family against a wide spectrum of infective microorganisms. Extracts from different parts of the plant, including leaves, stem, whole branch, fruits and wood have shown inhibitory effects to the growth of food-related microorganisms (Vesoul and Cock, 2011; Phan

Table 2.4. Antimicrobial activities of the extracts from different parts of *P. angustifolium*.

Species	Botanical tissues	Extraction	Methodology	Microorganisms	References
P. angustifolium	Whole branch, leaves, and stem	MeOH Water	Disc infusion assay	*C. albicans* *S. aureus*	Phan *et al.* (2020)
P. angustifolium	Fruits and leaves	Essential oils	*In vitro* minimum inhibit concentration (MIC) based on spectrophotometric method	*S. aureus* *S. epidermidis* *C. albicans*	Sadgrove and Jones (2013)
P. phylliraeoides (former of *P. angustifolium*)	Whole branch of plant material	MeOH Water Ethyl acetate Chloroform Hexane	Disc infusion assay MIC	*A. hydrophilia* *E. coli* *K. pneumonia* *P. mirabilis* *S. marcenscens* *S. sonnei* *S. aureus* *S. pyogenes* *A. niger*	Vesoul and Cock (2011)
P. angustifolium	Leaves and fruit	MeOH Water	MIC	*Proteus mirabilis*	Cock *et al.* (2015)
P. phylliraeoides	Fruit, wood, leaves	Ethanol	Ross River virus and polio assay	Ross River virus	Semple *et al.* (1998)

Note: MeOH – Methanol; MIC – minimum inhibition concentration.

et al., 2020), anti-proteus activity against *Proteus mirabilis* for treatment and prevention of rheumatoid arthritis (Cock *et al.*, 2015) and antiviral activity against Ross River virus (Semple *et al.*, 1998). These findings clearly indicate that the traditional medicinal knowledge of the Australian Indigenous people, including the antimicrobial properties of *P. angustifolium*, supports the scientific literature.

Antioxidant capacity

Radical scavenging activity is normally expressed as the percentage inhibition of DPPH (2,2-di phenyl-1-picrylhydrazyl), a stable free radical molecule, with a higher percentage of inhibition indicating a greater potential antioxidant capacity. Antioxidant capacity of the methanolic extract of *P. angustifolium* leaves, stem and whole branch have been assessed using EC_{50} value (i.e. the amount/concentration of the plant extract necessary to scavenge/reduce the concentration of initial DPPH radical by 50%; Phan *et al.*, 2020). The authors also reported that the methanolic extract of 0.5–1 g *P. angustifolium* leaf powder or 2 g stem powder could exhibit a similar DPPH scavenging activity as that of 0.1 mg ascorbic acid (99% purity). The antioxidant properties of *P. angustifolium* leaf and stem extracts could be related to the phytochemicals determined in this plant, discussed earlier.

Cytotoxicity

Traditional medicinal applications of the plants belonging to the genus *Pittosporum* have been reported for a long time. The cytotoxic effects of saponins extracted from *P. angustifolium* and other species have been reported using cell culture and animal studies. Isolated acylated triterpene saponins from *P. angustifolium* could inhibit the function of the human DNA-topoisomerase I (an essential enzyme in DNA metabolism) (Bäcker *et al.*, 2016). It has been reported that DNA-topoisomerases help to correct problems associated with DNA replication, transcription, recombination and chromatin remodelling – hence, targeting them could be an effective anti-cancer strategy. In fact, triterpene saponins from the ethanolic extracts of *P. angustifolium* seeds have shown antitumorigenic activity through exhibiting cytotoxicity against three different tumorigenic cell lines (MCF7-human breast cancer, 5637-human urinary bladder carcinoma and LN18-human glioblastoma). It is noteworthy that the extract also showed cytotoxicity to a non-tumorigenic cell line (HaCaT-human keratinocyte) (Bäcker *et al.*, 2014b).

For other species belonging to the Pittosporaceae family, the guaiane-type sesquiterpene glycoside extracted from the fruits of *P. undualatum* did not exhibit antiproliferative activity against two human cancer cell lines (lung large cell carcinoma COR-L23 and amelanotic melanoma C32 cells), despite clearly showing *in vitro* anti-inflammatory effects (Mendes *et al.*, 2013). Recent studies using animal models have shown the bioactivities of extracts of *P. mannii* Hook, including antispasmodic and spasmolytic effects (Joseph *et al.*, 2015). In addition, antinociceptive effects of the aqueous and methanolic extracts from the leaves of this plant were observed in a mouse model (Wandji *et al.*, 2016).

Conclusions and Future Trends

Generally, different parts of *P. angustifolium* contain a diverse range of phytochemicals, mainly polyphenols, saponins, essential oils and possibly tannins. Geographical locations, growing conditions and different botanical tissues are the main factors causing variations in the phytochemical composition of this plant and subsequently its bioactivity. Triterpene glycosides could be identified as the main saponin compounds, whereas flavonoids and phenolic acids are predominant polyphenols in *P. angustifolium*. Reported data on proximate, minerals and trace elements as well as vitamins indicated that this plant is a rich source of dietary fibre and contains a moderate amount of water-soluble vitamins (B and C).

Studies on the potential bioactivity of *P. angustifolium* using the extracts and essential oils of this plant have shown inhibitory effects against a wide spectrum of microorganisms including Gram-negative and Gram-positive bacteria, fungal species and Ross River virus. Isolated triterpene saponins extracted from the seeds and leaves of this plant have shown cytotoxic effects against important types of tumorigenic cell lines, suggesting potential applications of *P. angustifolium* as a functional (food) ingredient, natural antimicrobial and/or supplement.

However, there are still gaps in knowledge in the current literature in fully understanding the effects of *P. angustifolium* on human health. Further studies investigating the bioaccessibility/bioavailability of bioactive compounds in *P. angustifolim* leaves and stem using *in vitro* digestive model and well designed human studies are warranted to substantiate potential health benefits, including any medicinal and therapeutic properties of *P. angustifolium*.

Acknowledgements

The authors acknowledge the Traditional Owners of the lands on which the *P. angustifolium* grows and respect the knowledge and experience the Traditional Owners hold regarding the care, harvest and use of these plants.

The authors acknowledge the Commonwealth Department of Industry and Science via the Innovation Connection Grant Scheme, Mr Ken Murray (Gumby Gumby Australia) – our industry partner and The Australian Research Council (ARC) Industrial Transformation Training Centre for Uniquely Australian Foods (Grant number: IC180100045) for project funding.

References

Ahmad, R.S., Butt, M.S., Huma, N., Sultan, M.T., Arshad, M.U. *et al.* (2014) Quantitative and Qualitative Portrait of Green Tea Catechins (Gtc) Through Hplc. *International Journal of Food Properties* 17(7), 1626–1636. DOI: 10.1080/10942912.2012.723232.

AL-Hashimi, A.G. and Ali, S. (2016) The nutritional value and antioxidant activity of bay leaves (*Laurusnobilis* L.). *Basrah Journal of Veterinary Research* 15(2), 246–259. DOI: 10.33762/bvetr.2016.124310.

Amoo, S.O., Aremu, A.O., Moyo, M. and Van Staden, J. (2012) Antioxidant and acetylcholinesterase-inhibitory properties of long-term stored medicinal plants. *BMC Complementary and Alternative Medicine* 12, 87. DOI: 10.1186/1472-6882-12-87.

Australian Government (2014) Australia New Zealand Food Standards Code - Standard 1.2.8 - Nutrition Information Requirements. Available at: https://www.legislation.gov.au/Details/F2015C00968 (accessed 22 March 2022).

Bäcker, C., Jenett-Siems, K., Siems, K., Wurster, M., Bodtke, A. *et al.* (2013) Triterpene Glycosides from the Leaves of *Pittosporum angustifolium*. *Planta Medica* 79(15), 1461–1469. DOI: 10.1055/s-0033-1350806.

Bäcker, C., Jenett-Siems, K., Bodtke, A. and Lindequist, U. (2014a) Polyphenolic compounds from the leaves of *Pittosporum angustifolium*. *Biochemical Systematics and Ecology* 55, 101–103. DOI: 10.1016/j.bse.2014.02.015.

Bäcker, C., Jenett-Siems, K., Siems, K., Wurster, M., Bodtke, A. *et al.* (2014b) New Mono- and bisdesmosidic briterpene glycosides from *Pittosporum angustifolium* lodd. *Zeitschrift Für Naturforschung B* 69(9–10), 1026–1044. DOI: 10.5560/znb.2014-4143.

Bäcker, C., Jenett-Siems, K., Siems, K., Niedermeyer, T.H.J., Wurster, M. *et al.* (2015) Taraxastane-type triterpene saponins isolated from *Pittosporum angustifolium* Lodd. *Zeitschrift Für Naturforschung B* 70(6), 403–408. DOI: 10.1515/znb-2015-0005.

Bäcker, C., Drwal, M.N., Preissner, R. and Lindequist, U. (2016) Inhibition of DNA-Topoisomerase I by Acylated Triterpene Saponins from *Pittosporum angustifolium* Lodd. *Natural Products and Bioprospecting* 6(2), 141–147. DOI: 10.1007/s13659-016-0087-5.

Cayzer, L.W., Crisp, M.D. and Telford, I.R.H. (2000) Revision of *Pittosporum* (Pittosporaceae) in Australia. *Australian Systematic Botany* 13(6), 845. DOI: 10.1071/SB99021.

Cock, I.E., Winnett, V., Sirdaarta, J. and Matthews, B. (2015) The potential of selected Australian medicinal plants with anti-Proteus activity for the treatment and prevention of rheumatoid arthritis. *Pharmacognosy Magazine* 11(Suppl 1), S190-208. DOI: 10.4103/0973-1296.157734.

Cooper, R.C. (1956) The Australian and New Zealand Species of *Pittosporum*. *Annals of the Missouri Botanical Garden* 43(2), 87. DOI: 10.2307/2394673.

D'Acquarica, I., Di Giovanni, M.C., Gasparrini, F., Misiti, D., D'Arrigo, C. *et al.* (2002) Isolation and structure elucidation of four new triterpenoid estersaponins from fruits of *Pittosporum tobiraait*. *Tetrahedron* 58(51), 10127–10136. DOI: 10.1016/S0040-4020(02)01364-9.

D'Antuono, I., Carola, A., Sena, L.M., Linsalata, V., Cardinali, A. *et al.* (2018) Artichoke polyphenols produce skin anti-age effects by improving endothelial cell integrity and functionality. *Molecules* 23(11), 2729. DOI: 10.3390/molecules23112729.

David, G., Steven, W., Kim, M. and John, M. (2002) A new triterpene saponin from *Pittosporum viridiflorum* from the Madagascar rainforest. *Journal of Natural Products* 65(1), 65–68. DOI: 10.1021/np010327t.

El Dib, R.A., Eskander, J., Mohamed, M.A. and Mohammed, N.M. (2015) Two new triterpenoid estersaponins and biological activities of *Pittosporum tobira* "Variegata" (Thunb.) W. T. Aiton leaves. *Fitoterapia* 106, 272–279. DOI: 10.1016/j.fitote.2015.09.006.

Errington, S.G. and Jefferies, P.R. (1988) Triterpenoid sapogenins of *Pittosporum phillyraeoides*. *Phytochemistry* 27(2), 543–545. DOI: 10.1016/0031-9422(88)83138-8.

FSANZ (n.d.) Food Standards Australia and New Zealand. Online. Available at: https://www.foodstandards.gov.au/Pages/default.aspx (accessed 7 August 2020).

Jenkins, K.J. and Atwal, A.S. (1994) Effects of dietary saponins on fecal bile acids and neutral sterols, and availability of vitamins A and E in the chick. *The Journal of Nutritional Biochemistry* 5(3), 134–137. DOI: 10.1016/0955-2863(94)90084-1.

Joseph, N., Tom Esther, N.L., Télesphore Benoît, N., Paul Désiré, D.D., Oumarou Bibi-Farouck, A. *et al.* (2015) Effects of the aqueous extract of *Pittosporum mannii* Hook. f. (Pittosporaceae) stem barks on spontaneous and spasmogen-induced contractile activity of isolated rat duodenum. *Journal of Ethnopharmacology* 172, 1–9. DOI: 10.1016/j.jep.2015.05.047.

Khosravi-Largani, M., Pourvali-Talatappeh, P., Rousta, A.M., Karimi-Kivi, M., Noroozi, E. *et al.* (2018) A review on potential roles of vitamins in incidence, progression, and improvement of multiple sclerosis. *ENeurologicalSci* 10, 37–44. DOI: 10.1016/j.ensci.2018.01.007.

Lassack, E.V. and Mccarthy, T. (1983) *Australian Medicinal Plants*. Methuen Australia, Sydney.

Latz, P.K. (1995) *Bushfires and Bushtucker: Aboriginal Plant Use in Central Australia*. IAD Press, Alice Springs, Australia.

Madikizela, B. and McGaw, L.J. (2017) *Pittosporum viridiflorum* Sims (Pittosporaceae): A review on A useful medicinal plant native to South Africa and tropical Africa. *Journal of Ethnopharmacology* 205, 217–230. DOI: 10.1016/j.jep.2017.05.005.

Managa, M.G., Shai, J., Thi Phan, A.D., Sultanbawa, Y. and Sivakumar, D. (2020) Impact of Household Cooking Techniques on African Nightshade and Chinese Cabbage on Phenolic Compounds, Antinutrients, in vitro Antioxidant, and β-Glucosidase Activity. *Frontiers in Nutrition* 7. DOI: 10.3389/fnut.2020.580550.

Manase, M.J., Mitaine-Offer, A.-C., Miyamoto, T., Tanaka, C., Delemasure, S. *et al.* (2013) New triterpenoid estersaponins from the root barks of *Pittosporum verticillatum* subsp. verticillatum and evaluation of cytotoxicities. *Fitoterapia* 91, 231–235. DOI: 10.1016/j.fitote.2013.09.002.

Mani, B. and Dennis, T. (2014) Evaluation of the antioxidant potential of *Pittosporum dasycaulon* Miq. stem bark. *Food Science and Biotechnology* 23(2), 539–545. DOI: 10.1007/s10068-014-0074-8.

Mendes, S.A.C., Mansoor, T.A., Rodrigues, A., Armas, J.B. and Ferreira, M.-J.U. (2013) Anti-inflammatory guaiane-type sesquiterpenes from the fruits of *Pittosporum undulatum*. *Phytochemistry* 95, 308–314. DOI: 10.1016/j.phytochem.2013.06.019.

Moloto, M.R., Phan, A.D.T., Shai, J.L., Sultanbawa, Y. and Sivakumar, D. (2020) Comparison of phenolic compounds, carotenoids, amino acid composition, in vitro antioxidant and anti-diabetic activities in the leaves of seven cowpea (*Vigna unguiculata*) cultivars. *Foods* 9(9), 1285. DOI: 10.3390/foods9091285.

Negrelle, R.R.B., Mielke, E.C., Cuquel, F.L. and Pulido, E.E. (2018) *Pittosporum undulatum* Vent.: subsidies to the control and management. *Ornamental Horticulture* 24(4), 295–302. DOI: 10.14295/oh.v24i4.1457.

Olaokun, O.O., Mkolo, N.M., Mogale, M.A. and King, P.H. (2017) Phytochemical Screening, Antioxidant, Anti-inflammatory, and Glucose Utilization Activities of Three South African Plants Used Traditionally to Treat Diseases. *Biology and Medicine* 09(05), 2. DOI: 10.4172/0974-8369.1000412.

Otten, J., Hellwig, J. and Meyers, L. (2006) *Dietary Reference Intakes: The Essential Guide to Nutrient Requirements*. The National Academies Press, Washington, DC.

Phan, A.D.T., Chaliha, M., Hong, H.T., Tinggi, U., Netzel, M.E. *et al.* (2020) Nutritional Value and Antimicrobial Activity of *Pittosporum angustifolium* (Gumby Gumby), an Australian Indigenous Plant. *Foods* 9(7), 887. DOI: 10.3390/foods9070887.

Raes, K., Knockaert, D., Struijs, K. and Van Camp, J. (2014) Role of processing on bioaccessibility of minerals: Influence of localization of minerals and anti-nutritional factors in the plant. *Trends in Food Science & Technology* 37(1), 32–41. DOI: 10.1016/j.tifs.2014.02.002.

Rjeibi, I., Ncib, S., Ben Saad, A. and Souid, S. (2017) Evaluation of nutritional values, phenolic profile, aroma compounds and biological properties of *Pittosporum tobira* seeds. *Lipids in Health and Disease* 16(1), 206. DOI: 10.1186/s12944-017-0596-1.

Sadgrove, N.J. and Jones, G.L. (2013) Chemical and biological characterisation of solvent extracts and essential oils from leaves and fruit of two Australian species of *Pittosporum* (Pittosporaceae) used in aboriginal medicinal practice. *Journal of Ethnopharmacology* 145(3), 813–821. DOI: 10.1016/j.jep.2012.12.019.

Sadgrove, N.J. and Jones, G.L. (2016) Phytochemical variability of *Pittosporum angustifolium* Lodd. (*Pittosporaceae*): a traditional and contemporary Aboriginal Australian medicine. *Acta Horticulturae* (1125), 303–308. DOI: 10.17660/ActaHortic.2016.1125.39.

Saini, R.K., Nile, S.H. and Park, S.W. (2015) Carotenoids from fruits and vegetables: Chemistry, analysis, occurrence, bioavailability and biological activities. *Food Research International* 76(Pt 3), 735–750. DOI: 10.1016/j.foodres.2015.07.047.

Samtiya, M., Aluko, R.E. and Dhewa, T. (2020) Plant food anti-nutritional factors and their reduction strategies: an overview. *Food Production, Processing and Nutrition* 2(1), 6. DOI: 10.1186/s43014-020-0020-5.

Savage, G.P. (2003) Saponins. In: Caballero, B. (ed.) *Encyclopedia of Food Sciences and Nutrition*, 2nd edn. Academic Press, Oxford, UK.

Semple, S.J., Reynolds, G.D., O'Leary, M.C. and Flower, R.L. (1998) Screening of Australian medicinal plants for antiviral activity. *Journal of Ethnopharmacology* 60(2), 163–172. DOI: 10.1016/s0378-8741(97)00152-9.

Seo, Y., Berger, J.M., Hoch, J., Neddermann, K.M., Bursuker, I. *et al.* (2002) A new triterpene saponin from *Pittosporum viridiflorum* from the Madagascar rainforest. *Journal of Natural Products* 65(1), 65–68. DOI: 10.1021/np010327t.

Shieh, P., Govindasamy, A. and Kuo, D. (2009) Evaluation of antioxidant polyphenols in Taiwan's medicinal plants. *Asian Journal of Chemistry* 21, 5556–5562.

Shohag, M.J.I., Wei, Y., Yu, N., Zhang, J., Wang, K. *et al.* (2011) Natural variation of folate content and composition in Spinach (*Spinacia oleracea*) germplasm. *Journal of Agricultural and Food Chemistry* 59(23), 12520–12526. DOI: 10.1021/jf203442h.

Tabassum, N., Lee, J.-H., Yim, S.-H., Batkhuu, G.J., Jung, D.-W, *et al.* (2016) Isolation of 4, 5-O-dicaffeoylquinic acid as a pigmentation inhibitor occurring in *Artemisia capillaris* Thunberg and its validation in vivo. *Evidence-Based Complementary and Alternative Medicine* 2016, 7823541. DOI: 10.1155/2016/7823541.

Tajik, N., Tajik, M., Mack, I. and Enck, P. (2017) The potential effects of chlorogenic acid, the main phenolic components in coffee, on health: a comprehensive review of the literature. *European Journal of Nutrition* 56(7), 2215–2244. DOI: 10.1007/s00394-017-1379-1.

United States Department of Agriculture USDA (2019) USDA Food Composition Databases. Available at: https://wicworks.fns.usda.gov/resources/usda-food-composition-databases (accessed 22 March 2022).

Valentová, K., Vrba, J., Bancířová, M., Ulrichová, J. and Křen, V. (2014) Isoquercitrin: pharmacology, toxicology, and metabolism. *Food and Chemical Toxicology* 68, 267–282. DOI: 10.1016/j.fct.2014.03.018.

Vesoul, J. and Cock, I. (2011) An examination of the medicinal potential of *Pittosporum phylliraeoides*: Toxicity, antibacterial and antifungal activities. *Pharmacognosy Communications* 1(2), 8–17. DOI: 10.5530/pc.2011.2.3.

Wandji, B.A., Tatsinkou Bomba, F.D., Awouafack, M.D., Nkeng-Efouet, P.A., Kamanyi, A. *et al.* (2016) Antinociceptive effects of the aqueous and methanol extracts of the leaves of *Pittosporum mannii* Hook. F. (Pittosporaceae) in mice. *Journal of Ethnopharmacology* 187, 224–231. DOI: 10.1016/j.jep.2016.04.041.

Waseem, M., Akhtar, S., Manzoor, M.F., Mirani, A.A., Ali, Z. *et al.* (2021) Nutritional characterization and food value addition properties of dehydrated spinach powder. *Food Science & Nutrition* 9(2), 1213–1221. DOI: 10.1002/fsn3.2110.

Zhang, Y., Zhou, W.-E., Yan, J.-Q., Liu, M., Zhou, Y. *et al.* (2018) A review of the extraction and determination methods of thirteen sssential vitamins to the human body: an update from 2010. *Molecules* 23(6), 1484. DOI: 10.3390/molecules23061484.

3 Spider Plant (*Cleome gynandra*)

Eugénie Kayitesi* and Siphosanele Mafa Moyo
Department of Consumer and Food Sciences, University of Pretoria, Hatfield, Pretoria, South Africa

Abstract

Spider plant (*Cleome gynandra* L.) is an annual plant that grows as a weed or a volunteer crop and is used as a vegetable or traditionally as medicine. The nutritional value of raw spider plant consists of an average of 84.94% moisture, 5.88% ash, 7.58% fibre, 2.56% fat, 7.33% protein and 37.06% carbohydrates, and contributes to about 152.88 kcal of energy per 100 g portion. It contains essential amino acids such as valine, histidine, threonine, leucine, isoleucine, lysine, methionine, phenylalanine and tryptophan. Spider plant is a rich source of minerals such as phosphorus, potassium, sulfur, calcium, magnesium, iron, manganese, copper and zinc. It is also a good source of vitamins such as the provitamin A carotenoids, vitamin E, vitamin C and some B vitamins. In addition to its nutritional value, spider plant consists of phytochemicals such alkaloids, steroids, flavonoids, saponins, phenolic acids, tannins and cardiac glycosides known to exert health-promoting properties. The nutritional value and phytochemical profile of the vegetable suggests it can be used to reduce nutritional deficiencies, contribute towards daily dietary nutrient requirements, and provide health benefits such as antidiabetic, anticancer, antiarthritic and hepatoprotective properties. These health benefits are attributable to its bioactive mechanisms such as antioxidant and anti-inflammatory properties. Furthermore, the plant's antimicrobial properties point to its use as an antimicrobial preservative ingredient in food processing or as an antimicrobial agent in drug use.

Introduction

Cleome gynandra L. (Fig. 3.1) is a plant native to Asia and sub-Saharan Africa (Wu *et al.*, 2018) and is present in tropical and subtropical areas all over the world (Adhikari and Paul, 2018). It is usually referred to as cats' whiskers, African spider flower or spider plant (Chweya and Mnzava, 1997). Synonyms include *Gynandropsis gynandra* (L.) Briq., *Gymnogonia pentaphylla* (L.) R. Br. ex Steud., *Gynandropsis pentaphylla* Blanco., *Cleome pentaphylla* L., *Cleome pentaphylla* var. glabra Kuntze., *Pedicellaria pentaphylla* (L.) Schrank., *Pedicellaria pentaphylla* (L.) Schrank., *Pedicellaria triphylla* (L.) Pax., *Gynandropsis pentaphylla* (L.) DC., *Podogyne pentaphylla* (L.) Hoffmanns., *Gynandropsis heterotricha* DC., *Gynandropsis glandulosa* C. Presl. and *Cleome acuta* Schumach (Adhikari and Paul, 2018). The plant is not formally cultivated but semi-domesticated. It grows as a weed or a volunteer crop on disturbed soils and fertile land near homesteads and home gardens (Nemahunguni *et al.*, 2020). It is mainly used as a medicine or food in most communities (Onyango *et al.*, 2013). Spider plant is specially known for its C_4 photosynthetic pathway which is an adaptational mechanism to thrive in semi-arid zones, and so it is classified as a drought tolerant but not a drought resistant plant (Chweya and Mnzava, 1997). The biodiversity conservation of the plant and its contribution to food

*Corresponding author: eugenie.kayitesi@up.ac.za

DOI: 10.1079/9781789248067.0003

Fig. 3.1. Spider plant (*Cleome gynandra* L.).

security is related to its economic potential, its high nutritional value and phytochemical content that is linked to its antioxidant, antimicrobial, anti-inflammatory and antimalarial properties (Onyango *et al.*, 2013). However, the nutritional value and phytochemical profile may vary according to production techniques used, environmental growing conditions, soil fertility and plant phase of growth at harvest (Chweya and Mnzava, 1997).

As it is not formally cultivated, the production of the vegetable mainly depends on the rain, and manure is used as fertilizer. Furthermore, spider plant has a short growth period and is usually harvested within three to five weeks of germination (Oshingi *et al.*, 2019). During harvesting, an entire plant can be collected or individual leaves with stems are selected daily or every third day (Directorate of Plant Production, 2010; Oelofse and Averbeke, 2012). Like most green leafy vegetables, the main constraint to large-scale production, marketing and consumption of the plant is high perishability in the fresh form (Maseko *et al.*, 2017). Gao *et al.* (2011) isolated polyphenol oxidase from the leaves of spider plant to enhance understanding of the enzyme that will provide useful information to prevent enzymatic browning of the leaves during processing and storage. Leaves and young shoots are cooked as a leafy vegetable or sun dried for later consumption (Maroyi, 2013). Usually, the leaves are boiled for about one to two hours with the option to add groundnut butter, salt, onion or tomatoes. The prepared vegetable may be consumed as an accompaniment to maize meal porridge (Fig. 3.2) or sorghum porridge. Dried leaves are generally stored for three to four months, or up to a year (Oelofse and Averbeke, 2012). Some studies also show the use of spider plant as a fermented and canned product (Ahmad, 2012; Onyeoziri *et al.*, 2017; Muhialdin *et al.*, 2018).

Fig. 3.2. Dried and cooked spider plant (*Cleome gynandra* L.) served as an accompaniment to maize meal porridge and beef stew.

Botanical Characteristics

Spider plant is indexed under the genus *Cleome* that exists within the Cleomaceae family which was previously included in the family of Capparaceae (Uzilday *et al.*, 2012; El zayat *et al.*, 2020). The plant grows to about 1.5 m in height as an annual upright or semi-upright herb that can develop stout and branched stems as well as a long tap root consisting of a few secondary roots (Chweya and Mnzava, 1997). The stems can be of green or violet to purple pigmentation and of great or weak glandular pubescence (Anbazhagi *et al.*, 2009; Kiebre *et al.*, 2015; Omondi *et al.*, 2017). The leaves of spider plant are oppositely arranged on the stem and the attached petioles are 3–23 cm long (Anbazhagi *et al.*, 2009). The leaves are palmate compound with an average of five leaflets that have a shape that varies from ovate-elliptic to lanceolate-oblanceolate and displaying a base that is cuneate and apex acute to obtuse (Kwarteng *et al.*, 2018). Leaves can have clear-green or dark-green colour, finely toothed margins or round ends with a sparsely but distinctly hairy surface, surface area about 2–10 cm long and 2–4 cm wide (Chweya and Mnzava, 1997; Kiebre *et al.*, 2015). Both the stem and leaf petioles are thickly glandular and infrequently glabrous (Chweya and Mnzava, 1997).

Spider plant flowers can be unisexual (staminate) or bisexual (hermaphroditic). The hermaphrodite type consists of a stamen and carpel within the same flower, hence the synonym *Gynandropsis* taken from gynophore (female) and androphore (male) (Imanirampa and Alele, 2016). The staminate type consists of a reduced, non-functional or even absent female part of the flower (Sogbohossou *et al.*, 2018). The flower consists of four narrow petals which are either white, pale pink or lilac, four sepals and six stamens that have long purple filaments (Chweya and Mnzava, 1997; Bayat *et al.*, 2018). The whole inflorescence is about 30 cm in length while the flower is about 1–2.5 cm in diameter (Chweya and Mnzava, 1997). The fruit shape can be thin and rough or thin and smooth, measuring up to 12 cm long and 8–10 mm wide (Chweya and Mnzava, 1997; Kiebre *et al.*, 2015). The fruit is dry dehiscent and composed of two carpels; it is green and turns yellow when ripe to release many seeds which are subglobose, grey to black, 1.0–1.5 mm in diameter and irregularly ribbed (Chweya and Mnzava, 1997; Anbazhagi *et al.*, 2009). The seed optimally germinates at 31°C and at a low of 12°C and a high of 36°C (Oelofse and Averbeke, 2012).

Traditional Use

Spider plant is mainly consumed as a cooked vegetable and used in traditional folk medicine to treat certain ailments. Traditionally, the plant can be used as a disinfectant and in the treatment of microbial diseases by applying the leaf paste externally twice a day on wounds to avoid sepsis (Bala *et al.*, 2012; Shanmugam *et al.*, 2012; Lokesha, 2018). The decoction of leaves, whole plant and hypogeaus organs has been used to alleviate fever, headaches, body pain and malaria (Jeruto *et al.*, 2008; Urso *et al.*, 2016; Shah and Rahim, 2017). Furthermore, the plant can be used to facilitate childbirth as it has the ability to contract the uterus when taken as a decoction (Kamatenesi-Mugisha and Oryem-Origa, 2007; Jeruto *et al.*, 2008; Lokesha, 2018). It is also used to stimulate milk production in lactating mothers and help them in regaining their health (Kwarteng *et al.*, 2018). The plant is used in the treatment of earache, inflammation, bronchitis, constipation, stomach ache, pneumonia, rheumatism, uterine complaints as well as in the treatment of snake-bites (Jayanthi *et al.*, 2012; Lokesha, 2018).

Nutritional Composition

The proximate composition of spider plant is summarized in Table 3.1. Fresh leaves of the spider plant constitute an average 84.94% moisture. Moisture content may vary depending on the maturity of the plant, time of harvest and whether leaves and petioles were combined in analysis or only the leaves were used (Schönfeldt and Pretorius, 2011). The protein content varies between 3.9–16.04% dry matter (DM). The variation may be influenced by the presence of a petiole which forms part of the edible portion of spider plant. Furthermore, the nitrogen levels in

Table 3.1. Proximate composition of raw spider plant.

		%		
		Average	Range	Reference
Moisture	FW	84.94	79.3–88.8	(Hassan *et al*., 2005; Schönfeldt and Pretorius, 2011; van Jaarsveld *et al*., 2014; Koua *et al*., 2015; Matenge *et al*., 2017; Harris *et al*., 2017; Ochieng *et al*., 2018)
Ash	DM	5.88	1.46–18.77	(Hassan *et al*., 2005; Schönfeldt and Pretorius, 2011; van Jaarsveld *et al*., 2014; Koua *et al*., 2015; Harris *et al*., 2017; Matenge *et al*., 2017)
	FW	2.82		(Ochieng *et al*., 2018)
Crude fibre	DM	7.58	1.8–22.10	(Hassan *et al*., 2005; van Jaarsveld *et al*., 2014; Koua *et al*., 2015; Matenge *et al*., 2017; Harris *et al*., 2017)
	FW	1.66		(Ochieng *et al*., 2018)
Lipids	DM	2.56	0.3–6.50	(Hassan *et al*., 2005; Schönfeldt and Pretorius, 2011; van Jaarsveld *et al*., 2014; Koua *et al*., 2015; Matenge *et al*., 2017)
	FW	2.43		(Ochieng *et al*., 2018)
Proteins	DM	7.33	3.9–16.04	(Hassan *et al*., 2005; Schönfeldt and Pretorius, 2011; van Jaarsveld *et al*., 2014; Koua *et al*., 2015; Harris *et al*., 2017; Matenge *et al*., 2017)
	FW	5.24		(Ochieng *et al*., 2018)
Carbohydrates	DM	37.06	13.1–59.28	(Hassan *et al*., 2005; van Jaarsveld *et al*., 2014; Koua *et al*., 2015)
Energy (kcal/100 g)	DM	152.88	46.61–280.32	(Hassan *et al*., 2005; Schönfeldt and Pretorius, 2011; van Jaarsveld *et al*., 2014; Koua *et al*., 2015)

FW, fresh weight; DM, dry matter.

the soil may have an impact on the protein concentrations reported (Schönfeldt and Pretorius, 2011). According to van Jaarsveld *et al*. (2014), spider plant provides about 19 and 11% of the protein recommended dietary allowance (RDA) for children aged 4–8 years and women aged 19–30 year bs, respectively, which is higher than the percentage protein RDA contribution of other vegetables such as tsamma melon (*Citrullus lanatus* (Thunb.), Chinese cabbage (*Brassica rapa* L. subsp. *chinensis*), pigweed (*Amaranthus cruentus* L.), cowpea (*Vigna unguiculata* L. Walp.), Jew's mallow (*Corchorus olitorius* L.), pumpkin (*Cucurbita maxima* Duchesne) and black nightshade (*Solanum retroflexum* Dun.). The nutrient contribution was calculated from an average portion size of 90 g for children (4–8 years) and 130 g for women (19–30 years). The lipid content is low in the range of 0.3–6.50% DM. Other leafy vegetables such as swiss chard and spinach have a lipid content of 0.2 and 0.39% DM respectively (USDA, 2019). The ash content, which represents total mineral content of spider plant is in the range of 1.8–18.77% DM. Carbohydrates are present at about 13.1–59%. The crude fibre content of spider plant is in a range of 1.8–22%. Spider plant generally provides about 152.88 Kcal/100 g energy.

Table 3.2. Mineral content of spider plant.

		mg/100 g		
		Average	Range	Reference
Ca	DM	1533.0	425.2–2800.0	(Hassan *et al*., 2005; Koua *et al*., 2015; Omondi *et al*., 2017; Moyo *et al*., 2018)
	FW	197.6	126–260.1	(Jiménez-Aguilar and Grusak, 2015; Ochieng *et al*., 2018; Gowele *et al*., 2019)
Mg	DM	1166.7	208.7–4500.0	(Hassan *et al*., 2005; Koua *et al*., 2015; Omondi *et al*., 2017; Moyo *et al*., 2018)
	FW	96.68	67.63–148.0	(Jiménez-Aguilar and Grusak, 2015; Ochieng *et al*., 2018; Gowele *et al*., 2019)
P	DM	692.7	250.2–1100.0	(Hassan *et al*., 2005; Koua *et al*., 2015; Omondi *et al*., 2017; Moyo *et al*., 2018)
	FW	127	110–144	(Jiménez-Aguilar and Grusak, 2015)
K	DM	2033.0	114.4–3900	(Hassan *et al*., 2005; Koua *et al*., 2015; Omondi *et al*., 2017; Moyo *et al*., 2018)
	FW	593	457–729	(Jiménez-Aguilar and Grusak, 2015)
Na	DM	86.0	30.50–141.40	(Hassan *et al*., 2005; Koua *et al*., 2015)
	FW	12.5	5.1–22.7	(Jiménez-Aguilar and Grusak, 2015; Ochieng *et al*., 2018)
Fe	DM	161.81	27.0–598.2	(Hassan *et al*., 2005; Koua *et al*., 2015; Omondi *et al*., 2017; Moyo *et al*., 2018)
	FW	11.3	1.67–39.0	(Jiménez-Aguilar and Grusak, 2015; Ochieng *et al*., 2018; Gowele *et al*., 2019)
Zn	DM	16.22	4.4–42.30	(Hassan *et al*., 2005; Koua *et al*., 2015; Omondi *et al*., 2017; Moyo *et al*., 2018)
	FW	1.104	0.46–1.57	(Jiménez-Aguilar and Grusak, 2015; Gowele *et al*., 2019)
Cu	DM	1.76	0.8–3.45	(Hassan *et al*., 2005; Koua *et al*., 2015; Omondi *et al*., 2017; Moyo *et al*., 2018)
	FW	0.178	0.142–0.214	(Jiménez-Aguilar and Grusak, 2015)
Mn	DM	25.85	4.7–66.2	(Hassan *et al*., 2005; Omondi *et al*., 2017; Moyo *et al*., 2018)
	FW	0.514	0.402–0.626	(Jiménez-Aguilar and Grusak, 2015)

FW, fresh weight; DM, dry weight.

Spider plant is a rich source of macro minerals such as phosphorus, potassium, sulfur, calcium, magnesium, iron, manganese, copper and zinc as shown in Table 3.2. The most abundant minerals are calcium, magnesium and potassium at an average of 1533.0mg/100 g DM, 1166.7mg/100 g DM and 2033.0 mg/100 g DM, respectively. Spider plant contributes about 20 and 29% of calcium RDA for children aged 4–8 years and women aged 19–30 years, respectively. Whereas, for magnesium, it contributes 39% (children aged 4–8 years) and 23% (women aged 19–30 years) of the RDA (van Jaarsveld *et al*., 2014). The iron levels reported in spider plant (27.0–598.2 mg/100 g DM) are higher than those of spinach – *Spinacia oleracea* L. (4–35 mg/100 g) (Murcia *et al*., 2020). For iron, spider plant contributes about 14% (children aged 4–8 years) and 11% (women aged 19–30 years) of the RDA (van Jaarsveld *et al*., 2014). Table 3.3 summarizes the vitamins found in spider plant which include high concentrations of lipophilic vitamin A (434 μg

Table 3.3. Vitamin content of spider plant.

			mg/100 g		
			Average	Range	Reference
Vitamin A	RAE	FW	0.434	-	(van Jaarsveld *et al.*, 2014)
Provitamin A	α-carotene	FW	0.05	-	(Gowele *et al.*, 2019)
	β-carotene	FW	3.51	1.8–5.9	(Schönfeldt and Pretorius, 2011; van Jaarsveld *et al.*, 2014; Gowele *et al.*, 2019; Sogbohossou *et al.*, 2019)
		DM	4.59	0.4–8.7	(Mibei and Ojijo, 2011; Neugart *et al.*, 2017)
	β-cryptoxanthin	FW	0.13	-	(Gowele *et al.*, 2019)
	Total carotenoids	DM	104.67	77.2–162.3	(van der Walt *et al.*, 2009; Neugart *et al.*, 2017)
Vitamin E	α-tocopherol	FW	1.3	0.02–3.72	(Gowele *et al.*, 2019; Sogbohossou *et al.*, 2019)
	γ-tocopherol	FW	0.017	0.002–0.04	(Gowele *et al.*, 2019; Sogbohossou *et al.*, 2019)
Vitamin B	Thiamine	FW	0.06	-	(van Jaarsveld *et al.*, 2014)
	Riboflavin	FW	0.145	0.08–0.21	(Schönfeldt and Pretorius, 2011; van Jaarsveld *et al.*, 2014)
	Folate	FW	0.121	-	(van Jaarsveld *et al.*, 2014)
Vitamin C	Total ascorbic acid	FW	92.13	2–140	(van Jaarsveld *et al.*, 2014; Jiménez-Aguilar and Grusak, 2015; Koua *et al.*, 2015; Gowele *et al.*, 2019)
		DM	183.48	18–319	(Chipurura, 2010; Mibei and Ojijo, 2011; Harris *et al.*, 2017)
	Reduced ascorbic acid	FW	108.14	94–119	(Jiménez-Aguilar and Grusak, 2015)

FW, fresh weight; DM, dry matter; RAE, retinol activity equivalents.

retinol activity equivalents (RAE)/100 g FW) and provitamin A carotenoids such as β-carotene (1.18–5.9 mg/100 g FW and 5.7–8.7 mg/100 g DM), α-carotene (0.05 mg/100 g FW), β-cryptoxanthin (0.13 mg/100 g FW) (Mibei and Ojijo, 2011; Schönfeldt and Pretorius, 2011; van Jaarsveld *et al.*, 2014; Neugart *et al.*, 2017; Gowele *et al.*, 2019; Sogbohossou *et al.*, 2019). Spider plant provides 72% of vitamin A RDA for children between the ages of 4–8 years and 59% for women between the ages of 19–30 years (van Jaarsveld *et al.*, 2014). Lipophilic vitamin E tocopherols are present such as α-tocopherol (0.2–37.2 μg/g FW) and γ-tocopherols (0.02–0.4 μg/g FW) (Gowele *et al.*, 2019; Sogbohossou *et al.*, 2019). Water soluble vitamins such as thiamine, riboflavin, folate and ascorbic acid were also quantified.

Fatty acids present in spider plant leaves make up about 1% of the dry weight (Glew *et al.*, 2009). However, it has a high concentration of essential fatty acid α-linolenic acid at 58% but a low concentration of linoleic acid at 7.80% (Table 3.4). Saturated fatty acids were also quantified, with palmitic acid being present in the highest concentration among the fatty acids at 662.0 mg/100 g DM and 22.1%.

Table 3.5 summarizes amino acids found in spider plant. Essential amino acids such as valine, histidine, threonine, leucine,

Table 3.4. Fatty acid content of spider plant.

	(mg/100 g DM) (van der Walt *et al.*, 2008)	% (Glew *et al.*, 2009)
Myristic acid (14:0)	8.6 ± 0.3	1.03 ± 0.03
Palmitic acid (16:0)	662.0 ± 21.2	22.1 ± 0.4
Stearic acid (18:0)	127 ± 2.5	3.31 ± 0.21
Arachidic acid (20:0)	38.3 ± 3.6	-
Behenic acid (22:0)	35.6 ± 3.0	1.91 ± 0.01
Lignoceric acid (24:0)	38.0 ± 1.9	0.28 ± 0.49
Palmitoleic acid (16:1 n-7)	91.4 ± 0.5	0.53 ± 0.07
Linoleic acid (18:2 n-6)	385.7 ± 9.7	7.80 ± 0.22
Linolenic acid (18:2 n-3)	1554.3 ± 59.4	-
α-linolenic acid (18:3 n-3)	-	58.2 ± 1.0
Total fatty acids	2941.6	950 mg/100 g DM

DM, dry matter.

Table 3.5. Amino acid content of spider plant.

	mg/g DM (Glew *et al.*, 2009)	mg/100g FW (Ochieng *et al.*, 2018)
Alanine	20.9 ± 0.5	-
Arginine	14.7 ± 0.1	0.48 ± 0.001
Aspartate	28.4 ± 0.4	0.58 ± 0.001
Cysteine	4.04 ± 0.11	-
Glutamate	37.1 ± 0.8	4.56 ± 0.002
Glycine	14.7 ± 0.2	-
Histidine	5.19 ± 0.10	-
Isoleucine	13.4	0.54 ± 0.002
Leucine	22.3 ± 0.4	1.02 ± 0.002
Lysine	15.5 ± 0.20	0.28 ± 0.002
Methionine	1.27 ± 0.13	0.25 ± 0.001
Phenylalanine	15.7 ± 0.3	0.75 ± 0.002
Proline	22.7 ± 0.5	-
Serine	11.5 ± 0.1	-
Threonine	13.0 ± 0.1	0.40 ± 0.002
Tyrosine	6.42 ± 0.02	-
Valine	17.6 ± 0.3	0.06 ± 0.002
Tryptophan	-	0.08 ± 0.002

isoleucine, lysine, methionine, phenylalanine and tryptophan, as well as non-essential amino acids such as alanine, aspartic acid, cysteine, glutamate, glycine, arginine, proline, serine and tyrosine are present. Glutamate was the most abundant amino acid at 37.1 mg/g DM or 4.56 mg/100 g FW.

Phytochemical Quality

The qualitative phytochemical screening of spider plant shows the presence of alkaloids, steroids, flavonoids, anthraquinones, saponins, phenols, tannins and cardiac glycosides (Mibei and Ojijo, 2011; Sango *et al.*, 2016). Within the class of phytochemicals, phenolic compounds and glucosinolates have been quantified. In spider plant, the total phenolic content (TPC) varies from 188.2 to 1923.9 mg gallic acid equivalents (GAE)/100 g DM (van der Walt *et al.*, 2009; Moyo *et al.*, 2013, 2016, 2018; Konan *et al.*, 2014; Matenge *et al.*, 2017) while flavonoids are present in the range of 71.55–565 mg quercetin equivalents (CE)/100 g DM (Moyo *et al.*, 2013, 2016, 2018). The phenolic acid profile of spider plant shows the presence of chlorogenic acid, ferulic acid, protocatechuic acid, p-hydroxybenzoic acid, salicylic acid, caffeic acid, p-coumaric acid, sinapic acid, ferulic acid, vanillic acid, caffeic acid glucoside, coumaric acid glucoside, dimethoxycinnamoyl coumaric acid, dimethoxycinnamoyl glucuronide and esculetin glucoside (Matenge *et al.*, 2017; Moyo *et al.*, 2018; Chandradevan *et al.*, 2020). Flavonoids such as quercetin rutinoside, quercetin glucoside, kaempferol rutinoside, quercetin 7-rhamnoside, quercetin 3-rutinoside, quercetin 7-rutinoside, quercetin 3,7-dirhamnoside, quercetin, quercetin-3-glucoside, kaempferol-3-o-glucoside, kaempferol 7-rhamnoside, kaempferol 3-rutinoside, kaempferol and kaempferol-3-glucoside-7-rhamnoside are present (Kasem and Fatahy, 2016;

Chandradevan *et al.*, 2020). Glucosinolates such as aliphatic 3-hydroxypropyl glucosinolate, aliphatic 2-propenyl glucosinolate, indole 3-indolylmethyl glucosinolate and 4-hydroxy-3- indolylmethyl have been identified (Omondi *et al.*, 2017). Other compounds such as α-amyrin, α-amyrin acetate, stigmasetrol and sitosterol have also been isolated from spider plant (Ranjitha *et al.*, 2009). The phytochemical profile of spider plant may differ according to the environmental growth conditions and extraction solvents used in the quantification of these compounds.

Antinutritional Compounds

Despite the nutritional value of spider plant, a major limitation to bioaccessibility and bioavailability of the important nutrients such as minerals and protein may be because of the presence of antinutritional compounds. These include cyanogenic glycosides, oxalates, phytates, nitrates and tannins widely reported in leafy vegetables (Etong and Abbah, 2014; Natesh *et al.*, 2017). These compounds may inhibit the absorption of minerals such as copper, calcium, iron and zinc through the formation of insoluble complexes, particularly when present in high concentrations, and also reduce digestibility of protein (Udousoro and Akpan, 2014; Sango *et al.*, 2016). Oxalates particularly bind to calcium (Ca) to form complexes (calcium oxalate crystals) that prevent Ca from being absorbed by the body as well as cause irritation of the gut lining. Table 3.6 summarizes some of the antinutritive compounds found in raw spider plant. The quantity of oxalates ranges from 5.12 to 511.50mg/100g DM. According to Koua *et al.* (2015), the 511.50mg/100g oxalate quantified in spider plant may not impair the absorption of calcium as the ratio of oxalates to calcium (0.70) is below the critical level of 2.5 that is known to reduce calcium bioavailability (Koua *et al.*, 2015). Phytates were quantified at a range of 0.3 to 70.14mg/100g DM, with calculated ratio to calcium at 0.95, meaning the levels may not hinder calcium absorption (Koua *et al.*, 2015). However, the quantity of oxalates and phytates found in spider plant may affect the bioavailability of iron as the ratio calculated was above the critical level of 0.4 known to impair iron absorption (Koua *et al.*, 2015).

Nitrates are in the range 64.3 to 81.2mg/100g DM. The consumption of nitrates per serving of vegetables is not toxic, however, the byproducts and metabolites (nitrite, nitric oxide and N-nitrous compounds) from nitrates are of health concern (Natesh *et al.*, 2017). The acceptable daily nitrate intake is set at 3.7mg/kg body weight and at 0.06mg/kg body weight for nitrite by the World Health Organization (Natesh *et al.*, 2017). Other antinutrients of concern are cyanogenic glycosides and tannins which may affect protein digestibility (Natesh *et al.*, 2017; Ram *et al.*, 2020). Condensed tannins present in spider plant and related to the bitter taste of the vegetable have been quantified in a range of 0.23–0.49mg/g (Kutsukutsa *et al.*, 2014). Cyanogenic glycoside lotaustralin was identified in a range of 7.3–1500μg/kg in some of the samples of spider planted tested, however, the levels were lower than the regulatory limit of 10mg/kg hydrogen cyanide (HCN) equivalents (Misihairabgwi *et al.*, 2019). Cyanogenic glycoside presence in vegetables is also related to the poisonous nature of HCN (Natesh *et al.*, 2017) which is produced as a result of cyanogenesis (Ram *et al.*, 2020).

Table 3.6. Antinutritive compounds found in spider plant (mg/100g dry matter).

Oxalates	Phytates	Nitrate	References
511.50 ± 23.33	70.14 ± 0.07	-	(Koua *et al.*, 2015)
38.4 ± 1.7	0.02 ± 0.0	81.2 ± 1.6	(Mwanri *et al.*, 2018)
48.3 ± 1.5	0.3 ± 0.0	78.6 ± 0.8	(Mwanri *et al.*, 2018)
30.5 ± 0.5	0.3 ± 0.0	64.3 ± 3.7	(Mwanri *et al.*, 2018)
5.12 ± 0.9	-	-	(Wakhanu *et al.*, 2016)
10.17 ± 0.09	-	-	(Agbo *et al.*, 2014)

Implications of Food Processing on Nutrients and Phytochemical Quality

After harvesting, as with most leafy vegetables, spider plant is processed to increase palatability or reduce concentrations of plant toxins, extend shelf life through various preservation techniques and enhance the digestibility of essential nutrients present in the plant material (Masarirambi *et al.*, 2010). Spider plant leaves are usually consumed as a potherb, as flavouring in sauces and as a fermented product. Preparation may involve boiling the leaves and soaking them in milk for about two to three days (Mishra *et al.*, 2011). Alternatively, the leaves are boiled in salty water for two hours, and tomatoes, onions, oil, full cream or groundnut butter may be added (Directorate of Plant Production, 2010; Oelofse and Averbeke, 2012). For preservation purposes, the leaves are washed, blanched and dried, either in the shade or sun and stored for three to four months or a year. Other preservation techniques employed include freezing, canning and pickling (Oelofse and Averbeke, 2012; Muhialdin *et al.*, 2018). However, processing techniques used have implications on nutritional value and phytochemical quality of spider plant.

Boiling for 5 and 23 minutes seems to drastically reduce the proximate composition of spider plant leaves as summarized in Table 3.7. The loss of nutrients was greater with a longer cooking time. The decrease in nutritional value is influenced by leaching of water-soluble compounds such as free amino acids, nucleic acids or nucleotides as constituents of protein, or calcium and magnesium as constituents of ash (Njoroge *et al.*, 2016; Ochieng *et al.*, 2018). The decrease in crude fibre composition could be due to loss of hemicellulose branched structure through the breakage of glycosidic bonds. The protein content of spider plant was also reduced due to blanching and drying processes, however the blanching at 80°C combined with hot air drying at 65°C for seven hours or solar drying at 60°C for two days better retained nutrients than blanching at 90°C or 100°C combined with open sun drying or solar drying for two days at 60°C (Njoroge *et al.*, 2016). Therefore, mild blanching temperatures are recommended to reduce the loss of protein after drying. Fermentation processes also reduced the nutritional value of spider plant due to lactic acid bacteria proteolytic activity, utilization of carbohydrates as a source of energy and bioconversion lignocelluloses (Muhialdin *et al.*, 2018). Minerals such as iron were positively affected by fermentation and boiling for 23 and 90 minutes. However, blanching and drying processes investigated by Njoroge *et al.* (2016) caused extreme losses as high as 95 and 97% for iron and calcium, respectively, as shown in Table 3.8. For the amino acids, minimal losses of less than 5% occurred across all processing methods investigated as summarized in Table 3.9 with the exception of tryptophan with losses of about 19% after blanching at 90°C for five seconds and oven drying at 40°C for eight hours.

Most vitamins tested were negatively affected by processing methods as summarized in Table 3.9. Vitamins B and C are known to be water soluble compounds which might leach into cooking water during boiling and blanching. Vitamin C is also known to be heat sensitive. Using steam to blanch spider plant seems to preserve more vitamins than using water, since the vegetable material is not in contact with the cooking liquid, hence preventing leaching. Furthermore, it is noted that, during storage of up to six weeks, further losses of vitamins can occur (Khumbo and Kingsley, 2014). Information on the implications of processing on phytochemical content is scarce. Moyo *et al.* (2016) showed that of all the processing methods investigated (i.e. boiling, blanching, steaming and drying), steaming retained the most total phenolic and total flavonoid contents in spider plant. Furthermore, an increase in compounds such as quercetin rutinoside and quercetin rhamnoside occurred after the cooking and drying methods investigated (Moyo, 2016). The increase in these compounds was attributed to their release from the disintegration of cell wall structures and cellular constituents by heat and water. Loss of antinutrients such as oxalates has been reported after fermentation, boiling, sautéing and cooking of spider plant in lye and milk. Oxalates are known to be water soluble compounds, hence blanching or boiling of leafy vegetables high in antinutrients is advised.

Table 3.7. Implications of processing on proximate composition of spider plant.

Processing method	Processing conditions	Nutrient	% Change#	Effect	Reference
Boiling	23 min, 460 g vegetable to 500 ml water	Protein	21.65	Decrease	(Schönfeldt and Pretorius, 2011)
		Fat	58.3	Decrease	
		Ash	29.60	Decrease	
	5 min, 1 kg vegetable in 1 l water at 95°C	Protein	5.24[a]	Decrease	(Ochieng *et al.*, 2018)
		Fat	13.17	Decrease	
		Ash	23.76	Decrease	
		Fibre	28.96	Decrease	
Blanching and drying	Blanched at 80°C for 10 min and hot air dried at 65°C for 7 hours	Protein	33.7	Decrease	(Njoroge *et al.*, 2016)
		Fibre	19.05	Decrease	
	Blanched at 80°C for 10 min and solar dried at 60°C for 2 days	Protein	33.73	Decrease	
		Fibre	2.38	Decrease	
	Blanched at 90°C for 5 min and solar dried at 60°C for 2 days	Protein	43.37	Decrease	
		Fibre	2.3	Increase	
	Blanched at 100°C for 30 min and open sun dried for 3 days	Protein	84.33	Decrease	
		Fibre	14.29	Decrease	
	Blanched at 90°C for 5 min and hot air dried at 65°C for 7 hours	Protein	40.96	Decrease	
		Fibre	7.14	Decrease	
	Blanched at 90°C for 5 s and oven dried at 40°C for 8 hours	Protein	20.80	Decrease	(Ochieng *et al.*, 2018)
		Fibre	28.31	Decrease	
		Fat	9.87	Decrease	
		Ash	4.61	Increase	
Fermentation	Spontaneous fermentation in 10% brine and incubated at 28°C for 3 days. Fermented leaves were dried in the convection dryer at 45°C for 48 hours.	Protein	≈55	Decrease	(Muhialdin *et al.*, 2018)
		Ash	≈25	Increase	
		Fibre	≈55	Decrease	
		Fat	-	No change	
		Carbohydrates	≈44	Decrease	
	Blanched at 90°C for 5 s and fermented in 5% glucose brine for 48 hours at 40°C to <4.5 pH	Protein	27.4	Decrease	(Ochieng *et al.*, 2018)
		Ash	10.3	Decrease	
		Fibre	11.4	Decrease	
		Fat	11.1	Decrease	

[a] Change not significant; # in comparison to the raw untreated vegetable.

Bioactive Properties

Compounds such as organosulfur, zinc, iron, carotenoids, phenolics, tocopherols and phytosterols contribute to the bioactive properties of food. The beneficial effects of these compounds occur through various mechanisms of action, such as reducing the synthesis of proinflammatory cytokines, increasing the expression of antioxidant enzymes, or participating in different intracellular signalling pathways, regulating oxidative processes and increasing the expression of neurotrophic factors (Martín Ortega and Segura Campos,

Table 3.8. Implications of food processing on mineral content of spider plant.

Processing method	Processing conditions	Nutrient	% Change[#]	Effect	Reference
Boiling	90 mins, 100 g vegetable to 200 ml water	Aluminium	18	Decrease	(Ouédraogo *et al.*, 2013)
		Calcium	35.9	Decrease	
		Copper	15.38	Decrease	
		Iron	16.5	Increase	
		Potassium	51.04	Decrease	
		Magnesium	54.54	Decrease	
		Manganese	26.67	Decrease	
		Sodium	69.70	Decrease	
		Phosphorus	26.93	Decrease	
		Zinc	26.66	Increase	
	23 min, 460 g vegetable to 500 ml water	Iron	1.40	Increase	(Schönfeldt and Pretorius, 2011)
		Zinc	-	No change	
		Magnesium	37.33	Decrease	
		Calcium	32.57	Decrease	
		Phosphorus	25.17	Decrease	
	5 min, 1 kg vegetable in 1 l water at 95°C	Magnesium	3.4	Decrease	(Ochieng *et al.*, 2018)
		Sodium	7.74	Decrease	
		Iron	5.2	Decrease	
		Calcium	4.61	Decrease	

Continued

Table 3.8. Continued

Processing method	Processing conditions	Nutrient	% Change[#]	Effect	Reference
Blanching and drying	Blanched at 80°C for 10 min and hot air dried at 65°C for 7 hours	Iron	94.2	Decrease	(Njoroge *et al.*, 2016)
		Calcium	97.30	Decrease	
	Blanched at 90°C for 5 min and hot air dried at 65°C for 7 hours	Iron	95.7	Decrease	
		Calcium	97.43	Decrease	
	Blanched at 80°C for 10 min and solar dried at 60°C for 2 days	Iron	93.33	Decrease	
		Calcium	97.47	Decrease	
	Blanched at 90°C for 5 min and solar dried at 60°C for 2 days	Iron	95.7	Decrease	
		Calcium	97.62	Decrease	
	Blanched at 100°C for 30 min and open sun dried for 3 days	Iron	94.8	Decrease	
		Calcium	97.51	Decrease	
	Blanched at 90°C for 5 s and oven dried at 40°C for 8 hours	Magnesium	12.18	Increase	(Ochieng *et al.*, 2018)
		Sodium	6.47	Increase	
		Iron	30.96	Increase	
		Calcium	1.95	Decrease	
Fermentation	Blanched at 90°C for 5 s and fermented in 5% glucose brine for 48 hours at 40°C to <4.5 pH	Magnesium	0.82	Decrease	(Ochieng *et al.*, 2018)
		Sodium	7.17	Decrease	
		Iron	11.57	Increase	
		Calcium	3.62	Decrease	

Table 3.9. Implications of processing on amino acids and vitamins of spider plant.

Processing method	Processing conditions	Nutrient	% Change#	Effect	Reference
Boiling	5 min, 1 kg vegetable in 1 l water at 95°C	Valine	0.48	Decrease	(Ochieng *et al.*, 2018)
		Threonine	0.50	Increase	
		Leucine	-	No change	
		Isoleucine	0.74	Increase	
		Aspartic acid	0.35	Decrease	
		Lysine	-	No change	
		Glutamic acid	-	No change	
		Methionine	2.42	Increase	
		Phenylalanine	0.54	Decrease	
		Arginine	2.10	Decrease	
		Tryptophan	13.9	Decrease	
	23 min, 460 g vegetable to 500 ml water	Vitamin B2	25.0	Decrease	(Schönfeldt and Pretorius, 2011)
		β-carotene	46.9	Increase	
Blanching	Steam blanched with the lemon juice treated water and blanched for 4–5 min	Vitamin C	19.9	Decrease	(Khumbo and Kingsley, 2014)
	Steam blanched for 4–5 min	Vitamin C	39.95	Decrease	
	Blanched with lemon juice treated boiling water for 2–3 min	Vitamin C	59.95	Decrease	
	Blanched in boiling water for 2–3 min	Vitamin C	80.0	Decrease	
Blanching and drying	Steam blanched with the lemon juice treated water and blanched for 4–5 min, sun dried for 1–2 days and stored for 6 weeks	Vitamin C	95.51	Decrease	(Khumbo and Kingsley, 2014)
	Steam blanched for 4–5 min, sun dried for 1–2 days and stored for 6 weeks	Vitamin C	96.23	Decrease	

Continued

Table 3.9. Continued

Processing method	Processing conditions	Nutrient	% Change[#]	Effect	Reference
	Blanched with lemon juice treated boiling water for 2–3 min, sun dried for 1–2 days and stored for 6 weeks	Vitamin C	97.0	Decrease	
	Blanched in boiling water for 2–3 min, sun dried for 1–2 days and stored for 6 weeks	Vitamin C	97.78	Decrease	
	Blanched at 80°C for 10 min and hot air dried at 65°C for 7 hours	Vitamin C	53.43	Decrease	(Njoroge *et al.*, 2016)
		β-carotene	64.07	Decrease	
	Blanched at 90°C for 5 min and hot air dried at 65°C for 7 hours	Vitamin C	55.62	Decrease	
		β-carotene	6.96	Decrease	
	Blanched at 80°C for 10 min and solar dried at 60°C for 2 days	Vitamin C	41.74	Decrease	
		β-carotene	66.02	Decrease	
	Blanched at 90°C for 5 min and solar dried at 60°C for 2 days	Vitamin C	52.81	Decrease	
		β-carotene	64.08	Decrease	
	Blanched at 100°C for 30 min and open sun dried for 3 days	Vitamin C	55.62	Decrease	
		β-carotene	69.90	Decrease	
	Blanched at 90°C for 5 s and oven dried at 40°C for 8 hours	Valine	-	No change	(Ochieng *et al.*, 2018)
		Threonine	-	No change	
		Leucine	0.29	Decrease	
		Isoleucine	0.37	Increase	
		Aspartic acid	0.17	Decrease	
		Lysine	1.42	Decrease	
		Glutamic acid	4.06	Decrease	
		Methionine	1.61	Increase	
		Phenylalanine	0.134	Decrease	

Continued

Table 3.9. Continued

Processing method	Processing conditions	Nutrient	% Change[#]	Effect	Reference
		Arginine	0.84	Decrease	
		Tryptophan	18.99	Decrease	
Fermentation	Blanched at 90°C for 5 s and fermented in 5% glucose brine for 48 hours at 40°C to <4.5 pH	Valine	0.48	Decrease	(Ochieng *et al.*, 2018)
		Threonine	0.25	Increase	
		Leucine	0.196	Decrease	
		Isoleucine	0.556	Decrease	
		Aspartic acid	0.346	Decrease	
		Lysine	1.068	Decrease	
		Glutamic acid	0.044	Increase	
		Methionine	-	No change	
		Phenylalanine	0.401	Increase	
		Arginine	1.89	Decrease	
		Tryptophan	7.59	Decrease	

2018). Spider plant has interesting bioactive properties such as antioxidant (Bala *et al.*, 2012; Meda *et al.*, 2013; Konan *et al.*, 2014; Koua *et al.*, 2015; Moyo *et al.*, 2018; Muhialdin *et al.*, 2018; Chandradevan *et al.*, 2020), antibacterial (Ranjitha *et al.*, 2009; Sridhar *et al.*, 2014; Moyo *et al.*, 2016), antifungal (Ranjitha *et al.*, 2009; Imanirampa and Alele, 2016) and anti-inflammatory (Narendhirakannan *et al.*, 2007a; Bala *et al.*, 2013; Chandradevan *et al.*, 2020).

Spider plant extract *in vitro* free radical scavenging ability/antioxidant activity has been confirmed through various assays such as the 2,2-diphenyl-1-picrylhydrazyl (DPPH) scavenging assay, ferric reducing antioxidant (FRAP) assay, 2,2-azinobis (3-ethylbenzothiazoline-6-sulfonic acid (ABTS) assay and oxygen radical absorbance capacity (ORAC). The *in vitro* anti-inflammatory potential has been tested by nitric oxide (NO) inhibition activity assay. These assays test the ability of extracts to scavenge free radicals through the donation of an electron to a free radical ($R^{\cdot}$) or through the homolytic rupture of the O-H bond and consequent donation of a hydrogen atom to the free radical ($R^{\cdot}$) by the bioactive compound (Leopoldini *et al.*, 2011). Free radicals are of concern as they can form reactive oxygen/nitrogen species (ROS/RNS) which can overwhelm an antioxidant system and induce oxidative stress in living biological systems or in foods that are rich in fats (Fernandes *et al.*, 2018). Hence, dietary plant phenolics can assist to lower oxidative stress by scavenging free radicals and/or promoting endogenous defences against oxidative stress (Barrajón-Catalán *et al.*, 2014). To further ascertain the bioactive properties of spider plant, animal models have been used to illustrate the anti-inflammatory activity and radical scavenging activity (Narendhirakannan *et al.*, 2005a, b, 2007b).

The Minimum inhibitory concentration (MIC) method was used to confirm antimicrobial activity of spider plant against bacterial strains such as *Enterococcus aerogenes*, *Enterrococcus cloacae*, *Escherichia coli*, *Pseudomonas aeruginosa*, *Staphylococcus aureus*, *Bacillus subtilis* and fungal strains such as *Candida albicans*, *Candida glabrata*, *Microsporum canis*, *Trichophyton rubrum*, *Trichophyton mentagrophytes*, *Aspergillus flavus*, *Aspergillus niger* and *Fusarium axisporum*. Bioactive compound antimicrobial action could be due to the coagulation of cell composition or disruption of the cytoplasmic membrane, proton motive force, electron flow and active transport mechanisms of the microbes (Omojate *et al.*, 2014).

Health Benefits

Research on spider plant extracts shows its potential in the prevention of non-communicable diseases such as diabetes, cancer and cardiovascular disease. Oxidative stress, mitochondrial alterations as well as inflammation are major factors influencing the onset and development of these diseases (Camps and García-Heredia, 2014). The environment, genetics, nutrition and lifestyle factors are suggested as influencing the pro-oxidative and pro-inflammatory conditions in the body (Camps and García-Heredia, 2014). Therefore, regular consumption of foods comprising of bioactive compounds confers beneficial effects to health such as control and prevention of metabolic diseases, protection against cardiovascular disease, prevention and treatment of cancer, as well as protection against neurodegenerative diseases (Martín Ortega and Segura Campos, 2018). Table 3.10 summarizes the health benefits of spider plant (i.e., anticancer, immunosuppressive, immunostimulant, hepatoprotective, hypolipidemic and antidiabetic properties). The anti-diabetic and anti-cancer properties of spider plant were hypothesized to be influenced by the presence of flavonoids in the extract exerting radical scavenging/chelation activities (Bala *et al.*, 2010; Ravichandra *et al.*, 2014).

Conclusions and Future Trends

The use of spider plant to meet the nutritional requirements of individuals and contribute to the improvement of human health should be widely advocated. The vegetable is a rich source of nutrients such as protein and fibre at an average of 7.33 and 5.85%, respectively. It also has appreciable amounts of vitamin C and

Table 3.10. Health benefits of spider plant.

Health benefit tested	Extraction solvent	Animal model used	Experimental outcome	References
Anticancer	Methanol	Swiss albino male mice	Lifespan of EAC tumour bearing mice elevated Significant decrease in ($p<0.01$) in tumour volume, viable cell count and tumour weight Red blood cells, haemoglobin, white blood cells and lymphocyte count reverted to normal level in spider plant extract treated mice	(Bala *et al.*, 2010)
Immunosuppressive and immunostimulant	Aqueous & ethanol	Albino male or female rats	Extracts studied on humoral and cell mediated immunity in normal and cyclophosphamide-induced immunosuppressed rats Significant immunosuppressant activity from ethanol extract Ethanol extract decreased the number of white blood cells and splenic lymphocytes Phagocytic index, cellular and humoral antibody response decreased	(Rastogi *et al.*, 2009)
Hepatoprotective	70% ethanol	Swiss albino mice, male or female	Ethanolic extract treatment at 300 mg kg^{-1} for 10 days and subsequent tumour inoculation significantly lowered the impact of EAC on hepatocytes The extract also reduced the effects of EAC on liver tissue near to normal	(Dash *et al.*, 2017)
Hypolipidemic	70% ethanol	Albino Wistar male rats	Hydroalcoholic extract normalized the altered lipid profile in dexamethasone induced hyperlipidemia rats Decreased the atherogenic index as in dexamethasone induced hyperlipidemia rats	(Mallikarjuna *et al.*, 2018)
Estrogenic property	Methanol	Albino female mice	Effect of extract on the 17β estradiol induced folliculogenesis of female albino mice was studied Extracts administered reduced follicular development of female mice with respect to control	(Deka and Kalita, 2016)

Continued

Table 3.10. Continued

Health benefit tested	Extraction solvent	Animal model used	Experimental outcome	References
Antidiabetic	Ethanol	Wistar white male rats	Diabetes mellitus induced by alloxan in rats Reduced serum glucose, elevated dyslipidaemia levels, SGOT and SGPT levels occurred in metformin treated group, alloxan induced group, and ethanolic extract treated groups Increased serum HDL status occurred in all ethanolic extract treated groups as compared to normal control and diabetic control	(Ravichandra *et al.*, 2014)

Ehrlich ascites carcinoma (EAC), serum glutamic oxaloacetic transaminase (SGOT), serum glutamic pyruvic transaminase (SGPT).

provitamin A carotenoids. Cooking methods such as steaming or shorter boiling times are recommended to preserve the nutritional value of the vegetable while getting rid of the non-beneficial antinutrients. Its importance is further strengthened by its health benefits such as its anticancer and antidiabetic properties. However, the full utilization of the vegetable requires further investigation into the effects of processing, especially on the phytochemical value as this information is scare. Furthermore, bioavailability and bioaccessibility studies of spider plant compounds need to be addressed. The enhanced utilization of the vegetable also depends on studies that aim to develop novel products, reduce the bitter taste, improve post-harvest handling, marketing and seed distribution of the vegetable.

References

Adhikari, P.P. and Paul, S.B. (2018) Medicinally important plant *Cleome gynandra*: A phytochemical and pharmacological explanation. *Asian Journal of Pharmaceutical and Clinical Research* 11(1), 21. DOI: 10.22159/ajpcr.2018.v11i1.22037.

Agbo, A.E., Christophe, K., Anin, O.L.A., Soro, L.C., N'Zi, J.C. *et al.* (2014) Seasonal variation in nutritional compositions of spider plant (*Cleome gynandra* L.) in south Côte d'Ivoire. *International Journal of Agricultural Policy and Research* 2, 406–413.

Ahmad, A.F.H.B. (2012) The effect of pickling on total phenolic contents and colour *Cleome gynandra*, Honours thesis. Universiti Teknologi MARA, Selangor, Malaysia.

Anbazhagi, T., Kadavul, K., Suguna, G. and Petrus, A.J.A. (2009) Studies on the pharmacognostical and *in vitro* antioxidant potential of *Cleome gynandra* Linn. leaves. *Natural Product Radiance* 8, 151–157.

Bala, A., Kar, B., Haldar, P.K., Mazumder, U.K. and Bera, S. (2010) Evaluation of anticancer activity of *Cleome gynandra* on Ehrlich's ascites carcinoma treated mice. *Journal of Ethnopharmacology* 129(1), 131–134. DOI: 10.1016/j.jep.2010.03.010.

Bala, A., Kar, B., Karmakar, I., Kumar, R.B.S. and Haldar, P.K. (2012) Antioxidant activity of Cat's whiskers flavonoid on some reactive oxygen and nitrogen species generating inflammatory cells is mediated by scavenging of free radicals. *Chinese Journal of Natural Medicines* 10(5), 321–327. DOI: 10.1016/S1875-5364(12)60065-X.

Bala, A., Chetia, P., Dolai, N., Khandelwal, B. and Haldar, P.K. (2013) Cat's whiskers flavonoid attenuated oxidative DNA damage and acute inflammation: its importance in lymphocytes of patients with rheumatoid arthritis. *Inflammopharmacology* 22(1), 55–61. DOI: 10.1007/s10787-013-0193-5.

Barrajón-Catalán, E., Herranz-López, M., Joven, J., Segura-Carretero, A., Alonso-Villaverde, C. *et al.* (2014) Molecular promiscuity of plant polyphenols in the management of age-related diseases: far beyond their antioxidant properties. In: Camps, J. (ed.) *Advances in Experimental Medicine and Biology*. Springer, London, United Kingdom, pp. 141–159. DOI: 10.1007/978-3-319-07320-0_11.

Bayat, S., Schranz, M.E., Roalson, E.H. and Hall, J.C. (2018) Lessons from *Cleomaceae*, the Sister of Crucifers. *Trends in Plant Science* 23(9), 808–821. DOI: 10.1016/j.tplants.2018.06.010.

Camps, J. and García-Heredia, A. (2014) Introduction: oxidation and inflammation, a molecular link between non-communicable diseases. In: Camps, J. (ed.) *Advances in Experimental Medicine and Biology*, Vol. 824. Springer, London, United Kingdom, pp. 1–4. DOI: 10.1007/978-3-319-07320-0_1.

Chandradevan, M., Simoh, S., Mediani, A., Ismail, N.H., Ismail, I.S. *et al.* (2020) UHPLC-ESI-Orbitrap-MS Analysis of Biologically Active Extracts from *Gynura procumbens* (Lour.) Merr. and *Cleome gynandra* L. Leaves. *Evidence-Based Complementary and Alternative Medicine* 2020, 3238561. DOI: 10.1155/2020/3238561.

Chipurura, B. (2010) Nutritional content, phenolic compounds composition and antioxidant activities of selected indigenous vegetables of Zimbabwe. Masters thesis. University of Zimbabwe, Harare, Zimbabwe.

Chweya, J.A. and Mnzava, N.A. (1997) Cat's Whiskers. Cleome gynandra L. In: Heller, J., Engels, J., and Hammer, K. (eds) *Promoting the Conservation and Use of Underutilized and Neglected Crops*. Institute of Plant Genetics and Crop Plant Research/International Plant Genetic Resources Institute, Rome/Gatersleben, Italy/Germany, pp. 1–53.

Dash, J., Preetha, S.P., Ramesh, S., Parthiban, M., Sriram, P. *et al.* (2017) Hepatoprotective effect of *Cleome gynandra* on Ehrlich ascites carcinoma in mice. *International Journal of Chemical Studies* 5, 565–568.

Deka, J. and Kalita, J.C. (2016) Traditional medicine & clinical naturopathy effects of *Cleome gynandra* Linn. leaf extract on ovarian folliculogenesis of albino mice. *Journal of Traditional Medicine & Clinical Naturopathy* 5, 3–6.

Directorate of Plant Production (2010) *Cleome: Production Guideline*. Department of Agriculture, Forestry and Fisheries, Pretoria.

El zayat, M.A.S., Ali, M.E.S. and Amar, M.H. (2020) A systematic revision of Capparaceae and Cleomaceae in Egypt: An evaluation of the generic delimitations of Capparis and Cleome using ecological and genetic diversity. *Journal of Genetic Engineering and Biotechnology* 18(1), 1–15. DOI: 10.1186/s43141-020-00069-z.

Etong, D.I. and Abbah, E.A. (2014) Estimation of nutrients and anti-nutrients components of three Nigerian leafy vegetables. *Food Science and Quality Management* 33, 10–21.

Fernandes, R.D.P.P., Trindade, M.A. and de Melo, M.P. (2018) Natural antioxidants and food applications: healthy perspectives. In: Holban, A.M. and Grumezescu, A.M. (eds) *Alternative and Replacement Foods*. Academic Press, London, pp. 31–64.

Gao, Z.J., Liu, J.B. and Xiao, X.G. (2011) Purification and characterisation of polyphenol oxidase from leaves of *Cleome gynandra* L. *Food Chemistry* 129(3), 1012–1018. DOI: 10.1016/j.foodchem.2011.05.062.

Glew, R.S., Amoako-Atta, B., Ankar-Brewoo, G., Presley, J., Chuang, L.T. *et al.* (2009) Non-cultivated plant foods in west Africa: Nutritional analysis of the leaves of three indigenous leafy vegetables in Ghana. *Food* 3, 39–42.

Gowele, V.F., Kinabo, J., Jumbe, T., Kirschmann, C., Frank, J. *et al.* (2019) Provitamin A carotenoids, tocopherols, ascorbic acid and minerals in indigenous leafy vegetables from Tanzania. *Foods* 8(1), 35. DOI: 10.3390/foods8010035.

Harris, J., Beatrice, M. and Eric, C. (2017) Nutrient composition of cats whiskers (*Cleome gynandra* L.) from different agro ecological zones in Malawi. *African Journal of Food Science* 11(1), 24–29. DOI: 10.5897/AJFS2016.1478.

Hassan, L.G., Umar, K.J. and Gwaram, N.S. (2005) Nutritional composition of the leaves and stems of *Gynandropsis gynandra* L. *Biological and Envionmental Sciences Journal for the Tropics* 2, 112–119.

Imanirampa, L. and Alele, P.E. (2016) Antifungal activity of *Cleome gynandra* L. aerial parts for topical treatment of Tinea capitis: An *in vitro* evaluation. *BMC Complementary and Alternative Medicine* 16, 1–8. DOI: 10.1186/s12906-016-1187-9.

Jayanthi, P., Aravindhan, V. and Rajendran, A. (2012) Phytotherapeutic plants of madukkarai hills in the southern western ghats of coimbatore district, Tamil Nadu, India. *International Journal Of Ayurvedic And Herbal Medicine* 2, 807–906.

Jeruto, P., Lukhoba, C., Ouma, G., Otieno, D. and Mutai, C. (2008) An ethnobotanical study of medicinal plants used by the Nandi people in Kenya. *Journal of Ethnopharmacology* 116(2), 370–376. DOI: 10.1016/j.jep.2007.11.041.

Jiménez-Aguilar, D.M. and Grusak, M.A. (2015) Evaluation of minerals, phytochemical compounds and antioxidant activity of Mexican, Central American, and African green leafy vegetables. *Plant Foods for Human Nutrition* 70(4), 357–364. DOI: 10.1007/s11130-015-0512-7.

Kamatenesi-Mugisha, M. and Oryem-Origa, H. (2007) Medicinal plants used to induce labour during childbirth in western Uganda. *Journal of Ethnopharmacology* 109(1), 1–9. DOI: 10.1016/j.jep.2006.06.011.

Kasem, W.T. and Fatahy, S. (2016) Flavonoids and isoenzymes as chemotaxonomic markers in *Cleome* L. (*Cleomaceae Bercht* and J. Presl). *Current Botany* 7, 11. DOI: 10.19071/cb.2016.v7.3026.

Khumbo, T.M. and Kingsley, G.M. (2014) Effect of lemon juice treatment and sun drying on vitamin C retention in three steam and water blanched indigenous vegetables over six weeks storage period. *African Journal of Food Science* 8(6), 316–321. DOI: 10.5897/AJFS2014.1167.

Kiebre, Z., Bationo/kando, P., Sawadogo, N., Sawadogo, M., Zongo, J.-D. *et al.* (2015) Selection of phenotypic interests for the cultivation of the plant *Cleome gynandra* L. in the vegetable gardens in Burkina Faso. *Journal of Experimental Biology and Agricultural Sciences* 3(3), 288–297. DOI: 10.18006/2015.3(3).288.297.

Konan, Y., Witabouna, K.M., Bassirou, B. and Kagoyire, K. (2014) Antioxidant activity and total phenolic content of nine plants from Cote d'Ivoire (West Africa). *Journal of Applied Pharmaceutical Science* 4, 36–41.

Koua, G.A.Y., Zoue, L.T., Ouattara, S. and Niamke, S.L. (2015) Studies on nutritive value and antioxidant properties of three neglected leafy vegetables of Cote d'Ivoire. *Pakistan Journal of Nutrition* 14(12), 877–884. DOI: 10.3923/pjn.2015.877.884.

Kutsukutsa, R.T., Gasura, E., Mabasa, S. and Ngadze, E. (2014) Variability in condensed tannins and bitterness in spider plant genotypes. *African Crop Science Journal* 22, 275–280.

Kwarteng, A.O., Abogoom, J., Adu Amoah, R., Nyadanu, D., Ghunney, T. *et al.* (2018) Current knowledge and breeding perspectives for the spider plant (*Cleome gynandra* L.): A potential for enhanced breeding of the plant in Africa. *Genetic Resources and Crop Evolution* 65(5), 1529–1550. DOI: 10.1007/s10722-018-0626-5.

Leopoldini, M., Russo, N. and Toscano, M. (2011) The molecular basis of working mechanism of natural polyphenolic antioxidants. *Food Chemistry* 125(2), 288–306. DOI: 10.1016/j.foodchem.2010.08.012.

Lokesha, R. (2018) The African spider plant (*Gynandropsis pentaphylla* – DC. Syn. *Cleome gynandra* Linn.) (*Capparaceae*): Phytochemistry, pharmacological and biotechnological properties - A review. *International Journal of Pharmacognosy and Chinese Medicine* 2, 1–14.

Mallikarjuna, G., Shireesha, T., Suma, S.V., Shalini, P.M., Ramulu, N.M. *et al.* (2018) Evaluation of hypolipidemic activity of *Cleome gynandra* L. against dexamethasone induced hyperlipidemia in rats. *International Journal of Pharmaceutical Sciences Review and Research* 50, 47–52.

Maroyi, A. (2013) Use of weeds as traditional vegetables in Shurugwi District, Zimbabwe. *Journal of Ethnobiology and Ethnomedicine* 9(1), 1–10. DOI: 10.1186/1746-4269-9-60.

Martín Ortega, A.M. and Segura Campos, M.R. (2018) Bioactive compounds as therapeutic alternatives. In: Segura Campos, M.R. (ed.) *Bioactive Compounds: Health Benefits and Potential Applications*. Woodhead Publishing, Duxford, UK, pp. 247–264.

Masarirambi, M.T., Mavuso, V., Songwe, V.D., Nkambule, T.P. and Mhazo, N. (2010) Indigenous post-harvest handling and processing of traditional vegetables in Swaziland: A review. *African Journal of Agricultural Research* 5, 3333–3341.

Maseko, I., Mabhaudhi, T., Tesfay, S., Araya, H., Fezzehazion, M. *et al.* (2017) African leafy vegetables: A review of status, production and utilization in South Africa. *Sustainability* 10(2), 16. DOI: 10.3390/su10010016.

Matenge, S.T.P., Li, J., Apau, S. and Tapera, R. (2017) Nutritional and phytochemical content of indigenous leafy vegetables consumed in Botswana. *Frontiers in Food & Nutrition Research* 3, 1–7.

Meda, N.T.R., Bangou, M.J., Bakasso, S., Millogo-Rasolodimby, J. and Nacoulma, O.G. (2013) Antioxidant activity of phenolic and flavonoid fractions of *Cleome gynandra* and *Maerua angolensis* of Burkina faso. *Journal of Applied Pharmaceutical Science* 3, 36–42.

Mibei, E.K. and Ojijo, N.K.O. (2011) Effects of processing on chemical composition of four African leafy vegetables. *Electronic Journal of Environmental, Agricultural and Food Chemistry* 10, 3121–3131.

Mishra, S.S., Moharana, S.K. and Dash, M.R. (2011) Review on cleome gynandra. *International Journal of Research in Pharmacy and Chemistry* 1(3), 681–689.

Misihairabgwi, J.M., Ishola, A., Sulyok, M. and Krska, R. (2019) Mycotoxin and cyanogenic glycoside assessment of the traditional leafy vegetables mutete and omboga from Namibia. *Food Additives & Contaminants. Part B, Surveillance* 12(4), 245–251. DOI: 10.1080/19393210.2019.1616829.

Moyo, M., Amoo, S.O.O., Ncube, B., Ndhlala, A.R.R., Finnie, J.F.F. *et al.* (2013) Phytochemical and antioxidant properties of unconventional leafy vegetables consumed in southern Africa. *South African Journal of Botany* 84, 65–71. DOI: 10.1016/j.sajb.2012.09.010.

Moyo, M., Amoo, S.O., Aremu, A.O., Gruz, J., Šubrtová, M. *et al.* (2018) Determination of mineral constituents, phytochemicals and antioxidant qualities of *Cleome gynandra*, compared to *Brassica oleracea* and *Beta vulgaris*. *Frontiers in Chemistry* 5, 1–9. DOI: 10.3389/fchem.2017.00128.

Moyo, S.M. (2016) Effects of cooking and drying on the phenolic compounds, antioxidant activity and antibacterial activity of *Cleome gynandra* (spider plant). Mtech thesis, University of Johannesburg, Johannesburg, South Africa.

Moyo, S.M., Kayitesi, E., Mavumengwana, V. and Madala, N.E. (2016) Effects of cooking and drying on the total phenolic, total flavonoid content, antioxidant and antibacterial activity of *Cleome gynandra* (spider plant). In: *International Conference on Advances in Science Engineering, Technology & Natural Resources (ICASETNR-16)*, (pp. 136–141).

Muhialdin, B.J., Sukor, R., Ismail, N., Ahmad, S.W., Me, N. *et al.* (2018) The effects of fermentation process on the chemical composition and biological activity of spider flower (*Gynandropsis gynandra*). *Journal of Pure and Applied Microbiology* 12(2), 497–504. DOI: 10.22207/JPAM.12.2.08.

Murcia, M.A., Jiménez-Monreal, A.M., Gonzalez, J. and Martínez-Tomé, M. (2020) Spinach. In: Jaiswal, A.K. (ed.) *Nutritional Composition and Antioxidant Properties of Fruits and Vegetables*. pp. 181–195.

Mwanri, A.W., Mamboleo, T.F., Msuya, J.M. and Gowele, V.F. (2018) Oxalate, phytate and nitrate content in African nightshade, spider plant and amaranths at different stages of maturity. *African Journal of Food Science* 12(11), 316–322. DOI: 10.5897/AJFS2018.1735.

Narendhirakannan, R.T., Kandaswamy, M. and Subramanian, S. (2005a) Anti-inflammatory activity of *Cleome gynandra* L. on hematological and cellular constituents in adjuvant-induced arthritic rats. *Journal of Medicinal Food* 8(1), 93–99. DOI: 10.1089/jmf.2005.8.93.

Narendhirakannan, R.T., Subramanian, S. and Kandaswamy, M. (2005b) Free radical scavenging activity of *Cleome gynandra* L. leaves on adjuvant induced arthritis in rats. *Molecular and Cellular Biochemistry* 276(1–2), 71–80. DOI: 10.1007/s11010-005-3234-6.

Narendhirakannan, R.T., Subramanian, S. and Kandaswamy, M. (2007a) Anti-inflammatory and lysosomal stability actions of *Cleome gynandra* L. studied in adjuvant induced arthritic rats. *Food and Chemical Toxicology* 45(6), 1001–1012. DOI: 10.1016/j.fct.2006.12.009.

Narendhirakannan, R.T., Subramania, S. and Kandaswamy, M. (2007b) Evaluation of anti-inflammatory activity of Cleome gynandra L. leaf extract on acute and chronic inflammatory arthritis studied in rats. *Journal of Pharmacology and Toxicology* 2(1), 44–53. DOI: 10.3923/jpt.2007.44.53.

Natesh, H.N., Abbey, L. and Asiedu, S.K. (2017) An overview of nutritional and anti nutritional factors in green leafy vegetables. *Horticulture International Journal* 1(2), 1–9. DOI: 10.15406/hij.2017.01.00011.

Nemahunguni, N.K., Gupta, S., Kulkarni, M.G., Finnie, J.F. and Van Staden, J. (2020) The effect of biostimulants and light wavelengths on the physiology of *Cleome gynandra* seeds. *Plant Growth Regulation* 90(3), 467–474. DOI: 10.1007/s10725-019-00546-7.

Neugart, S., Baldermann, S., Ngwene, B., Wesonga, J. and Schreiner, M. (2017) Indigenous leafy vegetables of Eastern Africa - A source of extraordinary secondary plant metabolites. *Food Research International* 100(Pt 3), 411–422. DOI: 10.1016/j.foodres.2017.02.014.

Njoroge, E.W., Matofari, J.W., Mulwa, R.M.S. and Anyango, J.O. (2016) Effects of blanching time/ temperature combination coupled with solar-drying on the nutritional and microbial quality of indigenous leafy vegetables in Kenya. *African Journal of Food Science and Technology* 6, 209–219.

Ochieng, B.D., Owaga, E.E. and Njoroge, D.M. (2018) Effect of selected processing methods on the nutritional and anti-nutritional properties of spider plant (*Gynandropsis gynandra*). *Journal of Agriculture and Food Technology* 8, 1–9.

Oelofse, A. and Averbeke, W.V. (2012) *Nutritional Value and Water Use of African Leafy Vegetables for Improved Livelihoods*, 1st edn. Water Research Commission and Department of Agriculture, Forestry & Fisheries, Pretoria, South Africa.

Omojate, G.C., Enwa, F.O., Jewo, A.O. and Eze, C.O. (2014) Mechanisms of antimicrobial actions of phytochemicals against enteric pathogens – A review. *Journal of Pharmaceutical, Chemical and Biological Sciences* 2, 77–85.

Omondi, E.O., Engels, C., Nambafu, G., Schreiner, M., Neugart, S. *et al.* (2017) Nutritional compound analysis and morphological characterization of spider plant (*Cleome gynandra*) - an African indigenous leafy vegetable. *Food Research International* 100(Pt 1), 284–295. DOI: 10.1016/j.foodres.2017.06.050.

Onyango, C.M., Kunyanga, C.N., Ontita, E.G., Narla, R.D. and Kimenju, J.W. (2013) Current status on production and utilization of spider plant (*Cleome gynandra* L.) an underutilized leafy vegetable in Kenya. *Genetic Resources and Crop Evolution* 60(7), 2183–2189. DOI: 10.1007/s10722-013-0036-7.

Onyeoziri, I.O., Kinnear, M. and de Kock, H.L. (2017) Relating sensory profiles of canned amaranth (*Amaranthus cruentus*), cleome (*Cleome gynandra*), cowpea (*Vigna unguiculata*) and Swiss chard (*Beta vulgaris*) leaves to consumer acceptance. *Journal of the Science of Food and Agriculture* 98(6), 2231–2242. DOI: 10.1002/jsfa.8710.

Oshingi, S., Fekadu, F.D., Emmanuel, O.O., Traud, W. and Mary, O.A.O. (2019) *Cleome gynandra* L. origin, taxonomy and morphology: A review. *African Journal of Agricultural Research* 14(32), 1568–1583. DOI: 10.5897/AJAR2019.14064.

Ouédraogo, I.W.K., Tranchant, C. and Bonzi-Coulibaly, Y.L. (2013) Evaluation of mineral contents in *Cleome gynandra* leaves and stalks from Burkina Faso. *Journal of Cameroon Academy of Sciences* 11, 49–53.

Ram, S., Narwal, S., Gupta, O.P., Pandey, V. and Singh, G.P. (2020) Anti-nutritional factors and bioavailability: approaches, challenges, and opportunities. In: Gupta, O.P., Pandey, V., Narwal, S., Sharma, P., and Ram, S. (eds) *Wheat and Barley Grain Biofortification*. Woodhead Publishing, Duxford, UK, pp. 101–128.

Ranjitha, J., Bakiyalakshmi, K., Anand, M. and Sudha, P.N. (2009) Phytochemical investigation of n-hexane extract of leaves of *Cleome gynandra*. *Asian Journal of Chemistry* 21, 3455–3458.

Rastogi, B., Tiwari, U., Dubey, A., Bawra, B., Nandini, D. *et al.* (2009) Immunomodulating activity of *Cleome gynandra*. *Pharmacologyonline* 2, 151–169.

Ravichandra, B., Saketh Ram, P., Saritha, C.H. and Shankaraia, P. (2014) Anti-diabetic and Anti dyslipidemia activities of *Cleome gynandra* in alloxan induced diabetic rats. *Journal of Pharmacology and Toxicology* 9(1), 55–61. DOI: 10.3923/jpt.2014.55.61.

Sango, C., Marufu, L. and Zimudzi, C. (2016) Phytochemical, anti-nutrients and toxicity evaluation of *Cleome gynandra* and *Solanum nigrum*: Common indigenous vegetables in Zimbabwe. *British Biotechnology Journal* 13(3), 1–11. DOI: 10.9734/BBJ/2016/25164.

Schönfeldt, H.C. and Pretorius, B. (2011) The nutrient content of five traditional South African dark green leafy vegetables—A preliminary study. *Journal of Food Composition and Analysis* 24(8), 1141–1146. DOI: 10.1016/j.jfca.2011.04.004.

Shah, A. and Rahim, S. (2017) Ethnomedicinal uses of plants for the treatment of malaria in Soon Valley, Khushab, Pakistan. *Journal of Ethnopharmacology* 200, 84–106. DOI: 10.1016/j.jep.2017.02.005.

Shanmugam, S., Rajendran, K. and Suresh, K. (2012) Traditional uses of medicinal plants among the rural people in Sivagangai district of Tamil Nadu, Southern India. *Asian Pacific Journal of Tropical Biomedicine* 2(1), S429-S434. DOI: 10.1016/S2221-1691(12)60201-9.

Sogbohossou, E.O.D., Achigan-Dako, E.G., Maundu, P., Solberg, S., Deguenon, E.M.S. *et al.* (2018) A roadmap for breeding orphan leafy vegetable species: a case study of *Gynandropsis gynandra* (Cleomaceae). *Horticulture Research* 5(1), 2. DOI: 10.1038/s41438-017-0001-2.

Sogbohossou, E.O.D., Kortekaas, D., Achigan-Dako, E.G., Maundu, P. and Stoilova, T. (2019) Association between vitamin content, plant morphology and geographical origin in a worldwide collection of the orphan crop *Gynandropsis gynandra* (Cleomaceae). *Planta* 250(3), 933–947. DOI: 10.1007/s00425-019-03142-1.

Sridhar, N., Surya Kiran, B.V.V.S., Tharaka Sasidhar, D. and Kanthal, L.K. (2014) *In vitro* antimicrobial screening of methanolic extracts of *Cleome chelidonii* and *Cleome gynandra*. *Bangladesh Journal of Pharmacology* 9, 161–166. DOI: 10.3329/bjp.v9i2.17759.

Udousoro, I.I. and Akpan, E.B. (2014) Changes in anti-nutrients contents of edible vegetables under varied temperature and heating time. *Current Research in Nutrition and Food Science Journal* 2(3), 146–152. DOI: 10.12944/CRNFSJ.2.3.06.

Urso, V., Signorini, M.A., Tonini, M. and Bruschi, P. (2016) Wild medicinal and food plants used by communities living in Mopane woodlands of southern Angola: Results of an ethnobotanical field investigation. *Journal of Ethnopharmacology* 177, 126–139. DOI: 10.1016/j.jep.2015.11.041.

USDA (2019) FoodData Central. Available at: https://fdc.nal.usda.gov/fdc-app.html#/food-details/168462/nutrients (accessed 5 November 2020).

Uzilday, B., Turkan, I., Sekmen, A.H., Ozgur, R. and Karakaya, H.C. (2012) Comparison of ROS formation and antioxidant enzymes in *Cleome gynandra* (C_4) and *Cleome spinosa* (C_3) under drought stress. *Plant Science* 182, 59–70. DOI: 10.1016/j.plantsci.2011.03.015.

van der Walt, A.M., Ibrahim, M.I.M., Bezuidenhout, C.C. and Loots, D.T. (2008) Linolenic acid and folate in wild-growing African dark leafy vegetables (morogo). *Public Health Nutrition* 12(4), 525–530. DOI: 10.1017/S1368980008002814.

van der Walt, A.M., Loots, D.T., Ibrahim, M.I.M. and Bezuidenhout, C.C. (2009) Minerals, trace elements and antioxidant phytochemicals in wild African dark-green leafy vegetables (morogo). *South African Journal of Science* 105, 444–448.

van Jaarsveld, P., Faber, M., van Heerden, I., Wenhold, F., Jansen van Rensburg, W. *et al.* (2014) Nutrient content of eight African leafy vegetables and their potential contribution to dietary reference intakes. *Journal of Food Composition and Analysis* 33(1), 77–84. DOI: 10.1016/j.jfca.2013.11.003.

Wakhanu, J. A., Kimiywe, J. and Hudson, N. (2016) Oxalate levels in selected African indigenous vegetable recipes from the Lake Victoria Basin, Kenya. International Journal of Agriculture and Environmental Research, 01, 88–105.

Wu, T. hor, Solberg, S.O., Yndgaard, F. and Chou, Y.Y. (2018) Morphological patterns in a world collection of Cleome gynandra. *Genetic Resources and Crop Evolution* 65(1), 271–283. DOI: 10.1007/s10722-017-0529-x.

4 Amaranth (*Amaranthus*)

Nadaraj Govender[1]* and Himansu Baijnath[2]
[1]University of KwaZulu-Natal, School of Education, Edgewood Campus, Pinetown, Durban, South Africa; [2]University of KwaZulu-Natal, School of Life Sciences, Westville Campus, Durban, South Africa

Abstract

The amaranth plant has been known for centuries. It is an annual leafy vegetable plant found globally, but used mostly in Africa and in Asian countries as a natural food and traditional medicine in both rural and urban communities. On the American continent, it is also grown as a grain. It is a C_4 plant, so it reduces CO_2 levels and so helps to address climate change. The leaf and stem are valuable sources of dietary fibre, proteins, phytochemicals and vitamins for human and animal consumption. Recent studies indicate several benefits for food processing of its nutrients and phytochemicals. Due to its variability, hardiness in different climates and nutritional benefits, it is now being considered for the status of a famine food. While studies on amaranth leaf have generally focused on the northern hemisphere, this chapter includes its use in the southern hemisphere.

Introduction

The genus *Amaranthus* L. has been cultivated as a leafy vegetable crop worldwide since ancient times (Stallknecht and Schulz-Schaeffer, 1993). Mesoamerican people living in Central and South America have used grain amaranth as a staple in their diets from ancient times. We focus here on leaf amaranth (Fig. 4.1). People in early civilizations were aware by experience of its nutritional and medicinal benefits. Recent biochemical research confirms the value of amaranth for its protein, fibres and phytochemicals (Stallknecht and Schulz-Schaeffer, 1993). In Africa, most amaranth species are harvested in the wild as a food resource, but some species such as *Amaranthus cruentus* L. are cultivated for sale as a leafy vegetable at markets (Achigan-Dako *et al.*, 2014). It is highly resistant to drought, heat, diseases and pests (Joshi and Verma, 2020). The value of amaranth leaf as a dietary supplement, sometimes combined with starchy staples, is becoming recognized, especially in poverty stricken and rural areas of Africa, Asia and other parts of the world. Currently the leaves and stems are being rediscovered as feedstock for farm animals. Other promising uses of the crop include use as biofuel (ŢÎŢEI, 2020), fish food (Ngugi *et al.*, 2017; Gołaś *et al.*, 2020) and for phytoremediation in extracting heavy metals (Ramanlal *et al.*, 2020). Due to its hardiness and nutritional value, improved varieties are being sought and hybridized for commercial cultivation (Dinssa *et al.*, 2019).

Distribution and Availability

Amaranth species occur across the Americas, Australia, Africa, Asia and Europe, where they are either grown commercially as food or occur

*Corresponding author: Govendern37@ukzn.ac.za

 Handbook of Phytonutrients in Indigenous Fruits and Vegetables (eds D. Sivakumar *et al.*)
DOI: 10.1079/9781789248067.0004

Fig. 4.1. Amaranth leafy vegetable with flower clusters.

as weeds in the wild or in cultivated fields. Among the over 70 known amaranth species is the extensive distribution of species native to America, where they are cultivated largely as a grain crop, and species that are grown as leafy vegetable crops across a wide variety of climates, from the hot humid regions of the Asian subcontinent, the Caribbean, the south Pacific islands, and across Africa to the tropical lowlands to mountainous regions, as well as in China, Russia, the Mediterranean lands and Australia (Ebert *et al.*, 2011). The rapid growth and high yield of *Amaranthus* species is due to its characteristic good tolerance of poor, saline and acidic soils and drought resistance.

Amaranth species grown as vegetables are represented by, among others, *Amaranthus tricolor* L., *Amaranthus dubius* L., *A. cruentus* L., *Amaranthus edulis* L., *Amaranthus retroflexus L.*, *Amaranthus viridis* L., *Amaranthus hybridus* L., *Amaranthus gracilis* Desf., *Amaranthus gangeticus* L., *Amaranthus blitum* L. and *Amaranthus palmeri* S.Wats. (Bressani, 2003; Dmitrieva and Ivanov, 2020). Several domestic ornamental or vegetable varieties have been developed from *A. tricolor* L., which most likely originated in India and was then introduced to South America (Rastogi and Shukla, 2013) and southern China. The amaranth vegetable types appear to have originated in southern Asia and then spread throughout the tropics and temperate areas (Grubben and Van Sloten, 1981). Other common garden ornamentals include *Amaranthus caudatus* L. and *Amaranthus hypochondriacus* L. There are more than 20 species of amaranth cultivated or wild that occur in India (Rastogi and Shukla, 2013) and southern Africa (Van Wyk and Gericke, 2017). Specific species found in Africa are *Amaranthus graezianus* L. and *Amaranthus thunbergii* Moq. (Alemayehu *et al.*, 2015). Several other species, such as *A. spinosus* L. and *A. retroflexus* L., are considered to be weeds and in particular, *A. palmeri* S.Wats., *A. hybridus* L. and *A. retroflexus* L. are considered by crop farmers to be invasive and there have been systematic attempts to remove them manually or chemically. However, in southern Africa, *A. hybridus* L. occurs both as a weed and a commonly cultivated herb. Other species grown as green vegetables include *A. thunbergii* Moq., *Amaranthus spinosus* L. and *A. retroflexus* L. (Van Wyk and Gericke, 2017).

To conserve the germplasm of traditional vegetables like amaranth, there is a drive for gene banks to collect germplasm from gardeners and farmers in Africa and Asia before these are replaced by modern varieties (Ebert, 2020). The major germplasm collections, comprising several thousand germplasm accessions, are at 1) Rodale Research Institute, USA; 2) USDA North Central Regional Plant Introduction Station, Iowa State University, Ames; 3) Universidad del Cuzco, Cuzco, Peru; 4) National Bureau of Plant Genetic Resources, Shimla, India; 5) World Vegetable Centre (WVC), previously known as the Asian Vegetable Research and Development Centre (AVRDC), Tainan, Taiwan, where the South-East Asian accessions are kept, and Arusha, Tanzania, where Nigerian and African cultivars are stored; 6) National Horticultural Research Institute (NHR) at Ibadan, Nigeria and 7) ARC–Roodeplaat Vegetable and Ornamental Plant Institute, South Africa.

Botanical Description

Taxonomy

The genus *Amaranthus* is as a dicotyledonous herbaceous species with three subgenera, namely, *Acnida* (L.) *Aellen* ex K.R. Robertson,

Amaranthus L. and *Albersia (Kunth)* Gren. and Godr. as noted by Costea and DeMason (2001) and Thapa and Blair (2018). A taxonomic key for the cultivated species of amaranth has been developed by Feine-Dudley (Grubben and Van Sloten, 1981; Stallknecht and Schulz-Schaeffer, 1993). Amaranth is classified by means of the shape and dimensions of leaves, and the flower structure and inflorescence type. The small number of suitable traits, phenotypic plasticity, gene flow and hybridization made it difficult to establish the taxonomy and phylogeny of the whole genus, despite various studies using molecular markers (Costea and DeMason, 2001; Stetter and Schmid, 2017). Wild species have been noted to have longer bracteoles in comparison to those in cultivation (Costea and DeMason, 2001). Of the almost 100 species (The Plant List, 2013), 17 are grown for their edible leaves and three cultivated as food grain (Sarker *et al.*, 2020a). There is a distinct group of ten species of amaranth that are all native to North America, which are all dioecious, while the rest of the monoecious species are endemic (Steckel, 2007).

Amaranth is an annual or short-lived perennial. Plants are herbaceous, may be low lying, spreading or erect, tall and bushy with green, purple or red leaves on slender stalks. The plant may reach up to a height of 3 m with a succulent, fibrous stem that is hollow with grooves and bracteoles when mature. The stem diameter can range from 4 to 15 mm depending upon plant stand density and available soil moisture. The leaves of different amaranth species vary in shape; they may be elliptic, lanceolate, ovate, rhombic or rhombic-ovate with acute, obtuse or acuminate leaf tips. The leaves are alternate, with petioles being up to 1.5 cm in length. Leaves have wavy margins and hairs along the veins of the lower leaf surfaces. The floral buds arise directly in the leaf axils. The unisexual or bisexual brightly coloured flowers produced in summer or autumn are purple, orange, red or gold. They are developed on branched, densely packed, catkin-like or ament. These flower clusters are referred to as glomerules. A glomerule (Fig. 4.1) may be a diachsial cyme or biparous cyme (where the terminal peduncle gives rise to a flower, and two lateral branches also have terminal flowers that form large flowering panicles).

Physiology

Amaranth is a C_4 plant, making effective use of sunlight with relative low uptake of water and the ability to withstand stress even in saline conditions. It photosynthesizes via the Hatch-Slack pathway – one of three known processes for carbon fixation. The combination of anatomical features in amaranth, such as Kranz anatomy, a well-developed root system, stomatal conductance, and maintenance of leaf area and C_4 metabolism, results in increased efficiency in using CO_2 under diverse environmental conditions of temperature and available moisture, which contributes to the wide geographic distribution of the genus (Khatri-Chhetri *et al.*, 2020).

The flower heads can each produce up to 60,000 seeds. Seed heads can vary in length from 10 to 60 cm and the shape can range from pendulous to spike. The fruit is a single seeded utricle, which may be classified as dehiscent, semi-dehiscent or indehiscent (Stallknecht and Schulz-Schaeffer, 1993). The seeds are quite small (0.9 to 1.7 mm diameter) and weights vary from 1000 to 3000 seeds/g. The vegetable plant seeds are usually shiny black or brown while the grain-bearing plant seeds are white, beige or gold. The flowers are either wind or insect pollinated.

Uses of the leafy plant

Depending on the locations and habitats, the leaves and young shoots are utilized in various ways. Harvested material may be consumed as cooked or boiled and also as a supplement to cereals. Livestock is allowed to forage on the plants or material may be gathered and used as fodder. Some species are also used as traditional medicine.

As food

With its delicate spinach-like flavour, leaves are eaten raw or in salads. Leaves may be prepared alone or in a mixture with other vegetables, more especially fried or steamed with potatoes. In southern Africa leaves may be traditionally dried into balls and stored away for later use with maize porridge (Akinola *et al.*, 2020). In the arid regions, dried leaves are powdered and mixed with sauces in the dry season (Jansen, 2004). The ash of *A. spinosus* L. is useful to tenderize tough vegetables for cooking and is also used as a condiment, like salt (Alegbejo, 2013). In the early days, native Americans like the Hopi people boiled meat with amaranth leaves while the Cocopa, Mohave and Pima group baked *A. palmeri* S. Wats. (Steckel, 2007).

As forage and fodder

The leaf of amaranth is used as fresh forage, or dried or ensiled fodder for many animals, including cattle, goats, chickens, pigs and rabbits (Peiretti, 2018). Amaranth species such as *A. spinosus* L. are browsed by sheep and goats, and are reported to increase the milk yield in cattle (CABI, 2020). It is also used as traditional veterinary medicine for cattle (Vincent *et al.*, 2015). However, the high oxalate content can affect ruminant health and can lead to hypocalcemia or kidney failure of ruminants (Jansen, 2004; Peiretti, 2018). Ensiling the forage has been suggested as a way of reducing nitrate and oxalate concentration and improving digestibility (Peiretti, 2018).

As fuel

Amaranth may also be utilized as a fuel, which is pertinent in many African regions, where increasing forest decimation results in local inhabitants having to seek firewood (Alemayehu *et al.*, 2015). In India the smoke from dried amaranth stems which are burnt is used to break rock boulders without difficulty (Topwal, 2019).

As a medicinal herb

As shown in Table 4.1, Indigenous communities in different countries use amaranth to treat a range of ailments. Different parts of the plant are used as a diuretic, a remedy for coughs and colds, as a treatment for throat and urinary infections, to address gastric problems, dermatological problems, high blood pressure or to remove tapeworms. Research by Sharma *et al.* (2014) reported that *A. spinosus* has wound healing, anti-inflammatory, antimicrobial and antioxidant properties. Recent *in vivo* studies on mice (Nataru *et al.*, 2014; Saravanan, 2016) revealed that ethanolic extracts of leaves of *A. spinosus* L. enhanced the recovery of a tumour-bearing specimen, thus suggesting its potential as an anti-tumour agent.

Production and Preparation

There is no data on global amaranth production (D'Amico and Schoenlechner, 2017) but in Indonesia it is the third most produced vegetable with annual production being about 100,000 to 150,000 tons (Sulaiman and Andini, 2020). Amaranth leaf is usually cultivated as a monocrop but may also be used for intercropping in small-scale cultivation. In commercial planting, the plants are usually spaced in rows, about one metre apart. Harvesting is usually repeated every ten days. While the leaves of all species are edible, the main leafy cultivars are *A. tricolor* L. (mostly in Asia), *A. blitum* L. and *A. spinosus* L. (both in tropical Africa), *A. creuntus* L. and *A. hybridus* L. (in Africa and Latin America) as well as *A. dubius* Mart.ex Thell. and *A. lividus* L., a synonym of *A. blitum* subsp. *oleraceus* (L.) Costea (Jansen, 2004). Most amaranth cultivars grow rapidly and may be harvested from six to eight weeks. Some amaranth leaf is sold directly from the farm to individual consumers and traders, with a proportion being lost in pre- and post-harvest phases. The plants have a high potential for biomass yield (Corke *et al.*, 2016) of about 30–60 tons per hectare (Van Wyk and Gericke, 2017). Yields of dry matter of whole plants are 1 to 11 tons per hectare at four weeks, while yields of edible leaves are 0.3 to 0.7 tons per hectare. High levels of nitrogen are required for optimal yields. Most species are adapted for growth during short day-lengths. While amaranth is grown year-round, it could be grown in the gap periods between winter and hot summer.

Amaranth is not well established as a major food source and there are presently limited market outlets in the developing world (Aderibigbe *et al.*, 2020). Amaranth is also produced for animal feed. For fodder, the fresh plant can be provided for cattle and pigs, or it can be dehydrated or made into pellets to feed farm animals (Bressani, 2003). In China, it is largely produced as forage to feed hogs (Rastogi and

Table 4.1. Traditional medicinal uses of *Amaranthus* species.

Medicinal use	Plant part/s	Species	Country	References
Diuretic	Root	*A. spinosus* L.	Philippines, India, Thailand and Indo-China	Lemmens and Bunyapraphatsara, 1999
Treatment of gonorrhoea	Decoction of the root	*A. spinosus* L.	Philippines, India, Thailand and indo-China	Lemmens and Bunyapraphatsara, 1999
Emollient applied externally for earache and to burns, wounds, boils and haemorrhoids	Bruised leaves	*A. spinosus* L.	Indonesia and many countries	Lemmens and Bunyapraphatsara, 1999
Expectorant and relief of breathing problems in acute bronchitis	-	*A. spinosus* L.	Malaysia	Lemmens and Bunyapraphatsara, 1999
Haemorrhoids		*A. spinosus* L.	Africa	Lemmens and Bunyapraphatsara, 1999
Constipation	-	-	Benin	Akubugwo *et al*., 2007
Fever, haemorrhage, anaemia, kidney complaints	Leaves	-	India Benin	Akubugwo *et al*., 2007; Joshi and Verma, 2020
Pain in limbs	Macerated plants		Ghana	Jansen, 2004
Tumours	Leaves (heated)	-	Gabon	Jansen, 2004
Abortifacient	Leaves	A.*caudatus L.* *A. spinosus* L.	India	Lemmens and Bunyapraphatsara, 1999; Van Wyk and Gericke, 2017
Extract used as eyewash to treat ophthalmic infections and for convulsions in children	Plant sap	-	-	Joshi and Verma, 2020
Stimulate contractions in labour	Infusion	*A. thunbergii* Moq.	Africa	Van Wyk and Gericke, 2017
Decoction for treating eczema and cobra bites (roots) and for diarrhoea (tender leaves)	Roots and leaves	*A. viridis* L.	West Nepal	Kunwar *et al*., 2012.
Juice of leaves is applied topically on eczema or boils	Paste of leaves mixed with turmeric powder	*A. spinosus* L.	Uttarakhand, northern India	Sharma *et al*., 2014
Bleeding during menstruation, diarrhoea, fever, malaria	Leaves and roots	*A. spinosus* L.	Bangladesh	Jannat *et al*., 2020

Continued

Table 4.1. Continued

Medicinal use	Plant part/s	Species	Country	References
To treat constipation (laxative), stomach ache, wounds, boils, diarrhoea, diuresis, gonorrhoea, eczema, leucorrhoea	Leaves and roots	*A. spinosus* L.	Rajastan, India	Jeph and Khan, 2020
Stem juice to treat kidney stones	Dried stems	*A. spinosus* L.	Uttarakhand, Northern India	Topwal, 2019
A tea for relief of pulmonary problems and haemorrhoids. Used to purify blood, as a diuretic, to treat swellings and digestive upsets	Leaves - boiled	*A. caudatus* L.	Tamilnadu and Kerala hills, India	Topwal, 2019

Shukla, 2013). In South Africa, the leaf is produced commercially for processing – in canning or being mixed with maize. In East Africa, the amaranth value chain is relatively underdeveloped, but consumers prefer amaranth leaves over other plant parts (Nzomo *et al.*, 2015). For rural women in Africa amaranth leafy vegetable production can be a profitable venture, providing employment opportunities, but its production is still dominated by men and smallholder farmers (Adeagbo and Adejumo, 2020).

Successful plant growth and nutritional quality are dependent on season, geographical location, temperature, soil type and local and commercial utilization. Jimoh *et al.* (2020) observed that particle size and physicochemical characteristics of the soil affected the range of minerals found in plant tissues. Also, as the plant aged, the nutritional properties changed. They noted that for an optimal yield of vitamins A and E, the clayey loam soil was the most suitable, especially when *A. caudatus* plants were harvested prior to flowering. For vitamin C, the sandy-clayey loam was best at the same flowering stage. Also, loam and clayey loam soils produced the highest proximate compositions of yield of vitamins at the pre-flowering and flowering stages. However, mineral elements (micro and macro) were highest in the control (unfractionated mixture of silt, loam and clay) and loam soils (Jimoh *et al.*, 2020).

Adding poultry manure and urea-N to amaranth prolonged the vegetative stage and increased foliage production (Shiyam and Binang, 2011). Depending on the season, the yield of *Amaranthus* species is affected (Dinssa *et al.*, 2019). Recently, Hoidal *et al.* (2020a) reported that the nutritional value changed greatly throughout development, with the highest concentration of iron around halfway through the vegetative growth phase and the highest levels of vitamin A, copper and magnesium at the end of the vegetative growth phase. They noted that palatability was most marked in the young leaves and dropped as plants aged. Furthermore, growers gathering both seeds and foliage can effectively harvest up to 50% of leaves at any time during vegetative growth or bud formation while still maintaining seed yield.

Nutritional Composition

Chemical composition of vegetable amaranth

The chemical composition of the vegetative part of amaranth is variable, since it depends on the species, geographical climate, plant nutrition and the growth stage of the plant taken for analysis. Tender stems and leaves of green and red morph genotypes of amaranth are a good source of protein, vitamins and dietary fibre (Joshi and Verma, 2020). Amaranth is also rich in Mg and polyphenols; the latter give it its astringent taste and have antioxidant properties (Kyriacou *et al.*, 2019). However, different antioxidant activities are obtained in different genotypes and recently reported in a review study of 20 different species, noting the scavenging activity of red colour amaranth (*A. tricolor*) genotypes were found to be excellent compared to green colour (*A. lividus*) genotype of amaranth (Park *et al.*, 2020; Sarker *et al.*, 2020a). In a study, for other genotypes, *A. spinosus*, *A. dubius*, *A. viridis* and *A. tricolor* were evaluated to scavenge 1,1-diphenyl-2-picrylhydrazyl (DPPH), and the radical scavenging activity was found to be higher for *A. spinosus* (House *et al.*, 2020). The study also noted higher levels of phenolic content in *A. spinosus* and high flavonoid content in *A. dubius*, corroborating to their high antioxidant potential and among all showed better anti-inflammatory and antiproliferative activities.

On a fresh weight (FW) basis, amaranth contain moisture (84%, wet basis), protein (3200 mg/100 g), fat (3200 mg/100 g), carbohydrates (7000 mg/100 g) and vitamin C (0.85 mg/100 g); and on a dry weight (DW) basis, potassium (27 mg/100 g), calcium (20 mg/ 100 g), magnesium (27 mg/100 g) and iron (70 mg/100 g) (Mahangade *et al.*, 2020) (Table 4.2). The total lipids for the vegetable is 3.8 g/100 g, the crude fibre is 14.9 g/100 g and the ash content is 17.7 g/100 g (Bressani, 2003).

The mineral content of amaranth is quite high with no significant difference between raw and cooked amaranth leaves (Table 4.2), however the cooking process does decrease the nutritional value for other nutrients. Zn content may also vary with harvest time. Makobo *et al.* (2010) reported that, in the same genotype, Zn content in amaranth leaves harvested three weeks after emergence is 85% higher than the content in leaves harvested eight weeks after emergence. Amaranth is also a great source of vitamins B2, B3, B5, B6, C and vitamin E, which is a very important vitamin for heart health (Tang and Tsao, 2017).

In terms of protein constituents, analysis of leaf extracts of amaranth confirms the presence of several essential amino acids (in mg of amino acid per g of N) such as lysine, (300), methionine (78), leucine (421), isoleucine (327), cysteine (24), valine (355), tyrosine (232), threonine (356), tryptophan (80) and phenylalanine (314) (Bressani, 2003; Nwaogu *et al.*, 2006). Amaranth is particularly high in the amino acid lysine, which is largely lacking in maize and

Table 4.2. Chemical composition of vegetable amaranth in comparison with cowpea (*Vigna unguiculata*).

	Amaranth vegetable[1] mg/100 g	Raw amaranth[2] mg/100 g	Cooked amaranth[2] mg/100 g	[2]Cowpea raw leaves mg/100 g
Moisture	84%, (FW)	-	-	89.8%
Protein	3200	-	-	4100
Carbohydrates	7000			
Vitamin C	0.85	43.3	41.1	36
Fat	300	-	-	-
Potassium	27 (DW)	611	641	455
Calcium	20	215	209	63
Magnesium	27	55	55	43
Iron	70	2.32	2.26	1.92
Lipids	3.8	-	-	250
Crude fibre	14.9	-	-	-
Ash	17.7	-	-	1050

[1] The % daily value (DV) is how much a nutrient in a serving of food contributes to a daily diet;. 2000 calories a day is used for general nutrition advice. [2] USDA SR Legacy (2018) values. FW, fresh weight; DW, dry weight.

Table 4.3. Comparison of amino acid in leaves of amaranth and cowpea (Nielsen *et al.*, 1997).

Amino acid	Amaranth mg of amino acid/g of N	Cowpea mg/g N
Lysine	300	594
Methionine	78	313
Leucine	421	838
Isoleucine	327	413
Cysteine	24	-
Valine	355	381
Tyrosine	232	281
Threonine	356	413
Tryptophan	80	-
Phenylalanine	314	381

wheat. Both amaranth and cowpea leafy vegetables are used as traditional food in southern Africa and Asia. Cowpea (*Vigna unguiculata* L.) is also a nutrition-rich leafy vegetable and can supplement food nutrients such as protein sources. In sub-Saharan Africa and Asia, it grows in semi-arid areas with low rainfall and in sandy soils. Table 4.2 confirms that the protein content in cowpea leaf is much higher than amaranth leaf but the calcium content is significantly higher in amaranth. The amino acids of some cowpea cultivars (Table 4.3) were shown to be sufficient to fulfil some of the daily essential amino acid requirements (Moloto *et al.*, 2020). Leucine was the predominant essential amino acid in both cowpea and amaranth.

Amaranth microgreens have generated interest as a novel culinary ingredient due to the health benefits associated with the density of secondary metabolites and vitamins, which protect the human body against oxidative stress, being greater than in sprouts or seeds (Ebert *et al.*, 2014). For example, β-carotene is a precursor to vitamin A, which is essential to cellular health and vision and is also a powerful antioxidant that may reduce the risk of cancer. Fenugreek leaves had the highest content of β-carotene (9.15 mg/100 g), followed by amaranth (8.17 mg/100 g), carrot (8.14 mg/100 g) and pumpkin (1.90 mg/100 g) (Veda *et al.*, 2006). As vitamin A % RDA includes β-carotene which is not directly available, the % RDA upper recommended limit of β-carotene is 4.8 mg/100 g, thus for amaranth it is 170%. It is of great interest that food preparation techniques such as stir-frying in the presence of a small quantity of oil led to an enormous increase in the bioaccessibility of β-carotene from these vegetables, the increase being 263% (fenugreek leaves), 192% (amaranth leaves), 63% (carrot) and 53% (pumpkin) (Veda *et al.*, 2006).

Phytochemicals

Phytochemicals are bioactive chemicals produced by plants which help them to flourish and protect themselves. Thus far, over 25,000 phytochemicals have been isolated. In general phytochemicals are associated with the colourful organs of the plant which include nuts, grains and legumes. Some phytochemicals have toxic properties while others are used for traditional medicine. Species in the genus *Amaranthus* are a rich source of phytochemicals, including betacarotene, betacyanin, betaxanthin, chlorophyll, flavonoids, isoquercitrin, nicotiflorins, phenolic compounds, rutin and vitamin C (Sarker *et al.*, 2020a). These phytochemicals, because of their potential health benefits, are used in the food industry and in medicine (Joshi and Verma, 2020). Flavonoids have been noted to show antimalignant and antimutagenic properties. In testing for the inhibition of lipid peroxidation, the extracts of boiled amaranth displayed relatively high levels of antioxidants (Lindsey *et al.*, 2002). The red forms of amaranth are a rich source of betacyanin (a-cyanin) which can be used as a natural colourant in the food industry (Corke *et al.*, 2016).

Recently, Sarker *et al.* (2020a) evaluated 17 genotypes of stem amaranth (*A. lividus* L.) and found significant amounts of copper, zinc, phytopigments such as chlorophyll *a*, chlorophyll *ab* chlorophyll *b*, betalain, betaxanthin, betacyanin and carotenoids in the leaves (Table 4.4). These chemicals, namely, phenolics, phytopigments, flavonoids and vitamins, obtained from stem amaranth, contribute to dietary antioxidant sufficiency in humans.

However, environmental stresses, such as drought due to climatic change, cause oxidative damage in the plant. As a defence, these plants

Table 4.4. Amaranth genotypes (17 from D 25 to D 38) and mean nutritional components of *A. lividus* L. (Sarker *et al*., 2020a).

Nutrient	Mean of nutritional components
	mg/ 100 g FW
Protein	3460
Dietary fibre	78,890
Fat	290
Carbohydrates	7240
	mg/ 100 g FW
Chlorophyll *a*	27.76
Chlorophyll *b*	14.30
Chlorophyll *ab*	42.06
	µg/ 100 g FW
Betacyanin	31.12
Betaxanthin	31.81
Betalain	62.92
Total carotenoids	1244.34
	mg/g DW
K	9.72
Ca	24.47
Mg	29.73
Fe	1138.89
Mn	273.92
Cu	25.99
Zn	1016.62

have developed protection systems to overcome the oxidative damage by synthesis of phenolics and flavonoids. These compounds can detoxify reactive oxygen species (ROS) in plants, and also have the capacity to cure human diseases caused by oxidative damage and ageing. The concentrations of these phenolic acids and flavonoids were found to be remarkably increased with the severity of drought stress of amaranth leafy vegetable genotype VA3. Sarker and Oba (2018) reported that in this genotype the most abundant phenolic compounds were found consisting of five hydroxybenzoic acids (gallic acid, vanilic acid, *p*-hydroxybenzoic acid, salicylic acid (phenolic phytohormone) and ellagic acid) and six hydroxycinnamic acids (caffeic acid, chlorogenic acid, *trans*-cinnamic acid, *p*-coumaric acid, *m*-coumaric acid and sinapic acid). These were remarkably increased with the increment of the severity of drought stress (DS) in the order: Control <LDS (low) <MDS (moderate) <SDS (severe). In LDS, MDS and SDS, these phenolic acids and flavonoid concentrations were increased by (19, 4, 7, 9 6, 8, 4, 8, 13, 31 and 22%); (42, 32, 23, 19, 42, 26, 27, 105, 109, 47 and 50%) and (69, 59, 54, 66,111, 65, 72, 41, 115, 45, 154, 65 and 60%) respectively (Sarker and Oba, 2018). Such increased concentrations is positive news in times of climate change and global warming, because it means that farmers in semi-arid and dry areas of the world would be able to grow amaranth as a substitute crop. Plants can be protected with the production of phenols, flavonoids and antioxidants which are of value for human wellbeing.

Generally, plant phenolics protect plants against ultraviolet radiation or predators, aggressive pathogens and parasites, in addition to providing plant colour. Plant phenolics can act as antioxidants in various ways. Conforti *et al*. (2011) found that amaranth leaf extracts contained phenolics which prevented nitric oxide production and scavenged free radicals. Phenolic hydroxyl groups are a good source of hydrogen. These hydrogen-donating antioxidants can take up reactive oxygen and nitrogen species in a termination reaction, thus breaking the cycle of the generation of new free radicals. Among the phenolic compounds identified in amaranth stems, leaves and flowers, Neugart *et al*. (2017) recorded benzoic acid, hydroxycinnamic acids, and flavonols and their glycosides.

The main betacyanins identified by Li *et al*. (2015) in amaranth samples were amaranthine and isoamaranthine. Further extractions from different organs of amaranth revealed 11 phenolic compounds (chlorogenic acid, 2,4-dihydroxybenzoic acid, ellagic acid, ferulic acid, gallic acid, genistic acid, kaempferol-3-rutinoside, protocatechuic acid, quercetin, rutin and salicylic acid). The leaves revealed anticancer properties by inhibiting the proliferation of breast, colon and liver cancer lines and the researchers concluded that leaves are a good source of antioxidants. These properties were also reported in a review by Park *et al*. (2020) who confirmed the antioxidant properties and noted that amaranth chemo-preventative value may be exploited as an anti-tumor agent. Research on *A. caudatus* L. (Karamać *et al.*, 2019) recorded the presence of rutin, flavonols and hydroxycinnamic acid at different growth stages with the flowering one showing the highest antioxidant activity.

Postharvest handling and processing

Crop gathering in developing countries is achieved manually while in developed parts, machinery is used. Harvested mature leaves or young, tender uprooted plants are bundled, washed and packed to be transported to the site of sale. Leaves are sold by weight or volume and the young plants as bunches. Smallholder farmers in developing parts of the world experience challenges in extending the shelf life of produce as cold storage is unaffordable and electricity supply often interrupted. In stores and markets leaves can be kept fresh by sprinkling water on them to maintain freshness. Bunches of young plants may be kept fresh by keeping the roots immersed in water. Leaf powder can also be prepared for sale following sun drying. Despite these efforts vegetable crop loss following harvesting can in some cases exceed 25% (Chakraborty *et al.*, 2018). Nevertheless, proper postharvest handling methods such as modified atmosphere packaging (MAP) together with temperature control management can improve the quality and shelf life of vegetables, thus reducing losses and increasing availability and marketing of the product (Nyaura, 2020).

Several studies have explored extending the shelf life of amaranth while maintaining nutritional quality and improving safety, thus preventing postharvest losses (Kosgey, 2021). Nutrient loss is quite high with fresh amaranth losing over 80% of its initial β-carotene and vitamin C content. The effect of minimal processing such as slicing, blanching and drying techniques impacts on the nutritional quality and year-round properties of amaranth. Slicing and cooking of amaranth leaves causes a decrease in the mineral content (K, P, Mg) compared with sun-dried amaranth. Freeze-drying is used to extend the shelf life of foods by retarding lipid oxidation as well as preventing microbial growth (Marques *et al.*, 2009). In addition, it is used for industrial-scale preservation in the long-term storage of foods (Pérez-Gregorio *et al.*, 2011). In a study of African indigenous vegetables (AIVs), freeze-dried unsliced blanched had a significant difference on both β-carotene and vitamin C retention as the contents in blanched samples were lower by about 10–15% as compared to fresh leaf samples (Kosgey, 2021). The study further indicated that MAP in plastic bags, stage of maturity and blanching all had a significant effect on the stability of β-carotene and vitamin C of amaranth and other AIVs. The results of this study will be useful to AIV farmers in reducing postharvest losses.

Processing methods of amaranth, such as solar drying and cooking, alter antinutrients. Solar drying of amaranth lowers oxalic acid, phytates and polyphenols. According to Chege *et al.* (2014), solar drying concentrate per unit nutrients of amaranth leaves of ß-carotene, iron and zinc retention of 77.5%, 94.3% and 95.4% respectively. Drying leads to loss of water-soluble vitamins while fat-soluble vitamins are well retained. For example, vitamin C is greatly affected by processing methods including oxidative, photodegradation and enzymatic activities. Of note is the fact that drying significantly increases the carbohydrate and protein content in *Amaranthus hybridus* (66.74 and 12.09%), compared to blanching, boiling and the control (Kosgey, 2021). Of the three methods, drying is the most effective processing method in terms of retaining nutrients and is a worthwhile preservation method for amaranth leaves, but it takes longer to cook the dried matter (Kosgey, 2021).

Antinutritive compounds

Studies report that leaf amaranth contains antinutrient or toxic compounds which, if consumed in excess, could limit micronutrient uptake and affect their use as food. Some of the reported antinutrients in amaranth are phytic acid (PA), saponins and phenolic compounds, tannins, oxalates, protease inhibitors, lectins, alkaloids, polyphenolics and nitrates. Some of the phytochemicals are discussed below.

Phytates

Phytate in amaranth was found to be 1.32 mg/100 g (Akubugwo *et al.*, 2007) and spinach, 3660 mg/100 g (Sotelo *et al.*, 2010). Phytic acid, while it contains antinutrients, has been shown to have many important health benefits, including the prevention of cancer, heart disease, diabetes, kidney stones and osteoporosis. Phytic acid is also considered to be a natural antioxidant and is suggested to have potential functions of reducing lipid peroxidation and as a preservative in food. Phytates can also affect the endogenous pool of zinc which is critical to the human immune system (Cuadrado *et al.*, 2018).

Nitrates

Amaranth has a tendency to accumulate nitrates, with the value range of 0.307–0.733 mg in vegetable amaranth on a dried weight basis (DWB) (Bressani, 2003; Onyango *et al.*, 2008). The stems contain twice as much nitrate as the leaves. Application of nitrate-containing fertilizers increases the nitrate content of the plant. However, because of its low concentration and high solubility, it is leeched out during cooking so it is not problematic to humans. However, amaranth grown under dry conditions can be detrimental to livestock. Nitrates can build up in these plants under drought stress (Steckel, 2007). When these stressed plants are eaten by cattle, nitrate is converted to nitrite in their stomachs, which in high enough concentrations can be poisonous to them (Burrouis and Tyrl, 2001).

Oxalates and other phytochemicals

Oxalate consumption of 4–5 g in adults can result in death in humans and serious poisoning or death in domesticated animals and high toxicity and/or death in pastoral animals. However, in most AIVs the oxalate levels are far lower, although consumption of large amounts of sorrel and rhubarb can be dangerous and small cooked quantities are recommended. High oxalates have been known to have a negative effect of Ca and Fe absorption. Oxalates in plants may act as a buffer and protection in disease and pest resistance (Noonan and Savage, 1999). The highest level of oxalates in mg/100 g FW is: amaranth (10,050), taro (278–574) yam (486–781), beetroot (300–920), sweet potato (470), coriander leaf (1268), rhubarb (raw, 275–1336, cooked, 260), moringa (2754) (Falowo *et al.*, 2018), watercress (cooked, 1100) (Sotelo *et al.*, 2010) and spinach (varying from 400 to 900 mg/100 g). Sotelo *et al.* (2010) cite 7810 mg/100 g for spinach. Purple and green amaranth showed total oxalate of 8,106 and 10,056 mg/100 g DM (Radek and Savage, 2008). Oxalate accumulates during dry conditions, for example, spinach from 560 (winter) to 740 (summer) mg/100 g (Noonan and Savage, 1999). Leaf extracts in a quantitative phytochemical analysis revealed the presence of saponins (3%), tannins (7.5%), oxalates (10.3%) and cyanogenic glycosides (0.022%) but no alkaloids (Nwaogu *et al.*, 2006). Oxalates are important as they bind divalent minerals such as Ca, making them nutritionally unavailable. Their concentrations are 1–8% DWB (3.5 to 6.0 g in plants), their stems having a lower amount than the leaves (Bressani, 2003; Onyango *et al.*, 2008). During boiling, the amount of oxalate is reduced and drops to a non-toxic level (0.12–0.6 g), which is below toxic levels for humans of 2–5 g per day (Bressani, 2003). The application of humic acid (HA) via foliage spray or soil surface has shown to increase the betacyanin (an antioxidant) content and reduce the oxalate content in amaranth (Lestari and Dewi, 2020).

It can be seen in Table 4.5 that moringa (*Moringa oleifera* Lam.) has low antinutritive compounds compared to the other AIVs listed. Spinach has high oxalate with amaranth having a middle to high range. Nitrates are low and seldom cause a problem for humans as they are leached out in cooking in the AIVs listed (Bressani, 2003). Saponins are low for the AIVs listed. They do have antibiotic activity and are toxic to fish and insects but they appear to be practically non-toxic to humans (Oakenfull, 1981).

Novel Products/Formulations

Different organs or whole plants of selected *Amaranthus* species are used to prepare natural dyes which include red, yellow or green colours (Rastogi and Shukla, 2013). Plant dyes are selected because of the carcinogenic concerns

Table 4.5. Comparative value of antinutritive compounds of AIVs (mg/100g, DW).

AIV	Saponins	Phytates	Oxalate	Tannins	Nitrates	References
Moringa (control)	Low	Low	Low 1050–2574	Low 2070	Low 1700	Teixeira *et al.*, 2014
Amaranth	Low	1.32	3500–10 050	41.26	307–733	Radek and Savage, 2008 Falowo *et al.*, 2018; Jimoh *et al.*, 2020 Kundub *et al.*, 2021
Spinach	-	206	High 938–12570	62.5	High 11–165	Jaworska and Kmiecik, 1999; Radek and Savage, 2008 Iammarino *et al.*, 2014; Shankar *et al.*, 2014
Cowpea	0.06	2.90	1.42	2.75	-	Chikwendu *et al.*, 2014

around the use of synthetic dyes as food colourants. Red plant pigments are favoured for use in foods, medicines and beverages.

The leaves of amaranths are rich in calcium and magnesium and hence are useful in the treatment and prevention of osteoporosis and other ailments that are related to a deficiency of calcium. Moreover, magnesium is favoured for regulating blood sugar levels as it assists gluconeogenesis and as such can prevent diabetes (Harinarayan and Akhila, 2019). Hence, there is ongoing interest in the food and pharmaceutical industries in the mandatory fortification of foods with amaranth leaf and its products. This approach is supported by major concerns over the impact of fast foods on diabetes, obesity and other food-related health problems.

Future Trends

Driven by the need to reduce malnutrition and promote healthy indigenous foods, there has been increased research and development concerning production, establishing food supply and processing chains and chemical studies on the genus *Amaranthus*. However, there remains a need for greater commercialization, industrialization and food-processing technologies for amaranth products (Aderibigbe *et al.*, 2020). In Africa, the traditional methods of shade and sun-dried or freezer storage (where electricity and costs are affordable) are used in postharvesting. In a study determining the effect of blanching, boiling and drying on the nutritional composition of fresh and processed leaf samples, the results revealed that all the vegetables analysed contained an appreciable amount of moisture, ash, protein, fibre, carbohydrate, β-carotene and vitamins (Abdulrasheed-Adeleke *et al.*, 2021). While the results also revealed that all the processing methods employed significantly reduced the nutritional composition of the vegetables studied, drying in the shade retained more nutrients compared to the other processing methods. However, while conventional sun-drying is the cheapest and most accessible method for food preservation, β-carotene in amaranth was reduced more in open sun-dried vegetables as compared to solar-dried vegetables (Mulokozi and Svanberg, 2003). Good results for the stability of β-carotene content were observed for processed product in low temperature solar and cabinet dryers compared to shade and sun and this had the least drastic effect on the β-carotene, ascorbic acid and chlorophyll content of the processed product (Negi and Roy, 2000).

Several studies were conducted regarding solar- and freeze-drying techniques to promote the shelf-life of amaranth in Africa. For example, the installation of cost-effective solar dryers in place of traditional drying methods enables subsistence farmers to perform the postharvest drying at 50°C and this controlled drying method prevents the depletion of vitamins and functional compounds. In the case of freezer storage of amaranth microgreens, Rocchetti *et al.* (2020) investigated the combined effect of storage at 4°C for ten days on the phytochemical profile of amaranth microgreens. *In vitro* digestion found a significant increase in total phenolic content (36–88%), antioxidants (6–43%) and total betalains (41–57%), with maximum levels observed when the material was stored for ten days prior to digestion. Also, during the freeze-storage, the quality of the vegetables in terms of ascorbic acid, β-carotene and total phenolics is maintained when the vegetable was stored at 4°C for a period of up to four days (Onyango, 2010).

Evaporative cooling from 4–10°C is a low-cost temporary storage technology that also offers smallholder vegetable farmers an alternative to expensive cold rooms with minimal weight loss, better vegetable quality preservation and higher vitamin C preservation (51% loss at room temperature) compared to vegetables stored at ambient room conditions (Ambuko *et al.*, 2017).

Crop production technologies must be implemented to reduce the nitrate concentration of amaranth. More studies should highlight how the genetic background of amaranth microgreens and their production under effective light management can facilitate a product of superior quality (Kyriacou *et al.*, 2019).

Fast foods, such as instant noodles, eaten by many Asians and Africans, are a quick meal but they lack essential nutrients and are high in salt and carbohydrates which may cause obesity and cardio-related health risks (Huh *et al.*, 2017). However, noodles fortified with amaranth (morogo) leaf are nutritious and available on the market – launched in 2015 with Nestlé in partnership with the Nestlé Research Center in Switzerland, the South African Council for Scientific and Industrial Research and the Agricultural Research Council in South Africa (Nestlé, 2015). Renewed and sustained interest in this traditional indigenous food would also help reduce unemployment among the youth in rural populations through stimulating cultivation of amaranth and associated agricultural chains. More such product development is recommended with other AIVs (Qumbisa *et al.*, 2020).

A novel development has been the increased demand for amaranth leaf in aquaculture; it provides a cheaper alternative to fishmeal, which is expensive because of its protein ingredients. Amaranth supplemented fishmeal is characterized by a low content of antinutritional factors, mainly saponins and phytic acid, relative to other plants used in feeds (Ngugi *et al.*, 2017; Gołaś *et al.*, 2020). Collecting organic waste of amaranths from local vegetable markets for recycling on fish farms could provide employment. Further, betalain-rich vegetable amaranth (*A. tricolor* L.) extract blended into quaternary ammonium chitosan (QC)/fish gelatin (FG) films could be used as an antioxidant and antimicrobial food packaging material (Hu *et al.*, 2020).

The antifungal properties of amaranth leaf have been investigated as an alternative to artificial fungicides, which pose threats to the environment and human health through food contamination (Bahrami-Teimoori *et al.*, 2017). Ethyl acetate extract from *A. viridis* L. leaves contains potent antifungal constituents, which can be investigated in terms of designing commercial but ecofriendly antifungal compounds effective against a number of plant pathogens (Akbar *et al.*, 2020).

Further developmental research could enhance amaranth use as an energy crop for second-generation biofuels (Jamwal *et al.*, 2020). It is estimated that the total green protein supply collected from 1 hectare of amaranth plants could lead to an additional 6400–12,400 tons of bioethanol. This yield is ten-fold higher than that of soybean seed proteins and chemical surfactant (Madadi *et al.*, 2020).

There is current interest in highly pigmented salt-tolerant vegetable amaranth genotypes growing in high saline coastal belts around the globe, because they contain nutraceuticals, antioxidant phytochemicals and phenolics, suitable for making healthy drinks (Sarker *et al.*, 2020b). However, further detailed experimentation is required to standardize and stabilize the functional components of vegetable amaranth for extraction of juice.

There is potential for research into selective breeding of amaranth to improve horticultural varieties and seed yield, and macro- and micronutrients of cultivated amaranth species, including vitamin A, iron and zinc content. Finally, due to their harvesting high-leaf yield, ability to thrive under intolerant conditions, high nutritional content and minimal antinutrient compounds, as well their variety of possible uses, amaranth leaf species are likely to receive increased attention.

References

Aderibigbe, O.R., Ezekiel, O.O., Owolade, S.O., Korese, J.K., Sturm, B. *et al.* (2020) Exploring the potentials of underutilized grain amaranth (*Amaranthus* spp.) along the value chain for food and nutrition security: a review. *Critical Reviews in Food Science and Nutrition* 62(3), 656–669. DOI: 10.1080/10408398.2020.1825323.

Abdulrasheed-Adeleke, T., Shaibu, S.O.O., Agboola, A.R., Hamzah, R.U. and Garba, M.H. (2021) Comparative studies on the effect of processing methods on the nutritional value of *Basella alba*, *Talinum triangulare*, *Celosia argentea*, *Amaranthus hybridus* and *Gnetum africanum* leaves. *GSC Advanced Research and Reviews* 6(3), 044–052. DOI: 10.30574/gscarr.2021.6.3.0025.

Achigan-Dako, E.G., Sogbohossou, O.E. and Maundu, P. (2014) Current knowledge on *Amaranthus* spp.: research avenues for improved nutritional value and yield in leafy amaranths in sub-Saharan Africa. *Euphytica: Netherlands Journal of Plant Breeding* 197(3), 303–317. DOI: 10.1007/s10681-014-1081-9.

Adeagbo, O.A. and Adejumo, O.O. (2020) Economic analysis of dry season vegetable production in Ogun State, Nigeria. *African Journal of Economic and Management Studies* 11(3), 427–441. DOI: 10.1108/AJEMS-03-2019-0125.

Akbar, M., N. Sherazi, I., Khalil, T., S. Iqbal, M., Akhtar, S. *et al.* (2020) Identification of antifungal compounds from slender amaranth. *Planta Daninha* 38. DOI: 10.1590/s0100-83582020380100063.

Akinola, R., Pereira, L.M., Mabhaudhi, T., de Bruin, F.-M. and Rusch, L. (2020) A Review of indigenous food crops in Africa and the implications for more sustainable and healthy food systems. *Sustainability* 12(8), 3493. DOI: 10.3390/su12083493.

Akubugwo, I., Obasi, N., Chinyere, G. and Ugbogu, A. (2007) Nutritional and chemical value of *Amaranthus hybridus* L. leaves from Afikpo, Nigeria. *African Journal of Biotechnology* 6(24), 2833–2839. DOI: 10.5897/AJB2007.000-2452.

Alegbejo, J.O. (2013) Nutritional value and utilization of Amaranthus (*Amaranthus* spp.)–a review. *Bayero Journal of Pure and Applied Sciences* 6(1), 136–143. DOI: 10.4314/bajopas.v6i1.27.

Alemayehu, F.R., Bendevis, M.A. and Jacobsen, S.E. (2015) The potential for utilizing the seed crop Amaranth (*Amaranthus* spp.) in East Africa as an alternative crop to support food security and climate change mitigation. *Journal of Agronomy and Crop Science* 201(5), 321–329. DOI: 10.1111/jac.12108.

Ambuko, J., Wanjiru, F., Chemining'wa, G., Owino, W. and Mwachoni, E. (2017) Preservation of postharvest quality of leafy amaranth (*Amaranthus* spp.) vegetables using evaporative cooling. *Journal of Food Quality* 1–6. DOI: 10.1155/2017/5303156.

Bahrami-Teimoori, B., Nikparast, Y., Hojatianfar, M., Akhlaghi, M., Ghorbani, R. *et al.* (2017) Characterisation and antifungal activity of silver nanoparticles biologically synthesised by *Amaranthus retroflexus* leaf extract. *Journal of Experimental Nanoscience* 12(1), 129–139. DOI: 10.1080/17458080.2017.1279355.

Bressani, R. (2003) Amaranth. In: Caballero, B.T., Luiz, C. and Finglas, P.M. (eds) *Encyclopedia of Food Sciences and Nutrition*. Elsevier, Amsterdam, Netherlands, pp. 166–173.

Burrouis, G. and Tyrl, R. (2001) *Toxic Plants of North America*. Iowa State University, Ames, IA.

CABI (2020) *Amaranthus spinosus (spiny amaranth)*. Available at: https://www.cabi.org/isc/datasheet/46 53 (accessed 25 March 2022).

Chakraborty, I., Chattopadhyay, A., Chakraborty, I. and Chattopadhyay, A. (2018) Pre-and post-harvest losses in vegetables. *Advances in Postharvest Technologies of Vegetable Crops; CRC Press: Boca Raton, FL, USA* 1431(1), 25–87. DOI: 10.1111/nyas.13872.

Chege, P., Kuri, E., Kimiywe, J. and Nyambaka, H. (2014) Changes in nutrient content for β-carotene, iron and zinc in solar dried and stored *Amaranthus cruentus* vegetables. *International Journal of Agriculture Innovations and Research* 3, 880–882.

Chikwendu, J., Igbatim, A. and Obizoba, I. (2014) Chemical composition of processed cowpea tender leaves and husks. *International Journal of Scientific and Research Publications* 4, 1–5.

Conforti, F., Marrelli, M., Carmela, C., Menichini, F., Valentina, P. *et al.* (2011) Bioactive phytonutrients (omega fatty acids, tocopherols, polyphenols), *in vitro* inhibition of nitric oxide production and free radical scavenging activity of non-cultivated Mediterranean vegetables. *Food Chemistry* 129(4), 1413–1419. DOI: 10.1016/j.foodchem.2011.05.085.

Corke, H., Cai, Y. and Wu, H. (2016) Amaranth: Overview. In: G., W.S. (ed.) *Reference Module in Food Science*. Elsevier, Amsterdam, Netherlands, pp. 1–6.

Costea, M. and DeMason, D.A. (2001) Stem morphology and anatomy in *Amaranthus* L. (Amaranthaceae), taxonomic significance. *Journal of the Torrey Botanical Society* 128(3), 254–281. DOI: 10.2307/3088717.

Cuadrado, C., Takacs, K., Szabó, E. and Pedrosa, M. (2018) Non-nutritional factors: Lectins, Phytic Acid, Proteases Inhibitors, Allergens. In: Martín-Cabrejas, M.Á. (ed.) *Legumes*. Royal Chemistry Society, London, UK, pp. 152–176. DOI: 10.1039/9781788015721.

Dinssa, F.F., Hanson, P., Ledesma, D.R., Minja, R., Mbwambo, O. *et al*. (2019) Yield of vegetable amaranth in diverse Tanzanian production environments. *HortTechnology* 29(4), 516–527. DOI: 10.21273/HORTTECH04374-19.

Dmitrieva, O. and Ivanov, S. (2020) Comparative study of amaranth species (*Amaranthus* spp.) in the temperate continental climate of Russian Federation. *Acta Agriculturae Slovenica* 115(1), 15–24. DOI: 10.14720/aas.2020.115.1.1281.

D'Amico, S. and Schoenlechner, R. (2017) Amaranth: its unique nutritional and health-promoting attributes. In: *Gluten-Free Ancient Grains*. Elsevier, pp. 131–159.

Ebert, A.W. (2020) The role of vegetable genetic resources in nutrition security and vegetable breeding. *Plants* 9(6), E736. DOI: 10.3390/plants9060736.

Ebert, A., Wu, T. and Wang, S. (2011) Vegetable amaranth (*Amaranthus* L.). International Cooperators' Guide. AVRDC Publication Number: 11-754. Shanhua, Tainan, Taiwan. AVRDC -The World Vegetable Center, p. 9.

Ebert, A., Wu, T. and Yang, R. (2014) Amaranth sprouts and microgreens – a homestead vegetable production option to enhance food and nutrition security in the rural-urban continuum. In: *Proceedings of the Regional Symposium on Sustaining Small-Scale Vegetable Production and Marketing Systems for Food and Nutrition Security (SEAVEG 2014)*, Bangkok, Thailand, p. 227.

Falowo, A.B., Mukumbo, F.E., Idamokoro, E.M., Lorenzo, J.M., Afolayan, A.J. *et al*. (2018) Multi-functional application of *Moringa oleifera* Lam. in nutrition and animal food products: A review. *Food Research International* 106, 317–334. DOI: 10.1016/j.foodres.2017.12.079.

Gołaś, I., Potorski, J., Woźniak, M., Niewiadomski, P., Aguilera-Arreola, M.G. *et al*. (2020) Amaranth Meal and Environmental *Carnobacterium maltaromaticum* Probiotic Bacteria as Novel Stabilizers of the Microbiological Quality of Compound Fish Feeds for Aquaculture. *Applied Sciences* 10(15), 5114. DOI: 10.3390/app10155114.

Grubben, G. and Van Sloten, D. (1981) *Genetic Resources of Amaranth - a Global Plan of Action*. IBPGR, Rome.

Harinarayan, C.V. and Akhila, H. (2019) Modern India and the Tale of Twin Nutrient Deficiency-Calcium and Vitamin D - Nutrition Trend Data 50 Years - Retrospect, Introspect, and Prospect. *Frontiers in Endocrinology* 10, 493. DOI: 10.3389/fendo.2019.00493.

Hoidal, N., Jacobsen, S.E., Odone, A. and Alandia, G. (2020a) Defoliation timing for optimal leaf nutrition in dual-use amaranth production systems. *Journal of the Science of Food and Agriculture* 100(13), 4745–4755. DOI: 10.1002/jsfa.10533.

House, N.C., Puthenparampil, D., Malayil, D. and Narayanankutty, A. (2020) Variation in the polyphenol composition, antioxidant, and anticancer activity among different *Amaranthus* species. *South African Journal of Botany* 135, 408–412. DOI: 10.1016/j.sajb.2020.09.026.

Hu, H., Yao, X., Qin, Y., Yong, H. and Liu, J. (2020) Development of multifunctional food packaging by incorporating betalains from vegetable amaranth (*Amaranthus tricolor* L.) into quaternary ammonium chitosan/fish gelatin blend films. *International Journal of Biological Macromolecules* 159, 675–684. DOI: 10.1016/j.ijbiomac.2020.05.103.

Huh, I.S., Kim, H., Jo, H.K., Lim, C.S., Kim, J.S. *et al*. (2017) Instant noodle consumption is associated with cardiometabolic risk factors among college students in Seoul. *Nutrition Research and Practice* 11(3), 232–239. DOI: 10.4162/nrp.2017.11.3.232.

Iammarino, M., Di Taranto, A. and Cristino, M. (2014) Monitoring of nitrites and nitrates levels in leafy vegetables (spinach and lettuce): a contribution to risk assessment. *Journal of the Science of Food and Agriculture* 94(4), 773–778. DOI: 10.1002/jsfa.6439.

Jamwal, V.L., Kapoor, N. and Gandhi, S.G. (2020) Biotechnology of biofuels: historical overview, business outlook and future perspectives. In: *Biotechnology Business-Concept to Delivery*. Springer, pp. 109–127.

Jannat, K., Jahan, R., Rahman, T., Hossan, M.S., Shova, N.A. *et al.* (2020) Some plants used as phytomedicine by tribal healers of Chittagong Hill Tracts, Bangladesh. In: Mahendra Rai, S.B. and Chistiane, M.F. (eds) *Wild Plants: The Treasure of Natural Healers*. CRC Press, New York, p. 90.

Jansen, P. (2004) *Amaranthus hypochondriacus* L. (PROTA - Plant Resources of Tropical Africa), Issue. Available at: http://www.prota4u.org/search.asp

Jaworska, G. and Kmiecik, W. (1999) Content of selected mineral compounds, nitrates III and Vand oxalates in spinach [*Spinacia oleracea* L.] and New Zealand spinach [*Tetragonia expansa* Murr.] from spring and autumn growing seasons. *Electronic Journal of Polish Agricultural Universities. Series Food Science and Technology* 2(2). Available at: http://www.ejpau.media.pl/articles/volume2/issue 2/food/art-03.pdf

Jeph, A. and Khan, J. (2020) Ethnomedicinal study in reserve forest area of Jhunjhunu District, Rajasthan, India. *The Journal of the Society for Tropical Plant Research* 7(2), 379–387. DOI: 10.22271/tpr.2020.v7.i2.044.

Jimoh, M.O., Afolayan, A.J. and Lewu, F.B. (2020) Nutrients and antinutrient constituents of *Amaranthus caudatus* L. Cultivated on different soils. *Saudi Journal of Biological Sciences* 27(12), 3570–3580. DOI: 10.1016/j.sjbs.2020.07.029.

Joshi, N. and Verma, K.C. (2020) A review on nutrition value of amaranth (*Amaranthus caudatus* L.): the crop of future. *Journal of Pharmacognosy and Phytochemistry* 9, 1111–1113.

Karamać, M., Gai, F., Longato, E., Meineri, G., Janiak, M.A. *et al.* (2019) Antioxidant activity and phenolic composition of amaranth (*Amaranthus caudatus*) during plant growth. *Antioxidants* 8(6), 173. DOI: 10.3390/antiox8060173.

Khatri-Chhetri, A., Regmi, P.P., Chanana, N. and Aggarwal, P.K. (2020) Potential of climate-smart agriculture in reducing women farmers' drudgery in high climatic risk areas. *Climatic Change* 158(1), 29–42. DOI: 10.1007/s10584-018-2350-8.

Kosgey, G.C. (2021) *Nutritional Quality and Anti-Oxidant Properties of Minimally Processed African Indigenous Leafy Vegetables* [MSc, JKUAT-AGRICULTURE]. Kenya. Available at: http://ir.jkuat.ac.ke /handle/123456789/5430 (accessed 20 May 2020).

Kundub, P., Sethia, N., Guptab, P. and Bhattic, J.S. (2021) Evaluation of phytochemicals and anti-nutritional profile in underutilised green leafy vegetables. *European Journal of Molecular & Clinical Medicine* 8, 443–455.

Kunwar, R.M., Mahat, L., Sharma, L.N., Shrestha, K.P., Kominee, H. *et al.* (2012) Underutilized plant species in far west Nepal. *Journal of Mountain Science* 9(5), 589–600. DOI: 10.1007/s11629-012-2315-8.

Kyriacou, M.C., El-Nakhel, C., Pannico, A., Graziani, G., Soteriou, G.A. *et al.* (2019) Genotype-specific modulatory effects of select spectral bandwidths on the nutritive and phytochemical composition of microgreens. *Frontiers in Plant Science* 10, 1501. DOI: 10.3389/fpls.2019.01501.

Lemmens, R. and Bunyapraphatsara, N. (1999) *Amaranthus spinosus* L. In: Padua, L.S. d., Bunyapraphatsara, N., and Lemmens, R.H.M.J. (eds) *Plant Resources of South-East Asia*. Backhuys Publishers, pp. 110–113.

Lestari, F. and Dewi, K. (2020) Effects of humic acid on vegetative growth, yield, oxalic acid and betacyanin content of red amaranth (*Amaranthus tricolor* L.). 2260(1). In: *AIP Conference Proceedings*, AIP Publishing LLC, p030011. DOI: 10.1063/5.0015774.

Li, H., Deng, Z., Liu, R., Zhu, H., Draves, J. *et al.* (2015) Characterization of phenolics, betacyanins and antioxidant activities of the seed, leaf, sprout, flower and stalk extracts of three *Amaranthus* species. *Journal of Food Composition and Analysis* 37, 75–81. DOI: 10.1016/j.jfca.2014.09.003.

Lindsey, K., Motsei, M.L. and Jager, A.K. (2002) Screening of South African food plants for antioxidant activity. *Journal of Food Science* 67(6), 2129–2131. DOI: 10.1111/j.1365-2621.2002.tb09514.x.

Madadi, M., Wang, Y., Xu, C., Liu, P., Wang, Y. *et al.* (2020) Using *Amaranthus* green proteins as universal biosurfactant and biosorbent for effective enzymatic degradation of diverse lignocellulose residues and efficient multiple trace metals remediation of farming lands. *Journal of Hazardous Materials* 406, 124727. DOI: 10.1016/j.jhazmat.2020.124727.

Mahangade, P.S., Mani, I., Beaudry, R., Müller, N. and Chopra, S. (2020) Using amaranth as a model plant for evaluating imperfect storages: assessment of solar-refrigerated and evaporatively-cooled structures in India. *HortScience 1*(aop)(11), 1–7. DOI: 10.21273/HORTSCI15249-20.

Makobo, N., Shoko, M. and Mtaita, T. (2010) Nutrient content of Amaranth (*Amaranthus cruentus* L.) under different processing and preservation methods. *World Journal of Agricultural Sciences* 6(6), 639–643.

Marques, L.G., Prado, M.M. and Freire, J.T. (2009) Rehydration characteristics of freeze-dried tropical fruits. *LWT - Food Science and Technology* 42(7), 1232–1237. DOI: 10.1016/j.lwt.2009.02.012.

Moloto, M.R., Phan, A.D.T., Shai, J.L., Sultanbawa, Y. and Sivakumar, D. (2020) Comparison of phenolic compounds, carotenoids, amino acid composition, in vitro antioxidant and anti-diabetic activities in the leaves of seven cowpea (*Vigna unguiculata*) cultivars. *Foods* 9(9), 1285. DOI: 10.3390/foods9091285.

Mulokozi, G. and Svanberg, U. (2003) Effect of traditional open sun-drying and solar cabinet drying on carotene content and vitamin A activity of green leafy vegetables. Plant foods for human nutrition 58(3), 1–15. DOI: 10.1023/B:QUAL.0000041153.28887.9c.

Nataru, S., Pulicherla, Y. and Gaddala, B. (2014) A review on medicinal plants as A potential source for cancer. *International Journal of Pharmaceutical Sciences Review and Research* 26, 235–248.

Negi, P.S. and Roy, S.K. (2000) Effect of blanching and drying methods on β-carotene, ascorbic acid and chlorophyll retention of leafy vegetables. *LWT - Food Science and Technology* 33(4), 295–298. DOI: 10.1006/fstl.2000.0659.

Nestlé (2015) *Nestlé in society Creating Shared Value and meeting our commitments 2015*. Available at: https://www.google.com/search?client=firefox-b-d&q=Nestl%C3%A9+in+society+Creating+Shared+Value+and+meeting+our+commitments+2015

Neugart, S., Baldermann, S., Ngwene, B., Wesonga, J. and Schreiner, M. (2017) Indigenous leafy vegetables of Eastern Africa - A source of extraordinary secondary plant metabolites. *Food Research International* 100, 411–422. DOI: 10.1016/j.foodres.2017.02.014.

Ngugi, C.C., Oyoo-Okoth, E., Manyala, J.O., Fitzsimmons, K. and Kimotho, A. (2017) Characterization of the nutritional quality of amaranth leaf protein concentrates and suitability of fish meal replacement in Nile tilapia feeds. *Aquaculture Reports* 5, 62–69. DOI: 10.1016/j.aqrep.2017.01.003.

Nielsen, S., Ohler, T. and Mitchell, C. (1997) Cowpea leaves for human consumption: production, utilization and nutrient composition. In: Singh, B., Mohan Raj, D., Dashiell, K., and Jackai, L. (eds) *Adv Cowpea Res*, Vol. 326. Corporativo, IITA, pp. 326–332.

Noonan, S.C. and Savage, G.P. (1999) Oxalate content of foods and its effect on humans. *Asia Pacific Journal of Clinical Nutrition* 8(1), 64–74.

Nwaogu, L., Ujowundu, C. and Mgbemena, A. (2006) Studies on the nutritional and phytochemical composition of *Amaranthus hybridus* leaves. *Bio-research* 4(1), 28–31. DOI: 10.4314/br.v4i1.28608.

Nyaura, J.A. (2020) *Shelf Life Stability and Nutrient Content of Vegetable Amaranth*. MSc, JKUAT-IBR. Available at: http://ir.jkuat.ac.ke/handle/123456789/5241 (accessed 20 May 2020).

Nzomo, E., Ariyawardana, A., Sila, D. and Sellahewa, J. (2015) Reaping the potential benefits of amaranth: Value chain challenges ahead for Kenya. *Acta Horticulturae* 1102(1102), 191–198. DOI: 10.17660/ActaHortic.2015.1102.24.

Oakenfull, D. (1981) Saponins in food—a review. *Food Chemistry* 7(1), 19–40. DOI: 10.1016/0308-8146(81)90019-4.

Onyango, C. (2010) *Preharvest and postharvest factors affecting yield and nutrient contents of vegetable amaranth (Var. Amaranthus hypochondriacus)*, PhD, Wageningen University, Wageningen, Netherlands. Available at: https://library.wur.nl/WebQuery/wurpubs/fulltext/136215 (accessed 20 May 2020).

Onyango, C.M., Shibairo, S.I., Imungi, J.K. and Harbinson, J. (2008) The physico-chemical characteristics and some nutritional values of vegetable amaranth sold in Nairobi-Kenya. *Ecology of Food and Nutrition* 47(4), 382–398. DOI: 10.1080/03670240802003926.

Park, S.-J., Sharma, A. and Lee, H.-J. (2020) A Review of Recent Studies on the Antioxidant Activities of A Third-Millennium Food: *Amaranthus* spp. *Antioxidants* 9(12), 1236. DOI: 10.3390/antiox9121236.

Peiretti, P. (2018) Amaranth in animal nutrition: a review. *Livestock Research for Rural Development* 30(5). Available at: http://www.lrrd.org/lrrd30/5/peir30088.html (accessed 20 May 2020).

Pérez-Gregorio, M., Regueiro, J., González-Barreiro, C., Rial-Otero, R. and Simal-Gándara, J. (2011) Changes in antioxidant flavonoids during freeze-drying of red onions and subsequent storage. *Food Control* 22(7), 1108–1113. DOI: 10.1016/j.foodcont.2011.01.006.

Qumbisa, N.D., Ngobese, N., Kolanisi, U, *et al*. (2020) Potential of using amaranthus leaves to fortify instant noodles in the South African context: A review. *African Journal of Food, Agriculture, Nutrition and Development* 20(4), 16099–16111. DOI: 10.18697/ajfand.92.18690.

Radek, M. and Savage, G.P. (2008) Oxalates in some Indian green leafy vegetables. *International Journal of Food Sciences and Nutrition* 59(3), 246–260. DOI: 10.1080/09637480701791176.

Ramanlal, D.B., Kumar, R.N., Kumar, J.N. and Thakkar, R. (2020) Assessment of phytoremedial potential of invasive weeds *Acalypha indica* and *Amaranthus viridis*. *Environmental Sustainability* 3(4), 415–425. DOI: 10.1007/s42398-020-00129-7.

Rastogi, A. and Shukla, S. (2013) Amaranth: a new millennium crop of nutraceutical values. *Critical Reviews in Food Science and Nutrition* 53(2), 109–125. DOI: 10.1080/10408398.2010.517876.

Rocchetti, G., Tomas, M., Zhang, L., Zengin, G., Lucini, L. *et al.* (2020) Red beet (*Beta vulgaris*) and amaranth (*Amaranthus* sp.) microgreens: Effect of storage and in vitro gastrointestinal digestion on the untargeted metabolomic profile. *Food Chemistry* 332, 127415. DOI: 10.1016/j.foodchem.2020.127415.

Saravanan, R. (2016) *Evaluation of In-Vitro and In-Vivo Anticancer Activity of Leaf Extracts of Amaranthus Spinosus Linn*, MSc, College of Pharmacy Madras Medical College, Chennai.

Sarker, U. and Oba, S. (2018) Drought stress enhances nutritional and bioactive compounds, phenolic acids and antioxidant capacity of *Amaranthus* leafy vegetable. *BMC Plant Biology* 18(1), 258. DOI: 10.1186/s12870-018-1484-1.

Sarker, U., Hossain, M.M. and Oba, S. (2020a) Nutritional and antioxidant components and antioxidant capacity in green morph *Amaranthus* leafy vegetable. *Scientific Reports* 10(1), 1–10. DOI: 10.1038/s41598-020-57687-3.

Sarker, U., Hossain, M.N., Iqbal, M.A. and Oba, S. (2020b) Bioactive Components and Radical Scavenging Activity in Selected Advance Lines of Salt-Tolerant Vegetable Amaranth. *Frontiers in Nutrition* 7, 587257. DOI: 10.3389/fnut.2020.587257.

Shankar, K.S., Suma, K., Chary, K.B., Kumar, A.P., Grover, M. *et al.* (2014) Improving phytochemical and nutritional quality of spinach (*Spinacia oleracea*) through phosphate solubilizing bacteria. *Indian Journal of Dryland Agricultural Research and Development* 29(2), 104. DOI: 10.5958/2231-6701.2014.01224.X.

Sharma, J., Gairola, S., Sharma, Y.P. and Gaur, R. (2014) Ethnomedicinal plants used to treat skin diseases by Tharu community of district Udham Singh Nagar, Uttarakhand, India. *Journal of Ethnopharmacology* 158, 140–206. DOI: 10.1016/j.jep.2014.10.004.

Shiyam, J. and Binang, W. (2011) Effect of poultry manure and urea-n on flowering occurrence and leaf productivity of amaranthus cruentus. *Journal of Applied Sciences and Environmental Management* 15(1), 13–15.

Sotelo, A., González-Osnaya, L., Sánchez-Chinchillas, A. and Trejo, A. (2010) Role of oxate, phytate, tannins and cooking on iron bioavailability from foods commonly consumed in Mexico. *International Journal of Food Sciences and Nutrition* 61(1), 29–39. DOI: 10.3109/09637480903213649.

Stallknecht, G. and Schulz-Schaeffer, J. (1993) Amaranth rediscovered. In: Janick, J. and Simon, J.E. (eds) *New Crops*. Wiley, Oxford, UK, pp. 211–218.

Steckel, L.E. (2007) The dioecious *Amaranthus* spp.: here to stay. *Weed Technology* 21(2), 567–570. DOI: 10.1614/WT-06-045.1.

Stetter, M.G. and Schmid, K.J. (2017) Analysis of phylogenetic relationships and genome size evolution of the *Amaranthus* genus using GBS indicates the ancestors of an ancient crop. *Molecular Phylogenetics and Evolution* 109, 80–92. DOI: 10.1016/j.ympev.2016.12.029..

Sulaiman, M.I. and Andini, R. (2020) Potential of Amaranth in Alleviating Malnutrition in Indonesia. In: Amaranth, N.V. o (ed.) *Nutritional Value of Amaranth*. IntechOpen, pp. 1–12. DOI: 10.5772/intechopen.88889.

Tang, Y. and Tsao, R. (2017) Phytochemicals in quinoa and amaranth grains and their antioxidant, anti-inflammatory, and potential health beneficial effects: a review. *Molecular Nutrition & Food Research* 61(7), 1–16. DOI: 10.1002/mnfr.201600767.

Teixeira, E.M.B., Carvalho, M.R.B., Neves, V.A., Silva, M.A. and Arantes-Pereira, L. (2014) Chemical characteristics and fractionation of proteins from *Moringa oleifera* Lam. leaves. *Food Chemistry* 147, 51–54. DOI: 10.1016/j.foodchem.2013.09.135.

Thapa, R. and Blair, M.W. (2018) Morphological assessment of cultivated and wild amaranth species diversity. *Agronomy* 8(11), 272. DOI: 10.3390/agronomy8110272.

The Plant List (2013) *Version 1.1*. Available at: http://www.theplantlist.org (accessed 20 May 2020).

ŢÎŢEI, V. (2020) The quality of green mass and silage from Amaranthus hypochondriacus growing under the conditions of the Republic of moldova. *Studii Şi Comunicări, Ştiinţele Naturii* 36, 51–56.

Topwal, M. (2019) Review on amaranth: nutraceutical and virtual plant for providing food security and nutrients. *Acta Scientific Agriculture* 3, 9–15.

USDA SR Legacy (2018) U.S. Department of Agriculture. Available at: https://fdc.nal.usda.gov/faq.html (accessed 20 May 2020).

Van Wyk, B.-E. and Gericke, N. (2017) *People's Plants: A Guide to Useful Plants of Southern Africa*, 2nd edn. Briza Publications.

Veda, S., Kamath, A., Platel, K., Begum, K. and Srinivasan, K. (2006) Determination of bioaccessibility of β-carotene in vegetables by in vitro methods. *Molecular Nutrition & Food Research* 50(11), 1047–1052. DOI: 10.1002/mnfr.200600076.

Vincent, L., Sivaraj, N., Anushma, P., Ganeshan, S. and Rajasekharan, P. (2015) Diversity, distribution, collection and conservation of amaranth germplasm from Andhra Pradesh. In: *III International Symposium on Underutilized Plant Species 1241*.

5 African Pumpkin (*Momordica balsamina* L.)

Petunia Mashiane[1]* and Dharini Sivakumar[2]
[1]Department of Horticulture, Tshwane University of Technology, Pretoria, South Africa; [2]Phytochemical Food Network Research Group, Tshwane University of Technology, Pretoria, South Africa

Abstract

African pumpkin (*Momordica balsamina* L.) leaves have been used as food and in traditional folk medicine from ancient times due to the presence of phytochemicals and proteins. Research has shown African pumpkin leaves have the potential to manage chronic diseases such as diabetes, cancer, ulcers and hypertension. It is evident from the literature that African pumpkin leaves contain high amounts of protein as compared to other indigenous and exotic leafy vegetables such as lettuce, spinach, moringa and amaranthus. This chapter includes the health/traditional uses, nutritional values/promoting compounds, antioxidant activity and biological properties of African pumpkin leaves, and describes the changes in phytochemical components during preharvest and postharvest processing, to give insight to the content at consumption stage. The information presented in this chapter will benefit food producers and consumers.

Introduction

African pumpkin leaves possess nutraceutical properties and have the potential to be included as part of sustainable food security strategies (Odhav *et al.*, 2007; Jibril *et al.*, 2018). Young African pumpkin leaves contain momordicin (3, 7, 23-trihydroxycucurbitan-5, 24-dien-19-al), an alkaloid mainly responsible for their medicinal properties (Karumi *et al.*, 2004). African pumpkin has been used as a food and as a remedy for various types of disease such as diabetes (Dias, 2019; Kabir *et al.*, 2019). Thakur *et al.* (2009) regarded this vegetable as a 'hidden gift of nature' in his monograph. African pumpkin is known in South Africa as Inkakha (Tsonga), Intshungu (IsiZulu), Balsamina apple, Balsamina pear and Ikakha (Isindebele); in the USA as southern balsam pear (or balsampear) and wonder-apple; in Spain as balsamina; and in Germany as Balsamapfel, Gemeiner Balsamapfel and Gemeiner Balsamkuerbisand Wunderapfel (Spengler *et al.*, 2009; CABI, 2020). It is also known as African cucumber, balsam apple and southern balsam pear (CABI, 2020).

African pumpkins have long been part of traditional diets in southern African communities (Matenge *et al.*, 2017). They are cheap and contain readily available sources of several important micronutrients (Bvenura and Afolayan, 2015). Therefore, inclusion of this leafy vegetable in the daily diet needs to be promoted as a practical and sustainable way to benefit the health of consumers (Bvenura and Afolayan, 2015; Mncwango

*Corresponding author: MashianeP@tut.ac.za

DOI: 10.1079/9781789248067.0005

et al., 2020). In developing countries, starch-based staple foods provide energy, but protein deficiency still prevails among populations; therefore, identifying traditional leafy vegetables that are rich in protein can help to alleviate this. Authors have reported that African pumpkin leaves could be promoted as a protein supplement for cereal-based diets in poor rural communities (Hassan and Umar, 2006; Jibril *et al.*, 2018; Kadiri *et al.*, 2020). Furthermore, African pumpkin leaves contain high amounts of potassium, which is a good source for the management of hypertension and other cardiovascular conditions (Hassan and Umar, 2006). Traditional vegetables play a major role in diet diversification strategies for the sub-Saharan African population to combat hidden hunger (Burchi *et al.*, 2011). Rural communities use cooked African pumpkin leaves as a vegetables, consumed with the staple starch maize meal porridge. Alternatively the leaves and green fruits are cooked together with crushed ground nuts as a gravy (Mncwango *et al.*, 2020). Some African tribes (Tonga) use the boiled plant, cooked with nuts or maize meal, and then discard the boiled water (Mncwango *et al.*, 2020). Such traditional practices can bring about changes in the phytonutritional content of African pumpkin leaves. These changes can be positive or negative to health. Traditional vegetables are generally consumed as stir-fries or cooked as curries. They are also boiled, steamed and drunk as a tea (Mashiane *et al.*, 2021).

Traditional vegetables were an integral part of traditional agricultural systems in the past, but due to rapid urbanization and the introduction to fast food chains, the notion of 'low-status food' adversely affected the consumption of these vegetables (Faber *et al.*, 2010; Fabbri and Crosby, 2016). The consumption of African pumpkin leaves as a vegetable has declined compared to other commonly consumed traditional leafy vegetables such as *C. olitorius*, *C. gynandra*, pumpkin leaves and *Amaranthus* species (Bvenura and Afolayan, 2015). These traditional leafy vegetables are currently sold in supermarkets and traditional food markets while African pumpkin leaves are only sold on the streets in informal markets and in small quantities, despite the fact that they are high in proteins compared to other conventional leafy vegetables such as cabbage, spinach, lettuce and kale (Bvenura and Afolayan, 2015).

In addition to the changes in the phytonutritional compounds of the leaves during maturity, postharvest handling and food processing techniques such as cooking and drying practices can affect health benefits (Nagarani *et al.*, 2014; Essack *et al.*, 2017). In the literature, African pumpkin leaves have been well documented for their medicinal properties, however there are still gaps in the information regarding phytonutritional properties, preservation techniques during offseason and bioactive properties. Therefore, this chapter highlights traditional uses, nutrition and phytochemical composition, antinutritive compounds and changes in health-promoting compounds during postharvest processing (cooking), giving an insight to content at the consumption stage, along with biological properties. Furthermore, we attempt to identify areas of weakness in the current data and existing gaps that need attention to maximize the use of African pumpkin leaves as a food source with great health benefits.

Botany

African pumpkin (*Momordica balsamina* L.) is indigenous to South Africa and tropical Africa, tropical Asia, Arabia, India and Australia (CABI, 2020). The genus name *Momordica* – from the Latin *mordicus*, biting – alludes to the sculptured seed surfaces and margins, giving a bitten appearance (Singh and Devi, 2018). The specific epithet *balsamina*, from the Latin *balsamum*, refers to one of the medicinal uses of this plant (CABI, 2020). It is a trailing or climbing annual or perennial tendril-bearing herb (Thakur *et al.*, 2009; Jibril *et al.*, 2018; CABI, 2020; Kadiri *et al.*, 2020). The leaf-lamina (Fig. 5.1) is 1–9 cm long and 1.2–12 cm in width, broadleafed and deeply five-lobed to seven-lobed, lobes sinuate-toothed to lobed, acute; petiole 1.5 cm long (CABI, 2020; Kadiri *et al.*, 2020). Species of this genus are characterized by seeds that are always enveloped in a bright red pulp and male flowers are often subtended by prominent bracts (Idris *et al.*, 2018; CABI, 2020).

Fig. 5.1. *Momordica balsamina* Linn., African pumpkin.

Traditional Uses

African pumpkin is used as a food and a bitter flavouring as well as for medicinal purposes (Hassan and Umar, 2006). Traditional uses include: paste or hot water extract to reduce or control the symptoms of various diseases such as leprosy, piles and jaundice (Muronga *et al.*, 2020); the treatment of ringworm, bowel movement, cough, congestion and chest pain (Thakur *et al.*, 2009); treatment of boils, burns, external sores, frostbite, haemorrhoids, headache; and as a purgative for children (Mshelia and Karumi, 2013; Kamalaja *et al.*, 2018). The leaves have been used as an antiemetic against vomiting, nausea, malaria fever (Muronga *et al.*, 2020; Olalere and Gan, 2020), syphilis, rheumatism, hepatitis, stomach and intestinal disorders (Bello *et al.*, 2018). In Nigeria, the leaves are cooked with other green vegetables as a soup especially for lactating mothers (Hassan and Umar, 2006). The leaf sap of African pumpkin is a useful metal cleaner and has been used as a soap (CABI, 2020). African pumpkin leaves are also used for the treatment of excessive uterine bleeding and moist chopped leaves are placed on the navel wounds of newly born babies as a folklore practice in Nigeria (Bello *et al.*, 2018). Thakur *et al.* (2009) also reported that leaves have been used in reducing and relieving menstrual pains for young girls. The leaves can also be used as a tea for blood liver deficiencies (Thakur *et al.*, 2009) and digestion disorders (Muronga *et al.*, 2020).

Phytonutritional Composition

Macro- and micronutrients

The proximate composition of African pumpkin leaves has been reported by two research groups from Africa (Hassan and Umar, 2006; Kadiri *et al.*, 2020). The protein content of African pumpkin leaves (28.77 g/100 g) is very high compared to commonly consumed vegetables such as lettuce (0.9 g/100 g), spinach (2.35 g/100 g), Swiss chad (1.8 g/100 g) and other indigenous plants (moringa 9.4 g/100 g), amaranthus (2.46 g/100 g), pumpkin leaves (3.15 g/100 g), sweet potato leaves (2.49 g/100 g) (USDA Food data, 2020) and African nightshade (*Solanum retroflexum*) leaves (6.0 g/100 g) (Yuan *et al.*, 2018), but lower than cowpea leaves (cultivars VOP3 30.2 g/100 g and VOP4 31.3 g/100 g) (Moloto *et al.*, 2020). Proteins are large complex molecules composed of amino acids and play a vital role in cellular function and structure, and in regulation of metabolic activities in all living organisms. Hence, they are of primary importance in the daily human diet (Natesh *et al.*, 2017). Differences in the agroclimatic conditions may account for variations in protein content (Nagarani *et al.*, 2014). Protein deficient diets are almost certain to be generally nutrient-poor ones. The potentially adverse health outcomes of protein deficiency (e.g. poor bone health) (WHO/FAO/UNU, 2007) include significant stunting and wasting. In South Africa about 16% of children are underweight for their age and one in five are stunted (Schönfeldt and Gibson Hall, 2012). Hassan and Umar (2006), Jibril *et al.* (2018) and Kadiri *et al.* (2020) have reported that African pumpkin leaves could be promoted as a protein supplement for cereal-based diets in poor rural communities. Safe protein intake was set at 0.8 g/kg per day for both men and women, determined by the WHO/FAO/UNU (2007) guidelines. Therefore, 56 g of protein intake per day for an average body weight of 70 kg for a man or woman is recommended. Thus, African pumpkin leaves contribute 51.2%

of the recommended daily allowance (RDA) for a man or woman with this average weight. RDA is an average daily dietary intake level sufficient to meet the nutrient requirements for nearly all (97–98%) healthy individuals in a particular life stage and gender group (Van Jaarsveld *et al.*, 2014).

Macro elements are minerals that are necessary for the human body in large quantities, while trace elements are needed in small quantities. Different minerals play different roles in human health. Calcium and magnesium are important for human bones and tooth development, muscle function and metabolic processes (Kadiri *et al.*, 2020). Table 5.1 shows the mineral composition of African pumpkin leaves as compared to other leafy vegetables (indigenous and exotic). The data shows that African pumpkin leaves contain higher levels of potassium (1, 320 mg/100 g) as compared to other leafy vegetables. Lettuce contains the lowest amounts of potassium (141 mg/100 g). African pumpkin leaves contain the highest amounts of calcium as compared to other leafy vegetable (941 mg/100 g), while lettuce contains the lowest amounts of calcium (18 mg/100 g) (Hassan and Umar, 2006). Overall, African pumpkin leaves contain high amounts of minerals as compared to other leafy vegetables (Table 5.1). African pumpkin leaves contribute the highest percentage RDA (94.1%) for calcium as compared to other conventional leafy vegetables, followed by spinach with an RDA of 94% calcium, while lettuce contains the lowest calcium RDA percentage (1.8%). Potassium and magnesium levels range from 1% > of RDA for all the different traditional leafy vegetables, with lettuce having the lowest RDA value of 3% for potassium and 1.75% of magnesium. Zinc content showed a large variation, with lettuce having the lowest (<10% of the RDA).

Flyman and Afolayan (2007) studied the effect of plant maturity on the mineral content of African pumpkin leaves (Table 5.2). Their results showed that zinc content was higher at the initial stage and decreased continuously until the final stage. The iron content in the leaves increased between stage 1 and stage 2, then increased from stage 4 to 5 before decreasing until the final stage. For manganese, there were some significant fluctuations between stages 1 and 3, then the concentrations increased significantly until stage 6, followed by a decrease between the final two stages.

Comparing the nutrient content with RDA, a 100 g daily serving of African pumpkin leaves alone would provide at least 200% of adult male (31–50 years) requirements for iron at all stages of maturity. For females (31–50 years) a similar portion would only provide at least 90% of iron requirements at stage 1; however, both vegetables would make an excess contribution of this mineral from stage 2 until stage 7. As for zinc, the percentage contribution would progressively decline with advancing maturity. For example, a 100 g serving of African pumpkin leaves would exceed adult female requirements at stage 1, but the contribution declines to 97% at stage 4 and 28% at stage 7. It is important to note that the actual bioavailability of either iron or zinc at any maturity stage of the leaves will be affected by the interactions between both zinc and iron. It has been reported that iron inhibits the absorption of zinc when the iron to zinc ratio is greater than 2.

Amino acids such as glutamine, glycine, lysine, alanine, leucine, valine, proline, serine, isoleucine, phenylalanine, tryptophan, arginine, histidine, threonine, methionine and asparagine are reported in African pumpkin leaves (Kadiri *et al.*, 2020). The leaves contain the following essential amino acids: phenylalanine (4.43 g/100 g), valine (3.98 g/100 g), threonine (3.27 g/100 g), tyrosine (3.10 g/100 g), isoleucine (3.34 g/100 g), methionine (1.12 g/100 g), cysteine (0.97 g/100 g), leucine (7.94 g/100 g) and lysine (4.03 g/100 g). The non-essential amino acids proline (3.25 g/100 g), arginine (4.13 g/100 g), histidine (2.17 g/100 g), alanine (4.17 g/100 g), glutamic acid (11.66 g/100 g), glycine (3.99 g/100 g), serine (3.73 g/100 g) and aspartic acid (8.25 g/100 g) were reported in African pumpkin leaves (Hassan and Umar, 2006; Kadiri *et al.*, 2020).

The presence of essential amino acids in the leaves may reduce depletion of glutathione in the body, thereby contributing to general wellness through this antioxidant effect (Hassan and Umar, 2006). WHO/FAO/UNU (2007) guidelines suggest that a 19.67 g portion of African pumpkin leaves is able to fulfil the daily requirement of phenylalanine, leucine, lysine and threonine for adults with 70 kg body weight.

Table 5.1. Mineral composition of different leafy vegetables per mg/100 g dry weight (DW) (values for African pumpkin adapted from Hassan and Umar, 2006, values for other species adopted from the USDA food data 2020 (USDA Food data, 2020).

			Exotic vegetable (mg)			Indigenous vegetable (mg)			
Mineral content	RDA% African pumpkin	African pumpkin	Lettuce	Spinach	Swiss chard	Moringa	Amaranthus	Pumpkin leaves	Sweet potato leaves
Calcium	94.1%	941	18	94	51	185	215	39	78
Potassium	28.1%	1320	141	553	379	337	611	436	508
Magnesium	55.5%	222	7	76	81	42	55	38	70
Iron	753.75%	60.30	0.41	2.71	1.8	4.0	2.32	2.22	0.97
Sodium	8.2%	122.49	10	76	213	9.0	20	11	6.0
Zinc	39.75%	3.18	0.15	-	0.36	0.6	0.9	0.2	-
Manganese	504.3%	11.60	-	-	0.37	1.06	0.89	0.36	-

Table 5.2. The effect of maturity stages on the mineral content of African pumpkin leaves per mg/kg FW (Flyman and Afolayan, 2007).

Maturity stage	Age in days	Manganese (mg/kg)	Iron (mg/kg)	Zinc (mg/kg)	Potassium (%)	Calcium (%)	Magnesium (%)	Sodium (%)
1	21	150.3	160	389.09	2.16	0.22	0.35	0.01
2	28	174.76	248	161.53	2.27	0.23	0.45	0.13
3	35	164.82	251.76	85.83	2.62	4.08	1.45	0.07
4	44	176.07	260.13	77.83	3.22	2.61	1.29	0.03
5	50	186.92	320.98	82.63	3.12	1.76	1.13	0.01
6	57	194.24	277.53	47.04	3.03	0.94	0.82	0.0
7	64	167.04	248.04	22.50	2.74	1.04	0.84	0.0

Antinutritive compounds

It is of importance to note that while there are many groups of chemical compounds in plants that have health benefits, some can be very toxic and harmful to human health when consumed (Natesh *et al.*, 2017). Some of the harmful organic compounds found in plants are oxalates, tannins, nitrates and saponins, referred to as antinutrients (Natesh *et al.*, 2017). Antinutrients reduce the bioavailability of nutrients such as minerals and vitamins (Ogbuji and Ndulaka, 2016). This, in turn affects the nutritional value of the plants (Patel *et al.*, 2018). Anti-nutrients comprise of amino acids to proteins, simple amines to alkaloids, glycosides and phenolic compounds (Essack *et al.*, 2017). African pumpkin leaves contain glycosides, oxalates, tannins, phytates, alkaloids and hydrocyanic acid (Natesh *et al.*, 2017; Patel *et al.*, 2018). Tannins are plant polyphenols that form complexes with zinc and macro-molecules like proteins and polysaccharides, reducing the absorption of these nutrients as well as forming complexes with protein and thereby inhibiting digestion and absorption (Ogbuji and Ndulaka, 2016; Essack *et al.*, 2017). Tannins are also responsible for protein deficiency syndrome known in Sub-Saharan Africa as 'kwashiorkor' (Mshelia *et al.*, 2017). Phytic acid is the major phosphorus storage compound in traditional leafy vegetables. It has been recorded to inhibit the absorption of minerals and reduce the bioavailability of metal ions like zinc ion as well as affect protein and starch digestion (Essack *et al.*, 2017). Phytate intake of 4–9 mg/100 g decreases iron absorption in humans four- or fivefold. An over-rich phytate diet is associated with nutritional diseases such as rickets and osteomalacia in children and adults (Sobowale *et al.*, 2011). Oxalic acid exists in many leafy vegetables, depending on the species, and can occur as soluble salts of potassium or sodium, as insoluble salts of calcium, magnesium or iron or as a combination of soluble and insoluble salts (Mshelia *et al.*, 2017; Natesh *et al.*, 2017). Oxalates in foods can affect the bioavailability of minerals like calcium. They form strong chelates with dietary calcium inhibiting its absorption (Natesh *et al.*, 2017). Oxalic acids can decrease calcium absorption and aid the formation of kidney stones (Sobowale *et al.*, 2011; Mshelia *et al.*, 2017). The adverse effect of calcium absorption is higher when the rate of oxalate/calcium is more than 9/4. Besides dietary intake, oxalates are also formed in the human body as a by-product of ascorbic acid and glyoxylate metabolism (Natesh *et al.*, 2017). Oxalates also bind with calcium to form calcium oxalate crystals which prevent the absorption and utilization of calcium by the body thereby causing diseases such as rickets and osteomalacia (Ogbuji and Ndulaka, 2016). High calcium, affected by saponin levels, binds with exogenous and endogenous proteins including enzymes of the digestive tract, thereby affecting the utilization of protein (Sobowale *et al.*, 2011).

Different postharvest processing techniques are often utilized to reduce antinutritional factors in vegetables (Mshelia *et al.*, 2017; Singh *et al.*, 2018). African pumpkin and *Amaranthus* hybrids were boiled at a temperature of 95°C at different times, for 5 and 15 minutes respectively. Raw African pumpkin leaves contain high amounts of oxalic acid (11.1 mg/100 g) as compared to *Amaranthus* hybrids (4.65 mg/100 g). Boiling for 15 minutes significantly reduced the oxalate content for African pumpkin leaves to 2.2 mg/100 g. Moreover, boiling reduced the phytic acid content of the leaves, because of disruption of the cell wall due to the thermal treatment which releases soluble phytic acid into the boiling water (Essack *et al.*, 2017). Consumption of large amounts of oxalic acid causes corrosive gastroenteritis, shock, convulsive symptoms, low plasma calcium, high plasma oxalates and renal damage. A daily intake of 50 mg oxalate is recommended. A lethal dose of oxalate has been reported to be 3–5 g for humans (Ogbuji and Ndulaka, 2016). Tannins bind to both proteins and carbohydrates. The presence of tannins causes browning and other pigmentation problems in both fresh and processed leafy vegetables. Tannins can precipitate protein in the gut, reducing digestibility or inhibiting the digestive enzymes and microorganisms. Tannins can inhibit the activities of some enzymes such as amalyse, trypsin and lipase (Ogbuji and Ndulaka, 2016). Raw *Amaranthus* leaves contained the highest tannin content (0.022 mg/mL) (Table 5.3) as compared to the other traditional leafy vegetables, followed by fat hen (wild

Table 5.3. Antinutritional compounds of different traditional leafy vegetables subjected to boiling for five minutes at a temperature of 95°C (Essack *et al*., 2017).

Antinutritional compound mg/ml	African pumpkin	Amaranthus	Black jack	Black nightshade	Fat hen	Devil's thorn	Matweed
Tannin content	0.000 ± 0.007[a]	0.020 ± 0.006	0.002 ± 0.001	0.000 ± 0.001	0.003 ± 0.001	0.000 ± 0.002	0.000 ± 0.001
Phytic acid content	0.050 ± 0.035	0.010 ± 0.004	0.040 ± 0.015	0.070 ± 0.055	0.031 ± 0.002	0.010 ± 0.000	0.010 ± 0.006
Oxalic acid content	10.2 ± 0.6	3.92 ± 0.5	3.3 ± 0.2	6.7 ± 0.0	4.25 ± 0.74	2.5 ± 0.5	2.5 ± 0.5

[a] Data presented as mean ± standard deviation, *n*=3

goosefoot) (0.11 mg/ml) and African pumpkin with 0.01 mg/ml. Raw matweed has the lowest tannin content (0.005 mg/ml). Raw black nightshade leaves contain the highest amounts of phytate (0.14 mg/ml) as compared to the other traditional leafy vegetables, *Amaranthus* the lowest (0.01 mg/ml). Raw African pumpkin leaves contain the highest level of oxalate (11.1 mg/ml), while matweed contains the lowest. Table 5.3 indicates the changes in antinutrient content after traditional leafy vegetables were subjected to boiling for five minutes at 95°C. The antinutritional compounds were reduced for all the traditional leafy vegetables, therefore Essack *et al.* (2017) concluded that boiling for this time can be regarded as a standard to reduce antinutritional compounds.

Phytonutrional compounds

Fresh African pumpkin leaves contain 30.07 mg/100 g total chlorophyll and 4.54 mg/100 g carotenoid (Mashiane *et al.*, 2021). Phenolic compounds are a group of secondary metabolites commonly present in the epidermis of plant leaves and are categorized as flavonoids and non-flavonoids. They possess benzene rings with at least one hydroxyl substituent. The following phenolic compounds were reported in African pumpkin leaves: 3-coumaroylquinic acid (16.40 μg/g); 5-caffeoylquinic acid (36.64 μg/g); catechin (79.21 μg/g); quercetin-3-O-glucoside (111.70 μg/g); 3,5-dicaffeoylquinic acid (24.68 μg/g); 4,5-dicaffeoylquinic acid (68.93 μg/g); quercetin (15.94 μg/g) (Madala *et al.*, 2016; Moyo, 2019). Madala *et al.* (2016) also reported the presence of rutin in African pumpkin leaves. The total phenolic compounds was reported as 64.53 mg GAE/g t, DW basis (Moyo *et al.*, 2013).

Implications of postharvest processing on phytonutrients

Cooking is a common form of processing to enhance taste and palatability and causes changes in the phytochemical components of African pumpkin leaves (Moyo, 2019). Cooking also causes a number of physical and chemical changes in vegetables (Mashiane *et al.*, 2021). It affects the physical and chemical properties of plant cell walls by breaking the weak bonds between polysaccharides and the cleavage of glycosidic linkages, resulting in solubilization of the dietary fibre (Uusiku *et al.*, 2010). Thus, the stability of phenolic compounds is affected by postharvest processing and food preparation methods and the activity of phenolic compounds may be affected by the processing method (Moyo, 2019). The successful effects of these phenolic compounds on health depends on the amount adsorbed during *in vitro* digestion in the gastrointestinal tract from the food matrix (bioavailable) and then absorbed into the circulatory system. Household boiling was found to increase extricable phenolics from the cell matrix of African pumpkin leaves (Moyo *et al.*, 2020).

Moyo (2019) showed that more polar phenolic compounds such as quercetin-3-O-robinoside, rutin, quercetin, quercetin-3-O-glucoside, catechin and caftaric acid leached into the boiling water as a result of disruption of the cell wall. The researcher reported the presence of fewer phenolic compounds in the digested extracts but 4-caffeoylquinic acid remained at a high level. The findings of Moyo (2019) illustrate that there was no significant difference observed in the concentrations of the total phenolic compounds of the oral, gastric and duodenal digestion phases, their concentrations were between 57-69% (bioavailable). However, the bioaccessibility of the flavonoid was improved in the raw vegetable and 72.57% was noted after duodenal digestion. Boiling African pumpkin leaves and retaining the boiled water increased the flavonoid content (148.78 ± 16 μM TE/g DW) (Moyo, 2019). Therefore, Moyo, 2019 suggested that boiled water must be retained with the leaves to improve the antioxidant capacity of African pumpkin leaves during household cooking. These findings demonstrated positive correlations between total phenols, total flavonoids and TEAC antioxidant activity in African pumpkin leaves. Non-digested and digested extracts protected DNA, L929 and Caco-2 cells from 2,2'-azobis(2-methylpropionamide) dihydrochloride (AAPH)-induced oxidative damage. No extracts of African pumpkin or spinach could protect low-density lipoprotein (LDL) from oxidative damage while the majority of African pumpkin and spinach extracts (non-digested and digested) protected DNA,

L929 and Caco-2 cells from AAPH induced oxidative damage. Moyo (2019) emphasized that the consumption of African pumpkin could positively aid in reducing the conditions for cancer risk. In another study, the influence of different household cooking methods such as boiling, microwaving, stir-frying and steaming on colour properties, antioxidants, antioxidant activity, FRAP activity and the sensory quality of African pumpkin leaves was investigated (Mashiane *et al.*, 2021). The findings showed steaming increased the total carotenoid content by 22.65%, reduced the loss of total phenolic content and retained the antioxidant activity. Stir-fried leaves of African pumpkin were mostly preferred, then the steamed leaves, then boiled or microwaved leaves. Mashiane *et al.* (2021) suggested that since steaming improved the antioxidant components, steamed leaves could be used as functional ingredients in food recipes.

Postharvest Handling and Shelf Life

Postharvest handling refers to the processes performed immediately after removing a plant or part of a plant from its growth media until the produce reaches the final consumer in the desired form. Handling, including packaging methods, affects quantity, quality and price. Improving postharvest handling and processing of vegetables is one way of overcoming perishability constraints and ensuring a continued high quality food supply. Postharvest handling and shelf life for traditional leafy vegetables have been previously documented (Moyo *et al.*, 2018), however there is limited information pertaining to the postharvest handling and shelf life of African pumpkin leaves. Masarirambi *et al.* (2010) reported the traditional ways of harvesting African pumpkin leaves and of drying them. The harvesting technique is to cut part of the vine from the mother plant, taking care not to kill the mother plant. Tender pumpkin leaves and vine tips are picked. Water can be periodically misted onto the leaves to prolong shelf life. African pumpkin leaves are highly perishable and fresh-cut leaves have a limited shelf life of up to four days at 2°C (Sorifa, 2018). Storage of dehydrated leaves is recommend at low temperatures to reduce the degradation of ascorbic acid and browning (Uusiku *et al.*, 2010).

Traditionally, African pumpkin leaves are sun-dried, however the leaves are blanched before drying. Masarirambi *et al.* (2010) recommend that the leaves should be immersed in boiling or near boiling water and steamed for a very short time (usually less than three minutes) to minimize the loss of nutrients. The blanched leaves are spread evenly on a flat tray and placed on a rock surface or rooftop, directly in the sun. The blanched produce is occasionally stirred to facilitate even drying. During the night the vegetables are taken indoors to prevent moisture reabsorption from dew or rain. The blanched leaves are regarded as dry when they are hard and brittle. Using traditional methods, product quality is usually relatively low, drying is uneven and often caramelized and crusted pieces are produced. Direct exposure to the sun often destroys colour, vitamins and phytonutrients in the leaves. Further research on the different drying techniques of African pumpkin leaves is recommended to improve postharvest handling and reduce losses.

Bioactive Properties and Health Benefits

Bioactive compounds such as dietary fibre, vitamins, minerals and phytochemicals are responsible for mitigating some human diseases (Faujdar *et al.*, 2013). All these different compounds may contribute to overall health (Dias, 2019). Some phytochemicals (glucosinolates, thiosulfates, polyphenols, bioactive peptides, etc.) have positive effects on human health (Souda *et al.*, 2018). They are strong antioxidants and reduce the risk of chronic disease by protecting against free radical damage (Faujdar *et al.*, 2013) by modifying metabolic activation and detoxification of carcinogens, or by influencing mechanisms that alter the course of tumour cells (Dias, 2019). Dietary fibre is one of the major constituents of traditional leafy vegetables. Dietary fibre and other bioactive compounds have been usually addressed separately in nutritional studies. However, dietary fibre transports a significant number of phytochemicals through the human gut – vitamins and minerals linked to the fibre matrix (Dias and Imai, 2017; Dias, 2019). Therefore the associated phytochemicals, vitamins and minerals of the whole food may contribute to the overall health benefits usually attributed to dietary fibre alone. African pumpkin is rich in carotenoids, tocopherols and vitamin C (Dias, 2019).

The quality and quantity of bioactive compounds present in African pumpkin leaves can be affected by various aspects such as harvest time, temperature, stage of maturity, postharvest storage (Sivakumar *et al.*, 2018), seasonal variation, fertilizers or chemicals applied to the plant, variety, geographic location and method used for analysis (Uusiku *et al.*, 2010; Sorifa, 2018). Furthermore, bioactive compound extraction from African pumpkin leaves may depend on temperature, length of extraction, particle size and number of extraction of samples, the water to powder ratio, the ratio of water and plant material and the type of solvents used (Sorifa, 2018).

The hypoglycemic effect of African pumpkin may also be due to the presence of phenols and flavonoids which serve as antioxidants to prevent toxic effects on tissue, along with some bioactive substance in the plant extract which has hypoglycemic potentials (Ludidi *et al.*, 2019). It is evident from previous studies conducted on African pumpkin leaves that consumption is very effective in lowering blood glucose level (Dias and Imai, 2017; Siboto *et al.*, 2018; Sani *et al.*, 2019) and there are positive effects on antidiabetic activity. Glycosides, saponins, alkaloids, triterpenes, momordica Ic, oleanolic acid 3-O-glucuronide, charantin, 3-O-monodesmoside, proteins, steroids and polysaccharides are found in African pumpkin leaves (Sani *et al.*, 2019), and these compounds have been reported to contain hypoglycemic activity (Abdoon *et al.*, 2019). African pumpkin leaves also contain syringic acid that has blood sugar regulation properties (Jibril *et al.*, 2018; Olalere and Gan, 2020). Syringic acid has been determined to inhibit the activity of α-glucosidase enzyme and when α-glucosidase is inhibited, fewer carbohydrates are converted to sugars and blood sugar is able to remain more steady (Dias, 2019; Ludidi *et al.*, 2019). In a study by Jibril *et al.* (2018) an extract of the African pumpkin leaves introduced into the bloodsteam of Japanese quails was found to reduce the blood glucose level, this was attributed to the α-glucosidase inhibitory activity of the African pumpkin leaf extract (Jibril *et al.*, 2018).

Several researchers have showed that there are various medicinal properties of African pumpkin leaves such as anti-HIV (Mónico *et al.*, 2019), antitumour, antimutagenic, antibacterial and anticancer (Olalere and Gan, 2020). The therapeutic agent 'momordin' is capable of inhibiting the growth of HIV and other viruses (Kamalaja *et al.*, 2018). African pumpkin leaves contains high amounts of potassium, which is a good source for the management of hypertension and other cardiovascular conditions (Thakur *et al.*, 2009; Muronga *et al.*, 2020). They offer protection against cardiovascular disease (Olalere and Gan, 2020) because they are free of saturated fat, trans fat and cholesterol and rich in bioactive compounds such as carotenoids, flavonoids, dietary fibre and sterols (Dias, 2019). In a study by Abdoon *et al.* (2019), it was found found that the leaves of African pumpkin can help boost the immune system. Souda *et al.* (2018) demonstrated the methanol extract of African pumpkin leaves showed antifungal activity against *C. albacans* species at a concentration (MIC – 0.938 mg/ml and MBC-1875 mg/ml).

Conclusions and Future Trends

African pumpkin leaves are rich in phytonutrients and minerals. The adoption of different food processing methods (such as cooking) and an understanding of the variation of mineral content with maturity stage can reduce the loss of valuable phytonutrients in processed foods and further promote bioavailability of these nutrients to consumers. Since African pumpkin are a good source of carotenoids and protein, the leaves can be included in national food databases to improve the nutritional status of rural dwellers. In addition, the processed leaves can be used as a functional ingredient to improve the diets of consumers to suit the modern lifestyle. African pumpkin leaves have been widely researched for medicinal purposes, therefore further research on postharvest handling is recommended to encourage commercialization of the plant.

References

Abdoon, I.H., Osman, B., Mukhtar, M.M. and Elsheikh, H. (2019) Modulatory effects of *Momordica balsamina* on Th1/Th2 cytokine profiles in immune-challenged rats. *Journal of Medicinal Plants Research* 13(15), 343–350. DOI: 10.5897/JMPR2019.6788.

Bello, A., Muhammad, F., Kankara, S.S., Abdulkadir, B. and Buhari, Y. (2018) Antimicrobial activity of Balsam Apple (*Momordica balsamina* L.). *Journal of Microbiology Research* 3(1), 24–29.

Burchi, F., Fanzo, J. and Frison, E. (2011) The role of food and nutrition system approaches in tackling hidden hunger. *International Environmental Research Public Health* 8(2), 358–378. DOI: 10.3390/ijerph8020358.

Bvenura, C. and Afolayan, A.J. (2015) The role of wild vegetables in household food security in South Africa: A review. *Food Research International* 76(4), 1001–1011. DOI: 10.1016/j.foodres.2015.06.013.

CABI (2020) *Momordica balsamina* (common balsamina apple). Available at: https://www.cabi.org/isc/datasheet/34677 (accessed 9 May 2020).

Dias, J.S. (2019) Nutritional quality and effect on disease prevention of vegetables. *Food and Nutrition Sciences* 10(4), 369–402. DOI: 10.4236/fns.2019.104029.

Dias, J.S. and Imai, S. (2017) Vegetables consumption and its benefits on diabetes. *Journal of Nutritional Therapeutics* 6(1), 1–10. DOI: 10.6000/1929-5634.2017.06.01.1.

Essack, H., Odhav, B. and Mellem, J.J. (2017) Screening of traditional South African leafy vegetables for specific anti-nutritional factors before and after processing. *Food Science and Technology* 37(3), 462–471. DOI: 10.1590/1678-457x.20416.

Fabbri, A.D.T. and Crosby, G.A. (2016) A review of the impact of preparation and cooking on the nutritional quality of vegetables and legumes. *International Journal of Gastronomy and Food Science* 3, 2–11. DOI: 10.1016/j.ijgfs.2015.11.001.

Faber, M., Oelofse, A., Van Jaarsveld, P.J., Wenhold, F.A.M. and Van Rensburg, W.S. (2010) African leafy vegetables consumed by households in the Limpopo and KwaZulu-Natal provinces in South Africa. *South African Journal of Clinical Nutrition* 23(1), 30–38. DOI: 10.1080/16070658.2010.11734255.

Faujdar, S., Nain, S. and Kalia, A.N. (2013) Antioxidant activity and estimation of total phenolic content of *Momordica balsamina* L. *International Journal of Advanced Research* 2(2), 233–238.

Flyman, M.V. and Afolayan, A.J. (2007) Effect of plant maturity on the mineral content of the leaves of *Momordica balsamina* L. and *Vigna Unguiculata subsp. Sesquipedalis* (L,) Verdc. *Journal of Food Quality* 31, 661–671.

Hassan, L.G. and Umar, K.J. (2006) Nutritional value of Balsam Apple (*Momordica balsamina* L.) leaves. *Pakistan Journal of Nutrition* 5(6), 522–529. DOI: 10.3923/pjn.2006.522.529.

Idris, S.B., Sani, A.A. and Bobi, S.S. (2018) Phytochemical and anti-oxidant properties of *Momordica balsamina* L. *Journal of Veterinary Science* 13, 71–75.

Jibril, A., Musa, U., Saidu, B., Ajape, A., Maina, I. *et al.* (2018) Effects of powdered Balsam Apple (*Momordica balsamina* L.) leafy supplement on blood glucose level and the haematological parameters of Japanese Quails. *Journal of Livestock Research* 8(2), 1–7. DOI: 10.5455/ijlr.20170817041629.

Kabir, N., Umar, I., James, D., Inuwa, H., Atiku, M. *et al.* (2019) Antidiabetic Potentials of Aqueous Leaf Extracts of Momordica balsamina Linn and Leptadenia hastata (Pers) Decne Alone and in Combination in Streptozotocin-Induced Diabetic Rats. *Tropical Journal of Natural Product Research* 3(1), 10–16. DOI: 10.26538/tjnpr/v3i1.3.

Kadiri, O.J., Okafor, S.I. and Ogaji, J.I. (2020) Safety and health benefits profile studies of leaf extracts of *Momordica balsamina* Linn (Cucurbitaceae) found in North Central Nigeria. GSC. *Biological and Pharmaceutical Sciences* 11(2), 278–286. DOI: 10.30574/gscbps.2020.11.2.0039.

Kamalaja, T., Prashanthi, M. and Rajeswari, K. (2018) Effect of cooking methods on bioactive compounds in vegetables. *International Journal of Chemical Studies* 6(4), 3310–3315.

Karumi, Y., Onyeyili, P.A. and Ogugbuaja, V.O. (2004) Identification of active principles of *M. balsamina* (Balsam Apple) leaf extract. *Journal of Medicinal Science* 4(3), 179–182.

Ludidi, A., Baloyi, M.C., Khathi, A., Sibiya, N.H. and Ngubane, P.S. (2019) The effects of Momordica balsamina methanolic extract on haematological function in streptozotocin-induced diabetic rats: Effects on selected markers. *Biomedicine & Pharmacotherapy = Biomedecine & Pharmacotherapie* 116, 108925. DOI: 10.1016/j.biopha.2019.108925.

Madala, N.E., Piater, L., Dubery, I. and Steenkamp, P. (2016) Distribution patterns of flavonoids from three Momordica species by ultra-high performance liquid chromatography quadrupole time of flight mass spectrometry: a metabolomic profiling approach. *Revista Brasileira de Farmacognosia* 26(4), 507–513. DOI: 10.1016/j.bjp.2016.03.009.

Masarirambi, M.T., Mavuso, V., Songwe, V.D., Nkambule, T.P. and Mhazo, N. (2010) Indigenous post-harvest handling and processing of traditional leafy vegetables in Swaziland: A review. *African Journal of Agricultural Research* 5(24), 3333–3341.

Mashiane, P., Mashitoa, F.M., Slabbert, R.M. and Sivakumar, D. (2021) Impact of household cooking techniques on colour, antioxidant and sensory properties of African pumpkin and pumpkin leaves. *International Journal of Gastronomy and Food Science* 23, 100307. DOI: 10.1016/j.ijgfs.2021.100307.

Matenge, S.T.P., Apau, S. and Tapera, R. (2017) Nutritional and phytochemical content of indigenous leafy vegetables consumed in Botswana. *Food and Nutrition Research* 3(1), 1–7.

Mncwango, N.C., Mavengahama, S., Ntuli, N.R. and Van Jaarsveld, C.M. (2020) Diversity, consumption dynamics and ethnomedicinal claims of traditional leafy vegetables consumed by a rural community in the KwaMbonambi area, Northern KwaZulu-Natal, South Africa. *Biodiversitas* 21(3), 1201–1207. DOI: 10.13057/biodiv/d210347.

Moloto, R.M., Phan, A.D.T., Shai, J.L., Sultanbawa, Y. and Sivakumar, D. (2020) Comparison of phenolic compounds, carotenoids, amino acid composition, in vitro antioxidants and anti-diabetic activities in the leaves of seven Cowpea (*Vigna unguiculata*) cultivars. *Foods (Basel, Switzerland)* 9(1285), 1–23. DOI: 10.3390/foods9091285.

Mónico, A., Ramalhete, C., André, V., Spengler, G., Mulhovo, S. *et al.* (2019) Cucurbalsaminones A-C, Rearranged Triterpenoids with 5/6/3/6/5-Fused pentacyclic carbon skeleton from *Momordica balsamina* L. as multidrug resistance reversers. *Journal of Natural Products* 82(8), 2138–2143. DOI: 10.1021/acs.jnatprod.9b00019.

Moyo, S.M. (2019) Influence of In vitro digestion on the bioaccessibility and bioactivity of phenolic compounds from raw and boiled indigenous green leafy vegetables. PhD thesis. University of Johannesburg. Available at: https://hdl.handle.net/10210/418370 (accessed 6 September 2022).

Moyo, M., Amoo, S.O., Ncube, B., Ndhlala, A.R., Finnie, J.F. *et al.* (2013) Phytochemical and antioxidant properties of unconventional leafy vegetables consumed in South Africa. *South African Journal of Botany* 84, 65–71. DOI: 10.1016/j.sajb.2012.09.010.

Moyo, S.M., Mavumengwana, V. and Kayitesi, E. (2018) Effects of cooking and drying on phenolic compounds and antioxidant activity of African green leafy vegetables. *Food Reviews International* 34(3), 248–264. DOI: 10.1080/87559129.2017.1289384.

Moyo, S.M., Serem, J.C., Bester, M.J., Mavumengwana, V. and Kayitesi, E. (2020) African green leafy vegetables health benefits beyond nutrition. *Food Reviews International* 37(6), 1–9. DOI: 10.1080/87559129.2020.1717519.

Mshelia, S.H. and Karumi, Y. (2013) Antiulcerogenic effect of extracts of *Momordica balsamina* L. against experimentally induced gastric ulceration in rats. *International Journal of Medicinal Plants Research* 6(4), 332–336.

Mshelia, H.S., Karumi, Y. and Dibal, N.I. (2017) Therapeutic effect of *Momordica balsamina* leaf extract on ethanol-induced gastric ulcer in Wistar rats. *Annals of Research Hospitals* 1(3), 1–5. DOI: 10.21037/arh.2017.04.03.

Muronga, M., Tshikhudo, P.P., Msagati, T.A.M. and Mudau, F.N. (2020) Three selected edible crops of the genus *Momordica* as a potential source of phytochemicals: from production to pharmaceutical usage. Available at: Preprints (www.preprints.org) online (accessed 25 October 2020).

Nagarani, G., Abirami, A. and Siddhuraju, P. (2014) Food prospects and nutraceutical attributes of *Momordica* species: A potential tropical bioresources – A review. Food science and human wellness 3(4), 117–126. DOI: 10.1016/j.fshw.2014.07.001.

Natesh, N.H., Asiedu, S.K. and Abbey, L. (2017) An overview of nutritional and anti nutritional factors in green leafy vegetables. *Horticulture International Journal* 1(2), 58–65. DOI: 10.15406/hij.2017.01.00011.

Odhav, B., Beekrum, S., Akula, U.S. and Baijnath, H. (2007) Preliminary assessment of nutritional value of traditional leafy vegetables in KwaZulu-Natal, *South Africa. Journal of Food Composition and Analysis* 20(5), 430–435. DOI: 10.1016/j.jfca.2006.04.015.

Ogbuji, C.A. and Ndulaka, J.C. (2016) Evaluation of the antinutrient composition of the component herbs of a traditional antidiabetic formulation. *International Journal of Biomedical and Health Sciences* 12(2), 39–45.

Olalere, O.A. and Gan, C.Y. (2020) Multi-step reflux extraction of bio-pharmaceutical phenolic bioactives from balsam apple (*Momordica balsamina* L.). *Journal of Taibah University for Science* 14(1), 227–234. DOI: 10.1080/16583655.2020.1721722.

Patel, K., Neeharika, B., Jessie, S.W., Kumari, B.A., Prabharkar, N.B. *et al.* (2018) Effect of blanching on anti-nutritional factors of Bethua leaves. *The Pharma Innovation Journal* 7(4), 214–216.

Sani, A.A., Idris, S.B. and Yakubu, S. (2019) Anti-diabetic activity of methanolic extract of *Momordica balsamina* L. in rabbits. *Scientific Research Journal* 1–6. DOI: 10.31364/SCIRJ/v7.i6.2019.P0619657.

Schönfeldt, H.C. and Gibson Hall, N. (2012) Dietary protein quality and malnutrition in Africa. *The British Journal of Nutrition* 108 Suppl 2, S69–76. DOI: 10.1017/S0007114512002553.

Singh, B., Singh, S. and Koley, T.K. (2018) Advances in post-harvest technologies of vegetable crops. Post-harvest Biology and Technology. Apple Academic Press 9, Spinnaker Way, Waretown USA. DOI: 10.1201/9781315161020.

Siboto, A., Sibiya, N., Khathi, A. and Ngubane, P. (2018) The Effects of *Momordica balsamina* Methanolic Extract on Kidney Function in STZ-Induced Diabetic Rats: Effects on Selected Metabolic Markers. *Journal of Diabetes Research* 2018, 1–8. DOI: 10.1155/2018/7341242.

Singh, S. and Devi, B. (2018) Phytopharmacological evaluation of *Momordica balsamina* L. From Southern Haryana, India. *Kenkyu Journal of Pharmacy Practice & Health Care* 4(4), 17–34. DOI: 10.31872/2018/KJPHC-100112.

Sivakumar, D., Chen, L. and Sultanbawa, Y. (2018) A comprehensive review on beneficial dietary phytochemicals in common traditional Southern African leafy vegetables. *Food Science & Nutrition* 6(4), 714–727. DOI: 10.1002/fsn3.643.

Sobowale, S.S., Olatidoye, O.P., Olorode, O.O. and Akinlotan, J.V. (2011) Nutritional potentials and chemical value of some tropical leafy vegetables consumed in South West Nigeria. *Journal of Sciences and Multidisciplinary Research* 3, 55–65.

Sorifa, A.M. (2018) Nutritional compositions, health promoting phytochemicals and value added products of bitter gourd: a review. *International Food Research Journal* 25(5), 1763–1772.

Souda, S., George, S., Mannathoko, N., Goercke, I. and Chabaesele, K. (2018) Antioxidant and antibacterial activity of methanol extract of *Momordica balsamina* L. *International Journal of Applied Sciences* 10(2), 7–17. DOI: 10.21013/jas.v10.n2.p1.

Spengler, G., Ramalhete, C., Martins, M., Martins, A., Serly, J. *et al.* (2009) Evaluation of cucurbitane-type triterpenoids from *Momordica balsamina* L. on P-Glycoprotein (ABCB1) by flow cytometry and real-time fluorometry. *Anticancer Research* 29(10), 3989–3993.

Thakur, G.S., Bag, M., Sanodiya, B.S., Bhadouriya, P., Debnath, M. *et al.* (2009) *Momordica balsamina*: a medicinal and neutraceutical plant for health care management. *Current pharmaceutical biotechnology* 10(7), 667–682. DOI: 10.2174/138920109789542066.

USDA Food data (2020). Available at: https://www.fns.usda.gov/usda-food-patterns

Uusiku, N.P., Oelotse, A., Duodu, K.G., Bester, M.J. and Faber, M. (2010) Nutritional value of leafy vegetables of Sub-Saharan Africa and their potential contribution to human health: A review. *Journal of Food Composition and Analysis* 23(6), 499–509. DOI: 10.1016/j.jfca.2010.05.002.

Van Jaarsveld, P., Faber, M., Van Heerden, I., Wenhold, F., Van Rensburg, W.J *et al.* (2014) Nutrient content of eight African leafy vegetables and their potential contribution to dietary reference intakes. *Journal of Food Composition and Analysis* 33(1), 77–84. DOI: 10.1016/j.jfca.2013.11.003.

World Health Organization (WHO) and Food and Agriculture Organization (FAO) of the United States Nation (UN) (2007) Fruit and vegetables for health. Available at: http://www.fao.org/ag/magazine/fao-who-fv.pdf (accessed 18 September 2020).

Yuan, B., Byrnes, D., Giurleo, D., Villani, T., Simon, J.E. *et al.* (2018) Rapid screening of toxic glycoalkaloids and micronutrients in edible nightshades (Solanum spp). *Journal of Food and Drug Analysis* 26, 751–760.

6 Pumpkin Leaves (*Cucurbita maxima*, *Cucurbita moschata* and *Cucurbita pepo*)

Florence M. Mashitoa and Dharini Sivakumar*
Phytochemical Food Network Research Group, Department of Crop Sciences, Tshwane University of Technology, Pretoria, South Africa

Abstract
Pumpkin leaf accessions are more nutritious than most exotic and African leafy vegetables (ALVs). Pumpkin leaves are rich in iron content which is vital to the proper function of haemoglobin, which reduces the chances of anaemia. Pumpkin leaves also have the ability to reduce incidences of non-communicable chronic diseases such as type 2 diabetes. However, there is limited information on the dietary phytochemicals of pumpkin leaves. This review summarizes information on the nutrients of pumpkin leaf accessions, postharvest processing, antinutritive compounds and antidiabetic activity along with the problematic aspects of the common antinutritive compounds of pumpkin leaves, in that they reduce bioavailability of minerals such as iron and zinc.

Introduction

Pumpkins belong to the family Cucurbitaceae. This family is classified depending on the texture and shape of the stems, such as in *Cucurbita pepo*, *Cucurbita moschata* and *Cucurbita maxima* (Uusiku *et al.*, 2010). Nowadays, the market offers a wide variety of vegetables, with pumpkins being one of them because of their many applications for nutrition (Rodríguez *et al.*, 2006). Pumpkin leaves are an essential part of the pumpkin plant, and have long been used in traditional medicine to treat common illnesses. In addition, in numerous delicious vegetarian recipes, pumpkin leaves can be prepared to complement a well balanced, nutritious diet (Sharma *et al.*, 2019). They can be harvested or purchased from ethnic markets and wholesale vendors (Vorster *et al.*, 2002). In South Africa, members of the Cucurbitaceae family are very common leafy vegetables and one of the few cultivated types in Africa. Pumpkins are often grown in South Africa as a minor crop with maize covering the soil surface which helps to suppress weeds (Vorster *et al.*, 2002). In the wild, some cucurbits grow from where they are harvested. The leaves, flowers and young fruits are picked and cooked when harvested as a leafy vegetable (Hart and Vorster, 2006). The leaves (*Cucurbita maxima*, *Cucurbita moschata* and *Cucurbita pepo*) (Fig. 6.1) of the ordinary pumpkin are known in Sepedi as *mophotse*, *mphodi* and *monyaku*. Virtually all parts of the plant can be used, including the leaves, shoots, flowers and immature fruits, depending on the species (Zanish *et al.*, 2013). The leaves are regularly harvested as the plant grows (Onyeike and Omubo-Dede, 2002).

Pumpkin leaves are a rich source of minerals such as magnesium (38 g/100 g), calcium (39 g/100 g) and protein (3 g/100 g), but are

*Corresponding author: SivakumarD@tut.ac.za

DOI: 10.1079/9781789248067.0006

Fig. 6.1. Leaf types of different pumpkin species.

low in cholesterol (Odhav *et al.*, 2007). Their mineral and vitamin content exceeds levels found in exotic vegetables like spinach and cabbage (Odhav *et al.*, 2007). Pumpkin leaves contain polyphenols which are scientifically evident to be antidiabetic (Shi *et al.*, 2005). As a good source of iron, pumpkin leaves are comparable to other dark-green, leafy vegetables like spinach, collard greens and kale (van Rensburg *et al.*, 2007). According to Trumbo *et al.* (2001), the leaves of *C. maxima* are richer in iron content (9.2 mg/100 g) than black nightshade leaves (7.2 mg/100 g), Chinese cabbage leaves (1.4 mg/100 g), Jew's mallow leaves (3.6 mg/100 g) and cowpea leaves (4.7 mg/100 g).

Pumpkin leaves in the Vhembe district of Limpopo Province are among the highly favoured ALV species (Van Rensburg *et al.*, 2005). Pumpkin leaves as ALV are an essential food, consumed with starch staples in South African rural communities (Lewu and Mavengahama, 2010). There is limited information on the processing of pumpkin leaves in South Africa. Most of the information is on the fruits and seeds of pumpkin accessions.

Botany

Pumpkin leaves have a width of 20 to 30 cm. In hot climates with heavy rainfall, pumpkins usually grow at low altitudes but some varieties have been found above 2200 m (Mabberley, 2008). Groups of *C. moschata* include cheese, crookneck (*C. moschata*), and bell (Heistinger, 2013). The leaves of *C. moschata* often have white spots near the veins. The typical cultivated *Cucurbita* species has five-lobed or palmately divided leaves with long petioles, with the leaves alternately arranged on the stem (Heistinger, 2013). The stems in some species are angular. All of the above-ground parts may be hairy with various types of trichomes, which are often hardened and sharp (van Rensburg *et al.*, 2007). Spring-like tendrils grow from each node and are branching in some species. *C. argyrosperma* has ovate-cordate (egg-shaped to heart-shaped) leaves. *C. moschata* plants can have light or dense pubescence. The leaves of all three of these species may or may not have white spots (Heistinger, 2013). Cucurbitaceae are almost all vine-like, annual, herbaceous plants. The leaves and stems of some species are covered in sharp, stiff translucent hairs that can irritate human skin. The leaves vary considerably in shape and size from almost the entire leaves to deeply lobed leaves that vary from dark to light green (van Rensburg *et al.*, 2007). Flowers are monoecious, yellow or white and vary in colour. The shape of *C. pepo* leaves differs widely. *Cucurbita pepo* is an annual plant with leaves; it has a climbing stem of up to 12 m long and fruit with round fibrous flesh (Dietmar *et al.*, 2005). There is wide variation in fruit size, fruit weight, shape and rind colour, vine length and branching, leaf size, quality of fruit and seed size. Some are among the largest fruits produced by any plant group. The pumpkin plant is a tropical vegetable widely used in many Latin American and Caribbean dishes (Ruth *et al.*, 1998).

Traditional use

Pumpkin leaves are an important part of the pumpkin plant, and have long been used in traditional medicine to cure common ailments. Additionally, pumpkin leaves can be prepared in a variety of delicious veggie recipes to complement a well-balanced, nutritional diet (Van den Heever and Venter, 2007). They can be harvested or purchased from certain ethnic markets and wholesale vendors (van Rensburg *et al.*, 2007). Fresh pumpkin leaves start to decrease in quality and nutritional value almost immediately after harvest. However, they can be frozen and stored for up to three or four weeks in plastic freezer bags by separating each row with paper towels (Ojieh *et al.*, 2007). Additionally, it is a popular custom to sun-dry pumpkin leaves for later medicinal and culinary use throughout many regions of Africa, particularly South Africa, where the dried leaves are harvested and packaged with the pumpkins' flowers for added decoration (Ojieh *et al.*, 2007).

Different forms of processing are used based on the different cultures in sub-Saharan Africa and their resources (Vorster *et al.*, 2005). These include boiling, cooking, blanching and sun-drying. All through the African region, the leaves, flowers and even immature fruits of pumpkin leaves are boiled and consumed as a potherb or as a relish, stew or side accompaniment to a dish (Hart and Vorster, 2006).

The leaves of the pumpkin show neuroprotective effects. Ethyl acetate extract from the flowers of the pumpkin reduce the respiratory rate, diarrhoea and exophthalmos. Pumpkin-supplemented food has many health benefits since it has many functional compounds such as squalene in the seeds. Squalene showed positive effects on the treatment of certain cancers. Raw pumpkin seed tocopherol isomers (α, β, γ, and δ) have benefits in the treatment of diabetes due to their antioxidant properties (Caili *et al.*, 2006). In a study in South Africa, it was revealed that arthritis can be treated with orally taken *Cucurbita pepo* leaves (Asowata-Ayodele *et al.*, 2016). The extract of the *Cucurbita maxima* leaves, fruits and flowers is used for pharmacological and microbiological studies (Peter *et al.*, 2014). Noumedem *et al.* (2013) also revealed that the extract from *C. pepo* leaves showed the best antimicrobial activity.

The fresh leaves are cooked or consumed raw in a number of healthy salads. For an added energy boost, fresh pumpkin leaves can be blended into delicious fruit and vegetable breakfast smoothies. Pumpkin leaves can be added to a number of hearty soup recipes (Oloyede, 2012). For best results, they can be prepared with beans and healthy oils and fats. Pumpkin leaves can also be used for medicinal purposes by boiling the leaves in hot water over medium heat for half an hour and the warmed liquid consumed twice daily. This preparation is used for reducing fever and treating intestinal parasites (Noumedem *et al.*, 2013).

Nutrient comparison of exotic leafy vegetables and pumpkin leaves

Unlike exotic species, wild fruits and vegetables are well adapted to the harsh local climatic conditions (Bvenura and Sivakumar, 2017). The indigenous methods of processing and preserving food become vital to ensure food availability for rural households whose main meal consists mainly of a starch (i.e. porridge), indigenous vegetables (merogo) and legumes (Oloyede, 2012). The cultural practice of eating indigenous leafy vegetables and legumes adds vital nutritional elements to the traditional African diet, since good vegetables are a source of vitamins A and C with protein derived from the legumes, such as bambara, ground nuts and

Table 6.1. Estimated nutrient contribution of an average portion of pumpkin leaves to recommended dietary allowance (RDA %) for adults.

	Pumpkin leaves	RDA (%)
Minerals		
Calcium	177	22
Magnesium	67	21
Phosphorus	102	13
Sodium	12	1
Potassium	351	7
Copper	0.21	22
Zinc	0.75	9
Iron	9.2	49
Manganese	0.54	30
Thiamine	0.04	3
Riboflavin	0.10	8
Folic acid	47	8
Vitamin C	2	1
Vitamin A	325	44

cowpeas (Khumbane, 2004). Nutrient content of raw plant foods varies widely and is affected by factors such as variety or cultivar; the part of the plant consumed; the stage of maturity; the geographic site of production or the climate; the harvesting and postharvest handling conditions; and the method of storage (Van Jaarsveld *et al.*, 2014). Comparing nutrient content of leaves from different data sources is therefore challenging. Common errors introduced during sampling procedures, sample preparation and analytical methods can also affect the reported nutrient values (Van Jaarsveld *et al.*, 2014).

In Table 6.1, the estimated amount of nutrients retained after preparation and taking the yield 366 factor into account are expressed as a percentage of the recommended daily allowance (RDA) for adults. The RDA is the average daily dietary intake level (370) that is sufficient to meet the nutrient requirement for nearly all (97 to 98%) healthy individuals in a particular life stage and gender group (Otten *et al.*, 2006; Ross *et al.*, 2011). ALVs are low in energy (<5% of 376, the estimated requirement). Together with the high fibre and low-fat content, these characteristics of ALVs are potentially important elements of a dietary pattern to treat and prevent obesity (Van Jaarsveld *et al.*, 2014). The calculated estimates showed that ALVs provide 10% or less of the RDA for adults for protein, zinc, thiamin, riboflavin and vitamin C (Van Jaarsveld *et al.*, 2014). Phosphorus, copper and folate content ranged from 4 to 26% of RDA. Calcium and magnesium content for pumpkin leaves was above 10%. Iron content showed almost 50% RDA (Van Jaarsveld *et al.*, 2014).

The composition of vitamins and minerals in pumpkin leaves and exotic leafy vegetables varies (Table 6.2). Among the vegetables, the highest zinc, iron and copper were in pumpkin leaves followed by asparagus with copper, green cauliflower with zinc and spinach with iron. The concentrations of phosphorus in pumpkin leaves are higher and lower in spinach (Table 6.2). Phosphorus may lower blood pressure and normalize heart rhythm in humans (Soetan *et al.*, 2010). The calcium composition of *Cucurbita maxima* leaves reported by Trumbo *et al.* (2001) is lower compared to the composition reported by Gupta *et al.* (2005). Gupta *et al.* (2005) reported protein content of *C. maxima* higher than that of Trumbo *et al.* (2001). The magnesium content in pumpkin leaves (67 mg/100 g) is higher than that of kale (33 mg/100 g), asparagus (14 mg/100 g) and cauliflower (20 mg/100 g) but lower than that of spinach (77 mg/100 g). Pumpkin leaves are poor in vitamin C when compared to exotic vegetables but rich in folic acid (Table 6.2). Folic acid helps the body to produce and maintain new cells and also helps in preventing changes to DNA that may lead to cancer (Peter *et al.*, 2014). Pumpkin leaves are lower in sodium (12 mg/100 g) than spinach (79 mg/100 g), kale (53 mg/100 g) and cauliflower (23 mg/100 g) but higher than asparagus (2 mg/100 g) (Table 6.2).

The leaves of the pumpkin are rich in protein (9%), fat (18%) and vitamins (20%) (Sharma *et al.*, 2019). Pumpkin leaves have the highest iron content, and thus appear to be a very good source of non-haeme iron compared to the other ALVs (Van Jaarsveld *et al.*, 2014). The iron content of raw pumpkin leaves, for example, in the SAFOODS database, is 2.2 mg/100 g (Kruger *et al.*, 1998). The leaf extract is a good source of iron that increases the haemoglobin level in the blood. It is accepted that glycerolipids make an essential contribution to the structure of biological membranes

Table 6.2. Comparison of mineral and vitamin concentrations in pumpkin leaves and exotic vegetables (USDA, 2007).

	Unit	Pumpkin leaves	Spinach	Kale	Cauliflower	Asparagus
Proximate analysis						
Moisture	g	85.6	91.4	89.6	89.79	93.22
Protein	g	2.9	2.86	2.92	2.95	2.2
Ash	g	1.5			0.88	0.58
Fat	g	0.2	0.39	1.49	0.3	0.12
Energy	kJ	222	96	146	130	85
Dietary fibre	g	3.0 g	2.2 g	4.1 g	3.2	2.1
Carbohydrates	g	9.8	3.63	4.42	6.09	3.88
Minerals						
Calcium	mg	177	99	254	33	24
Magnesium	mg	67	79	33	20	14
Phosphorus	mg	102	49	55	62	52
Sodium	mg	12	79	53	23	2
Potassium	mg	351	558	348	300	202
Copper	mg	0.21	0.13	0.053	0.041	0.189
Zinc	mg	0.75	0.53	0.39	0.64	0.54
Iron	mg	9.2	2.71	1.6	0.73	2.14
Manganese	mg	0.54			0.247	0.158
Vitamins						
Thiamine	mg	0.04	0.078	0.113	0.08	0.143
Riboflavin	mg	0.10	0.189	0.347	0.102	0.141
Folic acid	μg	47	0	0	0	0
Vitamin C	mg	2	28.1	93.4	88.1	5.6
Total β-carotene	μg	4247	5626	2870	93	449
All-trans-β-carotene	μg	3547				
Vitamin A	μg RAE	325	469	241	8	38

and a-linolenic acid, the major fatty acid in leaves, is used in photosynthetic electron transport, confectionary products, alcohol, sweets, candy and fermented into beverages (Anju and Bhat, 2013) while the immature fruit is used as a vegetable. All pumpkin plant parts, including edible protein-rich seed, fruit and greens have wide nutritional aspects and food usage. It is useful as a food additive because of its high amount of beta-carotene and moderate amount of carbohydrate, vitamins such as B6, K, thiamine and riboflavin, and minerals such as potassium, phosphorus, magnesium, iron and selenium. It is used in making juice, pomade, pickles and dried products (Rakcejeva *et al.*, 2011). Past work on the nutrient composition of *C. pepo* leaves reveals that the leaf has 43.8% protein, which is similar to that of soybean (Oloyede, 2012).

Dietary fibre is an important constituent in pumpkin and other vegetables, helping to reduce serum cholesterol level, risk of coronary heart disease, and contributing to prevent colon and breast cancers and hypertension (Rodríguez *et al.*, 2006). The fibre level in pumpkin leaves and fruits is sufficiently high for this (Ajala, 2009). The observed value may imply that pumpkin leaves have a short shelf life since microorganisms that cause spoilage thrive due to high moisture content (Emebu and Anyika, 2011). The protein values of pumpkin, as observed, confer on it the advantage of being a rich source of vegetable protein. Incorporating both the leaves and fruits of the pumpkin in the diet can provide sufficient amounts of protein, which in turn provides several benefits, such as provision of imperative body constituents, maintenance of fluid balance, formation of hormones and enzymes and contributions to immune function (Bahru *et al.*, 2018). Furthermore the composition of minerals and vitamins in comparison with the other exotic vegetables are shown in Table 6.2. Table 6.3 demonstrates the compositions of proximate analysis, minerals and vitamins in pumpkin leaf accessions.

Antinutritive compounds

Undesirable chemical substances, referred to as antinutrients, are abundant in both cultivated and wild plant species (Dagostin, 2017). The level of adversity is largely dependent on the diet pattern and method of processing involved before the consumption of the specific plant food. When a plant food is consumed as a nutritional source, antinutrients are consumed and pose a health risk to the consumer (Ugwu and Oranye, 2006). Some of the antinutrients commonly found in pumpkin leaves are nitrates, oxalates, tannins, phytates and cyanogenic glycosides (Natesh *et al.*, 2017).

Tannins are identified as plant polyphenols that are capable of forming complexes with metals ions and macro-molecules like proteins and polysaccharides (Olawoye and Gbadamosi, 2017). Tannins affect the nutritional value of food products by chelating metals like iron and zinc and reducing the absorption of these nutrients as well as forming complexes with protein, thereby inhibiting their digestion and absorption (Olawoye and Gbadamosi, 2017). Tannins are responsible for a protein deficiency syndrome that can cause a decrease in growth rate as well as having a non-palatable taste (Essack *et al.*, 2017).

Phytic acid is the major phosphorus storage compound in traditional leafy vegetables. Phytic acid has been recorded to inhibit the absorption of minerals and reduce the bioavailability of metal ions like zinc and iron as well as affect protein and starch digestion (Dagostin, 2017). A phytic acid intake of 4–9 mg/100 g decreases iron absorption in humans four- to fivefold. An excessively phytate rich diet is associated with nutritional diseases such as rickets and osteomalacia in children and adults (Unuofin *et al.*, 2018). The levels of phytic acid and oxalic acid can be effectively reduced by cooking and blanching methods (Oboh, 2005). However, these practices can also lead to the leaching out of nutritional elements (Mepba *et al.*, 2007).

Oxalic acid exists in many leafy vegetables and plant foods (Akindahunsi and Salawu, 2005). Depending on species, oxalic acid can occur as soluble salts of potassium and sodium or as insoluble salts of calcium, magnesium or iron. It can also occur as a combination of soluble and insoluble salts (Natesh *et al.*, 2017). This forms strong chelates, with dietary calcium inhibiting its absorption (Akwaowo *et al.*, 2000). It's well known that oxalic acid and its salts can have a deleterious effect on human nutrition and health, mostly by decreasing calcium absorption and aiding the formation of kidney stones (Olawoye and Gbadamosi, 2017).

Blanching causes a significant reduction in value of all the antinutritional factors analysed (p>0.05). The tannin, phytate and oxalate contents of *C. pepo* leaves were observed to be lower than those reported for various vegetables by Onyeka and Nwambekwe (2007). The significant reduction in the oxalate value of a blanched sample may have been part of the reason for the increase in calcium content in the sample (Onyeka and Nwambekwe, 2007).

Antidiabetic activity

Diabetes, a metabolic disease, is a serious problem in modern society due to the severe health complications linked with it (Martha and Gutierrez, 2016). Numerous plants are used in

Table 6.3. Compositions of proximate analysis, minerals and vitamins in pumpkin leaf accessions.

Species (leaves)	Micronutrients	References
C. maxima	Moisture	
C. pepo	87.30 g/100 g FW	Schönfeldt and Pretorius, 2011
	85 g/100 g FW	Gupta *et al.*, 2005
	10.17 g/100 g FW	Bahru *et al.*, 2018
C. maxima	Protein	
C. pepo	249 g/100 g FW	Schönfeldt and Pretorius, 2011
	4.10 g/100 g FW	Gupta *et al.*, 2005
	26.31 g/100 g FW	Bahru *et al.*, 2018
C. maxima	Fat	Schönfeldt and Pretorius, 2011
C. pepo	0.12 g/100 g FW	Bahru *et al.*, 2018
	1.97 g/100 g FW	
C. maxima	Ash	
C. pepo	3.23 g/100 g FW	Schönfeldt and Pretorius, 2011
	3.04 g/100 g FW	Gupta *et al.*, 2005
	10.96 g/100 g FW	Bahru *et al.*, 2018
C. maxima	Carbohydrates	Schönfeldt and Pretorius, 2011
C. pepo	0.80 g/100 g FW	Bahru *et al.*, 2018
	28.15 g/100 g FW	
C. pepo	Fibre	Bahru *et al.*, 2018
	17.55 g/100 g FW	
C. pepo	Vitamin C	Bahru *et al.*, 2018
	37 mg/100 g FW	
C. pepo	Vitamin B1	Bahru *et al.*, 2018
	0.20 mg/100 g FW	
C. maxima	Iron	
C. pepo	15.9 mg/100 g FW	Schönfeldt and Pretorius, 2011
	4.38 mg/100 g FW	Gupta *et al.*, 2005
	4.38 mg/100 g FW	Bahru *et al.*, 2018
C. maxima	Zinc	
C. pep	0.7 mg/100 g FW	Schönfeldt and Pretorius, 2011
	0.62 mg/100 g FW	Gupta *et al.*, 2005
	0.62 mg/100 g FW	Bahru *et al.*, 2018
C. maxima	Calcium	Gupta *et al.*, 2005
C. pepo	302 mg/100 g FW	Bahru *et al.*, 2018
	383 mg/100 g FW	
C. maxima	Phosphorus	Gupta *et al.*, 2005
C. pepo	34 mg/100 g FW	Bahru *et al.*, 2018
C. maxima	119 mg/100 g FW	
		Gupta *et al.*, 2005
	Potassium	
	368 mg/100 g FW	Gupta *et al.*, 2005
	Sodium	
	12 mg/100 g FW	Gupta *et al.*, 2005
	Magnesium	
	150 mg/100 g FW	

FW, fresh weight.

South Africa for the treatment of type 2 diabetes, and among them is *C. pepo*, which was evaluated using *in vitro* α-glucosidase and α-amylase activity (Martha and Gutierrez, 2016). The results demonstrated inhibitory activity against both enzyme systems, adding evidence to its traditional use in the management of type 2 diabetic post-prandial hyperglycemia (Neel *et al.*, 2017). Antidiabetic activity studies are mostly done on pumpkin seeds and fruits/pulp.

In vitro cellular assays can be used to determine the short term (i.e. few hours) and long term (i.e. few days) effects of antioxidants (Kwon *et al.*, 2008). Furthermore, these cell culture systems can also be used to determine whether any cytotoxic molecules are present in the plant extracts. In these assays, the protective effects of antioxidants take place predominantly through radical scavenging activity and metal ion chelation, either extracellularly or intracellularly depending on the solubility of the extracts (Melidou *et al.*, 2005). The dichlorofluorescein (DCF) assay is used to measure the ability of plant extracts to protect cells against oxidative damage.

Antioxidant activity

The antioxidant activity of plant extracts mainly depends on the phenolic content of the extract (Salah *et al.*, 1995). Generally, plant extracts that contain a high amount of polyphenols exhibit high antioxidant activity (Wong and Kitts, 2006). Different assays such as 2,2-azobis-(3- ethyl-benzothiazoline-6-sulfonic acid) diammonium salt (ABTS), 2,2-diphenyl-2-picrylhydrazyl (DPPH), ferric reducing ability of plasma (FRAP) and oxygen radical antioxidant capacity (ORAC) have been used to determine the radical scavenging capacity and antioxidant content of African leafy vegetables and/or spinach. The analysis of pumpkin leaf (*C. maxima*) ethanol extract showed the presence of alkaloids, flavonoids, saponins, triterpens and alkaloids. Moreover, Biswas *et al.* (2013) showed that the ethanolic extract of *Cucurbita maxima* leaves showed the presence of reducing sugars, alkaloids, saponins, tannins, steroid and glycosides. Mavhungu (2011) revealed that pumpkin leaves had the highest amounts of total flavonoids (12.01 mg/g dry weight (DW) or 1.36 mg/g DW), while jute mallow had the least amounts with 0.18 mg/g DW (0.02 mg/g FW). According to Mavhungu (2011), pumpkin leaves showed a higher total phenolic content (2.7 mg GAE/g) than cowpea leaves (2.4 mg GAE/g) and amaranth leaves (2.1 mg GAE/g). Pumpkin leaves had the highest antioxidant activities ABTS (62.1 μmol TEAC/g), DPPH (83.6 μmol TEAC/g) and ORAC (36.0 μmol TEAC/g) when compared to amaranth leaves (Mavhungu, 2011).

Processing

Drying

Drying is a way of processing leafy vegetables to make them available during periods of short supply (Van Jaarsveld *et al.*, 2014). Drying is a postharvest process that can increase the availability of leafy vegetables to farmers, especially those who cannot afford packaging. Drying reduces microbiological activity through reduced moisture content in food.

Different processing techniques are often utilized in order to reduce antinutritional factors (Chen, 2007). Some processing techniques are performed at a household level (domestically) and others are performed on a larger scale in industry (Raes *et al.*, 2014). Drying is a food preservation method that has been used for many centuries and is beneficial in reducing microbial decay, ensuring physiochemical stability, reducing weight and transport costs as well as improving handling and storability (Kumar *et al.*, 2010). Vorster *et al.* (2008) reported that there are two main methods for preserving indigenous leafy vegetables in South Africa. These are sun-drying of fresh leaves and sun-drying of blanched leaves. Both these methods transform the leafy vegetable into a dry product that has a long shelf life (Vorster *et al.*, 2005). Van Averbeke *et al.* (2012) reported that electrification of rural areas has introduced the new preservation technology of freezing leaves. However, there are limited published data on the effect of various drying methods on the quality of pumpkin leaves in South Africa. Such information is necessary to establish suitable drying methods for cultivated pumpkin leaves in the country.

Blanching

Blanching is a thermal treatment commonly applied in a variety of vegetable preservation treatments and is particularly important in freezing because of its very considerable influence on quality. The primary objective is to inactivate enzymes responsible for alterations in sensory quality attributes (off-flavours and odours) and in nutritional value (loss of vitamins) during storage (Oulai *et al.*, 2014). Kunyanga *et al.* (2012) revealed that blanching showed significant loss of phenolics in *Cucurbita maxima* leaves by 43%. Like the minerals, blanching significantly led to a reduction in ascorbic acid content of *Cucurbita pepo* leaves (Oulai *et al.*, 2014). This reduction could be attributed to the solubility of ascorbic acid in water and a consequent loss in the blanching medium.

The wet matter moisture and carbohydrate content of blanched portions are higher than the air-dried portions, however, the dry matter moisture, protein, fats, crude fibre and ash content are lower in blanched than in air-dried portions of *C. pepo* (Oulai *et al.*, 2014). The high moisture content indicates short shelf-life and high vulnerability to microbial attack. Blanching caused a significant increase in the moisture content of *C. pepo* which could be as a result of water adsorption by the leaf. The fat content of *C. pepo* was low when compared with other vegetables such as *M. oleifera* (2.98 g/100 g), *G. senegalensis* (1.97 g/100 g) and *C. esculentus* (2.67 g/100 g) on a fresh weight basis (Nkafamiya *et al.*, 2010). This suggests that *C. pepo* leaves, in addition to the pulp and seed, are low lipid-containing foods. The ash content of raw *C. pepo* leaf was 2.9 g/100g and reduced to 1.68 g/100g when blanched. This value is lower than 10.92 g/100g reported in pumpkin leaf extract (Nworgu *et al.*, 2007). This suggests that the leaf in the study was lower in mineral composition than pumpkin leaf extract (Omotoso, 2005; Nnamani *et al.*, 2009), thus indicating that the vegetable may be beneficial to patients who experience inability to excrete minerals, such as renal patients.

The carbohydrate content of the vegetable increased from 1.50 g/100g in a raw sample to 2.73 g/100g in a blanched sample. This is contrary to the findings of Nkafamiya *et al.* (2010), in which the carbohydrate composition declined as a result of blanching. This could be attributed to the breakdown of ketogenic amino acids to glucose thereby increasing the carbohydrate content of the vegetable. *Cucurbita pepo* contributes about 2.85 and 1.42% RDA for sodium in the raw and blanched forms respectively. The potassium content adds about 14.04 and 5.10% RDA to the diet of a 70 kg adult male when 100g fresh weight (FW) of this vegetable is served in or with the diet (Nnamani *et al.*, 2009).

Cooking

Pumpkin leaves are rarely consumed in the raw form; they are usually cooked (Funke, 2011). Cooking methods such as boiling, microwaving and stir-frying are generally used to prepare such dishes. It is well known that cooking induces profound changes in chemical composition, affecting bioavailability and the amount of bioactive compounds in vegetables (Wachtel-Galor *et al.*, 2008). Cooking induces significant changes in chemical composition affecting concentration and nutrient bioavailability (Yuan *et al.*, 2009). Various cooking methods are used based on convenience and taste preference rather than nutrient retention. Some cooking methods may oxidize antioxidants (Funke, 2011) and affect nutrient retention. It is therefore important to choose a cooking method that leads to optimal nutrient retention and bioavailability (Shahnaz *et al.*, 2003).

The contents of epicatechin, rutin, myricetin, quercetin, kaempferol, luteolin and apigenin in pumpkin leaves was shown to decrease significantly after boiling (Mavhungu, 2011). The content of total flavonoids of extracts from boiled pumpkin leaves was also less than that of raw samples. Boiling decreases the phenolic and antioxidant content of *C. maxima* (Mavhungu, 2011). The decrease may have resulted from leaching of vegetable antioxidants into the cooking medium (Roy *et al.*, 2007; Wachtel-Galor *et al.*, 2008) or may be due to oxidation of polyphenol components by polyphenol oxidase in vegetables (Yamaguchi *et al.*, 2003). Yamaguchi *et al.* (2001) found both decreases and increases in flavonoid content after cooking. An increase in antioxidant activity, as was observed for pumpkin leaves at 60 minutes, may have resulted from liberation of high amounts of antioxidant components due to thermal destruction of vegetable cell wall structures and subcellular compartments (Mavhungu, 2011), the production of redox active secondary plant metabolites or breakdown products from intracellular compartments, matrix modifications and more efficient release of antioxidants during homogenization of samples (Wachtel-Galor *et al.*, 2008).

Conclusion

The seed, flowers, fruit pericarp and leaves of the pumpkin have been used as base materials for innovative functional foods, nutraceutical and pharmaceutical preparations. All parts of the pumpkin are cheap and easily available in developing countries. Thus, *Cucurbita* can be used to make different and value-added products, so that the consumer may receive the health benefits through those food products.

References

Ajala, L. (2009) The effect of boiling in the nutrients and anti-nutrients in two non-conventional vegetables. *Pakistan Journal of Nutrition* 8(9), 1430–1433. DOI: 10.3923/pjn.2009.1430.1433.

Akindahunsi, A.A. and Salawu, S.O. (2005) Antioxidant indices of some green leafy vegetables. *Tropical Sciences* 45(1), 33–35. DOI: 10.1002/ts.43.

Akwaowo, E.U.A., Ndon, B.A. and Etuk, E.U. (2000) Minerals and antinutrients in fluted pumpkin (*Telfairia occidentalis Hook f.*). *Food Chemistry* 70(2), 235–240. DOI: 10.1016/S0308-8146(99)00207-1.

Anju, B. and Bhat, M.A. (2013) Study on physico-chemical characteristics of pumpkin blended cake. *Journal of Food Processing & Technology* 4(9), 1000262. DOI: 10.4172/2157-7110.1000262.

Asowata-Ayodele, A.M., Afolayan, A.J. and Otunola, G.A. (2016) Ethnobotanical survey of culinary herbs and spices used in the traditional medicinal system of Nkonkobe Municipality, Eastern Cape, South Africa. *South African Journal of Botany* 104, 69–75. DOI: 10.1016/j.sajb.2016.01.001.

Bahru, T.B., Hailemariam Barkea, T. and Sali, A.A. (2018) Investigation of Primary Metabolites in Young Leaf and Fruit of Three Varieties of Pumpkin (Cucurbita pepo) from Gurage Zone, Ethiopia. *Journal of Analytical & Bioanalytical Techniques* 09(4), 1000407. DOI: 10.4172/2155-9872.1000407.

Biswas, N.N., Bishwajit, B., Md, S.R., Md, S.M. and Rahman, S.E. (2013) Phytochemical and pharmacological evaluation of *Cucurbita maxima* Duchesne and *Euphorbia royleana Boiss*. *Khulna University Studies* 11, 26–35.

Bvenura, C. and Sivakumar, D. (2017) The role of wild fruits and vegetables in delivering a balanced and healthy diet. *Food Research International* 99(Pt 1), 15–30. DOI: 10.1016/j.foodres.2017.06.046.

Caili, F.U., Huan, S.H. and Quanhong, L.I. (2006) A review on pharmacological activities and utilization technologies of pumpkin. *Plant Foods Human Nutrition* 61(2), 70–77. DOI: 10.1007/s11130-006-0016-6.

Chen, Q. (2007) Postharvest technologies for fresh leafy vegetables in Yunnan, China. In: *RETA 6376 Workshop on Best Practices in Postharvest Management of Leafy Vegetables in GMS Countries*, Hanoi, Vietnam, pp. 25–27.

Dagostin, J.L.A. (2017) Use of blanching to reduce anti-nutrients, pesticides, and microorganisms [New perspectives on food blanching]. In: Richter Reis, F. (ed.) *New Perspectives on Food Blanching*. Springer, USA, pp. 61–94.

Dietmar, F., Winkler, C., Wirleitner, B., Schroecksn, K. and Schennach, H. (2005) Extract of pumpkin seeds suppresses stimulated peripheral blood mononuclear cell *In vitro*. *American Journal of Immunology* 1(1), 6–11. DOI: 10.3844/ajisp.2005.6.11.

Emebu, P.K. and Anyika, J.U. (2011) Proximate and mineral composition of Kale (*Brassica oleraceae*) grown in Delts State, Nigeria. *Pakistan Journal of Nutrition* 10(2), 190–194. DOI: 10.3923/pjn.2011.190.194.

Essack, H., Odhav, B. and Mellem, J.J. (2017) Screening of traditional South African leafy vegetables for specific anti-nutritional factors before and after processing. *Food Science and Technology* 37(3), 462–471. DOI: 10.1590/1678-457x.20416.

Funke, O.M. (2011) Evaluation of nutrient contents of amaranth leaves prepared using different cooking methods. *Food Science and Nutrition* 2(4), 249–252. DOI: 10.4236/fns.2011.24035.

Gupta, S., Lakshmia, A.J., Manjunathb, M.N. and Prakash, J. (2005) Analysis of nutrient and antinutrient content of underutilized green leafy vegetables. *LWT - Food Science and Technology* 38(4), 339–345. DOI: 10.1016/j.lwt.2004.06.012.

Hart, T.G.B. and Vorster, H.J. (2006) *Indigenous Knowledge on the South African Landscape – Potentials for Agricultural Development. Urban, Rural and Economic Development Programme*. Occasional paper No 1. HSRC Press, Cape Town, South Africa, p. 52.

Heistinger, A. (2013) The manual of seed saving: harvesting, storing and sowing techniques for vegetables, herbs and fruits. Timber Press.

Khumbane, T. (2004) Food security: traditional knowledge and permaculture. *South African Rural Development Quarterly* 2, 44–49.

Kruger, M., Sayed, N., Langenhoven, M. and Holing, F. (1998) *Composition of South African Foods, Vegetables and Fruit. Supplement to the MRC Food Composition Tables 1991*. Medical Research Council, Tygerberg, p. 160.

Kumar, V., Pandey, N., Mohan, N. and Singh, R.P. (2010) Antibacterial and antioxidant activity of different extract of *moringa oleifera* leaves -an in vitro study. *International Journal of Pharmaceutical Sciences* 10, 89–94.

Kunyanga, C.N., Imungi, J.K., Okoth, M., Momanyi, C., Biesalski, H.K. *et al.* (2012) Antioxidant and antidiabetic properties of condensed tannins in acetonic extract of selected raw and processed indigenous food ingredients from Kenya. *Journal of Food Science* 76(4), C560-C567. DOI: 10.1111/j.1750-3841.2011.02116.x.

Kwon, Y.I., Apostolidis, E. and Shetty, K. (2008) Inhibitory potential of wine and tea against α-Amylase and α-Glucosidase for management of hypeglycemia linked to type 2 diabetes. *Journal of Food Biochemistry* 32(1), 15–31. DOI: 10.1111/j.1745-4514.2007.00165.x.

Lewu, F. and Mavengahama, S. (2010) Wild vegetables in Northern Kwazulu Natal, South Africa: current status of production and research needs. *Scientific Research and Essays* 5(20), 3044–3048.

Mabberley, D.J. (2008) *Mabberley's Plant-Book: A Portable Dictionary of Plants, Their Classification and Uses*, 3rd edn. Cambridge University Press, pp. 1–1021.

Martha, R. and Gutierrez, P. (2016) Review of *Cucurbita pepo* (Pumpkin) its Phytochemistry and Pharmacology. *Medicinal Chemistry* 6(1), 12–21. DOI: 10.4172/2161-0444.1000316.

Mavhungu, N.P. (2011) *Antioxidants properties and cellular protective effects of selected African green leafy vegetables*. Unpublished PhD dissertation, University of Pretoria, Pretoria.

Melidou, M., Riganakos, K. and Galaris, D. (2005) Protection against nuclear DNA damage offered by flavonoids in cells exposed to hydrogen peroxide: the role of iron chelation. *Free Radical Biology and Medicine* 39(12), 1591–1600. DOI: 10.1016/j.freeradbiomed.2005.08.009.

Mepba, H., Eboh, L. and Banigo, D. (2007) Effects of processing treatments on the nutritive composition and consumer acceptance of some Nigerian edible leafy vegetables. *African Journal of Food, Agriculture, Nutrition and Development* 7, 1–18.

Natesh, N.H., Abbey, L. and Asiedu, S.K. (2017) An overview of nutritional and anti nutritional factors in green leafy vegetables. *Horticulture International Journal* 1(2), 58–65. DOI: 10.15406/hij.2017.01.00011.

Neel, R., Vandana, Md.N. and Md, I. (2017) A Review on *Cucurbita* pepo. *International Journal of Pharmacognosy and Phytochemical Research* 9(9), 1190–1194. DOI: 10.25258/phyto.v9i09.10305.

Nkafamiya, I.I., Oseameahon, S.A., Modibbo, U.U. and Haggai, D. (2010) Vitamins and effect of blanching on nutritional and antinutritional values of nonconventional leafy vegetables. *African Journal of Food Science* 4, 335–341.

Nnamani, C.V., Oselebe, H.O. and Agbatutu, A. (2009) Assessment of nutritional values of three underutilized indigenous leafy vegetables of Ebonyi State, Nigeria. *African Journal of Biotechnology* 8, 2321–2324.

Noumedem, J.A.K., Mihasan, M., Lacmata, S.T., Stefan, M., Kuiate, J.R. *et al.* (2013) Antibacterial activities of the methanol extracts of ten Cameroonian vegetables against Gram-negative multidrug-resistant bacteria. *BMC Complementary and Alternative Medicine* 13, 26. DOI: 10.1186/1472-6882-13-26.

Nworgu, F.C., Ogungbenro, S.A. and Solesi, K.S. (2007) Performance and some blood chemistry indices of broiler chicken served *Cucurbita* spp leaves extract supplement. *American-Eurasian Journal of Agriculture and Environmental Science* 2(1), 90–98.

Oboh, G. (2005) Effect of blanching on the antioxidant properties of some tropical green leafy vegetables. *LWT - Food Science and Technology* 38(5), 513–517. DOI: 10.1016/j.lwt.2004.07.007.
Odhav, B., Beekrum, S. and Akula, U. (2007) Preliminary assessment of nutritional value of traditional leafy vegetables in KwaZulu-Natal, South Africa. *Journal of Food Composition and Analysis* 20(5), 430–435. DOI: 10.1016/j.jfca.2006.04.015.
Ojieh, G., Oluba, O., Oguniowo, Y., Adebaesi, K., Eigdangbe, G. *et al.* (2007) Composition studies of *Citrullus lanatus* (egusi melon) seed. *International Journal of Nutrition and Wellness* 6, 19–26.
Olawoye, B.T. and Gbadamosi, S.O. (2017) Effect of different treatments on *in vitro* protein digestibility, antinutrients, antioxidant properties and mineral composition of *Amaranthus viridis* seed. *Cogent Food & Agriculture* 3(1), 1296402. DOI: 10.1080/23311932.2017.1296402.
Oloyede, F.M. (2012) Growth, yield and antioxidant profile of pumpkin (*Cucúrbita pepo* L.) leafy vegetable as affected by NPK compound fertilizer. *Journal of Soil Science and Plant Nutrition* 12(3), 379–388. DOI: 10.4067/S0718-95162012005000001.
Omotoso, O.T. (2005) Nutritional quality, functional properties and anti-nutrient compositions of the larva of *Cirina forda* (Westwood) (*Lepidoptera: Saturniidae*). *Journal of Zhejiang University. SCIENCE B* 7(1), 51–55. DOI: 10.1631/jzus.2006.B0051.
Onyeike, E.N. and Omubo-Dede, T.T. (2002) Effect of heat treatment on the proximate composition, energy values, and levels of some toxicants in African yam bean (*Sphenostylis stenocarpa*) seed varieties. *Plant Foods for Human Nutrition* 57(3–4), 223–231. DOI: 10.1023/a:1021833516234.
Onyeka, E.U. and Nwambekwe, I.O. (2007) Phytochemical profile of some green leafy vegetables in South East, Nigeria. *Nigerian Food Journal* 25(1), 67–76. DOI: 10.4314/nifoj.v25i1.33655.
Otten, J., Helllwig, J. and Meyers, L.D. (2006) *Dietary Reference Intakes: The Essential Guide to Nutrient Requirements*, 1st edn. National Academies Press.
Oulai, P., Lessoy, Z., Megnanou, R.M., Doue, R. and Niamke, S. (2014) Proximate composition and nutritive value of leafy vegetables consumed in Northern Cote d'Ivorie. *European Scientific Journal* 10(6), 1857–1881.
Peter, E.A., Hudson, N., Alice, O.N., Stanley, O., William, T. *et al.* (2014) Evaluation of micronutrients in seeds of pumpkin varieties grown by smallholder farmers in the Lake Victoria basin. *African Journal of Food Science and Technology* 4, 221–228.
Raes, K., Knockaert, D., Struijs, K. and Van Camp, J. (2014) Role of processing on bioaccessibility of minerals: Influence of localization of minerals and anti-nutritional factors in the plant. *Trends in Food Science & Technology* 37(1), 32–41. DOI: 10.1016/j.tifs.2014.02.002.
Rakcejeva, T., Galoburda, R., Cude, L. and Strautniece, E. (2011) Use of dried pumpkins in wheat bread production. *Procedure Food Science* 1, 441–447. DOI: 10.1016/j.profoo.2011.09.068.
Rodríguez, R., Jiménez, A., Fernández, B., Guillen, R. and Heredia, A. (2006) Dietary fibre from vegetable products as source of functional ingredients. *Trends in Food Science & Technology* 17(1), 3–15. DOI: 10.1016/j.tifs.2005.10.002.
Ross, A.C., Manson, J.E., Abrams, S.A., Aloia, J.F., Brannon, P.M. *et al.* (2011) The 2011 report on dietary reference intakes for calcium and vitamin D from the institute of medicine. *Journal of Clinical Endocrinology and Metabolism* 96(1), 53–58. DOI: 10.1210/jc.2010-2704.
Roy, M.K., Takenaka, M., Isobe, S. and Tsushida, T. (2007) Antioxidant potential, anti-proliferative activities, and phenolic content in water-soluble fractions of some commonly consumed vegetables: effects of thermal treatment. *Food Chemistry* 103(1), 106–114. DOI: 10.1016/j.foodchem.2006.08.002.
Ruth, L.M., Sanja, K.C. and Abdulahi, H.Z. (1998) Assessment and production of tropical vegetables. *Journal of Plant Science* 23, 453–468.
Salah, N., Miller, N.J., Paganga, G., Tijburg, L., Bolwell, G.P. *et al.* (1995) Polyphenolic flavanols as scavengers of aqueous phase radicals and as chain-breaking antioxidants. *Archives of Biochemistry and Biophysics* 322(2), 339–346. DOI: 10.1006/abbi.1995.1473.
Schönfeldt, H.C. and Pretorius, B. (2011) The nutrient content of five traditional South African dark green leafy vegetables—A preliminary study. *Journal of Food Composition and Analysis* 24(8), 1141–1146. DOI: 10.1016/j.jfca.2011.04.004.
Shahnaz, A., Khan, K.M., Sheikh, M.A. and Shahid, M. (2003) Effect of peeling and cooking on nutrients in vegetables. *Pakistan Journal of Nutrition* 2(3), 189–191. DOI: 10.3923/pjn.2003.189.191.
Sharma, P., Kaur, G., Kehinde, B.A., Chhikara, N., Panghal, A. *et al.* (2019) Pharmacological and biomedical uses of extracts of pumpkin and its relatives and applications in the food industry: a review. *International Journal of Vegetable Science* 26(1), 79–95. DOI: 10.1080/19315260.2019.1606130.

Shi, J., Nawaz, H., Pohorly, J., Mittal, G., Kakuda, Y. *et al.* (2005) Extraction of polyphenolics from plant material for functional foods-Engineering and technology. *Food Reviews International* 21(1), 139–166. DOI: 10.1081/FRI-200040606.

Soetan, KO., Olaiya, C.O. and Yewole, O.E. (2010) The importance of mineral elements for humans, domestic animals and plants: a review. *African Journal of Food Science* 4(5), 200–222.

Trumbo, P., Yates, A.A., Schlicker, S. and Poos, M. (2001) Dietary reference intakes: Vitamin A, vitamin K, arsenic, boron, chromium, copper, iodine, iron, manganese, molybdenum, nickel, silicon, vanadium, and zinc. *Journal of the American Dietary Association* 101(3), 294–301. DOI: 10.1016/S0002-8223(01)00078-5.

Ugwu, F.M. and Oranye, N.A. (2006) Effects of some processing methods on the toxic components of African breadfruit (*Treculia africana*). *African Journal of Biotechnology* 5(22), 2329–2333.

Unuofin, J.O., Otunola, G.A. and Afolayan, A.J. (2018) Polyphenolic content, antioxidant and antimicrobial activities of *Vernonia mespilifolia* Less. Used in folk medicine in the Eastern Cape Province, South Africa. *Journal of Evidence-Based Integrative Medicine* 23, 2515690X18773990. DOI: 10.1177/2515690X18773990.

USDA (2007) U.S. Department of Agriculture (USDA), Agricultural Research Service. (2007) USDA 630 Table of nutrient retention factors, Release 6. Retrieved March 27, 2012 from the Nutrient 631 Data Products and Services Page. Available at: http://www.ars.usda.gov/Services/docs.htm?docid=9448 (accessed 10 November 2020).

Uusiku, N.P., Oelofse, A., Duodu, K.G., Bester, M.J. and Faber, M. (2010) Nutritional value of leafy vegetables of sub-Saharan Africa and their potential contribution to human health: A review. *Journal of Food Composition and Analysis* 23(6), 499–509. DOI: 10.1016/j.jfca.2010.05.002.

Van Averbeke, W., Chabalala, M.P., Okorogbona, A.O.M., Rumania, T.D., Azeez, J.O. *et al.* (2012) Plant nutrient requirements of African leafy vegetables. In: Oelofse, A. and Van Averbeke, W. (eds) *Nutritional Value and Water Use of African Leafy Vegetables for Improved Livelihoods; WRC TT535/12*. Water Research Commission, Pretoria, South Africa.

Van den Heever, E. and Venter, S.L. (2007) Nutritional and medicinal properties of traditional leafy vegetables. *Acta Horticulturae (ISHS)* 752, 127–130.

Van Jaarsveld, P., Faber, M., van Heerden, I., Wenhold, F., Van Rensburg, W.J. *et al.* (2014) Nutrient content of eight African leafy vegetables and their potential contribution to dietary reference intakes. *Journal of Food Composition and Analysis* 33(1), 77–84. DOI: 10.1016/j.jfca.2013.11.003.

Van Rensburg, W.J., Vorster, I.H., Van Zij, J.J.B. and Venter, S.L. (2005) *Conservation of African Leafy Vegetables in South Africa*. Project: indigenous and traditional vegetables genebank, ARC – VOPI, Pretoria, South Africa, p. 92.

van Rensburg, W.S.J, Van Averbeke, W., Slabbert, R., Faber, M., Van Jaarsveld, P. *et al.* (2007) African leafy vegetables in South Africa. *Water SA* 33(3), 317. DOI: 10.4314/wsa.v33i3.180589.

Vorster, H.J., Jansen van Rensburg, W.S., Van Zijl, J.J.B. and Van den Heever, E. (2002) Germplasm management of African leafy vegetables for the nutritional and food security needs of vulnerable groups in South Africa. Progress Report. ARC – VOPI, Pretoria, South Africa, p. 130.

Vorster, I.H.J., Stevens, J.B. and Steyn, G.J. (2008) Production systems of traditional leafy vegetables, challenges for research and extension. *South African Journal of Agriculture and Extension* 37, 85–96.

Vorster, I.H.J., Jansen Van Rensburg, W.J., Vanzyl, J.J.B. and Venter, S. (2005) Recreating awareness of traditional leafy vegetables in communities. In: *Proceedings of the Regional Workshop on African Leafy Vegetables for Improved Nutrition*. IPGRI, Nairobi, Kenya, p. 6.

Wachtel-Galor, S., Wong, K.W. and Benzie, I.F.F. (2008) The effect of cooking on Brassica vegetables. *Food Chemistry* 110(3), 706–710. DOI: 10.1016/j.foodchem.2008.02.056.

Wong, P.Y.Y. and Kitts, D.D. (2006) Studies on the dual antioxidant and antibacterial properties of parsley (*Petroselinum crispum*) and cilantro (*Coriandrum sativum*) extracts. *Food Chemistry* 97(3), 505–515. DOI: 10.1016/j.foodchem.2005.05.031.

Yamaguchi, T., Mizobuchi, T., Kajikawa, R., Kawashima, H., Miyabe, F. *et al.* (2001) Radical-scavenging activity of vegetables and the effect of cooking on their activity. *Food Science and Technology Research* 7(3), 250–257. DOI: 10.3136/fstr.7.250.

Yamaguchi, T., Katsuda, M., Oda, Y., Terao, J., Kanazawa, K. *et al.* (2003) Influence of polyphenol and ascorbate oxidases during cooking process on the radical scavenging activity of vegetables. *Food Science and Technology Research* 9(1), 79–83. DOI: 10.3136/fstr.9.79.

Yuan, G., Sun, B., Yuan, J. and Wang, Q. (2009) Effects of different cooking methods on health-promoting compounds of broccoli. *Journal of Zhejiang University. Science. B* 10(8), 580–588. DOI: 10.1631/jzus.B0920051.

Zanish, A., Workneth, T.S. and Woldetsadk, K. (2013) Effects of accessions on the chemical quality of fresh pumpkin. *African Journal of Biotechnology* 12, 7092–7098.

7 African Nightshade (*Solanum nigrum* complex species)

Siphosanele Mafa Moyo* and Eugénie Kayitesi
Department of Consumer and Food Sciences, University of Pretoria, Hatfield, Pretoria, South Africa

Abstract

African nightshades belong to the *Solanum nigrum* complex species which is characterized by taxonomical complexity. Species such as *Solanum scabrum, Solanum villosum, Solanum americanum, Solanum eldorettii, Solanum chenopodioides, Solanum retroflexum, Solanum florulentum, Solanum grossidentatum, Solanum tarderemotum, Solanum physalifolium* and *Solanum nigrum* have been regarded as African nightshades due to their origin or naturalization in Africa. Some species grow as weeds and some are cultivated. They are sensitive to low soil moisture levels and are not considered to be drought tolerant. African nightshades can be used to contribute to daily nutrient requirements as they are good sources of proteins (2.5% to 39.7%), fibre (1.3% to 19%), minerals and vitamins. Furthermore, they can contribute up to about 1485.3 kJ of energy per 100 g of vegetable portion. These species are also beneficial beyond their nutritional value as they possess phytochemicals with antioxidant and anti-inflammatory bioactivities. Some species possess anti-cancer, anti-diabetic and hepatoprotective activities. The consumption of these species is hindered by the presence of antinutritional compounds such as phytates, oxalates, steroidal glycoalkaloids and saponins. However, processing methods such as boiling and blanching have been found to reduce the concentration of antinutrients to non-toxic levels.

Introduction

Solanum nigrum complex species, also known as *Solanum* section *Solanum* (black nightshade) is composed of about 30 morphologically diverse taxa (Ojiewo *et al.*, 2007; Mohy-Ud-din *et al.*, 2009). The taxa that have been naturalized or that originate in Africa are referred to as African nightshade (Mwai *et al.*, 2009). These include *Solanum scabrum, Solanum villosum, Solanum americanum, Solanum eldorettii, Solanum chenopodioides, Solanum retroflexum, Solanum florulentum, Solanum grossidentatum, Solanum tarderemotum, Solanum physalifolium* and *Solanum nigrum* (Mwai *et al.*, 2007, 2009; Mohy-Ud-din *et al.*, 2009; Viljoen, 2011; Ayua *et al.*, 2016; Ronoh *et al.*, 2018). Species such as *S. physalifolium, S. scabrum, S. americanum, S. nigrum, S retroflexum* and *S. villosum* are among the most consumed in sub-Saharan Africa (Van Wijk *et al.*, 2017; Yuan *et al.*, 2018). There are some morphological resemblances between some *S. nigrum* complex species, and species such as *S. villosum, S. americanum, S. chenopodioides* and *S. retroflexum* are possibly confused with *S. nigrum* L. and reported not to be consumed (Mohy-Ud-din *et al.*, 2009; Ojiewo *et al.*, 2013). The taxonomical complexity of *S. nigrum* complex species is due to discordant variation, genetic variation, historical reasons, phenotypic

*Corresponding author: siphosanele@gmail.com

DOI: 10.1079/9781789248067.0007

plasticity, polyploidy and natural hybridization (Edmonds and Chweya, 1997). While much of the complex taxonomy of *S. nigrum* complex has been clarified for Asia, Europe, Australia and the Americas, South-East Asia and Africa still lack comprehensive taxonomic clarification (Mwai *et al.*, 2007). By studying the flavonoid profiles of *S. nigrum* complex species, Mohy-Ud-din *et al.* (2009) suggested that *S. americanum*, *S. chenopodioides*, *S. nigrum* and *S. villosum* are distinct species as a significant distance was found between them, however, *S. retroflexum* did not show a significant distance from *S. nigrum* and hence can be considered as a subspecies or variety of *S. nigrum*.

African nightshades grow as weeds close to trees, on cultivated land or in the shade near buildings (Keller *et al.*, 2004). Some species such as *S. scabrum*, *S. villosum* and *S. retroflexum* are collected in cultivated form in West Africa, East Africa and South Africa, respectively (Muchiri *et al.*, 2005; Maseko *et al.*, 2017). The stems and leaves are used for various medicinal purposes, or as vegetables and herbs while the berries are utilized in the preparation of preserves or consumed as fruits (Jacoby *et al.*, 2003; Jagatheeswari *et al.*, 2013). These species grow well in soils rich in organic matter and require more than 500–1200 mm of rain/year (Keller *et al.*, 2004; Ojiewo *et al.*, 2013). They are sensitive to low soil moisture levels and therefore are not drought tolerant (Keller *et al.*, 2004). Hence, the species grow well during the rainfall season (Ojiewo *et al.*, 2013). The earliest young leaves of transplanted African nightshades can be harvested at about four to five weeks by cutting the stem at 15 cm distance from the ground. If shoots develop after the first harvest, they can be harvested again at 7- to 14-day intervals and this may involve uprooting of the whole plant (Keller *et al.*, 2004). It has been suggested that harvesting of nightshades between 60–90 days after planting is beneficial due to a high concentration of most nutrients and phytochemicals at that phase of plant maturity (Kirigia *et al.*, 2019). At that phase of maturity, it can be proposed to consume African nightshades within four days when stored at 5°C or soon after harvest as room temperature storage is not recommended (Kirigia *et al.*, 2019).

For consumption, African nightshades are prepared differently according to various communities, e.g. in Kenya, they are boiled and/or fried together with the addition of onions and tomatoes as well as milk and cream. Other ingredients such as potatoes, pumpkins, tender gourds and blood are added to alleviate bitterness of some African nightshade species (Ontita *et al.*, 2016). In South Africa, African nightshade can be boiled with the addition of tomatoes and groundnuts or stored in dried-cooked form for about 6–12 months (Nesamvuni *et al.*, 2001).

Botany

African nightshade species are from the Solanaceae family under the genus *Solanum* (Mohy-Ud-din *et al.*, 2009). These species display variation in their phenotype especially the vegetative characteristics, e.g. plant habit, stem winging and leaf form and size (Jagatheeswari *et al.*, 2013). These species are either herbs or shrubs with stems that can grow to about 2 m in height (Särkinen *et al.*, 2018). Stems can have smooth/inconspicuously dentate ridges (e.g. *S. americanum*, *S. chenopodioides*, and *S. retroflexum*) or prominently and dentate ridges (e.g. *S. scabrum*) (Edmonds and Chweya, 1997). The leaf structure can be highly variable within a single plant or species (Särkinen *et al.*, 2018). Leaves are usually ovate-lanceolate (e.g. *S. americanum*), elliptic-lanceolate (e.g. *S. chenopodioides*), rhomboidal to ovate-lanceolate (e.g. *S. retroflexum Solanum villosum*) or ovate (e.g. *S scabrum*) (Edmonds and Chweya, 1997). Leaf margins vary from entire (e.g. *S. americanum*, and *S. chenopodioides*), entire sinuate or sinuate-dentate finely lobed (e.g. *S. nigrum* L.) (Fig. 7.1), lobed dentate (e.g. *S. retroflexum*), entire to sinuate (e.g. *S. scabrum*) to entire to sinuate-dentate (e.g. *S. villosum*) (Mwai *et al.*, 2007). *S. scabrum* has distinctively large leaves of about 10.0–12.0 cm long and 6.0–8.0 cm broad while *S. americanum*, *S. chenopodioides*, *S. retroflexum* and *S. villosum* have smaller leaf surfaces of about 3.0–6.2 cm long × 1.1–4.0 cm broad, 2.7–5.4 cm long × 2.2–5.4 cm broad, 4.0–4.8 cm long × 3.0–4.2 cm broad and 2.0–7.0 cm long × 1.5–4.0 cm broad, respectively (Edmonds and Chweya, 1997).

Generally, for most species, the inflorescence consists of simple umbellate cymes (e.g. *S.*

Fig. 7.1. African nightshade (*Solanum nigrum* complex species).

americanum, *S. chenopodioides* and *S. retroflexum*) or forked lax often extended cymes (*S. scabrum*). The fruiting peduncle can be facing upward (e.g. *S. americanum*), deflexed, densely pubescent (e.g. *S. chenopodioides*), facing upward or horizontal, hairy surface could be sparse to dense (e.g. *S. nigrum* L.), facing upward, densely pubescent (e.g. *S. retroflexum*), facing upward, sparsely to densely pubescent (e.g. *S. scabrum*) or erect (e.g. *S. villosum* and *S. scabrum*) (Mwai *et al.*, 2007). The fruiting pedicels can be erect (e.g. *S. americanum*), reflexed (e.g. *S. scabrum*, *S. chenopodioides*, *S. retroflexum* and *S. nigrum* L.) or deflexed in fruit (e.g. *S. villosum*) (Mwai *et al.*, 2007). Species can be identified by the position of the calyx lobes in fruit, as some have sepals flexed away from the mature berries (e.g. *S. americanum*, *S. scabrum*, and *S. retroflexum*) or adherent to the base of the mature berry (e.g. *S. villosum* and *S. chenopodioides*) (Edmonds and Chweya, 1997). The corollas of *S. nigrum* complex species are variously stellate with the colour being generally white or pale violet tinged (Särkinen *et al.*, 2018). *S. retroflexum* usually consists of a distinct purple central streak on the exterior the individual petals (Edmonds and Chweya, 1997). The anthers are commonly yellow but otherwise brown or purplish/brown for *S. scabrum* (Edmonds and Chweya, 1997).

The mature berries can be globose, dark purple/black, extremely shiny, dropping from calyces after ripening (e.g. *S. americanum*), slightly flattened, black, dull and remaining on calyces when fully ripe (e.g. *S. chenopodioides*), slightly flattened to rounded, dark purple to black, dull, remaining on calyces when fully ripe (e.g. *S. nigrum* L.), slightly flattened, black to dark purple with dull exterior, staying on calyces when completely ripe (e.g. *S retroflexum*), slightly flattened, deep purple to black, shiny and staying on plant when entirely ripe (e.g. *S. scabrum*) or globose, red, orange or yellow and falling from calyces when ripe (e.g. *S. villosum*) (Mwai *et al.*, 2007). Seed size varies for *S. nigrum* complex species from 1 to 3 mm long (Särkinen *et al.*, 2018). Diploid types typically have smaller seeds as compared to polyploid species e.g. *S. americanum* (diploid) 1–1.5 mm, *S. chenopodioides* (diploid) 1.0–1.8 mm, *S retroflexum* (tetraploid) 1.5–2.3 mm, *S. villosum* (tetraploid) 1.6–2.2 mm and *S. scabrum* (hexaploid) is 2–2.8 mm (Edmonds and Chweya, 1997; Särkinen *et al.*, 2018). The seeds require an optimum germination temperature between 15–30°C and temperature fluctuation to break dormancy (Edmonds and Chweya, 1997).

Traditional Use

Much interest in African nightshades emanates from their long use as ingredients in traditional food preparation and in folk medicines. Species such as *S. scabrum* and *S. americanum* have been utilized in the treatment of gastric ulcers (Silva *et al.*, 2017; Mureithi *et al.*, 2020). Furthermore, several other uses have been reported for *S. americanum* in the treatment of skin diseases, gynaecological disorders, inflammation of bladder, spasm dysentery, fever, casing gas formation in the stomach, analgesic, sedative, cough, joint pains and protozoal infections (Silva *et al.*, 2017; Usman *et al.*, 2018; Vagula *et al.*, 2018). For *S. villosum*, its use in the treatment of sore eyes, liver and spleen enlargement in addition to inflammatory swelling has been reported (Chowdhury *et al.*, 2008; Abdel-Hamid *et al.*, 2017). *S. nigrum* has mainly been utilized to treat liver disorders, eye diseases, inflammation conditions, oedema, chronic skin conditions such as psoriasis and ringworm, painful periods, fevers, diarrhoea and peptic ulcers (Albouchi *et al.*, 2018; Krithika and Verma, 2019). The ash from the burnt leaves of *S. nigrum* is reported to be used to treat fungal skin infections (Ontita *et al.*, 2016). *S. retroflexum* has been reported to be used in the treatment of tuberculosis, haemorrhoids, bleeding haemorrhoids and headache (Hulley and Van Wyk, 2019).

Nutritional Composition

Nutrients such as carbohydrates, fats, proteins, vitamins and minerals are essential to maintain good health (Singh *et al.*, 2019). Their presence in plant foods varies according to genetic and environmental factors (e.g. amount of sunlight, drought stress and temperature), plant cultivar/variety, harvest, storage, or transportation conditions to market as well as processing conditions (Fahey, 2016). Nutritional information on some of the African nightshade species such as *S. villosum* and *S. chenopodioides* is lacking. Table 7.1 summarizes the proximate composition of African nightshades which displayed pronounced variability in their macronutrient content. All of these species have a high moisture content of about 85–92% water. The protein content of nightshade vegetables ranges from 2.5 to 39.7% dry matter (DM) content and is mainly related to the amino acid composition (Wiesler, 2011). Adults are generally required to consume about 0.8–1.0 g of proteins/body weight (kg)/day, with a limit of 0.66 g/kg/d to prevent protein deficiency (Sá *et al.*, 2020). A 70 g serving of *S. americanum* leaves can provide about 11% of the 50 g recommended daily intake of proteins in a 2000 caloric diet intake a day (Xu *et al.*, 2020).

The amino acid profiles for some African nightshades are shown in Table 7.2 and include both essential and non-essential amino acids. However, proteins from plants are usually deficient in at least one of the essential amino acids (Wiesler, 2011). Hassan and Umar (2008) found that leucine (8.25 mg/100 g protein) and glutamic acid (12.23 mg/100 g protein) were the most predominant amino acids in *S. americanum*.

In most instances, leafy vegetables possess a low lipid content of about 1.3–5.6% DM and these lipids are in the form of monounsaturated fatty acids, polyunsaturated fatty acids, saturated fatty acids and phytosterols (Aramrueang *et al.*, 2019). But high lipid values have been reported for *S. americium*, *S. scabrum* and *S. nigrum l. Var virginicum* at 20.1% DM, 7.1% DM and 4.6 % DM, respectively, as shown in Table 7.1. Furthermore, this high lipid content could be related to the high energy value provided by the vegetables (Table 7.1). Fatty acids such as palmitic acid (26.81%), linoleic acid (34.82%) and linolenic acid (25.07%) as well as cholesterol, stigmasterol and β-sitosterol were found in *S. villosum* (Abdel-Hamid, 2018). African nightshades contain fibre in the range of 1 to 20% (Table 7.1). The method of dietary fibre determination may cause large variations in values reported from different studies (Pinela *et al.*, 2020). About 30–38 g and 21–25 g of dietary fibre intake is recommended a day for men and women, respectively (Suitor and Murphy, 2013). Therefore, the leaves of African nightshades, especially *S. americanum* and *S. scabrum* could be beneficial sources of dietary fibre in the human diet.

Leafy green vegetables are generally a valuable sources of minerals (Aramrueang *et al.*, 2019). African nightshade mineral content is summarized in Table 7.1 as ash content in the range 1.0 to 17.4%. *S. americanum* shows the highest ash value at 17.4%. It contains a high concentration of minerals such as iron, potassium and calcium (Table 7.3). Potassium predominantly occurred in all the African nightshades shown in Table 7.3 except for *S. nigrum l. Var virginicum* and *S. nodiflorum*. The mineral content of vegetables is highly influenced by the prevailing environmental conditions and maturity at harvest (Odhav *et al.*, 2007). According to van Jaarsveld *et al.* (2014), *S. retroflexum* contributes about 97 and 115% of adequate intake values of magnesium for women (19–30 years) and children (4–8 years), respectively. Although these African nightshades are good sources of minerals, antinutrients such

Table 7.1. Proximate composition of African nightshades.

Species		Protein (%)	Fibre (%)	Carbohydrates (%)	Fat (%)	Ash (%)	Moisture (%)	Energy (kJ/100 g)	Reference
S. americanum	DM	11.3	19.9	31.8	20.1	17.4	85.3	1477.6	Hassan and Umar, 2008
S. nigrum	FW	3.3	1.4	2.3	0.6	1.23	91.6	89.5	Adeyeye *et al*., 2018
S. nigrum	FW	2.3	2.2	1.7	0.23	2.2	91.5	75.1	Kouadio *et al*., 2019
S. nigrum l. Var *virginicum*[a]	DM	24.9	6.8	53.5	4.6	10.8	84.7	1485.3	Akubugwo *et al*., 2007
S. nodiflorum[a]	FW	3.0	2.4	9.0	0.6	2.24	85.0	230.1	Odhav *et al*., 2007
S. nodiflorum[a]	FW	3.8	1.5	2.7	0.8	1.3	90.9	115.9	Adeyeye *et al*., 2018
S. retroflexum	DM	4.2	1.3	5.7	0.8	2.0	85.5	159.0	Maziya-Dixon *et al*., 2019
S. retroflexum	DM	2.5	2.2	4.1	0.2	1.0	92.2	120.0	van Jaarsveld *et al*., 2014
S. scabrum	DM	39.7	14.07	25.94	7.11	10.57	87.71	-	Traoré *et al*., 2017

[a] Synonym of *S. americanum*, DM, dry matter content; FW, fresh weight basis.

Table 7.2. Amino acid content of African nightshades (adapted from Chowdhury *et al.*, 2008; Hassan and Umar, 2008).

Amino acid	*S. villosum* (g/16 g nitrogen)	*S. americanum* (mg/100 g protein)
Essential		
Histidine	-	2.66
Isoleucine	4.78	3.46
Leucine	4.00	8.25
Lysine	1.62	4.22
Methionine	1.90	0.99
Phenylalanine	3.43	4.58
Threonine	9.14	3.02
Tryptophan	-	-
Valine	5.64	4.36
Non-essential		
Alanine	5.66	4.09
Arginine	5.74	5.19
Aspartic acid	32.99[a]	8.55
Cysteine	1.44	0.64
Glutamic acid	3.46[b]	12.23
Glycine	8.89	4.98
Proline	-	3.34
Serine	6.29	4.17
Tyrosine	5.02	2.77

[a] Includes asparagine; [b] Includes glutamine.

as phytates may interfere with their bioavailability (Odhav *et al.*, 2007).

The calcium content of nightshades is higher than that found in spinach, lettuce and swiss chard and lower than in kale. All the leafy vegetables contain high amounts of potassium in the range of 168 to 558 mg/100 g. Hence, the consumption of African nightshades should also be encouraged as they contain similar amounts of minerals as the other commonly consumed vegetables.

Fat-soluble vitamins such as D, E retinoids and K, as well as water-soluble vitamins such as pantothenic acid, folic acid, ascorbic acid, thiamine, riboflavin, cobalamin and biotin are crucial in the maintenance of physiological properties of the body but cannot be synthesized in sufficient amounts by humans, and hence should be provided from the diet (Wiesler, 2011). African nightshades are a good source of vitamin beta carotene at 7.56 mg/100 g FW (*S. retroflexum*), 7.84 mg/100 g DM (*S. nigrum*), 8.74 mg/100 g DM (*S. scabrum*), 9.58 mg/100 g DM (*S. americanum*) and 13.81 mg/100 g DM (*S. villosum*) (Nesamvuni *et al.*, 2001; Yuan *et al.*, 2018). Studies have also shown appreciable amounts of vitamin C and folate in *S. retroflexum* (Nesamvuni *et al.*, 2001; van Jaarsveld *et al.*, 2014), while *S. nigrum*, *S. scabrum*, *S. americanum* and *S. villosum* also show appreciable amounts of vitamin E (Yuan *et al.*, 2018). Xanthophylls such as zeaxanthin, lutein have also been detected in *S. scabrum* (Odongo et *al.*, 2018). The consumption and utilization of xanthophylls in the body is associated with lower risk of eye disease (Ma and Lin, 2010). About 57 and 70% of vitamin A recommended dietary allowance (RDA) is provided for by *S. retroflexum* for women aged 19–30 years and children aged 4–8 years, respectively (van Jaarsveld *et al.*, 2014). As is the case with other nutrients, the concentration of vitamins may be affected by plant cultivar, climatic conditions, harvest period, nutrition, processing, food handling, or preparation and storage (Wiesler, 2011). With daily servings of about one to two cups (180–270 g), African nightshades could significantly contribute to the daily requirements of nutrients to prevent deficiency in the body (Nesamvuni *et al.*, 2001).

Phytochemical Quality

Phytochemicals are non-nutritive compounds of plant origin that are reported to possess bioactive properties linked to the provision of health benefits and protection against diseases in the body (Thakur *et al.*, 2020). These include nitrogen-containing compounds, polyphenols, carotenoids, organosulfur compounds and alkaloids (Tiwari and Cummins, 2013). Table 7.4 summarizes the phytochemical compounds found in different African nightshade species. Generally, the most predominant phenolic compounds found in African nightshades are phenolic acids and derivatives of quercetin and kaempferol. The type and concentration of phytochemicals detected from plant extracts is highly influenced by extraction method and

Table 7.3. Mineral content of African nightshades (mg/100 g).

	S. americanum	*S. retroflexum*	*S. nigrum*	*S. nigrum l. Var virginicum*	*S. nodiflorum*	*S. nodiflorum*	*S. scabrum*
	DM	DM	FW	DM	DM	FW	FW
Calcium	4930.0	199	214.5	17.33	2067	201.5	136
Copper	19.9	0.16	0.15	-	6	0.145	0.132
Iron	688.0	7.2	3.82	13.01	85	3.69	1.24
Magnesium	210.6	92	25.0	247.59	277	36.5	107
Manganese	72.7	2.08	0.23	1.52	3	0.24	0.855
Potassium	11,600.0	257	414.0	42.89	nd	540.0	430
Phosphorus	110.8	36	nd	75.22	478	nd	58
Sodium	106.6	8	180	2.71	431	252.0	0.357
Zinc	10.3	0.56	0.147	0.07	23	0.128	0.488
Reference	Hassan and Umar, 2008	van Jaarsveld *et al.*, 2014	Adeyeye *et al.*, 2018	Akubugwo *et al.*, 2007	Odhav *et al.*, 2007	Adeyeye *et al.*, 2018	Jiménez-Aguilar and Grusak, 2015

Nd, not determined; DM, dry matter; FW, fresh weight basis.

Table 7.4. Phytochemical compounds detected in African nightshade.

Species	Phytochemical compounds	References
S. americum	Quercetin-G, quercetin-G-Rha-G, chlorogenic acid, kaempferol-G-G, quercetin-G-G, kaempferol-G-Xyl	Yuan *et al.*, 2018
S. scabrum	3-CQA, 5-CQA, 4-CQA (dimer), kaempferol-3-rhamnosylrhamnogalactoside isomer 2, caffeoylmalate isomer 1, caffeoylmalate isomer 2 (dimer), quercetin glycosides, coumaric acid, quercetin-3-glucosylrhamnogalcatoside, quercetin-3-rhamnogalactoside, quercetin-3-rhamnosylrhamnogalactoside isomer 1, quercetin-3-rhamnosylrhamnogalactoside isomer 2, kaempferol-3-rhamnosylrhamnogalactoside isomer 1, sinapoylmalate, kaempferol-3-diglucoside, quercetin-3-pentosylrutinoside, sinapic acid, quercetin-3-rutinoside, caffeic acid derivatives	Neugart *et al.*, 2017; Odongo *et al.*, 2018
S. nigrum	Quercetin-G-G, kaempferol-G-G, quercetin-G-rha, quercetin-G	Yuan *et al.*, 2018
S. retroflexum	Feruroylglycoside, ellagic acid, trans-4-CQA, cis-4-CQA, cis-5-CQA, trans-5-CQA glycoside, 3, 4-di CQA, cis-3, 5-di-CQA, trans-3, 5-di-CQA, trans-4, 5-di-CQA, cis-4, 5-di-CQA, 3-p-CoQA, trans-5-p-CoQA, cis-5-p-CoQA, kaempferol-3-glucoside, quercetin-7-glucoside,coumaroyl-hexose, trans-5-FQA, cis-5-FQA, 4-FQA, 3-FQA, quercetin-3-rutinoside, vanillic acid, kaempferol-3-0-rutinoside, 3-CQA, isorhamnetin-3-0-rutinoside, trans-5-CQA,caffeic acid, gallic acid, 4-p-CoQA, 2,5 dihydrobenzoic acid, p-coumaric acid, ferulic acid, cis-5-CQA glycoside, quercetin, quercetin-3-rhamnoside, catechin	Daji *et al.*, 2018; Degrain *et al.*, 2020
S. villosum	Phaseolic acid, quercetindihexoside, scopoletin, kaempferol dihexoside, isoquercetin	Abdel-Hamid *et al.*, 2017

G, mannosyl, glucosyl, galactosyl or other hexosyl, Xyl -Xylosyl; rha, rhamnosyl; FQA, feruloyquinic acids, di-CQA -di-caffeoylquinic acid, p-CoQA -p-coumaroylquinic acid, CQA -caffeoylquinic acid.

conditions such as temperature and time, the nature of solvent as well as the concentration and polarity of the solvent (Thakur *et al.*, 2020).

For *S. scabrum*, compounds such as quercetin-3-glucosylrhamnogalcatoside (185237.00 µg/g DM), caffeoylmalate (52628 µg/g DM) and 4-caffeoylquinic acid (30839 µg/g DM) were predominant (Odongo *et al.*, 2018). An accession of *S. nigrum* was found to contain chlorogenic acid (282 mg/100 g DM), quercetin (131 mg/100 g DM), kaempferol (17.1 mg/100 g DM), isorhamnetin (1.90 mg/100 g DM) and rhamnetin (351 mg/100 g DM). The concentrations of these compounds in *S. nigrum* are lower than those reported for *S. scabrum* at 287 mg/100 g DM, 511 mg/100 g DM, 66.6 mg/100 g DM and 12.1 mg/100 g DM for chlorogenic acid, quercetin, kaempferol and isorhamnetin, respectively (Yuan *et al.*, 2020). For *S. retroflexum*, phenolic acids were quantified at 4955.3 mg/kg, with 2,5 dihydroxybenzoic acid being the most predominant polyphenol at 1593 mg/kg. Furthermore, flavonoids such as catechin and quercetin were also quantified at 726.5 mg/kg and 326.0 mg/kg, respectively (Degrain *et al.*, 2020). Mohy-Ud-din *et al.* (2009) compared the flavonoid contents of five *S. nigrum* complex species. *S. americanum* exhibited the highest quercetin content at 6.46 mg/100 g, while *S. villosum*, *S. chenopodioides*, *S. retroflexum* and *S. nigrum* contained 4.91 mg/100 g, 3.18 mg/100 g, 3.18 mg/100 g and 3.06 mg/100 g of quercetin, respectively.

Antinutritive Compounds

Some phytochemicals produced by plants are termed antinutrients, as they may act as toxins or interfere with the digestibility of essential nutrients. These include alkaloids, saponins and several glycosides, lectins, phytates, oxalates, phyto-oestrogens and tannins (Essack *et al.*, 2017; Petroski and Minich, 2020). Oxalates may hinder calcium absorption and increase the risk of calcium kidney stone formation (Petroski and Minich, 2020). Tannins may form complexes with proteins, polysaccharides and metal ions (Essack *et al.*, 2017) while phytates may hinder the absorption of calcium, iron and zinc (Petroski and Minich, 2020).

Raw leaves of *S. retroflexum* were found to contain about 55.4 mg/100 g FW of tannins, 88 mg/100 g FW of oxalates and 88% FW of phytates (Managa *et al.*, 2020a). Alkaloids as high as 463.8% were quantified in *S. nigrum* together with 0.14 mg/l phytates, 6.8 mg/l oxalates, 0.006 mg/l tannins and 32.2 mg/g cyanogenic glycosides (Essack *et al.*, 2017). Furthermore, Ejoh *et al.* (2017) also reported the presence of phytates (0.03 mg/100g DM), oxalates (0.43 mg/100 g DM) and tannins (0.009 mg/100 g DM) in *S. nigrum*. Nightshade *S. villosum* was found to have oxalate, phytate and nitrates in the concentration of 28.7 mg/100 g DM, 0.04 mg/100 g DM and 66.8 mg/100 g DM (Mwanri *et al.*, 2018). Factors such as stages of maturity at the time of harvest may determine concentrations found in leafy vegetables. For instance, *S. villosum* showed increased oxalate, phytate and nitrate levels between 21–35 days of maturity (Mwanri *et al.*, 2018). Furthermore, the geographical area, cultivar, season and plant growing conditions may determine levels of antinutrients in a plant (Sinha and Khare, 2017; Petroski and Minich, 2020). To prevent toxicity, consumption of less than 2–5 g of oxalates a day is recommended (Ejoh *et al.*, 2017). In addition, the oxalic content ratio to calcium should be lower than 2.5 for human consumption (Pinela *et al.*, 2020). The negative impact of phytates is dependent on pH and the proportion of phytates to metal ions, e.g., the critical level in which phytates can reduce iron absorption is 1, whereas for zinc it is 15 and for calcium 0.17 (Petroski and Minich, 2020).

Other compounds of concern in imparting toxicity after consumption of vegetables are steroidal glycoalkaloids and saponins. These compounds are a common occurrence in plants of the nightshade family and were detected in *S. retroflexum* and *S. villosum* (Abdel-Hamid *et al.*, 2017; Daji *et al.*, 2018). Steroidal alkaloids such as dihydroxy solasodine solatrioside, dihydroxy solasodine chacotrioside, dihydroxy solasodine diglycoside, α-solasonine, α-solamargine, solasonine diglycoside, solasodine, 12-hydroxysolasonine, 3-O-chacotriosyl-21-hydroxysolasodine, hydroxy solasodine diglycoside and hydroxy solasodine were identified in *S. villosum*. In addition, several steroidal saponins such as spirostenolpentaglycoside, Δ5,6-dehydrotigonin, tigonin, diosgenintetraglycoside, tigogeninhexaglycoside, tigogenintetraglycoside, degalactotigonin, tigogenin pentaglycoside and tigogenintriglycosid were also identified in *S. villosum* (Abdel-Hamid *et al.*, 2017).

Mohy-Ud-Din *et al.* (2010) compared the concentration of steroidal glycoalkaloids in five species from the *S. nigrum* complex. *S. nigrum* exhibited the highest concentration of α-solamargine at 5.03 mg/g followed by *S. americanum* at 1.96 mg/g, *S. retroflexum* at 1.45 mg/g and *S. villosum* at 1.17 mg/g. Highest concentrations of α-solasonine were found in *S. retroflexum* (5.7 mg/g) followed by *S. nigrum* (4.7 mg/g), *S. americanum* (3.29 mg/g), *S. chenopodioides* (3.08 mg/g) and *S. villosum* (1.5 mg/g). Toxicity of glycoalkaloids is realized at concentrations higher than 5 mg/kg (Ronoh *et al.*, 2018). According to Yuan *et al.* (2018), the glycoalkaloid content in the four species (*S. americanum*, *S. nigrum*, *S. scabrum* and *S. villosum*) studied was found to be safe for consumption.

Implications of Food Processing on Nutrients and Phytochemical Quality

Thermal treatments such as steaming, boiling, microwaving and frying are commonly used in the domestic cooking of leafy vegetables (Mehmood and Zeb, 2020). For preservation purposes, leafy vegetables are usually boiled or blanched and dried by various methods. Although processing is necessary to enhance the palatability of food and render it safe for consumption, it may alter the concentration of nutrients and phytochemicals in food.

Table 7.5 summarizes the influence of boiling and blanching on some African

Table 7.5. Implications of boiling and blanching for the proximate composition of African nightshades.

Processing method	Species	Processing conditions	Nutrient	% Change	Effects of processing
Boiling	*S. nigrum*[a]	15 min, 500 g vegetable to 1 l of water	Protein	3.36	↓
			Carbohydrates	7.63	↓
			Ash	7.28	↓
			Fibre	13.77	Increase
			Lipids	2.49	Increase
		20 min, 500 g vegetable to 1 l of water	Protein	4.09	↓
			Carbohydrates	12.76	↓
			Ash	8.66	↓
			Fibre	22.37	Increase
			Lipids	3.14	Increase
	S. nigrum[b]	15 min, 500 g vegetable to 4 l of water	Protein	0.95	↓
			Ash	1.75	↓
			Fibre	0.2	↓
			Lipids	0.16	↓
	S. scabrum[c]	30 min, 30 g vegetable to 2 l of water	Protein	0.44	↓
			Carbohydrates	30.53	↓
			Ash	3.25	↓*
		60 min, 30 g vegetable to 2 l of water	Protein	7.55	↓
			Carbohydrates	23.6	↓
			Ash	1.77	↓ *
Blanching	*S. scabrum*[c]	90–92°C for 2 min, 30 g vegetable	Protein	0.77	Increase *
			Carbohydrates	33.75	Increase
			Ash	1.73	↓ *
			lipids	0.32	↓ *

[a] Kouassi Dappah *et al*., 2020; [b] Lola, 2009; [c] Traoré *et al*., 2017; * not significant.

nightshades. The process of boiling and blanching lowered the protein concentration of *S. scabrum* (7.55% after 60 minutes) and *S. nigrum* (4.09% after 20 minutes). Protein denaturation and solubilization into cooking water could have occurred during the cooking process to induce a reduction in protein measured. Furthermore, decrease in protein could have been influenced by chemical reactions leading to the formation of protein-fibre complexes (Arias-Rico *et al*., 2020). Nightshades *scabrum* and *nigrum* also experienced losses in carbohydrate concentration of 30.53% after 30 minutes and 12.76 after 20 minutes, respectively. Maillard reactions and solubilization of carbohydrates into cooking water could explain such losses. Non-significant losses in ash content were recorded for *S. scabrum*, however, significant losses (8.66%) were reported for *S. nigrum* after boiling for 20 minutes, probably due to leaching of minerals into the cooking water. The breakdown of cell walls to liberate cellulose could explain the increase in fibre content after cooking in *S. nigrum* as reported by Kouassi Dappah *et al.* (2020). However, a decrease in fibre content reported by Lola (2009) could be influenced by the conversion of insoluble dietary fibre into digestible carbohydrates (Farhath Khanum *et al.*, 2000).

Most African nightshades suffered losses in mineral concentrations after boiling, as shown in Table 7.6. Mineral elements may leach into the cooking water or form complexes with other constituents of the vegetable material. However, the addition of alkaline salt during the process of boiling appeared to have a positive effect on the concentration of sodium, zinc, iron, potassium, copper and magnesium in *S. nigrum*. This was attributed to the addition of alkaline salt which contains the forementioned minerals (Ejoh *et al.*, 2017). *S. nigrum* and *S. scabrum* suffered losses in vitamin content studied, i.e. vitamin C, β-carotene, zeaxanthin and lutein. Although β-carotene is a heat-stable nutrient, lengthy cooking methods could cause degradation and polymerization of its structure (Musotsi *et al.*, 2019). To minimize losses of nutrients during boiling it is advised that a minimal amount of water should be used to reduce the amount left over after cooking as most soluble nutrients remain in the cooking water that is normally discarded (Musotsi *et al.*, 2019).

Different cooking methods may cause changes in the phytochemical quality of leafy vegetables and therefore affect the bioactive and health properties of the vegetables (Essack *et al.*, 2017). The process of boiling enhanced most of the phytochemical content of *S. nigrum* as shown in Table 7.7. In addition, water blanching also improved the phytochemical contents of *S. retroflexum*. Increases in phytochemical content after boiling and water blanching could be related to the release of bound phenolics from cellular structures due to the degradation of cell walls and subcellular structures. However, due to different boiling conditions as well as different vegetable to water ratios, decreased amounts of phenolics after boiling have been reported. For *S. scabrum*, the concentration of most compounds shown in Table 7.7 was reduced after boiling. This could be attributed to leaching of phenolics into cooking water as they exhibit high water solubility (Murador *et al.*, 2018). The traditional blanching process of African nightshades often involves the addition of lime or lemon juice into the blanching water (Managa *et al.*, 2020a). Water and steam blanching with 5% lemon juice improved the phenolic compound content (neochlorogenic acid, chlorogenic acid, caffeoylmalic acid and rutin) of *S. retroflexum* (Managa *et al.*, 2020a). As a method used for preserving African nightshade, fermenting using laboratory strain 75 had a positive impact on the concentration of phenolics such as coumaric acid, gallic acid, vanillic acid, ellagic acids, catechin and quercetin (Degrain *et al.*, 2020).

As a constituent of plant phytochemicals, antinutrients that pose a threat to human health when consumed in large quantities can be reduced in food by different processing methods. For African nightshades *S. nigrum* and *S. retroflexum*, boiling and blanching reduced the concentration of cyanogenic glycosides, alkaloids, tannins, saponins, oxalates and phytates (Ejoh *et al.*, 2017; Essack *et al.*, 2017; Managa *et al.*, 2020a). Antinutrients such as tannins, phytates and oxalates are water soluble – therefore they can be lost in the cooking water through leaching (Ejoh *et al.*, 2017; Petroski and Minich, 2020).

Bioactive Properties

A wide range of bioactive compounds are present in foods and these offer multiple physiological properties (Hernández-Ledesma *et al.*, 2015). Such compounds include polyphenols and micronutrients and are recognized for their anti-inflammatory, antioxidant properties (Mattioli *et al.*, 2018). Inflammation and oxidative stress are major factors in the progression of ageing as well as age-related ailments, i.e., cancer, cardiovascular diseases (CVDs) and neurodegenerative diseases (Aleksandrova *et al.*, 2021). Oxidative stress occurs in cells and tissues when the endogenous antioxidant system is overwhelmed by the accumulated presence of reactive oxygen species (ROS), leading to oxidative damage and cellular ageing (Aleksandrova *et al.*, 2021). Furthermore, ROS may induce a chain of reactions that enhances proinflammatory gene expression in which this state of inflammation may further aggravate the production of ROS (Aleksandrova *et al.*, 2021). Therefore, to reduce the occurrence of these processes in the body, exogenous sources of food with antioxidant and anti-inflammatory activities are recommended.

The ability of African nightshades to act as antioxidants has been investigated through a variety of assays such as ferric reducing antioxidant power assay (FRAP), trolox

Table 7.6. Implications of boiling on mineral and vitamin composition of African nightshades.

Species	Processing conditions	Nutrient	% Change	Effects of processing	References
S. nigrum	15 min, 500 g vegetable to 1 l of water	Calcium	28.83	↓	Kouassi Dappah *et al.*, 2020
		Iron	26.29	↓	
		Magnesium	37.77	↓	
		Phosphorus	22.88	↓	
	20 min, 500 g vegetable to 1 l of water	Calcium	36.03	↓	
		Iron	39.20	↓	
		Magnesium	43.19	↓	
		Phosphorus	26.24	↓	
S. nigrum	10 min, 450 g vegetable to 1 l of water	Calcium	35.0	↓	Ejoh *et al.*, 2017
		Copper	0.48	↓	
		Iron	1.29	Increase	
		Magnesium	50.0	↓	
		Manganese	24.79	↓	
		Potassium	27.59	Increase	
		Sodium	42.86	↓	
		Zinc	43.83	↓	
	10 min, 450 g vegetable to 1 l of water containing 10 g of alkaline salt	Calcium	30.0	↓	
		Copper	50.0	Increase	
		Iron	28.0	Increase	
		Magnesium	1.89	Increase	
		Manganese	20.23	↓	
		Potassium	-	No change	
		Sodium	94.4	Increase	
		Zinc	16.49	Increase	
S. scabrum	30 min, 30 g vegetable to 2 l of water	β-carotene	67.24	↓	Traoré *et al.*, 2017
	60 min, 30 g vegetable to 2 l of water	β-carotene	85.97	↓	
S. nigrum	15 min, 500 g vegetable to 1 l of water	Vitamin C	70.25	↓	Kouassi Dappah *et al.*, 2020
	20 min, 500 g vegetable to 1 l of water	Vitamin C	70.46	↓	
S. scabrum	20 min,[a] vegetable to 0.1 l of water	β-carotene	-	No Change	Odongo *et al.*, 2018
		Zeaxanthin	18	↓	
		Lutein	44	↓	

[a] Amount not provided; * not significant.

Table 7.7. Effect of boiling on phenolic compounds of African nightshade.

S. scabrum[a]		*S. nigrum*[b]		*S. retroflexum*[c]	
Phenolic compound	Effect (% Change)	Phenolic compound	Effect (% Change)	Phenolic compound	Effect [!]
3-caffeoylquinic acid	↓ (5.47)	3-caffeoylquinic acid	↑ (98.16)	Cryptochlorogenic acid	↑ *
5-caffeoylquinic acid	↑ (12.63)	Catechin	↑ (97.87)	Neochlorogenic acid	↑
4-caffeoylquinic acid	↓ (50.26)	4-caffeoylquinic acid	↑ (98.98)	Chlorogenic acid	↑
Caffeoylmalate	↓ (2.25)	5-caffeoylquinic acid	↑ (98.58)	Caffeoylmalic acid	↑
Caffeoylmalate	↓ (29.28)	Myricitin	↑ (38.02) *	Rutin	↑
Quercetin-3-glucosylrhamnogalcatoside	↓ (10.31)	Quercetin-3-O-robinobioside	↑ (94.53)	Kaempferol- O-rhamnosyl hexoside	↑ *
Coumaric acid	↑ (8.00)	Rutin	↑ (95.32)		
Kaempferol-3-rhamnosylrhamnogalactoside	↓ (7.44)	Quercetin-3-O-glucoside	↑ (100)		
Quercetin-3-rutinoside	↓ (14.73)	3,4-Dicaffeoylquinic acid	↑ (88.93)		
		3,5-Dicaffeoylquinic acid	↑ (100)		
		4,5-Dicaffeoylquinic acid	↑ (63.61)		
		Quercetin	↓ (21.75) *		
Processing condition: boiled for 20 min, # vegetable to 0.1 l of water		**Processing condition:** boiled for 20 min, 500 g of vegetable in 1 l of water		**Processing condition:** blanched at 95°C for 5 min	

[a] Odongo *et al.*, 2018; [b] Moyo *et al.*, 2020a; [c] Managa *et al.*, 2020a; # amount not provided; * not significant; **!** effect determined from a graph, therefore exact % change cannot be provided

Table 7.8. Radical scavenging activities of African nightshades.

Species	Antioxidant assay	Units	Measured activity	References
S. americanum	ABTS	TE mg/g DM	24.81	Yuan *et al.*, 2018
S. nigrum	TEAC	μM TE/g DM	650.95	Moyo *et al.*, 2020a
	ORAC	μM TE/g DM	299.30	
S. nigrum	ABTS	TE mg/g DM	16.80	Yuan *et al.*, 2018
S. nigrum	DPPH	% inhibition*	75.23	Chester *et al.*, 2017
S. scabrum	AAPH	μmol TE/g FW	34.50	Jiménez-Aguilar and Grusak, 2015
S. scabrum	ABTS	TE mg/g DM	21.36	Yuan *et al.*, 2018
S. scabrum	ORAC	μM TE/μg	2675	Tufts *et al.*, 2015
S. retroflexum	FRAP	mg of TEAC/100 g	3.96	Managa *et al.*, 2020b
	DPPH	mg of TEAC/100 g	0.30	
	ABTS	mg of TEAC/100 g	1.93	
S. retroflexum	DPPH	IC_{50}, μg/ml	198.9	Kheoane *et al.*, 2016
S. villosum	ABTS	TE mg/g DM	20.99	Yuan *et al.*, 2018

[a] % inhibition at 125 μg/ml; DM, dry matter weight; FRAP, ferric reducing antioxidant power assay; TEAC, trolox equivalence antioxidant capacity; ORAC, oxygen radical antioxidant capacity; AAPH, 2, 2′-Azobis-(2-amidinopropane) dihydrochloride; DPPH, 2,2-diphenyl-1-picrylhydrazyl; IC_{50}, 50% inhibition of DPPH

equivalence antioxidant capacity (TEAC), 2,2-diphenyl-1-picrylhydrazyl (DPPH) and oxygen radical antioxidant capacity (ORAC) as summarized in Table 7.8. The antioxidant properties of African nightshades have been correlated to their bioactive constituents such as phenolics and carotenoids which can quench singlet oxygen, chelate metal ions, donate hydrogen atoms and scavenge reactive radical species (Chester *et al.*, 2017). These assays are based on the ability of extracts to scavenge free-radical species through the hydrogen-atom transfer (HAT) mechanism, i.e. donation of hydrogen atoms, by the single-electron transfer (SET) mechanism, i.e. transfer of a single electron to reduce the target compound, or through their capacity to attract pro-oxidative metal ions (Craft *et al.*, 2012). Cellular antioxidant assays have been also used to investigate the ability of *S. nigrum* extracts to prevent oxidative stress in a physiological environment. It was found that *S. nigrum* can prevent the development of oxidative stress in human colon adenocarcinoma (Caco-2) and mouse fibroblast (L929) cell lines induced by 2, 2′-Azobis-(2-amidinopropane) dihydrochloride (AAPH) (Moyo *et al.*, 2020a). Therefore, bioactive constituents may lower or prevent oxidative stress also through activation of an NRF2 pathway that induces the synthesis of antioxidant enzymes that can reduce overall cell oxidative stress (Odongo *et al.*, 2018; Moyo *et al.*, 2020b). To test the anti-inflammatory ability of African nightshades, *S. nigrum* methanolic extract was found to have a substantial exposure-dependent anti-inflammatory reaction in egg white and carrageenin induced hind paw oedema in rats at a concentration of 100 mg/kg and 200 mg/kg (Arunachalam *et al.*, 2009). However, *S. scabrum* was found to have proinflammatory effect (Tufts, 2012). African nightshades have also been shown to possess antimicrobial activity. *S. villosum* methanol extract was reported to have maximum inhibition of growth against *Klebsiella pneumonia* and *Escherichia coli* with less inhibition towards *Enterobacter aerogenes* and *Vibrio cholera* (AbdelGawwad *et al.*, 2020).

Furthermore, *S. villosum* was also found to have antimicrobial properties against *Pseudomonas aeruginosa*, *Escherichia coli*, *Staphylococcus aureus* and *Bacillus subtilis* (Chowdhury *et al.*, 2008). Antimicrobial action could be due to the alteration of cell membrane permeability and the impact of bioactive compounds in different intracellular functions caused by the presence of hydrogen binding the phenolics to enzymes (Bouarab-Chibane *et al.*, 2019).

Bioaccessibility and Bioavailability

The ingestion of nutrients is followed by both bioaccessibility and bioavailability. Bioaccessibility denotes the concentration of food component that is discharged from the food matrix and is present in the gut with the possibility of being absorbed into the intestinal lumen (Gutierrez-Grijalva *et al.*, 2016). Bioavailability in terms of nutrition may refer to a utilized nutrient or stored quantity of the overall absorbed nutrient while for non-nutrients it may refer to the compound absorbed and detected at the site of action (Shahidi and Peng, 2019).

Nutrient bioavailability may be influenced by the occurrence of other nutrients (Pressman *et al.*, 2017). For instance, if there are elevated zinc levels, iron and copper absorption may be decreased while the presence of vitamin C improves non-haem iron (Pressman *et al.*, 2017). The absorption of some minerals can be hindered by the presence of oxalates, phytates, phenolic compounds and fibre (Wiesler, 2011). There is a dearth of data regarding the bioaccessibility and bioavailability of nutrients and non-nutrients from African nightshades. Nevertheless, *S. americanum* was found to have 33% of β-carotene and 58% of lutein bioaccessible after *in vitro* digestion with only about 3 and 11% of β-carotene and lutein available after *in vitro* colonic fermentation respectively (Serrano *et al.*, 2005). About 17–52% of total phenolics and 31–72% of total flavonoids were bioaccessible after the *in vitro* duodenal digestion of *S. nigrum* (Moyo *et al.*, 2020a). Some phenolics e.g. 3,5-dicafeolyquinic acid, quercetin-3-O-robinoside, 4-cafeolyquinic acid, 3-cafeolyquinic acid, 5-cafeolyquinic acid, catechin, myricetin, 3,4-dicafeolyquinic acid and rutin were drastically diminished following *in vitro* duodenal digestion of boiled *S. nigrum* (Moyo *et al.*, 2020a). Phenolic compound bioaccessibility may be affected by the nature of the vegetable material (raw or cooked), as well as their sensitivity to the changing pH environment from the stomach phase to the small intestine and the colon.

Health Benefits

African nightshades show a wide array of health benefits related to their nutritive and bioactive properties. *S. villosum* was found to have antifibrotic and hepatoprotective properties in rats by the regulation of fibrogenic mediator's expression and improving antioxidant capacity (Abdel-Hamid *et al.*, 2017). The epithelial mesenchymal transition and function of mitochondria in MCF-7 breast cancer cells was suppressed by *S. nigrum* extracts, hence showing its anti-cancer activity (Lai *et al.*, 2016). Anti-cancer activities of *S. nigrum* were also established by Alajmi *et al.* (2018) as the extracts showed significant cytotoxic property against human liver cancer cell line (HepG2) at IC_{50}: 20.4 µg/ml and against MCF-7 cell line at IC_{50}: 30.1 µg/ml. *S. scabrum* and *S. villosum* were also demonstrated to have antiulcerogenic properties by inhibiting ulcer formation on ethanol-induced gastric lesions in rats by 76.37% and 63.30 %, respectively (Mureithi *et al.*, 2020). Antidiabetic efficacy of *S. villosum* was recognized in diabetes-induced mice by testing the hypoglycaemic effect of the extracts (Nyaga *et al.*, 2019).

Novel Products/Formulations

African nightshades value addition deserves more attention as these vegetables contain a wide array of bioactive compounds and nutrients that are essential for human health. Manhivi *et al.* (2020) developed a vegetable soup consisting of nightshade (*S. retroflexum*), Chinese cabbage and pumpkin leaves in a ratio of 1.1.1 (w/w). A 100 g of the soup powder contained about 1399.14 kJ energy, 5.4 g moisture, 6.4 g fat, 2.9 g saturated fat, 7.7 g protein, 51.7 g carbohydrate, 19 g fibre, 9.9g ash and 1347 mg of sodium. Furthermore, the soup contained phytochemicals such as protocatechuic acid, catechin, caffeic acid, epicatechin, vanillic acid, 2,4-dihydroxybenzoic acid, coumaric acid, ferulic acid, ellagic acid, quercetin, luteolin, apigenin, kaempferol and isorhamnetin.

Conclusions and Future Trends

The taxonomic complexity and numerous synonyms related to the *S. nigrum* complex species is a major concern as valuable data concerning the nutritional, phytochemical and bioactive properties of African nightshades might be difficult to trace. Nonetheless, African nightshades are valuable sources of nutrients and phytochemicals which can be affected by different processing methods such as boiling, blanching, steaming, frying and drying. The presence of antinutrients in African nightshades may be reduced by boiling and blanching as these compounds are highly heat-labile and soluble in water. Postharvest techniques such as blanching and freeze/solar drying as well as modified atmospheric packaging are recommended for extending the shelf life of African nightshades during marketing as they are highly perishable foods. To fully recognize the health potential of African nightshades, more research into bioaccessibility and bioavailability of nutrients and phytochemicals from raw and processed vegetables is required. Furthermore, the popularity of these nightshades may be enhanced by the development of novel products with high nutritional value.

References

AbdelGawwad, M.R., Mahmood, A., Al Farraj, D.A., El-Abedein, A.I.Z., Mahmoud, A.H. *et al.* (2020) In-vitro antimicrobial activities of *Solanum villosum* (L.) lam; crude extract solvent comparison. *Journal of King Saud University - Science* 32(3), 2129–2133. DOI: 10.1016/j.jksus.2020.01.035.

Abdel-Hamid, A.E.E. (2018) A phytochemical and biological study of *Solanum villosum* Mill. belonging to Family Solanaceae. Master's thesis, Cairo University, Cairo, Eygpt.

Abdel-Hamid, A.E., El Dine, R.S., Sendker, J., El Zalabani, S.M., Meselhy, M.R. *et al.* (2017) Metabolic profiling of *Solanum villosum* Mill subsp. *miniatum* (Bernh. ex Willd.): hepatoprotective and antifibrotic activity in a rat model of liver fibrosis. *Pharmacognosy Magazine* 13, 179–188.

Adeyeye, A., Ayodele, O.D. and Akinnuoye, G.A. (2018) Evaluation of the nutritional composition of some less common edible leafy vegetables in Nigeria. *American Journal of Food Science and Nutrition* 5, 26–31.

Akubugwo, I.E., Obasi, A.N. and Ginika, S.C. (2007) Nutritional potential of the leaves and seeds of Black nightshade - *Solanum nigrum* L. Var *virginicum* from Afikpo-Nigeria. *Pakistan Journal of Nutrition* 6(4), 323–326. DOI: 10.3923/pjn.2007.323.326.

Alajmi, M.F., Alam, P., Rehman, M.T., Husain, F.M., Khan, A.A. *et al.* (2018) Interspecies Anticancer and Antimicrobial Activities of Genus *Solanum* and Estimation of Rutin by Validated UPLC-PDA Method. *Evidence-Based Complementary and Alternative Medicine* 2018, 1, 13. DOI: 10.1155/2018/6040815.

Albouchi, F., Attia, M., Hanana, M. and Hamrouni, L. (2018) Ethnobotanical notes and phytopharmacologiques on *Solanum nigrum* Linn (Family: *Solanaceae*). *American Journal of Phytomedicine and Clinical Therapeutics* 6, 1–9.

Aleksandrova, K., Koelman, L. and Rodrigues, C.E. (2021) Dietary patterns and biomarkers of oxidative stress and inflammation: a systematic review of observational and intervention studies. *Redox Biology* 42, 101869. DOI: 10.1016/j.redox.2021.101869.

Aramrueang, N., Asavasanti, S. and Khanunthong, A. (2019) *Leafy Vegetables*. In: Pan, Z., Zhang, R., and Zicari, S. (eds) *Integrated Processing Technologies for Food and Agricultural By-Products*. Academic Press, London, United Kingdom, pp. 245–272.

Arias-Rico, J., Macías-León, F.J., Alanís-García, E., Cruz-Cansino, N.D.S., Jaramillo-Morales, O.A. *et al.* (2020) Study of edible plants: Effects of boiling on nutritional, antioxidant, and physicochemical properties. *Foods* 9(5), 1–14. DOI: 10.3390/foods9050599.

Arunachalam, G., Subramanian, N. and Perumal, G. (2009) Evaluation of anti-inflammatory activity of methanolic extract of *Solanum nigrum* (Solanaceae) 5, 151–156.

Ayua, E., Mugalavai, V., Simon, J., Weller, S., Obura, P. *et al.* (2016) Ascorbic acid content in leaves of nightshade (*Solanum spp.*) and spider plant (*Cleome gynandra*) varieties grown under different fertilizer regimes in western Kenya. *African Journal of Biotechnology* 15(7), 199–206. DOI: 10.5897/AJB2015.14936.

Bouarab-Chibane, L., Forquet, V., Lantéri, P., Clément, Y., Léonard-Akkari, L. *et al.* (2019) Antibacterial properties of polyphenols: characterization and QSAR (Quantitative structure-activity relationship) models. *Frontiers in Microbiology* 10, 1–23. DOI: 10.3389/fmicb.2019.00829.

Chester, K., Paliwal, S., Khan, W. and Ahmad, S. (2017) UPLC-ESI-MS/MS and HPTLC method for quantitative estimation of cytotoxic glycosides and aglycone in bioactivity guided fractions of *Solanum nigrum* L. *Frontiers in Pharmacology* 8, 1–15. DOI: 10.3389/fphar.2017.00434.

Chowdhury, N., Laskar, S. and Chandra, G. (2008) Mosquito larvicidal and antimicrobial activity of protein of *Solanum villosum* leaves. *BMC Complementary and Alternative Medicine* 8, 1–6. DOI: 10.1186/1472-6882-8-62.

Craft, B.D., Kerrihard, A.L., Amarowicz, R. and Pegg, R.B. (2012) Phenol-based antioxidants and the in vitro methods used for their assessment. *Comprehensive Reviews in Food Science and Food Safety* 11(2), 148–173. DOI: 10.1111/j.1541-4337.2011.00173.x.

Daji, G., Steenkamp, P., Madala, N. and Dlamini, B. (2018) Phytochemical composition of *Solanum retroflexum* analysed with the aid of ultra-performance liquid chromatography hyphenated to quadrupole-time-of-flight mass spectrometry (UPLC-qTOF-MS). *Journal of Food Quality* 2018, 1–8. DOI: 10.1155/2018/3678795.

Degrain, A., Manhivi, V., Remize, F., Garcia, C. and Sivakumar, D. (2020) Effect of lactic acid fermentation on color, phenolic compounds and antioxidant activity in African nightshade. *Microorganisms* 8(9), 1–12. DOI: 10.3390/microorganisms8091324.

Edmonds, J.M. and Chweya, J.A. (1997) Black Nightshades *Solanum nigrum* L and related species. In: Heller, J., Engels, J., and Hammer, K. (eds) *Promoting the Conservation and Use of Underutilized and Neglected Crops*. Institute of Plant Genetics and Crop Plant Research/International Plant Genetic Resources Institute, Rome, Italy/Gatersleben, Germany, pp. 1–115.

Ejoh, R., Djuikwo, N.R.V. and Mbofung, C.M. (2017) Mineral profile and the effect of processing of some leafy vegetables indigenous to Cameroon. *African Journal of Food, Agriculture, Nutrition and Development* 17(3), 12362–12376. DOI: 10.18697/ajfand.79.15410.

Essack, H., Odhav, B. and Mellem, J.J. (2017) Screening of traditional South African leafy vegetables for specific anti-nutritional factors before and after processing. *Food Science and Technology* 37(3), 462–471. DOI: 10.1590/1678-457x.20416.

Fahey, J.W. (2016) Brassica: characteristics and properties. In: Caballero, B., Finglas, P.M. and Toldrá, F. (eds) *Encyclopedia of Food and Health*. Academic Press, Oxford, UK, pp. 469–477.

Farhath Khanum, F., Siddalinga Swamy, M., Sudarshana Krishna, K.R., Santhanam, K. and Viswanathan, K.R. (2000) Dietary fiber content of commonly fresh and cooked vegetables consumed in India. *Plant Foods for Human Nutrition* 55(3), 207–218. DOI: 10.1023/a:1008155732404.

Gutierrez-Grijalva, E.P., Ambriz-Pere, D.L., Leyva-Lopez, N., Castillo-Lopez, R.I. and Heiedia, J.B. (2016) Review: dietary phenolic compounds, health benefits and bioaccessibility. *Archivos Latinoamericanos de Nutricion* 66(2), 87–100.

Hassan, L.G. and Umar, K.J. (2008) Nutritional value of nightshade (*Solanum americanum* L.) Leaves. *Electronic Journal of Food and Plant Chemistry* 3, 14–17.

Hernández-Ledesma, B., Hsieh, C.C. and Martínez-Villaluenga, C. (2015) Food bioactive compounds against diseases of the 21st century 2016. *BioMed Research International* 2015, 2–4. DOI: 10.1155/2015/241014.

Hulley, I.M. and Van Wyk, B.E. (2019) Quantitative medicinal ethnobotany of Kannaland (western Little Karoo, South Africa): non-homogeneity amongst villages. *South African Journal of Botany* 122, 225–265. DOI: 10.1016/j.sajb.2018.03.014.

Jacoby, A., Labuschagne, M.T. and Viljoen, C.D. (2003) Genetic relationships between Southern African *Solanum retroflexum* Dun. and other related species measured by morphological and DNA markers. *Euphytica* 132(1), 109–113. DOI: 10.1023/A:1024657827796.

Jagatheeswari, D., Bharathi, T., Sheik, H. and Ali, J. (2013) Black nightshade (solanum nigrum L.)-an updated overview. *International Journal of Pharmaceutical & Biological Archives* 4, 288–295.

Jiménez-Aguilar, D.M. and Grusak, M.A. (2015) Evaluation of minerals, phytochemical compounds and antioxidant activity of Mexican, Central American, and African green leafy vegetables. *Plant Foods for Human Nutrition* 70(4), 357–364. DOI: 10.1007/s11130-015-0512-7.

Keller, G.B., Mndiga, H. and Maass, B.L. (2004) Production and consumption issues of traditional vegetables in Tanzania from the farmers point of view. MSc thesis. Georg August University of Göttingen, Göttingen, Germany.

Kheoane, P.S., Tarirai, C., Gadaga, T.H., Leonard, C. and Nyanzi, R. (2016) Antioxidant and prebiotic activity of selected edible wild plant extracts. *Journal of Food Research* 6(1), 7–21. DOI: 10.5539/jfr.v6n1p7.

Kirigia, D., Winkelmann, T., Kasili, R. and Mibus, H. (2019) Nutritional composition in African nightshade (*Solanum scabrum*) influenced by harvesting methods, age and storage conditions. *Postharvest Biology and Technology* 153, 142–151. DOI: 10.1016/j.postharvbio.2019.03.019.

Kouadio, I.A., Koffi, M.K. and Konan, G.A. (2019) Nutritional intake of *Solanum nigrum* Linn. leaves fortified with peanut paste consumed at breakfast in rural zones in Côte d'Ivoire. *International Journal of Biological and Chemical Sciences* 13(2), 586–596. DOI: 10.4314/ijbcs.v13i2.1.

Kouassi Dappah, D., Yao N'Dri, D., Kouakou Kouassic, N. and N'Guessan Amani, G. (2020) Effect of boiling on chemicals, phytochemicals and nutritional composition of *Solanum nigrum* L. leaves harvested in Abidjan (Côte d'Ivoire). *International Journal of Biochemistry Research & Review* 28, 1–11. DOI: 10.9734/ijbcrr/2019/v28i430159.

Krithika, R. and Verma, R.J. (2019) *Solanum nigrum* confers protection against CCl4-induced experimental hepatotoxicity by increasing hepatic protein synthesis and regulation of energy metabolism. *Clinical Phytoscience* 5(1), 1–8. DOI: 10.1186/s40816-018-0096-5.

Lai, Y.-J., Tai, C.-J., Wang, C.-W., Choong, C.-Y., Lee, B.-H. *et al.* (2016) Anti-Cancer Activity of Solanum nigrum (AESN) through Suppression of Mitochondrial Function and Epithelial-Mesenchymal Transition (EMT) in Breast Cancer Cells. *Molecules (Basel, Switzerland)* 21(5), 1–11. DOI: 10.3390/molecules21050553.

Lola, A. (2009) The effect of boiling on the nutrients and anti-nutrients in two non conventional vegetables. *Pakistan Journal of Nutrition* 8(9), 1430–1433. DOI: 10.3923/pjn.2009.1430.1433.

Ma, L. and Lin, X.M. (2010) Effects of lutein and zeaxanthin on aspects of eye health. *Journal of the Science of Food and Agriculture* 90(1), 2–12. DOI: 10.1002/jsfa.3785.

Managa, M.G., Mpai, S., Remize, F., Garcia, C. and Sivakumar, D. (2020a) Impact of moist cooking methods on colour, anti-nutritive compounds and phenolic metabolites in African nightshade (*Solanum retroflexum* Dun.). *Food Chemistry* 325, 1–8. DOI: 10.1016/j.foodchem.2020.126805.

Managa, M.G., Sultanbawa, Y. and Sivakumar, D. (2020b) Effects of different drying methods on untargeted phenolic metabolites, and antioxidant activity in Chinese cabbage (*Brassica rapa* L. subsp. chinensis) and nightshade (*Solanum retroflexum* Dun.). *Molecules* 25(6), 1–23. DOI: 10.3390/molecules25061326.

Manhivi, V.E., Sultanbawa, Y. and Sivakumar, D. (2020) Enhancement of the phytonutrient content of a gluten-free soup using a composite of vegetables. *International Journal of Food Properties* 23(1), 1051–1065. DOI: 10.1080/10942912.2020.1778028.

Maseko, I., Mabhaudhi, T., Tesfay, S., Araya, H., Fezzehazion, M. *et al.* (2017) African leafy vegetables: a review of status, production and utilization in South Africa. *Sustainability* 10(2), 16. DOI: 10.3390/su10010016.

Mattioli, R., Mosca, L., Sánchez-Lamar, A., Tempera, I. and Hausmann, R. (2018) Natural bioactive compounds acting against oxidative stress in chronic, degenerative, and infectious diseases. *Oxidative Medicine and Cellular Longevity* 2018, 1–2. DOI: 10.1155/2018/3894381.

Maziya-Dixon, B., Mdziniso, P., Jele, B., Zwane-Mdaka, S., Ndwandwe, B. *et al.* (2019) *Eswatini Food Composition Table*. Ministry of Agriculture, Mbabane, Eswatini.

Mehmood, A. and Zeb, A. (2020) Effects of different cooking techniques on bioactive contents of leafy vegetables. *International Journal of Gastronomy and Food Science* 22, 1–8. DOI: 10.1016/j.ijgfs.2020.100246.

Mohy-Ud-Din, A., Khan, Z.U.D., Ahmad, M. and Kashmiri, M.A. (2010) Chemotaxonomic value of alkaloids in *Solanum nigrum* complex. *Pakistan Journal of Botany* 42, 653–660.

Mohy-Ud-din, A., Khan, Z.U.D., Ahmad, M., Kashmiri, M.A., Yasmin, S. *et al.* (2009) Chemotaxonomic significance of flavonoids in the Solanum nigrum complex. *Journal of the Chilean Chemical Society* 54(4), 486–490. DOI: 10.4067/S0717-97072009000400037.

Moyo, S.M., Serem, J.C., Bester, M.J., Mavumengwana, V. and Kayitesi, E. (2020a) The impact of boiling and *in vitro* human digestion of *Solanum nigrum* complex (Black nightshade) on phenolic compounds bioactivity and bioaccessibility. *Food Research International* 137, 1–10. DOI: 10.1016/j.foodres.2020.109720.

Moyo, S.M., Serem, J.C., Bester, M.J., Mavumengwana, V. and Kayitesi, E. (2020b) African green leafy vegetables health benefits beyond nutrition. *Food Reviews International* 00(6), 1–18. DOI: 10.1080/87559129.2020.1717519.

Muchiri, S.V., Agong, S.G., Kahangi, E.M., Masinde, P.W., Chadha, M.L. *et al.* (2005) Evaluation of the growth and nutritional quality of African nightshade accessions (*solanum* spp) at different developmental stages. In: Abukutsa-Onyango, M.O., Muriithi, A.N., Anjichi, V.E., Ngamau, K., and Agong, S.G. (eds) *Proceedings of the Third Horticulture Workshop on Sustainable Horticultural Production in the Tropics, 26th- 29th November 2003*. Department of Horticulture, Maseno University (MSU), Maseno, Kenya, pp. 1–15.

Murador, D., Braga, A.R., Da Cunha, D. and De Rosso, V. (2018) Alterations in phenolic compound levels and antioxidant activity in response to cooking technique effects: A meta-analytic investigation. *Critical Reviews in Food Science and Nutrition* 58(2), 169–177. DOI: 10.1080/10408398.2016.1140121.

Mureithi, P.M., Kavoi, B.M., Onyango, C.M. and Mathiu, M.P. (2020) Antiulcerogenic effects of selected African nightshades (Solanum nigrum Linn.) genotypes on the rat stomach: A morphologic and morphometric study. *International Journal of Morphology* 38(4), 940–946. DOI: 10.4067/S0717-95022020000400940.

Musotsi, A.A., Makokha, A. and Sheila, M. (2019) Quantitative changes of ascorbic acid and beta carotene in African nightshade (*Solanum nigrum*) and spider plant (*Cleome gynandra*) due to traditional cooking methods used in Western Kenya. *Journal of Food and Nutritional Sciences Research* 1, 51–63.

Mwai, G.N., Onyango, J.C. and Abukusta-onyango, M.O. (2007) Taxonomic identification and characterization of African nightshades (*Solanum* L. section *Solanum*). *African Journal of Food, Agriculture, Nutrition and Development* 7(15), 1–17. DOI: 10.18697/ajfand.15.IPGRI2-1.

Mwai, G.N., Ojiewo, C.O., Masinde, P.W. and Agong, S.G. (2009) Good agricultural practices for African indigenous vegetables. In: Nono-Womdim, R., Ojiewo, C., Abang, M., and Oluoch, M.O. (eds) *Proceedings of a Technical Consultation Workshop*, eds. International Society for Horticultural Science (ISHS), Gent-Oostakker, pp. 55–74.

Mwanri, A.W., Mamboleo, T.F., Msuya, J.M. and Gowele, V.F. (2018) Oxalate, phytate and nitrate content in African nightshade, spider plant and amaranths at different stages of maturity. *African Journal of Food Science* 12(11), 316–322. DOI: 10.5897/AJFS2018.1735.

Nesamvuni, C., Steyn, N.P. and Potgieter, M.J. (2001) Nutritional value of wild leafy plants consumed by the Vhavenda. *South African Journal of Science* 97, 51–54.

Neugart, S., Baldermann, S., Ngwene, B., Wesonga, J. and Schreiner, M. (2017) Indigenous leafy vegetables of Eastern Africa - A source of extraordinary secondary plant metabolites. *Food Research International* 100(Pt 3), 411–422. DOI: 10.1016/j.foodres.2017.02.014.

Nyaga, S.N., Mathiu, P.M., Onyango, C.M. and Areba, G.O. (2019) Antidiabetic properties of *Solanum villosum* and *Solanum nigrum* var *sarrachoides* in a streptozotocin-induced diabetic mice model. *International Journal of Basic & Clinical Pharmacology* 8(11), 2396–2402. DOI: 10.18203/2319-2003.ijbcp20194774.

Odhav, B., Beekrum, S., Akula, U. and Baijnath, H. (2007) Preliminary assessment of nutritional value of traditional leafy vegetables in KwaZulu-Natal, South Africa. *Journal of Food Composition and Analysis* 20(5), 430–435. DOI: 10.1016/j.jfca.2006.04.015.

Odongo, G.A., Schlotz, N., Baldermann, S., Neugart, S., Huyskens-Keil, S. *et al.* (2018) African nightshade (Solanum scabrum Mill.): Impact of cultivation and plant processing on its health promoting potential as determined in a human liver cell model. *Nutrients* 10(10), 1–20. DOI: 10.3390/nu10101532.

Ojiewo, C.O., Murakami, K., Masinde, P.W. and Agong, S.G. (2007) Floral genetics of African nightshade (*Solanum* section *Solanum*). *International Journal of Plant Developmental Biology* 1, 75–81.

Ojiewo, C.O., Mwai, G.N., Abukutsa-Onyango, M.O., Agong, S.G. and Nono-Womdim, R. (2013) Exploiting the genetic diversity of vegetable african nightshades. *Bioremediation, Biodiversity and Bioavailability* 7, 6–13.

Ontita, E., Onyango, C., Onwonga, R. and Nyamongo, D. (2016) Indigenous knowledge on the uses of African nightshades (*Solanum nigram* L.) species among three Kenyan communities. *Asian Journal of Agricultural Extension, Economics & Sociology* 14(3), 1–8. DOI: 10.9734/AJAEES/2016/31099.

Petroski, W. and Minich, D.M. (2020) Is there such a thing as "anti-nutrients"? A narrative review of perceived problematic plant compounds. *Nutrients* 12(10), 1–32. DOI: 10.3390/nu12102929.

Pinela, J., Carvalho, A.M. and Ferreira, I.C. (2020) Watercress. In: Jaiswal, A.K. (ed.) *Nutritional Composition and Antioxidant Properties of Fruits and Vegetables*. Academic Press, London, United Kingdom, pp. 197–219.

Pressman, P., Clemens, R.A. and Hayes, A.W. (2017) Bioavailability of micronutrients obtained from supplements and food: A survey and case study of the polyphenols. *Toxicology Research and Application* 1, 1–7.

Ronoh, R., Ekhuya, N.A., Linde, M., Winkelmann, T., Abukutsa-Onyango, M. *et al.* (2018) African nightshades: genetic, biochemical and metabolite diversity of an underutilised indigenous leafy vegetable and its potential for plant breeding. *Journal of Horticultural Science and Biotechnology* 93(2), 113–121. DOI: 10.1080/14620316.2017.1358112.

Sá, A.G.A., Moreno, Y.M.F. and Carciofi, B.A.M. (2020) Plant proteins as high-quality nutritional source for human diet. *Trends in Food Science & Technology* 97, 170–184. DOI: 10.1016/j.tifs.2020.01.011.

Särkinen, T., Poczai, P., Barboza, G.E., van der Weerden, G.M., Baden, M. *et al.* (2018) A revision of the old world black nightshades (Morelloid clade of *Solanum* L., *Solanaceae*). *PhytoKeys* 106(106), 1–223. DOI: 10.3897/phytokeys.106.21991.

Serrano, J., Goñi, I. and Saura-Calixto, F. (2005) Determination of β-carotene and lutein available from green leafy vegetables by an *in vitro* digestion and colonic fermentation method. *Journal of Agricultural and Food Chemistry* 53(8), 2936–2940. DOI: 10.1021/jf0480142.

Shahidi, F. and Peng, H. (2019) Bioaccessibility and bioavailability of phenolic compounds. *Journal of Food Bioactives* 4, 11–68. DOI: 10.31665/JFB.2018.4162.

Silva, E.L., Almeida-Lafetá, R.C., Borges, R.M. and Staerk, D. (2017) Dual high-resolution inhibition profiling and HPLC-HRMS-SPE-NMR analysis for identification of α-glucosidase and radical scavenging inhibitors in *Solanum americanum* Mill. *Fitoterapia* 118, 42–48. DOI: 10.1016/j.fitote.2017.02.002.

Singh, P.K., Singh, R.P., Singh, P. and Singh, R.L. (2019) Food Hazards: Physical, Chemical, and Biological. In: Singh, R.L. and Mondal, S. (eds) *Food Safety and Human Health*. Academic Press, London, UK, pp. 15–65.

Sinha, K. and Khare, V. (2017) Review on: antinutritional factors in vegetable crops. *The Pharma Innovation Journal* 6, 353–358.

Suitor, C.W. and Murphy, S.P. (2013) Nutrition guidelines to maintain health. In: Coulston, A.M., Rock, C.L., and Monsen, E.R. (eds) *Nutrition in the Prevention and Treatment of Disease*. Academic Press, San Diego, USA, pp. 231–247.

Thakur, M., Singh, K. and Khedkar, R. (2020) Phytochemicals: Extraction process, safety assessment, toxicological evaluations, and regulatory issues. In: Prakash, B. (ed.) *Functional and Preservative Properties of Phytochemicals*. Academic Press, London, United Kingdom, pp. 341–361.

Tiwari, U. and Cummins, E. (2013) Factors influencing levels of phytochemicals in selected fruit and vegetables during pre- and post-harvest food processing operations. *Food Research International* 50(2), 497–506. DOI: 10.1016/j.foodres.2011.09.007.

Traoré, K., Parkouda, C., Savadogo, A., Ba/Hama, F., Kamga, R. *et al.* (2017) Effect of processing methods on the nutritional content of three traditional vegetables leaves: Amaranth, black nightshade and jute mallow. *Food Science & Nutrition* 5(6), 1139–1144. DOI: 10.1002/fsn3.504.

Tufts, H.R. (2012) The anti-inflammatory and antioxidant properties of kenyan leafy green vegetables, wild fruits, and medicinal plants: relation of the health-promoting properties of biodiversity to contribute to kwashiorkor alleviation. Master's thesis, McGill University Montreal, Canada.

Tufts, H.R., Harris, C.S., Bukania, Z.N. and Johns, T. (2015) Antioxidant and anti-inflammatory activities of Kenyan leafy green vegetables, wild fruits, and medicinal plants with potential relevance for kwashiorkor. *Evidence-Based Complementary and Alternative Medicine* 2015, 1–9. DOI: 10.1155/2015/807158.

USDA (2019) FoodData Central. Available at: https://fdc.nal.usda.gov/fdc-app.html#/food-details/168462/nutrients (accessed 5 November 2020).

Usman, H., Victor, V. and Waziri, I. (2018) Qualitative phytochemical screening and in vitro antimicrobial activities of *Solanum americanum* Mill. *Journal of Engineering, Technology, and Environment* 14, 104–110.

Vagula, J.M., Rocha, B.A., Silva, A.R., Narain, N., Bersani-Amado, C.A. *et al.* (2018) Analysis of *Solanum americanum* Mill. by ultrafast liquid chromatography with diode array and time-of-flight mass spectrometry detection with evaluation of anti-inflammatory properties in rodent models. *Analytical Letters* 51(13), 1973–1985. DOI: 10.1080/00032719.2017.1399413.

van Jaarsveld, P., Faber, M., van Heerden, I., Wenhold, F., Jansen van Rensburg, W. *et al.* (2014) Nutrient content of eight African leafy vegetables and their potential contribution to dietary reference intakes. *Journal of Food Composition and Analysis* 33(1), 77–84. DOI: 10.1016/j.jfca.2013.11.003.

Van Wijk, Y., Tusenius, M.L., Rust, R., Cowling, R.M. and Wurz, S. (2017) Modern vegetation at the Klasies river archaeological sites, Tsitsikamma coast, South-Eastern Cape, South Africa: A reference collection. *Plant Ecology and Evolution* 150(1), 13–34. DOI: 10.5091/plecevo.2017.1286.

Viljoen, E. (2011) Morphology and genetic relationships in members of the *Solanum nigrum* L. complex used for jam production in the highveld of South Africa. Master's thesis, University of Pretoria, Pretoria, South Africa.

Wiesler, F. (2011) Nutrition and quality. In: Marschner, H. (ed.) *Marschner's Mineral Nutrition of Higher Plants*. Academic Press, London, United kingdom, pp. 271–282.

Xu, Y., Liang, D., Wang, G.T., Wen, J. and Wang, R.J. (2020) Nutritional and functional properties of wild food-medicine plants from the coastal region of South China. *Journal of Evidence-Based Integrative Medicine* 25, 1–13. DOI: 10.1177/2515690X20913267.

Yuan, B., Byrnes, D., Giurleo, D., Villani, T., Simon, J.E. *et al.* (2018) Rapid screening of toxic glycoalkaloids and micronutrients in edible nightshades (*Solanum spp.*). *Journal of Food and Drug Analysis* 26(2), 751–760. DOI: 10.1016/j.jfda.2017.10.005.

Yuan, B., Dinssa, F.F., Simon, J.E. and Wu, Q. (2020) Simultaneous quantification of polyphenols, glycoalkaloids and saponins in African nightshade leaves using ultra-high performance liquid chromatography tandem mass spectrometry with acid assisted hydrolysis and multivariate analysis. *Food Chemistry* 312, 1–9. DOI: 10.1016/j.foodchem.2019.126030.

8 Moringa (*Moringa oleifera*)

Stephen A. Akinola* and Dharini Sivakumar
Phytochemical Food Network Research Group, Department of Crop Sciences, Tshwane University of Technology, Pretoria, South Africa

Abstract
Moringa oleifera is a prized plant found in many countries of the world, laden with huge potential to be used as food, nutraceuticals and for medicinal purposes. It is highly nutritive and contains both the essential minerals and vitamins necessary for health, besides its high protein content. It has a good amino acid balance and contains diverse phenolic compounds. It has a water purifying ability and is rich in phenolic compounds such as quercetin, kaempferol, zeatin, caffeoylquinic acid and B-sitosterol. *Moringa oleifera* plants are used in the treatment of physiological and pathological diseases due to their antihypertensive, anticholesterolemic, hepatoprotective, antiulcer, antipyretic, antidiabetic, antitumour, anti-inflammatory, antibacterial, antifungal and antiepileptic activities. They are sources of antioxidants which could prevent oxidative stress, reduce ageing and prevent the development of carcinogens. *Moringa oleifera* could be a remedy against irritable bowel disease, inflammation and cardiovascular diseases owing to its antioxidant activities. Parts of the plant have been utilized in food formulations, cosmetics, pharmaceuticals and herbal medicine. However, the potential of *Moringa oleifera* leaf could be further extended in the field of dietectics and nutrigenomics. Further investigation on its bioactivies and the bioaccessibilities of it nutrients and functional compounds could be explored using clinical studies. This review focuses on the nutritional composition, phytochemical content, biological activities and processing implications of *Moringa oleifera*, coupled with its prospects.

Introduction

Moringa oleifera is generally known as a miracle plant, based on its use in diverse areas including nutrition and medicine. *M. oleifera* is a plant that is typical to the tropical and semi-arid parts of the world: South-East and South Asia, India, Saudi Arabia, Pakistan, Afganistan, Bangladesh, Pacific and Carribean islands and Africa (Sengupta and Gupta, 1970; Veeraragava, 1998; Fahey, 2005). It is often used as a food ingredient in areas where there is food insecurity and other foods are seasonal (Afzal *et al.*, 2020). *M. oleifera* belongs to the Moringaceae family and remains the widely cultivated species out of the 13 total species (Oyeyinka and Oyeyinka, 2018). Based on regions where it is cultivated, *M. oleifera* (Latin) is called by different names such as mother's best friend, horseradish tree, drumstick tree, Ben oil tree, magic tree, miracle tree (English), Benzolive, Angela (Spanish), Marango, Mulangay (Telugu), kelor, Sainjna (Hindu), Sahajan tree (Unami), Moringueiro (Portugese), La ken (Chinese) and Rawag (Arabian) (Fahey, 2005; Anwar *et al.*, 2007; Mishra *et al.*, 2011; Awodele *et al.*, 2012). In Nicaragua, *M. oleifera* was introduced as an ornamental plant used for permanent fencing. *M. oleifera* is widely researched and has gained application in diverse fields.

*Corresponding author: akinolastephen3@gmail.com

 Handbook of Phytonutrients in Indigenous Fruits and Vegetables (eds D. Sivakumar *et al.*)
DOI: 10.1079/9781789248067.0008

However, according to NRC (2006) it is still considered under-exploited, neglected and under-utilized; a 'lost crop', especially in Africa. Despite this, different parts of *M. oleifera* leaves have found applications in food formulations, especially in African countries, and have been used in the production of yoghurts (Hekmat *et al.*, 2015), soup mixes (Babayeju *et al.*, 2014), dough meal and confectionary (Kolawole *et al.*, 2013; Chinma *et al.*, 2014; Karim *et al.*, 2015). *M. oleifera* leaf has been used as an ingredient in food formulations and fortifications to manage malnutrition among children (Boateng *et al.*, 2019) and caregivers (Boateng *et al.*, 2018), in water purification and for medicinal purposes, all of which account for its description as a miracle plant or miracle tree (Fuglie, 1999; Oyeyinka and Oyeyinka, 2018). *M. oleifera*'s tree trunk has been used in the production of gums, its seeds for water purification, its flower nectar as food for bees in honey production and its leaves as forage in animal husbandry. *M. oleifera* seed powder is used in water treatments due to its ability to flocculate and coagulate suspended solids and pathogens in water due to its inherent bioactive protein compounds (Ghebremichael *et al.*, 2005). Extracts from *M. oleifera* leaves have been used as plant growth promoters. These have been developed into foliar sprays that help plants to retain firmness and increase resistance to diseases and pests. Studies have shown that plants sprayed with the growth hormones from *M. oleifera* leaves produce larger fruits compared to unsprayed plants, and have increased yields, shelf life, heavier leaves, shoots and sugar levels (Makkar and Becker, 1996; Foidl *et al.*, 2001). *M. oleifera* leaves have been used as a feed ingredient and forage for livestock due to their rich nutritional properties. The high productivity of *M. oleifera* plant per unit area makes its use versatile compared to other forage crops. *M. oleifera* is one of the global functional plants that have been shown to be of value to nutrition and health and has found applications in industries not limited to food and pharmacological settings. Hence, this review seeks to underscore its nutritional quality, phytochemical properties and biological activities in terms of its potential for increased utilization.

Botany

M. oleifera is a perennial tree (Fig. 8.1). The stems are seldom straight but short, with a height of 1.5 to 2 m prior to branching. Branches can extend into a disorganized mass that forms a canopy that is umbrella-shaped. The leaves often have long petioles of 8–10 pairs of pinnae, with petiole

Fig. 8.1. *Moringa oleifera* parts: (a) leaf; (b) leaf and branches; (c) seeds and flower; (d) tree.

bases and pinnae having glands. The leaves are 20–70 cm long and appear as greyish-downy in colour at mid-age and can be elliptic or obovate in arrangement. The flower petals are cream, white or yellow in colour and are 0.1 to 2.5 m long, 2.5 cm wide and have a pleasant smell. The petals are five and slender-spatulated while the sepals are linear-lanceolated (Foidl *et al.*, 2001). The fruit pod is three-lobed and spans about 0.2–0.6 m, containing up to 12 to 35 seeds. *M. oleifera* seeds have semi-permeable hulls, coupled with white wings and are brownish in colour. The seeds weigh on average 0.3 g per seed, pods 7.6 g and the average number of seeds/pod is 12 (Ferrao and Ferrao, 1970; Makkar and Becker, 1997). *M. oleifera* is a fast growing plant that can stretch to 4 m in height within the first year of life. The average height of a matured *M. oleifera* tree ranges from 7 to 12 m and 2 to 4 m in trunk diameter (Foidl *et al.*, 2001). Seed weight is an average of 0.3 g and its kernel to hull ratio is 75:25 (Makkar and Becker, 1997).

M. oleifera is a cultivated and versatile crop (Nouman *et al.*, 2014a) that is drought and disease resistant (Foidl *et al.*, 2001). There are about 13 indigenous species of *M. oleifera* of which 11 are domesticated in Africa (Leone *et al.*, 2015a) and two in Asia. *M. oleifera* grows well in different soil types except in heavy clay soils. It is unable to tolerate unceasing flooded soils nor stagnant water, hence it is suited to subtropical and tropical parts of the world (Anwar *et al.*, 2007; Stadtlander and Becker, 2017).

M. oleifera grows in different soil types with pH ranging from 4.5 to 9 (Foidl *et al.*, 2001). It thrives well in slightly acidic, nutrient rich and well drained soils. According to Oliveira *et al.* (2009), *M. oleifera* can tolerate salt water at an early stage of life but may not be able to do the same at maturity. It can be cultivated effectively at annual temperature conditions of 18.7 to 28.5°C. It is also capable of growing in harsh environments with temperatures of 48°C. An annual rainfall of 250 to 1500 mm is regarded as adequate for its growth and it can tolerate drought and moderate frost which accounts for its domestication in different regions of the world (Price, 1985, 2000; HDRA, 2002; Afzal *et al.*, 2020). *M. oleifera* can be cultivated through stem cutting or by seed planting (Palada and Chang, 2003). It regrows fast after pruning and has a yearly high biomass yield of about 580 tonnes per ha (Foidl *et al.*, 2001). It can be propagated by either asexual or sexual methods and can be easily managed due to the low water requirements for growth (Foidl *et al.*, 2001). The season, cultivar type, fertilizer application and weather conditions are factors that affect the productivity of *M. oleifera* trees (Palada *et al.*, 2006; Nouman *et al.*, 2016). *M. oleifera* can be harvested at any time of the year, however this is best done in the rainy season. Trimming of the tree back by half helps to rejuvenate the plant. Leaves are harvested by stripping off the stems, cleaned to remove dirt and soil. The leaves can be dried in the dark to prevent loss of vitamins (Doerr and Cameron, 2005).

Traditional Use

Moringa species are high in protein, lipids, sugar, starch and polypeptides (stem and leaves), and this has enhanced their use locally and industrially. *M. oleifera* is known as a miracle plant because of the versatile application of its parts in diverse fields (Stadtlander and Becker, 2017). *M. oleifera* has been applied traditionally in the treatment of genito-urinary disease, typhoid fever, dermatological disease, sores, malaria and arthritis (Abe and Ohtani, 2013; Yabesh *et al.*, 2014; Sivasankari *et al.*, 2014). Similarly, *M. oleifera* is recorded as being used as a diuretic, antispasmodic, abortifacient, expectorant and as a stimulant (Mishra *et al.*, 2011). *M. oleifera* functions as an antioxidant, anticancer, anti-inflammatory, neuroprotective, antidiabetic, antiviral, anti-obesity, cardioprotective, antifungal, antibacterial and hepatoprotective biological material (Singh *et al.*, 2020). Traditionally, it has been used in the treatment of infertility and reproductive diseases in women. It boosts lactation and repels immune suppressions (Thurber and Fahey, 2009; Stohs and Hartman, 2015). It has been used to treat poor vision, anaemia and joint pain (Abe and Ohtani, 2013; Popoola and Obembe, 2013). Different part of the *M. oleifera* plant have been used in the treatment of diverse diseases and deficiencies (Anwar *et al.*, 2007).

M. oleifera leaves, roots and stems have been used in various ways, such as food and in traditional medicines (Amaglo *et al.*, 2010). The bark of *M. oleifera* has been used as an antibacterial and antifungal agent, its flowers as stimulants and as diuretic medicines (Mishra *et al.*, 2011), while other applications include antiseptics and medicine to improve cardiovascular function in the body. According to Mishra *et al.* (2011), *M. oleifera* pods have anthelmintic and antipyretic properties and have been used to manage diabetes locally. Its roots have been used to treat epilepsy, nervous-system associated disorders, hepatomegaly, splenomegaly and inflammation. A decoction of *M. oleifera* is used to treat sore throat while its leaves are cooked and used to treat catarrh and influenza.

The leaves, fruits, flowers and immature pods (Fig. 8.1) have been eaten as vegetables in Africa, Asia, the Middle East and the USA (Hawaii) (Anwar *et al.*, 2007). In developing countries where malnutrition is endemic, *M. oleifera* leaf powder has been used as a food supplement to manage cases of malnutrition (Thurber and Fahey, 2009; Stadtlander and Becker, 2017). *M. oleifera* leaves have been applied in feed formulations for the livestock industry and for human nutrition due to the high protein content (Amaglo *et al.*, 2010). Authors have reported the ability of *M. oleifera* to boost meat quality and animal production (Qwele *et al.*, 2013; Nkukwana *et al.*, 2014). *M. oleifera* leaves have served several purposes such as a food, a manure, a natural coagulant in water treatment, a growth enhancer and a biopesticide (Bennett *et al.*, 2003; Anwar *et al.*, 2007; Nouman *et al.*, 2012). However, the utilization of *M. oleifera* is limited by the presence of antinutrients and toxins.

Nutritional Composition of Leaves

M. oleifera leaves are a nutritionally rich food source that could play an important role in human and animal nutrition. Studies have reported a high protein content in *M. oleifera* leaves compared to other edible plant leaves. The protein content of *M. oleifera* leaf ranges from 25 to 30.29 g/100 g as reported by Jongrungruangchok *et al.* (2010) while Richter *et al.* (2003) and Hlophe and Moyo (2014) reported 25 g/100 g and 32.1 g/100 g on a dry weight (DW) basis, respectively. Several variations exist in the nutrient composition of *M. oleifera* parts, as shown in Table 8.1. The protein content of *M. oleifera* varies according to the region in which it is cultivated and the part of the plant. *M. oleifera* seeds contains the highest content of protein followed by its kernel, leaf and pod (Table 8.1). *M. oleifera* seed cultivated in Nigeria, as reported by Olagbemide and Philip (2014), had higher crude protein and lipid content compared to seeds from other countries. About 70 g of *M. oleifera* leaves per day is adequate to meet the recommendaed daily allowance (RDA) for a healthy man (Asiedu-Gyekye *et al.*, 2014). The protein content of *M. oleifera* leaves cultivated in Africa has been reported to range from 25.4 to 32.1 g/100 g and is comparable to the protein content of other vegetables (Table 8.2) such as cowpea leaves (21.5–43.7 g/100 g), kale (36.8 g/100 g), spinach (38.2 g/100 g), mustard green (34g /100 g), sprout (34.1 g/100 g), carrot (18.38 g/100 g), okra (17.81 g/100 g), water leaf (32.35 g/100 g), rosselle calyx (5.76 g/100 g), jute mallow (16.26 g/100 g) and fluted pumpkin leaves (19.29 g/100 g) (Singh *et al.*, 2001; Kahlon *et al.*, 2007; Ndlovu and Afolayan, 2008; Oluwalana *et al.*, 2011; Ukegbu and Okereke, 2013; Chikwendu *et al.*, 2014; Gonçalves *et al.*, 2016; Amoasah *et al.*, 2018).

The mineral content of *M. oleifera* varies with region of cultivation, as shown in Table 8.3. *M. oleifera* leaf from Bangladesh was reported to have high potassium content (3250 mg/100 g) compared to samples from other countries (Ahmed *et al.*, 2018; Islam *et al.*, 2020). *M. oleifera* leaf from Malawi had the highest magnesium (879 mg/100 g) compared to samples from Nigeria (650 mg/100 g) and other countries (Banjo, 2012; Stadtlander and Becker, 2017; Ogbe and Affiku, 2020). The zinc content of *M. oleifera* leaf from Nigeria is higher than leaves from other countries and could help to boost cognitive development in children if used in the development of food products targeted at such functions (Ogbe and Affiku, 2020). *M. oleifera* is dense in minerals such as iron, calcium, potassium, magnesium and copper (Afuang *et al.*, 2003; Hekmat *et al.*, 2015) compared to common vegetables (Table 8.4). The flower contains potassium and calcium (Makkar and Becker, 1996). The iron and calcium content in *M. oleifera* leaf is higher than that in spinach, kale, cabbage, carrot, okra, cowpea, rosselle, waterleaf, jute mallow and fluted pumpkins, and exceeds the iron and calcium content RDA value of 1.8 mg/100 g (Table 8.4). *M. oleifera* leaf is

Table 8.1. Variation in the proximate composition of parts of *Moringa oleifera* plants (%).

Part	Protein (%)	CHO (%)	Lipids (%)	Moisture (%)	Fibre (%)	Ash (%)	Country	References
Leaves	27.7	44.69	13.4	6.22	8.07	11.24	Malawi	Stadtlander and Becker (2017); Chatepa and Mbewe (2018)
Leaves	25.4	nd	4.6	6.3	7.3	9.0	Nicaragua	Afuang *et al*. (2003)
Leaves	30.29	nd	6.5	9.53	11.4	7.64	South Africa	Moyo *et al*. (2011)
Leaves	30.29	63.11	2.11	3.21	5.14	7.93	Nigeria	Banjo (2012); Teixeira *et al*. (2014); Ogbe and Affiku (2020)
Leaves	29.0	nd	nd	68.82	19.1	nd	Bangladesh	Islam *et al*. (2020)
Kernel	36.7	nd	37.4	4.9	21.9	nd	Nicaragua	Stadtlander and Becker (2017)
Kernel	34.7	nd	28.5	6.1	nd	nd	Ethiopia	Stadtlander and Becker (2017)
Kernel	37.5	nd	37.0	4.8	nd	nd	India	Stadtlander and Becker (2017)
Kernel	23.29	nd	2.26	10.48	19.91	7.07	Thailand	Foidl *et al*. (2001)
Kernel	37.86	11.55	41.18	2.6	4.8	4.6	Malawi	Chatepa and Mbewe (2018)
Stem	13.2	18.5	12.9	4.2	44.2	6.7	Malaysia	Abdulkadir *et al*. (2016)
Pods	18.4	10.4	16.9	8.1	39.2	6.7	Malaysia	Abdulkadir *et al*. (2016)
Pods	2.1	8.53	0.20	nd	3.2	nd	Bangladesh	Taher *et al*. (2017)
Seed	26.5	27.5	20.1	4.0	21.0	6.3	Malaysia	Abdulkadir *et al*. (2016)
Seed	35.97	8.67	38.67	9.97	2.87	3.87	Nigeria	Olagbemide and Philip (2014)
Seed	28.54	23.27	34.91	3.17	7.90	5.37	Malawi	Chatepa and Mbewe (2018)

CHO, Carbohydrates, Nd, not determined.

Table 8.2. Comparative analysis of the proximate composition of *M. oleifera* and common vegetables on DW basis.

Parameters (g/100 g)	Protein	Lipids	CHO	Moisture	Fibre	Ash	References
Moringa *(Moringa oleifera)*	32.1	1.72	63.11	3.21	5.14	7.93	Banjo (2012); Hlophe and Moyo (2014); Teixeira *et al*. (2014); Ogbe and Affiku (2020)
African spinach (*Amaranthus hybridus*)	18.7	1.58	50.32	6.23	4.13	17.41	Kahlon *et al*. (2007); Ndlovu and Afolayan (2008); Ukegbu and Okereke (2013)
Cabbage (*Brassica oleracea* var. capitata)	5.3	1.2	78.6	91	0.79	14.06	Yanyang *et al*. (2004); Ndlovu and Afolayan (2008)
Kale (*Brassica oleracea*)	36.8	0.26	2.36	81.39	3.0	1.33	Kahlon *et al*. (2007); Emebu and Anyika (2011)
Carrot (*Daucus carota* L.)	18.38	5.69	6.86	95.4	14.81	19.69	Warman and Havard (1997); Singh *et al*. (2001); Chantaro *et al*. (2008)
Okra (*Abelmoschus esculentus*)	17.81	4.8	46.42	6.31	10.9	8.6	Ukegbu and Okereke (2013)
Cowpea (*Vigna unguiculata*)	39.24	1.31	30.39	7.35	14.26	14.80	Chikwendu *et al*. (2014)
Waterleaf (*Talium triangulare*)	32.35	5.03	28.86	6.14	6.15	21.47	Oluwalana *et al*. (2011)
Rosselle (*Hisbiscus sabdariffa L.)*	5.76	2.88	62.46	6.97	16.12	5.80	Amoasah *et al*. (2018)
Jute mallow *(Cochorus olitorius* L.)	16.26	1.72	69.5	30.90	2.03	10.52	Ndlovu and Afolayan (2008); Onwordi *et al*. (2009)
Fluted pumpkin (*Telfaira occidentalis*)	19.29	5.51	38.87	6.23	9.45	10.10	Ukegbu and Okereke (2013)

CHO, Carbohydrates.

Table 8.3. Mineral content of *M. oleifera* leaf with sources from different countries (g/kg dry matter).

Country	Fe (mg/100 g)	Ca (mg/100 g)	Zn (mg/100 g)	P (mg/100 g)	Cu (mg/100 g)	Na (mg/100 g)	K (mg/ 100 g)	Mg (mg/kg)	Mn (g/kg)	References
Thailand	13.2	2370	2.1	307	0.659	29	1930	369	7.94	Jongrungruangchok *et al*. (2010); Stadtlander and Becker (2017)
Nicaragua	22.9	2080	2.07	336	0.008	70	1880	369	7.14	Makkar and Becker (1996); Jongrungruangchok *et al*. (2010); Stadtlander and Becker (2017)
South Africa	4.9	3650	0.003	299	0.0008	164	1500	500	0.009	Moyo *et al*. (2011)
India	0.013	1100	0.003	632	0.007	67	2960	373	0.051	Ramachandran *et al*. (1980); Stadtlander and Becker (2017)
Bangladesh	0.28	350	0.46	34	nd	nd	3250	9	0.037	Ahmed *et al*. (2018); Islam *et al*. (2020)
Nigeria	0.007	90	0.65	524	0.0061	19	440	650	0.008	Banjo (2012); Okiki *et al*. (2015); Ogbe and Affiku (2020)
Niger	0.037	1440	0.0022	321	0.0079	48	912	385	0.57	Sena *et al*. (1998)
Malawi	1.24	2220	0.021	370	0.0008	60	1350	879	0.77	Stadtlander and Becker (2017)
Uganda	0.581	2160	0.017	259	0.0004	29	1730	420	1.16	Stadtlander and Becker (2017)

Fe, Iron; Ca, Calcium; Zn, Zinc; P, Phosphorus; Cu, Copper; Na, Sodium; K, Potassium; Mg, Magnesium; Mn, Manganese; nd, not determined.

Table 8.4. Comparative analysis of the mineral composition of common vegetables on DW basis.

Parameters (mg/100 g)	Zn	Ca	Fe	Na	Mg	P	Cu	K	Vit. C	B-carot	References
Moringa *(Moringa oleifera)*	2.07	2080	22.9	70	369	336	0.65	1880	36.08	0.79	Makkar and Becker (1996); Jongrungruangchok *et al.* (2010); Clement *et al.* (2017); Stadtlander and Becker (2017)
Spinach (*Amaranthus hybridus*)	0.106	607	0.411	298	194	500	1.9	373	36.8	4.0	Bhattacharjee *et al.* (1998); Singh *et al.* (2001); Ndlovu and Afolayan (2008)
Cabbage (*Brassica oleracea* var. capitata)	1.94	603	2.72	82	141	251	0.16	265.47	58.36	9.9	Warman and Havard (1997); Ndlovu and Afolayan (2008); Park *et al.* (2014)
Kale (*Brassica oleracea*)	2.16	4.05	8.94	4.69	6.69	167.46	1.51	7.03	104.31	4.24	de Azevedo and Rodriguez-Amaya (2005); Acikgoz (2011); Emebu and Anyika (2011)
Okra (*Abelmoschus esculentus*)	0.233	366	3.2	18.3	45.3	44.5	0.102	267	4.9	1.93	EL-Qudah (2009); dos Santos *et al.* (2013)
Cowpea (*Vigna unguiculata*)	1.66	1.4	7.5	10	459	135.6	0.7	3120	1.39	0.25	Imungi and Potter (1983); Chikwendu *et al.* (2014)
Water leaf (*Talium triangulare*)	5	74	1.2	80	81	198	2.44	85	48.95	0.26	Yusuf *et al.* (2003); Oluwalana *et al.* (2011); Ogbonnaya and Chinedum (2013)
Rosselle (*Hisbiscus sabdariffa L.)*	0.2	14.8	8.5	72.1	100.7	35.3	8	101.5	4.69	700	Atta *et al.* (2010); Luvonga *et al.* (2010); Al-Sayed *et al.* (2020)

Continued

Table 8.4. Continued

Parameters (mg/100 g)	Zn	Ca	Fe	Na	Mg	P	Cu	K	Vit. C	B-carot	References
Jute mallow (*Cochorus olitorius* L.)	0.051	0.347	0.228	-	0.560	0.258	2.48	21.79	101.7	262.6	Yusuf *et al*. (2003); Ndlovu and Afolayan (2008); Onwordi *et al*. (2009); Musa *et al*. (2010)
Fluted pumpkin (*Telfaira occidentalis*)	6.80	3.20	2.73	2.30	0.78	10.6	0.92	3.05	8.8	18.8	Yusuf *et al*. (2003); Onoja (2014)
Carrot (*Daucus carota* L.)	3.2	2300	5.46	830	310	210	0.94	3660	2.87	2.85	Warman and Havard (1997); Singh *et al*. (2001)
RDA	1.1	100	1.8	150	42	125	90	470			Adelakun *et al*. (2012)

Zn, Zinc; Ca, Calcium; Fe, Iron; Na, Sodium; Mg, Magnesium; P, Phosphorus; Cu, Copper; K, Potassium; Vit. C, Vitamin C; B-carot, Beta-carotene; nd, not determined; RDA, Recommended Daily Allowance.

high in zinc and magnesium compared to other vegetables as shown in Table 8.4, except kale, fluted pumpkin and carrot. *M. oleifera* leaves are rich in ascorbic acids (36.08 mg/100 g) and carotenoids (0.79 mg/100 g), greater than levels found in carrots (Makkar and Becker, 1996; Hekmat *et al.*, 2015). According to Fahey (2005), the leaf is richer in vitamin C than okra and cowpea, rosselle calyx and fluted pumpkin (Table 8.4). *M. oleifera* leaves are a rich source of essential vitamins and minerals. The bioavailability of folate was reported to be 83% in *M. oleifera* leaves fed rats thus implying the potential of *M. oleifera* as a source of dietary folate (Saini *et al.*, 2016).

M. oleifera has a high amino acid profile (Olugbemi *et al.*, 2010) (Table 8.5). It is rich in the sulfur-containing amino acids (methionine and cysteine) thus making it suitable to replace legumes with limiting sulfur-containing amino

Table 8.5. Amino acid profile of *M. oleifera* leaf and pods against daily requirements for adults (DRA)FAO/WHO/UNU (2007).

M. Oleifera	Raw leaf (mg/g DM)	Leaf extract (mg/g DM)	Dry leaf (mg/g DM)	Pods (mg/g DM)	DRA	References
Arginine	30.28	15.64	1325	360	nd	Moyo *et al.* (2011); Nouman *et al.* (2014b)
Serine	20.79	10.34	nd	nd	nd	Moyo *et al.* (2011); Nouman *et al.* (2014b)
Aspartic acid	46.11	22.16	nd	nd	nd	Nouman *et al.* (2014b)
Glutamic acid	50.85	25.65	nd	nd	nd	Nouman *et al.* (2014b)
Glycine	26.62	13.73	nd	nd	nd	Nouman *et al.* (2014b)
Threonine	21.97	11.70	1188	390	15	Fuglie (2000); Nouman *et al.* (2014b)
Alanine	28.67	18.37	nd	nd	nd	Nouman *et al.* (2014b)
Tyrosine	18.88	9.71	nd	nd	25	Nouman *et al.* (2014b)
Proline	25.75	13.63	nd	nd	nd	Nouman *et al.* (2014b)
Methionine	8.96	4.97	350	140	15	Fuglie (2000); Nouman *et al.* (2014b)
Valine	27.58	14.26	1063	540	26	Fuglie (2000); Nouman *et al.* (2014b)
Phenylalanine	27.14	15.51	1388	430	15	Fuglie (2000); Nouman *et al.* (2014b)
Isoleucine	22.53	11.30	825	440	20	Fuglie (2000); Nouman *et al.* (2014b)
Leucine	42.89	21.84	1950	650	39	Fuglie (2000); Nouman *et al.* (2014b)
Histidine	13.57	7.50	613	110	10	Freiberger *et al.* (1998); Fuglie (2000)
Cysteine	5.18	3.39	nd	nd	15	Nouman *et al.* (2014b)
Lysine	26.77	14.06	1125	150	30	Freiberger *et al.* (1998); Fuglie (2000); Nouman *et al.* (2014b)
Tryptophan	9.26	5.27	425	80	4	Freiberger *et al.* (1998); Fuglie (2000); Nouman *et al.* (2014b)

DRA, daily requirements for adults according to FAO/WHO/UNU (2007); nd, not determined.

acids (Ferreira *et al.*, 2008). The amino acid profile in *M. oleifera* leaf and pod are within the limits stipulated by FAO/WHO/UNU (2007) for healthy adults. The essential amino acids in *M. oleifera* dried leaf exceed the RDA values for adults, thus making it suitable to replace limiting amino acids in other staples and for application in developing targeted diets and managing malnutrition in areas where food insecurity is a challenge.

Antinutritive Compounds

M. oleifera, despite being a good source of both macro- and micronutrients, contains antinutrients which could impair the bioaccessibility and availability of the nutrients to man and animals after consumption. The detection of alkaloids, oxalates, saponins, hydrogen cyanide and trypsin inhibitors has been reported in *M. oleifera* leaves (Nweze and Nwafor, 2014; Ogbe and Affiku, 2020). A comparison of the phytochemical content of *M. oleifera* with some leafy vegetables is presented in Table 8.6. *M. oleifera* leaves contain higher alkaloids (307 mg/100 g) compared to African spinach (3.54 mg/100 g), fluted pumpkins (2.20 mg/100 g), nightshade (6.6 mg/100 g) and cowpea (9.61 mg/100 g) as shown in Table 8.6 (Kidmose *et al.*, 2006; Moyo *et al.*, 2011; Nweze and Nwafor, 2014; Ogbe and Affiku, 2020). The phytate (2.59%) and tannin content (0.21 mg/g) in the dried leaf (2.59 mg/100 g; 3.12 mg/g) was higher than that in African spinach (1.32 mg/100 g, 0.049 mg/g) and cabbage (0.18 mg/100 g, 0.31 mg/g) while higher values were reported in nightshade and cowpeas (Owolabi *et al.*, 2012; Omenna *et al.*, 2016; Udousoro *et al.*, 2018; Managa *et al.*, 2020a, b). *M. oleifera* leaf contains trypsin inhibitors (3.0%) which could impair its protein digestibility (Gidamis *et al.*, 2003). The total polyphenols (2.2 mg GAE/ 100 g), saponins (1.60 mg/100 g) and trypsin inhibitors (3.0%) were higher than values in African spinach. Phytates and tannins are antinutrients that could bind to minerals thus making them non-bioavailable (Makkar and Becker, 1996). The presence of antinutrients in *M. oleifera* highlights the need for appropriate processing and pretreatment to aid the release of essential nutrients during digestion.

Implications of Food Processing on Nutrients and Phytochemicals

Processing influences the nutrient and antinutrient content of foods and different processing methods have been used to reduce the antinutrient content of vegetables. Antinutrients are chemical compounds that can block the availability and accessibility of food nutrients in the gut. Cooking is a traditional method of food processing and can help to reduce antinutrients in foods. *M. oleifera* parts have been processed by cooking. Gidamis *et al.* (2003) reported the reduction of antinutrients (phytates, trypsin inhibitors and tannins) in *M. oleifera* leaves and pods after cooking in boiling water. Similarly, processing methods such as UV radiation, exposure to high and moderate temperatures, precipitation and solar radiation have been reported to positively influence the bioactive properties of *M. oleifera* leaves (Vázquez-León *et al.*, 2017).

Similarly, Mosha *et al.* (1995) observed the inhibition of amylase and protease enzymes and reduction of the bioavailability of minerals in phytin-phosphorus complexes in foods. About a 24% decrease in phytin-phophorus content was reported in cooked moringa leaf compared to a 2.4% decrease in *M. oleifera* pods (Gidamis *et al.*, 2003). However, cooking in boiling water results in leaching of essential nutrients that could negatively affect the β-carotene and ascorbic acid content of *M. oleifera* leaves. However, the cooking of *M. oleifera* leaves and pods in boiling water was reported to increase the content of reducing sugars (Gidamis *et al.*, 2003), possibly due to the breakdown of complex carbohydrates into simple sugars upon cooking. Boiling was reported to produce a higher decrease in antinutritional factors in *M. oleifera* leaves compared to blanching and simmering methods (Sallau *et al.*, 2012). Fermentation has been used as a strategy to improve the bioaccessibility of nutrients. Furthermore, fermentation of *M. oleifera* leaf causes a reduction in tannins, phytates, trypsin inhibitors, steroids, saponins, terpenoids and

Table 8.6. Phytochemical content of *M. oleifera* leaf and extracts in comparison with some vegetables.

	M. oleifera				African spinach	Cabbage	Fluted pumpkin	Nightshade	Cowpea
Parameters	Wet	Dried	Aqueous extract	Ethanolic extract	Dried	Dried	Dried	Dried	Dried
Condensed tannins (mg/g)	0.21	3.12	9.36	9.19	0.049	0.31	0.48	5.56	3.44
Total polyphenols (mgGAE/100 g)	2.2	10,504	4581	1600	0.35	750	13.6	600.7	88.6
Phytates (mg/100 g)	2.59	103.67	nd	nd	1.32	0.18	21.72	8.8	5.33
Flavonoids (mgQE/g)	nd	12.5	31.28	31.28	0.083	2.61	0.76	1.05	1250
Alkaloids (mg/100 g)	nd	358.33	307	226	3.54	nd	2.20	6.6	9.61
Anthocyanins (mg/100 g)	nd	nd	0.06	0.05	nd	nd	nd	nd	nd
Trypsin inhibitors (%)	3.0	1.76	nd	nd	nd	nd	nd	nd	19.81
Oxalates (%)	0.45	3.13	nd	nd	nd	27.28	0.03	8.80	6.21
Saponins (mg/100 g)	1.60	200.0	146	172	1.68	nd	0.60	nd	7.17
Cardiac glycosides (%)	nd	nd	0.36	0.19	nd	nd	2.54	nd	nd
Anthraquinone (%)	nd	nd	11.68	10.86	nd	nd	nd	nd	nd
Terpenoids (%)	nd	25.5	4.84	4.26	nd	nd	nd	nd	nd
Hydrogen cyanide (%)	0.1	nd	nd	nd	1.69	9.35	4.86	nd	0.37
References	Makkar and Becker (1996); Kidmose *et al.* (2006); Moyo *et al.* (2011); Ijarotimi *et al.* (2013); Nweze and Nwafor (2014); Ogbe and Affiku (2020)				Akubugwo *et al.* (2008)	Kim *et al.* (2006); Prokopov *et al.* (2015); Udousoro *et al.* (2018); Managa *et al.* (2020a)	Madukwe *et al.* (2013); Udousoro *et al.* (2018); Ademiluyi *et al.* (2019); Ebana *et al.* (2019)	Yang *et al.* (2008); Velasco *et al.* (2011); Managa *et al.* (2020b)	Onwuka (2006); Yang *et al.* (2008); Owolabi *et al.* (2012); Omenna *et al.* (2016)

GAE, Gallic acid equivalent; QE, Quercetin equivalent; nd, not determined; mg, milligrams.

alkaloids compared to blanched *M. oleifera* leaves (Ijarotimi *et al.*, 2013). Hence, fermentation could help to aid an increase in nutrient quality and reduce antinutrients.

Phytochemicals

M. oleifera leaf has been reported as a good source of phytochemicals such as vitamin C, carotenoids and tocopherols which are capable of performing as free radical scavengers (DanMalam *et al.*, 2001; Saini *et al.*, 2014). Owing to the phytochemical and nutritional content of *M. oleifera* leaves, ingestion could help to manage immunosuppression in humans (DanMalam *et al.*, 2001). *M. oleifera* leaves contain glucosinolates, carotenoids and phenolic compounds, giving them potential as health promoters (Leone *et al.*, 2015b). Table 8.7 presents the comparative assessment of glucosinolate profile in *M. oleifera* and common vegetable leaves. The glucosinolate profiles in *M. oleifera* consist of glucoiberin (0.05 μmol/ g DW) and glucoraphanin (0.22 μmol/ g DW) – lower levels than that in broccoli, kale, cabbage, cauliflower and sprouts (Kushad *et al.*, 1999; Velasco *et al.*, 2011; Groenbaek *et al.*, 2016; Nugrahedi *et al.*, 2016; Tetteh *et al.*, 2019; Oyeniran *et al.*, 2020). However, the total glucosinolate content (101.54 μmol/ g DW) in *M. oleifera* leaf (Tetteh *et al.*, 2019) was reported higher than that in the assessed brassicas. Notably, *M. oleifera* leaves contain glucomoringin and glucotropaeolin, glucosinalbin, acetyl-4-α-rhamnopyranosyloxy-benzyl glucosinolate isomer I, acetyl-4-α-rhamnopyranosyloxy-benzyl glucosinolate isomer II and acetyl-4-α-rhamnopyranosyloxy-benzyl glucosinolate isomer III which are not detected in broccoli, kale, cabbage, cauliflower and sprouts (Table 8.7).

Carotenoids play a key role in the protection of photosynthetic activities of the plant, preventing the triplet molecules from excitation and reducing the excited energy caused by oxidation to normal status (Vázquez-León *et al.*, 2017). Carotenoids are essential in plants as they protect the leaves from excess radiation caused by heat which could damage the plant cells (Hongbo *et al.*, 2015). The study by Vázquez-León *et al.* (2017) reported that the climatic conditions, the age of the plant and soil parameters influence the bioactive compounds and antiradical activity of *M. oleifera* leaves. The carotenoid profile of *M. oleifera* leaf was compared with that of African spinach, spinach, cowpea and cabbage (Table 8.8). The total carotenoid content of *M. oleifera* (104.56 mg/100 g) was higher than in cabbage and spinach, while its violaxanthin content (18.30 mg/100 g) was higher than all the vegetables considered (Lakshminarayana *et al.*, 2005; Kidmose *et al.*, 2006; Joshi and Mehta, 2010). *M. oleifera* leaf contains zeaxanthin (4.13 mg/100 g) at a higher level than cowpea and cabbage. The lutein (35.05 mg/100 g) content is higher than spinach and cabbage. *M. oleifera* leaf has beta-carotene content that is higher than spinach, cowpea and cabbage. Hence. *M. oleifera* is a good source of carotenoids which could function as antioxidants in foods.

The phytochemical composition of *M. oleifera* is influenced by the stage of its maturity, germplasm and climatic conditions (Nouman *et al.*, 2014a). Vázquez-León *et al.* (2017) reported the detection of total phenolics, gallic acid and total carotenoids in *M. oleifera* leaves. Datasets on the phenolic contents of *M. oleifera* leaves have been limited to the Ghanaian and Malaysian cultivars which have improved the plant's utilization in the development of food and pharmaceutics (Bennett *et al.*, 2003). Therefore, the phenolic profiles in other cultivars of *M. oleifera* will be critical to improve the plant's use in the pharmaceutical and food industries. *M. oleifera* leaves are known as a rich source of kaempferol, rutin, quercetin, flavonoids and phenolics (Dillard and German, 2000; Siddhuraju and Becker, 2003). The leaf of *M. oleifera* largely contains β-carotene, quercetin, rutin, kaempferol, ellagic acid, epicathechins and 3-O-p-coumaroylquinic acid among others (Kumar and Pandey, 2013; Tumer *et al.*, 2015; Tetteh *et al.*, 2019; Oyeniran *et al.*, 2020) as presented in Table 8.9. The total phenol (107 mg GAE/g) and flavonoid content (21.53 mg/QUE/kg) in *M. oleifera* leaf has been shown to be higher than in jute mallow, fluted pumpkin, cowpea leaves, cabbage and spinach.

Table 8.7. Comparative glucosinolate content of common vegetables.

Glucosinolate (µmol/ g DW)	*M. oleifera*	Broccoli	Brussels sprouts	Cauliflower	Cabbage	Kale
Glucoiberin	0.05	0.1	nd	nd	2.46	1.99
Glucomoringin (µmol/100 g)	0.75	nd	nd	nd	nd	nd
Glucotropaeolin	0.15	nd	nd	nd	nd	nd
Glucoraphanin	0.22	7.1	1.0	nd	4.75	0.73
Glucosinalbin	0.84	nd	nd	nd	nd	nd
Gluconapin	nd	1.0	6.9	0.3	0.42	1.0
Glucosoonjnain	1.04	nd	nd	nd	nd	nd
Glucobrassicin	nd	0.3	0.5	0.1	4.74	4.11
Sinigrin	nd	0.1	8.9	9.3	7.8	10.4
Sinapic acid	nd	nd	1.91	nd	1.80	2.42
4-Methoxy-glucobrassicin	nd	0.4	0.4	1.0	0.36	0.2
4-OH glucobrassicin	nd	0.2	0.6	1.6	0.33	0.1
1-sinapoyl-2-feruloylgentiobioside	nd	nd	nd	nd	3.26	4.51
acetyl-4-α-rhamnopyranosyloxy-benzyl glucosinolate isomer I	2.46	nd	nd	nd	nd	nd
acetyl-4-α-rhamnopyranosyloxy-benzyl glucosinolate isomer II	1.25	nd	nd	nd	nd	nd
acetyl-4-α-rhamnopyranosyloxy-benzyl glucosinolate isomer III	9.46	nd	nd	nd	nd	nd
Total glucosinolates	101.54	12.8	25.1	15.1	12.75	15.0
References	Bennett *et al*. (2003); Bajpai *et al*. (2005); Mukunzi *et al*. (2011); Maldini *et al*. (2014); Chodur *et al*. (2018); Tetteh *et al*. (2019); Oyeniran *et al*. (2020)	Kushad *et al*. (1999)	Kushad *et al*. (1999)	Kushad *et al*. (1999)	Kushad *et al*. (1999); Velasco *et al*. (2011); Nugrahedi *et al*. (2016); Park *et al*. (2019)	Kushad *et al*. (1999); Velasco *et al*. (2011); Groenbaek *et al*. (2016)

Table 8.8. Comparison of *M. oleifera* leaf carotenoid profiles with some common vegetables.

Carotenoid (mg/100 g)	Zeaxanthin	Lutein	Violaxanthin	Neoxanthin	β-carot	T-carot	References
M. oleifera	4.13	35.05	18.30	9.60	37.8	104.56	Lakshminarayana *et al*. (2005); Kidmose *et al*. (2006); Joshi and Mehta (2010)
Spinach *(Amaranthus hybridus)*	4.84	18.34	15.51	10.77	31.5	87.25	Bunea *et al*. (2008); Sarker and Oba (2020)
African spinach	nd	30.92	nd	nd	17.4	113.6	Van der Walt *et al*. (2009); Okpalanma and Ojimelukwe (2017); Schröter *et al*. (2018)
Cowpea	0.04	109.1	17.8	nd	3.2	222.7	Andarwulan *et al*. (2012); Moloto *et al*. (2020)
Cabbage	0.012	0.35	0.082	nd	0.5	0.74	Singh *et al*. (2006); Park *et al*. (2019)

T-carot, total carotenoids; β-carot, beta-carotenoids; nd, not detected.

Iso-quercitrin was the most abundant phenolic in *M. oleifera* leaf, however, its ferulic acid content (0.078 mg GAE/g) was lower than in jute mallow, cowpea, fluted pumpkin and spinach (Table 8.9). Similarly, the catechin content was lower than in fluted pumpkin and cowpea but higher than in cabbage. The quercitrin (63.19 mg GAE/g) and kaempferol (58.15 mg GAE/g) content of *M. oleifera* leaf was higher than in jute mallow, fluted pumpkin, cowpea leaves, cabbage and spinach, however, jute mallow had higher cirsiliol and naringic content than *M. oleifera* leaf (Maseko *et al*., 2019; Abiola *et al*., 2016). The flower of *M. oleifera* has been reported to contain palmitic acid and phytosterols (Inbathamizh and Padmini, 2012) while its seeds contain lectines, myricetin and tocopherols (Singh *et al*., 2013). *M. oleifera* flower is rich in polyphenols (Alhakmani *et al*., 2013) and could find application as a natural antioxidant to deter the initiation and progression of oxidative degenerative diseases. The flower is rich in alpha- and gamma-tocopherols, flavonoids (Makkar and Becker, 1996), rhamnetin, quercetin, isoquercitrin, kaempferat, kaempherol and kaempferitrin (Siddhuraju and Becker, 2003).

Bioactive Properties

Antioxidative properties

According to Moyo *et al*. (2012), *M. oleifera* is a known source of dietary antioxidants comparably higher than some vegetables. As shown in Table 8.10, the aqueous and ethanolic extract of *M. oleifera* leaf (39.2 mgTE/g FW) has higher DPPH (2,2-diphenyl-1-picrylhydrazyl) antioxidant power than extracts from jute mallow (*Cochorus olitorius* L.), water leaf (*Talium triangulare*), spinach (*Amaranthus hybridus*), fluted pumpkin (*Telfaira occidentalis*) and cabbage (*Brassica oleracea* var. capitata). Likewise, the ABTS (2,2'-azino-bis3-ethylbenzothiazoline-6-sulfonic acid) (6.72 and 37.17 μmol TE/g FW) and FRAP (Ferric ion reducing antioxidant power) (44.8 μmol TE/g FW; 58.26 μmol TE/g FW) activities of aqueous and ethanolic extract of *M. oleifera* leaf (Braham *et al*., 2020; Rocchetti *et al*., 2020) were higher than waterleaf, fluted pumpkin and cabbage extract, but not spinach and jute mallow leaves. Hence, *M. oleifera* leaf could deliver good antioxidant activity and could be used as a functional ingredient in food that assists in the scavenging of generated free radicals.

Table 8.9. Phenolic profiles of *M. oleifera* leaves in comparison with some common vegetables (mg/g).

Phenolic	*M. oleifera*	Jute mallow	Fluted pumpkin	Cowpea	Cabbage	Spinach
Gallic acid	1.034	0.083	22.19	6.7	nd	0.041
Ellagic acid	0.018	nd	nd	nd	nd	nd
Epicatechins	16.67	nd	nd	nd	0.034	nd
Ferulic acid	0.078	3.92	0.94	0.014	0.067	0.316
Quinic acid	nd	nd	nd	nd	nd	nd
Catechins	10.08	nd	29.17	36.86	0.011	nd
Caffeic acid	0.409	0.31	9.17	0.020	0.005	0.094
p-coumaric acid	nd	nd	nd	nd	nd	0.005
3-*O*-p-coumaroylquinic acid	6.457	2.71	nd	nd	nd	1.48
Myricetin	nd	nd	nd	nd	nd	0.076
Iso-quercitrin	61.52	0.045	nd	nd	nd	nd
Rutin	17.05	nd	0.463	nd	0.005	1.04
Chlorogenic acid	0.489	nd	0.482	0.042	0.037	0.26
Phlorodzin	nd	nd	nd	nd	nd	0.377
Gallotannins	nd	9.52	nd	14.22	nd	nd
3-caffeoyl-quinic acid	8.9	nd	nd	nd	nd	nd
5-caffeoyl-quinic acid	1.1	nd	nd	nd	nd	nd
Total phenols (mg GAE/g)	107	11.21	29.06	9.87	1.49	7.35
Flavonoids						
Naringin	nd	27.32	nd	nd	nd	nd
Cirsiliol	nd	183.30	nd	nd	nd	nd
Kaempferol	58.15	0.016	0.483	0.033	6.79	0.16
kaempferol3-O-Glc	0.06	nd	nd	nd	nd	nd
kaempferol3-O-(6′′-malonyl-Glc)	0.08	nd	nd	nd	nd	nd

Continued

Table 8.9. Continued

Phenolic	*M. oleifera*	Jute mallow	Fluted pumpkin	Cowpea	Cabbage	Spinach
Qercetin3-O-Glc	0.21	nd	nd	nd	nd	nd
Quercetin3-O-(6′′-malonyl-Glc)	0.25	nd	nd	nd	nd	nd
Quercitrin	63.19	3.13	nd	0.27	0.79	0.73
Total flavonoid (mg QUE/kg)	21.53	5.10	0.30	3.76	nd	4.57
References	Bennett *et al*. (2003); Bajpai *et al*. (2005); Mukunzi *et al*. (2011); Maldini *et al*. (2014); Chodur *et al*. (2018); Tetteh *et al*. (2019); Oyeniran *et al*. (2020)	Abiola *et al*. (2016); Maseko *et al*. (2019)	Adefegha and Oboh (2011); Salawu *et al*. (2013); Eseyin *et al*. (2018)	Van der Walt *et al*. (2009); Deng *et al*. (2013); Maseko *et al*. (2019)	Kim *et al*. (2006); Nugrahedi *et al*. (2016); Park *et al*. (2019)	Santos *et al*. (2014); Santiago-Saenz *et al*. (2018); Manyelo *et al*. (2020)

Nd, not detected.

Table 8.10. Comparative antioxidants properties of *M. oleifera* leaf compared to some vegetables.

	DPPH (mgTE/g FW)		ABTS (µmol TE/g FW)		FRAP (µmol TE/g FW)		
Vegetable	Aqueous extract	Ethanolic extract	Aqueous extract	Ethanolic extract	Aqueous extract	Ethanolic extract	References
Moringa (*M. oleifera*)	39.2	45.38	6.72	37.17	44.8	58.26	Braham *et al*. (2020); Rocchetti *et al*. (2020)
Spinach (*Amaranthus hybridus*)	0.423	0.138	144.24	336	2450	2320	Nana *et al*. (2012); Kraujalis *et al*. (2013); Jiménez-Aguilar and Grusak (2017)
Jute mallow (*Cochorus olitorius* L.)	0.0126	0.0486	139.55	33.19	27.52	13.16	Youssef *et al*. (2014); Abiola *et al*. (2016); Chigurupati *et al*. (2020)
Water leaf (*Talium triangulare*)	7.4	29	1.03	2.17	28.3	0.2	Oboh *et al*. (2008); Andarwulan *et al*. (2010); Adefegha and Oboh (2011); Liao *et al*. (2015); Abiola *et al*. (2016);
Fluted pumpkin (*Telfaira occidentalis*)	0.69	0.12	2.5	1.0	0.5	0.35	Oboh *et al*. (2008); Adefegha and Oboh (2011)
Cabbage (*Brassica oleracea* var. capitata)	0.25	0.31	0.35	0.45	0.91	2.0	Rubab *et al*. (2018); Oancea *et al*. (2019); Šola *et al*. (2020)

mg TE/g FW, milligram of trolox equivalent per gram fresh weight (FW); µmol TE/g FW, micromolar trolox equivalent per gram FW.

Antidiabetic Properties

The physiological conditions associated with the conversion of blood sugars to stored fat have increased in recent times among the population with cases rising yearly. Diabetes mellitus is an endocrinopathic condition characterized by hyperglycaemia as a result of defects in insulin secretion, resistance to insulin or destruction of beta cells in the pancreas (Yoon *et al.*, 2020). Diabetes is associated with kidney failure, cardiovascular disease, eye problems and stroke. It is a serious health problem that places a burden on the health and finances of patients and accounts for over 1 million deaths annually (Ramtahal *et al.*, 2015; Khan *et al.*, 2020). According to Khan *et al.* (2020), reporting on the epidemiology of diabetes, in 2017 alone over 462 million people were diabetic, with a projected increase to about 7079 persons per 100,000 population in 2030. Hence, there is a need to seek alternatives to manage this pathological condition. *M. oleifera* is a plant laden with potential to alter the effect of pathological conditions such as diabetes due to its phytochemical content.

Several studies have reported the antidiabetic properties of *M. oleifera* in diabetic adult rats and linked the biological activities to the antioxidative effect of the plant's phenolic compounds (Yassa and Tohamy, 2014). Similarly, a postprandial blood glucose level reduction was reported upon the oral administration of *M. oleifera* extract in diabetic rats (Jaiswal *et al.*, 2009). Khan *et al.* (2020) demonstrated the potential of aqueous *M. oleifera* leaf extract to cause a lowering of hyperglycaemia in patients using the metabolomics approach (Khan *et al.*, 2017). The antidiabetic potential of *M. oleifera* root and leaves was reported in an *in vitro* assay by evaluating alpha glucosidase and alpha amylase inhibitory activities. In their study, Tshabalala *et al.* (2020) reported a higher antidiabetic activity in *M. oleifera* leaves than its roots. The enzyme alpha amylase breaks down complex carbohydrates (starch) into simple sugar (glucose) which in turn enables glucose clearance from the bloodstream. However, when an inhibition of the alpha amylase occurs, it reduces the postprandial glycaemia (Sangeetha and Vedasree, 2012).

Hypolipidemic and Hepatoprotective Properties

M. oleifera has been reported to act against obesity in high fat fed rats (Nahar *et al.* (2016). Obesity is associated with diabetes and causes the accumulation of fat which can lead to the blockage of blood vessels (atherosclerosis), the build-up of tension in the blood vessels (hypertension) and cardiovascular disease (Azimi *et al.*, 2013). Organic extract of *M. oleifera* leaf hinders body and organ weight gain, and reduces glucose retention, triglycerides and total cholesterol levels in the blood. Chen *et al.* (2020) reported the hypoglycemic properties of its leaves. *M. oleifera* leaf was also reported to have a hepatoprotective effect against non-alcoholic fatty liver disease in adult obese rats (Asgari-Kafrani *et al.*, 2020) and to down-regulate the expression of adipogenesis-associated proteins such as leptin, adiponectin and resistin which are determinants for energy use and secretion of insulin in the body (Catalano *et al.*, 2010; Singh *et al.*, 2020), important for body weight loss. The study of Xie *et al.* (2018) reported the up-regulation of the lipolysis-associated genes found in the 3T3-L1 adipocyte. *M. oleifera* leaf has a hypolipidemic effect due to its functional phytochemical components (Chen *et al.*, 2020).

Antimicrobial Properties

The antimicrobial activity of *M. oleifera* plants has been reported against pathogens of clinical health importance. The study of Rahman *et al.* (2009) reported the antimicrobial activities of aqueous and leaf extracts of *M. oleifera* against bacterial pathogens such as *Streptococcus-B-haemolytica*, *Staphylococccus aureus*, *Bacillus subtilis*, *Shigella* spp., *Pseudomonas aeruginosa*, *Shigella sonnei*, *Sarcina lutea*, *Bacillus cereus* and *Bacillus megaterium*. *Salmonella typhi*, *Pseudomonas aeruginosa*, *Staphylococcus aureus*, *Enterobacter* spp. and *Escherichia coli* have been reported as sensitive to 200 mg/l aqueous extract of *M. oleifera* leaf (Nepolean *et al.*, 2009). The antifungal activity of *M. oleifera* leaf extract against anthracnose-causing fungal pathogens (*Colletotrichum gloesporioides*) growing on Maluma avocado was reported by Kubheka *et al.*

(2020). In their study, 10 and 15% of *M. oleifera* extract incorporated with gum arabic caused the inhibition of radial mycelial growth of *Collectrichum gloeosporoides*, thus suggesting the fungicidal potential of *M. oleifera* leaf extract. Likewise the antifungal activity of ethanolic extract of *M. oleifera* was reported against *Penicillum* spp., *Mucor* spp. and *Candida albicans* (Raj *et al.*, 2011), thus suggesting its potential to control some postharvest fungal pathogens and fruit diseases such as anthracnose which causes postharvest loses and production of mycotoxins. Therefore, *M. oleifera* plants have the potential to be used as biocontrol agents against notorious pathogens that threaten agricultural productivity, food security and safety.

Health Benefits

The study of Saini *et al.* (2014) reported the effectiveness of *M. oleifera* to manage iron deficiency compared to iron supplements. In their study, *M. oleifera* leaf was found to modulate the expression of iron-responsive genes. *M. oleifera* leaves increased the growth parameters, organ weight and physiological parameters in animals.

Besides improvement in growth, the leaf extract is effective against leukaemia and hepatocarcinoma and can ameliorate cadmiu-induced toxicities (Mallya *et al.*, 2017). *M. oleifera* leaves have been reported to have a nephroprotective and hepatoprotective effect against antibiotic-caused toxicities (Sharifudin *et al.*, 2013; El-Hadary and Ramadan, 2019). The study of Sharifudin *et al.* (2013) reported the ameliorating effect of *M. oleifera* leaves against acetaminophen-induced hepatotoxicity in rats while El-Hadary and Ramadan (2019) reported a similar observation against diclofenac sodium-induced liver toxicity in rats.

The neuroprotective effect of *M. oleifera* leaf extract is demonstrated by a clear lowering of oxidative induced stress (Stohs and Hartman, 2015). In the study of Mallya *et al.* (2017) conducted on Wistar rats, *M. oleifera* leaf extract showed a reversal effect of cadmium-induced toxicity on the erythropoiesis, liver necrosis and intracellular enzymes such as alanine aminotransferase (ALT) and aspartate aminotransferase (AST). The study revealed that *M. oleifera* had a therapeutic effect against hepatotoxicity that is often portrayed by the increase in plasma ALT and AST liver enzymes (Mallya *et al.*, 2017). Cadmium is capable of inhibiting the antioxidant activities of gluthatione peroxidase and superoxide dismutase enzymes but promotes lipid peroxidation which results in release of superoxides that cause oxidation (Amara *et al.*, 2011). Therefore abnormalities in the lipoprotein encourage the generation of cardiovascular diseases and atherosclerosis (Glew *et al.*, 2002). The administration of *M. oleifera* leaf extract was reported to reverse the cadmium-triggered hypothalamus-pituitary-gonadal axis destruction which often leads to damage in the sex organs (Stohs *et al.*, 2000; Mallya *et al.*, 2017).

Novel Products/Formulations

M. oleifera leaves are edible and can be eaten as salads and soups (Foidl *et al.*, 2001). The green pods can be boiled and eaten like green beans. However, the hulls are not desirable as food due to their bitter taste. The seeds are ground into sauce seasoning powders. The flowers can be eaten raw, blanched and used as an additive to salads. *M. oleifera* roots have been used in the production of pickles and sauces. *M. oleifera* leaf is used in the production of yoghurts (Hekmat *et al.*, 2015), smoothies (Aderinola, 2018), soup mixes (Babayeju *et al.*, 2014), dough meal (Karim *et al.*, 2015), confections (Kolawole *et al.*, 2013; Chinma *et al.*, 2014), medicinal cookies (Alam *et al.*, 2014) and weaning food (supplemented with *M. oleifera* leaf powder) (Boateng *et al.*, 2018, 2019). *M. oleifera* leaf has been used as an ingredient in food formulations to manage malnutrition among children (Boateng *et al.*, 2019) and caregivers (Boateng *et al.*, 2018). Badejo *et al.* (2014) reported the incorporation of *M. oleifera* leaf extract with *Cyperus esculentus* and *Hisbiscus sabdariffa* in the production of beverages. Similarly, *M. oleifera* leaf extract has been used to fortify South Indian instant soup mixes (Chan *et al.*, 2020).

Conclusions and Future Trends

M. oleifera is a versatile plant with great potential for use as a food and a nutraceutical. This review has reported the nutritive quality of *M. oleifera* leaf in comparison with common vegetables, its biological activities and its application in diverse fields (food, water treatment, medicinal use, etc.). However, *M. oleifera* is still under-exploited as a functional food. It could find application in nutrigenomics where genetic application is used to develop nutrition-based solutions to solve metabolism problems. Several researchers have focused on the nutritional content, bioactive properties and photochemistry. *M. oleifera* still lacks a standard preparation and standardization of dose to prevent toxicities that could emanate from its indiscriminate use. Hence, further studies to determine adequate doses and upscaling its processing will be important. Moreover, the commercialization of recipes for industrial-scale processing will be important to increase access to and utilization of *M. oleifera* in areas where it is not typically grown. Further studies are needed to understand the nutrient diversity of *M. oleifera* from different locations using molecular tools such as next generation sequencing (NGS). This will enable a better understanding of the genetic make-up of the plant, leading to effective genetic manipulation and utilization. The pharmacological properties of *M. oleifera* have been reported using both *in vitro* and *in vivo* methods, however, few studies have reported its use in clinical trials, hence knowledge can be advanced by conducting clinical trials using humans as subjects.

Acknowledgement

The authors acknowledge the support of the SARCHl Research Chair, Phytochemical Food Network to Improve Nutrition Quality for Consumer, under the National Research Foundation (Grant number 98352), Tshwane University of Technology, Pretoria, South Africa.

References

Abdulkadir, A.R., Zawawi, D.D. and Jahan, M.S. (2016) Proximate and phytochemical screening of different parts of *Moringa oleifera*. *Russian Agricultural Sciences* 42(1), 34–36. DOI: 10.3103/S106836741601002X.

Abe, R. and Ohtani, K. (2013) An ethnobotanical study of medicinal plants and traditional therapies on Batan Island, the Philippines. *Journal of Ethnopharmacology* 145(2), 554–565. DOI: 10.1016/j.jep.2012.11.029.

Abiola, M., Ighodaro, O.M., Aminu, A.O. and Ogunlana, A.I. (2016) The antioxidant and phenolic profiles of five green vegetables grown in Southern Nigeria. *Acta Scientiarum Polonorum. Technologia Alimentaria* 15(4), 391–397. DOI: 10.17306/J.AFS.2016.4.37.

Acikgoz, F.E. (2011) Mineral, vitamin C and crude protein contents in kale (*Brassica oleraceae* var. *acephala*) at different harvesting stages. *African Journal of Biotechnology* 10(75), 17170–17174.

Adefegha, S.A. and Oboh, G. (2011) Enhancement of total phenolics and antioxidant properties of some tropical green leafy vegetables by steam cooking. *Journal of Food Processing and Preservation* 35(5), 615–622. DOI: 10.1111/j.1745-4549.2010.00509.x.

Adelakun, O.E., Ade-Omowaye, B.I.O., Adeyemi, I.A. and Van de Venter, M. (2012) Mineral composition and the functional attributes of Nigerian okra seed (*Abelmoschus esculentus* Moench) flour. *Food Research International* 47(2), 348–352. DOI: 10.1016/j.foodres.2011.08.003.

Ademiluyi, A.O., Oyeniran, O.H., Jimoh, T.O., Oboh, G. and Boligon, A.A. (2019) Fluted pumpkin (*Telfairia occidentalis*) seed modulates some markers of erectile function in isolated rat's corpus cavernosum: Influence of polyphenol and amino acid constituents. *Journal of Food Biochemistry* 43(11), e13037. DOI: 10.1111/jfbc.13037.

Aderinola, T.A. (2018) Nutritional, antioxidant and quality acceptability of smoothies supplemented with *Moringa oleifera* leaves. *Beverages* 4(4), 104. DOI: 10.3390/beverages4040104.

Afuang, W., Siddhuraju, P. and Becker, K. (2003) Comparative nutritional evaluation of raw, methanol extracted residues and methanol extracts of moringa (Moringa oleifera Lam.) leaves on growth performance and feed utilization in Nile tilapia (Oreochromis niloticus L). *Aquaculture Research* 34(13), 1147–1159. DOI: 10.1046/j.1365-2109.2003.00920.x.

Afzal, S., Nawaz, M.F., Qadir, I., Gul, S., Yasin, G. *et al.* (2020) Variability in leaf mineral composition of *Moringa oleifera* in irrigated plains of Pakistan. *South African Journal of Botany* 129, 442–447. DOI: 10.1016/j.sajb.2019.11.025.

Ahmed, K.S., Jahan, I.A., Hossain, M.H., Ethane, N.J. and Saha, B. (2018) Mineral and trace element content in different parts of *Moringa oleifera* grown in Bangladesh. *Current Journal of Applied Science and Technology* 31(5), 1–10. DOI: 10.9734/CJAST/2018/45645.

Akubugwo, I.E., Obasi, N.A., Chinyere, G.C. and Ugbogu, A.E. (2008) Mineral and phytochemical contents in leaves of *Amaranthus hybridus* L and *Solanum nigrum* L. subjected to different processing methods. *African Journal of Biochemistry Research* 2(2), 040–044.

Al-Sayed, H.M., Hegab, S.A., Youssef, M.A., Khalafalla, M.Y., Almaroai, Y.A, *et al.* (2020) Evaluation of quality and growth of roselle (Hibiscus sabdariffa L.) as affected by bio-fertilizers. *Journal of Plant Nutrition* 43(7), 1025–1035. DOI: 10.1080/01904167.2020.1711938.

Alam, Md.A., Alam, M., Hakim, M., Huq, A.O. and Moktadir, S.G. (2014) Development of fiber enriched herbal biscuits: A preliminary study on sensory evaluation and chemical composition. *International Journal of Nutrition and Food Sciences* 3(4), 246–250. DOI: 10.11648/j.ijnfs.20140304.13.

Alhakmani, F., Kumar, S. and Khan, S.A. (2013) Estimation of total phenolic content, in-vitro antioxidant and anti-inflammatory activity of flowers of *Moringa oleifera*. *Asian Pacific Journal of Tropical Biomedicine* 3(8), 623–627. DOI: 10.1016/S2221-1691(13)60126-4.

Amaglo, N.K., Bennett, R.N., Lo Curto, R.B., Rosa, E.A.S., Lo Turco, V. *et al.* (2010) Profiling selected phytochemicals and nutrients in different tissues of the multipurpose tree *Moringa oleifera* L., grown in Ghana. *Food Chemistry* 122(4), 1047–1054. DOI: 10.1016/j.foodchem.2010.03.073.

Amara, S., Douki, T., Garrel, C., Favier, A., Ben Rhouma, K. *et al.* (2011) Effects of static magnetic field and cadmium on oxidative stress and DNA damage in rat cortex brain and hippocampus. *Toxicology and Industrial Health* 27(2), 99–106. DOI: 10.1177/0748233710381887.

Amoasah, B., Appiah, F. and Kumah, P. (2018) Effects of different drying methods on proximate composition of three accessions of roselle (*Hibiscus Sabdariffa*) Calyce. *International Journal of Plant & Soil Science* 21(1), 1–8. DOI: 10.9734/IJPSS/2018/38550.

Andarwulan, N., Batari, R., Sandrasari, D.A., Bolling, B. and Wijaya, H. (2010) Flavonoid content and antioxidant activity of vegetables from Indonesia. *Food Chemistry* 121(4), 1231–1235. DOI: 10.1016/j.foodchem.2010.01.033.

Andarwulan, N., Kurniasih, D., Apriady, R.A., Rahmat, H., Roto, A.V. *et al.* (2012) Polyphenols, carotenoids, and ascorbic acid in underutilized medicinal vegetables. *Journal of Functional Foods* 4(1), 339–347. DOI: 10.1016/j.jff.2012.01.003.

Anwar, F., Latif, S., Ashraf, M. and Gilani, A.H. (2007) *Moringa oleifera*: a food plant with multiple medicinal uses. *Phytotherapy Research* 21(1), 17–25. DOI: 10.1002/ptr.2023.

Asgari-Kafrani, A., Fazilati, M. and Nazem, H. (2020) Hepatoprotective and antioxidant activity of aerial parts of *Moringa oleifera* in prevention of non-alcoholic fatty liver disease in Wistar rats. *South African Journal of Botany* 129, 82–90. DOI: 10.1016/j.sajb.2019.01.014.

Asiedu-Gyekye, I.J., Frimpong-Manso, S., Awortwe, C., Antwi, D.A. and Nyarko, A.K. (2014) Micro- and macroelemental composition and safety evaluation of the nutraceutical *Moringa oleifera* leaves. *Journal of Toxicology* 2014, 1–13. DOI: 10.1155/2014/786979.

Atta, A., Diallo, A.B., Bakasso, Y., Sarr, B., Saadou, M. *et al.* (2010) Micro-element contents of roselle (*Hibiscus sabdariffa* L.) at different growth stages. *African Journal of Food, Agriculture, Nutrition and Development* 10(5). DOI: 10.4314/ajfand.v10i5.57898.

Awodele, O., Oreagba, I.A., Odoma, S., da Silva, J.A.T. and Osunkalu, V.O. (2012) Toxicological evaluation of the aqueous leaf extract of *Moringa oleifera* Lam. (Moringaceae). *Journal of Ethnopharmacology* 139(2), 330–336. DOI: 10.1016/j.jep.2011.10.008.

Azimi, A., Charlot, M.G., Torp-Pedersen, C., Gislason, G.H., Køber, L. *et al.* (2013) Moderate overweight is beneficial and severe obesity detrimental for patients with documented atherosclerotic heart disease. *Heart* 99(9), 655–660. DOI: 10.1136/heartjnl-2012-303066.

Babayeju, A., Gbadebo, C., Obalowu, M., Otunola, G., Nmom, I. *et al.* (2014) Comparison of organoleptic properties of egusi and efo riro soup blends produced with moringa and spinach leaves. *Food Science and Quality Management* 28, 15–18.

Badejo, A.A., Damilare, A. and Ojuade, T.D. (2014) Processing effects on the antioxidant activities of beverage blends developed from *Cyperus esculentus*, *Hibiscus sabdariffa*, and *Moringa oleifera* extracts. *Preventive Nutrition and Food Science* 19(3), 227–233. DOI: 10.3746/pnf.2014.19.3.227.

Bajpai, M., Pande, A., Tewari, S.K. and Prakash, D. (2005) Phenolic contents and antioxidant activity of some food and medicinal plants. *International Journal of Food Sciences and Nutrition* 56(4), 287–291. DOI: 10.1080/09637480500146606.

Banjo, O.S. (2012) Growth and performance as affected by inclusion of *Moringa oleifera* leaf meal in broiler chicks diet. *Growth* 2(9), 35–38.

Bennett, R.N., Mellon, F.A., Foidl, N., Pratt, J.H., Dupont, M.S. *et al.* (2003) Profiling glucosinolates and phenolics in vegetative and reproductive tissues of the multi-purpose trees *Moringa oleifera* L. (horseradish tree) and *Moringa stenopetala* L. *Journal of Agricultural and Food Chemistry* 51(12), 3546–3553. DOI: 10.1021/jf0211480.

Bhattacharjee, S., Dasgupta, P., Paul, A.R., Ghosal, S., Padhi, K.K. *et al.* (1998) Mineral element composition of spinach. *Journal of the Science of Food and Agriculture* 77(4), 456–458. DOI: 10.1002/(SICI)1097-0010(199808)77:4<456::AID-JSFA55>3.0.CO;2-M.

Boateng, L., Nyarko, R., Asante, M. and Steiner-Asiedu, M. (2018) Acceptability of complementary foods that incorporate *Moringa oleifera* leaf powder among infants and their caregivers. *Food and Nutrition Bulletin* 39(1), 137–148. DOI: 10.1177/0379572117708656.

Boateng, L., Nortey, E., Ohemeng, A.N., Asante, M. and Steiner-Asiedu, M. (2019) Sensory attributes and acceptability of complementary foods fortified with Moringa oleifera leaf powder. *Nutrition & Food Science* 49(3), 393–406. DOI: 10.1108/NFS-07-2018-0192..

Braham, F., Carvalho, D.O., Almeida, C.M.R., Zaidi, F., Magalhães, J.M.C.S. *et al.* (2020) Online HPLC-DPPH screening method for evaluation of radical scavenging phenols extracted from *Moringa oleifera* leaves. *South African Journal of Botany* 129, 146–154. DOI: 10.1016/j.sajb.2019.04.001.

Bunea, A., Andjelkovic, M., Socaciu, C., Bobis, O., Neacsu, M. *et al.* (2008) Total and individual carotenoids and phenolic acids content in fresh, refrigerated and processed spinach (*Spinacia oleracea* L.). *Food Chemistry* 108(2), 649–656. DOI: 10.1016/j.foodchem.2007.11.056.

Catalano, K.J., Stefanovski, D. and Bergman, R.N. (2010) Critical role of the mesenteric depot versus other intra-abdominal adipose depots in the development of insulin resistance in young rats. *Diabetes* 59(6), 1416–1423. DOI: 10.2337/db08-0675.

Chan, Y.K.K., Gurumeenakshi, G., Varadharaju, N., Cheng, Y.-L. and Diosady, L.L. (2020) Debittering moringa oleifera (lam.) leaves in fortified south indian instant soup. *Chemosensory Perception* 14(1), 11–18. DOI: 10.1007/s12078-020-09280-1.

Chantaro, P., Devahastin, S. and Chiewchan, N. (2008) Production of antioxidant high dietary fiber powder from carrot peels. *LWT - Food Science and Technology* 41(10), 1987–1994. DOI: 10.1016/j.lwt.2007.11.013.

Chatepa, L.E.C. and Mbewe, E.C. (2018) Proximate, physical and chemical composition of leaves and seeds of Moringa (*Moringa oleifera*) from Central Malawi: A potential for increasing animal food supply in the 21st century. *African Journal of Agricultural Research* 13(51), 2872–2880. DOI: 10.5897/AJAR2018.13535.

Chen, G.L., Xu, Y.B., Wu, J.L., Li, N. and Guo, M.Q. (2020) Hypoglycemic and hypolipidemic effects of *Moringa oleifera* leaves and their functional chemical constituents. *Food Chemistry* 333, 127478. DOI: 10.1016/j.foodchem.2020.127478.

Chigurupati, S., Aladhadh, H.S., Alhowail, A., Selvarajan, K.K. and Bhatia, S. (2020) Phytochemical composition, antioxidant and antidiabetic potential of methanolic extract from Corchorus olitorius Linn. grown in Saudi Arabia. *Medicinal Plants - International Journal of Phytomedicines and Related Industries* 12(1), 71. DOI: 10.5958/0975-6892.2020.00010.6.

Chikwendu, J.N., Igbatim, A.C. and Obizoba, I.C. (2014) Chemical composition of processed cowpea tender leaves and husks. *International Journal of Scientific and Research Publications* 4(5), 1–5.

Chinma, C.E., Abu, J.O. and Akoma, S.N. (2014) Effect of germinated tigernut and *Moringa* flour blends on the quality of wheat-based bread. *Journal of Food Processing and Preservation* 38(2), 721–727. DOI: 10.1111/jfpp.12023.

Chodur, G.M., Olson, M.E., Wade, K.L., Stephenson, K.K., Nouman, W. *et al.* (2018) Wild and domesticated *Moringa oleifera* differ in taste, glucosinolate composition, and antioxidant potential, but not myrosinase activity or protein content. *Scientific Reports* 8(1), 1–10. DOI: 10.1038/s41598-018-26059-3.

Clement, A., Olatunde, M., Patrick, O. and Joyce, O. (2017) Effect of drying temperature on nutritional content of *Moringa oleifera* leave. *World Journal of Food Science and Technology* 1(3), 93–96.

DanMalam, H., Abubakar, Z. and Katsayal, U.A. (2001) Pharmacognosostic studies on the leaves of *Moringa oleifera*. *Nigerian Journal of Natural Products and Medicine* 5(1), 45–49. DOI: 10.4314/njnpm.v5i1.11723.

de Azevedo, C.H. and Rodriguez-Amaya, D.B. (2005) Carotenoid composition of kale as influenced by maturity, season and minimal processing. *Journal of the Science of Food and Agriculture* 85(4), 591–597. DOI: 10.1002/jsfa.1993.

Deng, G.-F., Lin, X., Xu, X.-R., Gao, L.-L., Xie, J.-F. *et al.* (2013) Antioxidant capacities and total phenolic contents of 56 vegetables. *Journal of Functional Foods* 5(1), 260–266. DOI: 10.1016/j.jff.2012.10.015.

Dillard, C.J. and German, J.B. (2000) Phytochemicals: nutraceuticals and human health. *Journal of the Science of Food and Agriculture* 80(12), 1744–1756. DOI: 10.1002/1097-0010(20000915)80:12<1744::AID-JSFA725>3.0.CO;2-W.

Doerr, B. and Cameron, L. (2005) Moringa Leaf Powder. *ECHO Technical Note. USA*. ECHOcommunity.org. Available at: https://assets.echocommunity.org/publication_issue/6187fe3c-57be-4f54-815a-7998c2f78579/en/moringa-leaf-powder.pdf (accessed 29 March 2022).

dos Santos, I.F., dos Santos, A.M.P., Barbosa, U.A., Lima, J.S., dos Santos, D.C. *et al.* (2013) Multivariate analysis of the mineral content of raw and cooked okra (*Abelmoschus esculentus* L.). *Microchemical Journal* 110, 439–443. DOI: 10.1016/j.microc.2013.05.008.

Ebana, R.U.B., Edet, U.O., Anosike, K.I., Etok, C.A. and Kanu, T.O. (2019) Nutritional analysis and wine production potentials of *Telfairia occidentalis* (fluted pumpkin) leaves and *Cucumis sativus* L.(cucumber) using baker's and palm wine yeast strains. *World News of Natural Sciences* 22, 12–30.

El-Hadary, A.E. and Ramadan, M.F. (2019) Antioxidant traits and protective impact of *Moringa oleifera* leaf extract against diclofenac sodium-induced liver toxicity in rats. *Journal of Food Biochemistry* 43(2), e12704. DOI: 10.1111/jfbc.12704.

EL-Qudah, J.M. (2009) Identification and quantification of major carotenoids in some vegetables. *American Journal of Applied Sciences* 6(3), 492–497. DOI: 10.3844/ajassp.2009.492.497.

Emebu, P.K. and Anyika, J.U. (2011) Proximate and mineral composition of kale (*Brassica oleracea*) grown in Delta State, Nigeria. *Pakistan Journal of Nutrition* 10(2), 190–194. DOI: 10.3923/pjn.2011.190.194.

Eseyin, O.A., Sattar, M.A., Rathore, H.A., Aigbe, F., Afzal, S. *et al.* (2018) GC-MS and HPLC profiles of phenolic fractions of the leaf of *Telfairia occidentalis*. *Pakistan Journal of Pharmaceutical Sciences* 31(1), 45–50.

Fahey, J.W. (2005) *Moringa oleifera*: a review of the medical evidence for its nutritional, therapeutic, and prophylactic properties part 1. *Trees for Life Journal* 1(5), 1–15.

FAO/WHO/UNU (Food and Agriculture Organization of the United Nations, World Health Organization & United Nations University) (2007) Joint FAO/WHO/UNU Expert Consultation on Protein and Amino Acid Requirements in Human Nutrition (2002: Geneva, Switzerland). Protein and amino acid requirements in human nutrition: report of a joint FAO/WHO/UNU expert consultation. World Health Organization. Available at: https://apps.who.int/iris/handle/10665/43411 (accessed 29 March 2022).

Ferrao, A.M.B. and Ferrao, J.E. (1970) Fatty acids of the oil of moringueiro (*Moringa oleifera* [*M. pterygosperma*]). *Agron. Angolana* 30, 3–16.

Ferreira, P.M.P., Farias, D.F., Oliveira, J.T. de A. and Carvalho, A. de F.U. (2008) *Moringa oleifera*: bioactive compounds and nutritional potential. *Revista de Nutrição* 21(4), 431–437. DOI: 10.1590/S1415-52732008000400007.

Foidl, N., Makkar, H.P.S. and Becker, K. (2001) The potential of *Moringa oleifera* for agricultural and industrial uses. *What Development Potential for Moringa Products* 20.

Freiberger, C.E., Vanderjagt, D.J., Pastuszyn, A., Glew, R.S., Mounkaila, G. *et al.* (1998) Nutrient content of the edible leaves of seven wild plants from Niger. *Plant Foods for Human Nutrition* 53(1), 57–69. DOI: 10.1023/a:1008080508028.

Fuglie, L. (1999) *The Miracle Tree: Moringa Oleifera: Natural Nutrition for the Tropics*. Church World Service, Dakar, p. 172.

Fuglie, L.J. (2000) New Uses of Moringa Studied in Nicaragua. ECHO Development Notes #68, June, 2000. Available at: https://www.echocommunity.org/ht/resources/15db05d7-4693-425b-9d86-468143bd0ec9 (accessed 29 March 2022).

Ghebremichael, K.A., Gunaratna, K.R., Henriksson, H., Brumer, H. and Dalhammar, G. (2005) A simple purification and activity assay of the coagulant protein from *Moringa oleifera* seed. *Water Research* 39(11), 2338–2344. DOI: 10.1016/j.watres.2005.04.012.

Gidamis, A.B., Panga, J.T., Sarwatt, S.V., Chove, B.E. and Shayo, N.B. (2003) Nutrient and antinutrient contents in raw and cooked young leaves and immature pods Of *Moringa oleifera*, Lam. *Ecology of Food and Nutrition* 42(6), 399–411. DOI: 10.1080/03670240390268857.

Glew, R.H., Kassam, H.A., Bhanji, R.A., Okorodudu, A. and VanderJagt, D.J. (2002) Serum lipid profiles and risk of cardiovascular disease in three different male populations in northern Nigeria. *Journal of Health, Population, and Nutrition* 20(2), 166–174.

Gonçalves, A., Goufo, P., Barros, A., Domínguez-Perles, R., Trindade, H. *et al*. (2016) Cowpea (*Vigna unguiculata* L. Walp), a renewed multipurpose crop for a more sustainable agri-food system: nutritional advantages and constraints. *Journal of the Science of Food and Agriculture* 96(9), 2941–2951. DOI: 10.1002/jsfa.7644.

Groenbaek, M., Jensen, S., Neugart, S., Schreiner, M., Kidmose, U. *et al*. (2016) Nitrogen split dose fertilization, plant age and frost effects on phytochemical content and sensory properties of curly kale (*Brassica oleracea* L. var. sabellica). *Food Chemistry* 197(Pt A), 530–538. DOI: 10.1016/j.foodchem.2015.10.108.

HDRA (2002) Moringa oleifera - A Multipurpose Tree. Produced by the Tropical Advisory Service. HDRA - The Organic Organization. Available at: http://www.hdra.org.uk (accessed 26 November 2020).

Hekmat, S., Morgan, K., Soltani, M. and Gough, R. (2015) Sensory evaluation of locally-grown fruit purees and inulin fibre on probiotic yogurt in Mwanza, Tanzania and the microbial analysis of probiotic yogurt fortified with *Moringa oleifera*. *Journal of Health, Population, and Nutrition* 33(1), 60–67.

Hlophe, S.N. and Moyo, N.A.G. (2014) Replacing fishmeal with kikuyu grass and moringa leaves: effects on growth, protein digestibility, histological and haematological parameters in *Clarias gariepinus*. *Turkish Journal of Fisheries and Aquatic Sciences* 14, 795–806. DOI: 10.4194/1303-2712-v14_3_22.

Hongbo, C., Jiangbo, W., Xintian, D., Yan, H., Qiaoli, M. *et al*. (2015) Carotenoid accumulation affects redox status, starch. *BMC Plant Biology* 15(1), 1–16.

Ijarotimi, O.S., Fagbemi, T.N. and Osundahunsi, O.F. (2013) Comparative study of nutritional profiles and phytochemical components of raw, blanched and fermented flour from the leaves of *Moringa oleifera* Lam. *Malaysia Journal of Nutrition* 19(3), 371–382.

Imungi, J.K. and Potter, N.N. (1983) Nutrient contents of raw and cooked cowpea leaves. *Journal of Food Science* 48(4), 1252–1254. DOI: 10.1111/j.1365-2621.1983.tb09204.x.

Inbathamizh, L. and Padmini, E. (2012) Gas chromatography-mass spectrometric analyses of methanol extract of *Moringa oleifera* flowers. *International Journal of Chemical and Analytical Science* 2(5).

Islam, M.A., Sheikh, A., Waterman, C. and Hosenuzzaman, M. (2020) Morphology, pod yield and nutritional quality of two cultivars of Moringa (*Moringa oleifera*) in Bangladesh. *Indian Journal of Science and Technology* 13(37), 3893–3903. DOI: 10.17485/IJST/v13i36.1083.

Jaiswal, D., Kumar Rai, P., Kumar, A., Mehta, S. and Watal, G. (2009) Effect of *Moringa oleifera* Lam. leaves aqueous extract therapy on hyperglycemic rats. *Journal of Ethnopharmacology* 123(3), 392–396. DOI: 10.1016/j.jep.2009.03.036.

Jiménez-Aguilar, D.M. and Grusak, M.A. (2017) Minerals, vitamin C, phenolics, flavonoids and antioxidant activity of Amaranthus leafy vegetables. *Journal of Food Composition and Analysis* 58, 33–39. DOI: 10.1016/j.jfca.2017.01.005.

Jongrungruangchok, S., Bunrathep, S. and Songsak, T. (2010) Nutrients and minerals content of eleven different samples of *Moringa oleifera* cultivated in Thailand. *Journal of Health Research* 24(3), 123–127.

Joshi, P. and Mehta, D. (2010) Effect of dehydration on the nutritive value of drumstick leaves. *Journal of Metabolomics and Systems Biology* 1(1), 5–9.

Kahlon, T.S., Chapman, M.H. and Smith, G.E. (2007) In vitro binding of bile acids by okra, beets, asparagus, eggplant, turnips, green beans, carrots, and cauliflower. *Food Chemistry* 103(2), 676–680. DOI: 10.1016/j.foodchem.2006.07.056.

Khan, W., Parveen, R., Chester, K., Parveen, S. and Ahmad, S. (2017) Hypoglycemic potential of aqueous extract of *Moringa oleifera* leaf and In Vivo GC-MS metabolomics. *Frontiers in Pharmacology* 8, 577. DOI: 10.3389/fphar.2017.00577.

Karim, O., Kayode, R., Oyeyinka, S. and Oyeyinka, A. (2015) Physicochemical properties of stiff dough ‘amala’prepared from plantain (Musa paradisca) flour and moringa (moringa oleifera) leaf powder. *Hrana u Zdravlju i Bolesti: Znanstveno-Stručni Časopis Za Nutricionizam i Dijetetiku* 4(1), 48–58.

Khan, M.A.B., Hashim, M.J., King, J.K., Govender, R.D., Mustafa, H. *et al*. (2020) Epidemiology of type 2 diabetes - global burden of disease and forecasted trends. *Journal of Epidemiology and Global Health* 10(1), 107–111. DOI: 10.2991/jegh.k.191028.001.

Kidmose, U., Yang, R.Y., Thilsted, S.H., Christensen, L.P. and Brandt, K. (2006) Content of carotenoids in commonly consumed Asian vegetables and stability and extractability during frying. *Journal of Food Composition and Analysis* 19(6–7), 562–571. DOI: 10.1016/j.jfca.2006.01.011.

Kim, D.-O., Padilla-Zakour, O.I. and Griffiths, P.D. (2006) Flavonoids and antioxidant capacity of various cabbage genotypes at juvenile stage. *Journal of Food Science* 69(9), C685–C689. DOI: 10.1111/j.1365-2621.2004.tb09916.x.

Kolawole, F.L., Balogun, M.A., Opaleke, D.O. and Amali, H.E. (2013) An evaluation of nutritional and sensory qualities of wheat-moringa cake. *Agrosearch* 13(1), 87–94.

Kraujalis, P., Venskutonis, P.R., Kraujalienė, V. and Pukalskas, A. (2013) Antioxidant properties and preliminary evaluation of phytochemical composition of different anatomical parts of amaranth. *Plant Foods for Human Nutrition* 68(3), 322–328. DOI: 10.1007/s11130-013-0375-8.

Kubheka, S.F., Tesfay, S.Z., Mditshwa, A. and Magwaza, L.S. (2020) Evaluating the efficacy of edible coatings incorporated with Moringa leaf extract on postharvest of 'Maluma' avocado fruit quality and its biofungicidal effect. *HortScience: A Publication of the American Society for Horticultural Science* 55(4), 410–415. DOI: 10.21273/HORTSCI14391-19.

Kumar, S. and Pandey, A.K. (2013) Chemistry and biological activities of flavonoids: an overview. *ScientificWorldJournal* 2013, 162750. DOI: 10.1155/2013/162750.

Kushad, M.M., Brown, A.F., Kurilich, A.C., Juvik, J.A., Klein, B.P. *et al.* (1999) Variation of glucosinolates in vegetable crops of *Brassica oleracea*. *Journal of Agricultural and Food Chemistry* 47(4), 1541–1548. DOI: 10.1021/jf980985s.

Lakshminarayana, R., Raju, M., Krishnakantha, T.P. and Baskaran, V. (2005) Determination of major carotenoids in a few Indian leafy vegetables by high-performance liquid chromatography. *Journal of Agricultural and Food Chemistry* 53(8), 2838–2842. DOI: 10.1021/jf0481711.

Leone, A., Fiorillo, G., Criscuoli, F., Ravasenghi, S., Santagostini, L. *et al.* (2015a) Nutritional Characterization and Phenolic Profiling of *Moringa oleifera* Leaves Grown in Chad, Sahrawi Refugee Camps, and Haiti. *International Journal of Molecular Sciences* 16(8), 18923–18937. DOI: 10.3390/ijms160818923.

Leone, A., Spada, A., Battezzati, A., Schiraldi, A., Aristil, J. *et al.* (2015b) Cultivation, Genetic, Ethnopharmacology, Phytochemistry and Pharmacology of *Moringa oleifera* Leaves: An Overview. *International Journal of Molecular Sciences* 16(6), 12791–12835. DOI: 10.3390/ijms160612791.

Liao, D.Y., Chai, Y.C., Wang, S.H., Chen, C.W. and Tsai, M.S. (2015) Antioxidant activities and contents of flavonoids and phenolic acids of *Talinum triangulare* extracts and their immunomodulatory effects. *Journal of Food and Drug Analysis* 23(2), 294–302. DOI: 10.1016/j.jfda.2014.07.010.

Luvonga, W.A., Njoroge, M.S., Makokha, A. and Ngunjiri, P.W. (2010) Chemical characterisation of *Hibiscus sabdariffa* (Roselle) calyces and evaluation of its functional potential in the food industry. In: *Jomo Kenyatta University of Agriculture and Technology Annual Scientific Conference Proceedings*. pp. 631–638.

Madukwe, E.U., Nwabunze, A.M. and Onyibalu, C.L. (2013) Bioavailability of selected phytochemicals in the extracts of fluted pumpkin (*Telfairia occidentalis*), tomato (*Lycopersicum esculentum*) and eggplant (*Solanum melongena*). *International Journal of Basic and Applied Sciences* 2(3), 251. DOI: 10.14419/ijbas.v2i3.1027.

Makkar, H.P.S. and Becker, K. (1996) Nutrional value and antinutritional components of whole and ethanol extracted *Moringa oleifera* leaves. *Animal Feed Science and Technology* 63(1–4), 211–228. DOI: 10.1016/S0377-8401(96)01023-1.

Makkar, H.P.S. and Becker, K. (1997) Nutrients and antiquality factors in different morphological parts of the *Moringa oleifera* tree. *The Journal of Agricultural Science* 128(3), 311–322. DOI: 10.1017/S0021859697004292.

Maldini, M., Maksoud, S.A., Natella, F., Montoro, P., Petretto, G.L. *et al.* (2014) *Moringa oleifera*: study of phenolics and glucosinolates by mass spectrometry. *Journal of Mass Spectrometry* 49(9), 900–910. DOI: 10.1002/jms.3437.

Mallya, R., Chatterjee, P.K., Vinodini, N.A., Chatterjee, P. and Mithra, P. (2017) *Moringa oleifera* leaf extract: Beneficial effects on cadmium induced toxicities - A review. *Journal of Clinical and Diagnostic Research* 11(4), CE01–CE04. DOI: 10.7860/JCDR/2017/21796.9671.

Managa, M.G., Sultanbawa, Y. and Sivakumar, D. (2020a) Effects of different drying methods on untargeted phenolic metabolites, and antioxidant activity in Chinese cabbage (*Brassica rapa* L. subsp. chinensis) and nightshade (*Solanum retroflexum* Dun.). *Molecules (Basel, Switzerland)* 25(6), 1326. DOI: 10.3390/molecules25061326.

Managa, M.G., Mpai, S., Remize, F., Garcia, C. and Sivakumar, D. (2020b) Impact of moist cooking methods on colour, anti-nutritive compounds and phenolic metabolites in African nightshade (*Solanum retroflexum* Dun.). *Food Chemistry* 325, 126805. DOI: 10.1016/j.foodchem.2020.126805.

Manyelo, T.G., Sebola, N.A. and Mabelebele, M. (2020) Nutritional and phenolic profile of early and late harvested Amaranth leaves grown under cultivated conditions. *Agriculture* 10(10), 432. DOI: 10.3390/agriculture10100432.

Maseko, I., Mabhaudhi, T., Ncube, B., Tesfay, S., Araya, H.T. *et al.* (2019) Postharvest drying maintains phenolic, flavonoid and gallotannin content of some cultivated African leafy vegetables. *Scientia Horticulturae* 255, 70–76. DOI: 10.1016/j.scienta.2019.05.019.

Mishra, G., Singh, P., Verma, R., Kumar, S., Srivastav, S. *et al.* (2011) Traditional uses, phytochemistry and pharmacological properties of *Moringa oleifera* plant: an overview. *Der Pharmacia Lettre* 3(2), 141–164.

Moloto, M.R., Phan, A.D.T., Shai, J.L., Sultanbawa, Y. and Sivakumar, D. (2020) Comparison of phenolic compounds, carotenoids, amino acid composition, in vitro antioxidant and anti-diabetic activities in the leaves of Seven Cowpea (*Vigna unguiculata*) cultivars. *Foods*9(9), 1285. DOI: 10.3390/foods9091285.

Mosha, T.C., Gaga, H.E., Pace, R.D., Laswai, H.S. and Mtebe, K. (1995) Effect of blanching on the content of antinutritional factors in selected vegetables. *Plant Foods for Human Nutrition (Dordrecht, Netherlands)* 47(4), 361–367. DOI: 10.1007/BF01088275.

Moyo, B., Masika, B.J., Hugo, A. and Muchenje, V. (2011) Nutritional characterization of Moringa (*Moringa oleifera* Lam.) leaves. *African Journal of Biotechnology* 10(60), 12925–12933. DOI: 10.5897/AJB10.1599.

Moyo, B., Oyedemi, S., Masika, P.J. and Muchenje, V. (2012) Polyphenolic content and antioxidant properties of *Moringa oleifera* leaf extracts and enzymatic activity of liver from goats supplemented with *Moringa oleifera* leaves/sunflower seed cake. *Meat Science* 91(4), 441–447. DOI: 10.1016/j.meatsci.2012.02.029.

Mukunzi, D., Nsor-Atind, J., Xiaoming, Z., Gahungu, A., Karangwa, E. *et al.* (2011) Comparison of volatile profile of *Moringa oleifera* leaves from Rwanda and China using HS-SPME. *Pakistan Journal of Nutrition* 10(7), 602–608. DOI: 10.3923/pjn.2011.602.608.

Musa, A., Ezenwa, M.I., Oladiran, J.A., Akanya, H.O. and Ogbadoyi, E.O. (2010) Effect of soil nitrogen levels on some micronutrients, antinutrients and toxic substances in *Corchorus olitorius* grown in Minna, Nigeria. *African Journal of Agricultural Research* 5(22), 3075–3081.

Nahar, S., Faisal, F.M., Iqbal, J., Rahman, M.M. and Yusuf, M.A. (2016) Antiobesity activity of *Moringa oleifera* leaves against high fat diet-induced obesity in rats. *International Journal of Basic and Clinical Pharmacology* 5, 1263–1268. DOI: 10.18203/2319-2003.ijbcp20162427.

Nana, F.W., Hilou, A., Millogo, J.F. and Nacoulma, O.G. (2012) Phytochemical composition, antioxidant and xanthine oxidase inhibitory activities of *Amaranthus cruentus* L. and *Amaranthus hybridus* L. Extracts. *Pharmaceuticals* 5(6), 613–628. DOI: 10.3390/ph5060613.

National Research Council (2006) *Lost Crops of Africa: Volume II: Vegetables, Development, Security and Cooperation*. National Academy of Science, Washington, D.C, pp. 247–267.

Ndlovu, J. and Afolayan, A.J. (2008) Nutritional analysis of the South African wild vegetable *Corchorus olitorius* L. *Asian Journal of Plant Sciences* 7(6), 615–618. DOI: 10.3923/ajps.2008.615.618.

Nepolean, P., Anitha, J. and Emilin, R.R. (2009) Isolation, analysis and identification of phytochemicals of antimicrobial activity of *Moringa oleifera* Lam. *Current Biotica* 3(1), 33–37.

Nkukwana, T.T., Muchenje, V., Masika, P.J., Hoffman, L.C., Dzama, K. *et al.* (2014) Fatty acid composition and oxidative stability of breast meat from broiler chickens supplemented with *Moringa oleifera* leaf meal over a period of refrigeration. *Food Chemistry* 142, 255–261. DOI: 10.1016/j.foodchem.2013.07.059.

Nouman, W., Siddiqui, M.T. and Basra, S.M.A. (2012) *Moringa oleifera* leaf extract: An innovative priming tool for rangeland grasses. *Turkish Journal of Agriculture and Forestry* 36(1), 65–75.

Nouman, W., Basra, S.M.A., Siddiqui, M.T., Yasmeen, A., Gull, T. *et al.* (2014a) Potential of *Moringa oleifera* L. as livestock fodder crop: a review. *Turkish Journal of Agriculture and Forestry* 38(1), 1–14. DOI: 10.3906/tar-1211-66.

Nouman, W., Basra, S.M.A., Yasmeen, A., Gull, T., Hussain, S.B. *et al.* (2014b) Seed priming improves the emergence potential, growth and antioxidant system of *Moringa oleifera* under saline conditions. *Plant Growth Regulation* 73(3), 267–278. DOI: 10.1007/s10725-014-9887-y.

Nouman, W., Anwar, F., Gull, T., Newton, A., Rosa, E. *et al.* (2016) Profiling of polyphenolics, nutrients and antioxidant potential of germplasm's leaves from seven cultivars of *Moringa oleifera* Lam. *Industrial Crops and Products* 83, 166–176. DOI: 10.1016/j.indcrop.2015.12.032.

Nugrahedi, P.Y., Dekker, M., Widianarko, B. and Verkerk, R. (2016) Quality of cabbage during long term steaming; phytochemical, texture and colour evaluation. *LWT - Food Science and Technology* 65, 421–427. DOI: 10.1016/j.lwt.2015.08.034.

Nweze, N.O. and Nwafor, F. (2014) Phytochemical, proximate and mineral composition of leaf extracts of *Moringa oleifera* Lam. from Nsukka, South-Eastern Nigeria. *IOSR Journal of Pharmacy and Biological Sciences* 9(1), 99–103. DOI: 10.9790/3008-091699103.

Oancea, S., Mila, L. and Ketney, O. (2019) Content of phenolics, in vitro antioxidant activity and cytoprotective effects against induced haemolysis of red cabbage extracts. *Biotech. Lett* 24(1), 1–9. DOI: 10.25083/rbl/24.1/1.9.

Oboh, G., Raddatz, H. and Henle, T. (2008) Antioxidant properties of polar and non-polar extracts of some tropical green leafy vegetables. *Journal of the Science of Food and Agriculture* 88(14), 2486–2492. DOI: 10.1002/jsfa.3367.

Ogbe, A.O. and Affiku, J.P. (2020) Proximate study, mineral and anti-nutrient composition of *Moringa oleifera* leaves harvested from Lafia, Nigeria: potential benefits in poultry nutrition and health. *Journal of Microbiology, Biotechnology and Food Sciences* 9(4), 296–308.

Ogbonnaya, E.C. and Chinedum, E.K. (2013) Vitamin and carotenoid composition of raw and decoctions of water leaf (*Talinum triangulare*). *Biochemistry & Pharmacology* 02(3), 2167–0501. DOI: 10.4172/2167-0501.1000121.

Okiki, P.A., Osibote, I.A., Balogun, O., Oyinloye, B.E., Idris, O.O. *et al.* (2015) Evaluation of proximate, minerals, vitamins and phytochemical composition of *Moringa oleifera* Lam. cultivated in Ado Ekiti, Nigeria. *Advances in Biological Research* 9(6), 436–443.

Okpalanma, F.N. and Ojimelukwe, P.C. (2017) Effects of post-harvest storage and moist heat treatment on carotenoids and micronutrients in African Spinach (*Amaranthus hybridus*). *Asian Journal of Advances in Agricultural Research* 2(2), 1–11. DOI: 10.9734/AJAAR/2017/35356.

Olagbemide, P.T. and Philip, C.N.A. (2014) Proximate analysis and chemical composition of raw and defatted *Moringa oleifera* kernel. *Advances in Life Science and Technology* 24, 92–99.

Oliveira, F.R.A., Oliveira, F.A., Guimaraes, I.P., Medeiros, J.F., Oliveira, M.K.T. *et al.* (2009) Emergency of seedlings of *Moringa oleifera* Lam. irrigated with water of different levels of salinity. *Bioscience Journal* 25, 66–74.

Olugbemi, T.S., Mutayoba, S.K. and Lekule, F.P. (2010) Effect of Moringa (*Moringa oleifera*) inclusion in cassava based diets fed to broiler chickens. *International Journal of Poultry Science* 9(4), 363–367. DOI: 10.3923/ijps.2010.363.367.

Oluwalana, I.B., Ayo, J.A., Idowu, M.A. and Malomo, S.A. (2011) Effect of drying methods on the physicochemical properties of waterleaf (*Talinum triangulare*). *International Journal of Biological and Chemical Sciences* 5(3). DOI: 10.4314/ijbcs.v5i3.72167.

Omenna, E.C., Olanipekun, O.T. and Kolade, R.O. (2016) Effect of boiling, pressure cooking and germination on the nutritional and antinutrients content of cowpea (*Vigna unguiculata*). *ISABB Journal of Food and Agricultural Sciences* 6(1), 1–8. DOI: 10.23880/FSNT-16000104.

Onoja, I.U. (2014) The effect of different processing methods on the proximate, β-carotene and ascorbate composition of fluted pumpkin (Telfairia Occidentalis) leaves and its product, the leaf curd. *International Journal of Nutrition and Food Sciences* 3(5), 404. DOI: 10.11648/j.ijnfs.20140305.17.

Onwordi, C.T., Ogungbade, A.M. and Wusu, A.D. (2009) The proximate and mineral composition of three leafy vegetables commonly consumed in Lagos, Nigeria. *African Journal of Pure and Applied Chemistry* 3(6), 102–107.

Onwuka, G.I. (2006) Soaking, boiling and antinutritional factors in pigeon peas (*Cajanus cajan*) and cowpeas (*Vigna unguiculata*). *Journal of Food Processing and Preservation* 30(5), 616–630. DOI: 10.1111/j.1745-4549.2006.00092.x.

Owolabi, A.O., Ndidi, U.S., James, B.D. and Amune, F.A. (2012) Proximate, antinutrient and mineral composition of five varieties (improved and local) of cowpea, *Vigna unguiculata*, commonly consumed in Samaru community, Zaria-Nigeria. *Asian Journal of Food Science and Technology* 4(2), 70–72.

Oyeniran, O.H., Ademiluyi, A.O. and Oboh, G. (2020) Comparative study of the phenolic profile, antioxidant properties, and inhibitory effects of Moringa (*Moringa oleifera* Lam.) and Almond (*Terminalia catappa* Linn.) leaves on acetylcholinesterase and monoamine oxidase activities in the head region of Fruitfly (*Drosophila melanogaster* Meigen) in vitro. *Journal of Food Biochemistry* 45(3). DOI: 10.1111/jfbc.13401.

Oyeyinka, A.T. and Oyeyinka, S.A. (2018) *Moringa oleifera* as a food fortificant: Recent trends and prospects. *Journal of the Saudi Society of Agricultural Sciences* 17(2), 127–136. DOI: 10.1016/j.jssas.2016.02.002.

Palada, M.C. and Chang, L.C. (2003) *Suggested Cultural Practices For Moringa*. International co-operator's guide, Asian Vegetable Research and Development Center. AVRDC.

Palada, M.C., Chang, L., Yang, R. and Engle, L.M. (2006) Introduction and varietal screening of drumstick tree (*Moringa* spp.) for horticultural traits and adaptation in Taiwan. In: *International Conference on Indigenous Vegetables and Legumes. Prospectus for Fighting Poverty, Hunger and Malnutrition*, Vol. 752. pp. 249–253. DOI: 10.17660/ActaHortic.2007.752.40.

Park, S.-Y., Choi, S.R., Lim, S.-H., Yeo, Y., Kweon, S.J. *et al.* (2014) Identification and quantification of carotenoids in paprika fruits and cabbage, kale, and lettuce leaves. *Journal of the Korean Society for Applied Biological Chemistry* 57(3), 355–358. DOI: 10.1007/s13765-014-4081-5.

Park, C.H., Yeo, H.J., Park, S.Y., Kim, J.K. and Park, S.U. (2019) Comparative phytochemical analyses and metabolic profiling of different phenotypes of Chinese cabbage (*Brassica Rapa* ssp. Pekinensis). *Foods* 8(11), 587. DOI: 10.3390/foods8110587.

Popoola, J.O. and Obembe, O.O. (2013) Local knowledge, use pattern and geographical distribution of *Moringa oleifera* Lam. (Moringaceae) in Nigeria. *Journal of Ethnopharmacology* 150(2), 682–691. DOI: 10.1016/j.jep.2013.09.043.

Price, M.L. (1985) The Moringa Tree. ECHO Technical Note. ECHO, Myers, Florida.

Price, M.L. (2000) The Moringa Tree. Echo Technical Note. Online. Available at: http://www.echonet. org (accessed 3 December 2020).

Prokopov, T., Goranova, Z., Baeva, M., Slavov, A. and Galanakis, C.M. (2015) Effects of powder from white cabbage outer leaves on sponge cake quality. *International Agrophysics* 29(4), 493–500. DOI: 10.1515/intag-2015-0055.

Qwele, K., Hugo, A., Oyedemi, S.O., Moyo, B., Masika, P.J. *et al.* (2013) Chemical composition, fatty acid content and antioxidant potential of meat from goats supplemented with Moringa *(Moringa oleifera)* leaves, sunflower cake and grass hay. *Meat Science* 93(3), 455–462. DOI: 10.1016/j.meatsci.2012.11.009.

Rahman, M.M., Sheikh, M.M.I., Sharmin, S.A., Islam, MS., Rahman, M.A. *et al.* (2009) Antibacterial activity of leaf juice and extracts of *Moringa oleifera* Lam. against some human pathogenic bacteria. *Chiang Mai University Journal of Natural Sciences* 8(2), 219–227.

Raj, A.J., Gopalakrishnan, V.K., Yadav, S.A. and Dorairaj, S. (2011) Antimicrobial activity of *Moringa oleifera* (Lam.) root extract. *Journal of Pharmacy Research* 4(5), 1426–1427.

Ramachandran, C., Peter, K.V. and Gopalakrishnan, P.K. (1980) Drumstick (*Moringa oleifera*): A multipurpose Indian vegetable. *Economic Botany* 34(3), 276–283. DOI: 10.1007/BF02858648.

Ramtahal, R., Khan, C., Maharaj-Khan, K., Nallamothu, S., Hinds, A. *et al.* (2015) Prevalence of self-reported sleep duration and sleep habits in type 2 diabetes patients in South Trinidad. *Journal of Epidemiology and Global Health* 5(4 Suppl 1), S35-43. DOI: 10.1016/j.jegh.2015.05.003.

Richter, N., Siddhuraju, P. and Becker, K. (2003) Evaluation of nutritional quality of moringa (*Moringa oleifera* Lam.) leaves as an alternative protein source for Nile tilapia (*Oreochromis niloticus* L.). *Aquaculture* 217(1–4), 599–611. DOI: 10.1016/S0044-8486(02)00497-0.

Rocchetti, G., Pagnossa, J.P., Blasi, F., Cossignani, L., Hilsdorf Piccoli, R. *et al.* (2020) Phenolic profiling and *in vitro* bioactivity of *Moringa oleifera* leaves as affected by different extraction solvents. *Food Research International* 127, 108712. DOI: 10.1016/j.foodres.2019.108712.

Rubab, M., Chellia, R., Saravanakumar, K., Mandava, S., Khan, I. *et al.* (2018) Preservative effect of Chinese cabbage (*Brassica rapa* subsp. pekinensis) extract on their molecular docking, anti-oxidant and antimicrobial properties. *PloS One* 13(10), e0203306. DOI: 10.1371/journal.pone.0203306.

Saini, R.K., Shetty, N.P., Prakash, M. and Giridhar, P. (2014) Effect of dehydration methods on retention of carotenoids, tocopherols, ascorbic acid and antioxidant activity in *Moringa oleifera* leaves and preparation of a RTE product. *Journal of Food Science and Technology* 51(9), 2176–2182. DOI: 10.1007/s13197-014-1264-3.

Salawu, S.O., Sanni, D.M. and Akindahunsi, A.A. (2013) HPLC/DAD/MS phenolic profile, antioxidant activities and inhibitory action of struchium sparganophora (linn) and telfairia occidentalis (hook. F) against low density lipoprotein oxidation. *African Journal of Food Science and Technology* 4, 1–8.

Saini, R.K., Manoj, P., Shetty, N.P., Srinivasan, K. and Giridhar, P. (2016) Relative bioavailability of folate from the traditional food plant *Moringa oleifera* L. as evaluated in a rat model. *Journal of Food Science and Technology* 53(1), 511–520. DOI: 10.1007/s13197-015-1828-x.

Sallau, A.B., Mada, S.B., Ibrahim, S. and Ibrahim, U. (2012). Effect of boiling, simmering and blanching on the antinutritional content of *Moringa oleifera* leaves. International Journal of Food Nutrition and Safety, 2(1), 1-6.

Sangeetha, R. and Vedasree, N. (2012) In vitro α-amylase inhibitory activity of the leaves of *Thespesia populnea*. *ISRN Pharmacology* 2012, 515634. DOI: 10.5402/2012/515634.

Santiago-Saenz, Y.O., Hernández-Fuentes, A.D., Monroy-Torres, R., Cariño-Cortés, R. and Jiménez-Alvarado, R. (2018) Physicochemical, nutritional and antioxidant characterization of three vegetables (*Amaranthus hybridus* L., *Chenopodium berlandieri* L., *Portulaca oleracea* L.) as potential sources of phytochemicals and bioactive compounds. *Journal of Food Measurement and Characterization* 12(4), 2855–2864. DOI: 10.1007/s11694-018-9900-7.

Santos, J., Oliveira, M.B.P.P., Ibáñez, E. and Herrero, M. (2014) Phenolic profile evolution of different ready-to-eat baby-leaf vegetables during storage. *Journal of Chromatography. A* 1327, 118–131. DOI: 10.1016/j.chroma.2013.12.085.

Sarker, U. and Oba, S. (2020) Leaf pigmentation, its profiles and radical scavenging activity in selected *Amaranthus tricolor* leafy vegetables. *Scientific Reports* 10(1), 1–10. DOI: 10.1038/s41598-020-66376-0.

Schröter, D., Baldermann, S., Schreiner, M., Witzel, K., Maul, R. *et al.* (2018) Natural diversity of hydroxycinnamic acid derivatives, flavonoid glycosides, carotenoids and chlorophylls in leaves of six different amaranth species. *Food Chemistry* 267, 376–386. DOI: 10.1016/j.foodchem.2017.11.043.

Sena, L.P., Vanderjagt, D.J., Rivera, C., Tsin, A.T., Muhamadu, I. *et al.* (1998) Analysis of nutritional components of eight famine foods of the Republic of Niger. *Plant Foods for Human Nutrition* 52(1), 17–30. DOI: 10.1023/a:1008010009170.

Sengupta, A. and Gupta, M.P. (1970) Studies on the seed fat composition of Moringaceae family. *Fette, Seifen, Anstrichmittel* 72(1), 6–10. DOI: 10.1002/lipi.19700720103.

Sharifudin, S.A., Fakurazi, S., Hidayat, M.T., Hairuszah, I., Moklas, M.A.M. *et al.* (2013) Therapeutic potential of *Moringa oleifera* extracts against acetaminophen-induced hepatotoxicity in rats. *Pharmaceutical Biology* 51(3), 279–288. DOI: 10.3109/13880209.2012.720993.

Siddhuraju, P. and Becker, K. (2003) Antioxidant properties of various solvent extracts of total phenolic constituents from three different agroclimatic origins of drumstick tree (Moringa oleifera Lam.) leaves. *Journal of Agricultural and Food Chemistry* 51(8), 2144–2155. DOI: 10.1021/jf020444+.

Singh, G., Kawatra, A. and Sehgal, S. (2001) Nutritional composition of selected green leafy vegetables, herbs and carrots. *Plant Foods for Human Nutrition* 56(4), 359–364. DOI: 10.1023/A:1011873119620.

Singh, J., Upadhyay, A.K., Bahadur, A., Singh, B., Singh, K.P. *et al.* (2006) Antioxidant phytochemicals in cabbage (*Brassica oleracea* L. var. capitata). *Scientia Horticulturae* 108(3), 233–237. DOI: 10.1016/j.scienta.2006.01.017.

Singh, R.G., Negi, P.S. and Radha, C. (2013) Phenolic composition, antioxidant and antimicrobial activities of free and bound phenolic extracts of *Moringa oleifera* seed flour. *Journal of Functional Foods* 5(4), 1883–1891. DOI: 10.1016/j.jff.2013.09.009.

Singh, A.K., Rana, H.K., Tshabalala, T., Kumar, R., Gupta, A. *et al.* (2020) Phytochemical, nutraceutical and pharmacological attributes of a functional crop *Moringa oleifera* Lam: An overview. *South African Journal of Botany* 129, 209–220. DOI: 10.1016/j.sajb.2019.06.017.

Sivasankari, B., Anandharaj, M. and Gunasekaran, P. (2014) An ethnobotanical study of indigenous knowledge on medicinal plants used by the village peoples of Thoppampatti, Dindigul district, Tamilnadu, India. *Journal of Ethnopharmacology* 153(2), 408–423. DOI: 10.1016/j.jep.2014.02.040.

Šola, I., Vujčić Bok, V., Dujmović, M. and Rusak, G. (2020) Developmentally-related changes in phenolic and L-ascorbic acid content and antioxidant capacity of Chinese cabbage sprouts. *Journal of Food Science and Technology* 57(2), 702–712. DOI: 10.1007/s13197-019-04103-y.

Stadtlander, T. and Becker, K. (2017) Proximate composition, amino and fatty acid profiles and element compositions of four different Moringa species. *Journal of Agricultural Science* 9(7), 46. DOI: 10.5539/jas.v9n7p46.

Stohs, S.J. and Hartman, M.J. (2015) Review of the safety and efficacy of *Moringa oleifera*. *Phytotherapy Research* 29(6), 796–804. DOI: 10.1002/ptr.5325.

Stohs, S.J., Bagchi, D., Hassoun, E. and Bagchi, M. (2000) Oxidative mechanisms in the toxicity of chromium and cadmium ions. *Journal of Environmental Pathology, Toxicology and Oncology* 19(3), 201–213.

Taher, M.A., Nyeem, M.A.B., Ahammed, M.M., Hossain, M.M. and Islam, M.N. (2017) *Moringa oleifera* (Shajna): the wonderful indigenous medicinal plant. *Asian Journal of Medical and Biological Research* 3(1), 20–30. DOI: 10.3329/ajmbr.v3i1.32032.

Teixeira, E.M.B., Carvalho, M.R.B., Neves, V.A., Silva, M.A. and Arantes-Pereira, L. (2014) Chemical characteristics and fractionation of proteins from *Moringa oleifera* Lam. leaves. *Food Chemistry* 147, 51–54. DOI: 10.1016/j.foodchem.2013.09.135.

Tetteh, O.N.A., Ulrichs, C., Huyskens-Keil, S., Mewis, I., Amaglo, N.K. *et al*. (2019) Effects of harvest techniques and drying methods on the stability of glucosinolates in *Moringa oleifera* leaves during post-harvest. *Scientia Horticulturae* 246, 998–1004. DOI: 10.1016/j.scienta.2018.11.089.

Thurber, M.D. and Fahey, J.W. (2009) Adoption of *Moringa oleifera* to Combat Under-Nutrition Viewed Through the Lens of the "Diffusion of Innovations" Theory. *Ecology of Food and Nutrition* 48(3), 212–225. DOI: 10.1080/03670240902794598.

Tshabalala, T., Ndhlala, A.R., Ncube, B., Abdelgadir, H.A. and Van Staden, J. (2020) Potential substitution of the root with the leaf in the use of *Moringa oleifera* for antimicrobial, antidiabetic and antioxidant properties. *South African Journal of Botany* 129, 106–112. DOI: 10.1016/j.sajb.2019.01.029.

Tumer, T.B., Rojas-Silva, P., Poulev, A., Raskin, I. and Waterman, C. (2015) Direct and indirect antioxidant activity of polyphenol- and isothiocyanate-enriched fractions from *Moringa oleifera*. *Journal of Agricultural and Food Chemistry* 63(5), 1505–1513. DOI: 10.1021/jf505014n.

Udousoro, I.I., Udo, E.S., Udoh, A.P. and Udoanya, E.E. (2018) Proximate and antinutrients compositions, and health risk assessment of toxic metals in some edible vegetables. *Nigerian Journal of Chemical Research* 23(2), 51–62.

Ukegbu, P.O. and Okereke, C.J. (2013) Effect of solar and sun drying methods on the nutrient composition and microbial load in selected vegetables, African spinach (*Amaranthus hybridus*), fluted pumpkin (*Telferia occidentalis*), and okra (*Hibiscus esculentus*). *Sky Journal of Food Science* 2(5), 35–40.

Van der Walt, A.M., Loots, D.T., Ibrahim, M.I.M. and Bezuidenhout, C.C. (2009) Minerals, trace elements and antioxidant phytochemicals in wild African dark-green leafy vegetables (morogo). *South African Journal of Science* 105(11–12), 444–448. DOI: 10.1017/S1368980008002814.

Vázquez-León, L.A., Páramo-Calderón, D.E., Robles-Olvera, V.J., Valdés-Rodríguez, O.A., Pérez-Vázquez, A. *et al*. (2017) Variation in bioactive compounds and antiradical activity of *Moringa oleifera* leaves: influence of climatic factors, tree age, and soil parameters. *European Food Research and Technology* 243(9), 1593–1608. DOI: 10.1007/s00217-017-2868-4.

Veeraragava, T.D. (1998) *Drumstick (Moringa Oleifera* Lam.*). A Guide on Vegetable Culture (Eds. Veeraragava TD, Jawaharlal M, Seemanthini R)*. Suri Associates, Coimbatore, India, pp. 139–143.

Velasco, P., Francisco, M., Moreno, D.A., Ferreres, F., García-Viguera, C. *et al*. (2011) Phytochemical fingerprinting of vegetable *Brassica oleracea* and *Brassica napus* by simultaneous identification of glucosinolates and phenolics. *Phytochemical Analysis* 22(2), 144–152. DOI: 10.1002/pca.1259.

Warman, P.R. and Havard, K.A. (1997) Yield, vitamin and mineral contents of organically and conventionally grown carrots and cabbage. *Agriculture, Ecosystems & Environment* 61(2–3), 155–162. DOI: 10.1016/S0167-8809(96)01110-3.

Xie, J., Wang, Y., Jiang, W.-W., Luo, X.-F., Dai, T.-Y. *et al*. (2018) *Moringa oleifera* leaf petroleum ether extract inhibits lipogenesis by activating the AMPK signaling pathway. *Frontiers in Pharmacology* 9, 1447. DOI: 10.3389/fphar.2018.01447.

Yabesh, J.E.M., Prabhu, S. and Vijayakumar, S. (2014) An ethnobotanical study of medicinal plants used by traditional healers in silent valley of Kerala, India. *Journal of Ethnopharmacology* 154(3), 774–789. DOI: 10.1016/j.jep.2014.05.004.

Yang, R.Y., Lin, S. and Kuo, G. (2008) Content and distribution of flavonoids among 91 edible plant species. *Asia Pacific Journal of Clinical Nutrition* 17 Suppl 1, 275–279.

Yanyang, X., Min, Z., Mujumdar, A.S., Le-qun, Z. and Jin-cai, S. (2004) Studies on hot air and microwave vacuum drying of wild cabbage. *Drying Technology* 22(9), 2201–2209. DOI: 10.1081/DRT-200034275.

Yassa, H.D. and Tohamy, A.F. (2014) Extract of *Moringa oleifera* leaves ameliorates streptozotocin-induced Diabetes mellitus in adult rats. *Acta Histochemica* 116(5), 844–854. DOI: 10.1016/j.acthis.2014.02.002.

Yoon, S., Fleeman, L.M., Wilson, B.J., Mansfield, C.S. and McGreevy, P. (2020) Epidemiological study of dogs with diabetes mellitus attending primary care veterinary clinics in Australia. *The Veterinary Record* 187(3), e22–e22. DOI: 10.1136/vr.105467.

Youssef, K.M., Mokhtar, S. and Morsy, N. (2014) Effect of hot air drying variables on phytochemicals and antioxidant capacity of Jew's mallow (*Corchorus olitorius* L.) leaves. *Suez Canal University Journal of Food Sciences* 2(1), 11–18. DOI: 10.21608/scuj.2014.6667.

Yusuf, A.A., Arowolo, T.A. and Bamgbose, O. (2003) Cadmium, copper and nickel levels in vegetables from industrial and residential areas of Lagos City, Nigeria. *Food and Chemical Toxicology* 41(3), 375–378. DOI: 10.1016/s0278-6915(02)00223-5.

9 Sweet Potato (*Ipomoea batatas* [L.] Lam)

Mohammad Khairul Alam* **and Sheikh Nazrul Islam**
Institute of Nutrition and Food Science, University of Dhaka, Dhaka, Bangladesh

Abstract

Owing to its special nutritional and functional properties, sweet potato (*Ipomoea batatas* [L.] Lam) has become a research subject in recent years. Bioactive carbohydrates, proteins, lipids, carotenoids, anthocyanins, phenolic acids and minerals are the diverse nutrients present in the leaves of sweet potato. Different health benefits, such as antioxidant, cardioprotective, anti-inflammatory, anticancer, antidiabetic, antimicrobial and antiobesity, contribute to the specific composition of sweet potato leaves. Sweet potato leaves remain substantially underutilized in the commercial sense. To promote human health, sweet potatoes can be further grown as a sustainable crop for various nutritionally-enhanced and value-added food items. Therefore, the goal of this chapter is to give a broad view of the nutritional value, phytochemical composition and health properties of sweet potato leaves.

Introduction

Sweet potato (*Ipomoea batatas* [L.] Lam) belongs to the botanical family Convolvulaceae (Fig. 9.1). In many countries, its culture and production is essential, because it contributes to reducing food shortages in times of crisis (natural disasters or wars) and helps to sustain food security. With over 90 million tonnes of yearly output, it is one of the world's most valuable, versatile and untapped food crops, with Asian and African regions contributing the majority of its production (Fig. 9.2a) (FAOSTAT, 2020). It is also a major food crop in tropical and subtropical regions, so increasing production and consumption would benefit people in rural and urban areas (Alam *et al.*, 2016). Globally, sweet potatoes are an important crop, ranked sixth, and the fifth most important food crop in terms of fresh weight (FW) in developing countries, after rice, wheat, maize and cassava, with more than 100 countries producing sweet potato (CIP, 2019). Sweet potato is the fourth most important crop in Bangladesh after rice, wheat and potato. It is largely produced in a sporadic way in various river belts, charlands, deltas and seasonally inundated floodplains by marginal or subsistence farmers (BBS, 2019). The world production of sweet potatoes, as in Bangladesh, has seen a steady drop in the last decade (Fig. 9.2b,c).

Sweet potato covers about 24,000 ha crop area in Bangladesh, producing 246,000 tonnes annually with an average yield of 10.29 tonnes per ha (BBS, 2019). The average tuber yield in Bangladesh is very low as compared to many tropical and subtropical countries due to cultivation of local and poor quality native sweet potato varieties (BBS, 2019). But still Bangladesh is one of the leading producers of sweet potato in the world. In South Asia, Bangladesh is the second leading producer of

*Corresponding author: khairul.alam010@gmail.com

 Handbook of Phytonutrients in Indigenous Fruits and Vegetables (eds D. Sivakumar *et al.*)
DOI: 10.1079/9781789248067.0009

Fig. 9.1. Sweet potato (*Ipomoea batatas* [L.] Lam), belonging to the botanical family Convolvulaceae.

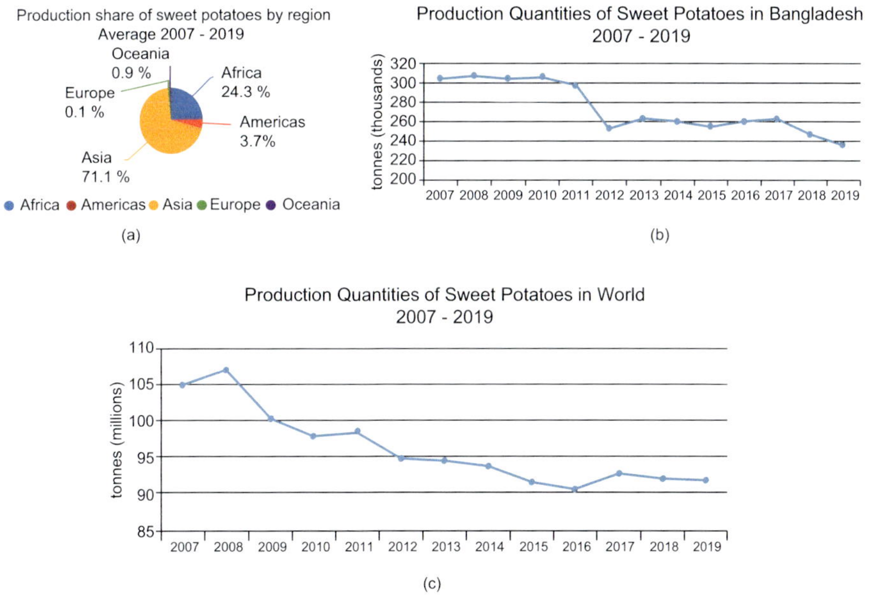

Fig. 9.2. Production quantities of sweet potato: (a) regional; (b) Bangladesh; (c) world share (FAOSTAT, 2020).

sweet potato after India and accounts for about 27 and 68% of the total production of South Asia respectively whereas Sri Lanka accounts for only 5% (FAOSTAT, 2020).

Almost all parts of the sweet potato (such as the stem, leaf and root) can be consumed, and the nutrients, bioactive compounds and antinutrients of these parts can differ. Sweet potato leaves, with the exception of a small amount of use as livestock feed, have largely gone unnoticed around the world. In countries where there are regular food shortages, using sweet potato leaves as a vegetable may significantly increase food supply. Sweet potato greens are consumed in limited quantities as a fresh vegetable in some parts of the world (Islam, 2006, 2014). Producers and researchers have become more focused on underutilized green leafy vegetables like sweet potato greens as a consequence of elevated food scarcity.

The sweet potato can be harvested many times during the year, yielding considerably more than other greens. In addition, it is one of the few vegetables that can be cultivated during the monsoon season, sometimes making sweet potato leaves the only greens available after a flood or typhoon in some countries. Sweet potato leaves are considered to be rich in protein, fibre, vitamin B, β-carotene and minerals such as iron, calcium, magnesium and zinc, and other dietary bioactive compounds (e.g. polyphenol compounds) (Islam, 2006, 2014; Wang *et al.*, 2016). The concentration of polyphenols from sweet potato leaves is seven to nine times as high as that of grape seeds (Xi *et al.*, 2015). Moreover, sweet potatoes are more resilient to diseases, pests and high moisture conditions as a crop than many other leafy vegetables grown in the tropical and subtropical regions.

Botany

The sweet potato is normally a dicotyledonous, herbaceous and perennial plant. However, by vegetative propagation, it is grown as an annual plant using either storage roots or stem cuttings. Moreover, due to the fast-growing nature of tops (leaves and tender vines), these can be harvested many times a year. In marginal lands the plant grows well – namely in char lands, homestead regions, low hill valleys, newly accreted lands and even problematic soils, such as saline belts, acidic soils and so on. A low quantity of fertilizers and agricultural inputs with low management practices can be enough to grow these plants. With a vine system that spreads horizontally and easily on the ground, the sweet potato growth habit is primarily prostrate. Erect, semi-erect, spreading and very spreading are the morphological characteristics of sweet potatoes (Huamari, 1992).

The systematic classification of sweet potato is as follows:

Family: Convolvulaceae

Genus: *Ipomoea*

Species: *Ipomoea batatas* [L.] Lam

The leaves of the sweet potato plant are either heart-shaped or palmately lobed with sympetalous medium-sized flowers and a smooth skin of yellow, orange, red, brown and purple colours. Due to anthocyanin pigmentation, the colour of the stem and leaves of the sweet potato plant ranges from green to purple (Li *et al.*, 2017; Chen *et al.*, 2018). Genetic variations may also influence the size, shape and colour of the leaves (Fig. 9.3) (Jackson *et al.*, 2020).

Traditional Use

In Bangladesh, sweet potato leaves are generally consumed as vegetables and served with rice in a meal. Besides the general use as vegetables, sweet potato leaves are also used in folk medicine to treat many illnesses, such as diabetes, hypertension, dysentery, diarrhoea, anaemia, constipation, fatigue, asthma, rheumatoid arthritis, meningitis, kidney disease and inflammation. They also have antimicrobial, analgesic, spasmolitic, spasmogenic, hypotensive, anticoagulant, psychotomimetic and anticancer activities. In addition, they are used for treating insect bites, burns, catarrh, dyslactea, fever, nausea, renosis, splenosis, stomach pain, mouth and throat tumours (Mohanraj and Sivasankar, 2014). Sweet potato leaf decoctions are used as bactericidal and fungicidal agents, laxatives and tonics (Mohanraj and Sivasankar, 2014). They are also used as fodder for livestock.

Nutritional Composition

Sweet potato leaves are rich in complex carbohydrates, dietary fibre, β-carotene (a precursor of vitamin A), vitamin C and thiamin (vitamin B1), riboflavin (vitamin B2), niacin (vitamin B3), lutein and other essential nutrients. They are also reported to contain essential mineral nutrients such as zinc, potassium, sodium, manganese, calcium, magnesium and iron (Islam, 2006, 2014; Sun *et al.*, 2014; Suárez *et al.*, 2020; Tang *et al.*, 2021). Table 9.1 summarizes the proximate and mineral composition of sweet potato leaves reported by several authors. Sweet potato leaves contain a significant amount of protein and have relatively higher content when compared to other commonly consumed leafy vegetables such as spinach. The leaf protein was also found to be of good quality considering the amino acid composition, and lysine was

Fig. 9.3. Genetical variations of sweet potato leaves (adapted from Jackson *et al.*, 2020).

reported to be the only or second limiting amino acid. This suggests the usefulness of sweet potato leaf as a protein resource (Ishida *et al.*, 2000; Suárez *et al.*, 2020). Table 9.2 shows the vitamin and essential amino acid composition of sweet potato leaves reported in the literature and Table 9.3 gives a comparison with other commonly consumed leafy vegetables. From Table 9.3, it can be seen that sweet potato leaves contain more proteins than other commercial leafy vegetables, however, in terms of amino acid composition, they contain similar amounts of most essential amino acids except for tryptophan and methionine. Sweet potato leaves are also considered to be comparable or superior in terms of vitamins and minerals when compared to spinach and other major and wild vegetables of Bangladesh (Islam, 2006, 2014; Rana *et al.*, 2019; Alam *et al.*, 2020a, b, d) (Table 9.3). Compared to traditional leafy vegetables, sweet potato leaves contain more minerals (calcium, potassium, magnesium, iron, zinc) and carotenoids (β-carotene and lutein) and therefore, could serve as an excellent source of essential dietary nutrients to prevent and/or manage vitamin A and other micronutrient-associated malnutrition (Ishida *et al.*, 2000; Islam, 2006, 2014; Alam *et al.*, 2016, 2020c; Li *et al.*, 2017; Suárez *et al.*, 2020; Tang *et al.*, 2021). However, it must be noted that variations in nutritional composition can occur and may be affected by different factors including genotype, cultural practices, maturity, harvesting, geography etc. (Sun *et al.*, 2014; Alam *et al.*, 2016, 2020c; Sams *et al.*, 2020).

Phytochemical Composition

Sweet potato leaves are reported to be a superior source of polyphenols, terpenoids, saponins, glycosides, alkaloids, steroids and other functional bioactive components (Wang *et al.*, 2016; Zhang *et al.*, 2020). Of these, the predominant bioactive components are phenolic compounds like phenolic acids (e.g. caffeic acid, monocafeoyl quinic (chlorogenic acid), caffeoylquinic acid (CQA) derivatives (mainly mono-CQA, di-CQA and 3,4,5-triCQA), p-coumaric acid, sinapic acid, hydroxybenzoic acids and p-anisic acids), flavonoids (e.g. quercetin, myricetin, luteolin, apigenin etc.), and anthocyanins (Wang *et al.*, 2016; Zhang *et al.*, 2020). The major phenolic

Table 9.1. Proximate and mineral composition of sweet potato leaves.

Nutrient	No. of cultivars studied	Reported content	Reference
Proximate composition (g/100 g DW)			
Protein	40	16.7–31.1	Sun *et al*. (2014)
	2	3.7–3.8	Ishida *et al*. (2000)
	1	26.2–30.8	Suárez *et al*. (2020)
	11	16.2–30.3	Tang *et al*. (2021)
Fat	40	2.1–5.3	Sun *et al*. (2014)
	2	0.3–1.0	Ishida *et al*. (2000)
	1	2.9–3.6	Suárez *et al*. (2020)
Ash	40	7.4–14.7	Sun *et al*. (2014)
	2	1.5–1.9	Ishida *et al*. (2000)
	1	10.39–11.30	Suárez *et al*. (2020)
Carbohydrate	40	42.0–61.3	Sun *et al*. (2014)
	2	0.9–2.0	Ishida *et al*. (2000)
Fibre	40	9.2–14.3	Sun *et al*. (2014)
	2	5.9–6.9	Ishida *et al*. (2000)
	1	49.8–51.8	Suárez *et al*. (2020)
	11	5.18–10.38	Tang *et al*. (2021)
Mineral composition (mg/100 g DW)			
Calcium	40	229.7–1958.1	Sun *et al*. (2014)
	2	174–187	Ishida *et al*. (2000)
	1	1318–1407	Suárez *et al*. (2020)
	11	500–1068	Tang *et al*. (2021)
Potassium	40	479.3–4280.6	Sun *et al*. (2014)
	2	357–639	Ishida *et al*. (2000)
	1	1963.8–3341.7	Suárez *et al*. (2020)
	11	4546–5966	Tang *et al*. (2021)
Magnesium	40	220.2–910.5	Sun *et al*. (2014)
	2	79–107	Ishida *et al*. (2000)
	1	455–513	Suárez *et al*. (2020)
	11	200–280	Tang *et al*. (2021)
Iron	40	1.92–21.77	Sun *et al*. (2014)
	2	5.43–5.54	Ishida *et al*. (2000)
	1	11.1–19.9	Suárez *et al*. (2020)
	11	8.82–18.44	Tang *et al*. (2021)
Zinc	40	1.20–3.23	Sun *et al*. (2014)
	2	0.60–0.86	Ishida *et al*. (2000)
	1	3.2–3.3	Suárez *et al*. (2020)
	11	2.80–4.84	Tang *et al*. (2021)

Continued

Table 9.1. Continued

Nutrient	No. of cultivars studied	Reported content	Reference
Copper	40	0.67–1.86	Sun *et al.* (2014)
	2	0.43–0.55	Ishida *et al.* (2000)
	1	1.35–1.52	Suárez *et al.* (2020)
	11	1.25–1.93	Tang *et al.* (2021)

compounds and anthocyanins reported in sweet potato leaves are presented in Table 9.4. The leaves also have higher levels of polyphenols than other commonly consumed vegetables, such as spinach, cabbage, aubergine and lettuce, and sweet potato stems and tubers (Islam, 2014). The concentration of polyphenols from sweet potato leaves was found to be seven to nine times higher than grape seeds (Xi *et al.*, 2015).

Antinutritive Compounds

Despite containing beneficial compounds, there are some antinutrients present in sweet potato leaves, such as phytic acid, cyanide, tannins, oxalates, alkaloids and anthraquinones (Islam, 2014; Wang *et al.*, 2016; Alam *et al.*, 2020a, b). However, their content is found to be higher than most commercial vegetables but reported to be less than one fifth that of spinach (Islam, 2014). Table 9.5 shows the content of antinutritive compounds in sweet potato leaves as well as commonly consumed leafy vegetables. From Table 9.5, it can be seen that spinach and green amaranth contain much higher amounts of oxalate than sweet potato leaves whereas the opposite trend is observed in terms of the content of tannins and phytic acid. However, due to scarce available data, such a comparison is hard to generalize. Moreover, cooking practices, such as boiling, can significantly lower, from 41 to 76%, the amount of antinutrients present in sweet potato leaves (Abong' *et al.*, 2020a). However, soluble oxalate content in spinach is reported to decrease significantly, by 87%, after boiling for 12 minutes (Chai and Liebman, 2005). Thus, it can be inferred from these findings that the reduction of antinutrients in spinach is more responsive to boiling than sweet potato leaves. However, it can also be inferred that the use of appropriate processing method (e.g. boiling) can render sweet potato leaves as suitable for consumption as a vegetable and that consumption should definitely be encouraged.

Bioactive Properties

As revealed *in vitro* and *in vivo*, specific chemical constituents allow sweet potato leaves to prevent and treat a number of disorders. Antioxidant, cardioprotective, anticancer, antidiabetic, antimicrobial and anti-inflammatory biological activities have been recently reported in the literature (Islam, 2006, 2014; Wang *et al.*, 2016). The presence of phenolic and anthocyanin compounds in sweet potato leaves has been attributed to the exertion of the above-mentioned biological activities. Phenolic compounds are naturally occurring secondary metabolites containing one or more phenol groups, which are well known for their antioxidant properties. Due to their unique and diverse structural configurations, they are regarded as potential therapeutic agents.

Health Benefits

In many parts of the globe, the sweet potato is a staple food and is a drought-tolerant crop that promises to boost food and nutritional security. Due to their abundant bioactive components and high nutritional value, sweet potatoes are particularly important. Sweet potatoes are, in essence, not only a healthy diversification of diet, but also a food substitute that ensures food security and promotes health (Mohanraj, 2018). The different benefits of sweet potato leaves for health are summarized in Fig. 9.4.

Table 9.2. Vitamin and essential amino acid composition of sweet potato leaves.

Nutrient	No. of cultivars studied	Reported content	Reference
Vitamin composition (mg/100 g)			
β-carotene	2	0.27–0.40	Ishida *et al.* (2000)
	14	35.21–52.01	Li *et al.* (2017)
Lutein	14	19.01–28.85	Li *et al.* (2017)
Vitamin C	2	62.7–81.0	Ishida *et al.* (2000)
	1	21.9–104	Suárez *et al.* (2020)
	11	99–511	Tang *et al.* (2021)
Vitamin B1	2	0.053–0.13	Ishida *et al.* (2000)
	1	0.49–0.62	Suárez *et al.* (2020)
Vitamin B2	2	0.248–0.254	Ishida *et al.* (2000)
	1	4.45–6.36	Suárez *et al.* (2020)
Vitamin B3	2	0.86–1.50	Ishida *et al.* (2000)
	1	0.45–0.54	Suárez *et al.* (2020)
Vitamin B6	2	0.12–0.33	Ishida *et al.* (2000)
Vitamin E	2	1.39–2.81	Ishida *et al.* (2000)
	1	3.24–5.80	Suárez *et al.* (2020)
Amino acid composition (mg/g DW)			
Isoleucine	2	9.60–10.62	Ishida *et al.* (2000)
	1	6.7–8.8	Suárez *et al.* (2020)
	11	25.6–44.8	Tang *et al.* (2021)
Leucine	2	18.3–20.3	Ishida *et al.* (2000)
	1	11.9–16.1	Suárez *et al.* (2020)
	11	45.3–79.1	Tang *et al.* (2021)
Lysine	2	11.8–12.9	Ishida *et al.* (2000)
	1	10.0–13.1	Suárez *et al.* (2020)
	11	31.6–54.7	Tang *et al.* (2021)
Methionine	2	3.67–4.10	Ishida *et al.* (2000)
	1	0.6–0.9	Suárez *et al.* (2020)
	11	4.1–8.0	Tang *et al.* (2021)
Phenylalanine	2	12.1–13.4	Ishida *et al.* (2000)
	1	8.8–11.1	Suárez *et al.* (2020),
	11	28.0–49.9	Tang *et al.* (2021)
Threonine	2	10.4–11.4	Ishida *et al.* (2000)
	1	7.0–8.9	Suárez *et al.* (2020)
	11	24.8–43.8	Tang *et al.* (2021)
Tryptophan	2	2.9–3.6	Ishida *et al.* (2000)
	11	2.6–4.0	Tang *et al.* (2021)

Continued

Table 9.2. Continued

Nutrient	No. of cultivars studied	Reported content	Reference
Valine	2	12.3–13.4	Ishida *et al.* (2000)
	1	8.8–11.4	Suárez *et al.* (2020)
	11	32.4–56.5	Tang *et al.* (2021)

Table 9.3. Comparison of selected nutrients of sweet potato leaves with traditional leafy vegetables.

Nutrient	Sweet potato leaves	Spinach[5]	Red Amaranth[5]	Green Amaranth[5]	Radish[5]
Protein (g/100 g)	16.2–30.3[1]	2.14	3.93	3.29	2.22
Amino Acids (mg/g)					
Isoleucine	25.6–44.8[1]	43.1	43.1	43.9	35.9
Leucine	45.3–79.1[1]	87.0	82.0	80.04	68.9
Lysine	31.6–54.7[1]	46.8	55.8	54.8	53.4
Methionine	4.1–8.0[1]	13.3	16.3	13.8	10.1
Phenylalanine	28.0–49.9[1]	53.8	47.7	56.8	60.0
Threonine	24.8–43.8[1]	47.0	40.4	41.6	39.9
Tryptophan	2.6–4.0[1]	15.3	12.0	10.1	13.9
Valine	32.4–56.5[1]	58.3	52.4	50.3	42.6
Minerals (mg/100 g)					
Calcium	229.7–1958.1[2]	82.29	245	330	234
Potassium	479.3–4280.6[2]	625	564	572	304
Magnesium	220.2–910.5[2]	86.97	177	194	57.96
Iron	1.92–21.77[2]	2.95	7.25	4.64	3.82
Zinc	1.20–3.23[2]	0.46	1.37	0.86	0.49
Vitamins (mg/100 g)					
Vitamin C	99–511[1]	30.28	86.20	83.54	65.76
Vitamin B2	4.45–6.36[3]	0.10	0.269	0.19	0.13
Vitamin E	3.24–5.80[3]	1.29	0.46	0.44	0.08
β-carotene	35.21–52.01[4]	2.65	8.46	8.55	2.59
Lutein	19.01–28.85[4]	3.85	7.44	8.40	1.74

[1] Tang *et al.* (2021); [2] Sun *et al.* (2014); [3] Suárez *et al.* (2020); [4] Li *et al.* (2017); [5] Longvah *et al.* (2017).

Antioxidant activity

Several biological effects including antioxidant activity have been documented from phytochemicals such as flavonoids and related phenolic compounds that are typically present in sweet potato leaves. Sweet potato leaves, as a source of antioxidant molecules, serve as scavengers for the reactive oxygen species within the cell (Islam, 2006, 2014; Sun *et al.*, 2014; Abong' *et al.*, 2020b; Makori *et al.*, 2020; Suárez *et al.*, 2020). A range of evidence indicates that degenerative diseases such as cancer, asthma, diabetes, cardiovascular disease, inflammation, senile dementia and eye disease are caused by the damaging effect of free radical reactions (Islam, 2006, 2014; Sun *et al.*, 2014; Abong' *et al.*, 2020b; Alam *et al.*, 2020a; Makori *et al.*, 2020; Suárez *et al.*, 2020).

Leaves from sweet potato cultivars have been reported to be effective in scavenging

Table 9.4. Phenolic acids, anthocyanins and flavonoids identified in sweet potato leaves.

Phenolic acids (Zhang *et al.*, 2020; Makori *et al.*, 2020; Suárez *et al.*, 2020; Tang *et al.*, 2021)	Anthocyanins (Islam, 2006; Makori *et al.*, 2020)	Flavonoids (Makori *et al.*, 2020; Suárez *et al.*, 2020)
5-caffeoylquinic acid	Cyaniding derivatives	Myricetrin
3-caffeoylquinic acid	Peonidin derivatives	Luteolin
4-caffeoylquinic acid	Pelargonidin derivatives	Isoquercitrin
3,4-dicaffeoylquinic acid		Astragalin
3,5-dicaffeoylquinic acid		Quercitrin
Caffeic acid		Tiliroside
4,5-dicaffeoylquinic acid		Quercetin
3,4,5-tricaffeoylquinic acid		Apigenin
		Kaempferol

Table 9.5. Common antinutritive compounds found in sweet potato leaves and traditional leafy vegetables.

Antinutritive compound (g/100 g)	Sweet potato leaves	Spinach	Drumstick leaves	Green amaranth	Radish
Oxalate	0.51–1.62[1]	1.145–12.576[2,3]	2.754[3]	3.38–4.33[6]	0.209[4]
Tannins	0.87–5.05[1]	-	0.018[5]	0.0004[7]	-
Phytic acid	0.49–0.58[1]	0.129[4]	0.273[5]	0.21–0.25[7]	-

[1] Abong' *et al.* (2020a, b); [2] Chai and Liebman (2005); [3] Radek and Savage (2008); [4] Yadav and Sehgal (2003); [5] Saha *et al.* (2015); [6] Mziray *et al.* (2001); [7] Sivakumar *et al.* (2018); - data not available.

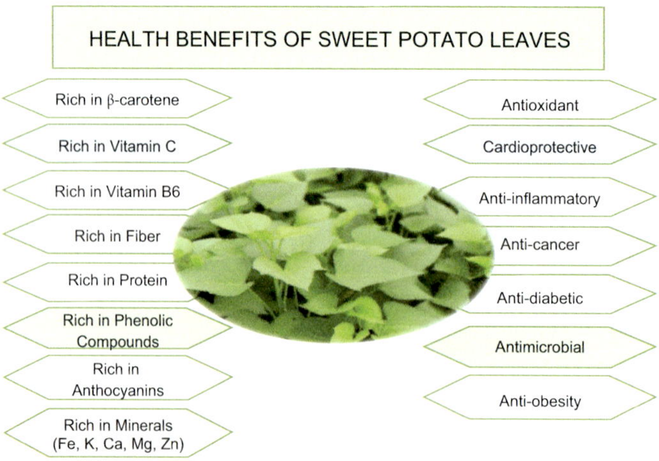

Fig. 9.4. Health benefits of sweet potato leaves.

free radicals by varying degrees (Sun *et al.*, 2014; Makori *et al.*, 2020; Suárez *et al.*, 2020). Compared to major commercial vegetables such as spinach, amaranth, drumstick leaves, cauliflower, aubergine etc., sweet potato leaf contains high concentrations of phenolics (Table 9.6). The scavenging ability of free radicals is associated with the content of total polyphenols and carbohydrates. This suggests that carbohydrates may be involved in the prevention of polyphenol oxidation (Sun *et al.*, 2014). In a study by Makori *et al.* (2020), leaf of sweet potato was found to be more effective in scavenging free radicals than the stalk, stem, skin and flesh. However, due to the different quantification methods used for assessing antioxidant activity in each study, comparison of antioxidant properties of leaves throughout the studies is to some extent difficult.

Antidiabetic effect

Studies have suggested that sweet potato leaf has the ability to reduce the blood glucose level. *In vivo* studies have documented the efficiency of sweet potato leaf extract in controlling blood sugar levels and reducing insulin resistance. Sweet potato leaf hot water extract was found to reduce the fasting blood glucose levels in both healthy and STZ-induced diabetic rats (Olowu *et al.*, 2011). Similarly, Nagamine *et al.* (2014) reported that diet containing sweet potato leaf extract can lessen hyperglycemia in KK-Ay mice, via increased glucagon-like peptide-1 (GLP-1) secretion. Generally, the antidiabetic property of sweet potato leaf has been attributed to its phytochemical content such as phenolic acids and anthocyanins (Nagamine *et al.*, 2014). Moreover, extract of sweet potato leaves can inhibit the α-amylase and α-glucosidase enzymes, which are responsible for the production of monosaccharides by breaking down starch and oligosaccharides (Zengin *et al.*, 2017).

Anticancer potential

Methanolic extracts of sweet potato leaves have been shown to effectively inhibit cell proliferation in all the studied human prostate cancer cell lines (LNCaP, DU145, PC-3, C4-2, C4-2B) with IC_{50} values in the 145–315 μg/ml range. Additionally, these extracts from leaves marginally affected the normal prostate epithelial cells (PrEC and RWPE-1) with IC_{50} values of 1000 and 1250 μg/ml for PrEC and RWPE-1 cells, respectively (Karna *et al.*, 2011). The molecular mechanism depicting the anticancer effect of leaf extracts in PC-3 prostate cancer cells, a representative of cancer cells, was linked to their cell cycle progression interference, clonogenic survival decreases, cell cycle modulations and regulatory molecules for apoptosis, and apoptosis induction (Karna *et al.*, 2011).

In different cancer cells, anthocyanins or anthocyanin-rich extracts have shown an inhibitory effect on cancerous cell development. Caffeoylquinic acid derivatives, 3,4,5-tricaffeoylquinic acid, in particular, have been shown to be effective in inhibiting the proliferation of human stomach Kato III cancer cells, colon cancer DLD-1 cells and promyelocytic leukemia HL-60 cells, dose-dependently and cell-line-dependently (Kurata *et al.*, 2007). The effectiveness of 3,4,5-tricaffeoylquinic acid is hyphothesized to be mediated with apoptosis at the molecular level by stimulating the activity of caspase-3 and the expression of c-Jun (Kurata *et al.*, 2007). Moreover, the reverse mutations induced by Trp-P-1 on *Salmonella typhimurium* TA were inhibited effectively by caffeoylquinic acid derivatives (Islam, 2006). There is a need for more effort to recognize key bioactive or bioactive mixtures with cancer prevention potential. The sweet potato-induced molecular mechanism of growth suppression in cancer cells remains to be explained.

Cardioprotective effect

Low-density lipoprotein oxidation can cause complications that can lead to cardiovascular disease caused by atherosclerosis (Wang *et al.*, 2016). Sweet potato leaf extract was capable of suppressing *in vitro* low-density lipoprotein oxidation, and this suppression was due to the antioxidant function of leaf phytochemicals in human subjects (Nagai *et al.*, 2011). The

Table 9.6. Total phenolic content and antioxidant activity of sweet potato leaves and traditional vegetables.

Total phenolic content and antioxidant activity	Sweet potato leaves							
	Suárez et al. (2020)[1]	Makori et al. (2020)[2]	Spinach[3]	Red amaranth[4]	Green amaranth[5]	Drumstick leaves[3]	Cauliflower[3]	Eggplant[3]
Total Phenolic content	9100 mg GAE/100 g	440–12080 mg CAE/100 g	258 mg GAE/100 g	17.223 mg GAE/100 g	1.483 mg GAE/100 g	43 mg GAE/100 g	896 mg GAE/100 g	59 mg GAE/100 g
Antioxidant activity								
DPPH	7.4g AAE/100 g	2.580–2.920g AAE/100 g	171 µmol TE/100 g	1997 µg TE/100 g	1374 µg TE/100 g	134 µmol TE/100 g	74 µmol TE/100 g	90 µmol TE/100 g
ABTS	10.6g AAE/100 g	3.950–5.300g AAE/100 g	-	3909 µg TE/100 g	2590 µg TE/100 g	-	-	-
FRAP	0.617 µmol TE/100 g	440–550 mg TE/100 g	-	-	-	-	-	-

[1] Suárez *et al.* (2020); [2] Makori *et al.* (2020); [3] Shaheen *et al.* (2014); [4] Sarker and Oba (2019); [5] Sarker *et al.* (2020); GAE: gallic acid equivalent; CAE: chlorogenic acid equivalents; AAE: ascorbic acid equivalent; TE: trolox equivalent; - data not available.

abundant phytochemical anthocyanin in sweet potato leaf has been reported to be capable of reducing the risk of coronary artery disease (Islam, 2006; Wang *et al.*, 2016).

Many studies have shown that sweet potato leaves are an excellent source of B6 vitamins, which facilitate homocysteine breakdown. The fact that homocysteine leads to the hardening of blood vessels and arteries is noteworthy. Sweet potato leaves play a major role in improving heart health as an exemplary source of potassium that reduces blood pressure and preserves fluid balance (Mohanraj, 2018).

The administration of aqueous sweet potato tuber extracts, containing tannin, saponin, flavonoids, terpenoids, alkaloids, anthraquinones, reducing sugar and cardiac glycosides, has resulted in a decrease in serum creatine and lactate dehydrogenase activity. The findings obtained indicate a possible cardioprotective effect of sweet potato aqueous tuber extracts (Shafe *et al.*, 2016). There are no studies reporting the inhibitory effect of sweet potato leaves on creatine and lactate dehydrogenase activity, however the bioactive compounds present in the leaves can be expected to possess this ability.

For the treatment of hypertension and congestive heart failure, an angiotensin-converting enzyme (ACE) inhibitor is mainly used. Freeze-dried powders of sweet potato tips comprising leaves and stem have been found to inhibit ACE activity (Cui *et al.*, 2011). The ACE inhibitory activity of sweet potato tips might be exerted by the caffeoylquinic acid (Ishiguro *et al.*, 2007).

Angiogenesis is a dynamic process where existing blood vessels form new blood vessels and could lead to the development and progression of many diseases, including atherosclerosis. This process is a consequence of the disintegration of extracellular matrices, endothelial cell migration and proliferation, and new blood vessel maturation (tube formation). The *in vitro* evaluation of purple sweet potato leaves on angiogenesis indices exhibited encouraging findings. In human umbilical vascular endothelial cells, sweet potato leaves were found to reduce cell proliferation and migration, tube formation and activity of MMP-22 secreted in a dose-dependent manner (Chen *et al.*, 2011).

Immunomodulatory effect

Some studies have concluded that sweet potato leaf extracts have modulatory effects on the immune system and health. In human aortic endothelial cells, purple sweet potato leaf extract was able to significantly inhibit TNF-α-induced monocyte-endothelial cell adhesion and attenuate VCAM-1, IL-8 and CD40 expression (Chao *et al.*, 2013). The anti-inflammatory effects of purple sweet potato leaf extract are mediated through modulation of NFXB and MAPK signalling (Chao *et al.*, 2013). Similarly, purple sweet potato leaf extract has shown to down-regulate the expression of pro-inflammatory molecules, such as TNF-α and IL-6 in adipocytes (Lee *et al.*, 2015). The presence of phenolic acids, anthocyanins and flavonoids in leaf extracts may be ascribed to the anti-inflammatory ability. As a rich source of manganese, phosphorus, magnesium and vitamins A and E, sweet potato leaves also promote a strong immune system and are important antioxidants that combat diseases (Johnson and Pace, 2010). Moreover, in a study involving 16 healthy human adults, the intake of purple sweet potato leaves was able to modulate T-lymphocyte functions, natural killer cell lytic activity and antibody production (Chen *et al.*, 2005).

Antiobesity effect

Obesity is considered to be a high risk factor to the development of type 2 diabetes, high blood pressure, heart disease, stroke, cancer and arthritis (Johnson and Pace, 2010). It has been found that sweet potato leaves possess antiobesity potential. In Sprague-Dawley rats fed a high-fat diet supplemented with sweet potato leaves, the gain in weight and adipose tissue was lower than in those not fed sweet potato leaves. In rats fed sweet potato leaves, the plasma triglyceride and total cholesterol and hepatic cholesterol levels were significantly lower relative to rats fed the high-fat diet alone (Kurata *et al.*, 2017). Additionally, administration of purple sweet potato leaves induced a dose-dependent decrease in preadipocyte 3T3-L1 cell proliferation, but did not alter the viability of the cells, and rendered ERK inactivation. Furthermore, pre-treatment with purple sweet potato leaves of confluent 3T3-L1 cells led to decreased lipid accumulation in differentiated 3T3-L1 cells. The inhibition of lipogenesis may be exerted through down-regulation of the expression of the transcription factors C/EBPI and SREBP-1 during 3T3-L1 adipocyte differentiation (Lee *et al.*, 2015).

Antimicrobial effect

Several foodborne pathogens, including *Escherichia coli* O157:H7, *Bacillus cereus* and *Staphylococcus aureus*, were effectively suppressed by lyophilized leaf powder from three sweet potato cultivars (Islam, 2008). Leaf extracts have shown limited effects on health-promoting bifidobacterium growth (*Bifidobacterium adolescentis*, *Bifidobacterium bifidum*, *Bifidobacterium breve*, *Bifidobacterium longum* and *Bifidobacterium infuntis*). Polysaccharides in lyophilized leaf extracts are known to be major antibacterial agents (Islam, 2008). The antimicrobial properties of proteins (10 and 22 kDa) in sweet potato leaves have also been claimed previously (Wang *et al.*, 2003). Mbaeyi-Nwa and Emejulu (2013) tested peptone, water and ethanol extracts of sweet potato leaves against *Escherichia coli*, *Salmonella typhi*, *Staphylococcus aureus*, *Aspergillus niger*, *Penicillium* spp., *Pseudomonas aeroginosa* and *Klebsiella pneumonia* for antimicrobial activity. They documented that by inhibiting the growth of all the organisms except *Escherichia coli* and *Penicillium* spp., the water extract exhibited excellent antimicrobial activity at different concentrations. However, 70% ethanolic sweet potato leaf extract has minimally affected the growth of *Streptococcus mutans*, *Streptococcus mitis*, *Staphylococcus aureus* and *Candida albicans* in both agar disk diffusion and agar well diffusion experiments (Pochapski *et al.*, 2011). Then, extraction solvents are a factor affecting the antibacterial ability of sweet potato leaves. Novel food products containing sweet potatoes with antibacterial properties against foodborne spoils and pathogenic bacteria are suggested to improve food quality and protection.

Other benefits

Sweet potato leaves, likely because of the sugar and other nutrients that influence body temperature, are thought to be capable of warming up the body. This property, along with its potent impact on congestion, is also beneficial for people suffering from bronchitis. To treat bronchitis, the concentration of vitamin C, iron and other nutrients helps (Nagai *et al.*, 2011).

The high fibre content of sweet potatoes aids in preserving water in the body. Magnesium, which is found in sweet potato leaves, helps with digestion as well. For the stomach and intestines, they have a soothing effect. In treating stomach ulcers, the B-complex vitamins, vitamin C, β-carotene, potassium and calcium found in sweet potato leaves are helpful. In addition, fibre in sweet potato leaves alleviates constipation (Ruhee *et al.*, 2020) and the associated formation of acid, which decreases the risk of ulcers (Mohanraj, 2018).

In herbal medicine, sweet potato leaf is used as a treatment for anaemia due to its hematinous effects (Osime *et al.*, 2009). In a recent study by Montejo *et al.* (2015), a powder diet of sweet potato leaves increased the amount of packed cells, haemoglobin and red blood cell count in mice. Similarly, an earlier study reported a substantial increase in the amount of packaged cells, white blood cells and platelets in rabbits fed with sweet potato leaf extract (Osime *et al.*, 2009).

Novel Products/Formulations

The application of sweet potato in developing functional foods is promising. Food producers have begun to look at utilizing sweet potato for formulating juice, pasta, ice cream and other novel food products. To date, few products have actually been formulated with sweet potato and this requires further research for marketing opportunities. Utilization of sweet potato in dairy products (e.g. curd and yogurt, acidophilus milk, etc.), noodles, bread, alcoholic and non-alcoholic beverages has been previously reported (Wang *et al.*, 2016). Thus, there exists a broad opportunity for developing sweet potato-based functional foods such as tea, confections etc., as nutritional supplements to convey beneficial polyphenolic compounds.

Conclusions and Future Trends

Sweet potato is a universal food crop that, for its nutritional and medicinal value, can be explored further. Sweet potato is a rich source of dietary fibre, protein, vitamins and minerals. Its leaves contain higher concentrations of polyphenolics than several commercially available vegetables. These polyphenolics are claimed to have antioxidant, anticancer, antidiabetic, cardioprotective, antimicrobial, immune enhancing and hepatoprotective properties. Future research is required into the functional properties and potential for the application of isolated bioactive compounds from sweet potato leaves. Usage of *in vivo* techniques such as molecular marker analysis of bioactive compound-governed modulations should be included in future studies. The effect of various cooking methods on the retention of phenolics and carotenoids must also be considered, as well as the necessity to standardize the proper cooking methods. Additionally, the standardization of methods for quantification, assessment of physiological activities, bioaccessibility and bioavailability of polyphenolics deserves subsequent consideration. Proper evaluation methods for exploring the mechanisms of action behind the interactions of different sweet potato bioactives and their use in the management of human diseases are necessary. To evaluate the safety of the bioactive constituents of sweet potato leaf, comprehensive toxicological studies should be performed. This will reinforce research interest in promoting the medicinal potential of sweet potato beyond its value in food supplies.

Acknowledgement

Authors are thankful to Professor Mohammad Akhtaruzzaman, Institute of Nutrition and Food Science, University of Dhaka, for his useful discussion.

References

Abong', G.O., Muzhingi, T., Okoth, M.W., Ng'ang'a, F., Emelda Ochieng, P. *et al.* (2020a) Processing methods affect phytochemical contents in products prepared from orange-fleshed sweetpotato leaves and roots. *Food Science & Nutrition* 9(2), 1070–1078. DOI: 10.1002/fsn3.2081.

Abong', G.O., Muzhingi, T., Okoth, M.W., Ng'ang'a, F., Ochieng', P.E. *et al.* (2020b) Phytochemicals in leaves and roots of selected Kenyan Orange Fleshed Sweet Potato (OFSP) varieties. *International Journal of Food Science* 2020, 3567972. DOI: 10.1155/2020/3567972.

Alam, M.K., Rana, Z.H. and Islam, S.N. (2016) Comparison of the proximate composition, total carotenoids and total polyphenol content of nine orange-fleshed sweet potato varieties grown in Bangladesh. *Foods* 5(3), 64. DOI: 10.3390/foods5030064.

Alam, M.K., Rana, Z.H., Islam, S.N. and Akhtaruzzaman, M. (2020a) Comparative assessment of nutritional composition, polyphenol profile, antidiabetic and antioxidative properties of selected edible wild plant species of Bangladesh. *Food Chemistry* 320, 126646. DOI: 10.1016/j.foodchem.2020.126646.

Alam, M.K., Sams, S., Khan, M.S., Azmir, J., Ahsan, M. *et al.* (2020b) Profiling of minerals, water soluble vitamins and carotenoid in selected unconventional leafy and non-leafy vegetables of bangladesh. *Natural Product Research*. DOI: 10.1080/14786419.2020.1849198.

Alam, M.K., Sams, S., Rana, Z.H., Akhtaruzzaman, M. and Islam, S.N. (2020c) Minerals, vitamin C, and effect of thermal processing on carotenoids composition in nine varieties orange-fleshed sweet potato (*Ipomoea batatas* L.). *Journal of Food Composition and Analysis* 92, 103582. DOI: 10.1016/j.jfca.2020.103582.

Alam, M.K., Tuli, R., Khan, M.S., Abdullah, A.T.M., Khatun, M. *et al.* (2020d) Chromatographic assessment of polyphenolic profile and total phenolic content and antioxidant activity of common leafy vegetables in Bangladesh. *Current Chromatography* 7(1), 40–50. DOI: 10.2174/221324060799920 0421144940.

BBS (2019) *Yearbook of Agricultural Statistics-2018, 30th series*. Bangladesh Bureau of Statistics, Statistics Division, Ministry of Planning, Government of the People's Republic of Bangladesh, Dhaka, Bangladesh, p. 323.

Chai, W. and Liebman, M. (2005) Effect of different cooking methods on vegetable oxalate content. *Journal of Agricultural and Food Chemistry* 53(8), 3027–3030. DOI: 10.1021/jf048128d.

Chao, P.Y., Huang, Y.P. and Hsieh, W.B. (2013) Inhibitive effect of purple sweet potato leaf extract and its components on cell adhesion and inflammatory response in human aortic endothelial cells. *Cell Adhesion & Migration* 7(2), 237–245. DOI: 10.4161/cam.23649.

Chen, Y.C., Chang, H.S., Lai, H.M. and Jeng, S.T. (2005) Characterization of the wound-inducible protein ipomoelin from sweet potato. *Plant, Cell and Environment* 28(2), 251–259. DOI: 10.1111/j.1365-3040.2005.01271.x.

Chen, C.-M., Li, S.-C., Chen, C.-Y.O., Au, H.-K., Shih, C.-K. *et al.* (2011) Constituents in purple sweet potato leaves inhibit in vitro angiogenesis with opposite effects ex vivo. *Nutrition* 27(11–12), 1177–1182. DOI: 10.1016/j.nut.2011.01.005.

Chen, S.-P., Wang, S.-Y., Huang, M.-Y., Lin, K.-H., Hua, S.-M. *et al.* (2018) Physiological and molecular analyses of chlorophyllase in sweet potatoes with different-colored leaves. *South African Journal of Botany* 114, 272–279. DOI: 10.1016/j.sajb.2017.11.021.

CIP (2019) *Institutional Brochure*. International Potato Center, Lima, Peru, p. 4. Available at: https://hdl.handle.net/10568/99442 (accessed 29 March 2022).

Cui, L., Liu, C., Li, D. and Song, J. (2011) Effect of processing on taste quality and health-relevant functionality of sweet potato tips. *Agricultural Sciences in China* 10(3), 456–462. DOI: 10.1016/S1671-2927(11)60025-4.

FAOSTAT (2020) Statistics Division of Food and Agriculture Organization of the United Nations. Available at: http://www.fao.org/faostat/en/#data (accessed 29 March 2022).

Huamari, Z. (1992) *Systematic botany and morphology of the sweet potato plant*. Technical Information Bulletin 25. International Potato Centre, Lima, Peru, p. 22.

Ishida, H., Suzuno, H., Sugiyama, N., Innami, S., Tadokoro, T, *et al.* (2000) Nutritive evaluation on chemical components of leaves, stalks and stems of sweet potatoes (*Ipomoea batatas* Poir). *Food Chemistry* 68(3), 359–367. DOI: 10.1016/S0308-8146(99)00206-X.

Ishiguro, K., Yoshimoto, M., Tsubata, M. and Takagaki, K. (2007) Hypotensive effect of sweetpotato tops. *Nippon Shokuhin Kagaku Kogaku Kaishi* 54(1), 45–49. DOI: 10.3136/nskkk.54.45.

Islam, S. (2006) Sweetpotato (*Ipomoea batatas* L.) leaf: its potential effect on human health and nutrition. *Journal of Food Science* 71(2), R13–R121. DOI: 10.1111/j.1365-2621.2006.tb08912.x.

Islam, S. (2008) Antimicrobial activities of *Ipomoea batatas* (L.) leaf. *Journal of Food, Agriculture and Environment* 6(1), 14–17.

Islam, S. (2014) *Nutritional and Medicinal Qualities of Sweetpotato Tops and Leaves*. Cooperative Extension Service, University of Arkansas, Fayetteville, Arkansas, FSA6135.

Jackson, D.M., Harrison, H.F., Jarret, R.L. and Wadl, P.A. (2020) Phenotypic variation in leaf morphology of the USDA, ARS sweetpotato (*Ipomoea batatas*) germplasm collection. *HortScience: A Publication of the American Society for Horticultural Science* 55(4), 465–475. DOI: 10.21273/HORTSCI14703-19.

Johnson, M. and Pace, R.D. (2010) Sweet potato leaves: properties and synergistic interactions that promote health and prevent disease. *Nutrition Reviews* 68(10), 604–615. DOI: 10.1111/j.1753-4887.2010.00320.x.

Karna, P., Gundala, S.R., Gupta, M.V., Shamsi, S.A., Pace, R.D. *et al.* (2011) Polyphenol-rich sweet potato greens extract inhibits proliferation and induces apoptosis in prostate cancer cells in vitro and in vivo. *Carcinogenesis* 32(12), 1872–1880. DOI: 10.1093/carcin/bgr215.

Kurata, R., Adachi, M., Yamakawa, O. and Yoshimoto, M. (2007) Growth suppression of human cancer cells by polyphenolics from sweetpotato (*Ipomoea batatas* L.) leaves. *Journal of Agricultural and Food Chemistry* 55(1), 185–190. DOI: 10.1021/jf0620259.

Kurata, R., Kobayashi, T., Ishii, T., Niimi, H., Niisaka, S. *et al.* (2017) Influence of sweet potato (*Ipomoea batatas* L.) leaf consumption on rat lipid metabolism. *Food Science and Technology Research* 23(1), 57–62. DOI: 10.3136/fstr.23.57.

Lee, S.L., Chin, T.Y., Tu, S.C., Wang, Y.J., Hsu, Y.T. *et al.* (2015) Purple sweet potato leaf extract induces apoptosis and reduces inflammatory adipokine expression in 3T3-L1 differentiated adipocytes. *Evidence-Based Complementary and Alternative Medicine* 2015, 126302. DOI: 10.1155/2015/126302.

Li, M., Jang, G.Y., Lee, S.H., Kim, M.Y., Hwang, S.G. *et al.* (2017) Comparison of functional components in various sweet potato leaves and stalks. *Food Science and Biotechnology* 26(1), 97–103. DOI: 10.1007/s10068-017-0013-6.

Longvah, T., Ananthan, R., Bhaskarachary, K. and Venkaiah, K. (2017) *Indian Food Composition Tables*. National Institute of Nutrition, Indian Council of Medical Research, Hyderabad, India.

Makori, S.I., Mu, T.H. and Sun, H.N. (2020) Total polyphenol content, antioxidant activity, and individual phenolic composition of different edible parts of 4 sweet potato cultivars. *Natural Product Communications* 15(7), 1934578X2093693. DOI: 10.1177/1934578X20936931.

Mbaeyi-Nwa, I.E. and Emejulu, V.N. (2013) Evaluation of phytochemical composition and antimicrobial activity of sweet potato (*Ipomoea batatas*) leaf. *Pakistan Journal of Nutrition* 12(6), 575–586. DOI: 10.3923/pjn.2013.575.586.

Mohanraj, R. (2018) Sweet potato: bioactive compounds and health benefits. In: Mérillon, J.-M. and Ramawat, K.G. (eds) *Bioactive Molecules in Food. Reference Series in Phytochemistry*. Springer, Cham, Switzerland, pp. 919–934. DOI: 10.1007/978-3-319-78030-6.

Mohanraj, R. and Sivasankar, S. (2014) Sweet potato (*Ipomoea batatas* [L.] Lam)- A valuable medicinal food: A review. *Journal of Medicinal Food* 17(7), 733–741. DOI: 10.1089/jmf.2013.2818.

Montejo, J.F., Mondonedo, J.A.B., Lee, M.G.A., Ples, M.B. and Vitor, R.J.S. (2015) Hematological effects of *Ipomoea batatas* (camote) and *Phyllanthus niruri* (sampa-sampalukan) from Philippines in the ICR mice (*Mus musculus*). *Asian Pacific Journal of Tropical Biomedicine* 5(1), 29–33. DOI: 10.1016/S2221-1691(15)30166-0.

Mziray, R.S., Imungi, J.K. and Karuri, E.G. (2001) Nutrient and antinutrient in contents of raw and cooked Amaranthus hybridus. *Ecology of Food and Nutrition* 40(1), 53–65. DOI: 10.1080/03670244.2001.9991637.

Nagai, M., Tani, M., Kishimoto, Y., Iizuka, M., Saita, E. *et al.* (2011) Sweet potato (*Ipomoea batatas* L.) leaves suppressed oxidation of low density lipoprotein (LDL) in vitro and in human subjects. *Journal of Clinical Biochemistry and Nutrition* 48(3), 203–208. DOI: 10.3164/jcbn.10-84.

Nagamine, R., Ueno, S., Tsubata, M., Yamaguchi, K., Takagaki, K. *et al.* (2014) Dietary sweet potato (*Ipomoea batatas* L.) leaf extract attenuates hyperglycaemia by enhancing the secretion of glucagon-like peptide-1 (GLP-1). *Food & Function* 5(9), 2309–2316. DOI: 10.1039/c4fo00032c.

Olowu, A., Adeyemi, O. and Adeneye, A. (2011) Hypoglycaemic effect of *Ipomoea batatas* aqueous leaf and stem extract in normal and streptozotocin-induced hyperglycaemic rats. *Journal of Natural Pharmaceuticals* 2(2), 56. DOI: 10.4103/2229-5119.83951.

Osime, E., Ediale, G., Omoti, C. and Famodu, A. (2009) Effect of sweetpotato leaf (*Ipomoea Batatas*) extract on some haematological parameters using rabbits. *Journal of Medicine and Biomedical Research* 7(1–2), 12–15. DOI: 10.4314/jmbr.v7i1-2.44540.

Pochapski, M.T., Fosquiera, E.C., Esmerino, L.A., Dos Santos, E.B., Farago, P.V. *et al.* (2011) Phytochemical screening, antioxidant, and antimicrobial activities of the crude leaves' extract from *Ipomoea batatas* (L.) Lam. *Pharmacognosy Magazine* 7(26), 165–170. DOI: 10.4103/0973-1296.80682.

Radek, M. and Savage, G.P. (2008) Oxalates in some Indian green leafy vegetables. *International Journal of Food Sciences and Nutrition* 59(3), 246–260. DOI: 10.1080/09637480701791176.

Rana, Z.H., Alam, M.K. and Akhtaruzzaman, M. (2019) Nutritional composition, total phenolic content, antioxidant and α-amylase inhibitory activities of different fractions of selected wild edible plants. *Antioxidants* 8(7), 203. DOI: 10.3390/antiox8070203.

Ruhee, R.T., Sams, S., Begum, A., Alam, M.K., Begum, P. *et al.* (2020) Dietary fibre content in ethnic and unconventional vegetables and fruits growing in Bangladesh. *Potravinarstvo Slovak Journal of Food Sciences* 14, 33–41. DOI: 10.5219/1202.

Saha, J., Biswal, A.K. and Deka, S.C. (2015) Chemical composition of some underutilized green leafy vegetables of Sonitpur district of Assam, India. *International Food Research Journal* 22(4), 1466–1473.

Sams, S., Alam, M.K., Ahsan, M. and Islam, S.N. (2020) Total carotenoids content and assessment of carotene profile by high-performance liquid chromatography in selected vegetables of Bangladesh with special reference to some unconventional varieties. *Current Chromatography* 7(1), 24–31. DOI: 10.2174/2213240607999200731010929.

Sarker, U. and Oba, S. (2019) Protein, dietary fiber, minerals, antioxidant pigments and phytochemicals, and antioxidant activity in selected red morph Amaranthus leafy vegetable. *PloS One* 14(12), e0222517. DOI: 10.1371/journal.pone.0222517.

Sarker, U., Hossain, M.M. and Oba, S. (2020) Nutritional and antioxidant components and antioxidant capacity in green morph amaranthus leafy vegetable. *Scientific Reports* 10(1), 1336. DOI: 10.1038/s41598-020-57687-3.

Shafe, M.O., Eze, E.D., Ubhenin, A.E. and Tende, J.A. (2016) Effects of aqueous tuber extract of *ipomea batatas* on cardiac enzymes, lipid profile and organ weights in wistar rats. *Journal of Basic Sciences and Applied Research* 4, 414–417.

Shaheen, N., Rahim, A.T.M., Mohiduzzaman, M., Banu, C.P., Bari, M.L. *et al.* (2014) *Food Composition Table for Bangladesh*, 2nd edn. Institute of Nutrition and Food Science, University of Dhaka, Dhaka, Bangladesh.

Sivakumar, D., Chen, L. and Sultanbawa, Y. (2018) A comprehensive review on beneficial dietary phytochemicals in common traditional Southern African leafy vegetables. *Food Science & Nutrition* 6(4), 714–727. DOI: 10.1002/fsn3.643.

Suárez, S., Mu, T., Sun, H. and Añón, M.C. (2020) Antioxidant activity, nutritional, and phenolic composition of sweet potato leaves as affected by harvesting period. *International Journal of Food Properties* 23(1), 178–188. DOI: 10.1080/10942912.2020.1716796.

Sun, H., Mu, T., Xi, L., Zhang, M. and Chen, J. (2014) Sweet potato (*Ipomoea batatas* L.) leaves as nutritional and functional foods. *Food Chemistry* 156, 380–389. DOI: 10.1016/j.foodchem.2014.01.079.

Tang, C.-C., Ameen, A., Fang, B.-P., Liao, M.-H., Chen, J.-Y. *et al.* (2021) Nutritional composition and health benefits of leaf-vegetable sweet potato in South China. *Journal of Food Composition and Analysis* 96, 103714. DOI: 10.1016/j.jfca.2020.103714.

Wang, Y.L., Wu, C.Y., Chang, C.T. and Sung, H.Y. (2003) Invertase inhibitors from sweet potato (*Ipomoea batatas*): purification and biochemical characterization. *Journal of Agricultural and Food Chemistry* 51(16), 4804–4809. DOI: 10.1021/jf026107i.

Wang, S., Nie, S. and Zhu, F. (2016) Chemical constituents and health effects of sweet potato. *Food Research International* 89(Pt 1), 90–116. DOI: 10.1016/j.foodres.2016.08.032.

Xi, L., Mu, T. and Sun, H. (2015) Preparative purification of polyphenols from sweet potato (*Ipomoea batatas* L.) leaves by AB-8 macroporous resins. *Food Chemistry* 172, 166–174. DOI: 10.1016/j.foodchem.2014.09.039.

Yadav, S.K. and Sehgal, S. (2003) Effect of domestic processing and cooking on selected antinutrient contents of some green leafy vegetables. *Plant Foods for Human Nutrition* 58(3), 1–11. DOI: 10.1023/B:QUAL.0000040359.40043.4f.

Zengin, G., Locatelli, M., Stefanucci, A., Macedonio, G., Novellino, E. *et al.* (2017) Chemical characterization, antioxidant properties, anti-inflammatory activity, and enzyme inhibition of *Ipomoea batatas* L. leaf extracts. *International Journal of Food Properties* 20(sup2), 1–13. DOI: 10.1080/10942912.2017.1357127.

Zhang, C., Liu, D., Wu, L., Zhang, J., Li, X. *et al.* (2020) Chemical characterization and antioxidant properties of ethanolic extract and its fractions from sweet potato (*Ipomoea batatas* L.) leaves. *Foods* 9(1), 15. DOI: 10.3390/foods9010015.

10 Coral Tree (*Erythrina variegata* L.)

Mohammad Khairul Alam* and Mohammad Akhtaruzzaman
Institute of Nutrition and Food Science, University of Dhaka, Dhaka, Bangladesh

Abstract

Underutilized and/or traditional plants as a medium of herbal medicines have received greater attention in recent years in terms of boosting health status due to their potential pharmacological properties. One such underutilized plant is *Erythrina variegata* L., which is known as coral tree in English and Mandar in Bengali. The nutritive values of the leaves of *Erythrina variegate* L. contain important constituents such as protein, fibre and minerals (calcium, phosphorus, magnesium, potassium, sodium, zinc, copper and iron) and the seed protein is considered to be of good quality considering the amino acid composition. This plant also contains dietary bioactive compounds, such as alkaloids, flavonoids, triterpenoids and many others, which exert a large variety of health potentials including antidiabetic properties. Diabetes mellitus (DM) is a chronic metabolic condition that affects the endocrine gland and, due to poor insulin secretory responses and beta cell dysfunction, is manifested by hyperglycemia and glucose intolerance. Each year, this disorder causes millions of deaths worldwide and is one of the leading causes of death. Despite the tremendous advances made in the production of modern antidiabetic drugs, due to the various side effects associated with them these drugs are mostly deemed not to be effective in the management and treatment of type 2 DM (T2DM). Thus, since they have little to no side effects, exploration of natural products as therapeutic sources for the treatment of T2DM is encouraged. Moreover, utilization of underutilized and/or traditional plants would allow the population to consume more of these plants, increase their production and hence food security, and contribute to preserving biodiversity as well as dietary diversification. Therefore, the potential of the *Erythrina variegata* L. plant as a source of bioactive compounds in the management of T2DM is highlighted in this chapter.

Introduction

Ethnobotanical science has become more diverse in recent years, not only at the national level but also at the international level. Some of the literature has revealed that there is a substantial difference between ethnomedicine's theoretical validation and its uses. Nutritionally useful herbal medicines are now being used to boost health status due to their pharmacological properties (Ghani, 2003; Andarwulan *et al.*, 2012; Alam *et al.*, 2019, 2020a). Plants are therefore used concurrently as both food and medicine (Ghani, 2003; Andarwulan *et al.*, 2012; Alam *et al.*, 2019, 2020a). Owing to the large variety of plants around the world, the possibility of using plants as a medicine is immense. Cultural and geographical variables have also stimulated the use of traditional or unconventional plants for the treatment of several diseases in various formulations, such as crude extracts, whole plants, plant paste,

*Corresponding author: khairul.alam010@gmail.com

 Handbook of Phytonutrients in Indigenous Fruits and Vegetables (eds D. Sivakumar *et al.*)
DOI: 10.1079/9781789248067.0010

Fig. 10.1. Coral tree (*Erythrina variegata* L.).

infusions etc. (Ghani, 2003; Dewick, 2009). Herbal medicines are gaining popularity all over the world due to their production from natural sources and their having no and/or fewer toxic effects (Andarwulan *et al.*, 2012; Shori, 2015; Alam *et al.*, 2019, 2020a, b; Rana *et al.*, 2019). Recently, a number of studies have shown that traditional or unconventional plants are used to treat chronic disorders, such as diabetes, cardiovascular disease, inflammation-related diseases and as well as in many other applications.

Erythrina variegata L. (Fig. 10.1) is one such unconventional plant that can have potential application in treating and/or managing chronic disorders. This plant belongs to the family Fabaceae (Kumar *et al.*, 2010; Lim, 2014; Tabassum *et al.*, 2019) and is a deciduous tree with thorny stems and branches, thick scarlet or crimson clusters and showy red flowers with black seeds. It grows all over Bangladesh (Rahman *et al.*, 2010; Tabassum *et al.*, 2019). The genus *Erythrina* consists of more than 110 species and is native to the tropical and subtropical regions (Kumar *et al.*, 2010; Lim, 2014; Tabassum *et al.*, 2019). *Erythrina variegata* L. is known as Palitamadar, Paniamadar, Raktamadar, Mandar or Madar in Bengali and coral tree or tiger's claw around the world.

Bangladesh is the fertile Bengal region of South Asia that became independent in 1971. The nation is situated between 20° and 27° N (latitude) and 88° and 93° E (longitude). This positions it in one of the most fertile areas, with alluvial soil which leads to the growing of diversified flora, such as *Erythrina variegata* L. This area has become an important part of the public health service (Rahman *et al.*, 2012) due to the available nature of conventional herbal medicines. Bangladesh's geographical and cultural similarities to India have influenced local people's engagement with Indian Ayurveda and Unani medicine.

Botany

Erythrina variegata L. is an upright, highly branched, medium-sized, deciduous tree up to 25 m tall and 8–12 m spread. Leaves are alternate, trifoliate, 20–30 cm long, 6–15 cm long petioles, the highest terminal; glandular leaflet-stalk. Leaflets, with acuminate tips and obtuse bases, are triangular to narrowly rhomboidovate, shining green. Inflorescence in thick axillary and terminal racemes appear before the leaves; tomentose rachis; small bracts; bright red flowers; tubular calyx, minutely five-toothed; long corolla, wide normal, ovate-elliptical, short-clawed and 7–9 cm long, subequal wings and keels; stamens ten, essentially connate, exserted; glabrous-style multi-ovulated ovary, incurved. Greyish smooth stem with dispersed broad spaced conical spikes on the stem and branch. Pods are long, up to 38 cm, typically black, thin-walled and constricted around the seeds. There are five to ten seeds in each pod (Lim, 2014).

The systematic classification of *Erythrina variegata* L. is given below:

Family: Fabaceae
Genus: *Erythrina*
Species: *Erythrina variegata* L.

Traditional Use

In various countries of the South Asian region (Ghani, 2003; Kumar *et al.*, 2010; Tabassum *et al.*, 2019), different parts of the *Erythrina variegata* L. plant have historically been used as folk medicine for various ailments. Uses include as a nervous sedative, febrifuge, laxative, anti-asthmatic and antiepileptic (Kumar *et al.*, 2010; Tabassum *et al.*, 2019). The plant is used nationwide in Bangladesh and is sometimes used as an astringent as well as in fever and leprosy, while the leaf paste is used externally to alleviate joint pain and treat inflammation (Tabassum *et al.*, 2019). Leaf juice is used for earache, toothache, constipation and cough, and is also known to induce menstruation and lactation (Rahman *et al.*, 2007, 2010). Potential remedial effects for the treatment of some diseases have been demonstrated by the plant, including bacterial infection, convulsion, cuts and wounds, helminthiasis, cough and insomnia (Kumar *et al.*, 2010; Lim, 2014; Tabassum *et al.*, 2019). The plant is also useful in ophthalmia and skin diseases. Moreover, young and old leaves of *Erythrina variegata* L. can be eaten with rice or mixed with other vegetables after being steamed or stewed. The flowers are also edible after boiling and seeds can be consumed after boiling or roasting.

Nutritional Composition

The literature concerning the nutritional composition of different parts of the *Erythrina variegata* L. plant is scant. However, some aspects related to the nutritional composition of *Erythrina variegata* L. have been reported by some authors (Samanta and Laskar, 2009; Lim, 2014; Alam *et al.*, 2020a). Table 10.1 and Table 10.2 summarize the nutritional composition of leaves and the amino acid composition of seed protein of *Erythrina variegata* L., respectively, and show a

Table 10.1. Nutritional value of *Erythrina variegata* L. leaves, spinach and amaranth.

	Erythrina variegata L. leaves			
Nutrients	Alam et al. (2020a)[1]	Lim (2014)[2]	Spinach[3]	Amaranth[3]
Proximate composition (g/100 g)				
Moisture	85.7	78.1	91.8	88.8
Protein	21.12	5.0	2.4	4.5
Fat	1.55	0.7	0.3	0.3
Fibre	17.55	3.0	2.2	4.2
Ash	20.15	2.6	1.2	1.7
Carbohydrate	39.63	10.6	2.1	0.5
Minerals (mg/100 g)				
Sodium	351.09	-	69	59
Potassium	2266.56	-	187	261
Magnesium	218.67	-	179	181
Calcium	292.11	639.1	11	256
Iron	22.19	-	2.2	6
Zinc	4.15	4.1	0.35	0.96
Copper	0.49	-	0.06	0.22
Ascorbic acid (mg/100 g)	70.43	-	51.8	42

-: not reported; [1] dry weight (DW) basis; [2] fresh weight (FW) basis; [3] Shaheen *et al.* (2014) FW basis.

Table 10.2. Amino acid composition of *Erythrina variegata* L. seeds, rice and lentils.

Amino acids (mg/100 g)	*Erythrina variegata* L. seed[1]	Rice[2]	Lentils[2]
Essential			
Isoleucine	3720	291	1048
Leucine	12,200	556	2017
Lysine	7160	243	2118
Methionine	290	159	147
Phenylalanine	6030	359	1431
Threonine	3980	241	1032
Tryptophan	-	78	257
Valine	8040	410	1366
Non-essential			
Serine	7270	4354	1515
Glycine	5630	307	1150
Histidine	4340	159	633
Arginine	2600	560	2265
Alanine	4100	390	1917
Proline	7360	316	1075
Tyrosine	7890	225	836
Cysteine	1320	138	216

-: not reported; [1] Samanta and Laskar (2009) DW basis; [2] Shaheen *et al.* (2014) FW basis.

comparison with commonly consumed vegetables, cereal and pulses.

Phytochemicals

Diverse types of alkaloid such as erythaline, erythratine, erysothrine, hypaphorine, spiroamine, carboxylated indole-3-alkylamines, tetracyclic alkaloids, scoulerine, (+) coreximine, L-reticuline, erybidine, water-soluble bases, such as erysovine and stachydrine, spiroamine alkaloids, isoquinoline (erythritol) and isococcolinine alkaloids have been found to be present in different parts of *Erythrina variegata* L. (Kumar *et al.*, 2010; Lim, 2014; Tabassum *et al.*, 2019). Besides alkaloids, isoflavone derivatives, triterpenoids (oleanolic acid, β-sitosterol, stigmasterol, lupeol etc.), glycosides and others have also been extracted from different parts of *Erythrina variegata* L. (Kumar *et al.*, 2010; Lim, 2014; Tabassum *et al.*, 2019). The leaves and seeds have been reported to contain lectin, a potential bioactive compound (Lim, 2014). The leaves also contain phenolic compounds such as gallic acid, catechin and epicatechin, vanillic acid, caffeic acid, rutin hydrate, ellagic acid, myricetin, kaempferol and quercetin (Muthukrishnan *et al.*, 2016; Alam *et al.*, 2020a). Chemical structures of commonly found phytochemicals in *Erythrina variegata* L. are presented in Fig. 10.2.

Antinutritive Compounds

The leaves and seeds (when eaten raw) of *Erythrina variegata* L. contain poisonous alkaloids (erythrinine), that could affect the central nervous system (Lim, 2014). Antinutritional compounds such as tannins have also been reported in the leaves of *Erythrina variegata* L. (Verdecia *et al.*, 2019).

Bioactive Properties

The extracts of different parts of *Erythrina variegata* L. are claimed to possess year-round, antidiabetic, anti-inflammatory, analgesic, antibacterial, antihelmintic, antitumour, hypotensive, β-glucosidase inhibitory and antiosteoporotic activities (Kumar *et al.*, 2010, 2011; Lim, 2014; Alam *et al.*, 2020a). *Erythrina variegata* L. also exhibits several other characteristic bioactive functions such as neuromuscular blocking, smooth muscle relaxant, CNS depressant and hydrocholeretic action (Kumar *et al.*, 2010). Alkaloids, flavonoids, isoflavonoids and other phenolic compounds present in *Erythrina variegata* L. are normally considered to exert the above mentioned bioactive properties.

Health Benefits: Antidiabetic Effect

T2DM is a metabolic condition demonstrated by hyperglycemia (high blood glucose levels) and glucose sensitivity, leading to improper secretion of insulin or the action of insulin to improve glucose uptake. The combination of various factors, such as obesity, stress, sedentary

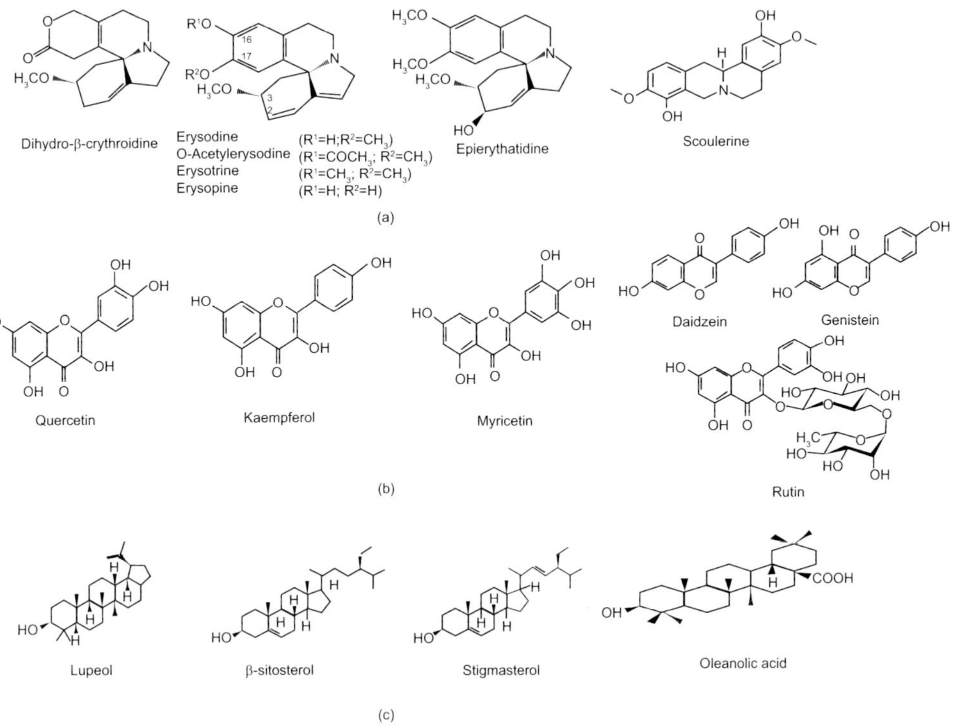

Fig. 10.2. Chemical structures of phytochemicals commonly found in *Erythrina variegata* L.: (a) alkaloids; (b) flavonoids; and (c) triterpenoids (Lim, 2014; Habtemariam, 2019; Tabassum *et al.*, 2019).

lifestyle and genetic factors related to insulin secretion or resistance, can contribute to the development of T2DM (Kaku, 2010; McCarthy, 2010; Alam *et al.*, 2020a). Owing to its high rate of morbidity, death and higher health costs for management and care, this disease is considered a burden worldwide. According to the 2019 International Diabetes Federation report, 450 million adults are living with diabetes worldwide, with 700 million cases expected by 2045 (International Diabetes Federation, 2019). At the global level, with almost 50% of cases undiagnosed, this disease is more common in low-income and middle-income countries. According to the report by World Health Organization (2016), 8% (12.88 million) of Bangladesh's total population was affected by diabetes while 3% of total all-age deaths occurred due to diabetes. Over time, a growing rate of DM prevalence among the Bangladeshi population has also been observed and the trend is still continuing (Talukder and Hossain, 2020).

To date, a cure for diabetes has not been achieved, in spite of great scientific advances. However, from lifestyle modifications to therapies, there are many strategies for its management. Oral antidiabetic medications are the most widely used treatments. Sulfonylureas, biguanides, thiazolidinediones, meglitinides, inhibitors of α-glucosidase and α-amylase, inhibitors of dipeptidyl peptidase 4 (DPP4), incretin mimetics and inhibitors of sodium-glucose cotransporter-2 (SGLT2) are some of them (Frandsen *et al.*, 2016; Habtemariam, 2019). However, the response rates of commercially available synthetic antidiabetic drugs are poor and frequently associated with serious adverse effects, as the mechanism of this condition is very complex. In addition to side effects, high synthetic drug prices and unavailability in remote areas remain a major obstacle for diabetes management. Therefore, because of their high effectiveness as natural antidiabetic agents, the search for traditional plants on antidiabetic activity has earned a

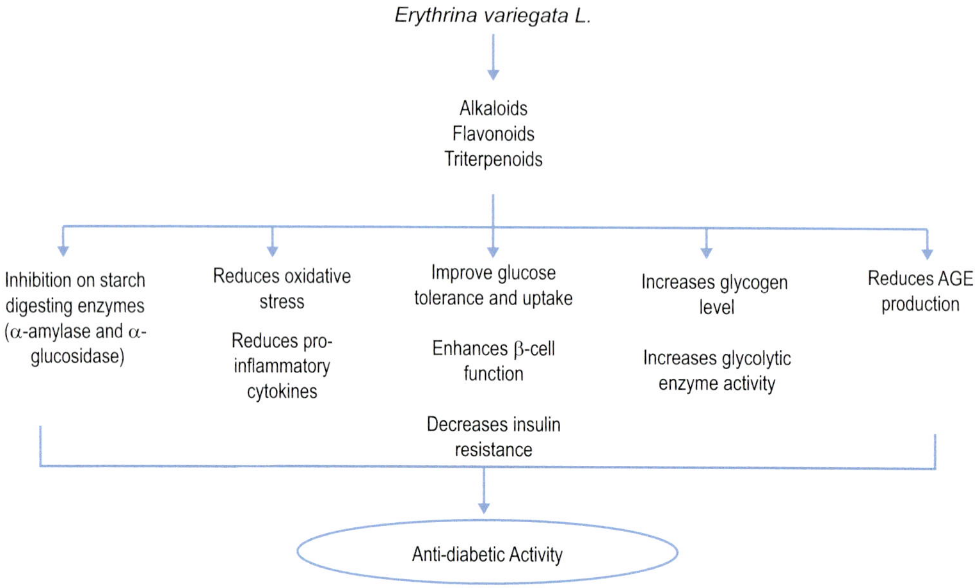

Fig. 10.3. Proposed antidiabetic effect of *Erythrina variegata* L.

lot of attention in recent years (Rana *et al.*, 2019; Alam *et al.*, 2020a). Plants accomplish this via the action of phytochemicals such as flavonoids, saponins, alkaloids, glycosides, terpenes etc. (Aba and Asuzu, 2018). These phytochemicals exert antidiabetic activity through different modes of action, such as insulin mimicking activity, acting on beta cells to produce insulin, enzyme inhibiting activities or activities to alter glucose utilization (Fig. 10.3) (Oboh *et al.*, 2015). As stated previously, the *Erythrina variegata* L. plant contains significant amount of alkaloids, flavonoids, glycosides, terpenes etc. and thus can be effectively utilized as a potential agent for managing T2DM. The antidiabetic potential of several bioactive compounds commonly found in *Erythrina variegata* L. plant are discussed below.

Alkaloids

In the plant kingdom, alkaloids are a widely distributed and structurally diverse group of secondary metabolites (Christodoulou *et al.*, 2018). They have a bitter taste, and over recent decades it is estimated that more than 12,000 plant alkaloids (PAs) have been found (Kutchan *et al.*, 2008; Ashihara *et al.*, 2013). About 14–20% of plant species have been found to possess alkaloids. These compounds are also relatively common in nature, and in the biosynthetic routes of PAs, amino acids are pivotal intermediaries. Different forms of these compounds have been identified, including glycosylated, methylated and hydroxylated compounds. A characteristic feature of these compounds is the presence of a N-atom in their chemical amine-base backbone (Kutchan *et al.*, 2008; Ashihara *et al.*, 2013). Toxicity is found in these secondary metabolites for herbivores, insects, animals and, in some cases, humans (Kutchan *et al.*, 2008; Ashihara *et al.*, 2013). For plant alkaloids, various types of classification systems have been identified based on their chemical structures and biosynthetic pathways. For example, they are split into three major groups, namely real, pseudo- and proto-alkaloids, based on their chemical structures. Specific biological properties have been unravelled by each class of alkaloids in this classification and, thus, the complexity of their backbone can decide or influence their biological activities (Kutchan *et al.*, 2008; Ashihara *et al.*, 2013).

The mechanism by which alkaloids aid in managing and/or treating T2DM is still not clear. However, it is thought that the antioxidant potential of alkaloids may be attributed to this activity as one of the main determinants of the development of diabetes has been linked to oxidative stress (Evans *et al.*, 2002; Bahadoran *et al.*, 2013). The reactive oxygen species (ROS) overwhelming the body's antioxidant mechanism facilitates the pathogenesis of diabetes and that is why in diabetic subjects more oxidative cells are found than in healthy subjects, i.e. an elevated level of ROS production (Martín-Gallán *et al.*, 2003; Rains and Jain, 2011). Several studies have also shown that oxidative stress and T2DM are closely related because of increased oxidative damage to essential macromolecules. There is an increased level of protein carbonylation and nitrosylation in insulin-sensitive tissues observed in T2DM, according to studies by Grimsrud *et al.* (2008) and Muellenbach *et al.* (2008).

Research findings have also shown a clear correlation between increased oxidative stress and protein unfolding leading to observation of protein functionality loss in several animal model studies (Pierce *et al.*, 2008; Pérez-Matute *et al.*, 2009). Oxidative stress triggers the alteration of two main pathways in diabetic patients, which are insulin resistance and insulin secretion. Adipocytokine dysregulation and suppression of insulin signals are triggered by oxidative stress, thus leading to insulin resistance. Furthermore, it has been shown that the increase in ROS production in T2DM subjects activates harmful pathways, such as glucosamine pathways, advanced glycation end (AGE) products, increased activity of aldose reductase, and PKCβ1/2 kinase (Brownlee, 2005).

Crude extracts of *Erythrina variegata* L. leaves and bark containing alkaloids have been reported to exhibit antioxidant activities (Rahman *et al.*, 2010; Sakat and Juvekar, 2010; Tabassum *et al.*, 2019) and inhibit aldose reductase activity and formation of AGE products (Gaviraj *et al.*, 2019). Other mechanisms induced by alkaloids are proposed to be improvement of GLUT 4, glucokinase activity, peroxisome PPARγ and hepatic glycogen content, reductions in total cholesterol, triglyceride, phosphoenol pyruvate carboxykinase, and attenuation of glucose-6-phosphatase activity (Sharma *et al.*, 2010). Moreover, in streptozotocin-induced diabetic rats, crude extract of *Erythrina variegata* L. leaf exhibited promising hypoglycemic action by significantly reducing the level of glycosylated haemoglobin (HbA1C), total cholesterol, low-density lipoprotein cholesterol and triglycerides while increasing the level of serum high-density lipoprotein cholesterol (Kumar *et al.*, 2011). The leaf extract was also found to normalize the activity of enzymes such as alanine transaminase, aspartate transaminase, and alkaline phosphatase in diabetic rats (Kumar *et al.*, 2011). These results suggest the potential antidiabetic application of *Erythrina variegata* L. plant.

Flavonoids

Flavonoids are a group of polyphenolic compounds that are widely present in the plant kingdom and are the most studied class of polyphenolic compounds. They consist of two benzene rings linked by a short three-carbon chain (Manach *et al.*, 2004; Vinayagam and Xu, 2015). Depending on the chemical structure, flavonoids can be categorized into six subclasses: anthocyanidins; flavan-3-ols; flavonols; flavones; flavanones and isoflavones (Manach *et al.*, 2004; Xiao *et al.*, 2013).

Different parts of *E. variegata* L. are found to contain significant amount of isoflavones (4,5,7-trihydroxy-8-prenyl isoflavone, alpinum isoflavone, 6-hydroxygenistein, indicanine C, scandenone, etc.), flavan-3-ols (catechin, epicatechin etc.), flavonols (kaempferol, myricetin, quercetin, rutin etc.) and many other secondary metabolites (Rahman *et al.*, 2010; Lim, 2014; Tabassum *et al.*, 2019; Alam *et al.*, 2020a). The antidiabetic potential of these flavonoids is accredited to their antioxidant capacities (Rahman *et al.*, 2010; Tabassum *et al.*, 2019; Alam *et al.*, 2020a), their ability to modulate inflammatory markers (Thongmee and Itharat, 2016) and their ability to act on starch-digesting enzymes such as α-amylase (Alam *et al.*, 2020a). Kaempferol, quercetin and rutin are responsible for inhibition of apoptosis, decreased production of caspase-3 in beta cells, improved signalling of cAMP, inhibition of AGE products, improved insulin synthesis and secretion, and decreased absorption of intestinal glucose by inhibition of GLUT2. They also have the ability to restore oxidant status and decrease IL-1β, IL-6, TNF-α, lipid peroxidation, nitrite and glycosylated haemoglobin production (Coskun *et al.*, 2005; Kobori *et al.*, 2009; Hao *et al.*, 2012; Abo-Salem, 2014; Niture *et al.*, 2014; Al-Numair *et al.*, 2015).

Triterpenoids

Triterpenes are a class of hydrocarbon compound consisting of three terpenes and six units of isoprene, such as squalene. They are categorized according to the number of rings they contain, such as pentacyclic and tetracyclic. Triterpenoids are called 'functional metabolites' of triterpenes. Oleanolic acid, stigmasterol, lupeol, β-sitosterol, polypodatetraene, malabaricane, lanostane, hopane, oleanane and so on are other examples of triterpenes (Lyu *et al.*, 2016). Their antidiabetic properties have been the focus of a large number of studies. Oleanolic acid, stigmasterol, lupeol and β-sitosterol are types of triterpenoid reported to be present in different parts of *Erythrina variegata* L. and these bioactive triterpenoids have been claimed to possess antidiabetic activity (Rahman *et al.*, 2010; Lim, 2014; Tabassum *et al.*, 2019).

Stigmasterol has been shown to decrease insulin resistance in high-fat KK-Ay mice in this respect (Wang *et al.*, 2017). When 50 mg/kg/day of this compound was given to these mice, an increase in glucose tolerance was observed after four weeks. This positive influence on diabetes could have been due to the ability of stigmasterol to normalize islet forms and avoid a shrinkage in islet size of the pancreas, as well as increasing levels of GLUT4 expression in skeletal muscle and white adipose tissue. In addition, improved oral glucose tolerance, decreased fasting glucose levels and increased insulin secretion from isolated rat pancreatic islet cells were shown in rats given β-sitosterol (Gupta *et al.*, 2011; Ponnulakshmi *et al.*, 2019). Further studies have confirmed that these effects are mediated by PI3K/Akt, insulin receptor and GLUT4 activation and regulation (Chai *et al.*, 2011; Ponnulakshmi *et al.*, 2019). A significant level of lipolytic effect was commonly exerted by the triterpenoids at lower concentrations, but gradually decreases with increasing concentrations. These findings were compatible with the results of real-time polymerase chain reaction (RT-PCR): the expression of the GLUT4 and insulin receptor genes were upregulated, while the PI3KA and AKT genes were downregulated (Chai *et al.*, 2011; Ponnulakshmi *et al.*, 2019).

Furthermore, 30 days of oral administration of β-sitosterol (20 mg/kg) to high-fat diet and T2DM rats, induced sucrose stabilized plasma insulin deficiency and blood glucose levels (Gupta *et al.*, 2011). Subsequently, the ability of triterpenoids to improve glucose homeostasis has been confirmed by various *in vitro* studies. In particular, an increasing amount of evidence shows that β-sitosterol and stigmasterol are likely to improve adipocyte glucose uptake and GLUT4 translocation and expression in rat skeleton (L6) cells (Wang *et al.*, 2017), and immortalized myoblast cells of the rat skeleton (L6) with endogenous GLUT4 expression and elevated myotube fusion capacity. AMPK mediated pathway tends to facilitate this beneficial effect of triterpenoids on glucose metabolism (Hwang *et al.*, 2008). Moreover, *in vitro* administration of oleanolic acid together with ursolic acid was found to inhibit α-amylase activity potentially with an IC_{50} value of 4 μM. Oleanolic acid was also found to be effective in inhibiting α-amylase at various concentrations (1–5 μg/ml) whereas lupeol, at 5 μg/ml, exhibited inhibitory effect against α-amylase (Ali *et al.*, 2006).

Cumulatively, these results indicate that triterpenoids can target the primary cause of insulin resistance and thus contribute to an overall improvement in the key metabolic elements of T2DM. The information gleaned in humans, however, is limited and more in-depth studies are needed to gain more insight into the dynamic mechanisms of triterpenoid action.

Conclusions and Future Trends

A possible source of knowledge on *Erythrina variegata* L. as a rich source of bioactive compounds such as alkaloids, flavonoids, triterpenoids etc. is given in this chapter, which highlights its potential in management of T2DM. Several plants with bioactive compounds as their main components are used as conventional antidiabetic medicines in different countries. As possible pharmaceutical agents in the treatment of diabetes, a significant number of studies have reported their positive effects. Natural bioactives (alkaloids, flavonoids, triterpenoids etc.) show a large range of pharmacological activities toward DM. Multi-target pharmaceutical effects have also been shown by some triterpenoids and they could simultaneously stimulate the secretion of insulin and reverse insulin resistance. Their possible antidiabetic potential contains several pathways, including insulin signalling pathway activation, α-glucosidase and α-amylase inhibition, and AMPK and PPAR activation. On the other hand, alkaloids and flavonoids as source of antioxidants could modulate the inflammatory process and thus lead to the management of T2DM. The diverse structures and roles of *Erythrina* bioactives makes them ideal chemical scaffolds for novel therapeutic agents, in various physiological processes. The authors of this chapter conclude that the potential of this genus has not been fully explored and that further therapeutic studies are needed to elucidate and validate the *in vivo* active compounds and comprehensive molecular mechanisms of antidiabetic agents. Before they can be used as likely targets for novel drug discovery, the toxicity of these compounds warrants further investigation.

Acknowledgement

Authors are thankful to Professor Sheikh Nazrul Islam, Institute of Nutrition and Food Science, University of Dhaka, for his useful discussion.

References

Aba, P.E. and Asuzu, I.U. (2018) Mechanisms of actions of some bioactive anti-diabetic principles from phytochemicals of medicinal plants: a review. *Indian Journal of Natural Products and Resources* 9, 85–96.

Abo-Salem, O.M. (2014) Kaempferol attenuates the development of diabetic neuropathic pain in mice: possible anti-inflammatory and anti-oxidant mechanisms. *Open Access Macedonian Journal of Medical Sciences* 2(3), 424–430. DOI: 10.3889/oamjms.2014.073.

Al-Numair, K.S., Chandramohan, G., Veeramani, C. and Alsaif, M.A. (2015) Ameliorative effect of kaempferol, a flavonoid, on oxidative stress in streptozotocin-induced diabetic rats. *Redox Report* 20(5), 198–209. DOI: 10.1179/1351000214Y.0000000117.

Alam, M.K., Rana, Z.H., Islam, S.N. and Akhtaruzzaman, M. (2019) Total phenolic content and antioxidant activity of methanolic extract of selected wild leafy vegetables grown in Bangladesh: A cheapest source of antioxidants. *Potravinarstvo Slovak Journal of Food Sciences* 13(1), 287–293. DOI: 10.5219/1107.

Alam, M.K., Rana, Z.H., Islam, S.N. and Akhtaruzzaman, M. (2020a) Comparative assessment of nutritional composition, polyphenol profile, antidiabetic and antioxidative properties of selected edible wild plant species of Bangladesh. *Food Chemistry* 320, 126646. DOI: 10.1016/j.foodchem.2020.126646.

Alam, M.K., Sams, S., Khan, M.S., Azmir, J., Ahsan, M. *et al.* (2020b) Profiling of minerals, water soluble vitamins and carotenoid in selected unconventional leafy and non-leafy vegetables of Bangladesh. *Natural Product Research* 1–5. DOI: 10.1080/14786419.2020.1849198.

Ali, H., Houghton, P.J. and Soumyanath, A. (2006) Alpha-Amylase inhibitory activity of some Malaysian plants used to treat diabetes; with particular reference to *Phyllanthus amarus*. *Journal of Ethnopharmacology* 107(3), 449–455. DOI: 10.1016/j.jep.2006.04.004.

Andarwulan, N., Kurniasih, D., Apriady, R.A., Rahmat, H., Roto, A.V. *et al.* (2012) Polyphenols, carotenoids, and ascorbic acid in underutilized medicinal vegetables. *Journal of Functional Foods* 4(1), 339–347. DOI: 10.1016/j.jff.2012.01.003.

Ashihara, H., Yokota, T. and Crozier, A. (2013) Biosynthesis and Catabolism of Purine Alkaloids. In: Giglioli-Guivarc'h, N. (ed.) *Advances in Botanical Research*, Vol. 68. pp. 111–138. Available at: http://www.sciencedirect.com/science/article/pii/B9780124080614000043 (accessed 24 January 2021).
Bahadoran, Z., Mirmiran, P. and Azizi, F. (2013) Dietary polyphenols as potential nutraceuticals in management of diabetes: a review. *Journal of Diabetes and Metabolic Disorders* 12(1), 43. DOI: 10.1186/2251-6581-12-43.
Brownlee, M. (2005) The pathobiology of diabetic complications: a unifying mechanism. *Diabetes* 54(6), 1615–1625. DOI: 10.2337/diabetes.54.6.1615.
Chai, J.W., Lim, S.L., Kanthimathi, M.S. and Kuppusamy, U.R. (2011) Gene regulation in β-sitosterol-mediated stimulation of adipogenesis, glucose uptake, and lipid mobilization in rat primary adipocytes. *Genes & Nutrition* 6(2), 181–188. DOI: 10.1007/s12263-010-0196-4.
Christodoulou, M.-I., Tchoumtchoua, J., Skaltsounis, A.-L., Scorilas, A. and Halabalaki, M. (2018) Natural alkaloids intervening the insulin pathway: new hopes for anti-diabetic agents? *Current Medicinal Chemistry* 26(32), 5982–6015. DOI: 10.2174/0929867325666180430152618.
Coskun, O., Kanter, M., Korkmaz, A. and Oter, S. (2005) Quercetin, a flavonoid antioxidant, prevents and protects streptozotocin-induced oxidative stress and β-cell damage in rat pancreas. *Pharmacological Research* 51(2), 117–123. DOI: 10.1016/j.phrs.2004.06.002.
Dewick, P.M. (2009) *Medicinal Natural Products: A Biosynthetic Approach*, 3rd edn. John Wiley & Sons Ltd, Chichester, UK. DOI: 10.1002/9780470742761.
Evans, J.L., Goldfine, I.D., Maddux, B.A. and Grodsky, G.M. (2002) Oxidative stress and stress-activated signaling pathways: a unifying hypothesis of type 2 diabetes. *Endocrine Reviews* 23(5), 599–622. DOI: 10.1210/er.2001-0039.
Frandsen, C.S., Dejgaard, T.F. and Madsbad, S. (2016) Non-insulin drugs to treat hyperglycaemia in type 1 diabetes mellitus. *The Lancet. Diabetes & Endocrinology* 4(9), 766–780. DOI: 10.1016/S2213-8587(16)00039-5.
Gaviraj, E.N., Ramarao, A., Veeresham, C., Shivakumar, B., Kalyane, N.V. *et al.* (2019) Inhibitory activities of some folklore remedies on aldose reductase of rat lens and generation of advanced glycation end products. *Research Journal of Pharmacy and Technology* 12(4), 1947. DOI: 10.5958/0974-360X.2019.00326.3.
Ghani, A. (2003) *Medicinal Plants of Bangladesh: Chemical Constituents and Uses*, 2nd edn. Asiatic Society of Bangladesh, Dhaka, Bangladesh.
Grimsrud, P.A., Xie, H., Griffin, T.J. and Bernlohr, D.A. (2008) Oxidative stress and covalent modification of protein with bioactive aldehydes. *The Journal of Biological Chemistry* 283(32), 21837–21841. DOI: 10.1074/jbc.R700019200.
Gupta, R., Sharma, A.K., Dobhal, M.P., Sharma, M.C. and Gupta, R.S. (2011) Antidiabetic and antioxidant potential of β-sitosterol in streptozotocin-induced experimental hyperglycemia. *Journal of Diabetes* 3, 29–37. DOI: 10.1111/j.1753-0407.2010.00107.x.
Habtemariam, S. (2019) *Medicinal Foods as Potential Therapies for Type-2 Diabetes and Associated Diseases: The Chemical and Pharmacological Basis of Their Action*, 1st edn. Academic Press, Cambridge, Massachusetts.
Hao, H., Shao, Z., Tang, D., Lu, Q., Chen, X. *et al.* (2012) Preventive effects of rutin on the development of experimental diabetic nephropathy in rats. *Life Sciences* 91(19–20), 959–967. DOI: 10.1016/j.lfs.2012.09.003.
Hwang, S.-L., Kim, H.-N., Jung, H.-H., Kim, J.-E., Choi, D.-K. *et al.* (2008) Beneficial effects of β-sitosterol on glucose and lipid metabolism in L6 myotube cells are mediated by AMP-activated protein kinase. *Biochemical and Biophysical Research Communications* 377(4), 1253–1258. DOI: 10.1016/j.bbrc.2008.10.136.
International Diabetes Federation (2019) *IDF Diabetes Atlas*, 9th edn. International Diabetes Federation, Brussels, Belgium. Available at: https://www.diabetesatlas.org/en/ (accessed 30 March 2022).
Kaku, K. (2010) Pathophysiology of type 2 diabetes and its treatment policy. *Japan Medical Association Journal* 53, 41–46.
Kobori, M., Masumoto, S., Akimoto, Y. and Takahashi, Y. (2009) Dietary quercetin alleviates diabetic symptoms and reduces streptozotocin-induced disturbance of hepatic gene expression in mice. *Molecular Nutrition & Food Research* 53(7), 859–868. DOI: 10.1002/mnfr.200800310.
Kumar, A., Lingadurai, S., Jain, A. and Barman, N.R. (2010) *Erythrina variegata* Linn: A review on morphology, phytochemistry, and pharmacological aspects. *Pharmacognosy Reviews* 4(8), 147–152. DOI: 10.4103/0973-7847.70908.

Kumar, A., Lingadurai, S., Shrivastava, T.P., Bhattacharya, S. and Haldar, P.K. (2011) Hypoglycemic activity of *Erythrina variegata* leaf in streptozotocin-induced diabetic rats. *Pharmaceutical Biology* 49(6), 577–582. DOI: 10.3109/13880209.2010.529615.

Kutchan, T.M., Frick, S. and Weid, M. (2008) Engineering plant alkaloid biosynthetic pathways: progress and prospects. In: Bohnert, H.J., Nguyen, H., and Lewis, N.G. (eds) *Advances in Plant Biochemistry and Molecular Biology*, Vol. 1. Pergamon Press, Oxford, UK, pp. 283–310.

Lim, T.K. (2014) *Erythrina variegata*. In: Lim, T.K. (ed) *Edible Medicinal and Non-Medicinal Plants: vol. 7, Flowers*. Springer Netherlands, Dordrecht, Netherlands, pp. 788–805.

Lyu, H., Chen, J. and Li, W.-L. (2016) Natural triterpenoids for the treatment of diabetes mellitus: a review. *Natural Product Communications* 11(10), 1579–1586. DOI: 10.1177/1934578X1601101037.

Manach, C., Scalbert, A., Morand, C., Rémésy, C. and Jiménez, L. (2004) Polyphenols: food sources and bioavailability. *The American Journal of Clinical Nutrition* 79(5), 727–747. DOI: 10.1093/ajcn/79.5.727.

Martín-Gallán, P., Carrascosa, A., Gussinyé, M. and Domínguez, C. (2003) Biomarkers of diabetes-associated oxidative stress and antioxidant status in young diabetic patients with or without subclinical complications. *Free Radical Biology & Medicine* 34(12), 1563–1574. DOI: 10.1016/s0891-5849(03)00185-0.

McCarthy, M.I. (2010) Genomics, type 2 diabetes, and obesity. *The New England Journal of Medicine* 363(24), 2339–2350. DOI: 10.1056/NEJMra0906948.

Muellenbach, E.A., Diehl, C.J., Teachey, M.K., Lindborg, K.A., Archuleta, T.L. *et al.* (2008) Interactions of the advanced glycation end product inhibitor pyridoxamine and the antioxidant α-lipoic acid on insulin resistance in the obese Zucker rat. *Metabolism* 57(10), 1465–1472. DOI: 10.1016/j.metabol.2008.05.018.

Muthukrishnan, S., Palanisamy, S., Subramanian, S., Selvaraj, S., Mari, K.R. *et al.* (2016) Phytochemical profile of *Erythrina variegata* by using high-performance liquid chromatography and gas chromatography-mass spectroscopy analyses. *Journal of Acupuncture and Meridian Studies* 9(4), 207–212. DOI: 10.1016/j.jams.2016.06.001.

Niture, N.T., Ansari, A.A. and Naik, S.R. (2014) Anti-hyperglycemic activity of rutin in streptozotocin-induced diabetic rats: an effect mediated through cytokines, antioxidants and lipid biomarkers. *Indian Journal of Experimental Biology* 52(7), 720–727.

Oboh, G., Agunloye, O.M., Adefegha, S.A., Akinyemi, A.J. and Ademiluyi, A.O. (2015) Caffeic and chlorogenic acids inhibit key enzymes linked to type 2 diabetes (*in vitro*): a comparative study. *Journal of Basic and Clinical Physiology and Pharmacology* 26(2), 165–170. DOI: 10.1515/jbcpp-2013-0141.

Pérez-Matute, P., Zulet, M.A. and Martínez, J.A. (2009) Reactive species and diabetes: counteracting oxidative stress to improve health. *Current Opinion in Pharmacology* 9(6), 771–779. DOI: 10.1016/j.coph.2009.08.005.

Pierce, A., Mirzaei, H., Muller, F., De Waal, E., Taylor, A.B. *et al.* (2008) GAPDH is conformationally and functionally altered in association with oxidative stress in mouse models of amyotrophic lateral sclerosis. *Journal of Molecular Biology* 382(5), 1195–1210. DOI: 10.1016/j.jmb.2008.07.088.

Ponnulakshmi, R., Shyamaladevi, B., Vijayalakshmi, P. and Selvaraj, J. (2019) In silico and in vivo analysis to identify the antidiabetic activity of beta sitosterol in adipose tissue of high fat diet and sucrose induced type-2 diabetic experimental rats. *Toxicology Mechanisms and Methods* 29(4), 276–290. DOI: 10.1080/15376516.2018.1545815.

Rahman, M.Z., Sultana, S.J., Faruquee, C.F., Ferdous, F., Rahman, M.S. *et al.* (2007) Phytochemical and biological investigations of *Erythrina variegata*. *Saudi Pharmaceutical Journal* 15, 140–145.

Rahman, M.Z., Rahman, M.S., Kaisar, A., Hossain, A. and Rashid, M.A. (2010) Bioactive isoflavones from *erythrina variegata* L. *Turkish Journal of Pharmaceutical Sciences* 7, 21–28.

Rahman, S.A., Kielmann, T., McPake, B. and Normand, C. (2012) Healthcare-seeking behaviour among the tribal people of Bangladesh: Can the current health system really meet their needs? *Journal of Health, Population, and Nutrition* 30(3), 353–365. DOI: 10.3329/jhpn.v30i3.12299.

Rains, J.L. and Jain, S.K. (2011) Oxidative stress, insulin signaling, and diabetes. *Free Radical Biology & Medicine* 50(5), 567–575. DOI: 10.1016/j.freeradbiomed.2010.12.006.

Rana, Z.H., Alam, M.K. and Akhtaruzzaman, M. (2019) Nutritional Composition, Total Phenolic Content, Antioxidant and α-Amylase Inhibitory Activities of Different Fractions of Selected Wild Edible Plants. *Antioxidants* 8(7), 203. DOI: 10.3390/antiox8070203.

Sakat, S.S. and Juvekar, A.R. (2010) Comparative Study of *Erythrina indica* Lam. (Febaceae) Leaves Extracts for Antioxidant Activity. *Journal of Young Pharmacists* 2(1), 63–67. DOI: 10.4103/0975-1483.62216.

Samanta, T.D. and Laskar, S. (2009) Chemical investigation of *Erythrina variegata* Linn. seed proteins. *Food Chemistry* 114(1), 212–216. DOI: 10.1016/j.foodchem.2008.09.040.

Shaheen, N., Rahim, A.T.M., Mohiduzzaman, M., Banu, C.P., Bari, M.L. *et al.* (2014) *Food Composition Table for Bangladesh*, 2nd edn. Institute of Nutrition and Food Science, University of Dhaka, Dhaka, Bangladesh.

Sharma, B., Salunke, R., Balomajumder, C., Daniel, S. and Roy, P. (2010) Anti-diabetic potential of alkaloid rich fraction from *Capparis decidua* on diabetic mice. *Journal of Ethnopharmacology* 127(2), 457–462. DOI: 10.1016/j.jep.2009.10.013.

Shori, A.B. (2015) Screening of antidiabetic and antioxidant activities of medicinal plants. *Journal of Integrative Medicine* 13(5), 297–305. DOI: 10.1016/S2095-4964(15)60193-5.

Tabassum, F., Hasan, C.M., Masud, M.M., Nur Manik, Md.I. and Ahsan, M. (2019) Isolation and Characterization of Secondary Metabilites and Evaluation of Antimicrobial, Antioxidant and Thrombolytic Potentials of *Erythrina variegata* L. Bark. *Asian Journal of Chemistry* 31(8), 1842–1846. DOI: 10.14233/ajchem.2019.21993.

Talukder, A. and Hossain, M.Z. (2020) Prevalence of diabetes mellitus and its associated factors in Bangladesh: Application of two-level logistic regression model. *Scientific Reports* 10(1), 10237. DOI: 10.1038/s41598-020-66084-9.

Thongmee, P. and Itharat, A. (2016) Anti-inflammatory activities of *Erythrina variegata* bark ethanolic extract. *Journal of the Medical Association of Thailand* 99(suppl 4), 166–171.

Verdecia, D.M., Herrera, R.S., Ramírez, J.L., Paumier, M., Bodas, R. *et al.* (2019) *Erythrina variegata* quality in the Cauto Valley, Cuba. *Agroforestry Systems* 94(4), 1209–1218. DOI: 10.1007/s10457-019-00353-z.

Vinayagam, R. and Xu, B. (2015) Antidiabetic properties of dietary flavonoids: a cellular mechanism review. *Nutrition & Metabolism* 12, 60. DOI: 10.1186/s12986-015-0057-7.

Wang, J., Huang, M., Yang, J., Ma, X., Zheng, S. *et al.* (2017) Anti-diabetic activity of stigmasterol from soybean oil by targeting the GLUT4 glucose transporter. *Food & Nutrition Research* 61(1), 1364117. DOI: 10.1080/16546628.2017.1364117.

World Health Organization (2016) WHO Diabetes country profiles, Bangladesh. Available at: https://www.who.int/diabetes/country-profiles/bgd_en.pdf (accessed 10 January 2021).

Xiao, J., Kai, G., Yamamoto, K. and Chen, X. (2013) Advance in dietary polyphenols as α-glucosidases inhibitors: a review on structure-activity relationship aspect. *Critical Reviews in Food Science and Nutrition* 53(8), 818–836. DOI: 10.1080/10408398.2011.561379.

11 Snake Melon (*Cucumis melo* var. *flexuosus*)

Riadh Ilahy[1], Imen Tlili[1], Amina Benabdallah[2], Zoltán Pék[3], Mohammed Wasim Siddiqui[4]*, Fozia Homa[4] and Helyes Lajos[3]

[1]National Agricultural Research Institute of Tunisia (INRAT), Laboratory of Horticulture, University of Carthage, Ariana, Tunisia; [2]BPE Laboratory, Faculty of Life and Natural Sciences, University of Chadli Bendjedid, El-Taref, Algeria; [3]Institute of Horticultural Science, Szent István Campus, Hungarian University of Agriculture and Life Sciences, Gödöllő, Hungary; [4]Bihar Agricultural University, Sabour, Bhagalpur, Bihar, India

Abstract

Snake melons (*Cucumis melo* var. *flexuosus* L. Naudin) are heirloom monoecious cucurbits that are mostly cultivated and distributed in the old world tropics and subtropics. Immature fruit are typically eaten raw or pickled, particularly in Mediterranean countries, where they have been prized for their crisp texture, refreshing qualities and non-sweet flavour since antiquity. These details were emphasized primarily by their depiction in wall paintings, statues and mosaics from Egyptian civilization and the Roman Empire, as well as other sources in ancient and medieval Islamic and Jewish texts.

There is a general trend in investing capital to increase the consumption of fresh fruit and vegetables in order to avoid significant chronic and age-related diseases, given the growing interest among growers and consumers in high quality, healthy foods/food items and the mortality due to low fruit and vegetable consumption. As a result, the levels of various bioactive compounds and antioxidant activity in fruit and vegetables are becoming an essential quality trait to analyse and quantify in terms of the nutritional quality of food so as to encourage the inclusion of certain still underutilized sources of bioactive compounds in the daily diet. This chapter concisely summarizes the important and valuable information about snake melon agronomic characteristics, functional consistency and therapeutic importance in a succinct manner. We've discussed how important snake melon is not only as a source of phytochemicals, but also for its healthy mineral composition, implying that this crop will become more important in the near future.

Introduction

Snake melon has different names in different parts of the world such as alficoz in Spain, fakous or fegous in Arabic Maghreb countries and northern Africa, agoor in Sudan, acur in Turkey, kakdi in India and the qishu'im in Hebrew (Pitrat, 2005; Paris *et al.*, 2012; Ilahy *et al.*, 2019). Snake melon is consumed at the green-immature stage of ripening and is similar to a cucumber, but longer and with a lighter green colour. It has been proven that snake

*Corresponding author: wasim_serene@yahoo.com

DOI: 10.1079/9781789248067.0011

melon was a widely spread and consumed vegetable across a wide geographic belt during the medieval period. Paris *et al.* (2012) reported that snake melon was depicted in ancient Egyptian and Roman mosaics.

Snake melons were very appreciated in the Jewish literature of the classical and early medieval period and the term qishu'im (or qishu'in), a synonym of snake melon, was widely reported in the Mishnah and the Tosefta book. Furthermore, the term quitha, another synonym of qishu'im, has been also reported in the Holy Koran (Chapter 1/Surat Al Baqara, Verse/Aya 61) as another name of the snake melon and it is proven that the Prophet Mohammad (Peace and Blessings of Allah be Upon Him) recommended the consumption of snake melon along with immature date fruit for various reported therapeutic values 14 centuries earlier (Ilahy *et al.*, 2019). These pleiotropic effects have been recently emphasized by Saddi *et al.* (2018).

Snake melon immature fruit are still an underestimated source of balanced minerals as well as an array of vitamins and moderate amounts of carotenoids, mainly the provitamin A, β-carotene and lutein, within green plastids. The fruit are stomachic and are traditionally used as a cooling light cleanser or moisturiser for the skin as well as various beneficial effects detailed in Ilahy *et al.* (2019).

Botany and Agricultural Traits

The consumption and production of cucurbits is significantly growing worldwide. In 2019, the area dedicated to this crop exceeded 1 million ha with a production attaining 28 million tons. In Tunisia, the area devoted to melon touched 9800 ha totalling a production exceeding 103,000 tons produced under open field and/or protected greenhouses (FAO, 2020). Off-season snake melon production represents a major income for a large number of smallholders particularly due to the elevated prices during this period.

Snake melon (*Cucumis melo* L. var. *flexuosus* Naudin, 2n=2x=24) (Fig. 11.1), is a Cucurbitaceae family annual plant widely grown in the old world tropical and subtropical regions, especially in the Mediterranean, the Near East, the Indian subcontinent and the Middle East (Fig. 11.1) (Chakravarty, 1966; Walters and Thieret, 1993; Pandey *et al.*, 2010; Ilahy *et al.*, 2019). Snake melon leaves have three to five lobes with whole to dentate margins and are ovate, base cordate, simple or palmate (Fig. 11.1a). Simple tendrils define the obtuse apex, which is synonymous with a pubescent petiole. Female flowers are single and axillary implanted, while male flowers are axillary clustered or solitary. Long to very long cylindrical, claviform fruits with a yellowish-green colour distinguish the *flexuosus* variety (Pitrat *et al.*, 2000; Burger *et al.*, 2010) (Fig. 11.1b). The fruit's skin is very thin, light green and bumpless, and the fruit is usually eaten without peeling.

Snake melon is an important revenue source for many farmers and smallholders in various Mediterranean countries based on the high price, particularly for off-season production. Genotypic differences in snake melon varieties affect several characteristics such as earliness, skin colour, fruit length and form (right or curved) (Fig. 11.1b), as well as the longitudinal fruit furrowing (Chakravarty, 1966; Pandey *et al.*, 2010; Ilahy *et al.*, 2020). Their susceptibility to powdery mildew is a limiting factor for snake melon production, particularly under glass (greenhouses). The skin colour varies from dark/light green to lined dark and light green (Pandey *et al.*, 2010; Ilahy *et al.*, 2020) and the furrowing can range from very shallow to very deep. Unlike melon, *flexuosus* do not accumulate sugars throughout ripening (Burger *et al.*, 2006).

Limited work has been focused on improving the productivity and horticultural performance of snake melon cultivars. Significant amounts of data exist concerning plant physiology and genetic characterization but very little on the optimization of good agricultural practices suitable for increasing yields and improving the quality. Previously, Jebari and Hamza (1990) evaluated the late season snake melon cropping under plastic greenhouses in the Tunisian coastal regions. The authors reported that late season production is economically interesting as it is characterized by short cycle, good yield and fruit quality. Recently, Ilahy *et al.* (2020) studied the variability affecting the main agronomic traits of various snake melon genotypes consisting of hybrids and breeding lines grown under greenhouse conditions. The authors revealed that the main traits shaping

(a) (b)

Fig. 11.1. (a) Snake melon (*Cucumis melo* var. *flexuosus*) leaves; (b) snake melon (*Cucumis melo* var. *flexuosus*) fruit.

the agronomic production varied significantly depending on the genotype. The prevailing skin colour determined visually was different among the considered genotypes. Skin colour varied between dark/light green, intense dark green and medium light green. Fruit shape also ranged from straight to elongate. Regarding fruit yield, it has been reported that the F1 hybrid (H5) was the most productive (3.6 kg/plant). This suggests the importance of such data for both consumers and breeders (Ilahy *et al.*, 2019, 2020).

Traditional Use

The fruits are stomachic, and traditionally used as a cooling light disinfectant or skin moisturiser, for injuries and burns, and for their claimed anti-inflammatory, anticancer, antioxidative stress, antibacterial and immune-modulatory, antidiabetic effects (Grover and Yadav, 2004; Kolawole and Ayankunle, 2012; Bao *et al.*, 2013; Joseph and Jini, 2013) as well as for treating kidney stones (Daraizadeh, 2014). The seeds are also antitussive, digestive, febrifuge and vermifuge.

Nutritional Composition and Phytochemicals

Most fruits and vegetables are year-round available and consumed fresh, cooked, hot or cold, canned, pickled, frozen or dried. Fruits and vegetables represent a significant source of selected minerals, vitamins and bioactive compounds with antioxidant activity. Furthermore, fruits and vegetables are low in calories and fat, rich in carbohydrates and fibres, containing moderate to high levels of ascorbic acid and carotenes depending on the fruit (Tlili *et al.*, 2011a, b; Ilahy *et al.*, 2019). In this context, snake melon fruits have attracted increasing attention due to their substantial nutrient content. Ilahy *et al.* (2019) reported that snake melon is rich in water (approximately 96% water), contains 1.2% protein, 4.5% carbohydrate, 0.7% fat, and 2% fibre. Snake melons can also be considered as a significant source of certain minerals, mainly potassium (264 mg/100 g), phosphorus (37 mg/100 g), calcium (33 mg/100 g), magnesium (27 mg/100g) and sodium (11 mg/100g) (Table 11.1). The content of these minerals is higher or similar to that provided by other cucurbits such as cucumber, pumpkin and melon (Table 11.2). Snake melon fruits harvested at an immature stage of ripeness (the desirable stage for consumption) offer a diverse mixture of nutrients as well as valuable bioactive secondary metabolites such as polyphenols. However, research in this area is very sparse and not complete (Henane *et al.*, 2013; Ilahy *et al.*, 2014, 2019, 2020). Snake melons are mostly consumed fresh and preserved as pickles for out of season use. Currently, with the growing awareness of the negative correlation between the consumption of fresh fruit and vegetables and the incidence of various chronic diseases (Erdman *et al.*, 2009; Fernández-García, 2014; Nahak *et al.*, 2014), snake melon consumption is on the increase. Since most studies have been focused on cucumber cultivars (Nasrin and Nahar, 2014; Shan *et al.*, 2020), we attempt here to critically examine some relevant studies conducted on the agricultural traits and nutritional quality of snake melon cultivars while integrating updated data from recent published works in this regard, aiming to highlight their importance for human health and increase their consumption worldwide.

Snake melons contain different carotenoid components such as lutein (165 mg/100 g) and β-carotene (144 mg/100 g) (Ilahy *et al.*, 2014, 2019). This data suggests the need for a comprehensive investigation of snake melon as an underestimated source of bioactive compounds and it seems likely that this crop will gain increasing interest as a valuable source of such compounds for human health. While the assessment of bioactive secondary metabolites and the antioxidant properties of various cucurbits such as watermelon (Tlili *et al.*, 2011a, b, 2013) and melon (Henane *et al.*, 2013) is well documented, studies are scarce and not complete for snake melon cultivars. Formerly, Mariod *et al.* (2009) comparatively evaluated the properties of six Sudanese cucurbit seeds and seed oils and found that flexuosus samples showed the highest content of fibre (>22%), important δ-tocopherol content (0.8 mg/100 g) and a total tocopherol content attaining 35 mg/100 g, higher than sesame oil (33.0 mg/100 g), which highlights the need to further investigate the possible industrial uses of such products.

Snake melon fruit are generally climacteric, consumed at the immature stage of ripening. Data regarding their nutritional composition is still scarce. Pandey *et al.* (2010) attempted to improve the nutritional quality of snake melon via the hybridization of the inbred snake melon Punjab Long Melon 1 (PLM1) with five genetically diverse inbred melons. The authors assessed the parents and the resulting hybrids in multi-location trials for various traits. Among the obtained hybrids, PLM1 × 04–02 and PLM1 × Punjab Wanga showed significant heterosis for ascorbic acid (until 11.6 mg/100 g FW) and carotenoid content (until 4470 μg/100 g FW) in marketable fruits, among others. Zinash *et al.* (2013) and Zhou *et al.* (2017) found ascorbic acid content ranging from 48 to 139 mg/kg FW in different pumpkin genotypes. As regard to carotenoid content, Souza *et al.* (2012) and de Carvalho *et al.* (2012) reported values ranging from 236.1 to 290.6 mg/kg in raw pumpkin (*C. moschata*) cultivars.

Previously, Ilahy *et al.* (2014) assessed total phenolic content and the hydrophilic antioxidant activity of six powdery mildew-resistant snake melon breeding lines bred in Tunisia (Fig. 11.2). The authors found that total phenolic content ranged from 42 to 52 mg GAE/100 g FW in L9P1 and L4P18 respectively whereas hydrophilic antioxidant activity ranged from 34 to 134 μM trolox/100 g FW in L9P1 and L4P18, respectively. In this study, the breeding line L4P18 exhibited concurrently the highest total phenolic content and hydrophilic antioxidant activity. Minor variations in total phenolic content were found among the studied snake melon lines ranging from 3 to 22% compared to

Table 11.1. Nutrient content of snake melon (per 100 g edible food) (TurKomp, http://www.turkomp.gov.tr).

Component	Unit	Average	Range
Energy	kJ	92	51–129
Water	g	93.81	91.48–95.6
Ash	g	0.58	0.42–0.78
Protein	g	0.65	0.42–1.19
Nitrogen	g	0.10	0.07–0.19
Total fat	g	0.47	0.12–0.71
Carbohydrates	g	3.03	0.69–4.51
Total dietary fibre	g	1.45	1.19–1.89
Water soluble fibre	g	0.42	0.35–0.59
Water insoluble fibre	g	1.03	0.84–1.36
Saccharose	g	0.00	0–0
Glucose	g	1.03	0.40–2.39
Fructose	g	1.26	0.47–2.57
Lactose	g	0.00	0–0
Maltose	g	0.00	0–0
Salt	mg	27	9–62
Iron, Fe	mg	0.49	0.17–0.84
Phosphorus, P	mg	32	27–37
Calcium, Ca	mg	24	8–33
Magnesium, Mg	mg	23	21–27
Potassium, K	mg	237	204–264
Sodium, Na	mg	11	4–25
Zinc, Zn	mg	0.32	0.26–0.43
Vitamin C	mg	8.7	2.1–21.3
Thiamin	mg	0.039	0.007–0.053
Riboflavin	mg	0.045	0.033–0.061
Niacin, performed	mg	0.500	0.443–0.581
Vitamin B6, total	mg	0.064	0.051–0.090
Folate, food	µg	48	28–66
Vitamin A	RE	9	8–12
Beta-carotene	µg	113	94–144
Lycopene	µg	0	0–0
Lutein	µg	88	24–165
Vitamin K-1	µg	22.8	11.9–42.5

the large variation observed in hydrophilic antioxidant activity (43%–300%). Consequently, the authors emphasized the importance of this data for consumers, breeders and food scientists hoping to improve the antioxidant content of snake melon cultivars. Further, they evidenced that this refreshing summer fruit is a valuable source of dietary antioxidants for the

Table 11.2. Comparison between different crops for nutrient content (per 100 g edible food) (TurKomp, http://www.turkomp.gov.tr).

Component	Snake melon	Cucumber	Pumpkin	Melon
Potassium	264	156	290	260
Phosphorus	37	27	19	18
Calcium	33	19	31	22
Magnesium	27	13	18	17
Sodium	11	3	3	12

Fig. 11.2. Snake melon breeding lines selected by the National Agricultural Research Institute of Tunisia.

Mediterranean population because of its availability and high consumption.

Ilahy *et al.* (2019) recently published a study that focused on the differences in the main antioxidant components and antioxidant behaviours of three F1 hybrids, six snake melon breeding lines, and the commonly cultivated and consumed cultivar Mornagui, grown in a greenhouse over two growing seasons. The functional quality of the cultivar was found to differ significantly. The total phenol content of the F1 hybrid LP5 was the highest (48 mg GAE/kg FW). L4P18 had the highest levels of carotenoids (36 mg-Ca. E/kg FW) and flavonoids (232 mg RE/kg FW), while L9P1 had the highest levels of ascorbic acid (146 mg/kg FW), dehydroascorbic acid (231 mg/kg FW) and total vitamin C (377 mg/kg FW). The hydrophilic and lipophilic antioxidant activity were higher in XP12 and L2P4 (215 M TROLOX/100 g FW), and L1P3 (458 M trolox/100 g FW) than in all other cultivars. Snake melon therefore exhibits lower carotenoid

Table 11.3. Comparison between different crops for their total carotenoids, total phenolics, flavonoids and total vitamin C content.

Component	Snake melon	Pumpkin	Melon
Total carotenoids	36 mg/kg FW	236.1–290.6 mg/kg FW	82.4–90.83 mg/kg FW
Total phenolics	42–52 mg GAE/100 g FW	50–453.7 mg GAE/100 g FW	117.66–184.33 mg GAE/kg FW
Flavonoids	377 mg/kg FW	0.51–8.23 mg QE/100 g FW	45.32–47.21 mg CE/kg FW
Total vitamin C	377 mg/kg FW	48–139 mg/kg FW	134.4–339.75 mg/kg FW
Reference	(Ilahy *et al*., 2019)	(Tlili *et al*., 2020)	(Tlili *et al*., 2019)

content but significantly higher levels of total phenolics, flavonoids and total vitamin C compared to other cucurbits such as pumpkin and melon (Table 11.3).

Antinutritive Compounds

Generally, various antinutritive compounds are likely to be detected in melon fruits such as phytate, tannins and saponins. These compounds hinder the uptake of various nutrients (Sarkiyayi and Agar, 2010). Phytate, oxalate and tannins use calcium, zinc, iron and copper to form insoluble complexes, decreasing severely the availability of such elements and leading to their deficiency. Following their intake, tannins can inhibit digestive enzymes and precipitate proteins. Saponins affect fruit taste by their bitterness, thereby limiting consumption. Nevertheless, it has been recently highlighted that such compounds are potent stimulators of the immune system and modulators of cholesterol levels, besides their anticancer properties. Seemingly, there is scarce data regarding the levels of such compounds in melon and snake melon, so further investigations should be planned in order to improve knowledge of such aspects, particularly for snake melon as it is consumed at the immature stage of ripening and it is supposed that the content of antinutrients decreases with ripening stages (Menon and Ramana Rao, 2012).

Bioactive Properties and Health Benefits

The scientific community is interested in expanding awareness about food functionality, bioactivity and its implications, with the overall goal of enhancing human health. Growing evidence suggests that consuming more fresh fruits and vegetables is linked to a lower risk of developing a number of common human diseases, including cardiovascular disease, prostate, liver and stomach cancers, osteoporosis and UV-related skin disorders. (Erdman *et al*., 2009; Fernández-García, 2014; Nahak *et al*., 2014).

The fruit cucumber (*Cucumis sativus*) has long been used in oriental medicine for its anti-inflammatory, antidiabetic and abortifacient effects. Most recent studies related to the bioactivity and therapeutic values have mainly been conducted on different cucumber, rather than snake melon, cultivars, and showed the presence of various bioactive compounds with particular bioactivities. In fact, antimicrobial sphingolipids were isolated from the stems of cucumber (Tang *et al*., 2010). Significant anti-inflammatory and analgesic activity with respect to diclofenac sodium and morphine were reported by Gill *et al*. (2010). Antidiarrhoeal activity of cucumber leaves was also reported by Nasrin and Nahar (2014). Gill *et al*. (2009) revealed that a methanolic extract of cucumber seeds has antiulcer properties. Patil *et al*. (2011) conducted a pharmacological assessment of the ameliorative impact of aqueous cucumber fruit extract on wound healing in Wistar rats, demonstrating that *C. sativus* could be responsible for wound contracting and an increased degree of epithelization in wound healing in laboratory animals due to its antioxidant property and the existence of flavonoids. According to Heidari *et al*. (2016), aqueous extract of *Cucumis sativus* serves as a hepato-protective and antioxidant agent against cumene hydroperoxide-induced hepatotoxicity, implying that the antioxidant and radical scavenging components of *Cucumis sativus* fruit extract will easily cross the cell membrane and deal with intracellular reactive oxygen species (ROS) formation. Mallik and Akhter (2012) studied the phytochemical composition of cucumber peel ethanol extracts, as well as their reducing ability, cytotoxicity and antifungal activities, and discovered that cucumber can be a source of biologically active compounds, acting as an antifungal, anticancer agent and pesticide. All of the above-mentioned beneficial effects and bioactivities associated with cucumber intake can also apply to snake melon, and future research initiatives should emphasize this.

Regarding snake melon, Ibrahim (2017) focused on the neuroprotective effect of snake melon leaf extract on the brains of rats with streptozotocin-induced diabetes. The authors concluded that snake melon leaf extract applied a neuroprotective effect against oxidative damage associated with diabetes.

Novel Products/Formulations

Although the data regarding snake melon is scarce, it has recently been highlighted that when snake melon and date fruit extracts were combined, pleiotropic protective effects were observed against various chronic diseases such inflammation, diabetic-associated heart tissue damage and apoptosis. This stresses the need to comprehensively focus on the profiles of different bioactives present in snake melon fruits and their behaviour during ripening, and to explore the effect of such combined extracts as an emerging novel therapy for the treatment of chronic and age-related diseases (Saddi *et al.*, 2018). This may be associated with the synergistic effects among different antioxidant classes (carotenoids, phenolics, flavonoids, vitamins and minerals) as well as the matrix, rather than the effect of a single compound.

Conclusions and Future Trends

Snake melon cultivars are a subject of growing interest based on the richness of their fruit in various compounds with beneficial effect on human health, such as vitamins, phenolics, flavonoids, carotenoids, a balanced mineral mixture as well as other compounds not fully revealed. These findings can be expected to contribute to forthcoming research regarding innovative and hitherto underestimated sources of functional compounds, and stress the need to focus on the profile of different compounds present in snake melon and understand their dynamic changes during the pre- and post-harvest periods. Additionally, the *in vivo* health benefits of snake melon bioactive molecules also requires further investigation.

References

Bao, B., Chen, Y.-G., Zhang, L., Na Xu, Y.L., Wang, X. *et al.* (2013) *Momordica charantia* (Bitter Melon) reduces obesity-associated macrophage and mast cell infiltration as well as inflammatory cytokine expression in adipose tissues. *PloS One* 8(12), e84075. DOI: 10.1371/journal.pone.0084075.

Burger, Y., Paris, H.S., Cohen, R., Katzir, N., Tadmor, Y. *et al.* (2010) Genetic diversity of *Cucumis melo*. *Horticultural Reviews* 36, 165–198. DOI: 10.1002/9780470527238.

Burger, Y., Sa'ar, U., Paris, H.S., Lewinsohn, E., Katzir, N. *et al.* (2006) Genetic variability for valuable fruit quality traits in Cucumis melo. *Israel Journal of Plant Sciences* 54(3), 233–242. DOI: 10.1560/IJPS_54_3_233.

Chakravarty, H. (1966) Monograph on the Cucurbitaceae of Iraq. *Dar Al-Jumhuriya: Ministry of Agriculture Technical Bulletin* 133, 107–118.

Daraizadeh, M. (2014) *Snake Melon, its Therapeutic Effects and its Beautiful Skin*, 2nd edn. Mehr Publication.

de Carvalho, L.M.J., Gomes, P.B., Godoy, R.L. de O., Pacheco, S. and do Monte, P.H.F. (2012) Total carotenoid content, α-carotene and β-carotene, of landrace pumpkins (*Cucurbita moschata* Duch): A preliminary study. *Food Research International* 47(2), 337–340. DOI: 10.1016/j.foodres.2011.07.040.

Erdman, J.W., Ford, N.A. and Lindshield, B.L. (2009) Are the health attributes of lycopene related to its antioxidant function? *Archives of Biochemistry and Biophysics* 483(2), 229–235. DOI: 10.1016/j.abb.2008.10.022.

Fernández-García, E. (2014) Skin protection against UV light by dietary antioxidants. *Food & Function* 5(9), 1994–2003. DOI: 10.1039/c4fo00280f.

Food and Agriculture Organization of the United Nations (2020) FAOSTAT statistics database. Available at: https://www.fao.org/faostat/en/#home (accessed 2 May 2020).

Gill, N.S., Garg, M., Bansal, R., Sood, S., Muthuraman, A. *et al.* (2009) Evaluation of antioxidant and antiulcer potential of *Cucumis sativum* L. seed extract in rats. *Asian Journal of Clinical Nutrition* 1(3), 131–138. DOI: 10.3923/ajcn.2009.131.138.

Gill, N.S., Sood, S., Muthuraman, A. and Garg, M. (2010) Antioxidant, anti-inflammatory and analgesic potential of *Cucumis sativus* seed. *Latin American Journal of Pharmacy* 29, 927–932.

Grover, J.K. and Yadav, S.P. (2004) Pharmacological actions and potential uses of *Momordica charantia*: a review. *Journal of Ethnopharmacology* 93(1), 123–132. DOI: 10.1016/j.jep.2004.03.035.

Heidari, H., Kamalinejad, M., Noubarani, M., Rahmati, M., Jafarian, I. *et al.* (2016) Protective mechanisms of Cucumis sativus in diabetes-related models of oxidative stress and carbonyl stress. *BioImpacts: BI* 6(1), 33–39. DOI: 10.15171/bi.2016.05.

Henane, I., Mezghani, N., Tlili, I., Rhim, T., Ghezal, I. *et al.* (2013) Agro-morphological characterization and assessment of variability in local germplasm of *Cucumis melo* L. in Tunisia. *Journal of Biodiversity and Environmental Sciences* 3, 198–207.

Ibrahim, D.S. (2017) Neuroprotective effect of *Cucumis melo* var. *flexuosus* leaf extract on the brains of rats with streptozotocin-induced diabetes. *Metabolic Brain Disease* 32(1), 69–75. DOI: 10.1007/s11011-016-9886-y.

Ilahy, R., Tlili, I., Hdider, C., Ben, A.A. and Jebari, H. (2014) Total phenolic content and hydrophilic antioxidant activity of different powdery mildew-resistant Fakous (*Cucumis melo* var. *flexuosus*) lines grown in Tunisia. In: *Proceedings of the 3rd International Congress on Bioactives Molecules*, Hammamet, Tunisia, 21-23 March 2014.

Ilahy, R., Tlili, I., Chikh Rrouhou, H., R'Him, T., Homa, F. *et al.* (2019) Functional quality traits of snake melon (*Cucumis melo* var. *flexuosus* L.) fruits as affected by genotypic differences. *Journal of Postharvest Technology* 7, 1–10.

Ilahy, R., Tlili, I., R'him, T., Chikh Rouhou, H., Siddiqui, M.W. *et al.* (2019) A first insight on snake melon (*Cucumis melo* var. *flexuosus*): an under estimated source of functional compounds. In: Mitra, S., Banik, A.K., Mani, A., Kuchi, V.S., and Meena, N.K. (eds) *Trends & Prospects in Post-Harvest Management of Horticultural Crops*. Today & Tomorrow's Printers and Publishers, New Delhi, India, pp. 529–542.

Ilahy, R., Tlili, I., Rouhou, H.C., Siddiqui, M.W., Mishra, P.M. *et al.* (2020) Determining the main agronomic traits of snake melon (*Cucumis melo* var. *flexuosus* L.) fruits as affected by genotypic differences. *Advances in Horticultural Science* 34, 113–119.

Jebari, H. and Hamza, N. (1990) Study of late season crops of muskmelon (*Cucumis melo* L.) and fakous (*Cucumis melo* var. *flexuosus*) in Tunisian Sahel. *Acta Horticulturae* 263, 47–52. DOI: 10.17660/ActaHortic.1990.263.4.

Joseph, B. and Jini, D. (2013) Antidiabetic effects of *Momordica charantia* (bitter melon) and its medicinal potency. *Asian Pacific Journal of Tropical Disease* 3(2), 93–102. DOI: 10.1016/S2222-1808(13)60052-3.

Kolawole, O.T. and Ayankunle, A.A. (2012) Seasonal variation in the anti-diabetic and hypolipidemic effects of *Momordica charantia* fruit extract in rats. *European Journal of Medicinal Plants* 2(2), 177–185. DOI: 10.9734/EJMP/2012/998.

Mallik, J. and Akhter, R. (2012) Phytochemical screening and in-vitro evaluation of reducing power, cytotoxicity and anti-fungal activities of ethanol extracts of *Cucumis sativus*. *International Journal of Pharmaceutical & Biological Archives* 3, 555–560.

Mariod, A.A., Ahmed, Y.M., Matthäus, B., Khaleel, G., Siddig, A. *et al.* (2009) A comparative study of the properties of six Sudanese cucurbit seeds and seed oils. *Journal of the American Oil Chemists' Society* 86(12), 1181–1188. DOI: 10.1007/s11746-009-1459-3.

Menon, S.V. and Ramana Rao, T.V. (2012) Nutritional quality of muskmelon fruit as revealed by its biochemical properties during different rates of ripening.

Nahak, G., Suar, M. and Sahu, R.K. (2014) Antioxidant potential and nutritional values of vegetables: a review. *Research Journal of Medicinal Plant* 8(2), 50–81. DOI: 10.3923/rjmp.2014.50.81.

Nasrin, F. and Nahar, L. (2014) Antidiarhoeal activity of cucumis sativus leaves. *International Journal of Pharmaceutics and Drug Analysis* 106–110.

Pandey, S., Dhillon, N.P.S., Sureja, A.K., Singh, D. and Malik, A.A. (2010) Hybridization for increased yield and nutritional content of snake melon (*Cucumis melo* L. var. *flexuosus*). *Plant Genetic Resources* 8, 127–131. DOI: 10.1017/S1479262110000067.

Paris, H.S., Daunay, M.C. and Janick, J. (2011) Occidental diffusion of cucumber (*Cucumis sativus*) 500-1300 CE: two routes to Europe. *Annals of Botany* 109(1), 117–126. DOI: 10.1093/aob/mcr281.

Patil, M.V., Bhise, S. and Kandhare, A. (2011) Pharmacological evaluation of ameliorative effect of aqueous extract of *Cucumis sativus* L. fruit formulation on wound healing in Wistar rats. *Chronicles of Young Scientists* 2(4), 207–213. DOI: 10.4103/2229-5186.93026.

Pitrat, M. (2005) Ressources génétiques, diversité et amélioration du melon. *Le Sélectionneur Français* 55, 3–12.

Pitrat, M., Hanelt, P. and Hammer, K. (2000) Some comments on infraspecific classification of cultivars of melon. In: *VII Eucarpia Meeting on Cucurbit Genetics and Breeding 510*, pp. 29–36.

Saddi, A.A., Mohamed, A.M. and Shaikh, A.M. (2018) Prophylactic mechanisms of *Cucumis melo* var. *flexuosus* and *Phoenix dactylifera* fruit extracts against diabetic cardiomyopathy in streptozotocin induced diabetic rats. *Pakistan Journal of Pharmaceutical Sciences* 31(Suppl 2), 699–707.

Sarkiyayi, S. and Agar, T.M. (2010) Comparative analysis on the nutritional and anti-nutritional contents of the sweet and bitter cassava varieties. *Advance Journal of Food Science and Technology* 2, 328–334.

Shan, N., Gan, Z., Nie, J., Liu, H., Wang, Z. *et al.* (2020) Comprehensive characterization of fruit volatiles and nutritional quality of three cucumber (*Cucumis sativus* L.) genotypes from different geographic groups after bagging treatment. *Foods* 9(3), 294. DOI: 10.3390/foods9030294.

Souza, C.D., Menezes, J.D.S., DCNo, R., Assis, J.G.A., Silva, S.R. *et al.* (2012) Carotenoides totais e vitamina A de cucurbitáceas do banco ativo de germoplasma da embrapa semiárido. *Ciênc Rural Santa Maria* 42, 926–933.

Tang, J., Meng, X., Liu, H., Zhao, J., Zhou, L. *et al.* (2010) Antimicrobial activity of sphingolipids isolated from the stems of cucumber (*Cucumis sativus* L.). *Molecules* 15(12), 9288–9297. DOI: 10.3390/molecules15129288.

Tlili, I., Hdider, C., Lenucci, M.S., Riadh, I., Jebari, H. *et al.* (2011a) Bioactive compounds and antioxidant activities of different watermelon (*Citrullus lanatus* (Thunb.) Mansfeld) cultivars as affected by fruit sampling area. *Journal of Food Composition and Analysis* 24(3), 307–314. DOI: 10.1016/j.jfca.2010.06.005.

Tlili, I., Hdider, C., Lenucci, M.S., Ilahy, R., Jebari, H. *et al.* (2011b) Bioactive compounds and antioxidant activities during fruit ripening of watermelon cultivars. *Journal of Food Composition and Analysis* 24(7), 923–928. DOI: 10.1016/j.jfca.2011.03.016.

Tlili, I., Hdider, C., Ilahy, R., R'him, T. and Jebari, H. (2013) Effect of growing period on the agronomic characteristics and phenolic content of different watermelon (*Citrullus lanatus* (Thunb.) Mansfeld) cultivars grown in Tunisia. In: Ilahy, R. (ed.) *Tunisian Fruits and Vegetables. Food* 7(Special Issue 1), 22–26.

Tlili, I., Henane, I., Ilahy, R., R'him, T., Siddiqui, M.W. *et al.* (2019) Melon fruit: Functional component and allied health claims. In: Mitra, S., Banik, A.K., Mani, A., Kuchi, V.S., and Meena, N.K. (eds) *Trends & Prospects in Post-Harvest Management of Horticultural Crop*. Today & Tomorrow's Printers and Publishers, New Delhi, India, pp. 543–560.

Tlili, I., Chikh-Rouhou, H., Ilahy, R., Jedidi, E., Bouhlel, R. *et al.* (2020) Pumpkin. In: Nayik, G.A. and Gull, A. (eds) *Antioxidants in Vegetables and Nuts - Properties and Health Benefits*. Springer, Cham, Switzerland, pp. 105–126.

Walters, T.W. and Thieret, J.W. (1993) Notes on economic plants. *Economic Botany* 47(1), 99–100. DOI: 10.1007/BF02862210.

Zhou, C.-L., Mi, L., Hu, X.-Y. and Zhu, B.-H. (2017) Evaluation of three pumpkin species: correlation with physicochemical, antioxidant properties and classification using SPME-GC-MS and E-nose methods. *Journal of Food Science and Technology* 54(10), 3118–3131. DOI: 10.1007/s13197-017-2748-8.

Zinash, A., Workneh, T.S. and Woldetsadik, K. (2013) Effect of accessions on the chemical quality of fresh pumpkin. *African Journal of Biotechnology* 12, 7092–7098.

12 Kakadu Plum (*Terminalia ferdinandiana*)

Saleha Akter[1,2], Anh Dao Thi Phan[1,2], Michael Netzel[1,2] and Yasmina Sultanbawa[1,2*]

[1]ARC Industrial Transformation Training Centre for Uniquely Australian Foods, Centre for Nutrition and Food Sciences, Queensland Alliance for Agriculture and Food Innovation, The University of Queensland, Coopers Plains, Queensland, Australia; [2]Centre for Nutrition and Food Sciences, Queensland Alliance for Agriculture and Food Innovation, The University of Queensland, St. Lucia, Queensland, Australia

Abstract

Terminalia ferdinandiana Exell is a native Australian fruit from the Combretaceae family. The use of *T. ferdinandiana* by the Indigenous people of Australia has a long history. The fruit is used as a customary food consumed by children and adults. On hunting trips, adults use *T. ferdinandiana* fruits for quick energy and refreshment. This fruit is regarded as a medicine by the tribes in central Arnhem Land, Northern Territory (NT). Research on this fruit has revealed that *T. ferdinandiana* fruits contain the highest levels of vitamin C naturally occurring in fruits around the world. This fruit also has a high level of ellagic acid, a health promoting phytocompound with diverse bioactivities. This chapter outlines information on the botany, traditional uses, nutritional and phytochemical compositions of *T. ferdinandiana* and its potential as a functional ingredient in the food industry. In addition, a detailed description is given of the bioactive properties of the plant, such as antioxidant, antimicrobial, anti-inflammatory and cytotoxic. *In vitro* bioaccessibility and absorption of the main phytochemicals of *T. ferdinandiana* are discussed along with the implications of food processing. The information provided will contribute to the emerging research on *T. ferdinandiana* to help improve the design and plan of future research. The chapter will significantly help the food industry to find future applications for *T. ferdinandiana* as a functional ingredient in food products.

Introduction

Terminalia ferdinandiana Exell is a native Australian fruit from the Combretaceae family and is commonly called Kakadu plum (KP), billy goat plum, gubinge or salty plum (Cunningham *et al.*, 2008), and is known by different Indigenous names such as Arangal, Gabiny (Yawuru community), Gubinge (Bardi community north of and around Broome), Kabinyn (Nyul Nyul), Madoorr (Bardi community near One Arm Point, Dampier Peninsula), Manmohban (in Kune language, in Maningrida area), Mi-marl-arl (by Wadeye community, NT) and Murunga (Eastern Arnhem Land, NT). This endemic Australian plant grows in the tropical grasslands of the Kimberley area of Western Australia, upper NT and in some parts of Queensland (Hegarty and Hegarty, 2001; Gorman *et al.*, 2006). Commercial interest in *T. ferdinandiana* as a natural source of ascorbic acid

*Corresponding author: y.sultanbawa@uq.edu.au

DOI: 10.1079/9781789248067.0012

arose over 30 years ago, however commercial harvesting only began after 1996 (Cunningham *et al.*, 2008). According to the private company Kakadu Plum Wild Harvest, *T. ferdinandiana* is an ingredient in new products being developed by over 17 health and cosmetic companies globally (Cunningham *et al.*, 2008). Several health and nutritional products containing *T. ferdinandiana* are available in Australia such as juice, sauces, jam, whey protein isolates, energy bars, probiotic punch, analgesic spray and dietary supplements (in both tablet and liquid form) (The Australian Superfood Co, 2015). Cosmetic products containing *T. ferdinandiana* include shampoo, conditioner, moisturiser, skin serum, soap, face mask, hand sanitizer and face masks (Nourished Life, The Benefits of Kakadu Plum, 2020). In some of the *T. ferdinandiana* products, fruits are directly dried and milled into powders with both pulps and seeds present. Growing commercial interest by local and international food industries has increased demand for *T. ferdinandiana* products, and consequently these products need to be stored for long periods. The quality and bioactivity of *T. ferdinandiana* products must therefore be ensured as safe and effective. A storage trial of *T. ferdinandiana* was carried out over an extended period demonstrating that it can retain its effectiveness over 18 months when frozen (−20°C). Other studies have also identified chemical markers to determine the quality of *T. ferdinandiana* fruit products during frozen storage (Sultanbawa *et al.*, 2015). A recent report on *T. ferdinandiana* demonstrated a food safe enzyme-assisted extraction method to prepare bioactive compounds from *T. ferdinandiana* (Mridusmita *et al.*, 2017).

This chapter summarizes and discusses information regarding the nutritional values, proximate compositions and bioactivities of *T. ferdinandiana*. In addition, the health benefits of the phytochemicals present are discussed.

Botany

T. ferdinandiana is a semi-deciduous tree, small to moderate in size, which grows in the upper rainfall band of Australian wet/dry tropical savannas (Gorman *et al.*, 2019). The taxonomical classification is described in Table 12.1. The distribution of *T. ferdinandiana* in Australia is illustrated in

Table 12.1. Taxonomical classification of *Terminalia ferdinandiana*.

Scientific classification
Kingdom: Plantae
Phylum: Charophyta
Class: Equisetopsida
Subclass: Magnoliidae
Superorder: Rosanae
Order: Myrtales
Family: Combretaceae
Genus: *Terminalia*
Species: *T. ferdinandiana*
Binomial name: *Terminalia ferdinandiana* Exell

Fig. 12.1. *T. ferdinandiana* flowers from September to November and starts fruiting in January, through until June. Trees usually bear fruits in the fifth or sixth year. The fruits are yellow/green in colour with smooth skin and the size of a medium olive, with ovoid fleshy drupes and a short beak. Traditionally, KP trees grow in the wild, and are relatively free from insects and predators. A mature *T. ferdinandiana* tree and a fruiting branch are shown in Fig. 12.2. Currently, KP fruit can only be obtained through wild harvest, however some orchard plantations have recently begun. During the harvesting season, fruits are collected manually, and the harvested fruit are sorted and graded to remove any unripe or low-quality fruit. After sorting, fruit can be immediately frozen and sold as whole frozen fruit. Alternatively, sorted fruits are sanitized, de-stoned and processed into a purée which can be immediately frozen and sold. Otherwise, the purée can be freeze-dried and supplied as a powder. Variability exists among different species and subspecies of *T. ferdinandiana*, perhaps due to genetic diversity, ripening stage, soil, climate and storage conditions (Konczak *et al.*, 2014).

Traditional Uses

The use of KP as a part of the diet of the Indigenous Australians has a long history (Table 12.2). The fruit is not a staple food but is rather used as a customary food, though it is popular with children and adults. On hunting trips adults use these

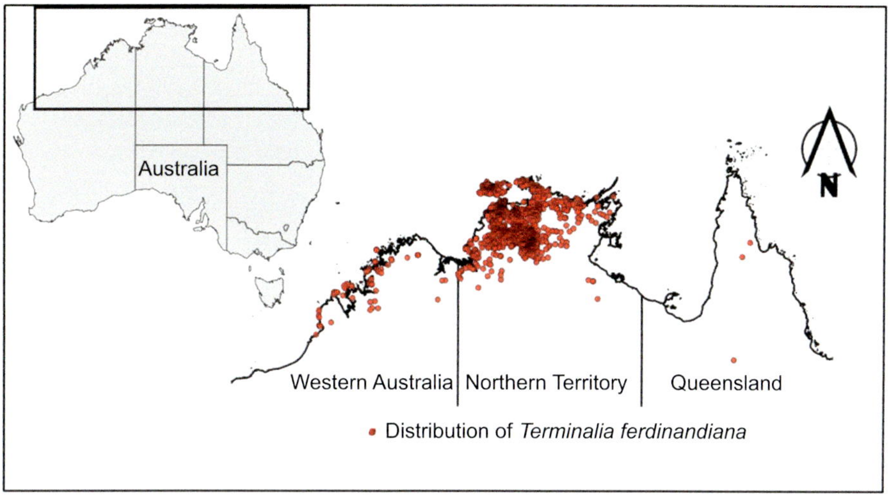

Fig. 12.1. Distribution of *Terminalia ferdinandiana* in Australia (Atlas of Living Australia, 2022).

Fig. 12.2. *Terminalia ferdinandiana* plant: (a) mature tree; (b) fruiting branch.

fruits for quick energy and refreshment. This fruit is regarded as a medicine rather than a food by the tribes in central Arnhem Land, NT (Konczak *et al.*, 2014). Modern use of the plum flesh includes sauces, jams and flavourings (Hegarty and Hegarty, 2001). The present use of KP includes food supplements such as beverages, capsules and powders; skin and care products; pharmacological products; and in gourmet bush foods such as chutneys, jams and pickles (Konczak *et al.*, 2014).

Nutritional Composition

Proximate composition

The results of the proximate composition of plant parts of *T. ferdinandiana* have been adapted from Akter *et al.* (2018, 2020a) and are shown in Table 12.3. Results are expressed as a mean of triplicate experiments on a dry weight (DW) basis

Table 12.2. Traditional uses of *T. ferdinandiana*.

Plant part	Traditional uses
Fruits and seeds	In Western Australia, refreshing drinks are made from fresh or dried fruits by soaking them in water for one or two days (Hegarty and Hegarty, 2001)
Kernels	The kernels (*T. grandiflora*) are used as food after breaking the thick seedcoat with stone (Kenneally *et al.*, 1996)
Gums	An edible jelly is prepared from the gums by soaking the pounded powder. They are also cooked in sand or can be chewed directly (Kenneally *et al.*, 1996)
Bark	The inner part of the bark is useful in the treatment of inflammatory conditions such as sores and boils as well as for ringworm infection and leprosy (Gorman *et al.*, 2006)

Table 12.3. Proximate composition of *T. ferdinandiana* plant parts (Akter *et al.*, 2018, 2020a).

	Plant part				
Proximate Profile		Fruits	Leaves	Seedcoats	Kernel
Protein	% (w/w)	4.7	7.1	2.3	32.0
Fat	% (w/w)	0.9	6.5	0.6	35.1
Saturated fat	% (w/w)	0.3	2.5	0.1	5.8
Monounsaturated fat	% (w/w)	<0.1	0.4	0.2	9.8
Polyunsaturated fat	% (w/w)	0.6	3.7	0.4	19.4
Trans fat	% (w/w)	<0.01	<0.01	<0.01	<0.01
Ash	% (w/w)	5.5	5.9	0.6	4.5
Moisture	% (w/w)	6.0	3.0	2.2	4.0
Dietary fibre (total)	% (w/w)	45.9	44.0	90.2	21.2
Energy	kJ/100 g	1110	1283	853	2065
Total sugar	g/100 g	2.3	2.5	<0.10	0.49
Available carbohydrate	%	37.0	33.4	4.0	3.2

(Akter *et al.*, 2018, 2020a). Protein, fat, ash and energy content were highest in the leaves, followed by the fruits and seedcoats. The seedcoats contain the highest amount of dietary fibre (90.2%). The moisture content of the fruit is higher than the leaves and seedcoats. Study on the proximate composition of the kernels of *T. ferdinandiana* showed that they contain 32% protein, 35% fat, and are a good source of minerals and nutrients.

Mineral composition

The mineral content in *T. ferdinandiana* kernels has been shown to be phosphorus at 872.8 mg/100 g DW and calcium at 538.5 mg/100 g DW, while sodium was 120.3 mg/100 g DW, potassium 669.3 mg/100 g DW and magnesium 421.1 mg/100 g DW. Important trace elements found in *T. ferdinandiana* kernels were zinc, manganese, copper and iron at levels of 6, 9.1, 2.5 and 6.1 mg/100 g, respectively. The authors suggested that *T. ferdinandiana* kernels can be a valuable dietary source of these trace elements (Akter *et al.*, 2018). Calcium content was found to be higher in the leaves (1343 mg/100 g on DW basis) compared to the fruits (295 mg/100 g DW) and seedcoats (131 mg/100 g DW). Potassium content in the fruits was found to be 2718 mg/100 g DW, phosphorus 73 mg/100 g DW, iron 1.7 mg/100 g DW and zinc 2.2 mg/100 g DW. Manganese levels in the leaves are much higher (25.5 mg/100 g DW) compared to the fruits (5.1 mg/100 g DW) and seedcoats

(1.3 mg/100 g DW) (Akter *et al.*, 2020a). The authors have suggested that *T. ferdinandiana* fruit presents no known risk for heavy metal toxicity if 1 g freeze dried powder is consumed per day.

Fatty acid composition

The fatty acid composition of the kernels consisted of omega-6 fatty acid, linoleic acid (50.2%), monounsaturated oleic acid (29.3%) and two saturated fatty acids, namely palmitic acid (12.0%) and stearic acid (7.2%). The authors have suggested that *T. ferdinandiana* kernels have the potential to be utilized as a novel protein source for dietary purposes and as a non-conventional supply of linoleic, palmitic and oleic acids (Akter *et al.*, 2018).

Vitamin C

T. ferdinandiana fruit has been reported to have high vitamin C content, though the level reported by different research groups has varied. Williams and colleagues have reported that total vitamin C level in whole fruit was 18.16 g/100g DW (Williams *et al.*, 2016) and in composite fruit extracts 14.04 g/100g DW (Williams *et al.*, 2014). Later, Akter and colleagues reported that the contents of L-AA, DHAA and total vitamin C in freeze-dried *T. ferdinandiana* fruits was 16.4, 4.7, and 21.1 g/100g DW, respectively (Akter *et al.*, 2020a). This variation could be due to wild harvesting of fruits occurring in different geographical locations, climatic and soil conditions, and seasonality.

Phytochemicals

T. ferdinandiana has a rich composition of phytochemicals with many of these compounds providing health benefits. Wild harvest practices in the northern region of Australia and parts of Western Australia, along with the natural abundance of ascorbic acid in *T. ferdinandiana* fruit, have contributed to increased research efforts and the growing commercial value of *T. ferdinandiana* (Gorman *et al.*, 2006; Cunningham *et al.*, 2008). Studies are still underway to determine the various biological activities attributed to phytochemicals isolated from this plant. Reports of the phytochemicals identified in *T. ferdinandiana* are summarized in Table 12.4 while chemical structures of selected phytochemicals are illustrated in Fig. 12.3. In addition to numerous beneficial phytochemicals present in *T. ferdinandiana* fruits, a novel pectin has also recently been identified (Konczak and Roulle, 2011; Chaliha *et al.*, 2018).

Tannins, ellagitannins and ellagic acid

T. ferdinandiana has been reported to contain ellagitannins and ellagic acid. Courtney *et al.* (2015) reported that ethyl acetate leaf extracts of *T. ferdinandiana* contain ETs, corilagin, castalagin and punicalin, along with chebulagic acid, ellagic acid dehydrate and trimethyl ellagic acid. Sirdaarta *et al.* (2015) have reported the presence of tannins, such as ethylgallate isomers, gallocatechin, trimethyl ellagic acid, chebulic acid and corilagin. Additionally, stilbenes, like resveratrol, combretastatin, diethyl stilbestrol monosulfate and resveratrol glucoside have been found in methanolic extracts of *T. ferdinandiana* fruits using non-targeted HPLC-MS QTOF analysis. Previous report on the ellagic acid (EA) content in *T. ferdinandiana* fruits ranged from 3050 to 14020 mg/100 g DW (Konczak *et al.*, 2014). However, Akter *et al.* (2020a) have reported that fruits contain 1400 mg/100 g DW free ellagic acid and 2760 mg/100 g DW total ellagic acid; leaves contain 2250 mg/100 g DW free ellagic acid and 9540 mg/100 g DW total ellagic acid; bark contains 94 mg/100 g DW free ellagic acid and 1373 mg/100 g DW total ellagic acid; and seedcoats contain 84 mg/100 g DW free ellagic acid and 88 mg/100 g DW total ellagic acid. The punicalagin content of *T. ferdinandiana* fruits and leaves was found to be 74 and 49 mg/100 g DW, respectively (Akter *et al.*, 2020a). Castalagin content was reported by the same research group as 1.1 mg/100 g DW for fruits and 0.8 mg/100 g DW for leaves (Akter *et al.*, 2020a).

Table 12.4. Different classes of compounds identified from *T. ferdinandiana*.

Plant part	Compound group	Name of compound	References
Fruit	Vitamin C	Ascorbic acid	(Netzel *et al.*, 2007; Williams *et al.*, 2014)
	Vitamin E components	α-tocopherol, γ-tocopherol, lutein, chlorophyll a, chlorophyll b	(Konczak and Roulle, 2011)
	Polyphenols/phenolic compounds	Ellagic acid, ethyl gallate, gallocatechin, trimethyl ellagic acid, chebulic acid, corilagin, gallic acid	(Sirdaarta *et al.*, 2015)
	Aromatic acids or phenylpropanoids	Hydroxy cinnamic acid, proanthocyanidins	
	Minerals	Ni, Cd, Pb, Al, Co, Na, Fe, Zn, Mo, Cu, Mn, Ca, Mg, K, P, Se,	(Konczak and Roulle, 2011)
	Stilbenes	Resveratrol, combretastatin, combretastatin A-1, diethylstilbestrol monosulphate, reseveratrol glucoside (piceid/polydatin)	(Courtney *et al.*, 2015)
	Monoterpenoids	Cineole (eucalyptol), cuminol, camphor, isomenthol, amyrintriterpenoid (3β,4α,16α,21β,22α) olean-12-ene-3,16,21,22,23,28-hexol	(Sirdaarta *et al.*, 2015)
	Flavonoids	Appanone A-7 methyl ether, xanthotoxin, phthalane, Eujavanoic acid, Rebamipide	(Sirdaarta *et al.*, 2015)
	Propionic acid derivative	Carpofen	(Sirdaarta *et al.*, 2015)
Leaf	Tannins	Exifone (4-galloyl pyrogallol), ellagic acid dehydrate, trimethyl ellagic acid, chebulic acid, punicalin, corilagin, castalagin, chebulagic acid	(Courtney *et al.*, 2015)

Antinutritive Compounds

Several reports are available on the presence of toxic chemicals in *Terminalia* genus despite the presence of a plethora of biologically important and efficacious phytochemicals in these plants. In 2001, Hegarty reported an antinutritive property of *T. ferdinandiana* even though no toxic or unsafe compounds were identified in *T. ferdinandiana* frozen fruits and fresh seeds. In this report, oxalic acid (0.24 g/ 100 g) and cyanogens (<0.1 mg HCN/ 100 g) on a fresh weight (FW) basis were identified, however saponins and alkaloids were not detected (Hegarty and Hegarty, 2001). Oxalic acid, a known antinutritional compound, has also been reported to be present in *T. ferdinandiana* (Williams *et al.*, 2016). Total oxalate content in fruits is 1420 mg/100 g DW, in leaves 1133 mg/100 g DW and in seedcoats 327 mg/100 g DW (Akter *et al.*, 2020a). The phytate levels of *T. ferdinandiana* plant parts show that the leaves contain the highest amount of phytate (121.72 mg/100 g DW) compared to fruits (37.6 mg/100 g DW) and seedcoats (8.44 mg/100 g DW) (Akter *et al.*, 2020a).

Implications of Food Processing on Nutrients and Phytochemicals

Phan *et al.* (2021) have recently reported the effects of drying methods and maltodextrin on the vitamin C content and quality of *Terminalia ferdinandiana* fruit powder. Freeze-drying and oven-drying were the two drying methods tested. Freeze-drying was a better dehydration

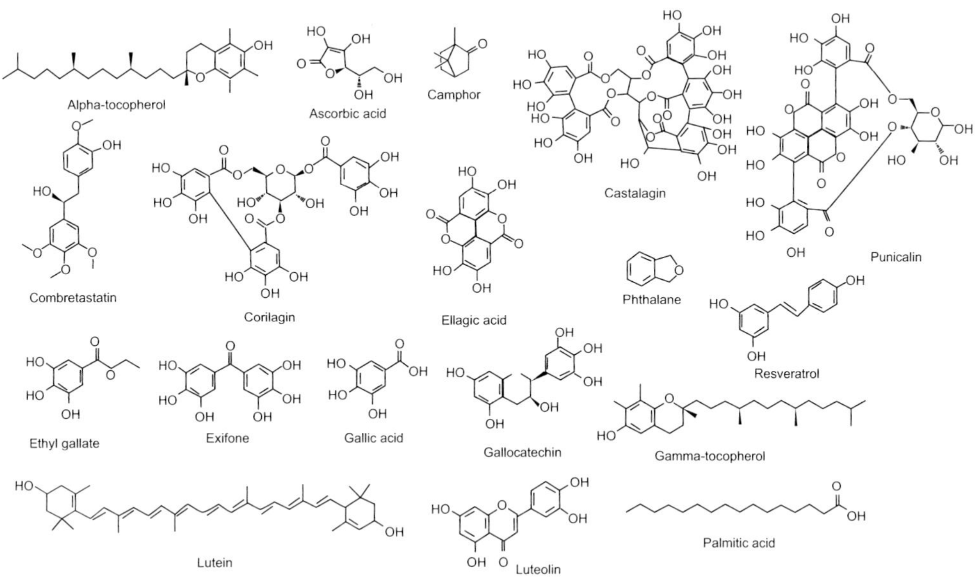

Fig. 12.3. Chemical structures of selected compounds identified from *T. ferdinandiana*.

technique than oven-drying in retaining vitamin C, reducing the formation of non-enzymatic browning and oxidation products and improving powder colour. Furoic acid and furfural have been reported as the main C5-compounds that could be generated during the degradation of AA in the presence or absence of oxygen in a low acidic environment (KP fruit purée with pH 3.37). Furoic acid (81–32 mg kg−1 DW) was produced at much higher levels than furfural (11–19 mg kg−1 DW). Compared with the control samples, adding maltodextrin (P<0.05) significantly reduced the percentage loss of ascorbic acid (AA) from 8.1 to 3.3% in the freeze-dried samples and from 18.9 to 14.4% in the oven-dried samples, particularly at maltodextrin levels of 10–15%. The results demonstrated the potential of utilizing maltodextrin as an AA stabilizer, contributing to the improvement of colour appearance of the KP powdered samples processed by freeze-drying and oven-drying (Phan *et al.*, 2021). Cozzolino and colleagues successfully applied mid-infrared (MIR) and near-infrared (NIR) spectroscopy for the measurement of vitamin C in KP plum powder samples. They calibrated IR spectroscopic instruments for the measurement and showed that IR spectroscopy (MIR, NIR and Fourier transform near-infrared (FTNIR)) models are capable of quantification of vitamin C in KP. They have suggested that high-throughput IR spectroscopy could be used as a routine technology to determine vitamin C in plant-based foods and derived products (Cozzolino *et al.*, 2021).

Bioactive Properties

Antioxidant activity

T. ferdinandiana is well known for its extremely high content of ascorbic acid (Williams *et al.*, 2016). Several studies have focused on the antioxidant potential of *T. ferdinandiana*. These studies have investigated various extraction and purification methods for the identification of phytochemicals that contribute to the antioxidant activity of this fruit (Netzel *et al.*, 2007; Konczak *et al.*, 2010, 2014; Tan *et al.*, 2010, 2011a, b, c; Konczak and Roulle, 2011). A bioassay guided fractionation of the *T. ferdinandiana* extracts was screened for antioxidant activity using oxygen radical absorbance capacity (ORAC) and cellular antioxidant activity (CAA) in Hep G2 cells. The authors reported that *T. ferdinandiana* extracts had strong antioxidant activity with an ORAC value of 3776 ± 603 μmol trolox/g DW and a CAA value of 52.2 ± 8.6 µmol quercetin equivalents/g DW (Tan *et al.*, 2011c). A lyophilized polyphenolic-rich extract of *T. ferdinandiana* exhibited the greatest cellular antioxidant activity (CAA assay) of 71.5 µmol QE/g (Tan *et al.*, 2011b). Antioxidant study was again performed by the same research group with 45 accessions of *T. ferdinandiana* and they reported that it contained high levels of total phenolics (505.2 to 376.1 mg GA E/g DW) and vitamin C (322.2 to 173.5 mg/g DW), with pronounced antioxidant capacities (FRAP: 5030.5 to 4244.9 µmol Fe2+/g DW; ORAC: 3861.5 to 2985.6 µmol trolox E/g DW) (Konczak *et al.*, 2014). An accelerated solvent extraction system was applied by another research group who reported the antioxidant potential of the *T. ferdinandiana* fruits with methanol extracts containing the highest TPC, and DPPH-free radical scavenging capacity (12.2 g/100 g DW and 93%, respectively) among the solvents used for the extraction (Akter *et al.*, 2019).

Anti-inflammatory activity

The anti-inflammatory activity of *T. ferdinandiana* was reported earlier. *T. ferdinandiana* alleviates hydrogen peroxide induced oxidative stress and inflammation *in vitro* (Chaliha and Sultanbawa, 2019). Caco-2 and KERTr cell lines were treated with an aqueous food grade EA enriched (EAE) extract of *T. ferdinandiana* or pure EA followed by the exposure to hydrogen peroxide (H_2O_2). Levels of reactive oxygen species (ROS) production and gene expression of molecular markers associated with oxidative stress and inflammation were monitored in that study. The authors reported that significant reduction in ROS production was observed in both cell types treated with 100 or 200µg/ml EA or EAE. Treatment of cells with EAE or EA showed up-regulation of mRNA expression of the antioxidative gene superoxide dismutase (SOD)-2, and down-regulation of the expression of inducible nitric oxide synthase (iNOS), soluble cell adhesion molecule (sICAM) and cyclooxygenase (COX)-2. Neither EAE nor EA had any effect on the constitutively expressed COX1 (Chaliha and Sultanbawa, 2019). Anti-inflammatory activity was also tested on a bioassay guided fractionation of the *T. ferdinandiana* extracts using nitrite concentration and prostaglandin E2 release in lipopolysaccharide (LPS)-activated murine macrophages (RAW 264.7 Cells). *T. ferdinandiana* extracts demonstrated an upregulation of the Nrf2/Keap1 ratio in Hep G2 cells. The extracts also inhibited the expression of COX-2 and iNOS in LPS-activated murine macrophages, potentially through the NF-κB, p44/42 mitogen activated protein kinase and Akt pathways (Tan *et al.*, 2011c). *T. ferdinandiana* has also been reported to have anti-inflammatory activity in autoimmune disorders (Courtney *et al.*, 2015; Sirdaarta *et al.*, 2015).

Cytotoxic activity

Extracts of *T. ferdinandiana* fruits and seedcoats were tested for the *in vitro* cell viability of undifferentiated and differentiated Caco-2, HT29-MTXE12 and HepG2 cells using the CyQUANT NF cell proliferation assay. Changes to cell viability produced IC_{50} values between 3650 and 14400 µg/ml for all extracts and cell lines tested with Hep G2 cells were impacted the most by *T. ferdinandiana* extracts, followed by HT29-MTX-E12 cells, and undifferentiated and differentiated Caco-2 cells. Extracts from seedcoats impacted cell viability less than fruit extracts (Akter *et al.*, 2021).

Antimicrobial activity

Numerous studies have been performed in the last decade to evaluate the pharmacological potential of *T. ferdinandiana*. Various research groups have revealed potential uses for *T. ferdinandiana* by identifying antibacterial activity (Cock and Mohanty, 2011; Sultanbawa *et al.*, 2015; Akter *et al.*, 2019) and antigiardial activity (Rayan *et al.*, 2015). Other reports have also indicated that *T. ferdinandiana* fruit and leaf extracts exhibit strong activity against antibiotic-resistant pathogens such as extended spectrum β-lactamase and methicillin-resistant *Staphylococcus aureus* (Cheesman *et al.*, 2019), and potent growth inhibitory properties against fish spoilage bacteria, notably *Shewanella* spp. (Wright *et al.*, 2019). A recent report has shown the suitability and effectiveness of high throughput screening methods for the determination of antimicrobial activity of *T. ferdinandiana* extracts (Akter *et al.*, 2021). Table 12.5 summarizes the published antimicrobial work on *T. ferdinandiana*.

Antiproliferative activity

A bioassay guided fractionation of the *T. ferdinandiana* extracts was reported to induce apoptosis and DNA damage in HT-29 cells determined by the cytokinesis block micronucleus cytome assay (Tan *et al.*, 2011c). Antiproliferative activities against CaCo-2 and HeLa cancer cells were determined by an MTS-based cell proliferation assay. The high antioxidant methanolic and aqueous extracts of *T. ferdinandiana* were reported to be potent inhibitors of cell proliferation, with IC_{50} values generally below 1000 μg/ml (Jamieson *et al.*, 2014). Fruit and leaf extracts were also reported to inhibit proliferation and induce apoptosis in Caco-2, HeLa, Jeg-3, JAR, MC3T3-E1, and MG63 cancer cell lines (Shalom and Cock, 2018). Solvent extracts prepared from *T. ferdinandiana* kinos and their antiproliferative activities against *G. duodenalis* were determined by direct enumeration. The antiproliferative activities against HeLa and Caco-2 cancer cell lines were determined using an MTS-based cell

Table 12.5. Antimicrobial activity of *T. ferdinandiana*.

Botanical tissue	Extraction	Methodology	Microorganisms	References
Fruit pulp and leaves	Methanol, deionized water, ethyl acetate, chloroform, hexane	Disc diffusion and microplate liquid dilution MIC assay	Fish spoilage bacteria *Shewanella putrefaciens* strain 200, *Shewanella baltica* strain OS155, *Shewanella frigidimarina* strain NCIMB 400 and *Shewanella loihica* strain PV-4	(Wright *et al.*, 2019)
Fruit pulp and leaves	Methanol, deionized water, ethyl acetate, chloroform or hexane	Disc diffusion and microplate liquid dilution MIC assay	*Bacillus anthracis*	(Wright *et al.*, 2016)
Fruits, leaves, seedcoats and barks	Methanol, ethanol, acetone, hexane, and distilled water	Disc diffusion, MIC and MBC	*Staphylococcus aureus* (NCTC 6571), Methicillin resistant *Staphylococcus aureus* (MRSA) clinical isolates, *Bacillus cereus*, *Listeria monocytogenes*, *Pseudomonas aeruginosa*, *Pseudomonas aeruginosa* clinical isolates	(Akter *et al.*, 2019)
Fruit pulp and leaves	Methanol, deionized water, ethyl acetate, chloroform or hexane	Disc diffusion, microplate liquid dilution MIC assay and disc diffusion MIC assay	*Brevibacterium linens* (ATCC9172), *Corynebacterium jeikeium* (ATCC43734) and *Propionibacterium acnes* (ATCC6919)	(McManus *et al.*, 2017)
Fruit pulp	Methanol and deionized water	Direct enumeration of parasite numbers	*Giardia duodenalis* S-2 (sheep strain 2) trophozoite strain	(Rayan *et al.*, 2015; Cock and Rayan, 2020)

assay. Methanolic and aqueous *T. ferdinandiana* kino extracts were strong inhibitors of Caco-2 and HeLa cell proliferation, with IC_{50} values substantially below 1000 µg/ml (Shalom *et al.*, 2018). *T. ferdinandiana* fruit and leaf extracts have been reported to have antiproliferative and apoptotic activities against a panel of human carcinoma cell lines. All extracts inhibited Caco-2, HeLa, Jeg-3, JAR, MC3T3-E1 and MG63 proliferation. The leaf ethyl acetate extract was the most potent inhibitor of proliferation (MC3T3-E1 IC_{50} of 6 µg/ml, Caco-2 IC_{50} of 102 µg/ml). Furthermore, IC50s < 500 µg/ml were determined against all cell lines tested against that extract. The methanolic leaf extract was also a potent inhibitor of cell proliferation (Jeg-3 IC_{50} of 147 mg/ml and MC3T3-E1 IC_{50}D 40 µg/ml). The fruit extracts were also good inhibitors of carcinoma cell proliferation. Caspase 3 activity was significantly elevated in Caco-2 cells exposed to these extracts, indicating that apoptosis was induced. The leaf ethyl acetate extract contained a high diversity and relative abundance of tannins and flavonoids (Shalom and Cock, 2018).

Cytoprotective activity

In vitro cytoprotective activity (Tan *et al.*, 2011b) and *in vitro* anticancer activity (Tan *et al.*, 2011a, c) were reported earlier on *T. ferdinandiana*. Polyphenolic-rich extracts of *T. ferdinandiana* extracts efficiently protected RAW 264.7 cells against hydrogen peroxide induced apoptosis in a dose-dependent manner (Tan *et al.*, 2011b). The authors suggested that the polyphenolic-rich extracts increased the Nrf2/Keap1 ratio by activation of the antioxidant response element (ARE) through the Nrf2/Keap1 complex.

Bioaccessibility and Bioavailability

The bioaccessibility and absorption of the bioactive compounds from *T. ferdinandiana* was evaluated by utilizing the Infogest *in vitro* digestion model and Caco-2-HT29-MTX-E12 co-culture cell absorption/transport model. The results of the study indicate that there were no significant changes (P>0.05) in the total ascorbic acid content throughout *in vitro* digestion, suggesting that ascorbic acid was not degraded during this process *in vitro*. The initial bioaccessibility was observed at 93% at the beginning of the gastric phase. Approximately 90% of AA was released from the solid phase into the liquid phase following digestion with pepsin for 60 minutes. Digestion with pancreatin and bile in the small intestinal phase resulted in 93, 100, and 79% bioaccessibility after 30, 60 and 120 minutes, respectively. Bioaccessibility of ellagic acid significantly increased from 33% at the beginning of the gastric phase to 48% at the end of the intestinal phase (I120). The increase in EA bioaccessibility may be due to the release of ellagic acid from ellagitannins and other ellagic acid derivatives produced during digestion. Bioaccessibility of total oxalic acid was estimated to be 72% at the beginning of the gastric digestion phase before significantly increasing to 98% in the intestinal phase. At the end of the intestinal phase the oxalic acid bioaccessibility was decreased to 89%. The release of calcium at the beginning of the gastric phase was 67% and remained steady throughout the gastric phase. Bioaccessibility of calcium significantly increased to 90% after 30 and 60 minutes in the intestinal phase before significantly decreasing to 79% at the end of the intestinal phase. The intestinal transport of ascorbic acid, ellagic acid, oxalic acid and calcium from digested *T. ferdinandiana* powder was measured using Caco-2-HT29-MTX-E12 co-culture cell monolayers. Intestinal digesta samples collected at different time points (I30, I60, and I120) throughout the *in vitro* digestion were applied to the co-culture transwell model for 120 minutes before apical and basolateral samples were collected. The amount of ascorbic acid, ellagic acid, oxalic acid and calcium was measured in the digesta initially applied to the apical chamber of the transwell model, and in samples collected from the basolateral chamber. Apparent permeability (*Papp*) was calculated with statistical significance observed (P≤0.05) in *Papp* values obtained from the transported I30, I60, and I120 digesta samples. Overall, Papp of the measured AA, EA and OA was significantly higher in the transported I30 digesta compared to the transported I60 and I120 digesta. However, the transport of Ca^{2+} did not significantly change when comparing transport of any digesta. The *Papp* values obtained for ellagic acid were higher than those calculated for ascorbic acid, oxalic acid and calcium. From the *Papp* observed in the study, it is apparent that ascorbic acid, ellagic acid, oxalic acid and calcium demonstrated significant intestinal transport. The findings from the study presented here indicate that ascorbic acid, ellagic acid, oxalic acid and calcium released from digested *T. ferdinandiana* powder are bioaccessible and somewhat resistant to human gastrointestinal conditions, potentially making these compounds available for absorption in the small intestine. Apparent permeability of the investigated compounds also indicated that a large proportion of these compounds may be transported through the intestinal barrier providing insights into the way these compounds behave *in vivo*.

Health Benefits

EA, a dietary polyphenol, and vitamin C or ascorbic acid (AA), a well known antioxidant, are found in *Terminalia ferdinandiana* and have attracted research interest due to the potential health benefits and vast range of pharmacological activities associated with these compounds (Williams *et al.*, 2016). Pharmacological properties of EA have been investigated including antioxidant, antibacterial, antiparasitic, antimutagenic, antiatherosclerotic, anticarcinogenic and antiapoptotic activity (Aguilera-Carbo *et al.*, 2008; Okuda *et al.*, 2009; Assunção *et al.*, 2017). While the clinical research is limited at this point, the clinical trials that have been conducted suggest that certain *Terminalia* plant extracts have demonstrable health-promoting qualities, certainly in terms of antimicrobial activity in oral applications. While traditional use and anecdotal evidence is abundant, further clinical trials are required to demonstrate the broader health-promoting activities of this interesting genus of plants. Further, as currently there have been no clinical trials conducted involving KP, much work is required in this area for this plant to fulfil its potential as a dietary supplement for the prevention of chronic disease.

Novel Products/Formulations

Due to increased awareness of the potential health benefits of *T. ferdinandiana*, there has been an increased interest both nationally and internationally to incorporate it in food, nutraceutical, pharmaceutical or cosmeceutical products. The fresh fruit has an aroma of stewed apples and pears, with some cooked citrus, pickled and fermented notes, and a tart-bitter taste. Due to its highly fibrous content it is often incorporated into products as an ingredient in powdered form (freeze-dried purée powder). Currently, a number of *T. ferdinandiana* foods (juices, sauces, jams and whey product isolates), nutraceuticals (energy bars, probiotics with Manuka honey, probiotic punch with cranberry, analgesic sprays, dietary supplement tablets) and cosmetic products (shampoos, conditioners, body cleansers, hydrating creams, hand balms, body butters, moisturisers, skin serums, face scrubs, soaps, exfoliators, fragrances, cleansing oils, face masks and hand sanitizers) are commercially available.

Conclusions and Future Trends

The development of novel functional ingredients from *T. ferdinandiana* fruits has been an important research focus that will empower Indigenous communities by developing Indigenous owned and controlled enterprises. For these enterprises to be successful, scientific evidence must be available regarding nutritional, antinutritional, functional and bioactive properties, to determine the safe use of *T. ferdinandiana* functional ingredients. Science should support traditional Indigenous knowledge that has a long history of customary *T. ferdinandiana* use for food and medicinal purposes. Such research will greatly support the use of native foods as novel functional ingredients in the food and beverage industries. Studies into the potential use of *T. ferdinandiana* by-products are lacking. This study has identified *T. ferdinandiana* kernels as a good source of fat (35%) and protein (32%) on a DW basis. The kernels also serve as a rich source of essential minerals (potassium, calcium and iron) and have high energy content, dietary fibre and linoleic acid (omega-6 fatty acid). Industrial processing of *T. ferdinandiana* fruits into purée generates seeds as a by-product, thus seedcoats are another by-product of *T. ferdinandiana* processing that could provide value-adding opportunities. The seedcoats contain 90% dietary fibre supporting this by-product for potential inclusion in diets as high fibre functional ingredients. Separation of the kernels from the seedcoat at scale, and further processing of the kernels and seedcoats, could play a role in the development of novel functional ingredients targeting the food and nutraceutical industries. Polyphenol-rich extracts were used in different studies with the resulting bioactive properties (antioxidant and antimicrobial) indicating that *T. ferdinandiana* fruit is not the only tissue with functional properties, but that the leaves, seedcoats and bark also have bioactive potential. Significant advances in the understanding of the biological properties of tannins and ellagitannins have been made due to disease prevention being attributed to tannin intake. Recent research has identified two notable ellagitannins, punicalagin and castalagin, in *T. ferdinandiana* along with high ellagic acid content. The presence of punicalagin in *T. ferdinandiana* can help promote this fruit as a novel functional ingredient with health-promoting value. Understanding the *in vitro* biological effects of phytocompounds is also important for assessing and predicting adverse health effects in humans. For example, it has been suggested that *T. ferdinandiana* seeds contain toxins and therefore should be avoided. However, nutritional composition along with the cell viability data obtained in this study suggests that seedcoats have potential to be included in the food industry as a high fibre functional ingredient. *In vitro* bioaccessibility and intestinal transport of the different compounds investigated showed variable release from *T. ferdinandiana* powder and apparent intestinal permeability. The different affinity of phytocompounds and minerals is well documented. Minerals are known to be chelated to the organic acids in plant food. The calcium and oxalic acid present in *T. ferdinandiana* fruit therefore might play an important role in protection against chronic diseases.

However, future research should be designed to fully understand the effects of *T. ferdinandiana* on human health as well as its potential as a functional food ingredient in the food industry. Complete polyphenol profiling should be performed on *T. ferdinandiana* fruits to better understand how fruit composition contributes to bioactivity. The pharmacokinetic profile, absorption rate, tissue distribution, intestinal and liver metabolism, and excretion of *T. ferdinandiana* polyphenols should also be assessed to determine their safety in the human body. Metabolites and breakdown products of *T. ferdinandiana* polyphenols should also be investigated in relation to bioavailability and bioactivity. A dynamic *in vitro* digestion model mimicking all the digestion steps could be applied to further determine the metabolic pathway of phytocompounds and vitamins. Detailed metabolite profiling and characterization of *T. ferdinandiana* from different geographical locations should be conducted to explore the phytochemical diversity in this fruit. Metabolic enzymes and pathways modulated by the structurally diverse phytochemicals of *T. ferdinandiana* also warrant further investigation.

Acknowledgement

The authors acknowledge the AgriFutures Australia (Grant 201430161) for project funding. The authors also acknowledge the Traditional Owners of the lands on which the *Terminalia ferdinandiana* grows and respect the knowledge and experience the Traditional Owners hold regarding the care, harvest and use of these plants.

References

Aguilera-Carbo, A., Augur, C., Prado-Barragan, L., Aguilar, C., Favela-Torres, E. *et al.* (2008) Extraction and analysis of ellagic acid from novel complex sources. *Chemical Papers* 62(4), 440–444. DOI: 10.2478/s11696-008-0042-y.

Akter, S., Netzel, M.E., Fletcher, M.T., Tinggi, U. and Sultanbawa, Y. (2018) Chemical and nutritional composition of *Terminalia ferdinandiana* (Kakadu Plum) Kernels: A novel nutrition source. *Foods (Basel, Switzerland)* 7(4), 60. DOI: 10.3390/foods7040060.

Akter, S., Netzel, M.E., Tinggi, U., Osborne, S.A., Fletcher, M.T. *et al.* (2019) Antioxidant rich extracts of *Terminalia ferdinandiana* inhibit the growth of foodborne bacteria. *Foods* 8(8), 281. DOI: 10.3390/foods8080281.

Akter, S., Netzel, M., Tinggi, U., Fletcher, M., Osborne, S. *et al.* (2020a) Interactions between phytochemicals and minerals in *Terminalia ferdinandiana* and implications for mineral bioavailability. *Frontiers in Nutrition* 7, 598219. DOI: 10.3389/fnut.2020.598219.

Akter, S., Addepalli, R., Netzel, M., Fletcher, M., Sultanbawa, Y. *et al.* (2021) Impact of polyphenol-rich extracts of *Terminalia ferdinandiana* fruits and seeds on viability of human intestinal and liver cells in vitro. *Food Chemistry* 2, 100024. DOI: 10.1016/j.fochms.2021.100024.

Akter, S., Sultanbawa, Y. and Cozzolino, D. (2021) High throughput screening to determine the antibacterial activity of *Terminalia ferdinandiana* (Kakadu plum): A proof of concept. *Journal of Microbiological Methods* 182, 106169. DOI: 10.1016/j.mimet.2021.106169.

Assunção, P.I.D., da Conceição, E.C., Borges, L.L. and de Paula, J.A.M. (2017) Development and validation of a HPLC-UV method for the evaluation of ellagic acid in liquid extracts of *Eugenia uniflora* L. (Myrtaceae) leaves and its ultrasound-assisted extraction optimization. *Evidence-Based Complementary and Alternative Medicine* 2017, 1501038. DOI: 10.1155/2017/1501038.

Atlas of Living Australia (2022) Distribution of Terminalia Ferdinandiana. Available at: https://www.ala.org.au/ (accessed 1 April 2022).

Chaliha, M. and Sultanbawa, Y. (2019) *Terminalia ferdinandiana*, a traditional medicinal plant of Australia, alleviates hydrogen peroxide induced oxidative stress and inflammation, in vitro. *Journal of Complementary & Integrative Medicine* 17(1). DOI: 10.1515/jcim-2019-0008.

Chaliha, M., Williams, D., Smyth, H. and Sultanbawa, Y. (2018) Extraction and characterization of a novel *Terminalia* pectin. *Food Science and Biotechnology* 27(1), 65–71. DOI: 10.1007/s10068-017-0201-4.

Cheesman, M.J., White, A., Matthews, B. and Cock, I.E. (2019) *Terminalia ferdinandiana* fruit and leaf extracts inhibit methicillin-resistant *Staphylococcus aureus* growth. *Planta Medica* 85(16), 1253–1262. DOI: 10.1055/a-1013-0434.

Cock, I.E. and Mohanty, S. (2011) Evaluation of the antibacterial activity and toxicity of *Terminalia ferdinandia* fruit extracts. *Pharmacognosy Journal* 3(20), 72–79. DOI: 10.5530/pj.2011.20.14.

Cock, I.E. and Rayan, P. (2020) Ascorbic acid potentiates the *Giardia duodenalis* growth inhibitory activity of pure *Terminalia ferdinandiana* Exell compounds. *Parasitology Research* 119(3), 1125–1137. DOI: 10.1007/s00436-019-06579-1.

Courtney, R., Sirdaarta, J., Matthews, B. and Cock, I.E. (2015) Tannin components and inhibitory activity of Kakadu plum leaf extracts against microbial triggers of autoimmune inflammatory diseases. *Pharmacognosy Journal* 07(1), 18–31. DOI: 10.5530/pj.2015.1.2.

Cozzolino, D., Phan, A.D.T., Netzel, M.E., Smyth, H. and Sultanbawa, Y. (2021) The use of vibrational spectroscopy to predict vitamin C in Kakadu plum powders (*Terminalia ferdinandiana* Exell, Combretaceae). *Journal of the Science of Food and Agriculture* 101(8), 3208–3213. DOI: 10.1002/jsfa.10950.

Cunningham, A.B., Garnett, S., Gorman, J., Courtenay, K. and Boehme, D. (2008) Eco–enterprises and *Terminalia ferdinandiana*: "Best Laid Plans" and Australian policy lessons. *Economic Botany* 63(1), 16–28. DOI: 10.1007/s12231-008-9055-2.

Gorman, J.T., Griffiths, A.D. and Whitehead, P.J. (2006) An analysis of the use of plant products for commerce in remote aboriginal communities of northern Australia. *Economic Botany* 60(4), 362–373. DOI: 10.1663/0013-0001(2006)60[362:AAOTUO]2.0.CO;2.

Gorman, J.T., Wurm, P.A.S., Vemuri, S., Brady, C. and Sultanbawa, Y. (2019) Kakadu Plum (*Terminalia ferdinandiana*) as a sustainable indigenous agribusiness. *Economic Botany* 74(1), 74–91. DOI: 10.1007/s12231-019-09479-8.

Hegarty, M.P. and Hegarty, E.E. (2001) *Food Safety of Australian Plant Bushfoods. RIRDC Publication No 01/28. IRDC Project No AGP-1A*. Rural Industries Research and Development Corporation, Barton, Australia.

Jamieson, N., Sirdaarta, J. and Cock, I. (2014) The anti-proliferative properties of Australian plants with high antioxidant capacities against cancer cell lines. *Pharmacognosy Communications* 4(4).

Kenneally, K.F., Edinger, D.C. and Willing, T. (1996) *Broome and beyond: Plants and People of the Dampier Peninsula, Kimberley, Western Australia*. Dept of Conservation and Land Management, Como, Western Australia.

Konczak, I. and Roulle, P. (2011) Nutritional properties of commercially grown native Australian fruits: Lipophilic antioxidants and minerals. *Food Research International* 44(7), 2339–2344. DOI: 10.1016/j.foodres.2011.02.023.

Konczak, I., Zabaras, D., Dunstan, M. and Aguas, P. (2010) Antioxidant capacity and hydrophilic phytochemicals in commercially grown native Australian fruits. *Food Chemistry* 123(4), 1048–1054. DOI: 10.1016/j.foodchem.2010.05.060.

Konczak, I., Maillot, F. and Dalar, A. (2014) Phytochemical divergence in 45 accessions of *Terminalia ferdinandiana* (Kakadu plum). *Food Chemistry* 151, 248–256. DOI: 10.1016/j.foodchem.2013.11.049.

McManus, K., Wood, A., Wright, M.H., Matthews, B., Greene, A.C. *et al.* (2017) *Terminalia ferdinandiana* Exell. Extracts inhibit the growth of body odour-forming bacteria. *International Journal of Cosmetic Science* 39(5), 500–510. DOI: 10.1111/ics.12403.

Mridusmita, C., David, W., David, E., Sharon, P., Heather, S. *et al.* (2017) Bioactive rich extracts from *Terminalia ferdinandiana* by enzyme-assisted extraction: A simple food safe extraction method. *Journal of Medicinal Plants Research* 11(5), 96–106. DOI: 10.5897/JMPR2016.6285.

Netzel, M., Netzel, G., Tian, Q., Schwartz, S. and Konczak, I. (2007) Native Australian fruits — a novel source of antioxidants for food. *Innovative Food Science & Emerging Technologies* 8(3), 339–346. DOI: 10.1016/j.ifset.2007.03.007.

Nourished Life, The Benefits of Kakadu Plum (2020) The Benefits of Kakadu Plum. Nourished Life. Available at: https://www.nourishedlife.com.au/article/165358/benefits-of-kakadu-plum.html (accessed 23 April 2020).

Okuda, T., Yoshida, T., Hatano, T. and Ito, H. (2009) Ellagitannins Renewed the Concept of Tannins. In: Quideau, S. (ed.) *Chemistry and Biology of Ellagitannins*. pp. 1–54. DOI: 10.1142/9789812797414_0001.

Phan, A.D.T., Adiamo, O., Akter, S., Netzel, M.E., Cozzolino, D. *et al.* (2021) Effects of drying methods and maltodextrin on vitamin C and quality of *Terminalia ferdinandiana* fruit powder, an emerging Australian functional food ingredient. *Journal of the Science of Food and Agriculture* 101(12), 5132–5141. DOI: 10.1002/jsfa.11159.

Rayan, P., Matthews, B., McDonnell, P.A. and Cock, I.E. (2015) *Terminalia ferdinandiana* extracts as inhibitors of *Giardia duodenalis* proliferation: a new treatment for giardiasis. *Parasitology Research* 114(7), 2611–2620. DOI: 10.1007/s00436-015-4465-4.

Shalom, J. and Cock, I.E. (2018) Terminalia ferdinandiana Exell. Fruit and leaf extracts inhibit proliferation and induce apoptosis in selected human cancer cell lines. *Nutrition and Cancer* 70(4), 579–593. DOI: 10.1080/01635581.2018.1460680.

Shalom, J., Rayan, P., Courtney, R., McDonnell, P.A. and Cock, I.E. (2018) Terminalia ferdinandiana Exell. Kino extracts have anti-giardial activity and inhibit Caco2 and HeLa cancer cell proliferation. *Pharmacognosy Communications* 8(2), 60–65. DOI: 10.5530/pc.2018.2.13.

Sirdaarta, J., Matthews, B. and Cock, I.E. (2015) Kakadu plum fruit extracts inhibit growth of the bacterial triggers of rheumatoid arthritis: Identification of stilbene and tannin components. *Journal of Functional Foods* 17, 610–620. DOI: 10.1016/j.jff.2015.06.019.

Sultanbawa, Y., Williams, D. and Smyth, H. (2015) *Changes in Quality and Bioactivity of Native Food During Storage. RIRDC Final Report 15-010 (15/010)*. Rural Industries Research and Development Corporation (RIRDC), Barton, Australia.

Tan, A.C., Konczak, I., Sze, D.M.-Y. and Ramzan, I. (2010) Towards the discovery of novel phytochemicals for disease prevention from native Australian plants: an ethnobotanical approach. *Asia Pacific Journal of Clinical Nutrition* 19(3), 330–334.

Tan, A.C., Konczak, I., Ramzan, I. and Sze, D.M.-Y. (2011a) Native Australian fruit polyphenols inhibit cell viability and induce apoptosis in human cancer cell lines. *Nutrition and Cancer* 63(3), 444–455. DOI: 10.1080/01635581.2011.535953.

Tan, A.C., Konczak, I., Ramzan, I. and Sze, D.M.-Y. (2011b) Antioxidant and cytoprotective activities of native Australian fruit polyphenols. *Food Research International* 44(7), 2034–2040. DOI: 10.1016/j.foodres.2010.10.023.

Tan, A.C., Konczak, I., Ramzan, I., Zabaras, D. and Sze, D.M.-Y. (2011c) Potential antioxidant, antiinflammatory, and proapoptotic anticancer activities of Kakadu plum and Illawarra plum polyphenolic fractions. *Nutrition and Cancer* 63(7), 1074–1084. DOI: 10.1080/01635581.2011.596646.

The Australian Superfood Co (2015) Kakadu Plum. Available at: https://austsuperfoods.com.au/kakadu-plum/ (accessed 1 April 2022).

Williams, D.J., Edwards, D., Pun, S., Chaliha, M., Sultanbawa, Y. *et al.* (2014) Profiling ellagic acid content: The importance of form and ascorbic acid levels. *Food Research International* 66, 100–106. DOI: 10.1016/j.foodres.2014.09.003.

Williams, D.J., Edwards, D., Pun, S., Chaliha, M., Burren, B. *et al.* (2016) Organic acids in Kakadu plum (*Terminalia ferdinandiana*): The good (ellagic), the bad (oxalic) and the uncertain (ascorbic). *Food Research International* 89(Pt 1), 237–244. DOI: 10.1016/j.foodres.2016.08.004.

Wright, M.H., Sirdaarta, J., White, A., Greene, A.C. and Cock, I.E. (2016) GC-MS headspace analysis of *Terminalia ferdinandiana* fruit and leaf extracts which inhibit *Bacillus anthracis* growth. *Pharmacognosy Journal* 9(1), 73–82. DOI: 10.5530/pj.2017.1.14.

Wright, M.H., Shalom, J., Matthews, B., Greene, A.C., Cock, I.E. *et al.* (2019) *Terminalia ferdinandiana* Exell: Extracts inhibit *Shewanella* spp. growth and prevent fish spoilage. *Food Microbiology* 78, 114–122. DOI: 10.1016/j.fm.2018.10.006.

13 Lillypilly/Riberry (*Syzygium* spp.)

Maral Seididamyeh[1,2], Anh Dao Thi Phan[1,2], Michael E. Netzel[1,2] and Yasmina Sultanbawa[1,2*]

[1]ARC Industrial Transformation Training Centre for Uniquely Australian Foods, Queensland Alliance for Agriculture and Food Innovation, The University of Queensland, Coopers Plains, Queensland, Australia; [2]Centre for Nutrition and Food Sciences, Queensland Alliance for Agriculture and Food Innovation, The University of Queensland, St. Lucia, Queensland, Australia

Abstract

Evergreen *Syzygium* plants have long been used in traditional medicine mainly due to their unique antimicrobial properties. This chapter encompasses a general description on the growth and botany of the two species, *S. luehmannii* and *S. aqueum*, as well as the reported data on their phytochemical, bioactive and health-related properties. The unique flavouring and colouring characteristics of these two fruits are attributed to their rich phytochemical content and mainly to the high amounts of anthocyanins such as cyaniding-3-galactoside. The notable antimicrobial properties of the fruit are promising to be used as a natural preservative in manufactured foods. However, there is still an urgent need for a detailed and thorough study of their phytochemicals, especially of *S. aqueum* as an underutilized fruit, together with their biological properties to be used in manufacturing novel health-promoting functional foods.

Introduction

The genus of evergreen flowering plants *Syzygium* Gaertn. belongs to Myrtaceae family, which is distributed in tropical and subtropical regions and encompasses about 1110–1200 species (Soh, 2017). It is indigenous to sub-Saharan Africa, South and South-East Asia, the Indo-Pacific region and South America, with a rich variety observed in south-eastern Asia and north-eastern Australia (Kress *et al.*, 2003; Craven and Biffin, 2010; Soh, 2017). The species in the genus *Syzygium* are commonly known as lillypillies and are used as a fresh fruit, spice and ornamental plant. Roughly 60 species of lillypilly are found in Australia having edible fruit, however, some are not consumed directly due to high levels of sourness or astringency (Gorst, 2002). A few *Syzygium* species have gained enough attention to be commercially cultivated such as *S. luehmannii*, for its fruit (used fresh or as a spice) or its timber (used in construction). Furthermore, the fruit, leaves and bark are used in traditional medicine, and as fabric dye (Soh, 2017).

Syzygium luehmannii (F. Muell.) L.A.S. Johnson, known as riberry (Fig. 13.1a), is native to the east coast of Australia (New South Wales and Queensland) and New Guinea (Goodenough Island). It is commonly grown in subtropical rainforests, littoral and riverine areas (Lim, 2012). The industry body, Australian Native Foods and Botanicals (ANFAB), has selected it as one of the

*Corresponding author: y.sultanbawa@uq.edu.au

 Handbook of Phytonutrients in Indigenous Fruits and Vegetables (eds D. Sivakumar *et al.*)
DOI: 10.1079/9781789248067.0013

key native crops (Clarke, 2012), due to its suitability for large plantations and commercialization.

The tree is fairly salt tolerant and grows better in moist and well-drained soils, however its adaptability to most conditions has been demonstrated (Robinson *et al.*, 1994). It produces pink to red edible fruit that are tart and slightly acidic, having a cranberry-like flavour with a hint of clove/cinnamon (Konczak *et al.*, 2010a; Lim, 2012). According to Hardwick (1996), riberries bruise easily and can resist freezing and cold storage temperatures (Hardwick, 1996).

Another species of interest to this chapter is *Syzygium aqueum* (Burm. F.) Alston, known as watery rose-apple (also as water jambu, water apple or water guava) (Fig. 13.1b) that originated in tropical Asia (Malaysia and Indonesia) (Manaharan *et al.*, 2012), New Guinea and north Queensland (Lim, 2012; ALA, 2021). Watery rose-apple is grown in fairly moist tropical regions (low altitude areas up to 1600 m above the sea level) (Eseyin *et al.*, 2007), i.e. wet, warm and humid conditions with well distributed annual rainfall. Nevertheless, with a supplemental water supply, it can also grow in the dry season (Lim, 2012). The tree is commonly cultivated in India, South-East Asia and the Pacific Islands (Lim, 2012) for its wood or edible fruit, which have red or yellow flesh (Eseyin *et al.*, 2007), a slight sweetness, an apple-like flavour and a crisp watery texture (Janick and Paull, 2008). The leaves and bark of the tree are traditionally used as herbal medicine, and the leaves are also used as food wrappers (Janick and Paull, 2008). The fruit, on the other hand, are generally consumed raw, and have thirst-relieving properties (Morton, 1987).

Ample opportunities in terms of developing new food products present themselves by exploring native edible plants as sources of health-promoting active compounds. Recently, more attention has been given to Australian native foods, which have been incorporated into various food products, in turn enriching the daily diet of Australian consumers. Clove (*S. aromaticum*), wax jambu (*S. samarangensis*) and jambolan (*S. cuminni*) are well known *Syzygium* species that enjoy extensive attention. Having said that, the majority of *Syzygium* species are still underutilized and their chemical and biological properties are yet to be explored. Some of the commercialized species are shown in Table 13.1, however, a fairly comprehensive list of *Syzygium* species can be found elsewhere (Craven and Biffin, 2010).

Botany

Syzygium trees generally have elliptic, glossy, dark-green, turpentine-scented tapering or blunt leaves with pointing apexes. The trunk is mostly dark grey, rough, cracked and flaking at the lower parts, turning smoother and lighter at the upper parts (Ahmad *et al.*, 2016). The fruit vary in size (i.e. <1 cm–7 cm in diameter), colour (i.e. cream to red) and flavour (i.e. acidic to fairly sweet). They contain a single oblong-shaped seed of green or brown colour (Gorst, 2002; Ahmad *et al.*, 2016).

(a) (b)

Fig. 13.1. (a) *Syzygium luehmannii* (IndiGrow, 2018); (b) *Syzygium aqueum* (Vinayaraj, 2008).

Table 13.1. Examples of *Syzygium* species.

Plant	Origin	Common names
S. anisatum (Vickery) Craven & Biffin	Australia	Anise myrtle, aniseed myrtle, ringwood
S. australe (J.C. Sendl. Ex Link) B. Hyland	Eastern Australia	Brush cherry, lilly pilly
S. fibrosum (F.M. Bailey) T.G. Hartley & L.M. Perry	Australia	Rainberry
S. jambos (L.) Alston	Australia	Rose apple
S. samarangense (Blume) Merr. & L.M. Perry	Sout-East Asia; Australia	Java apple; wax jambu
S. cumini (L.) Skeels	South-East Asia	Java plum, Jambolan
S. javanicum	South-East Asia	Water apple
S. aromaticum (L.) Merr. & L.M. Perry	South-East Asia	Clove
S. malaccensis (L.) Merr. & L.M. Perry	South-East Asia	Malay Rose Apple

Table 13.2. Botanical characteristics of *Syzygium luehmannii* and *Syzygium aqueum*.

Species	Botanical features	Reference
S. luehmannii	Medium to large trees (4–30 m) with smooth to slightly flaky bark and dense canopy Smooth, glossy, ovate to lanceolate leaves with 1–3 cm width and 2.5–7.5 cm length Cream to white flowers with free and spreading petals Pink to red pyriform (i.e. pear-shaped) or obovoid, either single seed or seedless fruits with 6–10 mm diameter and 9–12 mm length November–December is the flowering season and December–February is the fruit maturation period	(Konczak *et al.*, 2010a; Lim, 2012)
S. aqueum	Short and twisted trunk (3–10 m height) with brown flaky bark and branches, branching near the base Smooth, elliptic-cordate to obovate-oblong leaves with 2.5–16 cm width and 7–25 cm length Yellowish-white or pinkish flowers with spathulate petals White to red, glossy, waxy, bell-shaped fruits with white, spongy or crispy, juicy and slightly aromatic flesh, round and small seeds, may be seedless Fruits are 1.27 cm in width and 2.54 cm in length February–March is the flowering season and May–June is the fruit maturation period	(Morton, 1987; Verdcourt, 2001; Lim, 2012; Sonawane, 2018)

The botanical characteristics of *S. luehmannii* and *S. aqueum* are shown in Table 13.2.

Traditional Uses

Indigenous Australians traditionally foraged the fruit of these species, which were mostly eaten fresh. Based on anecdotal evidence, various Australian native plants including *Syzygium* species have been used by Indigenous people as traditional medicine (Cock, 2011; Richmond *et al.*, 2019). *S. luehmannii* has been used as a cosmetic or herbal medicine to treat different diseases such as skin fungal infections (Cock and Cheesman, 2019), and the pulp of the fruit has been used to treat earache (Lassak and McCarthy, 2011; Rayan *et al.*, 2016).

S. luehmannii is also used to treat gastrointestinal disorders, dysentery, diarrhoea, tuberculosis and respiratory complaints (Cock, 2011). The fruit are commonly used in jams, syrups, confections, ice creams and sauces served with meat (Lim, 2012).

Traditional utilization of different parts of *S. aqueum* has been mainly due to antibiotic properties (Panggabean, 1992). In Malaysian traditional medicine, various preparations of *S. aqueum* fresh fruit, bark, roots and ground dried leaves are used to relieve pain in childbirth (Panggabean, 1992; Lim, 2012), to treat thrush (Lim, 2012), to relieve itching and lessen swelling, and to treat cracked tongue (mouth ulcers) (Morton, 1987; Osman *et al.*, 2009), respectively. The fruit are eaten fresh as a fruit salad (rujak) in Indonesia and Malaysia, or are pickled (asinan) in Indonesia. The young leaves are also edible and Indonesians use them to wrap snacks made of fermented sticky rice (Lim, 2012).

Nutritional Composition

Native Australian fruit are rich in essential minerals and vitamins, providing the Indigenous people with the nutritional sources needed to survive throughout tens of thousands of years (Cherikoff, 1993). The reported proximate data of *S. luehmannii* and *S. aqueus* is given in Table 13.3. Higher protein content is usually found in Australian native fruit compared to European fruit. Riberry, for example, contains higher protein (0.8 g/100 g) compared to cherry tomato and blueberry (both 0.5 g/100 g) (NUTTAB, 2010).

Table 13.3. Macronutrient composition of *S. luehmannii* and *S. aqueum* fruit (g/100 g DW).

Property	*S. luehmannii*	*S. aqueum*
Energy (kJ)	325	68
Moisture	82	90–95
Protein	0.81–0.96	0.12–0.80
Fat	0.4	0.10–0.37
Carbohydrate	0.36	3–8.49
Sugars	0.36–0.42	ND
Fibre	6.8	0.86–1.37
Ash	0.8	0.33–1.24
References	(Miller *et al.*, 1993; Konczak *et al.*, 2009; RIRDC, 2012)	(Dignan *et al.*, 1994; Lim and Rabeta, 2013; Sonawane, 2018)

Human health is significantly dependent on ingesting the essential minerals needed for physiological processes such as protein functions (Bertini and Sigel, 2001). In order to enjoy the nutritional benefits of any native fruit such as *Syzygium*, investigating mineral composition is of importance. This is mostly to avoid the excessive consumption of particular elements, which may give rise to toxic effect exertion through free radical generation – for example, by redox-active elements such as manganese, iron and copper (Kozlowski *et al.*, 2009), which are also found in the *S. luehmannii* and *S. aqueum* fruit. The reported nutritional data for *S. luehmannii* and *S. aqueum* is summarized in Table 13.4, but is still incomplete and needs further study. *S. luehmannii* is found to be high in biologically essential minerals such as potassium, calcium, magnesium, sulfur, phosphorus and sodium (Table 13.4), for which the recommended daily intake (RDI) is more than 100 mg per day (Konczak and Roulle, 2011). It is comparable to European fruit and vegetables (Ekholm *et al.*, 2007). While the reported data on *S. aqueum* is scarce, it has been shown that higher concentrations are still less than the RDI of calcium, potassium, sodium, and phosphorus found in the fruit, compared to other elements (Table 13.4).

Physiologically essential trace elements (RDI less than 100 mg per day) (Konczak and Roulle, 2011) such as manganese, iron, zinc, copper, nickel, molybdenum and cobalt were also found mostly in *S. luehmannii* and fairly in *S. aqueum* fruit (Table 13.4), with concentrations well below 100 mg/100 g dry weight (DW) and comparable to European fruit (Ekholm *et al.*, 2007). The content of calcium, magnesium and zinc has been reported to be 307.7, 189.0 and 1.315 mg/100 g DW respectively in *S. luehmannii* (Konczak and Roulle, 2011), and 63.6 mg calcium and 2.0 mg zinc in 100 g of dry watery rose-apple (Lim and Rabeta, 2013). These concentrations are structurally required for DNA self-repair systems, and act as cofactors in enzymatic processes (Fenech, 2005). The content of iron, with its major function

Table 13.4. Mineral composition of *S. luehmannii* and *S. aqueum* fruit (mg/100 g DW).

Mineral*	*S. luehmannii*	*S. aqueum*
K	250–1715.7	48
Ca	100–307.7	2–63.6
Mg	48–189.0	ND
S	137	ND
P	49–118.8	13
Na	11–47.1	1–19.6
Mn	22.75	3.3
Fe	0.9–4.320	0.2–3.7
Al	1.665	ND
Zn	0.2–1.315	2.0
Cu	0.3–1.135	ND
Pb	0.208	ND
Ni	0.128	3.9
Cd	0.0245	4.6
Mo	0.0107	ND
Co	0.008	ND
References	(Miller *et al.*, 1993; Konczak and Roulle, 2011)	(Dignan *et al.*, 1994; Lim and Rabeta, 2013)

* K, potassium; Ca, calcium; Mg, Magnesium; S, Sulfur; P, Potassium; Na, Sodium; Mn, Manganese; Fe, Iron; Al, Aluminium; Zn, Zinc; Cu, Copper; Pb, Lead; Ni, Nickel; Cd, Cadmium; Mo, Molybdenum; Co, Cobalt.

in haemoglobin in red blood cells, has been reported to be 4.320 and 3.7 mg/100 g DW in *S. luehmannii* and *S. aqueum* fruit (Konczak and Roulle, 2011; Lim and Rabeta, 2013), which is comparable to European fruit (Ekholm *et al.*, 2007). Copper, with a concentration of 1.135 mg/100 g DW in *S. luehmannii*, is another trace element essential to suppress oxidative status by scavenging free radicals (Linder and Hazegh-Azam, 1996), which has not been reported in *S. aqueum*. The copper content in *S. luehmannii* is higher than in common European fruit (Ekholm *et al.*, 2007). Furthermore, it was shown that *S. luehmannii* contains 1.665, 0.208 and 0.0245 mg of aluminium, lead and cadmium (in 100 g DW) (Konczak and Roulle, 2011), while *S. aqueum* possesses 4.6 mg of cadmium in 100 g DW (Lim and Rabeta, 2013). These elements are required in minute levels by humans, and high concentrations are toxic (Konczak and Roulle, 2011).

Phytonutrients

Phytochemicals (bioactive plant chemicals) accumulate in various parts of plants such as the fruit and not only contribute to the plant's defensive mechanisms against disease and oxidative damage but also to its colour, aroma and flavour (Koche *et al.*, 2016). Phenolic acids, terpenoids and alkaloids are the three main groups of phytochemicals, with polyphenols the largest group (Tsao, 2010). There is however scarce information in the literature on *S. aqueum* fruit, while a few studies have been conducted on *S. luehmannii* fruit. The identified compounds in *S. luehmannii* fruit and the essential oils of *S. aqueum* fruit are given in Table 13.5.

The analysis of solvent-based extracts of *S. luehmannii* fruit resulted in identification of mainly phenolic acids (benzoic and hydroxycinnamic) and flavonoids (flavonols, flavanones and anthocyanins) (Konczak *et al.*, 2010a), which have antioxidant properties (Kähkönen *et al.*, 1999). Riberry extract is

Table 13.5. Identified phytochemicals in *S. luehmannii* and *S. aqueum* fruit.

Fruit	Major compounds identified	References
S. luehmannii	Malic acid; oxalic acid; α-tocopherol (vitamin E); vitamin C; cyanidin-3-galactoside; cyanidin-3-glucoside; cyanidin-3,5-diglucoside; quercetin rutinoside (rutin); quercetin hexoside; quercetin rhamnoside; kaempferol; myricetin hexoside; kaempferol/luteolin rutinoside; folates	(Konczak, 1988; Netzel *et al.*, 2007; Konczak *et al.*, 2010a; Konczak and Roulle, 2011)
***S.S. aqueum* (essential oil)**	Hexanal; (Z)-hex-3-enal; γ-terpinene; hex-1-ol; (Z)-hex-3-en-1-ol; p-cymene; terpinolene; β-caryophyllene; terpinen-4-ol; caryophyllene- oxide	(Wong and Lai, 1996)

rich in anthocyanins consisting of cyanidin 3-galactoside (0.25 ± 0.02 mg/g fresh weight –FW), cyanidin 3-glucoside (0.04 ± 0.006 mg/g FW) and cyanidin 3,5-diglucoside (0.02 ± 0.001 mg/g FW) (Konczak *et al.*, 2010a). Other compounds detected in *S. luehmannii* fruit are presented in Table 13.5. One of the prominent lipid-soluble antioxidants with protective effects for both human and plant cells is vitamin E (Havaux *et al.*, 2005), which was found to be 0.230 ± 0.040 mg/100 g FW in riberry with α-tocopherol as the main component (0.229 ± 0.044 mg/100 g FW) (Konczak and Roulle, 2011). Furthermore, the content of an essential group of water-soluble vitamins known as vitamin B9 (also as folate or folic acid) in *S. luehmannii* fruit has been reported to be 9 µg/100 g FW (Konczak *et al.*, 2009), which is comparable to that in blueberry (6 µg/100 g FW) (USDA, 2020). It is noteworthy that folate deficiency may give rise to an increased risk of dementia, cardiovascular disease, anaemia and birth defects (Duthie *et al.*, 2002; Quinlivan *et al.*, 2002). Another group of phytonutrients in fruit that not only influences the sensorial quality (such as flavour and colour) but also intermediates the plant metabolic processes such as senescence, is organic acids. Riberry was found to contain mainly malic acid (17.2 ± 0.34 mg/g FW) with small amounts of oxalic acid (0.23 ± 0.01 mg/g FW) (Konczak *et al.*, 2010a). The malic acid content of riberry is much higher than that of kiwi fruit (1.9 mg/g FW), apple (4.12 mg/g FW) and strawberry (1.11 mg/g FW) (Pérez *et al.*, 1997). Further studies are needed to identify and confirm the phytochemicals present not only in *S. luehmannii* and *S. aqueum* fruit, but also other *Syzygium* species.

Bioactive Properties

The observed bioactive properties of fruit are mainly due to the presence of secondary metabolites, also known as bioactives, which contribute to a plant's defensive mechanisms to protect it from predators as well as from external stressors including climate and soil condition variability, ultraviolet radiation and pathogenic organisms (Taiz *et al.*, 2015). Phenolic compounds and pigments constitute the majority of fruit bioactives, which are found in Australian native fruits such as *Syzygium*. The investigated bioactive properties of the *S. luehmannii* and *S. aqueum* fruits are presented in Table 13.6.

It has been reported that *S. luehmannii* fruits are rich in anthocyanins, proanthocyanidins, phenolics and flavonoids, with the total phenolic content ranging from 14 to 24 mg gallic acid equivalent (per g DW) (Konczak *et al.*, 2010a, b; Konczak and Roulle, 2011). The high antioxidant properties observed in native Australian fruits such as *S. luehmannii* are due to both hydrophilic (e.g. phenolics, flavonoids) and lipophilic (e.g. tocopherols) compounds (Babbar *et al.*, 2015). It is noteworthy to mention that 30% of the total antioxidant activity of *S. luehmannii* fruit is caused by lipophilic antioxidants, compared to ca. 0.5% in blueberry. One of the unique features of native Australian food (e.g. *S. luehmannii*) is the lipophilic antioxidant content (such as tocopherols) which has an equally essential contribution to a healthy diet as water-soluble phenolics (Konczak and Roulle, 2011). The potent antimicrobial activity of different parts of the *Syzygium* species (e.g. bark, leaves, fruit) such as *S. aromaticum* (Arora and Kaur, 2007; Pinto *et al.*, 2009; Batiha *et al.*, 2019) and *S. cumini* (Yadav *et al.*, 2017; Asghar *et al.*, 2020) is well known. The antifungal activity of *S. luehmannii* fruit and leaf extracts against the pathogens causing dermatophytic infections such as tinea pedis, tinea corpis, tinea cruris, onchomyosis, ringworm and candiadisis has been reported by Noé *et al.* (2019). Higher antifungal activity was observed in leaf extracts compared to the fruit. The authors also reported a high therapeutic index for all the investigated solvent extracts from both leaves and fruit of *S. luehmannii*, showing their promising therapeutic potential (Noé *et al.*, 2019).

As can be seen in Table 13.4, the fruit extracts of *S. luehmannii* showed cytotoxicity in *Artemia franciscana* nauplii bioassay (Sautron and Cock, 2014; Wright *et al.*, 2016; Murhekar *et al.*, 2017; Wood *et al.*, 2017; Noé *et al.*, 2019). However, fatality of nauplii and in turn high toxicity results can be influenced by the antioxidant capacity or the acidic pH of the studied extracts, due to the high sensitivity of brine shrimp nauplii used in this assay (Kwast

Table 13.6. Bioactive properties of *S. luehmannii* and *S. aqueum* fruit.

Fruit (Origin)	Solvent extract	Bioactive properties	Reference
***S. luehmannii* (Australia)**	Aqueous methanol (acidified)	**Antioxidant** 28.07 ± 2.45 μmol TE/g FW	(Netzel *et al*., 2007)
***S. luehmannii* (Australia)**	Aqueous methanol (acidified)	**Antioxidant** 7.5 ± 0.7 μmol GAE/gFW compared to 26.5 ± 0.6 μmol GAE/gFW of blueberry 33.2 ± 1.9 μmol Fe^{2+}/g FW 49.9 ± 6.4 μmol TE/ g FW	(Konczak *et al*., 2010a)
***S. luehmannii* (Australia)**	Acetone	**Antioxidant** 72.04 μM TE/g FW	(Konczak and Roulle, 2011)
***S. luehmannii* (Austrlia)**	Methanol, water, ethyl acetate, chloroform, hexane	**Antimicrobial** Polar extracts showed higher activity compared to the non-polar ones Inhibition zones ranging from 6 to 13 mm and MIC values of 0.1–44.2 mg/ml against *Alcaligenes faecalis*, *Aeromona hydrophillia*, *Escherichia coli*, *Klebsiella pneumonia*, *Pseudomonas fluorescens*, *Proteus mirabilis*, *Serratia marcenscens*, *Salmonella newport*, *Shigella sonnei*, *Bacillus cereus*, *Staphylococcus aureus*, *Staphylococcus epidermidis*, *Streptococcus pyogenes*, *Saccharomyces cerevisae* **Toxicity** Non-polar extracts (LC_{50} of 4149 μg/ml for chloroform extract) were less toxic than the polar ones (LC_{50} of 181–478 μg/ml)	(Sautron and Cock, 2014)
***S. luehmannii* (Australia)**	Ethyl acetate, water, methanol	**Antioxidant** 1.5, 59.2, and 94.6 mg AA/g DW, for ethyl acetate, water, and methanolic extracts, respectively **Antimicrobial** 161–275 μg/ml, with ethyl acetate extract showing MIC over 10,000 μg/ml Inhibition zones of 7–13 mm against *Clostridium perfringens* using 10 μl of the extracts **Toxicity** Toxic, LC_{50} of 414–478 μg/ml through *Artemia franciscana* nauplii bioassay (LC_{50}<1000 μg/ml)	(Wright *et al*., 2016)

Continued

Table 13.6. Continued

Fruit (Origin)	Solvent extract	Bioactive properties	Reference
***S. luehmannii* (Australia)**	Ethyl acetate, water, methanol	**Antimicrobial** Inhibition zones of 7–8 mm against *Shewanella putrifaciens* and *S. frigidimarina*, 4–9 mm against *S. baltica* (except for water extract), 10 mm against *S. loihica* (only methanolic extract) MIC values of 317–3500 µg/ml against the studied *Shewanella* species **Toxicity** Toxic, LC_{50} values of 414–478 µg/ml through *Artemia franciscana* nauplii bioassay (LC_{50} <1000 µg/ml)	(Murhekar *et al.*, 2017)
***S. luehmannii* (Australia)**	Water, methanol, ethyl acetate	**Antimicrobial** Moderate activity against *Brevibacterium linens* and *Cornyebacterium jeikeium* (6–8 mm of inhibition zone) Water and methanol extracts showed MIC > 1000 ppm, ethyl acetate extract showed MICs of 203 and 571 ppm against *C. jeikeium* and *B. linens*, respectively **Toxicity** Toxic, LC_{50} of 414–478 µg/ml through *Artemia franciscana* nauplii bioassay (LC_{50} <1000 µg/ml)	(Wood *et al.*, 2017)
***S. luehmannii* (Australia)**	Water, methanol, ethyl acetate	**Antimicrobial** Inhibition against *Candida albicans*, *Epidermophyton floccosum* var. *nigricans*, *Microsporum gypseum*, *Trichophyton mentagrophytes* and *T. rubrum* with MICs ranging 39–1955 µg/ml, **Toxicity** Methanolic and aqueous extracts were toxic with LC_{50} <1000 mg/ml (*Artemia* nauplii bioassay) Methanolic and aqueous extracts were non-toxic against normal human primary dermal fibroblasts by having LC_{50} of 1500–5000 mg/ml (>200 mg/ml)	(Noé *et al.*, 2019)
***S. aqueum* (Malaysia)**	Acetone, water	**Antioxidant** 3545.56 and 847.22 µM Fe(II)/g DW for acetone and water extracts, respectively	(Lim and Rabeta, 2013)
***S. aqueum* (Indonesia)**	Citrate buffer (pH 4.2)	**Antioxidant** IC_{50} of 4.857 µg/ml in citrate buffer (pH 4.2)	(Saptarini and Herawati, 2017)

and Hand, 1996; Noé *et al.*, 2019). In order to verify the safety and confirm the efficiency of the *Syzygium* extracts in complex biological systems, further studies are required.

Furthermore, the presence of phytoactive secondary metabolites (i.e. flavonoids, ellagitannins, proanthocyanidins) resulting in notable bioactive properties such as antioxidant, antimicrobial, anti-inflammatory, antidiabetic, antinociceptive and hepatoprotective has been observed in the leaf extracts of *S. aqueum* (Osman *et al.*, 2009; Ling *et al.*, 2010; Manaharan *et al.*, 2013; Manaharan *et al.*, 2014; Sobeh *et al.*, 2018; Rocchetti *et al.*, 2019), and *S. leuhmannii* (Cock, 2012; Wright *et al.*, 2016; Murhekar *et al.*, 2017).

Health Benefits

The worldwide application of native plants in folk medicine is well known, and can also be seen in Indigenous Australian communities. Traditionally, Indigenous Australians have always benefited from the therapeutic properties of some Australian *Syzygium* species. In general, the application of *Syzygium* as a traditional medicine has been due to this species' reported antifungal, antibacterial and anaesthetic properties (Olukoya *et al.*, 1993; Guleria *et al.*, 2013).

The abundant phytochemicals generally found in fruit are phenolic compounds, mainly hydrophilic. Studies have shown the molecular structure affects their interaction with enzymes and absorption throughout the gastrointestinal tract (Konczak and Roulle, 2011). This, together with the high antioxidant phytochemicals present in these fruits, affects the health of consumers by, for example, reducing oxidative stress and in turn preventing carcinogenesis (Netzel *et al.*, 2007; Mani *et al.*, 2020). The potential of some *Syzygium* species such as *S. aromaticum* (Kumar *et al.*, 2014), *S. cumini* (Khodavirdipour *et al.*, 2020) and *S. alternifolium* (Yugandhar *et al.*, 2017) in inhibiting breast cancer cell growth has been investigated. Rabeta *et al.* (2013) observed the anti-breast cancer property of the aqueous and methanolic extracts of *S. aqueum* fruit, which was shown to be effective against MCF-7 breast cancer cell lines, but not against MDA-MB-231 (Rabeta *et al.*, 2013).

Various anthocyanins have been identified in *S. luehmannii* fruit (Table 13.5), and it has been reported that these compounds can contribute to metabolic function (Kalt *et al.*, 2008; Tsuda, 2008) as well as brain performance, i.e. memory and cognitive function (Casadesus *et al.*, 2004; Andres-Lacueva *et al.*, 2005; Kalt *et al.*, 2008). As is shown in Table 13.6, various bioactivities have been observed by different plant parts of *Syzygium* species, and therefore have been suggested as a remedy for autoimmune diseases due to their inhibitory effect on the microbial triggers (Chikowe *et al.*, 2013; Sautron and Cock, 2014). However, the antimicrobial activity of *S. luehmannii* and *S. aqueum* fruit against the microbial initiators specific to each autoimmune disease needs to be adequately investigated. For example, the efficacy of ethyl acetate fruit extract of *S. luehmannii* against *Streptococcus pyrogenes*, which initiates the autoimmune disease rheumatic fever, has been reported (Chikowe *et al.*, 2013; Sautron and Cock, 2014).

It has been suggested that chronic diseases such as diabetes can be prevented by consumption of phytochemicals such as quercetin, resveratrol, carotenoids, sulphoraphane, silymarine and indole-3-carbino, which are found in plants (e.g. fruit) (Gulati *et al.*, 2015). Interestingly, the health-related bioactivity of *S. aqueum* leaf extracts has been recently studied, showing moderate to potent cytotoxity against the investigated human cancer cell lines (Ling *et al.*, 2010; Rocchetti *et al.*, 2019; Sarathchandrian *et al.*, 2020) as well as strong anti-inflammatory activity through lipoxygenase inhibition assay (Sobeh *et al.*, 2018). Furthermore, the isolated flavonoids from the *S. aqueum* leaves have shown antidiabetic (Manaharan *et al.*, 2012; Palanisamy and Manaharan, 2015) and anticancer (Subarnas *et al.*, 2015) activities. However, the medicinal properties (e.g. antidiabetic) of Australian *Syzygium* species, including the fruit of *S. luehmannii* and *S. aqueum* have not yet been explored, but may be a promising natural alternative to chemical treatments.

Novel products/formulations

Lillypillies such as *S. luehmannii* and *S. aqueum*, with their unique and strong characteristics in fruit flavour and colour can be used in various food products such as jams, ice creams, chutneys, drinks (either soft or alcoholic), sauces and various types of desserts. Exploring their phytochemical profiles, nutritional and bioactive properties may lead the food industry to develop novel products using the authentic native ingredients as sources of flavours, preservatives (antioxidant or antimicrobial), colourants or functional nutrients, by either using the fruit/ fruit extract powders or isolated nutritive and bioactive ingredients (Richmond *et al.*, 2019). In this way, the food industry may be able to address consumer demand for natural and health-promoting ingredients in manufactured foods. The broad-spectrum antimicrobial activity of *Syzygium* species including *S. luehmannii* and *S. aqueum* against both food spoilage and disease-causing organisms makes them promising as a possible food preservatives. They could be utilized in developing active edible coatings, smart packaging or be used directly in various food products such as bakery, dairy, dressings, fruit and vegetable products, to not only act as a food preservative but also help in disease prevention.

Lillypillies, including *S. luehmannii* and *S. aqueum* fruit, are rich in anthocyanins accumulating in the flesh or skin, which are water-soluble natural pigments with pH-dependent colours of red, purple or blue (Netzel *et al.*, 2006; Chuen *et al.*, 2016; Yang *et al.*, 2019). Anthocyanins undergo reactions with other compounds known as co-pigments such as phenolic acids, including chlorogenic acid, forming intramolecular complexes leading to colour stability enhancement that is desirable in food manufacturing (Jensen *et al.*, 2011).

Conclusions and Future Trends

Syzygium species are rich in health-promoting and nutritive phytochemicals leading to their unique flavouring and colouring properties as well as bioactivities with potential medicinal properties such as antioxidant, antimicrobial, anticancer and antidiabetic. Besides the fresh consumption of the fruits, various parts of these plants including *S. luehmannii* and *S. aqueum* have been traditionally used for their antimicrobial properties in preventing and treating different health issues, which have been validated by the high antimicrobial and antioxidant activities observed in the literature. Hence, these native and underutilized fruit give rise to opportunities for the food manufacturing industry to not only benefit from their sensorial characteristics but also their natural food preservative properties, simultaneously.

Despite some studies being conducted on *S. luehmannii* fruit, information about the phytochemicals of *S. aqueum*, in particular their biological properties, remains scarce. Therefore, a detailed study of the bioactivity, and in turn the medicinal properties, of the major compounds is required. Having said that, the reported toxicity results, and the safety aspects, warrant further research and validation for safe application in food products. Furthermore, the antimicrobial properties against microorganisms of concern as well as the impact of storage conditions and food processing on the bioactive compounds needs to be investigated. Also, there is a need to explore the antiviral potential of the *Syzygium* species including *S. luehmannii* and *S. aqueum*, in light of the emergence of new viruses and their strains, as is the case for the current global spread of the SARS-CoV-2 causing Covid-19 disease.

Acknowledgement

The authors acknowledge the Traditional Owners of the lands on which the plant species grow, respect the knowledge and experience the Traditional Owners hold regarding the care, harvest and use of these plants. Maral Seididamyeh would also like to thank the Australian Government for awarding a Research Training Program Scholarship.

References

Ahmad, B., Baider, C., Bernardini, B., Biffin, E., Brambach, F. *et al.* (2016) *Syzygium* (Myrtaceae): monographing a taxonomic giant via 22 coordinated regional revisions. *PeerJ Preprints* April 5. DOI: 10.7287/peerj.preprints.1930v1.

ALA (2021) *Syzygium aqueum (Burm.f.) Alston*. Atlas of Living Australia. Available at: https://bie.ala.org.au/species/https://id.biodiversity.org.au/node/apni/2909237

Andres-Lacueva, C., Shukitt-Hale, B., Galli, R.L., Jauregui, O., Lamuela-Raventos, R.M. *et al.* (2005) Anthocyanins in aged blueberry-fed rats are found centrally and may enhance memory. *Nutritional Neuroscience* 8(2), 111–120. DOI: 10.1080/10284150500078117.

Arora, D.S. and Kaur, G.J. (2007) Antibacterial activity of some Indian medicinal plants. *Journal of Natural Medicines* 61(3), 313–317. DOI: 10.1007/s11418-007-0137-8.

Asghar, M.A., Zahir, E., Asghar, M.A., Iqbal, J. and Rehman, A.A. (2020) Facile, one-pot biosynthesis and characterization of iron, copper and silver nanoparticles using *Syzygium cumini* leaf extract: As an effective antimicrobial and aflatoxin B1 adsorption agents. *PloS One* 15(7), e0234964. DOI: 10.1371/journal.pone.0234964.

Babbar, N., Oberoi, H.S. and Sandhu, S.K. (2015) Therapeutic and nutraceutical potential of bioactive compounds extracted from fruit residues. *Critical Reviews in Food Science and Nutrition* 55(3), 319–337. DOI: 10.1080/10408398.2011.653734.

Batiha, G.E.-S., Beshbishy, A.M., Tayebwa, D.S., Shaheen, H.M., Yokoyama, N. *et al.* (2019) Inhibitory effects of *Syzygiumaromaticum* and *Camelliasinensis* methanolic extracts on the growth of Babesia and Theileria parasites. *Ticks and Tick-Borne Diseases* 10(5), 949–958. DOI: 10.1016/j.ttbdis.2019.04.016.

Bertini, I. and Sigel, A. (2001) *Handbook on Metalloproteins*. Marcel Dekker Inc, New York. DOI: 10.1201/9781482270822.

Casadesus, G., Shukitt-Hale, B., Stellwagen, H.M., Zhu, X., Lee, H.-G. *et al.* (2004) Modulation of hippocampal plasticity and cognitive behavior by short-term blueberry supplementation in aged rats. *Nutritional Neuroscience* 7(5–6), 309–316. DOI: 10.1080/10284150400020482.

Cherikoff, V. (1993) *The Bushfood Handbook: How to Gather, Grow, Process and Cook Australian Wild Foods*. Cherikoff Pty Ltd, Boronia Park, NSW, Australia.

Chikowe, G., Mpala, L. and Cock, I.E. (2013) Antibacterial activity of selected Australian Syzygium species. *Pharmacognosy Communications* 3, 77.

Chuen, T.L.K., Vuong, Q.V., Hirun, S., Bowyer, M.C., Predebon, M.J. *et al.* (2016) Antioxidant and anti-proliferative properties of Davidson's plum (*Davidsonia pruriens* F. Muell) phenolic-enriched extracts as affected by different extraction solvents. *Journal of Herbal Medicine* 6(4), 187–192. DOI: 10.1016/j.hermed.2016.08.005.

Clarke, M. (2012) *Australian Natives Food Industry Stocktake*. RIRDC, Rural Industries Research and Development Corporation, Barton, Australia.

Cock, I.E. (2011) Medicinal and aromatic plants – Australia. UNESCO Encyclopedia of Life Support Systems (EOLSS). Available at: https://www.eolss.net/ (accessed 1 April 2022).

Cock, I. (2012) Antimicrobial activity of *Syzygiumaustrale* and *Syzygiumleuhmannii* leaf methanolic extracts. *Pharmacognosy Communications* 2(2), 71–77. DOI: 10.5530/pc.2012.2.11.

Cock, I.E. and Cheesman, M. (2019) Plants of the genus *Syzygium* (Myrtaceae): A review on ethnobotany, medicinal properties and phytochemistry. In: Goyal, M.R. and Ayeleso, A.O. (eds) *Bioactive Compounds of Medicinal Plants: Properties and Potential for Human Health*. Apple Academic Press, Waretown, New Jersey, USA.

Craven, L.A. and Biffin, E. (2010) An infrageneric classification of *Syzygium* (Myrtaceae). *Blumea-Biodiversity, Evolution and Biogeography of Plants* 55, 94–99.

Dignan, C.A., Burlingame, B.A., Arthur, J.M., Quigley, R.J. and Milligan, G.C. (1994) *The Pacific Islands Food Composition Tables*. South Pacific Commission, Nouméa, New Caledonia.

Duthie, S.J., Narayanan, S., Brand, G.M., Pirie, L. and Grant, G. (2002) Impact of folate deficiency on DNA stability. *The Journal of Nutrition* 132(8 Suppl), 2444S-2449S. DOI: 10.1093/jn/132.8.2444S.

Ekholm, P., Reinivuo, H., Mattila, P., Pakkala, H., Koponen, J. *et al.* (2007) Changes in the mineral and trace element contents of cereals, fruits and vegetables in Finland. *Journal of Food Composition and Analysis* 20(6), 487–495. DOI: 10.1016/j.jfca.2007.02.007.

Eseyin, O.A., Ebong, P., Ekpo, A., Igboasoiyi, A. and Oforah, E. (2007) Hypoglycemic effect of the seed extract of *Telfairiaoccidentalis* in rat. *Pakistan Journal of Biological Sciences* 10(3), 498–501. DOI: 10.3923/pjbs.2007.498.501.

Fenech, M. (2005) The Genome Health Clinic and Genome Health Nutrigenomics concepts: diagnosis and nutritional treatment of genome and epigenome damage on an individual basis. *Mutagenesis* 20(4), 255–269. DOI: 10.1093/mutage/gei040.

Gorst, J.R. (2002) Indigenous Fruits of Australia. *Acta Horticulturae* 575(575), 555–561. DOI: 10.17660/ActaHortic.2002.575.65.

Gulati, V., Gulati, P., Harding, I.H. and Palombo, E.A. (2015) Exploring the anti-diabetic potential of Australian Aboriginal and Indian Ayurvedic plant extracts using cell-based assays. *BMC Complementary and Alternative Medicine* 15, 1–11. DOI: 10.1186/s12906-015-0524-8.

Guleria, S., Tiku, A.K., Singh, G., Koul, A., Gupta, S. *et al.* (2013) In vitro antioxidant activity and phenolic contents in methanol extracts from medicinal plants. *Journal of Plant Biochemistry and Biotechnology* 22(1), 9–15. DOI: 10.1007/s13562-012-0105-6.

Hardwick, P. (1996) Riberry is a flavour bomb. *Australian Rainforest Bushfood Industry Association Newsletter* 2, 6–12.

Havaux, M., Eymery, F., Porfirova, S., Rey, P. and Dörmann, P. (2005) Vitamin E protects against photoinhibition and photooxidative stress in *Arabidopsisthaliana*. *The Plant Cell* 17(12), 3451–3469. DOI: 10.1105/tpc.105.037036.

IndiGrow (2018) *Riberry - Syzygium luehmannii*. Available at: https://www.indigigrow.com.au/product-page/riberry-syzygium-luehmannii (accessed 1 April 2022).

Janick, J. and Paull, R.E. (2008) *The Encyclopedia of Fruit and Nuts*. CABI Publishing, Wallingford, UK. DOI: 10.1079/9780851996387.0000.

Jensen, M.B., López-de-Dicastillo Bergamo, C.A., Payet, R.M., Liu, X. and Konczak, I. (2011) Influence of copigment derived from Tasmannia pepper leaf on Davidson's plum anthocyanins. *Journal of Food Science* 76(3), C447–C453. DOI: 10.1111/j.1750-3841.2011.02077.x.

Kähkönen, M.P., Hopia, A.I., Vuorela, H.J., Rauha, J.P., Pihlaja, K. *et al.* (1999) Antioxidant activity of plant extracts containing phenolic compounds. *Journal of Agricultural and Food Chemistry* 47(10), 3954–3962. DOI: 10.1021/jf990146l.

Kalt, W., Blumberg, J.B., McDonald, J.E., Vinqvist-Tymchuk, M.R., Fillmore, S.A.E. *et al.* (2008) Identification of anthocyanins in the liver, eye, and brain of blueberry-fed pigs. *Journal of Agricultural and Food Chemistry* 56(3), 705–712. DOI: 10.1021/jf071998l.

Khodavirdipour, A., Zarean, R. and Safaralizadeh, R. (2020) Evaluation of the anti-cancer effect of *Syzygiumcumini* ethanolic extract on HT-29 colorectal cell line. *Journal of Gastrointestinal Cancer* 1–7.

Koche, D., Shirsat, R. and Kawale, M. (2016) An overerview of major classes of phytochemicals: their types and role in disease prevention. *Hislopia Journal* 9, 1–11.

Konczak, I. (1988) Health attributes of indigenous Australian plants. In: Bran-Miller, J.C. and Cherikoff, V. (eds) *Australian Aboriginal Bushfoods: Their Nutritive Value*.

Konczak, I. and Roulle, P. (2011) Nutritional properties of commercially grown native Australian fruits: Lipophilic antioxidants and minerals. *Food Research International* 44(7), 2339–2344. DOI: 10.1016/j.foodres.2011.02.023.

Konczak, I., Zabaras, D., Dunstan, M., Aguas, P., Roulfe, P. *et al.* (2009) *Health Benefits of Australian Native Foods: An Evaluation of Health-Enhancing Compounds*. Rural Industries Research and Development Corporation, Barton, Australia.

Konczak, I., Zabaras, D., Dunstan, M. and Aguas, P. (2010a) Antioxidant capacity and hydrophilic phytochemicals in commercially grown native Australian fruits. *Food Chemistry* 123(4), 1048–1054. DOI: 10.1016/j.foodchem.2010.05.060.

Konczak, I., Zabaras, D., Dunstan, M. and Aguas, P. (2010b) Antioxidant capacity and phenolic compounds in commercially grown native Australian herbs and spices. *Food Chemistry* 122(1), 260–266. DOI: 10.1016/j.foodchem.2010.03.004.

Kozlowski, H., Janicka-Klos, A., Brasun, J., Gaggelli, E., Valensin, D. *et al.* (2009) Copper, iron, and zinc ions homeostasis and their role in neurodegenerative disorders (metal uptake, transport, distribution and regulation). *Coordination Chemistry Reviews* 253(21–22), 2665–2685. DOI: 10.1016/j.ccr.2009.05.011.

Kress, W.J., Defilipps, R.A., Farr, E. and Kyi, D.Y.Y. (2003) *A Checklist of the Trees, Shrubs, Herbs, and Climbers of Myanmar*. Department of Systematic Biology - Botany. National Museum of Natural History, Washington, D.C.

Kumar, P.S., Febriyanti, R.M., Sofyan, F.F., Luftimas, D.E. and Abdulah, R. (2014) Anticancer potential of *Syzygiumaromaticum* L. in MCF-7 human breast cancer cell lines. *Pharmacognosy Research* 6(4), 350–354. DOI: 10.4103/0974-8490.138291.

Kwast, K.E. and Hand, S.C. (1996) Oxygen and pH regulation of protein synthesis in mitochondria from *Artemiafranciscana* embryos. *The Biochemical Journal* 313 (Pt 1), 207–213. DOI: 10.1042/bj3130207.

Lassak, E.V. and McCarthy, T. (2011) *Australian Medicinal Plants: A Complete Guide to Identification and Usage*. Reed New Holland Publishers, Sydney, Australia.

Lim, T.K. (2012) *Edible Medicinal and Non-Medicinal Plants*. Springer, Dordrecht, the Netherlands. DOI: 10.1007/978-94-007-1764-0.

Lim, A.S.L. and Rabeta, M.S. (2013) Proximate analysis, mineral content and antioxidant capacity of milk apple, malay apple and water apple. *International Food Research Journal* 20.

Linder, M.C. and Hazegh-Azam, M. (1996) Copper biochemistry and molecular biology. *The American Journal of Clinical Nutrition* 63(5), 797S-811S. DOI: 10.1093/ajcn/63.5.797.

Ling, L.T., Radhakrishnan, A.K., Subramaniam, T., Cheng, H.M. and Palanisamy, U.D. (2010) Assessment of antioxidant capacity and cytotoxicity of selected Malaysian plants. *Molecules* 15(4), 2139–2151. DOI: 10.3390/molecules15042139.

Manaharan, T., Appleton, D., Cheng, H.M. and Palanisamy, U.D. (2012) Flavonoids isolated from *Syzygiumaqueum* leaf extract as potential antihyperglycaemic agents. *Food Chemistry* 132(4), 1802–1807. DOI: 10.1016/j.foodchem.2011.11.147.

Manaharan, T., Ming, C.H. and Palanisamy, U.D. (2013) *Syzygiumaqueum* leaf extract and its bioactive compounds enhances pre-adipocyte differentiation and 2-NBDG uptake in 3T3-L1 cells. *Food Chemistry* 136(2), 354–363. DOI: 10.1016/j.foodchem.2012.08.056.

Manaharan, T., Chakravarthi, S., Radhakrishnan, A.K. and Palanisamy, U.D. (2014) In vivo toxicity evaluation of a standardized extract of *Syzygiumaqueum* leaf. *Toxicology Reports* 1, 718–725. DOI: 10.1016/j.toxrep.2014.09.006.

Mani, J.S., Johnson, J.B., Hosking, H., Ashwath, N., Walsh, K.B. *et al.* (2020) Antioxidative and therapeutic potential of selected Australian plants: A review. *Journal of Ethnopharmacology* 268, 113580. DOI: 10.1016/j.jep.2020.113580.

Miller, J.B., James, K.W. and Maggiore, P.M.A. (1993) *Tables of Composition of Australian Aboriginal Foods*. Aboriginal Studies Press, Canberra, Australia.

Morton, J.F. (1987) *Fruits of Warm Climates*. Florida Flair Books.

Murhekar, S., Wright, M.H., Greene, A.C., Brownlie, J.C. and Cock, I.E. (2017) Inhibition of *Shewanella* spp. growth by *Syzygiumaustrale* and *Syzygiumluehmannii* extracts: natural methods for the prevention of fish spoilage. *Journal of Food Science and Technology* 54(10), 3314–3326. DOI: 10.1007/s13197-017-2782-6.

Netzel, M., Netzel, G., Tian, Q., Schwartz, S. and Konczak, I. (2006) Sources of antioxidant activity in Australian native fruits. Identification and quantification of anthocyanins. *Journal of Agricultural and Food Chemistry* 54, 9820–9826. DOI: 10.1021/jf0622735.

Netzel, M., Netzel, G., Tian, Q., Schwartz, S. and Konczak, I. (2007) Native Australian fruits — a novel source of antioxidants for food. *Innovative Food Science & Emerging Technologies* 8(3), 339–346. DOI: 10.1016/j.ifset.2007.03.007.

Noé, W., Murhekar, S., White, A., Davis, C. and Cock, I.E. (2019) Inhibition of the growth of human dermatophytic pathogens by selected Australian and Asian plants traditionally used to treat fungal infections. *Journal de Mycologie Medicale* 29(4), 331–344. DOI: 10.1016/j.mycmed.2019.05.003.

NUTTAB (2010) Food Nutrient Database. Available at: https://www.foodstandards.gov.au/science/monitoringnutrients/ausnut/ausnutdatafiles/Pages/foodnutrient.aspx (accessed 1 April 2022).

Olukoya, D.K., Idika, N. and Odugbemi, T. (1993) Antibacterial activity of some medicinal plants from Nigeria. *Journal of Ethnopharmacology* 39(1), 69–72. DOI: 10.1016/0378-8741(93)90051-6.

Osman, H., Rahim, A.A., Isa, N.M. and Bakhir, N.M. (2009) Antioxidant activity and phenolic content of *Paederiafoetida* and *Syzygiumaqueum*. *Molecules* 14(3), 970–978. DOI: 10.3390/molecules14030970.

Palanisamy, U.D. and Manaharan, T. (2015) *Syzygiumaqueum* leaf extracts for possible antidiabetic Treatment. *International Symposium on Medicinal Plants and Natural Products* (1098), 13–22. DOI: 10.17660/ActaHortic.2015.1098.1.

Panggabean, G. (1992) *Syzygiumaqueum, Syzygiummalaccense & Syzygiumsamarangense Edible Fruits and Nuts*, 2nd edn. Prosea Foundation Bogor, Indonesia.

Pérez, A.G., Olías, R., Espada, J., Olías, J.M. and Sanz, C. (1997) Rapid determination of sugars, non-volatile acids, and ascorbic acid in strawberry and other fruits. *Journal of Agricultural and Food Chemistry* 45, 3545–3549.

Pinto, E., Vale-Silva, L., Cavaleiro, C. and Salgueiro, L. (2009) Antifungal activity of the clove essential oil from *Syzygiumaromaticum* on *Candida*, *Aspergillus* and dermatophyte species. *Journal of Medical Microbiology* 58(Pt 11), 1454–1462. DOI: 10.1099/jmm.0.010538-0.

Quinlivan, E.P., McPartlin, J., McNulty, H., Ward, M., Strain, J.J. *et al.* (2002) Importance of both folic acid and vitamin B12 in reduction of risk of vascular disease. *Lancet* 359(9302), 227–228. DOI: 10.1016/s0140-6736(02)07439-1.

Rabeta, M.S., Chan, S., Neda, G.D., Lam, K.L. and Ong, M.T. (2013) Anticancer effect of underutilized fruits. *International Food Research Journal* 20, 551.

Rayan, P., Matthews, B., Mc Donnell, P. and Edwin Cock, I. (2016) Phytochemical analysis of *Tasmannialanceolata* extracts and inhibition of *Giardiaduodenalis* proliferation. *Pharmacognosy Journal* 8(3), 291–299. DOI: 10.5530/pj.2016.3.19.

Richmond, R., Bowyer, M. and Vuong, Q. (2019) Australian native fruits: Potential uses as functional food ingredients. *Journal of Functional Foods* 62, 103547. DOI: 10.1016/j.jff.2019.103547.

RIRDC (2012) *A Focus on Riberry*. Rural Industries Research and Development Corporation. Available at: https://anfab.org.au/edit/research_publications/14-121_Riberry.pdf (accessed 1 April 2022).

Robinson, B., Delaney, M. and Hardwick, P. (1994) Growing Australian fruit and nut trees with horticultural potential. In: *Bush Food Seminar, Papers and Proceedings*. Greening Australia, Lismore, NSW.

Rocchetti, G., Lucini, L., Ahmed, S.R. and Saber, F.R. (2019) In vitro cytotoxic activity of six *Syzygium* leaf extracts as related to their phenolic profiles: An untargeted UHPLC-QTOF-MS approach. *Food Research International* 126, 108715. DOI: 10.1016/j.foodres.2019.108715.

Saptarini, N.M. and Herawati, I.E. (2017) Antioxidant activity of water apple (*Syzygiumaqueum*) fruit and fragrant mango (*mangiferaodorata*) fruit. *Asian Journal of Pharmaceutical and Clinical Research* 10(14), 54. DOI: 10.22159/ajpcr.2017.v10s2.19487.

Sarathchandrian, I., Monisha, P., Kavimani, S. and Selvasudha, N. (2020) Assessment of in-vitro cytotoxicity and in-vivo antitumor activity of *Syzygiumaqueum*. *International Journal of Research in Pharmaceutical Sciences and Technology* 2, 37–43.

Sautron, C. and Cock, I.E. (2014) Antimicrobial activity and toxicity of *Syzygiumaustrale* and *Syzygiumleuhmannii* fruit extracts. *Pharmacognosy Communication* 4(1), 53–60. DOI: 10.5530/pc.2014.1.8.

Sobeh, M., Mahmoud, M.F., Petruk, G., Rezq, S., Ashour, M.L. *et al.* (2018) Syzygiumaqueum: A polyphenol-rich leaf extract exhibits antioxidant, hepatoprotective, pain-killing and anti-inflammatory activities in animal models. *Frontiers in Pharmacology* 9, 566. DOI: 10.3389/fphar.2018.00566.

Soh, W.K. (2017) Taxonomy of *Syzygium*. In: Nair, K.N. (ed.) *The Genus Syzygium: Syzygium Cumini and Other Underutilized Species*. CRC Press, New York. DOI: 10.1201/9781315118772.

Sonawane, M.S. (2018) Dietary benefits of watery rose apple (*Syzygiumaqueum* (Burm. f.) Alston). *International Archive of Applied Sciences & Technology* 9, 126–129.

Subarnas, A., Diantini, A., Abdulah, R., Zuhrotun, A., Hadisaputri, Y.E. *et al.* (2015) Apoptosis induced in MCF-7 human breast cancer cells by 2',4'-dihydroxy-6-methoxy-3,5-dimethylchalcone isolated from *Eugeniaaquea* Burm f. leaves. *Oncology Letters* 9(5), 2303–2306. DOI: 10.3892/ol.2015.2981.

Taiz, L., Zeiger, E., Møller, I.M. and Murphy, A. (2015) *Plant Physiology and Development*. Sinauer Associates Incorporated, Sunderland, Massachusetts.

Tsao, R. (2010) Chemistry and biochemistry of dietary polyphenols. *Nutrients* 2(12), 1231–1246. DOI: 10.3390/nu2121231.

Tsuda, T. (2008) Regulation of adipocyte function by anthocyanins; possibility of preventing the metabolic syndrome. *Journal of Agricultural and Food Chemistry* 56(3), 642–646. DOI: 10.1021/jf073113b.

USDA National Nutrient Database for Standard Reference (2020). Available at: https://fdc.nal.usda.gov/fdc-app.html#/food-details/1102702/nutrients

Verdcourt, B. (2001) *Syzygium aqueum (Burm.f.) Alston [family Myrtaceae], Flora of Tropical East Africa*. JSTOR. Available at: https://plants.jstor.org/collection/FTEA (accessed 1 April 2022).

Vinayaraj (2008) *Syzygium aqueum*. Wikimedia Commons. Available at: https://commons.wikimedia.org/wiki/File:Syzygium_aqueum-2.jpg (accessed 1 April 2022).

Wong, K.C. and Lai, F.Y. (1996) Volatile constituents from the fruits of four *Syzygium* species grown in Malaysia. *Flavour and Fragrance Journal* 11(1), 61–66. DOI: 10.1002/(SICI)1099-1026(199601)11:1<61::AID-FFJ539>3.0.CO;2-1.

Wood, A.J., McManus, K., Wright, M.H., Greene, A.C. and Cock, I.E. (2017) Growth inhibitory activity of selected Australian Syzygium species against Malodour forming bacteria. *Pharmacognosy Communications* 7(3), 129–136. DOI: 10.5530/pc.2017.3.19.

Wright, M.H., Lee, C.J., Pollock, C.E., Greene, A.C. and Cock, I.E. (2016) Growth inhibitory activity of selected high antioxidant Australian *Syzygium* species against the food poisoning and tissue necrotic pathogen *Clostridium perfringens*. *Pharmacognosy Communications* 6(2), 93–99. DOI: 10.5530/pc.2016.2.7.

Yadav, A.K., Saraswat, S., Sirohi, P., Rani, M., Srivastava, S. *et al.* (2017) Antimicrobial action of methanolic seed extracts of *Syzygiumcumini* Linn. on *Bacillussubtilis*. *AMB Express* 7(1), 1–10. DOI: 10.1186/s13568-017-0500-4.

Yang, L., Rong-Rong, C., Ji-Li, F. and Ke, Y. (2019) Total anthocyanins and cyanidin-3-o-glucoside contents and antioxidant activities of purified extracts from eight different pigmented plants. *Pharmacognosy Magazine* 15, 124.

Yugandhar, P., Vasavi, T., Uma Maheswari Devi, P. and Savithramma, N. (2017) Bioinspired green synthesis of copper oxide nanoparticles from *Syzygiumalternifolium* (Wt.) Walp: characterization and evaluation of its synergistic antimicrobial and anticancer activity. *Applied Nanoscience* 7(7), 417–427. DOI: 10.1007/s13204-017-0584-9.

14 Australian Green Plum (*Buchanania obovata*)

Selina Fyfe[1], Horst Joachim Schirra[2,3,4], Michael Rychlik[1,5], Heather E. Smyth[1] and Yasmina Sultanbawa[1*]

[1]The University of Queensland, Queensland Alliance for Agriculture and Food Innovation, Brisbane, Queensland, Australia; [2]The University of Queensland, Centre for Advanced Imaging, Brisbane, Queensland, Australia; [3]School of Environment and Science, Griffith University, Nathan, Queensland, Australia; [4]Griffith Institute for Drug Discovery, Griffith University, Nathan, Queensland, Australia; [5]Chair of Analytical Food Chemistry, Technical University of Munich, Freising, Germany

Abstract

Green plum fruit grow on the tree *Buchanania obovata* in the Northern Territory and Western Australia and the flesh and seed of the fruit is harvested and eaten by Aboriginal Australians. Green plums are a small drupe that grow to about 10–17 mm in length and are part of the family Anacardiaceae. They provide valuable nutrients to the Aboriginal people who live in remote parts of Australia. Green plums are high in potassium and are a good source of magnesium, calcium, phosphorus and dietary fibre. They have promisingly high levels of folate, the highest being the vitamer 5-methyl-tetrahydrofolate (5-CH_3-H_4folate). Green plums contain other phenolic compounds and phytonutrients that may have pharmacological effects and health benefits to those who eat this fruit, including gallic acid, chlorogenic acid and ferulic acid. Green plums are a traditional food that could be promoted further for their health benefits and nutritional properties, especially to the Aboriginal people who value and eat them.

Introduction

The green plum is the fruit of the tree *Buchanania obovata* and the flesh and seed of the fruit are harvested and eaten by Aboriginal Australians. *B. obovata* trees are native to Australia and grow across the northern parts of the Northern Territory and Western Australia. They grow in Arnhem Land from Nhulunbuy across to Darwin, on Groote Eylandt, around the Daly River, Keep River and McArthur River regions in the Northern Territory, as well as in the Kimberley region of Western Australia (Atlas of Living Australia, 2020). Green plums ripen in the summer, when it is hot and humid and around the start of the first rains that occur in the monsoonal rain season of these tropical areas (Head, 2002).

The green plum fruit has historically been eaten and is still eaten by Aboriginal communities in Australia (Ferguson *et al.*, 2017). Each Aboriginal Australian language where the green plum is consumed has its own name for the fruit including *munydjutj* in Djambarrpuyngu,

*Corresponding author: y.sultanbawa@uq.edu.au

 Handbook of Phytonutrients in Indigenous Fruits and Vegetables (eds D. Sivakumar *et al.*)
DOI: 10.1079/9781789248067.0014

mulaagi in Burarra, *yulmuru* in Alawa, *bikabaji* in Garawa, *bigigee* in Yanyuwa and *mangkarrba* in Anindilyakwa (Barr *et al.*, 1993).

Green plums are eaten by Aboriginal Australians as a raw fruit when they are soft (Levitt, 1981). They are also eaten when they have been sun-dried on the tree or on the ground, and when they have been dried and covered in red ochre, wrapped in paperbark and stored in rock shelters to preserve them for later eating when they are no longer in season (Fox and Garde, 2018). The seed is sometimes pounded with the pulp and they can be eaten together (Head, 2002) or the pounded seed and flesh can be stored in paperbark for later use (Purdie *et al.*, 2018). Green plums are a very popular fruit, liked for their sweet and pleasant flavour (Head, 2002; Hiddins, 2003; Fox and Garde, 2018).

Other parts of the *B. obovata* plant have been used by Aboriginal Australians as bush medicines. The young stems and twigs, leaf petioles and bark are used to treat toothache, inflamed and sore eyes and skin infections (Levitt, 1981; Barr *et al.*, 1993). The tree and fruit are also a major sacred totem for the Bolmo people (Fox and Garde, 2018).

Fruit and vegetables are important sources of vitamins, minerals, fibre and phytochemicals in the diet (Slavin and Lloyd, 2012). These nutrients are essential for health and some can only be obtained through the diet as they are unable to be synthesized in the body. The Aboriginal Australians who have eaten the green plum have obtained valuable nutrition from this fruit. Understanding the nutritional value of the green plum and the phytonutrients it contains helps understanding of the role the fruit plays in diet and health for those who eat it.

Botany

The taxonomy of the green plum is shown in Table 14.1 (Atlas of Living Australia, 2020). The accepted name of the tree is *Buchanania obovata* Engl. (The Plant List, 2013) and it was first described in 1883 (Candolle and Candolle, 1883). The *B. obovata* tree is in the family Anacardiaceae (Atlas of Living Australia, 2020), which contains other fruit and nuts that are well known and good sources of nutrition including the mango (*Mangifera indica*), pistachio nut (*Pistacia vera*) and cashew nut (*Anacardium occidentale*). The Anacardiaceae family also contains other fruit that provide valuable nutrition to Indigenous communities around the world such as the marula fruit (*Sclerocarya birrea*), the sumac (genus *Rhus*), the yellow mombin (*Spondias mombin*) and the chironji (*Buchanania lanzan*) (Fyfe *et al.*, 2020a).

Table 14.1. Taxonomy of the green plum, *Buchanania obovata* (Atlas of Living Australia, 2020).

Class	Green plum classification
Kingdom	Plantae
Phylum	Charophyta
Class	Equisetopsida
Subclass	Magnoliidae
Superorder	Rosanae
Order	Sapindales
Family	Anacardiaceae
Genus	*Buchanania*
Species	*Buchanania obovata*

B. obovata trees can grow to 15 m high and have a rough brown or grey bark that is tesselated and can be scaly (Barr *et al.*, 1993; Fox and Garde, 2018). The trees are deciduous and are often scraggly looking with soft hairy branchlets. They are able to fruit even when very small with only a few leaves present (Fox and Garde, 2018). Figure 14.1 shows variation in tree size with all of the trees in the figure bearing fruit. The leaves are situated alternate and are obovate to oblong-obovate in shape (Barr *et al.*, 1993). They are smooth, thick and leathery and have a prominently raised midvein and distinct venation (Fox and Garde, 2018). The flowers are bisexual and are a creamy-white colour; they grow to 5–7 mm in diameter (Barr *et al.*, 1993). There are numerous flowers growing on slender terminal panicles that are 100–180 mm long (Fox and Garde, 2018). The trees flower between July and November and fruit mostly between October and January.

The fruit is a smooth and fleshy ovoid shaped drupe that grows to between 10–17 mm long and each fruit contains a single hard-shelled seed (Barr *et al.*, 1993; Fox and Garde, 2018). The green plum grows to about 1 g in weight,

Fig. 14.1. *Buchanania obovata* trees with green plum fruit on them (all photographs by Selina Fyfe © The University of Queensland).

Fig. 14.2. Immature green plum fruit on the tree, and ripe fruit (all photographs by Selina Fyfe © The University of Queensland).

about the size of a blueberry or a small grape. When the fruit ripen they change colour from green to yellow-green (Fox and Garde, 2018). Figure 14.2 shows the immature fruit on the tree and the ripe fruit that have been harvested.

Green plums are harvested from the tree when ripe, in December and January, during the hot summer. Figure 14.3 shows the amount of water and solids from the flesh and seed of 1 kg of ripe green plums and the expected yield of flesh and freeze-dried flesh. This amount is expected from green plums that are harvested on the east side of the state of the Northern Territory, however, there is variation in yields especially between locations.

Nutritional Value

Proximates, minerals, metals and trace elements

The proximate analysis of green plums shows that they are a good source of dietary fibre in both the flesh and the seed (55.1 and

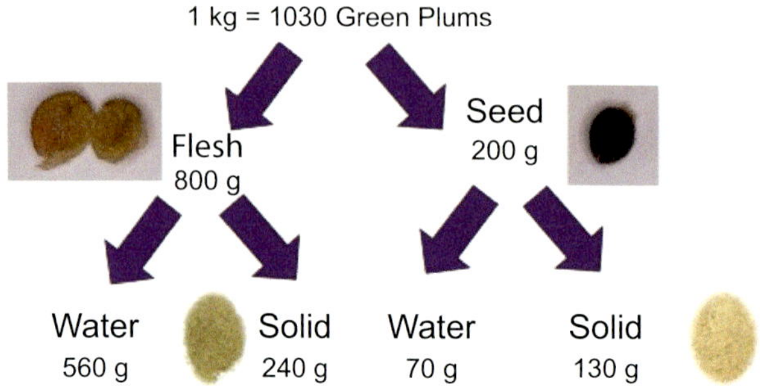

Fig. 14.3. The expected yield of water and solids from the flesh and seed of 1 kg of whole ripe green plums.

87.7 g/100 g dry weight (DW) respectively) (Fyfe *et al.*, 2018b). The pulp of the green plum, made up of both flesh and skin, is high in potassium (2274.7 mg/100 g DW) and a good source of magnesium (570.5 mg/100 g DW), calcium (426.0 mg/100 g DW) and phosphorus (216.8 mg/100 g DW) (Medley and Bollhöfer, 2016; Fyfe *et al.*, 2018b, 2020b). The green plum is a member of the Anacardiaceae family (Atlas of Living Australia, 2020) which contains other fruit that are also good dietary sources of potassium (Fyfe *et al.*, 2020a). Potassium intake is associated with blood pressure reduction and may be protective against age-related bone loss and reduction of kidney stones (Weaver, 2013). Magnesium plays a major role in the human body in controlling cardiac excitability, neuromuscular transmission and in helping regulate blood pressure. It is important for skeletal growth and development, helps prevent stroke and decreases the risk of chronic disease (Champagne, 2008). Calcium is required in the diet for bone health, acquiring bone mass and in soft tissues and body fluids, and works in muscle contraction, secretions of neurotransmitters, blood coagulation and digestion (Theobald, 2005). Phosphorus is essential for cellular signal transduction in every human body cell as well as in glycolysis, gluconeogenesis, skeletal mineralization and energy metabolism (Takeda *et al.*, 2012).

In remote locations the diet of Aboriginal people is deficient in potassium, calcium and magnesium when they rely only on store-bought food (Brimblecombe *et al.*, 2013). Traditional foods can therefore add more of these valuable nutrients back into the diet and increase diversity in dietary nutrients. The green plum contains other minerals and trace elements that are required by the human body for health, even at micronutrient levels. These include zinc (2.6 mg/100 g DW), iron (3.8 mg/100 g DW), selenium (0.022 mg/100 g DW) and manganese (1.5 mg/100 g DW) (Fyfe *et al.*, 2018b). Green plums ripen and are available on trees in the hot and humid summer (Head, 2002) when people are constantly sweating and losing fluid and electrolytes. The green plum may provide valuable electrolytes needed for rehydration during the fluid loss that occurs from sweating. Potassium is lost during sweating and its replacement may be important to achieve rehydration by helping to retain water in the intracellular spaces (Shirreffs *et al.*, 2004).

Vitamins

Vitamins are important phytonutrients in fruit and plant foods and are necessary and valuable for maintaining health and preventing disease. The green plum has not been extensively analysed for vitamin content; currently, only data on the vitamin C and folate content is published. These results show that green plums have promisingly high levels of folate (161 μg/100g FW) (Fyfe *et al.*, 2018b, 2020b). The folate levels are higher than in many other tropical fruit such as mango (55.8–74.5 μg/100g FW), guava (91.0 μg/100g FW), papayas (61.6–90.7 μg/100g FW) and dragonfruit (36.0 μg/100g FW), but are still within the natural range and lower than very high folate fruit such as yellow passionfruit (271 μg/100g

FW) and durian (175–440 μg/100g FW) (Striegel *et al.*, 2018, 2019).

Folate is a B vitamin and important for health. It is necessary for maintaining normal cell growth and division and for DNA synthesis, methylation and repair (Chango *et al.*, 2013; Dierkes and Nygard, 2013). Folate deficiency can lead to a number of clinically associated diseases including megaloblastic anaemia, increased blood homocysteine and an increased risk of neural tube defects and congenital disorders in developing foetuses (Chango *et al.*, 2013). Folate is not biosynthesized in the human body so must be obtained from the diet, through either natural foods such as fruit and plant leaves or through supplementation or fortification of other foods with synthetic folate (Chango *et al.*, 2013).

The folate vitamer 5-methyl-tetrahydrofolate (5-CH_3-H_4folate) is the most abundant folate in green plum (Fyfe *et al.*, 2018b). This vitamer is the main form of folate circulating in the human body and is able to cross the intestinal barrier into the bloodstream (Chango *et al.*, 2013). Approximately 98% of folate in human plasma is 5-CH_3-H_4folate and it is the preferred form of folate for transport into peripheral tissues (Pietrzik *et al.*, 2010). Folate in fruit such as the green plum is a valuable nutrient and important vitamin in the diet, and the consumption of this phytonutrient is necessary for maintaining health and wellbeing.

Folate is important in the diet of women prior to pregnancy as it is necessary for preventing neural tube defects in the fetus (Chango *et al.*, 2013). Folate deficiency and neural tube defects were higher in some Aboriginal populations in Australia compared with non-Aboriginal populations prior to mandatory fortification of folic acid in foods. In 2009, fortification with folic acid of wheat flour used for bread-making was made mandatory in Australia. Studies to understand the folate levels in the Aboriginal and non-Aboriginal populations in Western Australia and the effect on neural tube defects show that before fortification 10% of Aboriginal women and 26% of Aboriginal men had red blood cell folate concentrations below 250 ng/mL which is associated with folate deficiency (Maxwell *et al.*, 2013). Following fortification none of the participants were folate deficient and there was an increase in the mean red cell folate levels of 129 ng/ml for males and 186 ng/ml for females (Bower *et al.*, 2016). Prior to mandatory fortification, neural tube defects were 42% higher in Aboriginal births than in non-Aboriginal births in Australia (D'Antoine and Bower, 2019). Data from the Western Australia Register of Developmental Anomalies show that neural tube defects per 1000 births were 2.55 in 1980–1992 before the promotion of folic acid supplements; this level decreased to 2.43 in 1996–2006 when there was promotion of folic acid supplements and voluntary fortification of some foods, and then decreased again to 0.82 between 2010–2014 after mandatory fortification was implemented (Bower *et al.*, 2016). These studies show the effect that fortification had on the general Aboriginal population in Western Australia. However, these studies were researched on Aboriginal populations where the green plum does *not* grow. It is unknown if populations where the green plum grows had higher levels of folate prior to folic acid fortification in bread compared to the populations without them.

The green plum also contains low levels of vitamin C (170 mg/100g DW), similar to the amount in mangos (Fyfe *et al.*, 2018b). Vitamin C exists in plants as ascorbic acid and dehydroascorbic acid and is essential in the diet as humans are unable to synthesize it (Granger and Eck, 2018). It is essential for humans to eat it to prevent scurvy and other diseases including coronary heart disease, stroke and cancer (Granger and Eck, 2018). The green plum provides only low levels of vitamin C, but other bush foods such as the Kakadu plum (*Terminalia ferdinandiana*) provide much higher levels to the Aboriginal communities when in season (Konczak *et al.*, 2014).

Phenolic Compounds and Phytonutrients

The green plum also contains other phenolic compounds and phytonutrients that may have pharmacological effects and health benefits to those who eat it. Slightly underripe green plums were analysed for targeted chemical compounds that have functional and preservative effects in food products, such as antimicrobial and antioxidant properties (Fyfe *et al.*, 2018a). These compounds have been found in green plums but their pharmacological effects have not been tested on the green plum itself.

Green plums contain gallic acid, in both unbound and bound extractions of the flesh (unbound extraction of flesh 955.39 μg/g DW and bound extraction of flesh 3342.97 μg/g DW) and with lower levels in the seed (unbound extraction of seed 151.89 μg/g DW and bound extraction of seed 3.95 μg/g DW) (Fyfe *et al.*, 2018a). Gallic acid has a number of demonstrated pharmacological activities including antimicrobial activity, anticancer activity, protection of the gastrointestinal tract against disease, antioxidant activity that decreases oxidative consequences and anti-inflammatory activity (Kahkeshani *et al.*, 2019). Gallic acid from other fruits and plants has demonstrated antiobesity capabilities in experimental models by reducing excessive lipid storage in obese subjects and targeting adipose tissue to suppress lipogenesis, improve the signalling of insulin and simultaneously combating oxidative stress and increased pro-inflammatory responses (Dludla *et al.*, 2019). Gallic acid and its derivatives obtained from foods may be able to manipulate mitochondria to have a chemopreventive effect and protect against disease (Kosuru *et al.*, 2018).

Chlorogenic acid is found in the flesh (unbound extraction of flesh 91.83 μg/g DW and bound extraction of flesh 19.93 μg/g DW) and seed (unbound extraction of seed 695.19 μg/g DW and bound extraction of seed 45.71 μg/g DW) of the green plum (Fyfe *et al.*, 2018a). It is a dietary polyphenol with antioxidant activity, antimicrobial activity, anti-inflammatory activity, antiobesity activity, antidiabetic activity and with protective effects for the cardiac and hepatic body systems (Tajik *et al.*, 2017; Naveed *et al.*, 2018).

Green plums also contain ferulic acid, an antioxidant found in many plant foods (Fyfe *et al.*, 2018a). Ferulic acid has beneficial effects in health from anti-oxidant properties related to disorders including diabetes, cancer and neurodegenerative diseases (de Oliveira Silva and Batista, 2017). Ferulic acid is an efficient scavenger of reactive oxygen and nitrogen species and many of its protective effects occur because of its ability to scavenge free radicals (Mancuso and Santangelo, 2014). It has antimicrobial activity against yeast and bacteria, it decreases the level of some inflammatory mediators giving it anti-inflammatory properties, it neutralizes free radicals formed in diabetes and it has antioxidant and pro-oxidant properties against the reactive oxygen species produced in cancer cells (de Oliveira Silva and Batista, 2017). The green plum also contains ferulic acid precursors including p-coumaric acid and cinnamic acid (de Oliveira Silva and Batista, 2017; Fyfe *et al.*, 2018a). The green plum has levels of ferulic acid unbound in its flesh of 114.50 μg/g DW and unbound in the seed of 22.77 μg/g DW with lower levels of it bound in the flesh at 8.02 μg/g DW and in the seed 17.85 μg/g DW (Fyfe *et al.*, 2018a). In wheat grain, ferulic acid is the major contributor to *in vitro* antioxidant capacity and free or unbound ferulic acid has been found to have higher bioaccessibility than bound ferulic acid (Mateo Anson *et al.*, 2009).

Green plums also contain a number of other phytochemicals that may have health benefits including the flavonol quercetin (unbound flesh extraction 118.59 μg/g DW) and its derivatives quercetin 3-glucoside (unbound flesh extraction 874.17 μg/g DW, bound flesh extraction 39.64 μg/g DW, unbound seed extraction 13.19 μg/g DW) and quercetin 3,4'-diglucoside (bound seed extraction 63.37 μg/g DW). Myricetin is in the unbound flesh of the green plum at 180.96 μg/g DW. Green plums also contains unquantified amounts of salicylic acid, catechin, kaempferol, naringenin, other quercetin derivatives, cinnamic acid, vanillic acid and ellagic acid (Fyfe *et al.*, 2018a).

Potential for Health

The amount of bushfood eaten by Aboriginal populations in Australia and the role that nutrients from these bushfoods play in diet and health is largely unknown. Prior to colonization Aboriginals in Australia had a hunting and gathering diet. After colonization this transitioned to a 'station' diet and then changed to what is now a self-selected store diet that is essentially a western diet (Smith and Smith, 1999). The diets of people in remote Aboriginal communities are generally very poor quality with significant proportions of key nutrients obtained by processed foods that are nutrient fortified and poor quality (Brimblecombe *et al.*, 2013). In remote communities in the Northern Territory, white bread that has been fortified with fibre, folate and other micronutrients is a major source of vitamins and nutrients including iron, calcium, potassium, magnesium, sodium, protein fibre and folate (Brimblecombe *et al.*, 2013).

However, traditional bush foods are still eaten in remote Aboriginal communities and are available all year round. Of Aboriginal participants in one survey, 89% consumed some bush foods fortnightly and 71% consumed them at least weekly (Ferguson *et al.*, 2017). Bush foods are relied on by some members of remote Aboriginal communities as their food source during times of food insecurity (Ferguson *et al.*, 2017).

Traditional foods play a role in representing cultural identity in remote Aboriginal Australian populations. The knowledge of the traditional food system by the Aboriginal people made food choices within it seem simple when compared with a western diet that was not as well understood by them (Brimblecombe *et al.*, 2014). Supporting access to traditional foods and attending to past and present contexts of food in nutrition education can both be used to improve the diet and health outcomes for remote Aboriginal populations (Brimblecombe *et al.*, 2014).

Conclusion

The green plum has been and is still an important food in the diet of remote Aboriginal populations in Australia. It contains nutrients with health benefits that give vital phytonutrients including electrolytes for rehydration in the hot summer, folate for cell and DNA growth and repair and prevention of disease, and other valuable compounds with antioxidant, anti-inflammatory activity and other properties for maintaining a healthy body. Green plums are a traditional indigenous food that could be promoted further for its health benefits and nutritional properties especially to the Aboriginal people who value and eat it.

Acknowledgements

The authors acknowledge the Traditional Owners of the lands on which the green plums are harvested and respect the knowledge and experience the Traditional Owners hold regarding the care, harvest and use of these plants.

Funding is by an Australian Government Research Training Program Scholarship, The University of Queensland and the Australian Research Council Training Centre for Uniquely Australian Foods.

References

Atlas of Living Australia (2020) *Buchanania obovata Engl. Species page*. Available at: https://bie.ala.org.au/species/https://id.biodiversity.org.au/node/apni/2887105 (accessed 2 December 2020).

Barr, A., Chapman, J., Smith, N., Wightman, G., Knight, T. *et al.* (1993) *Traditional Aboriginal Medicines in the Northern Territory of Australia, by Aboriginal Communities of the Northern Territories*. Conservation Commission of the Northern Territory of Australia, Darwin, Northern Territory, Australia.

Bower, C., Maxwell, S., Hickling, S., D'Antoine, H. and O'Leary, P. (2016) Folate status in Aboriginal people before and after mandatory fortification of flour for bread-making in Australia. *The Australian & New Zealand Journal of Obstetrics & Gynaecology* 56(3), 233–237. DOI: 10.1111/ajo.12425.

Brimblecombe, J.K., Ferguson, M.M., Liberato, S.C. and O'Dea, K. (2013) Characteristics of the community-level diet of Aboriginal people in remote northern Australia. *The Medical Journal of Australia* 198(7), 380–384. DOI: 10.5694/mja12.11407.

Brimblecombe, J., Maypilama, E., Colles, S., Scarlett, M., Dhurrkay, J.G. *et al.* (2014) Factors influencing food choice in an Australian Aboriginal community. *Qualitative Health Research* 24(3), 387–400. DOI: 10.1177/1049732314521901.

Candolle, A. and Candolle, C. (1883) *Monographiæ phanerogamarum: Prodromi nunc continuatio*. G. Masson, Paris.

Champagne, C.M. (2008) Magnesium in hypertension, cardiovascular disease, metabolic syndrome, and other conditions: a review. *Nutrition in Clinical Practice* 23(2), 142–151. DOI: 10.1177/0884533608314533.

Chango, A., Watkins, D. and Abdennebi-Najar, L. (2013) The importance of folate in health. In: Preedy, V.R. (ed.) *B Vitamins and Folate: Chemistry, Analysis, Function and Effects*. Royal Society of Chemistry, Cambridge, UK.

de Oliveira Silva, E. and Batista, R. (2017) Ferulic acid and naturally occurring compounds bearing a feruloyl moiety: a review on their structures, occurrence, and potential health benefits. *Comprehensive Reviews in Food Science and Food Safety* 16(4), 580–616. DOI: 10.1111/1541-4337.12266.

Dierkes, J. and Nygard, O. (2013) B vitamins and disease. In: Preedy, V.R. (ed.) *B Vitamins and Folate*. The Royal Society of Chemistry, Cambridge, UK. DOI: 10.1039/9781849734714.

Dludla, P.V., Nkambule, B.B., Jack, B., Mkandla, Z., Mutize, T. *et al.* (2019) Inflammation and oxidative stress in an obese state and the protective effects of gallic acid. *Nutrients* 11, 23.

D'Antoine, H. and Bower, C. (2019) Folate status and neural tube defects in aboriginal Australians: the success of mandatory fortification in reducing a health disparity. *Current Developments in Nutrition* 3(8), nzz071. DOI: 10.1093/cdn/nzz071.

Ferguson, M., Brown, C., Georga, C., Miles, E., Wilson, A. *et al.* (2017) Traditional food availability and consumption in remote Aboriginal communities in the Northern Territory, Australia. *Australian and New Zealand Journal of Public Health* 41(3), 294–298. DOI: 10.1111/1753-6405.12664.

Fox, G. and Garde, M. (2018) *An-Me Arri-Ngun: The Food We Eat*. Gundjeihmi Aboriginal Corporation, Australia.

Fyfe, S.A., Netzel, G., Netzel, M.E. and Sultanbawa, Y. (2018a) *Buchanania obovata*: functionality and phytochemical profiling of the Australian native green plum. *Foods (Basel, Switzerland)* 7(5), E71. DOI: 10.3390/foods7050071.

Fyfe, S.A., Netzel, M.E., Tinggi, U., Biehl, E.M. and Sultanbawa, Y. (2018b) *Buchanania obovata*: An Australian indigenous food for diet diversification. *Nutrition & Dietetics* 75, 527–532.

Fyfe, S., Smyth, H.E., Schirra, H.J., Rychlik, M. and Sultanbawa, Y. (2020a) The nutritional potential of the native Australian green plum (Buchanania obovata) compared to other Anacardiaceae fruit and nuts. *Frontiers in Nutrition* 7, 600215. DOI: 10.3389/fnut.2020.600215.

Fyfe, S.A., Smyth, H.E., Schmitt-Kopplin, P., Rychlik, M. and Yasmina, S. (2020b) The nutritional potential of the Australian green plum (*Buchanania obovata*). In: *Metabolomics 2020: The 16th Annual Conference of the Metabolomics Society*. DOI: 10.3389/fnut.2020.600215.

Granger, M. and Eck, P. (2018) Dietary vitamin C in human health. *Advances in Food and Nutrition Research* 83, 281–310. DOI: 10.1016/bs.afnr.2017.11.006.

Head, L. (2002) Country and garden: ethnobotany, archaeobotany and Aboriginal landscapes near the Keep River, northwestern Australia. *Journal of Social Archaeology* 2, 173–196.

Hiddins, L. (2003) *Explore Wild Australia with the Bush Tucker Man*. Explore Australia, South Yarra, Victoria, Australia.

Kahkeshani, N., Farzaei, F., Fotouhi, M., Alavi, S.S., Bahramsoltani, R. *et al.* (2019) Pharmacological effects of gallic acid in health and diseases: A mechanistic review. *Iranian Journal of Basic Medical Sciences* 22(3), 225–237. DOI: 10.22038/ijbms.2019.32806.7897.

Konczak, I., Maillot, F. and Dalar, A. (2014) Phytochemical divergence in 45 accessions of *Terminalia ferdinandiana* (Kakadu plum). *Food Chemistry* 151, 248–256. DOI: 10.1016/j.foodchem.2013.11.049.

Kosuru, R.Y., Roy, A., Das, S.K. and Bera, S. (2018) Gallic acid and gallates in human health and disease: Do mitochondria hold the key to success? *Molecular Nutrition & Food Research* 62(1), 1700699. DOI: 10.1002/mnfr.201700699.

Levitt, D. (1981) *Plants and People: Aboriginal Uses of Plants on Groote Eylandt*. Australian Institute of Aboriginal Studies, Canberra, Australia.

Mancuso, C. and Santangelo, R. (2014) Ferulic acid: pharmacological and toxicological aspects. *Food and Chemical Toxicology* 65, 185–195. DOI: 10.1016/j.fct.2013.12.024.

Mateo Anson, N., van den Berg, R., Havenaar, R., Bast, A. and Haenen, G.R.M.M. (2009) Bioavailability of ferulic acid is determined by its bioaccessibility. *Journal of Cereal Science* 49(2), 296–300. DOI: 10.1016/j.jcs.2008.12.001.

Maxwell, S.J., Brameld, K.J., Bower, C., D'Antoine, H., Hickling, S. *et al.* (2013) Baseline investigations of folate status in Aboriginal and non-Aboriginal West Australians prior to the introduction of mandatory fortification. *The Australian & New Zealand Journal of Obstetrics & Gynaecology* 53(1), 26–31. DOI: 10.1111/j.1479-828X.2012.01484.x.

Medley, P. and Bollhöfer, A. (2016) Influence of group II metals on Radium-226 concentration ratios in the native green plum (*Buchanania obovata*) from the Alligator Rivers Region, Northern Territory, Australia. *Journal of Environmental Radioactivity* 151 Pt 3, 551–557. DOI: 10.1016/j.jenvrad.2015.07.013.

Naveed, M., Hejazi, V., Abbas, M., Kamboh, A.A., Khan, G.J. *et al.* (2018) Chlorogenic acid (CGA): A pharmacological review and call for further research. *Biomedicine & Pharmacotherapy = Biomedecine & Pharmacotherapie* 97, 67–74. DOI: 10.1016/j.biopha.2017.10.064.

Pietrzik, K., Bailey, L. and Shane, B. (2010) Folic Acid and L-5-Methyltetrahydrofolate. *Clinical Pharmacokinetics* 49(8), 535–548. DOI: 10.2165/11532990-000000000-00000.

Purdie, S., Patrick, P., Nyadbi, L., Thomas, P., Fletcher, D. *et al.* (2018) *Gija Plants and Animals: Aboriginal Flora and Fauna Knowledge from the East Kimberley, North Australia*. Northern Territory Government - Department of Environment and Natural Resources, Australia.

Shirreffs, S.M., Armstrong, L.E. and Cheuvront, S.N. (2004) Fluid and electrolyte needs for preparation and recovery from training and competition. *Journal of Sports Sciences* 22(1), 57–63. DOI: 10.1080/0264041031000140572.

Slavin, J.L. and Lloyd, B. (2012) Health benefits of fruits and vegetables. *Advances in Nutrition* 3(4), 506–516. DOI: 10.3945/an.112.002154.

Smith, P.A. and Smith, R.M. (1999) Diets in transition: Hunter-gatherer to station diet and station diet to the self-select store diet. *Human Ecology* 27(1), 115–133. DOI: 10.1023/A:1018709401639.

Striegel, L., Chebib, S., Dumler, C., Lu, Y., Huang, D. *et al.* (2018) Durian fruits discovered as superior folate sources. *Frontiers in Nutrition* 5, 114. DOI: 10.3389/fnut.2018.00114.

Striegel, L., Weber, N., Dumler, C., Chebib, S., Netzel, M.E. *et al.* (2019) Promising tropical fruits high in folates. *Foods* 8(9), 363. DOI: 10.3390/foods8090363.

Tajik, N., Tajik, M., Mack, I. and Enck, P. (2017) The potential effects of chlorogenic acid, the main phenolic components in coffee, on health: a comprehensive review of the literature. *European Journal of Nutrition* 56(7), 2215–2244. DOI: 10.1007/s00394-017-1379-1.

Takeda, E., Yamamoto, H., Yamanaka-Okumura, H. and Taketani, Y. (2012) Dietary phosphorus in bone health and quality of life. *Nutrition Reviews* 70, 311–321. DOI: 10.1111/j.1753-4887.2012.00473.x.

The Plant List (2013) *Version 1.1*. Available at: http://www.theplantlist.org/ (accessed 24 February 2021).

Theobald, H.E. (2005) Dietary calcium and health. *Nutrition Bulletin* 30(3), 237–277. DOI: 10.1111/j.1467-3010.2005.00514.x.

Weaver, C.M. (2013) Potassium and health. *Advances in Nutrition* 4(3), 368S-377S. DOI: 10.3945/an.112.003533.

15 Red Bush Apple (*Syzygium suborbiculare*)

Gabriele Netzel[1,2*], Olivia Wright[1,2,3], Yasmina Sultanbawa[1,2] and Michael E. Netzel[1,2]

[1]ARC Industrial Transformation Training Centre for Uniquely Australian Foods, Queensland Alliance for Agriculture and Food Innovation, The University of Queensland, Coopers Plains, Queensland, Australia; [2]Centre for Nutrition and Food Sciences, Queensland Alliance for Agriculture and Food Innovation, The University of Queensland, St. Lucia, Queensland, Australia; [3]School of Human Movement and Nutrition Sciences, The University of Queensland, St. Lucia, Queensland, Australia

Abstract

Australian native bush food is becoming increasingly popular, with hundreds of native edible plant species available throughout the country. *Syzygium suborbiculare*, also known as red bush apple or lady apple, can be found in northern Australia and is traditionally used by Indigenous Australians. There is also increasing attention in the local media about the 'culinary potential' of this native Australian fruit. However, there is little or no information available on the nutritional and phytochemical composition or the bioactive properties of red bush apple. Therefore, this chapter will give an overview of the available literature about the traditional use, nutritional and phytochemical composition, bioactive properties and potential health benefits of *Syzygium suborbiculare* and its close relatives, *Syzygium smarangense* and *Syzygium malaccense*. Recommendations for future research to elucidate the full nutritional and bioactive potential of red bush apple are given at the end of the chapter.

Introduction

Australian native bush food is becoming increasingly popular, with hundreds of native edible plant species available throughout the country (Ahmed and Johnson, 2000). The *Syzygium* genus (Mytracea family) is globally the largest genus with 1200–1800 species, most of them originating in Australia or South-East Asia. Some of the species like *Syzygium aromaticum* are already economically important and numerous studies have demonstrated their antimicrobial, anaesthetic and antifungal properties (Cock and Cheesman, 2018). There are also several reports on the antimicrobial properties and phytochemical composition of *Syzygium australe* and *Syzygium luehmannii* (Netzel *et al.*, 2006, 2007; Konczak *et al.*, 2010; Murhekar *et al.*, 2017). Another promising species is *Syzygium suborbiculare* (Fig. 15.1), also known as red bush apple or lady apple, which can be found in northern Australia. Red bush apple, usually wild harvested, is traditionally used by Indigenous Australians and more recently also commercially utilized by indigenous enterprises. There is also increasing attention in the local media about the 'culinary potential' of this native

*Corresponding author: g.netzel@uq.edu.au

 Handbook of Phytonutrients in Indigenous Fruits and Vegetables (eds D. Sivakumar *et al.*)
DOI: 10.1079/9781789248067.0015

Fig. 15.1. Red bush apple (*Syzygium suborbiculare*).

Australian fruit (Brann, 2018, 2019; Walker, 2021). However, there is little or no information available on the nutritional and phytochemical composition, bioactive properties or health benefits of *Syzygium suborbiculare*. Other *Syzygium* species such as *Syzygium malaccense* or *Syzygium samarangense* are already studied and commercially available in Thailand, Indonesia, Malaysia and Brazil (Shu *et al.*, 2011; Nunes *et al.*, 2016). *Syzygium samarangense* is one of the few *Syzygium* species which has been planted on a commercial scale and is used as an international trading commodity (Shu *et al.*, 2011; Supapvanich *et al.*, 2011). This chapter gives an overview of the available literature about the traditional use, nutritional and phytochemical composition, bioactive properties and potential health benefits of *Syzygium suborbiculare* and its close relatives, *Syzygium smarangense* and *Syzygium malaccense*.

Botany

Syzygium suborbiculare is a species of the *Syzygium* genus within the Myrtaceae family. Generally, Myrtaceae are trees and shrubs. Their flowers are radial symmetrical with a four or five parted calyx and corolla. They produce edible berries or loculicidal capsules. The most important genera include *Eugenia* and *Syzygium*. The taxonomic identification was very difficult due to the high morphological variation at all taxonomic levels. Hence, the *Eugenia* genus has been revised and now most of the species in the Oceanian area have been reclassified as *Syzygium*. *Syzygium suborbiculare*, formerly known as *Eugenia suborbiculare* or *Carea jambosoides* or commonly known as red bush apple or lady apple, is a small evergreen tree with opposite, simple, pinnately veined leaves. The trees are fairly drought- and salt-tolerant and grow in the open forests and woodlands of Western Australia, the Northern Territory, north-east Queensland and the Cape York Peninsula (Fig. 15.2). The tree blooms from June to November and produces fruit from October to February. The large, cream-coloured flowers have up to 5 cm long stamens which grow in dense clusters. The fruit usually measures between 30 × 35 mm and 70 × 90 mm and can be depressed globular, globular or ovoid. There is only one seed with a diameter of up to 55 mm, which can be easily removed from the pericarp (Fig. 15.1). The large edible apple-like fruit is eaten mainly raw by Indigenous Australians (Mitra, 2010; Lim, 2012).

Syzygium samarangense (syn. Java apple or wax apple) is native to the Philippines, Indonesia and Malaysia, but it also grows in moist tropical areas from sea level to an altitude of up to 1200 m. *Syzygium samarangense* blooms in the early to late dry season, having white to yellowish-white flowers with a diameter of 2.5 cm. The fruits are pear shaped and approximately 4–5.5 cm in diameter. The skin is described as very thin with a waxy glossy appearance. The fruits prefer warm temperatures for normal growth and ripening, and the fruit colour can vary from white to dark-green or pink to red. Interestingly, both lower and higher temperatures have an inhibitory effect on the development of the red fruit colour. Furthermore, the quality of the fruit depends on

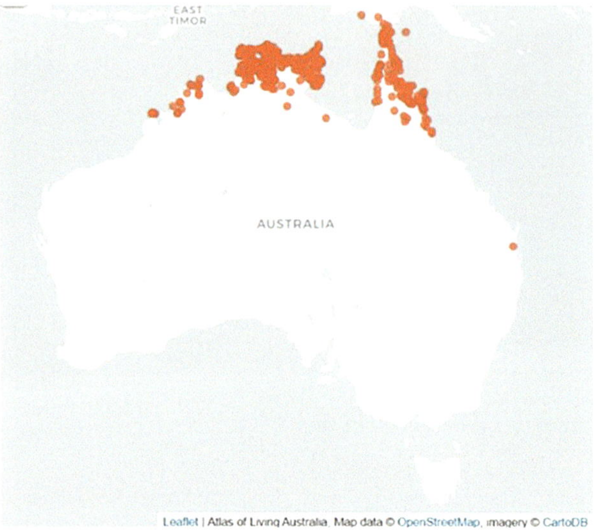

Fig. 15.2. Distribution of *Syzygium suborbiculare* in Australia (Atlas of Living Australia website. *Syzygium suborbiculare* (Benth.) T.G.Hartley and L.M.Perry).

the season and growing region. Fruits produced in the summer are bigger, whereas winter fruit have more total soluble solids and hardiness (Shu *et al.*, 2011). The pulp is described as apple-like, crisp and watery. The centre of the fruit is rich in fibre and usually seedless, although occasionally between three and six small seeds can be found (Supapvanich *et al.*, 2011; Rosnah *et al.*, 2012; Aung *et al.*, 2020).

Syzygium malaccense (syn. Malay apple, Malay rose apple, water apple) is a tree originating from South-East Asia, which grows up to 15 m. The tree bears fruit twice a year from February to March and from October to November. The succulent, pear-shaped fruits have a size of approximately 4.7 × 5.8 cm with a single seed. The smooth, waxy peel has an intense red colour when fully ripened. The soft, whitish flesh is described as juicy and sweet, similar to apples, with a characteristic rose flavour (Batista *et al.*, 2017; Fontan *et al.*, 2018).

Traditional Use

The fruits and juice of *Syzygium suborbiculare* have been used for the treatment of colds, chest congestions and earache. Stomach problems or diarrhoea are treated by infusions made from the bark or leaves. Heated leaves can also be used to stop the bleeding and swelling of wounds (Lim, 2012). Various parts of *Syzygium malaccense* have been used in traditional medicine as a remedy for cough, constipation, headache and diabetes. *Syzygium malaccense* is also known for its antibacterial, diuretic and abortifacient properties. It has been reported that an extract of *Syzygium malaccense* bark was used to treat diabetes mellitus (Bairy *et al.*, 2005), stomach ache as well as throat and mouth infections in infants (Itam and Anna, 2020). The fruit is usually eaten fresh or as hand-made products. Extracts of leaves and bark of *Syzygium samarangense* are traditionally used to treat diarrhoea, fever and rheumatic pains (Chattopadhyay *et al.*, 1998). In Malaysia, powdered dried leaves are used for the treatment of cracked tongue, whereas the roots are used to relieve itching and swelling. The fruits are usually eaten raw or cooked as a sauce (Khandaker *et al.*, 2015).

Nutritional Composition

Brand Miller (1993), the only scientific reference about the nutritional composition of *Syzygium suborbiculare* which could be found in the common scientific databases, reported only proximate levels, a few minerals/trace elements and three vitamins (Table 15.1). Nunes *et al.* (2016) published basic nutrient data of the peel and edible fruit parts of *Syzygium malaccense* grown at two different geographical locations in Brazil, whereas Batista *et al.* (2017) determined the nutritional composition of pulp, peel, seed and leaves of this species. Furthermore, the observed high ascorbic acid (vitamin C) content of up to 175 mg/100 g fresh wight (FW) in the edible parts of *Syzygium malaccense* is notable (Nunes *et al.*, 2016). Khandaker *et al.* (2015) found relatively high concentrations of trace elements in the extracts of different *Syzygium samarangense* parts. Nutritional data for the edible parts of *Syzygium suborbiculare*, *Syzygium malaccense* and *Syzygium samarangense* as reported in the literature are summarized in Table 15.1.

Phytochemicals

Most studies have investigated the leaves ('non-edible parts') and their extracts for phytochemicals. Myricetin and its glycosides, myricetin-3-rhamnoside, myricetin-3-glucoside and myricetin-arabinofuranoside have been found as the main phytochemicals/polyphenols in *Syzygium malaccense* leaves (Bairy *et al.*, 2005; Arumugam *et al.*, 2014, 2016). These phytochemicals were also found to be the main phenolic compounds in other *Syzygium* species and have shown strong antioxidant and antidiabetic properties (Tian *et al.*, 2011). However, only a few reports exist about the phytochemical composition in the edible fruits or edible fruit parts. Nunes *et al.* (2016) found cyanidin-3-glucoside as the main anthocyanin in *Syzygium malaccense*, followed by cyanidin-3,5-diglucoside and peonidin-3-glucoside (Table 15.2). Cyanidin-3-glucoside was also the main anthocyanin in *Syzygium samarangense* (Reynertson *et al.*, 2008). Anthocyanins, polyphenols like myricetin, quercetin and procyanidins (Table 15.2), have reported preventative effects against cardiovascular diseases, inflammatory processes, type 2 diabetes and cognitive decline (Wright *et al.*, 2021). Interestingly, Khandaker *et al.* (2015) found considerably more polyphenols in the fruit of *Syzygium samarangense* than in the leaves or bark. Phytochemicals and antioxidant capacity reported for the edible parts of *Syzygium malaccense* and *Syzygium samarangense* are summarized in Tables 15.2 and 15.3, respectively. However, no phytochemical or antioxidant capacity data for *Syzygium suborbiculare* could be found in the literature.

Health Benefits and Bioactive Properties

The leaf or bark extracts of *Syzygium malaccense* and *Syzygium samarangense* were used to investigate their bioactive properties and potential health benefits. Arumugam and co-workers studied the hypoglycaemic effect of *Syzygium malaccense* (Arumugam *et al.*, 2014, 2016) and could demonstrate that a myricetin-rich ethanolic leaf extract inhibited the activity of α-glucosidase similar to that of acarbose, which was used as a control. The same research group also investigated the effect of different *Syzygium malaccense* extracts on promoting adipogenesis using insulin as a positive control and showed that the myricetin-rich extract had a comparable effect to that of insulin. Myricetin and myricetin derivatives could be identified as the main phytochemicals in the leaves and leaf extracts. Chattopadhyay *et al.* (1998) reported a strong antimicrobial activity of *Syzygium samarangense* leaf extracts, whereas extracts from the stem bark only showed an inhibitory effect on *Streptococcus pyogens*. Khandaker *et al.* (2015) also investigated the antimicrobial activity of *Syzygium samarangense*, but used the edible fruits for their experiments. The red fruit varieties had a stronger antimicrobial activity than the pink or green varieties. No reports about the potential health benefits or antimicrobial activity of *Syzygium suborbiculare* were located in the literature.

Conclusions and Future Trends

Red bush apple (*Syzygium suborbiculare*) is an 'unknown' fruit regarding its nutritional composition, bioactive properties and potential health benefits. This is somewhat surprising due to its long history of traditional use by Indigenous

Table 15.1. Nutritional composition of *Syzygium suborbiculare*, *Syzygium malaccense* and *Syzygium samarangense*, data per 100 g FW (except pH).

Proximate	*Syzygium suborbiculare*	*Syzygium malaccense*		*Syzygium samarangense*				
Energy (kJ)	146							142
Moisture (g)	89.3	90.84 – 0.17	90.93 – 91.42	93.09 ± 0.24	90.66 – 92.41			90.6
Protein (g)	0.6	0.29 – 0.02	0.72 – 1.04	0.45 ± 0.014	4.00 – 5.12			0.5
Fat (g)	0.5	0.76 – 0.15	0.12 – 0.19	0.20 ± 0.008	0.1 – 0.08			0.2
Ash (g)	0.6	0.45 – 0.04	0.25 – 0.37	0.31 ± 0.006	2.77 – 3.26			0.2
Available carbohydrates (g)	10.9	4.91 – 0.52	0.52 – 0.56	2.25 ± 0.04	8.68 – 19.021	8.7 – 9.6		8.6
Total dietary fibre (g)	4.9	1.82 – 0.06	1.50 – 1.93	1.54 ± 0.02	3.92 – 5.21			1
Total Soluble Solids (°Brix)			5.46 – 6.02		5.2 – 9.0	8.43 – 8.76		
pH		3.53 – 0.03	3.75 – 3.82		4.44 – 4.63	2.89 – 3.65		
Minerals (mg)								
Ca	67						16 – 22	28
Cu	0.1						4.3 – 5.2	
Mg	32						15 – 22	13
Mn							1.8 – 2.2	
Fe	1.2						75 – 93	1
K	85						39 – 55	340
Na	48						38 – 50	25

Continued

Table 15.1. Continued

Proximate	*Syzygium suborbiculare*	*Syzygium malaccense*			*Syzygium samarangense*			
Zn	0.2						2.1 – 2.8	0.2
P	-							35
Vitamins (mg)								
Thiamine	0.04							0.02
Riboflavin	0.01							0.03
Niacin								0.03
Ascorbic acid	8		171.14 – 175.06				5.5 – 5.6	6
Reference	(Brand Miller, 1993)	(Fontan *et al.*, 2018)	(Nunes *et al.*, 2016)	(Batista *et al.*, 2017)	(Rosnah *et al.*, 2012)	(Moneruzzaman *et al.*, 2011)	(Khandaker *et al.*, 2015)	(Shu *et al.*, 2011)

[a] Sum of fructose, glucose and sucrose.

Table 15.2. Phytochemicals in *Syzygium malaccense* and *Syzygium samarangense*.

Compound	*Syzygium malaccense* (edible fruit parts; mg/g DW)	*Syzygium malaccense* (edible fruit pulp; mg/100 g FW)	*Syzygium samarangense* (edible fruit parts; mg/g DW)
Cyanidin 3,4 -diglucoside		5.519 ± 0.247	
Cyanidin 3-glucoside	$0.02 \pm 1.5 \times 10^{-3}$	42.77 ± 3.36	1.56 ± 0.01
Peonidin 3-glucoside		0.512 ± 0.025	
Isorhamnetin 3-glucoside		3.334 ± 0.205	
Isoquercitrin		0.549 ± 0.000	
Quercetin	<0.01 mg/g		$0.02 \pm 2.2 \times 10^{-3}$
Quercitrin	0.02 ± 0.01		$0.02 \pm 6.0 \times 10^{-3}$
Kaempferol 3-glucoside		1.315 ± 0.161	
Rutin	$0.02 \pm 1.4 \times 10^{-3}$	0.549 ± 0.100	<0.01 mg/g
(-)-Epicatechin gallate		1.374 ± 0.247	
(+)-Catechin		1.124 ± 0.087	
(-)-Epicatechin		1.374 ± 0.029	
Procyanidin A2		1.736 ± 0.246	
Procyanidin B1		1.311 ± 0.246	
Procyanidin B2		0.450 ± 0.001	
Ellagic acid	$0.01 \pm 1.7 \times 10^{-3}$		0.09 ± 0.01
	(Reynertson *et al.*, 2008)	(Batista *et al.*, 2017)	(Reynertson *et al.*, 2008)

FW: fresh weight; DW, dry weight.

Communities in Australia and growing attention to it as a native ingredient for Australian cuisine. Therefore, a systematic scientific approach is warranted to elucidate the full nutritional and bioactive potential of this native Australian fruit. A detailed compositional analysis including phytochemicals and potential antinutrients, *in vitro* bioaccessibility and bioactivity using state of the art gastrointestinal digestion models and cell culture assays as well as *in vivo* bioavailability and intervention studies with human subjects to substantiate bioactive properties are the suggested core activities for this approach. It should also be mentioned that a small volume of *Syzygium suborbiculare* fruits harvested in 2020 in the Northern Territory, would but for Covid, have undergone a detailed nutritional analysis as part of the research activities within the Australian Research Council funded Industrial Transformation Training Centre for Uniquely Australian Foods at the University of Queensland, Australia.

Table 15.3. Antioxidant capacity of *Syzygium malaccense* and *Syzygium samarangense*.

Antioxidant capacity assay	*Syzygium malaccense*			*Syzygium samarangense*			
DPPH	269 ± 7.66[b]	22.20–25.92[a]	627.5[b]	77.5–157[a]		283–358[b]	
TPC[c]	8.58 ± 0.12	7.64–8.03	0.159	18.0–23.8	2.85–3.27[d]	46.5–49.0	79.60 ± 2.83
FRAP[e]		0.09					
	(Reynertson *et al.*, 2008)	(Nunes *et al.*, 2016)	(Batista *et al.*, 2017)	(Reynertson *et al.*, 2008)	(Moneruzzaman *et al.*, 2011)	(Khandaker *et al.*, 2015)	(Esua *et al.*, 2017)

[a] μM trolox equivalents/g DW; [b] IC50 in μg/ml; [c] Total phenolic content in mg gallic acid equivalents (GAE)/g DW; [d] mg GAE/g FW; [e] ferric reducing antioxidant power in mM ferrous sulfate/g DW.

Acknowledgement

The authors acknowledge the Traditional Owners of the lands on which the Indigenous plant species were harvested and respect the knowledge and experience the Traditional Owners hold regarding the care, harvest, and use of these plants. This research is funded by the Australian Research Council (ARC) Industrial Transformation Training Centre (ITTC) for Uniquely Australian Foods (Grant number: IC180100045).

References

Ahmed, A.K. and Johnson, K.A. (2000) Horticultural development of Australian native edible plants. *Australian Journal of Botany* 48(4), 417. DOI: 10.1071/BT99042.

Arumugam, B., Manaharan, T., Heng, C.K., Kuppusamy, U.R. and Palanisamy, U.D. (2014) Antioxidant and antiglycemic potentials of a standardized extract of *Syzygium malaccense*. *LWT - Food Science and Technology* 59(2), 707–712. DOI: 10.1016/j.lwt.2014.06.041.

Arumugam, B., Palanisamy, U.D., Chua, K.H. and Kuppusamy, U.R. (2016) Potential antihyperglycaemic effect of myricetin derivatives from *Syzygium malaccense*. *Journal of Functional Foods* 22, 325–336. DOI: 10.1016/j.jff.2016.01.038.

Atlas of Living Australia (n.d.) *Syzygium suborbiculare* (Benth.) T.G.Hartley & L.M.Perry. Available at: https ://bie.ala.org.au/species/https://id.biodiversity.org.au/node/apni/2892768 (accessed 10 June 2021).

Aung, E.E., Kristanti, A.N., Aminah, N.S., Takaya, Y. and Ramadhan, R. (2020) Plant description, phytochemical constituents and bioactivities of *Syzygium* genus: A review. *Open Chemistry* 18(1), 1256–1281. DOI: 10.1515/chem-2020-0175.

Bairy, K.L., Sharma, A. and Shalini, A. (2005) Evaluation of the hypoglycemic, hypolipidemic and hepatic glycogen raising effects of *Syzygium malaccense* upon streptozotocin induced diabetic rats. *Journal of Natural Remedies* 5(1), 46–51.

Batista, Â.G., da Silva, J.K., Betim Cazarin, C.B., Biasoto, A.C.T., Sawaya, A.C.H.F. *et al.* (2017) Red-jambo (*Syzygium malaccense*): Bioactive compounds in fruits and leaves. *LWT - Food Science and Technology* 76, 284–291. DOI: 10.1016/j.lwt.2016.05.013.

Brand Miller, J. (1993) *Tables of Composition of Australian Aboriginal Foods / Janette Brand Miller, Keith W. James, Patricia M.A. Maggiore*. Aboriginal Studies Press, Canberra, Australia.

Brann, M. (2018) Indigenous apple crumble on the menu, as Arnhem Land communities start selling red bush apples. ABC Rural. Available at: https://www.abc.net.au/news/rural/2018-12-18/indigenous-harvest-of-red-bush-apple-maningrida-arnhem-land/10600936 (accessed 1 April 2022).

Brann, M. (2019) Indigenous community starts harvesting red bush apple that has high-end restaurants lining up. ABC Rural. Available at: https://www.abc.net.au/news/rural/2019-11-21/attica-restaurant-lines-up-for-taste-of-indigenous-bush-apple/11713876

Chattopadhyay, D., Sinha, B.K. and Vaid, L.K. (1998) Antibacterial activity of *Syzygium* species. *Fitoterapia* 69(4), 365–367.

Cock, I. and Cheesman, M. (2018) Plants of the genus *Syzygium* (Myrtaceae): A review on ethnobotany, medicinal properties and phytochemistry. In: *Bioactive Compounds of Medicinal Plants*. pp. 35–42.

Esua, J.O., Chin, N.L., Yusof, Y.A. and Sukor, R. (2017) Antioxidant bioactive compounds and spoilage microorganisms of wax apple (*Syzygium samarangense*) during room temperature storage. *International Journal of Fruit Science* 17(2), 188–201. DOI: 10.1080/15538362.2017.1285263.

Fontan, R.C.I., Sampaio, V.S., Souza, E.C., Pereira, R.G., Rodrigues, L.B. *et al.* (2018) Composition and thermophysical properties of malay rose apple pulp. *International Food Research Journal* 25(3), 1067–1073.

Itam, A. and Anna, L. (2020) Antioxidant activities, cytotoxic properties and total phenolic content of Syzygium malaccense (L.) Merr. & L.M. perry leaves extracts: A West Sumatera indonesian plant. *Pakistan Journal of Pharmaceutical Sciences* 33(1), 175–181.

Khandaker, M.M., Sarwar, M.J., Mat, N. and Boyce, A.N. (2015) Bioactive constituents, antioxidant and antimicrobial activities of three cultivars of wax apple (*Syzygium samarangense* L.) fruits. *Research Journal of Biotechnology* 10(1), 7–16. DOI: //WOS:000351641900002.

Konczak, I., Zabaras, D., Dunstan, M. and Aguas, P. (2010) Antioxidant capacity and hydrophilic phytochemicals in commercially grown native Australian fruits. *Food Chemistry* 123(4), 1048–1054. DOI: 10.1016/j.foodchem.2010.05.060.

Lim, T.K. (2012) *Syzygium suborbiculare*. In: *Edible Medicinal And Non Medicinal Plants: Volume 3, Fruits*. Springer, Dordrecht, the Netherlands, pp. 789–790.

Mitra, S.K. (2010) Important myrtaceae fruit crops. *Acta Horticulturae* 849(849), 33–38. DOI: 10.17660/ActaHortic.2010.849.2.

Moneruzzaman, K.M., Al-Saif, A.M., Alebidi, A.I., Hossain, A.B.M.S., Normaniza, O. *et al.* (2011) An evaluation of the nutritional quality evaluation of three cultivars of *Syzygium samarangense* under Malaysian conditions. *African Journal of Agricultural Research* 6(3). DOI: 10.5897/AJAR10.923.

Murhekar, S., Wright, M.H., Greene, A.C., Brownlie, J.C. and Cock, I.E. (2017) Inhibition of *Shewanella* spp. growth by *Syzygium australe* and *Syzygium luehmannii* extracts: natural methods for the prevention of fish spoilage. *Journal of Food Science and Technology* 54(10), 3314–3326. DOI: 10.1007/s13197-017-2782-6.

Netzel, M., Netzel, G., Tian, Q.G., Schwartz, S. and Konczak, I. (2006) Sources of antioxidant activity in Australian native fruits. Identification and quantification of anthocyanins. *Journal of Agricultural and Food Chemistry* 54(26), 9820–9826. DOI: 10.1021/jf0622735.

Netzel, M., Netzel, G., Tian, Q.G., Schwartz, S. and Konczak, I. (2007) Native Australian fruits — a novel source of antioxidants for food. *Innovative Food Science & Emerging Technologies* 8(3), 339–346. DOI: 10.1016/j.ifset.2007.03.007.

Nunes, P.C., Aquino, J. de S., Rockenbach, I.I. and Stamford, T.L.M. (2016) Physico-chemical characterization, bioactive compounds and antioxidant activity of Malay Apple [*Syzygium malaccense* (L.) Merr. & L.M. Perry]. *PloS One* 11(6), e0158134. DOI: 10.1371/journal.pone.0158134.

Reynertson, K.A., Yang, H., Jiang, B., Basile, M.J. and Kennelly, E.J. (2008) Quantitative analysis of antiradical phenolic constituents from fourteen edible Myrtaceae fruits. *Food Chemistry* 109(4), 883–890. DOI: 10.1016/j.foodchem.2008.01.021.

Rosnah, S., Wong, W.K., Noraziah, M. and Osman, H. (2012) Chemical composition changes of two water apple (*Syzygium samaragense*). *International Food Research Journal* 19(1), 167–174.

Shu, Z.H., Shiesh, C.C. and Lin, H.L. (2011) Wax apple (*Syzygium samarangense* (Blume) Merr. and LM Perry) and related species. In: Yahia, E.M. (ed.) *Postharvest Biology and Technology of Tropical and Subtropical Fruits, Vol 4: Mangosteen to White Sapote*. pp. 458–473.

Supapvanich, S., Pimsaga, J. and Srisujan, P. (2011) Physicochemical changes in fresh-cut wax apple (*Syzygium samarangenese* [Blume] Merrill & L.M. Perry) during storage. *Food Chemistry* 127(3), 912–917. DOI: 10.1016/j.foodchem.2011.01.058.

Tian, L.-W., Xu, M., Wang, D., Zhu, H.-T., Yang, C.-R. *et al.* (2011) Phenolic constituents from the leaves of *Syzygium forrestii* Merr. and Perry. *Biochemical Systematics and Ecology* 39(2), 156–158. DOI: 10.1016/j.bse.2011.01.014.

Walker, K. (2021) Red bush apples are trending in the NT, but what are they? SBS. Available at: https://www.sbs.com.au/food/article/2020/05/13/red-bush-apples-are-trending-nt-what-are-they (accessed 1 April 2022).

Wright, O.R.L., Phan, A.D.T., Hong, H.T., Netzel, G., Sultanbawa, Y. *et al.* (2021) Bioactive anthocyanins in selected fruits – a foodomics approach. In: Cifuentes, A. (ed.) *Comprehensive Foodomics*, Vol. 2. Elsevier, Oxford, UK, pp. 77–104.

16 Passion Fruit (*Passiflora edulis*)

H.A.C.O. Hettiarachchi, N.N.G. Chiranthika, G. Janarny and K.D.P.P. Gunathilake*

Department of Food Science and Technology, Faculty of Livestock, Fisheries and Nutrition, Wayamba University of Sri Lanka, Makandura, Gonawila, (NWP), Sri Lanka

Abstract

Passion fruit, *Passiflora edulis*, is a tropical and semi-tropical fruit with excellent biological properties such as flavour, nutritional value and pharmacological benefits. It has good potential for processing and value addition, due to its attractive properties and has been traditionally used in folk medicine to eliminate various diseases. The fruit is a good source of nutrients such as vitamins C and A, minerals, organic acid, soluble solid, dietary fibres and other biological compounds such as polyphenols and carotenoids. Tender leaves of the passion fruit are also edible and used as leafy greens, having good nutrient content such as minerals, vitamin C and polyphenols. Recent *in vitro*- and *in vivo*-based studies have shown that the fruits and leaves of the passion fruit have antioxidant, anti-inflammatory, antidiabetic, anticancer and antihypertensive properties. This chapter congregates the functional and medicinal properties of passion fruit.

Botany

Passiflora edulis, commonly known as passion fruit, grenadelle and weldodam belongs to the domain Eukaryota, kingdom Plantae (*clade* – angiosperm), order Malpighiales, family Passifloraceae, genus *Passiflora* and species *P. edulis*. There are two varieties, *Passiflora edulis* Sims (purple variety) and *Passiflora edulis* var. *Flavicarpa* Degenerer (yellow variety) (Taïwe and Kuete, 2017). *P. edulis* is a herbaceous perennial climber, originating in South America and commonly cultivated in tropical and subtropical regions. In Sri Lanka, the yellow variety is mostly grown in the lowland regions; the purple variety is grown in the hilly regions, upcountry, and a hybrid variety of both yellow and purple, known as 'Rahangala' hybrid is grown in common (Hapuarachchi, 2016). It is a tangled vine, with a stem up to 15 m in height, attaching with axillary tendrils (Fig. 16.1). Deeply three-lobed leaves are slightly leathery, having a glossy green above and paler green below. Solitary flowers are up to 7 cm in diameter, with white petals having five stamens and three styles with large stigmas (Taïwe and Kuete, 2017). Fruits are ovoid to spherical, with a diameter of 4–5 cm (Fig. 16.1). Immature fruits are greenish and once matured and ripened they become either purple or yellow in colour according to the variety. The flesh of the fruit is juicy, yellow to light orange and contains numerous seeds covered with two membranes in which the juice is contained. This plant is propagated either by seed or vegetatively (Hapuarachchi, 2016). The purple fruit is considered less acidic and having the best flavour compared to the yellow variety. The fast-growing vine can bear fruits after one to three years.

*Corresponding author: kdppgunathilake@yahoo.com

 Handbook of Phytonutrients in Indigenous Fruits and Vegetables (eds D. Sivakumar *et al.*)
DOI: 10.1079/9781789248067.0016

Fig. 16.1. *Passiflora edulis*: (a) flower having five stamens and three styles with large stigmas; (b) leaf; (c) fruit; (d) fruits hanging on the vine; (e) ripened fruit – yellow variety; f) dried seeds with fleshy arils removed (Fern, 2019).

Traditional Use

P. edulis is widely used as a folk medicine because of its sedative and tranquilizer activities (Barbosa *et al.*, 2008; Taïwe and Kuete, 2017). It is known to boost the iron content in the body since it contains a lot of iron and is used to treat iron deficiency and anaemia. It is used to prevent cancer owing to the rich antioxidants in the fruit. The pulp of the fruit acts as a digestive stimulant and tonic. Oral intake reduces the pain and stiffness of osteoarthritis in the knee (Bhat and Paliyath, 2016). Native Americans have used passion fruit flowers to cure wounds and bruises. The leaves of passion fruit are also used for injuries and boils (Wijeratnam, 2016). In ethnomedicine, it is recommended for laxative purposes. It relieves asthma and the high fibre content in the fruit protects the body from colon cancers and is highly beneficial for digestion. The ripened fruit is also used to increase urine acidity, thereby facilitating the body's metabolism. In traditional medicine, it is used in treating oedema, gout, diarrhoea, dysentery and ear diseases. Not only the fruits but also the leaves, stems and roots are used for various medical treatments such as inflammation, insomnia and pain. The leaves have exhibited antinociceptive, antitumour, antimicrobial and antioxidant effects (Patel, 2009; Taïwe and Kuete, 2017). *P. edulis* is commonly used in the treatment of anxiety and sleep disorders. The teas brewed using the leaves, roots and flowers of *P. edulis* have been used to settle nervousness and encourage good sleep (Wijeratnam, 2016). It is also used to cure many infectious disorders caused by bacteria, fungi, viruses, mycobacteria and protozoa. *P. edulis* is used to increase appetite and digestion. Further, *P. edulis* extracts are used in combination with other herbal medicines to treat depression. The therapeutic effects of *P. edulis* are are not limited to the above. The leaf extracts are used in treating symptoms of alcoholism, anxiety and migraine. A drink prepared from the flower has been shown to be effective in treating asthma, bronchitis and whooping cough. Passion fruit seed oil has been used as a stimulating lubricant and massage oil (Zibadi and Watson, 2004; Taïwe and Kuete, 2017). The leaves are boiled using a small amount of water and the collected extract drunk to treat dysentery. The traditional uses of *P. edulis* include treatment for skin inflammation and wrinkle prevention. In some countries, it is used in the treatment of dysmenorrhea, epilepsy, insomnia, neurosis and neuralgia. Many people in early civilizations used *P. edulis* to cure hysteria, neurasthenia, burns and haemorrhoids. It is even used to reduce menopausal symptoms and *Helicobacter pylori* infection (Barbosa *et al.*, 2008; Ingale and Hivrale, 2010; Taïwe and Kuete, 2017).

Nutritional Composition

Table 16.1 shows the nutritional composition of *P. edulis* per 100 g of the edible portion (in fresh weight – FW). Among total carbohydrates present in the fruit, dietary fibre accounts for 10.4 g/100 g of the edible portion (in FW) which is considered significantly high and therefore the fruit is used as a laxative to prevent constipation and also to treat colon cancers. Fibre is considered essential for the control of bowel movement to reduce the exposure time of the colon to any toxins (Wijeratnam, 2016). It is also rich in potassium which makes passion fruit effective in protecting against high blood pressure (Hapuarachchi, 2016). Furthermore, potassium is an essential constituent to relax the tension of blood vessels, acting as a vasodilator to promote blood flow. The amount of dietary fibre present in *P. edulis* pulp and seeds also helps to reduce low-density lipoprotein (LDL) cholesterol levels in the body and increase high-density lipoprotein (HDL) cholesterol levels. The high potassium content of *P. edulis* (7% of recommended daily allowance – RDA) will exert great cardiac benefits to the consumer. It has a high amount of iron and is a rich source of vitamin C. The iron present is in the form of non-haem, plant-based iron which is of great benefit for vegetarians who do not consume the proper amount of iron through animal food. The high amount of vitamin C facilitates iron absorption. Therefore, the iron absorption into the body is well boosted which is why *P. edulis* is commonly used for treating anaemia and iron deficiency. It is very high in vitamin A content, which helps to maintain cell growth, healthy skin and good eyesight.

Table 16.1. Nutritional composition of *P. edulis* (per 100 g of edible portion in FW) (Thokchom and Mandal, 2017).

Nutrients	Amount	% Daily Value
Energy	97 calories	5%
Proteins	2.2 g	4%
Total carbohydrates	23.38 g	18%
Dietary fibre	10.40 g	27%
Sugar	11 g	
Total fat	0.7 g	
Saturated fat	0.1 g	3%
Polyunsaturated fat	0.4 g	0%
Monounsaturated fat	0.1 g	
Cholesterol	0 mg	0%
Sodium	0 mg	0%
Potassium	348 mg	7%
Vitamin A	1274 IU	43%
Vitamin C	30 mg	50%
Vitamin E	0.02 µg	<1%
Vitamin K	0.7 mg	0.5%
Vitamin D	0 mg	0%
Folates	14 µg	3%
Niacin	1.500 mg	9%
Pyridoxine	0.100 mg	8%
Riboflavin	0.130 mg	10%
Thiamin	0.00 mg	0%
Calcium	12 mg	1.2%
Iron	1.60 mg	20%
Copper	0.086 mg	9.5%
Phosphorus	68 mg	10%
Selenium	0.6 µg	1%
Zinc	0.10 mg	1%
Magnesium	29 mg	7%

[a] per cent daily values are based on a 2000 calorie diet.

The fat-soluble carotenoids present in the fruit help the body to produce vitamin A. It also contains riboflavin and niacin along with minerals such as magnesium, copper and phosphorus. Magnesium is important for proper gene expression and carbohydrate metabolism. It is necessary to regulate heartbeat and nerve transmissions and for the development of bones. Phosphorus is also required for bone development and passion fruit contains a significant amount, which will assist in the prevention of bone deterioration with ageing (Wijeratnam, 2016).

Phytochemicals

The presence of phytochemically characterized primary constituents in *P. edulis* has been reported by many studies as indicated in Table 16.2. These constituents include flavonoids, phenols, cyanogenic compounds, alkaloids, vitamins, minerals, glycosides and terpenoid compounds. The esters listed in the table are responsible for the flavour and the appetizing smell of the fruit.

Antinutritive Compounds

Antinutritive compounds are chemical substances present in most food, and reduce the availability of macro- and micronutrients to function at their maximum; some are also poisonous to humans. However, low levels of some antinutritive compounds may provide health benefits such as reducing blood glucose, cholesterol and triglyceride levels (Natesh *et al.*, 2017). There are common antinutritive compounds in foods, such as phytates, tannins, oxalates, polyphenols, protease and cyanide. The antinutritive activity of tannins inhibits amylase, lipase, trypsin and chymotrypsin activities, reduces the protein quality of food and disturbs iron absorption. Phytates negatively affect mineral absorption and oxalates form insoluble oxalate 'stones' by binding with calcium, iron and magnesium. Cyanide is considered a neurotoxin that affects the heart and central nervous system (Wagasu *et al.*, 2016). Different parts of *P. edulis* contain antinutritive compounds at different levels. Tannins, polyphenols, phytates, oxalates, saponins, alkaloids and flavonoids were determined as antinutrients in the seeds, peel and pulp of *P. edulis* by Adeyeye and Aremu (2017). The highest levels of tested antinutritive compounds were found in the seeds, while the fruit pulp contained lower levels. Maniyan *et al.* (2015) studied the antinutritive compounds in fruit peels and reported that peel of *P. edulis* contains phytic acid, tannin and protease at the level of 0.1 μg/ml, 1.43 μg/ml and 0.12 mg/ml, respectively. Wagasu *et al.* (2016) reported that rind, seed and juice of *P. edulis var. flavicarpa* contain oxalate, tannin, cyanide and phytate as antinutrient factors in lower levels and the amounts were significantly different ($p<0.05$) in three parts.

Implications of Food Processing on Nutrients and Phytochemicals

Almost all foods are processed in different ways before they are consumed. Processing is necessary to eliminate undesirable microorganisms and extend the shelf life of the food. However, the nutritive value of the particular food is changed due to the processing as well as during harvesting and storing. *P. edulis* is a climacteric fruit. Common *P. edulis* postharvest quality losses are shrinkage and darkening of peel, moisture loss and microbial attacks resulting from higher respiration and transpiration after harvesting (Chen *et al.*, 2018). Further, it has been reported that storage temperature is the main factor that affects postharvest quality losses in *P. edulis* resulting in variations in physicochemical and sensory parameters (Kishore *et al.*, 2011). Kishore and others have studied the changes of physicochemical properties and sensory attributes with different storage temperatures and storage times. Physicochemical parameters such as weight loss, total soluble solids (TSS), titrable acidity, reducing and non-reducing sugars, and sensory parameters such as flavour intensity, sourness, sweetness, off flavour and overall acceptance were analysed and it was concluded that a storage temperature of 8°C showed optimum results for retaining initial

Table 16.2. Phytochemical constituents of *P. edulis*.

Phytochemical	Plant part	Reference
Flavonoids	Fruit pulp	Zeraik and Yariwake, 2010; Taïwe and Kuete, 2017
Isoorientin and total flavonoids		
C-glycosylflavones-schaftoside, isoschaftoside, orientin, isovitexin, luteolin-6-C-chinovoside, and luteolin-6-C-fucoside		Pereira and Vilegas, 2000; Dhawan *et al.*, 2004; Taïwe and Kuete, 2017
Apigenin, benzoflavone, homoorientin, kaempferol, lucenin, luteolin, passiflorine (named after the genus), quercetin ,rutin, saponarin, and vicenin and vitexin	Flower	Pereira and Vilegas, 2000; Dhawan *et al.*, 2004; Taïwe and Kuete, 2017
Chrysin		
Hydrocarbon nonacosane and the anthocyanidin pelargonidin-3-diglycoside		Pereira and Vilegas, 2000;Dhawan *et al.*, 2004; Taïwe and Kuete, 2017
Alkaloids		
Harmane, harmaline, harmalol, harmine, and harmol	Fruit pulp	Pereira and Vilegas, 2000; Dhawan *et al.*, 2004; Taïwe and Kuete, 2017
Terpenoids		
4-Hydroxy-β-ionol, 4-oxo-β-ionol, 4-hydroxy-7,8-dihydro-β-ionol, 4-oxo-7,8-dihydro-β-ionol, 3-oxo-α-ionol, isomeric 3-oxo retro-α-ionols, 3-oxo-7,8-dihydro-α-ionol, 3-hydroxy-1,1,6- trimethyl-1,2,3,4-tetrahydronaphthalene vomifoliol and dehydrovomi-foliol, terpene alcohols linalool, and α-terpeneol, terpenediols (E) and (Z)-2,6-dimethyl-octa-2,7-diene-1,6-diol, 2,6-dimethyl-octa-3,7-dien-2,6-diol, 2,6-dimethyl-1,8-octanediol, 2,6-dimethyl-octa-1,7-diene-3,6-diol, ionol derivatives oxygenated in position 3, and 2,5-dimethyl-4-hydroxy-3-(2H)-furanone (furaneol)	Fruit pulp	Chassagne *et al.*, 1996; Taïwe and Kuete, 2017
Ionones I and II		Näf *et al.*, 1977; Taïwe and Kuete, 2017
Glycosides		
A cyclopropanetriterpene glycoside, named passiflorine	Leaves	

Continued

Table 16.2. Continued

Phytochemical	Plant part	Reference
Flavonoid glycosides, namely, luteolin-6-C-chinovoside, luteolin-6-C-fucoside; cyclopentenoid cyanohydrin glycosides passicapsin and passibiflorin; cyanogenic glycosides passicoriacin, epipassicoriacin and epitetraphyllin B cyanogenic-β-rutinoside{(R)–mandelonitrile-α-l-rhamnopyranosyl-β-d-glucopyranoside} amygdalin, prunasin, man-delonitrilerhamnopyranosyl-β-d-glucopyranoside, sambunigrin; 6-O-α-l-arabinopy-ranosyl-β-d-glucopyranosides of linalool, benzyl alcohol, and 3 methyl-but-2en-1-ol; β-d-glucopyranoside and 6-O-α-l-rhamnopyranosyl-β-d-glucopyranoside of methyl salicy-late and β-d-glucopyranoside of eugenol		Dhawan *et al.*, 2004; Zucolotto *et al.*, 2012;Taïwe and Kuete, 2017
Polyphenolic compounds	Seeds	Wijeratnam, 2016
Piceatannol and scir-pusin B		
Miscellaneous and phytoconstituents		
Organic acids including formic, butyric, linoleic, linolenic, malic, myristic, oleic, and palmitic acids		
Phenolic compounds		
Amino acid α-alanine		
Esters like ethyl butyrate, ethyl caproate, n-hexyl butyrate, and n-hexyl caproate		
Sugars, - d-fructose, d-glucose, and raffinose		
Enzymes - catalase, pectin methylesterase, and phenolase		Duke, 2008; Ingale and Hivrale, 2010; Taïwe and Kuete, 2017
Coumarins (e.g., scopoletin and umbelliferone), maltol, phytosterols (e.g., lutenin), and cyanogenic glycosides		
Coumarins (e.g., scopoletin and umbelliferone), maltol, phytosterols (e.g., lutenin), and cyanogenic glycosides	Flower	Duke, 2008; Taïwe and Kuete, 2017
Cycloartanetriterpenoids and their saponins, including two new cycloartanetriterpenoidsapo-nins named cyclopassifloside XII (1) and XIII (2), together with six known cycloar-tanetriterpenoids, cyclopassifloic acids B and E, cyclopassiflosides II, VI, IX, and XI		Wang *et al.*, 2013; Taïwe and Kuete, 2017

physicochemical and sensory parameters of *P. edulis* up to 21 days. In addition to cold storage, the effect of high O_2 atmosphere packaging for storage on postharvest quality implications was studied by Chen *et al.* (2018). High O_2 atmosphere packaging was introduced as a proper and novel storage method for *P. edulis* since it prevented the respiration and shrinkage of peel, reduced the loss of vitamin C and TSS content while increasing the polyphenol (phenolic acid and flavonoid) content of the fruit.

Fruit juice, cordial, fruit concentrates, nectar, ice cream, yogurt, jelly and jam are some of the commercially available products from *P. edulis*. When processing these products, heat treatments (pasteurization) are applied. The temperatures are set at 80–95°C or beyond in different products. *P. edulis* is a good source of micronutrients and polyphenols. However, those vitamins and polyphenols are sensitive to heat and deterioration occurs during processing. Atzingen *et al.* (2013) reported that the most degraded vitamins due to heating are vitamin A, vitamin C, folic acid and thiamin. Capanoglu *et al.* (2013) conducted a study to evaluate polyphenol losses during the processing of grape juice concentrate. It was concluded that polyphenol losses occurred in every processing step (pressing, pasteurizing, clarifying, filtering and concentrating), and this resulted in a large loss of polyphenols in the final product. The total phenolic content of raw grapes was 1619 mg GAE/100 g dry weight (DW) and in the final product it was reduced up to 252 mg GAE/100 g DW, directly reducing the antioxidant activity. Gunathilake *et al.* (2018) studied the relationship between types of cooking methods and changes in polyphenols, carotenoids and antioxidant activity of some edible leaves including *P. edulis*. The reduction of polyphenol content in *P. edulis* leaves compared to raw leaves by the treatments of boiling, steaming and frying was 55.9%, 88.5% and 76.3%, respectively. Flavonoid content was also significantly ($p<0.05$) reduced by 83.5% after boiling and 71.2% after steaming when compared to flavonoid content in fresh *P. edulis* leaves. However, the reduction of nutrients and phenolic compounds in foods depends on the processing steps and the nature of the food and it is very important to investigate and apply processing methods that minimally affect the reduction of nutrient quality.

Bioactive Properties

Extracts of *P. edulis* from fruit, leaves or seed show a wide range of bioactive properties which promise to improve human health and well-being (Table 16.3).

Antioxidant activity

Antioxidant activity is defined as the ability to inhibit or eliminate free radical action and reduce oxidation stress in the human body (He *et al.*, 2020). There are large numbers of studies that prove the antioxidant activity of *P. edulis*. Rotta *et al.* (2019) reported that the fruit pulp of *P. edulis* is a good source of phenolic compounds such as vanillic acid and quercetin. Further *in vitro* antioxidant activity has been determined by 2,2'-azino-bis(3-ethylbenzothiazoline-6-sulfonic acid (ABTS) and 2,2-diphenyl-1-picrylhydrazyl (DPPH) radicals, and fruit pulp of *P. edulis* has potent antioxidant capacity and radical scavenging activity. Another study showed polyphenol content of *P. edulis* fruit as 286.6 mg gallic acid

Table 16.3. Bioactive properties from compounds extracted from *P. edulis*.

Bioactive property	Plant part	Compound	References
Antioxidant activity	Pulp	Phenolic compounds (vallinic acid, quercetin)	Septembre-Malaterre *et al.*, 2016; Rotta *et al.*, 2019
	Leaves	Polyphenols	Phamiwon and John, 2017; Thomas *et al.*, 2019
Antidiabetic activity	Leaves	Flavonoid glycosides	Cesar *et al.*, 2020
	Peel		Salgado *et al.*, 2010
Anti-inflammatory activity	Peel	Dietary fibre and polyphenols	Baú *et al.*, 2016
	Leaves		Beatriz *et al.*, 2007; Baú *et al.*, 2015
Antihypertensive activity	Rind	γ-aminobutyric acid	Ichimura *et al.*, 2006
Hypolipidemic activity	Rind and fruit juice		Barbalho *et al.*, 2011
Anticancer effect	Pulp and leaves	Bioactive polyphenols and polysaccharides	Aguillón *et al.*, 2018
	Leaves	Flavonoids, cardiotonic glycosides and alkaloids	(Ramirez *et al.*, 2019)
	Seed cake	Long and medium-chain fatty acids	Mota *et al.*, 2018

equivalent/100 g DW basis and 64% of DPPH, reducing antioxidant activity, resulting in a higher level compared to tested litchi, banana, papaya and mango samples (Septembre-Malaterre *et al.*, 2016). *P. edulis* leaves are also considered as important natural antioxidant sources and their antioxidant activity was reported as 26.23 µg/ml for DPPH assay, 20.27 µg/ml for ABTS assay and 711.01 µm/g in DW basis for FRAP by Phamiwon and John (2017). The study conducted by Thomas *et al.* (2019) stated that aqueous leaf extract of *P. edulis* had an IC50 value of 1185.54 µg/ml which showed a higher level of antioxidant activity.

Antidiabetic activity

The antidiabetic activity of *P. edulis* leaf extract has been determined by Cesar *et al.* (2020) using a rat model. They reported that leaf extract has an antihyperglycaemic activity which may have a therapeutic effect in preventing chronic type 2 diabetes mellitus. In addition, flavonoid glycosides in leaf extract were associated with that therapeutic property. Peel extract of *P. edulis* has been shown to produce a lowering of blood glucose by 59% and an increase in hepatic glycogen level by 71% (Salgado *et al.*, 2010).

Anti-inflammatory activity

A previous study determined the anti-inflammatory activity of *P. edulis* peel, using a rat model analysis and it was reported as the peel extract constituents affected to reduce the expression of cytokine in the pro-inflammatory stage in the rat intestine. Further, it was found that the peel dietary fibre and polyphenols were associated with the same therapeutic effect (Baú *et al.*, 2016). The anti-inflammatory activity of *P. edulis* leaf extract was studied by Baú *et al.* (2015) and Beatriz *et al.* (2007), who reported the extract could reduce colon pro-inflammatory condition in rats by decreasing the levels of IL-1βf and TNF.

Antihypertensive and hypolipidemic activity

Ichimura *et al.* (2006) revealed the antihypertensive activity of *P. edulis* rind extract and it has been concluded γ-aminobutyric acid (GABA) contained in peel extract might associate with that antihypertensive activity. The potential of preventing dyslipidemia (hypolipidemic activity) using *P. edulis* rind extract and fruit juice to maintain cholesterol levels, lower LDL and triglyceride levels and increase HDL levels has been evaluated by Barbalho *et al.* (2011).

Anticancer/antitumour activity

The anticancer activity of *P. edulis* fruit pulp and leaves has been studied. Both pulp and leaf extracts have been shown to have pro-apoptotic activity against liver cancer in an *in vitro* model since they contain bioactive polyphenols and polysaccharides (Aguillón-Osma *et al.*, 2019). Ramirez *et al.* (2019) studied the antitumour effect of *P. edulis* leaf extract and reported it had a chemopreventive effect against human colon cancer. They suggested that the major associating compounds were flavonoids, cardiotonic glycosides and alkaloids. Seed cake is a by-product of *P. edulis* and was subjected to a study of its anticancer properties by Mota *et al.* (2018). It was suggested that the presence of long-chain and medium-chain fatty acids in seed cake extract provide anti-tumour activity (tested *in vivo* using a rat model).

Bioaccessibility and Bioavailabilityng

Numerous phytonutrients, as well as trace elements in *P. edulis*, have been reported to deliver significant health benefits to consumers. However, these effects depend on the bioaccessibility and bioavailability of the bioactive compounds affected by the food matrix and physiological conditions, such as pH and enzymes (Campos-Vega *et al.*, 2015). Aguillón-Osma *et al.* (2019) reported that the phenolic content of *P. edulis* juice extract is affected greatly throughout the gut. It was also noted that ascorbic acid and kaempferol were bioaccessible in oral and gastrointestinal phases. Flavonoids such as quercetin were not available in the small intestinal stage, however lower content of chlorogenic acid was available in the intestinal phase.

Table 16.4. Compounds in *P. edulis* responsible for different bioactivities.

Bioactivity	Associated compounds	References
Antioxidant	Isoorientin, apigenin, vitexin, quercetin,	He *et al.*, 2020
Anti-inflammatory	Luteolin-8-*C*-*β*-digitoxopyranoside, apigenin, vitexin, quercetin	Hu *et al.*, 2018; He *et al.*, 2020
Nephroprotection	Saponins, tannins	Nerdy and Ritarwan, 2019
Neuroprotection	Thiol and thioester compounds, cycloartane triterpenoids	Tal *et al.*, 2016; He *et al.*, 2020
Antidepressant	Cycloartane triterpenoids	He *et al.*, 2020
Antiproliferative	Harmine	He *et al.*, 2020
Vasorelaxation	Piceatannol	He *et al.*, 2020

Health Benefits

P. edulis has traditionally been used by different communities to treat diverse conditions. *P. edulis* juice has been used as a cardiac tonic, to treat asthma, whooping cough and urinary infections, and to stimulate digestion. Recent research has shown that *P. edulis* is a good source of vitamin A, C, folic acid, niacin and minerals such as calcium, iron and potassium (Ng *et al.*, 2011). Traditional Chinese records (*ZhongHuaBenCao*/中华本草) recommend consumption of purple *P. edulis* at a dosage of 10 to 15 g could treat cough, insomnia, dysentery, dysmenorrhea and arthralgia (He *et al.*, 2020). Various researchers have investigated different health benefits that could be exerted by *P. edulis* and Table 16.4 lists some bioactive compounds that have exerted different bioactivities. As documented previously, *P. edulis* peel extract has demonstrated a dose-dependent hepatoprotective and nephroprotective activity in albino rats. These extracts were able to substantially reduce the level of the liver markers alanine transaminase and aspartame transaminase, and kidney biochemical markers urea and creatinine (Nerdy and Ritarwan, 2019). Aqueous extracts of *P. edulis* pulp showed strong dose-dependent neuroprotective effects in both *in vivo* and *in vitro* assays. This neuroprotective activity was expressed by controlling cell death induced by glutamate, mitochondrial depolarization and glutathione depletion, in the medium of cultured HT4 neurons. It was also reported that supplementation of fruit extract was able to improve the survival rate of dopaminergic neurons by 60% in mice injected with the neurotoxin (Tal *et al.*, 2016). Studies have also shown that these fruits are a good remedy to treat anxiety, headache, abdominal pain and epilepsy. Dry extracts have demonstrated the ability to control lipid and glucose levels. *P. edulis* has also been reported to alleviate inflammation by inhibiting pro-inflammatory cytokines and their mediators. Compounds such as the C-glycosyl flavonoids vicenin, orientin, chrysin, vitexin and kaempferol in *P. edulis* peel flour were able to reduce TNF-α, IL-1ß, IL-6, IL-12 and IL-17, and exhibit anti-inflammatory activity. The polyphenol piceatannol taken from extracts of *P. edulis* was able to induce apoptosis, alter the protein levels of p53, p21 and cyclin D1 and cause G0/G1 cell cycle arrest. These mechanisms greatly contributed to the anticancer activity in transgenic adenocarcinoma of the mouse prostate model and this revealed the health importance of *P. edulis* fruit (Kido *et al.*, 2020). In another study, *P. edulis* juice was able to efficiently reduce the number, size and invasiveness of transformed foci in a BALB/c 3T3 neoplastic transformation model and induce caspase-3 in the mammalian cell line MOLT-4, facilitating the changes in transformed foci by inducing apoptosis (Ng *et al.*, 2020). Passiflin, which is a bioactive protein extracted from seeds of *P. edulis*, specifically inhibits the fungus *Rhizoctonia solani* and exhibits antiproliferative activity against breast cancer cells. Administration of *P. edulis* extract for Alzheimer's disease in rats has improved brain function, with a reduction in the level of malondialdehyde in plasma and tissue. It has also improved neurobehavioural changes with the modulation of acetylcholinesterase in the serum, thus exerting an efficient neuroprotective mechanism. The *in vitro* and *in vivo* findings of Kumar *et al.* (2019) have revealed that *P. edulis* fruits are a good dietary source for the prevention of ulcers. This was evidenced by the inhibition of release of phosphate ions in the stomach and hydrogen peroxide-induced lipid peroxidation in gastrointestinal parietal cell homogenate. Studies also have reported that administration of *P. edulis* seed extracts to treat melanoma cells inhibited melanogenesis and increased the content of total soluble collagen in dermal fibroblast cells.

Novel Products/Formulations

Existing evidence for the health-promoting properties of different parts of *P. edulis* has led to the research and development of novel products in the food, pharmaceutical and cosmetic industries. He *et al.* (2020) reported that a *P. edulis* drink in combination with Chinese medicines such as *Lyciumbarbarum* and *Dendranthema morifolium* could act as an immunity booster and vitamin supplement. Researchers also suggest that *P. edulis* seed oil could be used for the development of tea cream with whitening and anti-wrinkle effects (He *et al.*, 2020). Polyphenol-rich tea drinks are susceptible to form precipitates, referred to as tea cream, caused by the interactions of catechin, proteins, and other polyphenols. This can also be used as functional food. *P. edulis* fibres added to skimmed yogurt increased the content of short-chain fatty acids and polyunsaturated fatty acids. These fibres were in turn able to increase the conjugated linolenic acid content. López-Vargas *et al.* (2014) recommended that the incorporation of 2.5% of *P. edulis* albedo to pork burgers could improve the cooking characteristics of the burger and make it a source of dietary fibre. Fermentation of *P. edulis* pulp with *Lactobacillus reuteri* and microencapsulation by spray-drying has resulted in the production of novel food products using powdered fruit pulp, rich in probiotics (Santos Monteiro *et al.*, 2020). The addition of *P. edulis* pulp as a flavour enhancer to yogurt has improved the nutritional value of the product by increasing the protein and mineral content along with a reduction in fat and carbohydrate. One study recommends the use of 10 to 20% of *P. edulis* pulp as a flavour enhancer for yogurt to produce a novel healthy dairy product commercially (Mbaeyi-Nwaoha and Ezeoke, 2019). Reports have also suggested the use of *P. edulis* peel flour in the formulation of bread which would enrich the fibre content. Gluten-free cookies have been developed using *P. edulis* flour combined with cassava flour. As reported by Garcia *et al.* (2019), the formula with the higher percentage of *P. edulis* flour has produced a cookie with higher dietary fibre content and thus a functional and nutritious product.

Conclusions and Future Trends

Passiflora edulis and its bioactives offer remarkable potential for research in the areas of chemistry, biotechnology and pharmacology. The fruit of *P. edulis* is rich in dietary fibre, potassium, iron, magnesium and phosphorus, as well as being a rich source of vitamins C and A, indicating the species exerts great cardiac benefits and may be an asset in the prevention of bone deterioration with ageing. Recent *in vitro-* and *in vivo*-based studies have shown that the fruits and leaves of *P. edulis* have antioxidant, anti-inflammatory, antidiabetic, anticancer and antihypertensive properties. *P. edulis* can also be used in traditional medicine systems as a sedative and anxiolytic. Based on various findings related to this species, its biological activities seem to be mainly related to the content of flavonoids, which are main secondary metabolites. *P. edulis* needs to be investigated for further possible exploitation in the nutraceutical and pharmaceutical industries regarding its health-promoting qualities or as a horticultural crop.

References

Adeyeye, E.I. and Aremu, M.O. (2017) Chemical composition of the raw fruit coat, seed and pulp of passion fruit (*Passiflora edulis*). *FUW Trends in Science & Technology* 2, 334–341.

Aguillón, J., Arango, S.S., Uribe, D.F., Loango, N. and Ramirez, V. (2018) Cytotoxic and apoptotic activity of extracts from leaves and juice of Passiflora edulis. *Liver Research, Disorders & Therapy* 4(2), 67–71. DOI: 10.15406/jlrdt.2018.04.00102.

Aguillón-Osma, J., Luzardo-Ocampo, I., Cuellar-Nuñez, M.L., Maldonado-Celis, M.E., Loango-Chamorro, N. *et al.* (2019) Impact of in vitro gastrointestinal digestion on the bioaccessibility and antioxidant capacity of bioactive compounds from Passion fruit (*Passiflora edulis*) leaves and juice extracts. *Journal of Food Biochemistry* 43(7), e12879. DOI: 10.1111/jfbc.12879.

Atzingen, M.C.B.C., Oliveira, L.S., Pinto-e-Silva, M.E.M. and Soares, R.A.M. (2013) Vitamin retention in vegetables submitted to heat treatment. *Nutrition & Food Science* 43(5), 505–512. DOI: 10.1108/NFS-12-2012-0126.

Barbalho, S.M., Damasceno, D.C., Spada, A.P.M., Lima, I.E.D.R.N., Araújo, A.C. *et al.* (2011) Effects of *Passiflora edulis* on the metabolic profile of diabetic Wistar rat offspring. *Journal of Medicinal Food* 14(12), 1490–1495. DOI: 10.1089/jmf.2010.0318.

Barbosa, P.R., Valvassori, S.S., Bordignon, C.L., Kappel, V.D., Martins, M.R. *et al.* (2008) The aqueous extracts of *Passiflora alata* and *Passiflora edulis* reduce anxiety-related behaviors without affecting memory process in rats. *Journal of Medicinal Food* 11(2), 282–288. DOI: 10.1089/jmf.2007.722.

Baú, C, Cazarin, B, B, Kelly, F.G.R., Cristina, T, Giovana, Â. *et al.* (2015) Intake of *Passiflora edulis* leaf extract improves antioxidant and anti-inflammatory status in rats with 2,4,6-trinitrobenzenesulphonic acid induced colitis. *Journal of Functional Foods* 17, 575–586. DOI: 10.1016/j.jff.2015.05.034.

Baú, C., Cazarin, B., Rodriguez-nogales, A., Algieri, F., Utrilla, M.P. *et al.* (2016) Intestinal anti-inflammatory effects of *Passiflora edulis* peel in the dextran sodium sulphate model of mouse colitis. *Journal of Functional Foods* 26, 565–576. DOI: 10.1016/j.jff.2016.08.020.

Beatriz, A., Maria, S., Paulo, E. and Fr, S. (2007) Evidence of anti-inflammatory effects of *Passiflora edulis* in an inflammation model. *Journal of Ethnopharmacology* 109, 281–288. DOI: 10.1016/j.jep.2006.07.031.

Bhat, R. and Paliyath, G. (2016) Fruits of tropical climates: dietary importance and health benefits. In: Caballero, B., Finglas, P.M., and Toldrá, F. (eds) *Encyclopedia of Food and Health*. Academic Press, Oxford, pp. 144–149.

Campos-Vega, R., Vázquez-Sánchez, K., López-Barrera, D., Loarca-Piña, G., Mendoza-Díaz, S. *et al.* (2015) Simulated gastrointestinal digestion and in vitro colonic fermentation of spent coffee (*Coffea arabica* L.): Bioaccessibility and intestinal permeability. *Food Research International* 77, 156–161. DOI: 10.1016/j.foodres.2015.07.024.

Capanoglu, E., de Vos, R.C.H., Hall, R.D., Boyacioglu, D. and Beekwilder, J. (2013) Changes in polyphenol content during production of grape juice concentrate. *Food Chemistry* 139(1–4), 521–526. DOI: 10.1016/j.foodchem.2013.01.023.

Cesar, B., Salles, C., Aparecido, M., Taniguthi, L. and Ferreira, J.N. (2020) *Passiflora edulis* leaf extract: Evidence of antidiabetic and antiplatelet effects in rats. *Biological & Pharmaceutical Bulletin* 43(1), 169–174. DOI: 10.1248/bpb.b18-00952.

Chassagne, D., Crouzet, J.C., Bayonove, C.L. and Baumes, R.L. (1996) Identification and quantification of passion fruit cyanogenic glycosides. *Journal of Agricultural and Food Chemistry* 44(12), 3817–3820. DOI: 10.1021/jf960381t.

Chen, F.P., Xu, X.Y., Luo, Z., Chen, Y. and Xiao, G. (2018) Effect of high O_2 atmosphere packaging on postharvest quality of purple passion fruit (passiflora edulis sims). *Journal of Food Processing & Preservation* 42, 1–7. DOI: 10.1111/jfpp.13749.

Dhawan, K., Dhawan, S. and Sharma, A. (2004) Passiflora: a review update. *Journal of Ethnopharmacology* 94(1), 1–23. DOI: 10.1016/j.jep.2004.02.023.

Duke, J.A. (2008) Dr. Duke's Phytochemical and Ethnobotanical Databases. Passiflora spp. Available at: https://phytochem.nal.usda.gov/phytochem/plants/show/1433?et= (accessed 5 April 2022).

Fern, K. (2019) *Passifloraedulis*. Useful Tropical Plants. Available at: http://tropical.theferns.info/viewtropical.php?id =Passiflora +edulis (accessed 10 November 2020).

Garcia, M.V., Milani, M.S. and Ries, E.F. (2019) Production optimization of passion fruit peel flour and its incorporation into dietary food. *Food Science and Technology International*. DOI: 10.1177/1082013219870011.

Gunathilake, K.D.P.P., Ranaweera, K.K.D.S. and Rupasinghe, H.P.V. (2018) Effect of different cooking methods on polyphenols, carotenoids and antioxidant activities of selected edible leaves. *Antioxidants* 7(9), 117. DOI: 10.3390/antiox7090117.

Hapuarachchi, S. (2016) Passionate about passion fruit. *Daily News*. Available at: http://www.dailynews.lk/2016/09/19/features/93430 (accessed 17 November 2020).

He, X., Luan, F., Yang, Y., Wang, Z., Zhao, Z. *et al.* (2020) *Passiflora edulis*: an insight into current researches on phytochemistry and pharmacology. *Frontiers in Pharmacology* 11, 617. DOI: 10.3389/fphar.2020.00617.

Hu, Y., Jiao, L., Jiang, M.-H., Yin, S., Dong, P. *et al.* (2018) A new *C*-glycosyl flavone and a new neolignan glycoside from *Passiflora edulis* Sims peel. *Natural Product Research* 32(19), 2312–2318. DOI: 10.1080/14786419.2017.1410809.

Ichimura, T., Yamanaka, A., Ichiba, T., Toyokawa, T., Kamada, Y. *et al.* (2006) Antihypertensive effect of an extract of *Passiflora edulis* rind in spontaneously hypertensive rats. *Bioscience, Biotechnology, and Biochemistry* 70(3), 718–721. DOI: 10.1271/bbb.70.718.

Ingale, A.G. and Hivrale, A.U. (2010) Pharmacological studies of *Passiflora* sp. and their bioactive compounds. *African Journal of Plant Science* 4, 417–426.

Kido, L.A., Hahm, E.-R., Kim, S.-H., Baseggio, A.M., Cagnon, V.H.A. *et al.* (2020) Prevention of prostate cancer in transgenic adenocarcinoma of the mouse prostate mice by yellow *P. edulis* extract and antiproliferative effects of its bioactive compound piceatannol. *Journal of Cancer Prevention* 25(2), 87–99. DOI: 10.15430/JCP.2020.25.2.87.

Kishore, K., Pathak, K.A., Shukla, R. and Bharali, R. (2011) Effect of storage temperature on physico-chemical and sensory attributes of purple passion fruit (*Passiflora edulis* Sims). *Journal of Food Science and Technology* 48(4), 484–488. DOI: 10.1007/s13197-010-0189-8.

Kumar, V.A., Ramkumar, M. and Kanthlal, S.K. (2019) The benefit of *P.edulis* as an anti-ulcerogenic diet: scientific evidence by *in vitro* and *in silico* H+/K+ATPase inhibitory activity assessment. *Current Computer-Aided Drug Design* 16(5), 555–563. DOI: 10.2174/15734099156661910251 21005.

López-Vargas, J.H., Fernández-López, J., Pérez-Álvarez, J.Á. and Viuda-Martos, M. (2014) Quality characteristics of pork burger added with albedo-fiber powder obtained from yellow passion fruit (*Passiflora edulis* var. flavicarpa) co-products. *Meat Science* 97(2), 270–276. DOI: 10.1016/j.meatsci.2014.02.010.

Maniyan, A., John, R. and Mathew, A. (2015) Evaluation of fruit peels for some selected nutritional and anti-nutritional factors. *Journal of Emergent Life Sciences Research* 1, 13–19.

Mbaeyi-Nwaoha, I.E. and Ezeoke, A.M. (2019) Production and quality evaluation of yoghurt enriched with two accessions of passion fruits (*Passiflora edulis* f. *flavicarpa*). *Asian Food Science Journal* 10, 1–24. DOI: 10.9734/afsj/2019/v10i130023.

Mota, N.S.R.S., Kviecinski, M.R., Zeferino, R.C., de Oliveira, D.A., Bretanha, L.C. *et al.* (2018) In vivo antitumor activity of by-products of Passiflora edulis f. flavicarpa Deg. Rich in medium and long chain fatty acids evaluated through oxidative stress markers, cell cycle arrest and apoptosis induction. *Food and Chemical Toxicology* 118, 557–565. DOI: 10.1016/j.fct.2018.06.010.

Näf, F., Decorzant, R., Willhalm, B., Velluz, A. and Winter, M. (1977) Structure and synthesis of two novel ionones identified in the purple passionfruit (Passiflora edulis Sims). *Tetrahedron Letters* 18(16), 1413–1416. DOI: 10.1016/S0040-4039(01)93059-X.

Natesh, H.N., Abbey, L. and Aseidu, S.K. (2017) An overview of nutritional and anti nutritional factors in green leafy vegetables. *Horticulture International Journal* 1(2), 58–65. DOI: 10.15406/hij.2017.01.00011.

Nerdy, N. and Ritarwan, K. (2019) Hepatoprotective activity and nephroprotective activity of peel extract from three varieties of the passion fruit (*Passiflora* Sp.) in the albino rat. *Open Access Macedonian Journal of Medical Sciences* 7(4), 536–542. DOI: 10.3889/oamjms.2019.153.

Ng, T.B., Lam, S.K., Cheung, R.C., Wong, J.H., Wang, H.X. *et al.* (2011) Antifungal protein from *P.edulis* (*Passiflora edulis*) Seeds. In: Preedy, V.R., Watson, R.R., and Patel, V.B. (eds) *Nuts and Seeds in Health and Disease Prevention*. Academic Press, Cham, Switzerland, pp. 865–871.

Patel, S.S. (2009) Morphology and pharmacology of passifloraedulis: a review. *Journal of Herbal Medicine and Toxicology* 3, 1–6.

Pereira, C.A.M. and Vilegas, J.H.Y. (2000) Chemical and pharmacological constituents of the genus *Passiflora*, with emphasis on P. alata dryander, P. edulis Sims and P. incarnata L. *Revista Brasileira de Plantas Medicinais* 3(1), 1–12.

Phamiwon, Z.A.S. and John, S. (2017) Phytochemical investigation and antioxidant activities of *Passiflora edulis* (passion fruit) leaves from the Ukhrul district, Manipur, India. *World Journal of Pharmaceutical Research* 6(14), 793–801.

Ramirez, V., Arango, S.S., Maldonado, M.E., Uribe, D., Aguillon, J. *et al.* (2019) Biological activity of Passiflora edulis f. Flavicarpa ethanolic leaves extract on human colonic adenocarcinoma cells. *Journal of Applied Pharmaceutical Science* 9(2), 64–71.

Ramirez, V., Arango, S.S., Uribe, D., Maldonado, M.E. and Aguillon, J. (2019) Cytotoxic and apoptotic activity of extracts from leaves and juice of *Passiflora edulis*. *Journal of Liver Research Disorders & Therapy* 4(2), 67–71.

Rotta, E.M., Rodrigues, C.A., Jardim, I.C.S.F., Maldaner, L. and Visentainer, J.V. (2019) Determination of phenolic compounds and antioxidant activity in passion fruit pulp (*Passiflora* spp.) using a modified QuEChERS method and UHPLC-MS/MS. *LWT* 100, 397–403. DOI: 10.1016/j.lwt.2018.10.052.

Salgado, J.M., Bombarde, T.A.D., Mansi, D.N., Piedade, S.M. de S. and Meletti, L.M.M. (2010) Effects of different concentrations of passion fruit peel (*Passiflora edulis*) on the glicemic control in diabetic rat. *Ciência e Tecnologia de Alimentos* 30(3), 784–789. DOI: 10.1590/S0101-20612010000300034.

Santos Monteiro, S., Albertina Silva Beserra, Y., Miguel Lisboa Oliveira, H. and Pasquali, M.A.D.B. (2020) Production of probiotic Passion fruit (*Passiflora edulis* Sims f. flavicarpa Deg.) drink using *lactobacillus reuteri* and microencapsulation via spray drying. *Foods* 9(3), 335. DOI: 10.3390/foods9030335.

Septembre-Malaterre, A., Stanislas, G., Douraguia, E. and Gonthier, M.-P. (2016) Evaluation of nutritional and antioxidant properties of the tropical fruits banana, litchi, mango, papaya, passion fruit and pineapple cultivated in Réunion French Island. *Food Chemistry* 212, 225–233. DOI: 10.1016/j.foodchem.2016.05.147.

Tal, Y., Anavi, S., Reisman, M., Samach, A., Tirosh, O. *et al.* (2016) The neuroprotective properties of a novel variety of passion fruit. *Journal of Functional Foods* 23, 359–369. DOI: 10.1016/j.jff.2016.02.039.

Taïwe, G.S. and Kuete, V. (2017) *Passiflora edulis*. In: Kuete, V. (ed.) *Medicinal Spices and Vegetables from Africa*. Academic Press, Cham, Switzerland, pp. 513–526.

Thokchom, R. and Mandal, G. (2017) Production preference and importance of passion fruit (*Passiflora edulis*): A review. *Journal of Agricultural Engineering and Food Technology* 4, 27–30.

Thomas, B., Vithiya, B.S.M., Prasad, T.A.A., Mohamed, S.B., Magdalane, C.M. *et al.* (2019) Antioxidant and photocatalytic activity of aqueous leaf extract mediated green synthesis of silver nanoparticles using *Passiflora edulis f. flavicarpa*. *Journal of Nanoscience and Nanotechnology* 19(5), 2640–2648. DOI: 10.1166/jnn.2019.16025.

Wagasu, R.S.U., Lawal, M., Amedu, A.M., Sbir, A.A., Kabir, S. *et al.* (2016) Comparative chemical analysis, phytochemical screening and antimicrobial activities of the rinds, seeds and juice of passion fruit (*Passiflora edulis* var. *flavicarpa*). *Journal of Natural Sciences Research* 6, 138–143.

Wang, C., Xu, F.-Q., Shang, J.-H., Xiao, H., Fan, W.-W. *et al.* (2013) Cycloartane triterpenoid saponins from water soluble of *Passiflora edulis* Sims and their antidepressant-like effects. *Journal of Ethnopharmacology* 148(3), 812–817. DOI: 10.1016/j.jep.2013.05.010.

Wijeratnam, S.W. (2016) Passion fruit. In: Caballero, B., Finglas, P.M., and Toldrá, F. (eds) *Encyclopedia of Food and Health*. Academic Press, Cham, Switzerland, pp. 230–234. DOI: 10.1016/B978-0-12-384947-2.00521-3.

Zeraik, M.L. and Yariwake, J.H. (2010) Quantification of isoorientin and total flavonoids in *Passiflora edulis* fruit pulp by HPLC-UV/DAD. *Microchemical Journal* 96(1), 86–91. DOI: 10.1016/j.microc.2010.02.003.

Zibadi, S. and Watson, R.R. (2004) Passion fruit (*Passiflora edulis*). *Evidence-Based Integrative Medicine* 1(3), 183–187. DOI: 10.2165/01197065-200401030-00005.

Zucolotto, S.M., Fagundes, C., Reginatto, F.H., Ramos, F.A., Castellanos, L. *et al.* (2012) Analysis of *C*-glycosyl flavonoids from South American *Passiflora* species by HPLC-DAD and HPLC-MS. *Phytochemical Analysis* 23(3), 232–239. DOI: 10.1002/pca.1348.

17 Baobab (*Adansonia digitata* L.)

Cyrielle Garcia[1,2]*, Vonimatra Razafindrazaka[3] and Charlotte Ralison[3]
[1]Qualisud, Univ La Réunion, CIRAD, Univ Montpellier, Montpellier SupAgro, Univ Avignon, Sainte Clotilde, France; [2]Qualisud, Univ Montpellier, CIRAD, Institut Agro, Avignon Université, Univ de La Réunion, ESIROI, Montpellier, France; [3]LABASAN (Laboratoire de Biochimie Appliquee aux Sciences de l'Alimentation et de Nutrition) Faculté des Sciences, Université d'Antananarivo, Madagascar

Abstract

Baobab (*Adansonia* sp.) is a majestic tree native to Africa, found throughout the dry regions of the continent such as sub-Saharan Africa, and in western Madagascar. It includes eight species, six of which are endemic to Madagascar island. The plant is well known to have multiple uses as food, for clothing and as medicine. The fruit is becoming increasingly recognized for its nutritional quality and antioxidant activity due to a particularly high vitamin C content, minerals and other bioactive molecules.

Baobab fruit use in strategies to increase mineral bioaccessibility or in bioactivity studies demonstrating antioxidant, antimicrobial and anti-inflammatory effects has been reported. Naturally occurring antioxidant compounds which are widely distributed in plants and present in baobab fruit pulp have stimulated research interest due to their wide range of health-promoting bioactivities. A growing scientific interest in the health-promoting effect of fruit pulp is notable, and lipid metabolism, myocardial and hepatoprotective effects are increasingly studied. Valorization and investigation of new ways of transforming this local resource should be encouraged to improve food security.

Introduction

Baobab (*Adansonia* sp.) is a large and iconic tree indigenous to Africa. Several names are used to describe the tree depending on its geographical location or physiological attributes including magic tree, chemist tree, symbol of the earth, upside-down tree, arbre a palabre, monkey bread of Africa, dead rat tree, cream of tartar or the small pharmacy tree among numerous others (De Smedt *et al.*, 2012; Fatema *et al.*, 2015; Rahul *et al.*, 2015; Asogwa *et al.*, 2021).

This symbolic, culturally significant and physically majestic tree is a keystone species in most landscapes of the drylands and has many uses. Every part of the plant is reported to be useful. It provides food, clothing and medicine as well as raw materials for many useful items (Rahul *et al.*, 2015). The fruit pulp, seeds, leaves, flowers, roots and bark of baobab are edible. In recent years, the global demand for baobab raw materials has grown (e.g. seed oil, fruit pulp) in the food and beverage, nutraceutical and cosmetic industries, leading to increased pressure on this wild harvest resource (Kamatou *et al.*, 2011). Indeed, baobab fruit has been referred to as a 'superfruit' based on its nutritional profile (e.g. vitamins and minerals) (Rahul *et al.*, 2015) and arouses much interest in terms of

*Corresponding author: cyrielle.garcia@univ-reunion.fr

 Handbook of Phytonutrients in Indigenous Fruits and Vegetables (eds D. Sivakumar *et al.*)
DOI: 10.1079/9781789248067.0017

commercialization and novel product development, generating increasing income perspectives in dryland regions of Africa.

Baobab fruit pulp is a mixture of carbohydrates, including mono- and disaccharides, pectin, proteins, lipids, fibre and amino acids. It is particularly rich in vitamins and minerals and has a well documented antioxidant capacity. Its composition allows baobab to be one of the top vitamin C providers. Bioactivity has been demonstrated for some of these constituents and baobab fruit pulp pretends to various health claims. In this chapter, available data on the uses of baobab fruit pulp, its composition and its health benefits are presented.

Botany and Traditional Use

The African baobab, *Adansonia* sp., belonging to the Bombaceae family and Malvales order, is a deciduous stem-succulent tree native to the dry regions of tropical Africa and is widely distributed in sub-Saharan Africa and western Madagascar (Diop *et al.*, 2006; De Smedt *et al.*, 2012).

The baobabs comprise eight species globally from three different taxonomic groups, each characterized following their floral morphologies (Baum, 1995). The *Adansonia* section includes the abundant African species (*A. digitata*), which is the most studied. The *Brevitubae* section includes two endemic species from Madagascar (*A. suarezensis* and *A. grandidieri*), while five species belong to the *Longitubae* section, four of which are from Madagascar (*A. rubrostipa*, *A. za*, *A. perrieri*, *A. madagascariensis*) and one from Australia (*A. gregorii*). The *A. digitata* species is found in many African countries throughout the savannas and savannah woodlands of southern Africa, for example in Malawi, Zimbabwe, Mozambique and South Africa, and especially in the warm parts of Limpopo Province. In West Africa, it is found in Mali, Benin, Senegal, the Ivory Coast, Cameroon and Burkina Faso, as well as Kenya, Uganda and Tanzania in East Africa (Kamatou *et al.*, 2011). Apart from *A. gregorii*, which is not endemic of the African continent, all species are found on the island of Madagascar, whose six are endemic to that region. The characterization of the population structure shows that baobab stands are denser with larger-sized and taller trees in riverine areas and natural forest compared to grazing lands, suggesting adverse effects of human disturbances on its populations (Birhane *et al.*, 2020).

The tree has a great ecological tolerance for both hot and dry cultivating conditions. This quality is attributed to the thick fire-resistant bark, shedding of leaves and an often hollow trunk that can absorb water in the rainy season and which contracts in the dry season (Sidibe *et al.*, 2002). Morphologies are variable according to the species. Trees have a large diameter of 10–12 m and a height of 5–30 m or more, with a cylindrical, spindle or bottle-shaped massive trunk (Fig. 17.1). The wood has a stratified structure and is globally soft and full of water (Baum, 1995). The tree and fruit morphology by species is presented in Fig. 17.1.

Baobab fruit are composed of a woody outer shell (epicarp) representing approximately 45% of the fruit mass, which contains the internal fruit pulp (endocarp) which is split in small floury, dehydrated and powdery slides that enclose multiple seeds and filaments (the red fibres) that subdivide the pulp into segments (Caluwé *et al.*, 2010). *A. digitata* L. fruit was physically characterized previously as a 17 cm long and 8 cm large fruit on average, weight of 150–300 g, of which 14–28% was pulp (Nour *et al.*, 1980; Diop *et al.*, 2006). We recently characterized fruit morphologies between different *Adansonia* species in Madagascar. It appeared that among fruits collected at the same maturity stage, *digitata* had heavier fruits than other species with 354.5 g on average, ranging from 150.6 to 602.1 g (n=43), whose pulp represented 21.5%. These values were much lower for *za* and *grandidieri* with, respectively, 195.4 g on average ranging from 49.6 to 437.2 g (n=112), whose pulp represented 19.8%, and 159.3 g on average, ranging from 37.9 to 534.9 g (n=131) whose pulp represented 19.3%. The minimum weight was observed for *rubrostipa* at 108.3 g on average, ranging from 42.6 to 245.1 g (n=80), whose pulp represented 15.3%. However, variabilities were observed within the same species in relation to the site of collection, which is presumed to impact fruit weight, given the importance of edapho-climatic factors on fruit maturation.

Fig. 17.1. Tree and fruit of the different *Adansonia* species.

The tree is well known in Africa for its significant role in international trade and as an important source of food and beverages, folk medicine and income for rural communities, especially in the dry season or at times of drought, with more than 300 uses reported for this species (De Smedt *et al.*, 2012). Baobab fruit pulp is traditionally used in ethnomedicine and pharmaceutically in treatment against malaria, fever, stomach-ache, diarrhoea, dysentery, giardiasis, scurvy, cough, smallpox, measles and pain after birth (Ghoneim *et al.*, 2016; Hussain *et al.*, 2019).

Various parts of the baobab – fruit pulp, seeds and leaves – are frequently consumed by rural people in both western and south-eastern Africa. By contrast, only fruit pulp and seeds are consumed in Madagascar (De Smedt *et al.*, 2012; Razafindralambo *et al.*, 2014). Traditionally, the dried pulp was artisanally ground to obtain a powder that could be directly used to produce a drink, with a characteristic sour flavour close to grapefruit or citrus. This can be consumed by itself or mixed with other juices. Milky drinks are also consumed by mixing pulp and the flour of a cereal to obtain 'gubdi' by mixing with acha, in Nigeria. The fruit pulp is used in sauces, or can be processed to a complete porridge preparation of maize flour (Obizoba and Anyika, 1994; Caluwé *et al.*, 2010; Asogwa *et al.*, 2021).

Composition

Nutritional compounds

Nutrient composition has been reported in multiple studies (Nour *et al.*, 1980; Odetokun, 1996; Glew *et al.*, 1997; Sena *et al.*, 1998; Murray *et al.*, 2001; Diop *et al.*, 2006; Caluwé *et al.*, 2010; Erwa *et al.*, 2019; Tsetegho Sokeng *et al.*, 2019) and is summarized in Table 17.1.

Baobab fruit pulp is characterized by a particularly high level of dry matter, with around 85% carbohydrate, particularly soluble sugars, representing the majority of compounds. The absence of starch, variability in fibre content and a richness in pectin have also been reported regarding the carbohydrate content (Caluwé *et al.*, 2010; Rahul *et al.*, 2015). Relatively low protein and fat content was observed, below 2 g/100 g, with the presence of all essential amino acids in the baobab fruit amino acid

Table 17.1. Nutritional composition of *A. digitata* L. fruit pulp (Diop *et al.*, 2006; Caluwé *et al.*, 2010; Tsetegho Sokeng *et al.*, 2019).

		Content			Proportion
Constituents	Moisture			Fruit %	4.7–25.9
	Dry matter	g/kg	662–953		85.5–89.5
	Metabolizable energy	kJ/100 g	810–1480		
Proximates	Protein	g/100 g	1.4–2.7	Pulp %	2.6–15.3
	Fat		0.1–4.1		0.2–4.3
	Ash		1.8–5.3		1.9–5.7
	Crude fibre		1.1–8.9		5.4–11.15
	Carbohydrate		44.4–80.4		23.2–79.4
	Soluble sugars		5.1–34.6		
Sugars	Sucrose	g/100 g	8.1–10.2	Sugars %	86
	α and β glucose		6.7–8.5		8
	β-fructopyranose		14.2–17.9		4
	Myoinositol				2
	Galacturonic acid				<1
Organic acids	Citric acid			Organic acids %	77.3
	Malic acid				17.6
	Ascorbic acid				3.2
	Gallic acid				1.3
	Other acids				0.34
	Acidity	mEq/kg	400–630		
Minerals	Calcium	mg/100 g	115.6–655		
	Copper		0.6–1.6		
	Iron		0.86–9.3		
	Potassium		726–2836.4		
	Magnesium		78–300		
	Manganese		0.8–2.1		
	Sodium		0.7–27.9		
	Phosphorus		35–1330		
	Zinc		0.47–3.18		
Vitamins	Ascorbic acid	mg/100 g	20–500		
	Thiamin (B1)		0.01–1		
	Riboflavin (B2)		0.06–0.09		
	Niacin (B3)		0.5–2.2		
	Pyridoxin (B6)		0.002–0.02		
	Pro-vitamin A (eq. retinol)		0.003–0.3		

profile. Concerning fat, the most represented fatty acids are palmitic and linolenic (Glew *et al.*, 1997). Interestingly, an exhaustive analysis of metabolite distribution by chemical classes showed a high number of organic metabolites, representing 67% of total metabolites detected by GC-MS within the pulp, whereas fatty acids and carbohydrates represented 23% and 10% of total metabolites, respectively (Khakimov *et al.*, 2016).

The pulp has an acidic nature due to high acidity in correlation with the elevated presence of organic acids such as citric acid, mainly 3300–3345 mg/100 g, which represents around 77.3% of organic acids, followed by malic, tartaric and succinic acid, with a pH around 3.3 (Tembo *et al.*, 2017; Adetola *et al.*, 2019). The combination of rich sugar and acid content results in a sweet, acidulous and refreshing taste to the pulp.

Vitamin profile is characterized by the vitamin C abundance in the baobab since the fruit is considered as one of the most important natural sources of ascorbic acid, with an average content estimated to be 290 mg/100 g (Chadare *et al.*, 2008), far higher than orange fruit and mango juice, containing 47.5 and 25 mg/100 g on average, respectively, and higher than guava juices containing 228 mg/100 g (Ciqual: Table de composition nutritionnelle des aliments, n.d.). This vitamin C content could be compared to other indigenous fruit like marula, containing 67–403 mg/100g (Hiwilepo-van Hal *et al.*, 2012). In regard to the recommended daily intake of vitamin C which is up to 75–90 mg/d for adults in Europe and the USA (EFSA, 2013; National Institutes of Health, 2019), these levels of vitamin C mean that only 25–31 g of baobab pulp would be sufficient to cover daily needs. The high vitamin C content might also have a possible influence on iron and zinc bioaccessibility, even if they are present in low amounts.

Reports of mineral composition show the presence of numerous compounds such as potassium, magnesium, phosphorus, copper, iron, sodium, zinc and manganese in descending order of importance, with some variability. A remarkable calcium amount of 302 mg/100 g ranging from 156–655 mg/100 g has also been described, making baobab fruit an attractive source of calcium for supplementation in populations with significant needs (Diop *et al.*, 2006; Chadare *et al.*, 2008; Caluwé *et al.*, 2010).

The antinutritional molecules detected in baobab fruit pulp – tannins (0.0051% and 0.0062%), hydrocyanic acid (0.0049% dry weight – DW) and total oxalate (0.0044% DW) – do not present any potential toxicity or harmful effects for consumers (Ghani and Agbejule, 1986).

Antioxidant compounds

Baobab fruit pulp has a well documented antioxidant capability. Studies on the antioxidant properties of *A. digitate* fruit constituents and leaves, estimated by the photochemiluminescence method, clearly indicate the superior antioxidant properties of the fruit. In particular the integral antioxidant capacity value of baobab fruit pulp was ten times higher than that of orange pulp (Vertuani *et al.*, 2002; Besco *et al.*, 2007). The richness in vitamin C clearly contributes to its overall antioxidant capacity, together with the many different phenolic compounds reported including procyanidin, epicatechin, gallic acid, hydroxycinnamic acid glycosides, tannins, phenols and flavonoids (Seukep *et al.*, 2013; Li *et al.*, 2017; Tembo *et al.*, 2017; Ismail *et al.*, 2019). Specifically, the content of flavonoids and procyanidins, in particular, (533 mg/100 g) was found to be higher than in many fruits, including apples (104.4 mg/100 g), blueberries (256 mg/100 g) or even cranberries (419 mg/100 g) (Ismail *et al.*, 2020). A combined HRMS, 1H and 13C NMR and 2D experimental analysis (COSY, HSQC, HMBC, and ROESY) allowed the identification of four hydroxycinnamic acid glycosides, six iridoid glycosides and three phenylethanoid glycosides from the dried baobab fruit pulp (Li *et al.*, 2017). An exhaustive LC-MS-QTOF analysis of baobab fruit pulp phytochemicals then discovered the presence of 46 phytochemicals (Ismail *et al.*, 2019). Among the phenolic acids, two different subclasses were detected which are mainly the derivatives of hydroxycinnamic acid and hydroxybenzoic acids. The interpretation of MS and MS/MS spectra among these subclasses led to the identification of chlorogenic acid, caffeic acid and p-hydroxycinnamic acid for the hydroxycinnamic acid derivatives, and protocatechuic acid, p-hydroxybenzoic acid and ellagic acid for hydroxybenzoic acid derivatives. Regarding the flavonoids, two classes were detected, and identified compounds were two flavanols (D-(+)-catechin, and (−)-epicatechin) and three flavonols (rutin, quercetin and quercetin-3-rhamnoside). Even allowing for the fact that the different solvent systems used for phenolic extraction had an effect on the estimated concentration, the most predominant phenolic compounds were concluded to be (−)-epicatechin, D-(+)-catechin, chlorogenic acid and p-hydroxybenzoic acid, reaching 1059.3, 805.5, 380.4 and 142.8 μg/g DW as maximal values, respectively. By comparison, a methanolic extraction led to the determination of procyanidin B2 and (−)-epicatechin as main flavan-3-ols in baobab fruit pulp (Tembo *et al.*, 2017).

The total phenolic (TPC) and flavonoid (TFC) content are related to the overall antioxidant capacity in a plant, giving redox potentials and the possibility to act as reducing agents, hydrogen donors and singlet oxygen quenchers. The determined values obtained after quantification of TPC, TFC and the antioxidant capacity of baobab fruit pulp are summarized in Table 17.2. The main *in vitro* radical scavenging activity assays were used to compare the antioxidant capacity of the phenolic extracts, DPPH (2, 2-dipheny-l-picrylhydrazyl), FRAP (ferric reducing antioxidant capacity), ABTS•+ (2,2'-azinobis-3-ethylbenzothiazoline-6-sulphonate) and ORAC (oxygen radical absorbance capacity).

Significant correlations were observed between phenolic levels and antioxidant activity by Ismail *et al.* (2019), whether it be measured by DPPH (TPC to DPPH: $r=0.707$, $p<0.01$; TFC to DPPH: $r=0.776$, $p<0.01$) or by FRAP (TPC to FRAP: $r=0.825$, $p<0.01$, TFC to FRAP: $r=0.700$, $p<0.01$). The TPC and TFC were also mentioned to be correlated ($r=0.754$, $p<0.01$), suggesting that baobab fruit pulp phenolic compounds contribute to antioxidant activity. Regarding the individual phenolic compounds, even if there is more variation on the correlation coefficients, it appears that all the phenolic compounds had a significant correlation with both TPC, TFC and FRAP, except caffeic acid, and with DPPH, except protocatechuic acid and caffeic acid, confirming their role in antioxidant activity.

Variation factors

Several studies have looked into *A. digitata* L. fruit nutrient content depending on the geographical location of the trees. Contrasting results were found: 1) significant variations were observed in pulp moisture, protein, fibre, ash and elemental content between Kenya and Malawi (Muthai *et al.*, 2017); 2) in pulp iron, potassium, total acidity and pH value among six locations of eastern and coastal Kenya (Stadlmayr *et al.*, 2020); while 3) fruit collection in each climatic zone of Benin did not reveal an influence of the agroclimatic zone on any part of the fruit biochemical composition, with the notable exception of vitamin C content (Assogbadjo *et al.*, 2012). However, it was observed that highly basic soils, rich in carbon, clay, fine silt and organic matter seemed to be correlated with some pulp biochemical parameters (Assogbadjo *et al.*, 2012), suggesting that the diversity of the soils where baobab trees are growing is an important parameter to consider for nutrient composition analysis of the fruits.

Regarding *Adansonia* species effect, high variations were found in the nutritional characteristics of five endemic species of Madagascar, namely *A. perrieri*, *za*, *madagascariensis*, *grandidieri* and *rubrostipa*, by comparison with the African Sudano-Sahelian species, *A. digitate* (Cisse *et al.*, 2013). Distinct biochemical profiles were found and allowed discrimination between the species of the three groups. *A. za* was separated from the other species and characterized by a high content of vitamin C and a low fat content. *A. digitata* and *A. grandidieri* showed low levels of sugars and calcium. Finally, the third group included *A. rubrostipa*, *A. perrieri* and *A. madagascariensis*, characterized by higher starch, ash and fat content.

Vitamin C is known to be a highly thermolabile compound that is easily degraded upon application of heat, and is also unstable in poor conditions. It is easily decomposed in fruit and vegetables during storage. Therefore, the preservation of vitamin C content is a major issue for baobab fruit pulp quality. Bioactive compounds

Table 17.2. TPC, TFC and antioxidant activities inventoried for *A. digitata* L. fruit pulp.

		A	B[1]	C	D	E	F	G
TPC	mg GA/100 g	1160.6	183.6–742.4	2190–289	329–1726	3518.3 (methanol extract) 4057.5 (acetone extract)		
TFC	mg RE/100 g	5872.4	1090.6–6832.5					
	mg QE/100 g					31.70 (methanol extract) 42.73 (acetone extract)		
DPPH	mg Trolox/100 g DW	2695.8	721.8–2501.5					
FRAP	mg Trolox/100 g DW	10,542.4	2242.4–5146.8					
	μmol/l			1666.8–2167.7				
	mmol AEAC/100 g					48.45		
ABTS•+	mg Trolox/100 g DW	3321.1						
ORAC	mmol TE/g				0.11–0.15			
PCL	mmol TE/g						0.24	11.1

AEAC, ascorbic acid equivalent; DW, dry weight; GA, gallic acid; PC, photochemiluminescence method; QE, quercitin equivalent; RE, rutin equivalent; TE, Trolox equivalent. [1] Ethyl-acetate extraction was excluded giving it was extracting very few phenolics according to the authors. Sources: A: Ismail *et al.*, 2020; B: Ismail *et al.*, 2019; C: Coe *et al.*, 2013; D: Cisse *et al.*, 2013; E: Lamien-Meda *et al.*, 2008; F: Besco *et al.*, 2007; G: Vertuani *et al.*, 2002.

of *A. digitata* pulp from Malawi and resulting juice were investigated after undergoing preservation treatment (Tembo *et al.*, 2017). Unlike microwaving, thermal pasteurization impacted on the quality of baobab juice. Conventional thermal pasteurization (72°C, 15 s) led to significantly higher vitamin C content in juices than longer or higher treatments and further showed extended half-life under the optimal refrigeration temperature of 6°C. Procyanidin B_2, (-)-epicatechin, TPC and antioxidant activity were also measured and showed fluctuation during thermal treatment and storage. Interestingly, it was reported that the addition of gum Arabic also contributed to the retention of vitamin C content in a baobab-based soft drink (Abdalla *et al.*, 2010).

Physiological Relevance

Bioaccessibility and bioavailability

Food-to-food fortification was investigated to increase mineral bioaccessibility using baobab fruit pulp as a facilitator, since mineral bioaccessibility from foods depends on modifiers of absorption as well as their mineral content. The high content in organic acids, especially ascorbic acid and citric acid, known as potent enhancers of mineral bioavailability in foods, was hypothesized to allow mineral chelation and to keep them in a soluble and absorbable form (Lönnerdal, 2000; Iyengar *et al.*, 2010). Ascorbic acid acts as a reducing agent for apical membrane-bound ferrireductase which reduces Fe^{3+} to Fe^{2+}, the only form in which non-haem iron is transported across the intestinal membrane (Mackenzie and Garrick, 2005).

Mineral bioaccessibility was observed on various fermented cereal grains commonly consumed in Zimbabwe (maize, red and white sorghum, pearl millet and red and white finger millet) (Gabaza *et al.*, 2018). This parameter depended on the type of cereal fermented, ranging from 7.08 to 22.7% (estimated by % dyalisate) for iron, with highest bioaccessibility observed for white sorghum and lowest for pearl millet. Variable effects on iron and zinc bioaccessibility were observed following baobab fortification. Iron bioaccessibility was increased for baobab-enriched cereals of maize, red sorghum and white finger millet, while no change was observed for other cereal fermented slurries, ranging from 8.42 to 44.9%. Otherwise, zinc bioaccessibility ranged from 0.74 to 14.4% for control cereals and 9.25 to 25.9% for baobab-enriched cereals. Notable bioaccessibility increase was observed for white sorghum, pearl millet and red finger millet, while a decrease was observed for maize and red sorghum, and no effect for white finger millet.

A targeted fortification of pearl millet porridge resulted in a 36% and a 12% increase of percentage iron and magnesium bioaccessibility, respectively, in millet plus baobab compared to millet alone (Adetola *et al.*, 2019). This higher bioaccessibility was likely mainly due to the high citric acid content, even if for iron the molar ratios of citric acid:iron (19:1) and ascorbic acid:iron (0.7:1) were below the efficient ratios described for optimal iron absorption in cereal food of 100:1 and 2-10:1, respectively (Hunt, 2005). However, in a similar way to moringa, fortification of pearl millet porridge with baobab decreased the percentage of zinc and calcium bioaccessibility, presumably due to their high content of phytates and phenolics, respectively.

Direct bioaccessibility and bioavailability of baobab fruit pulp nutrients for minerals had not been studied until recently. Despite a low content (1.67 mg/100 g), baobab pulp iron was demonstrated to be remarkably bioaccessible (99.7%) given it is in low-molecular-weight form in the pulp, but not bioavailable for uptake by Caco-2 cells (Khoja *et al.*, 2021). To date, there are no studies on the bioaccessibility and bioavailability of other baobab nutrients.

Bioactivity

Naturally occurring antioxidant compounds which are widely distributed in plants and present in baobab fruit pulp have stimulated research interest due to their wide range of health-promoting bioactivities such as antioxidant, antimicrobial and anti-inflammatory effects.

Polyphenols, as regard to their chemical structure, have been extensively investigated for antimicrobial properties. The presence of flavonols, in particular in baobab fruit pulp, is interesting in relation to their wide spectrum and high antimicrobial activity, and to the fact that most of them are able to suppress a number of microbial virulence factors (such as inhibition of biofilm formation, reduction of host ligands adhesion and neutralization of bacterial toxins), plus synergism with antibiotics (Daglia, 2012). The antimicrobial effect of *A. digitata* was tested from a powdered air-dried methanol extracted sample and compared to other Cameroonian dietary plants (Seukep *et al.*, 2013). The results indicated that the exhibited minimal inhibitory concentrations of baobab ranged from 256 to 1024 µg/ml. Baobab fruit pulp also displayed antimicrobial activities against a limited number of strains studied as it was active against 14.8% of the tested bacteria, however, this activity was the lower observed among the plant extracts tested. The results indicated a minimal inhibitory concentration, ranging from 256 to 1024 µg/ml.

Anti-inflammatory and analgesic profiles for the fibrous part of *A. digitata* fruit were investigated according to various extraction methods using microwave and either toluene, dichloromethane, pet-ether, methanol, water or

DMSO as extractive solvent (Fatema *et al.*, 2015). The anti-inflammatory activity performed by the carrageenan-induced rat paw oedema method, using ibuprofen as reference drug, and the analgesic profile, using aspirin as a standard reference drug, showed maximum profiles for the methanolic extract. The activities profile was observed to be good when extraction used polar solvent DMSO, by contrast with non-polar solvents like hexane, toluene and dichloromethane that showed poor percentage extraction along with poor anti-inflammatory or analgesic activity profiles, in the same way as the combination of water and methanol.

Different antioxidant extracts of *A. digitata* fruit pulp obtained by maceration, decoction and microwave techniques were evaluated for enzyme inhibition properties on enzymes related to neurodegenerative ailments (acetylcholinesterase and butyrylcholinesterase), diabetes (α-glucosidase and α-amylase) and skin hyperpigmentation (tyrosinase) (Hussain *et al.*, 2019). A variable enzyme inhibitory capacity of the fruit pulp extracts was observed according to the extraction method and polarity change in the solvent used. This could be attributed partly to the presence of phenolic compounds like chlorogenic acid, which have been reported as active inhibitors of α-amylase, α-glucosidase (Kalita *et al.*, 2018), caffeic, p-coumaric, ferulic, caffeic and sinapic acid that are known acetylcholinesterase and/or butyrylcholinesterase inhibitors (Szwajgier, 2013).

Additionally, various phenolic compounds demonstrated inhibitory activity on digestive enzymes, α-glucosidase and α-amylase notably, limiting or delaying starch digestion and absorption. These inhibitory properties seemed to be connected to the phenolic chelating, structure-altering, and biological function-limiting capacities and were connected to subsequently reduced postprandial glycaemic response *in vivo* (Garcia *et al.*, 2020).

These carbohydrate enzyme inhibition properties were explored for digestion control through baobab fruit extract baked into white bread at 1.25%, 1.88%, 2.50%, 3.13%, and 3.75% (Coe *et al.*, 2013). Assessment of glycaemic response showed a slower sugar release above 1.88% of baobab supplementation by comparison with white bread, either after 20 or 60 minutes of *in vitro* digestion. Similarly, the ingestion of randomized meals by nine healthy volunteers, with each participant serving as her/his own control on a crossover design, was tested. The results showed a dose-dependent reduction of the glycaemic response following the ingestion of white bread and a solution of baobab fruit extract, but no corresponding increase in satiety measured through the subjective feelings of hunger. Even if fibre might also play a role in delaying carbohydrate absorption, polyphenols were considered by the authors as the main contributing factor to reduced glucose release, as has been previously described in carbohydrate-rich foods like oats (Thondre *et al.*, 2011). Tests on another 17 healthy subjects consuming different beverages matched for carbohydrates and soluble fibre content also revealed a significant decrease for baobab-enriched milk drink glucose and insulin response by comparison with reference drinks (Ryan *et al.*, 2014).

Health benefits

As baobab fruit has been traditionally used as food and medicine by local sub-savannah residents in Africa throughout history due to its numerous properties, its effects on prevalent pathologies were investigated.

The efficiency of baobab fruit pulp was first tested as a solution on treatment of children with acute diarrhoea and resulting mild to moderate dehydration. The treatment was described as clinically equivalent to the World Health Organization (WHO) standard treatment but showed nutritional, economic and cultural advantages (Tal-Dia *et al.*, 1997).

To evaluate the effects of baobab fruit on lipid profile, dried fruit pulp was administered among a population of hyperlipidemic patients for four weeks. The study demonstrated the effectiveness of baobab fruit in lowering lipid levels in patients with dyslipidemia, with noticeable significant reduction in the levels of total cholesterol (120.06 vs. 49.06%) and triglycerides (78.13 vs. 57.44%) by comparison with the control group receiving atorvastatin only. These results were found to be in accordance with findings reported on the lipid-lowering effects of pectin and might be connected to a limitation of cholesterol reabsorption due to dietary fibre and their direct bile acid-binding capacity or their fermentation capacity leading to an increase in short chain fatty acid production in the large intestine, promoting stool excretion (Gadour *et al.*, 2017).

Several animal studies have revealed that treatment with oral baobab fruit pulp exerts significant health effects. Recently, high sugar/high fat diet-simulated metabolic syndrome in Wistar rats receiving an aqueous *A. digitata* fruit treatment resulted in a significant dose-dependent amelioration of the induced dyslipidemia and hyperglycaemia with a significant reduction in the Atherogenic Index of Plasma (Suliman *et al.*, 2020). The fruit normalized the elevated hepatic biomarkers. Creatinine urea and partial reduction in lesion intensity were observed in the hepatic, heart and kidney tissue. An anti-inflammatory effect was also observed since the inflammatory infiltration was eliminated, demonstrating the antimetabolic syndrome potential of baobab fruit pulp.

The effect of baobab fruit pulp was also evaluated in terms of myocardial function (Ghoneim *et al.*, 2016). An infarction rat model induced by isoproterenol-oxidative stress showed infiltration of inflammatory cells and congestion in the blood vessels at the histopathological level, while daily treatment with baobab fruit pulp brought these parameters to near normal level, along with normalization in levels of cardiac marker enzymes (creatine kinase MB, lactate dehydrogenase, aspartate aminotransferase, interlukin-1β, monocyte chemoattractant protein-1, myeloperoxidase, collagen and galectin-3), and with a concomitant decrease in the level of reduced glutathione and glutathione peroxidase activity in heart tissue. This study reveals that the efficiency of baobab treatment leading to a myocardial protective effect could be associated with the enhancement of the antioxidant defence system, through the activation of reduced glutathione and glutathione peroxidase in heart tissue, and the reduction of extensive myocardial damage by blocking the increase of cardiac marker enzymes.

A hepatoprotective effect was demonstrated on rats for fruit pulp aqueous (Al-Qarawi *et al.*, 2003) and methanolic extracts (Mohamed *et al.*, 2015; Hanafy *et al.*, 2016). The effects of liver injury by carbon tetrachloride (CCl_4) or paracetamol in animal model studies were demonstrated to be reversible by the administration of baobab fruit pulp extract. CCl_4 is a xenobiotic able to induce oxidative stress by initiating free radical mediated lipid peroxidation, leading to the accumulation of lipid-derived oxidation products that lead to liver injury and excess collagen deposition in the liver (Liu *et al.*, 2006). Paracetamol is a common analgesic-antipyretic drug, extensively metabolized into N-acetyl p-benzoquinoneimine after high dosage, which depletes GSH and leads to hepatotoxicity, and was shown to directly inhibit cellular proliferation and induce oxidative stress, resulting in lipid peroxidation, depletion of ATP levels and alteration of Ca^{++} homeostasis (Al-Qarawi *et al.*, 2003). Liver function tests were performed to reveal or confirm hepatocyte necrosis/damage since acute hepatitis generally produces markedly elevated levels of aminotransferases. Administration of baobab fruit extract led to the normalization of liver enzyme serum levels – alanine aminotransferase and aspartate aminotransferase notably – and to the regeneration of hepatocyte necrosis compared to CCl_4 or paracetamol intoxication (Al-Qarawi *et al.*, 2003; Mohamed *et al.*, 2015; Hanafy *et al.*, 2016). Additionally, the decreased levels of antioxidant enzymes such as superoxide dismutase, catalase and glutathione, observed in paracetamol treated rats and indicating tissue damage produced by free radicals, was reversed in liver tissue of the extract-treated animals, indicating a contribution to the antioxidant activity of the treatment (Hanafy *et al.*, 2016).

Food Product Opportunities

The numerous bioactive compounds and properties observed for baobab fruit pulp and the demonstration of favourable effects should promote interest for this food and its consumption, however, derived products from baobab are still limited for consumers.

The European Commission authorized baobab fruit pulp as a novel food ingredient in 2008, with 'generally recognized as safe' (GRAS) status. Approval as a food ingredient by the USA Food and Drug Administration was then acquired in 2009 (European Commission, 2008; US Food and Drug Administration, 2009), allowing baobab fruit pulp to enter the European and the US food markets and increasing demand for various products derived from it.

Common uses of baobab fruit pulp are related to its dissolution properties on aqueous solutions like water or milk, leading to use as a drink, alone or in combination with other fruit juices, as a sauce for food, as a fermenting agent in local brewing or as a substitute for cream in baking. The pulp is also a popular ingredient in ice products and baobab-based nectar, syrup and jam (Abdalla *et al.*, 2010; Caluwé *et al.*, 2010; Ryan *et al.*, 2014). Recently, baobab has been used as an ingredient in seasonings or for yogurts (Rahul *et al.*, 2015; Ismail *et al.*, 2019).

Considering food functions, the abundance of pectin and fibre in baobab fruit pulp would be an asset for thicker beverage development, such as smoothies, which have a high content of fruit pulp, and could be also used as a filler. The recent isolation of polysaccharides from Nigerian baobab fruits has opened new opportunities for valorization as a functional ingredient (Alba *et al.*, 2020). These xylogalacturonans are composed of around 69 mol% uronic acids and 13 mol% xylose and may have interesting colloidal properties which could be exploited in the areas of advanced texture modulation, next-generation biological emulsifiers or fibre-enriched foods, to act as prebiotics in helping to maintain microbiote balance and gut physiology.

Considering nutritional composition, the high mineral content, in particular of calcium and zinc, and the richness in antioxidant molecules has interesting potential in terms of improving local diets with locally available plants in developing countries where the risk of iron and zinc deficiencies still exists. The growing scientific interest in strategies of food-to-food fortification using vitamin C opulent baobab fruit pulp as a locally available food complement with cereals seems to be a very interesting and sustainable way to prevent mineral deficiencies in vulnerable populations. Several studies have already validated the concept of a baobab/cereal enrichment from a sensorial point of view for sorghum, rice cookies, maize and cowpea snack-bars (Mounjouenpou *et al.*, 2018; Netshishivhe *et al.*, 2019; Momanyi *et al.*, 2020). Enrichment of cereals with indigenous products has great potential in terms of improving macro- and micronutrient bioaccessibility in the resultant food products, thereby improving food security.

The bioactivity and health claims related to baobab fruit pulp make it a good candidate for a new generation of functional foods and drinks. However, further research in food processing, dietary formulation and nutrition is needed to diversify the range and maximize the potential of baobab derived food products.

Conclusions and Future Trends

Baobab is a long-used indigenous resource widely spread across Africa and every part of the plant is reported to be useful. The fruit pulp, mainly consumed as a drink, offers several health benefits in terms of qualitative nutritional composition and richness in bioactive molecules, even if further studies on bioaccessibility and bioavailability are necessary. The diversification of derived baobab pulp products would allow access to this local reservoir of bioactive compounds, in particular in populations suffering from nutritional deficiencies.

Acknowledgement

This work follows from the Interreg-V Qualinov 2 programme, a grant from the European Union and the Conseil Régional de La Réunion.

References

Abdalla, A., Yagoup, N.E.H. and Mudawi, H. (2010) Production and quality evaluation of baobab (*Adansonia digitata*) beverages. *Journal of Applied Sciences Research* 6, 729–741.

Adetola, O.Y., Kruger, J., White, Z. and Taylor, J.R.N. (2019) Comparison between food-to-food fortification of pearl millet porridge with moringa leaves and baobab fruit and with adding ascorbic and citric acid on iron, zinc and other mineral bioaccessibility. *LWT* 106, 92–97. DOI: 10.1016/j.lwt.2019.02.044.

Al-Qarawi, A.A., Al-Damegh, M.A. and El-Mougy, S.A. (2003) Hepatoprotective influence of *adansonia digitata* pulp. *Journal of Herbs, Spices & Medicinal Plants* 10, 1–6. DOI: 10.1300/J044v10n03_01.

Alba, K., Offiah, V., Laws, A.P., Falade, K.O. and Kontogiorgos, V. (2020) Baobab polysaccharides from fruits and leaves. *Food Hydrocolloids* 106, 105874. DOI: 10.1016/j.foodhyd.2020.105874.

Asogwa, I.S., Ibrahim, A.N. and Agbaka, J.I. (2021) African baobab: Its role in enhancing nutrition, health, and the environment. *Trees, Forests and People* 3, 100043. DOI: 10.1016/j.tfp.2020.100043.

Assogbadjo, A.E., Chadare, F.J., Kakaï, R.G., Fandohan, B. and Baidu-Forson, J.J. (2012) Variation in biochemical composition of baobab (*Adansonia digitata*) pulp, leaves and seeds in relation to soil types and tree provenances. *Agriculture, Ecosystems & Environment* 157, 94–99. DOI: 10.1016/j.agee.2012.01.021.

Baum, D.A. (1995) The comparative pollination and floral biology of Baobabs (*Adansonia*- Bombacaceae). *Annals of the Missouri Botanical Garden* 82(2), 322. DOI: 10.2307/2399883.

Besco, E., Braccioli, E., Vertuani, S., Ziosi, P., Brazzo, F. *et al.* (2007) The use of photochemiluminescence for the measurement of the integral antioxidant capacity of baobab products. *Food Chemistry* 102(4), 1352–1356. DOI: 10.1016/j.foodchem.2006.05.067.

Birhane, E., Asgedom, K.T., Tadesse, T., Hishe, H., Abrha, H. *et al.* (2020) Vulnerability of baobab (*Adansonia digitata* L.) to human disturbances and climate change in western Tigray, Ethiopia: Conservation concerns and priorities. *Global Ecology and Conservation* 22, e00943. DOI: 10.1016/j.gecco.2020.e00943.

Caluwé, E., Halamová, K. and Van Damme, P. (2010) *Adansonia digitata* L. – A review of traditional uses, phytochemistry and pharmacology. *Afrika Focus* 23(1). DOI: 10.21825/af.v23i1.5037.

Chadare, F.J., Linnemann, A.R., Hounhouigan, J.D., Nout, M.J.R. and Van Boekel, M.A.J.S. (2008) Baobab food products: a review on their composition and nutritional value. *Critical Reviews in Food Science and Nutrition* 49(3), 254–274. DOI: 10.1080/10408390701856330.

Ciqual: Table de composition nutritionnelle des aliments (n.d.). Available at: https://ciqual.anses.fr (accessed 1 April 2020).

Cisse, I., Montet, D., Reynes, M., Danthu, P., Yao, B. *et al.* (2013) Biochemical and nutritional properties of baobab pulp from endemic species of Madagascar and the African mainland. *African Journal of Agricultural Research* 8, 6046–6054. DOI: 10.5897/AJAR12.1231.

Coe, S.A., Clegg, M., Armengol, M. and Ryan, L. (2013) The polyphenol-rich baobab fruit (*Adansonia digitata* L.) reduces starch digestion and glycemic response in humans. *Nutrition Research* 33(11), 888–896. DOI: 10.1016/j.nutres.2013.08.002.

Daglia, M. (2012) Polyphenols as antimicrobial agents. *Current Opinion in Biotechnology* 23(2), 174–181. DOI: 10.1016/j.copbio.2011.08.007.

De Smedt, S., Cuní Sanchez, A., Van den Bilcke, N., Simbo, D., Potters, G. *et al.* (2012) Functional responses of baobab (*Adansonia digitata* L.) seedlings to drought conditions: Differences between western and south-eastern Africa. *Environmental and Experimental Botany* 75, 181–187. DOI: 10.1016/j.envexpbot.2011.09.011.

Diop, A.G., Sakho, M., Dornier, M., Cisse, M. and Reynes, M. (2006) Le baobab africain (*Adansonia digitata* L.): principales caractéristiques et utilisations. *Fruits* 61(1), 55–69. DOI: 10.1051/fruits:2006005.

EFSA (2013) Scientific opinion on dietary reference values for vitamin C. *EFSA Journal* 11, 3418. DOI: 10.2903/j.efsa.2013.3418.

Erwa, I., Shinger, M., Ishag, O., Ali, A., Ahmed, R. *et al.* (2019) Proximate and elemental composition of baobab fruit (*Adansonia digitata* L) pulp. *Journal of Chemical, Biological and Physical Sciences* 9(1). DOI: 10.24214/jcbps.B.9.1.04251.

European Commission (2008) Commission decision of 27 June 2008 authorising the placing on the market of Baobab dried fruit pulp as a novel food ingredient under Regulation. 2008/575/EC.

Fatema, S., Rode, P., Jadhav, S.B. and Farooqui, M. (2015) Anti-inflammatory and analgesic study of fibrous part of *Adansonia digitata* fruit using microwave extraction techniques. *Der Pharmacia Lettre* 7(5), 341–347.

Gabaza, M., Shumoy, H., Muchuweti, M., Vandamme, P. and Raes, K. (2018) Baobab fruit pulp and mopane worm as potential functional ingredients to improve the iron and zinc content and bioaccessibility of fermented cereals. *Innovative Food Science & Emerging Technologies* 47, 390–398. DOI: 10.1016/j.ifset.2018.04.005.

Gadour, M.O., Khidir, H.B., Adam, I. and Gasim, G.I. (2017) Effects of a powder of the fruit of *Adansonia digitata* (Tabaldia, Gongolase, or baobab tree) on serum lipids. *Journal of Herbal Medicine* 8, 14–16. DOI: 10.1016/j.hermed.2017.03.004.

Garcia, C., Guerin, M., Souidi, K. and Remize, F. (2020) Lactic fermented fruit or vegetable juices: Past, present and future. *Beverages* 6(1), 8. DOI: 10.3390/beverages6010008.

Ghani, A. and Agbejule, A.O. (1986) Pharmacognostic study of the fruits of *Adansonia digitata* L. In: *The State of Medicinal Plant Research in Nigeria*. University of Ife, Ife, Nigeria, pp. 181–185.

Ghoneim, M.A.M., Hassan, A.I., Mahmoud, M.G. and Asker, M.S. (2016) Protective effect of *Adansonia digitata* against isoproterenol-induced myocardial injury in rats. *Animal Biotechnology* 27(2), 84–95. DOI: 10.1080/10495398.2015.1102147.

Glew, R.H., Vanderjagt, D.J., Lockett, C., Grivetti, L.E., Smith, G.C. *et al.* (1997) Amino acid, fatty acid, and mineral composition of 24 indigenous plants of *Burkina Faso*. *Journal of Food Composition and Analysis* 10(3), 205–217. DOI: 10.1006/jfca.1997.0539.

Hanafy, A., Aldawsari, H.M., Badr, J.M., Ibrahim, A.K. and Abdel-Hady, S.E.-S. (2016) Evaluation of hepatoprotective activity of *Adansonia digitata* extract on acetaminophen-induced hepatotoxicity in rats. *Evidence-Based Complementary and Alternative Medicine* 2016, 4579149. DOI: 10.1155/2016/4579149.

Hiwilepo-van Hal, P., Bosschaart, C., van Twisk, C., Verkerk, R. and Dekker, M. (2012) Kinetics of thermal degradation of vitamin C in marula fruit (*Sclerocarya birrea* subsp. caffra) as compared to other selected tropical fruits. *LWT - Food Science and Technology* 49(2), 188–191. DOI: 10.1016/j.lwt.2011.12.038.

Hunt, J.R. (2005) Dietary and physiological factors that affect the absorption and bioavailability of iron. *International Journal for Vitamin and Nutrition Research* 75(6), 375–384. DOI: 10.1024/0300-9831.75.6.375.

Hussain, Z.T.E., Yagi, S., Mahomoodally, M.F., Mohammed, I. and Zengin, G. (2019) A comparative study of different solvents and extraction techniques on the anti-oxidant and enzyme inhibitory activities of *Adansonia digitata* L. (Baobab) fruit pulp. *South African Journal of Botany* 126, 207–213. DOI: 10.1016/j.sajb.2019.01.034.

Ismail, B.B., Pu, Y., Guo, M., Ma, X. and Liu, D. (2019) LC-MS/QTOF identification of phytochemicals and the effects of solvents on phenolic constituents and antioxidant activity of baobab (*Adansonia digitata*) fruit pulp. *Food Chemistry* 277, 279–288. DOI: 10.1016/j.foodchem.2018.10.056.

Ismail, B.B., Yusuf, H.L., Pu, Y., Zhao, H., Guo, M. *et al.* (2020) Ultrasound-assisted adsorption/desorption for the enrichment and purification of flavonoids from baobab (*Adansonia digitata*) fruit pulp. *Ultrasonics Sonochemistry* 65, 104980. DOI: 10.1016/j.ultsonch.2020.104980.

Iyengar, V., Pullakhandam, R. and Nair, K.M. (2010) Dietary ligands as determinants of Iron-Zinc interactions at the absorptive enterocyte. *Journal of Food Science* 75(8), H260–H264. DOI: 10.1111/j.1750-3841.2010.01796.x.

Kalita, D., Holm, D.G., LaBarbera, D.V., Petrash, J.M. and Jayanty, S.S. (2018) Inhibition of α-glucosidase, α-amylase, and aldose reductase by potato polyphenolic compounds. *PloS One* 13(1), e0191025. DOI: 10.1371/journal.pone.0191025.

Kamatou, G.P.P., Vermaak, I. and Viljoen, A.M. (2011) An updated review of *Adansonia digitata*: A commercially important African tree. *South African Journal of Botany* 77(4), 908–919. DOI: 10.1016/j.sajb.2011.08.010.

Khakimov, B., Mongi, R.J., Sørensen, K.M., Ndabikunze, B.K., Chove, B.E. *et al.* (2016) A comprehensive and comparative GC-MS metabolomics study of non-volatiles in Tanzanian grown mango, pineapple, jackfruit, baobab and tamarind fruits. *Food Chemistry* 213, 691–699. DOI: 10.1016/j.foodchem.2016.07.005.

Khoja, K.K., Aslam, M.F., Sharp, P.A. and Latunde-Dada, G.O. (2021) In vitro bioaccessibility and bioavailability of iron from fenugreek, baobab and moringa. *Food Chemistry* 335, 127671. DOI: 10.1016/j.foodchem.2020.127671.

Lamien-Meda, A., Lamien, C.E., Compaoré, M.M.Y., Meda, R.N.T., Kiendrebeogo, M. *et al.* (2008) Polyphenol content and antioxidant activity of fourteen wild edible fruits from Burkina Faso. *Molecules* 13(3), 581–594. DOI: 10.3390/molecules13030581.

Li, X.-N., Sun, J., Shi, H., Yu, L.L., Ridge, C.D. *et al.* (2017) Profiling hydroxycinnamic acid glycosides, iridoid glycosides, and phenylethanoid glycosides in baobab fruit pulp (*Adansonia digitata*). *Food Research International* 99(Pt 1), 755–761. DOI: 10.1016/j.foodres.2017.06.025.

Liu, J.-Y., Chen, C.-C., Wang, W.-H., Hsu, J.-D., Yang, M.-Y. *et al.* (2006) The protective effects of *Hibiscus sabdariffa* extract on CCl4-induced liver fibrosis in rats. *Food and Chemical Toxicology* 44(3), 336–343. DOI: 10.1016/j.fct.2005.08.003.

Lönnerdal, B. (2000) Dietary factors influencing Zinc absorption. *The Journal of Nutrition* 130(5S Suppl), 1378S-1383S. DOI: 10.1093/jn/130.5.1378S.

Mackenzie, B. and Garrick, M.D. (2005) Iron Imports. II. Iron uptake at the apical membrane in the intestine. *American Journal of Physiology. Gastrointestinal and Liver Physiology* 289(6), G981–G986. DOI: 10.1152/ajpgi.00363.2005.

Mohamed, M.A., Mohamed, I.A.A., El-Din, T., Hassan, A. and Hassan, M. (2015) Hepatoprotective effect of *Adansonia digitata* L. (baobab) fruits pulp extract on ccl4-induced hepatotoxicity in rats. *World Journal of Pharmaceutical Research* 4(8), 368–377.

Momanyi, D., Owino, W. and Makokha, A. (2020) Formulation, nutritional and sensory evaluation of baobab based ready-to-eat sorghum and cowpea blend snack bars. *Scientific African* 7, e00215. DOI: 10.1016/j.sciaf.2019.e00215.

Mounjouenpou, P., Ngono Eyenga, S.N.N., Kamsu, E.J., Bongseh Kari, P., Ehabe, E.E. *et al.* (2018) Effect of fortification with baobab (*Adansonia digitata* L.) pulp flour on sensorial acceptability and nutrient composition of rice cookies. *Scientific African* 1, e00002. DOI: 10.1016/j.sciaf.2018.e00002.

Murray, S.S., Schoeninger, M.J., Bunn, H.T., Pickering, T.R. and Marlett, J.A. (2001) Nutritional composition of some wild plant foods and honey used by Hadza Foragers of Tanzania. *Journal of Food Composition and Analysis* 14(1), 3–13. DOI: 10.1006/jfca.2000.0960.

Muthai, K.U., Karori, M.S., Muchugi, A., Indieka, A.S., Dembele, C. *et al.* (2017) Nutritional variation in baobab (*Adansonia digitata* L.) fruit pulp and seeds based on Africa geographical regions. *Food Science & Nutrition* 5(6), 1116–1129. DOI: 10.1002/fsn3.502.

National Institutes of Health (2019) Office of dietary supplements - vitamin C. U.S. Department of Health and Human Sciences - National Institutes of Health. Available at: https://ods.od.nih.gov/factsheets/VitaminC-Consumer (accessed 13 January 2021).

Netshishivhe, M., Omolola, A.O., Beswa, D. and Mashau, M.E. (2019) Physical properties and consumer acceptance of maize-baobab snacks. *Heliyon* 5(3), e01381. DOI: 10.1016/j.heliyon.2019.e01381.

Nour, A.A., Magboul, B.I. and Kheiri, N.H. (1980) Chemical composition of baobab fruit (*Adansonia digitata* L.). *Tropical Science* 22, 383–388.

Obizoba, I.C. and Anyika, J.U. (1994) Nutritive value of baobab milk (gubdi) and mixtures of baobab (*Adansonia digitata* L.) and hungry rice, acha (*Digitaria exilis*) flours. *Plant Foods for Human Nutrition* 46(2), 157–165. DOI: 10.1007/BF01088768.

Odetokun, S.M. (1996) The nutritive value of Baobab fruit (*Adansonia digitata*). *Rivista Italiana Delle Sostanze Grasse* 73, 371–373.

Rahul, J., Jain, M.K., Singh, S.P., Kamal, R.K., Anuradha A. *et al.* (2015) *Adansonia digitata* L. (baobab): a review of traditional information and taxonomic description. *Asian Pacific Journal of Tropical Biomedicine* 5(1), 79–84. DOI: 10.1016/S2221-1691(15)30174-X.

Razafindralambo, N.K., Lacoste, V., Olive, M.-M., Barivelo, T.A. and Reynes, J.-M. (2014) Anjozorobe hantavirus, a new genetic variant of Thailand virus detected in rodents from Madagascar. *Vector Borne and Zoonotic Diseases* 14(3), 212–219. DOI: 10.1089/vbz.2013.1359.

Ryan, L., Fraser, A. and Thondre, P.S. (2014) BAOBAB (*Adansonia digitata* L.) reduces the glycaemic and insulin response in healthy volunteers. *Journal of Nutrition & Intermediary Metabolism* 1, 44. DOI: 10.1016/j.jnim.2014.10.162.

Sena, L.P., Vanderjagt, D.J., Rivera, C., Tsin, A.T., Muhamadu, I. *et al.* (1998) Analysis of nutritional components of eight famine foods of the Republic of Niger. *Plant Foods for Human Nutrition* 52(1), 17–30. DOI: 10.1023/a:1008010009170.

Seukep, J.A., Fankam, A.G., Djeussi, D.E., Voukeng, I.K., Tankeo, S.B. *et al.* (2013) Antibacterial activities of the methanol extracts of seven Cameroonian dietary plants against bacteria expressing MDR phenotypes. *SpringerPlus* 2, 363. DOI: 10.1186/2193-1801-2-363.

Sidibe, M., Williams, J.T., Hughes, A., Haq, N. and Smith, R.W. (2002) *Baobab, Adansonia Digitata L.* International Centre for Underutilised Crops, Southampton, UK.

Stadlmayr, B., Wanangwe, J., Waruhiu, C.G., Jamnadass, R. and Kehlenbeck, K. (2020) Nutritional composition of baobab (*Adansonia digitata* L.) fruit pulp sampled at different geographical locations in Kenya. *Journal of Food Composition and Analysis* 94, 103617. DOI: 10.1016/j.jfca.2020.103617.

Suliman, H.M., Osman, B., Abdoon, I.H., Saad, A.M. and Khalid, H. (2020) Ameliorative activity of *Adansonia digitata* fruit on high sugar/high fat diet-simulated metabolic syndrome model in male Wistar rats. *Biomedicine & Pharmacotherapy = Biomedecine & Pharmacotherapie* 125, 109968. DOI: 10.1016/j.biopha.2020.109968.

Szwajgier, D. (2013) Inhibition of cholinesterases by phenolic acids detected in beer: A dose-response model approach. *African Journal of Biotechnology* 14, 1675–1681.

Tal-Dia, A., Toure, K., Sarr, O., Sarr, M., Cisse, M.F. *et al.* (1997) A baobab solution for the prevention and treatment of acute dehydration in infantile diarrhea. *Dakar Medical* 42(1), 68–73.

Tembo, D.T., Holmes, M.J. and Marshall, L.J. (2017) Effect of thermal treatment and storage on bioactive compounds, organic acids and antioxidant activity of baobab fruit (*Adansonia digitata*) pulp from Malawi. *Journal of Food Composition and Analysis* 58, 40–51. DOI: 10.1016/j.jfca.2017.01.002.

Thondre, P.S., Ryan, L. and Henry, C.J.K. (2011) Influence of polyphenols on *in vitro* starch digestibility of oat porridges. *Proceedings of the Nutrition Society* 70(OCE4). DOI: 10.1017/S0029665111001893.

Tsetegho Sokeng, A.J., Sobolev, A.P., Di Lorenzo, A., Xiao, J., Mannina, L. *et al.* (2019) Metabolite characterization of powdered fruits and leaves from *Adansonia digitata* L. (baobab): A multi-methodological approach. *Food Chemistry* 272, 93–108. DOI: 10.1016/j.foodchem.2018.08.030.

US Food and Drug Administration (2009) Agency Response Letter GRAS Notice no.GRN 000273. Available at: https://www.fda.gov/food/gras-notice-inventory/agency-response-letter-gras-notice-no-grn-000623 (accessed 6 April 2022).

Vertuani, S., Braccioli, E., Buzzoni, V. and Manfredini, S. (2002) Antioxidant capacity of *Adansonia digitata* fruit pulp and leaves. *Acta Phytotherapeutica* 5, 6.

18 Jatobá (*Hymenaea courbaril* L.)

André Gonzaga dos Santos*, Kátia Sivieri, Bárbara Paes Miglioli da Mata, Mateus Kawata Salgaço and Luis Vitor Silva do Sacramento
School of Pharmaceutical Sciences, São Paulo State University - Unesp, Araraquara city, São Paulo state, Brazil

Abstract

The fruit of *Hymenaea courbaril* L. is named jatobá in Brazil. The genus *Hymenaea* – Fabaceae-Caesalpinioideae – is common in Central and South American forests and Brazilian savannah. The floury pulp around the seeds is the main edible part, which is yellowish, sweet and has a characteristic odour. It is eaten fresh or used as an ingredient in traditional recipes. The nutritional composition of jatobá pulp highlights the fibres as the main component which may be related to several health-promoting effects. Jatobá pulp is also a source of protein and minerals. *H. courbaril* fruits present a diversity of secondary metabolites including sesquiterpenes, labdane- and clerodane-type diterpenes, carotenoids, biscoumarins and flavonoids. Phenolic compounds and carotenoids are important due to their known antioxidant activity. The pulp also contains labdane diterpenes which have showed anti-inflammatory action. The biscoumarins were isolated from the seeds and their toxicity must be further investigated. Besides the nutritional properties of the fruit, the different parts of the plant are also employed in traditional medicine in South America and its biological activities have been evaluated throughout the years. The fruits and other parts of the plant have presented antimicrobial, antileishmanial, antiplasmodial, myorelaxant, antioxidant and 5-lipoxygenase inhibition activities. Jatobá pulp consumption can promote nutritional and health benefits and the pulp can be processed by the food industry. The fruit can also be used for the production of food additives, pharmaceutical and cosmetic ingredients.

Introduction

Plants are sources of micronutrients and macronutrients necessary for good health. Brazil is considered privileged in the diversity of fruit species. One of these species is the *Hymenaea courbaril* L. – jatobazeiro, which is consumed and appreciated by the Brazilian local population (Arkcoll, 1984; Kinnup and Lorenzi, 2014). The species occurs in various areas of South America, mainly in the north of the Amazon. In Brazil, it occurs in Amazon rainforest, Caatinga, the central Brazilian savannah (Cerrado), Atlantic rainforest and Pantanal biomes (Oliveira *et al.*, 2018; Reflora, 2020).

The fruit of *H. courbaril* is named jatobá in Brazil (Fig. 18.1). The floury pulp around the seeds is the main edible part and is yellowish, sweet and has a characteristic odour. It is eaten fresh or used as an ingredient in porridges, breads, ice creams, regional dishes and beverages (Arkcoll, 1984; Kinnup and Lorenzi, 2014; Boniface *et al.*, 2017). Jatobá may be classified as a non-conventional edible plant (PANC) according to Kinnup and

*Corresponding author: andre.gonzaga@unesp.br

 Handbook of Phytonutrients in Indigenous Fruits and Vegetables (eds D. Sivakumar *et al.*)
DOI: 10.1079/9781789248067.0018

Fig. 18.1. *Hymenaea courbaril* L., tree and fruit.

Lorenzi (2014) because it is not part of the conventional agricultural matrix and is little-known by most people of certain regions or countries.

Besides the nutritional properties of the fruit, different parts of the plant are also employed in traditional medicine in South America and their biological activities have been evaluated throughout the years (Oliveira *et al.*, 2018).

Botany

The genus *Hymenaea* – Fabaceae-Caesalpinioideae – was transferred to Fabaceae-Detarioideae according to Azani *et al.* (2017). The *Hymenaea* species are common in Central and South American forests or the Cerrado – Brazilian savannah – landscapes (Rizzini, 2000; Ribeiro *et al.*, 2015; Zappi *et al.*, 2015). *Hymenaea* has resin-producing species and is traditionally employed as food and medicine (Kinnup and Lorenzi, 2014; Oliveira *et al.*, 2018).

Hymenaea courbaril L. – jatobazeiro – is a tree species reaching up to 20 m in height, with heliophyte characteristics. It blooms between the months of October and December and the fruits ripen from July (Lorenzi, 2014). In general, the leaves are alternate and bifoliolate, the corymbiform inflorescences have a tetramerous calyx, a pentamerous corolla, ten stamens, a glabrous or pubescent ovary and a hard, indehiscent fruit with a floury pulp (Pinto *et al.*, 2015). Botelho *et al.* (2000) carried out a morphological study of the fruit in which they reported that the floury pulp corresponds to the development of the seed aril.

H. courbaril has oblong to cylindrical fruits, which are 8–15 cm long and 4–6 cm in diameter. The fruits are green in colour when immature, dark brown when ripe and black when old. The epicarp is thick and dark brown, the endocarp thick and yellow and the pulp floury, sweet and yellowish. The fruits are composed of seeds (25 to 40% by weight), seed pods (50 to 70%) and pulp only 5 to 10% (Botelho *et al.*, 2000; Costa *et al.*, 2010; Lorenzi, 2014).

Traditional Use

H. courbaril has been used as a medicinal plant by the indigenous tribes of South America and also by non-indigenous people (Boniface *et al.*, 2017). All parts of *H. courbaril* are employed in traditional medicine, mainly bark, leaves, sap, resin and fruit (Barrett, 1994; Albuquerque *et al.*, 2007; Mendes and Carlini, 2007; Araújo *et al.*, 2008;

Chaves and Barros, 2012). It has been reported as being wound-healing, anti-inflammatory (Araújo *et al.*, 2008), a tonic during pregnancy (Coe and Anderson, 1996), a febrifuge and for the treatment of rheumatism (Cáceres *et al.*, 1987), leucorrhea, vaginitis (Cáceres *et al.*, 1993), malaria (Cartaxo *et al.*, 2010), gastric disorders, diarrhoea, dysentery and constipation, among other several uses (Boniface *et al.*, 2017).

Some articles cite the use of this species as a tonic, fortifying and energizing, acting to restore organic strength and to increase blood production, fighting weakness in children and weakness in general (Barrett, 1994; Mendes and Carlini, 2007). The bark tea and the ethanol bark extract ('vinho de jatobá') have been used for these purposes in Brazil (Boniface *et al.*, 2017).

In a study developed in a rural community of the Ceará state, Brazil, the local diversity of medicinal plants was recorded as 119 species used for the treatment of 92 diseases. Nineteen of these species had a great versatility of use, including *H. courbaril*. This study confirms the several uses of this species, reported by the population: anaemia, prostate disorders, kidney disorders, diluted blood, gastric disorders, influenza, bronchitis, expectorant, cancer (leukemia), lice, lung disorders, inflammation of the throat and cold sores (Coe and Anderson, 1996).

Nutritional Composition

The total fibres are the most important functional component in jatobá fruit pulp. They are the main component of the pulp (the edible part of the fruit) (Silva *et al.*, 2001). These fibres are important for the human body to stimulate the growth of the intestinal microbiota. More than half of the pulp is composed of total fibres (50.02%), followed by carbohydrates (32.20%), proteins (2.32%), ash (4.53%) and lipids (1.94%) (Dias *et al.*, 2013). Dietary fibre values in 75% of the samples were higher than 30% – very close to those described for the main source of fibre of the Brazilian population, black beans. The extremely high value of 30% dietary fibre in 100 g of jatobá pulp represents 200% of the recommended daily intake (RDI) (FDA, 2010). The consumption of dietary fibre has been related to several health-promoting effects, such as the prevention of cardiovascular diseases, diabetes, digestive tract diseases, overweight and obesity (Duranti, 2006).

In general, fruits are not great sources of protein, however, jatobá fruit pulp presents 6–12% protein (Table 18.1) and this value is relatively high when compared to other fruits such as lime orange (1.1%), banana nanica (1.4%) and fuji apple (0.3%) according to data from Taco (2011). Jatobá fruit pulp presents 2–8% of lipids and this value is relatively low since some fruit from Brazilian biomes, such as pequi (18%) and macaúba (40%), have higher values (Taco, 2011).

Another important macronutrient is carbohydrates, which supply the energy used in vital organic functions and for other activities in an organism (Ordóñez Pereda, 2005). In general, fruits of the Brazilian biome are rich in sugars as described by Martin *et al.* (2003), who cited baru (12.80%), mangaba (4.34%) and lobeira (5.74%), among others. When compared with jatobá the values obtained are lower since the total carbohydrates can vary from 3 to 34%.

Jatobá pulp is a rich source of minerals as indicated by its high ash content (Dias *et al.*, 2013). The mineral components of the pulp (mg/100 g), according to Dias *et al.* (2013), are sodium (468.15), potassium (229.40), phosphorus (137.60), magnesium (99.15), calcium (57.39), manganese (7.62), copper (2.07), iron (1.75), boron (1.32), zinc (1.14) and selenium (0.27). In addition to these macro and micronutrients, 100 g of jatobá fruit pulp contains 121.5 mg of vitamin C. In terms of energy, 100 g of jatobá pulp has 115 kcal (Schwartz, 2018).

The nutritional composition of jatobá seed is shown in Table 18.1. Most of the components surveyed presented high levels when considering RDI. Due to their balanced composition, rich in proteins, oil and digestible carbohydrates, the seeds showed high energy values, which ranged from 556 to 1804 kJ 100 g^{-1}, and the values of digestible carbohydrates varied from 3 to 34%. These energy values are comparable to those reported for black beans (1413 kJ 100 g^{-1}), and based on FDA regulations these values can be considered as good for high energy sources, representing 7–22% of RDI (FDA, 2010).

The seeds presented high levels (mg/100 g) of seed flour – potassium 139.65–366.0,

Table 18.1. Nutritional composition of *Hymnaea courbaril* L. pulp and seeds from Brazil compared to RDI standards based on an intake of 2000 kcal (8368 kJ).

Nutrient	Pulp: Dias *et al.* (2013)	Pulp: SiBBr (2018)	Seeds: Dias *et al.* (2013)	Seeds: Carvalho *et al.* (2011)	Reference Daily Intake (RDI), FDA (2010)
Lipid (g/100 g)	1.94	2.53	5.78	8.0	65 (g day^{-1})
Ash (g/100 g)	4.53	nd	1.95	2.5	-
Protein (g/100 g)	12.32	5.65	11.32	10.9	50 (g day^{-1})
Crude fibre (g/100 g)	50.02	51.0	72.14	77.8	25 (g day^{-1})
Carbohydrate (g/100 g)	32.20	25.21	8.81	nd	300 (g day^{-1})
Calories (kJ /100 g)	481	1026	556	1804	8368 (kJ day^{-1})
Cu (μg/100 g)	nd	nd	3350	1040	2000 (μg day^{-1})
Mn (mg/100 g)	99.15	117	3.90	2.5	2.0 (mg day^{-1})
Fe (mg/100 g)	0	0	3.43	3.8	18 (mg day^{-1})
Zn (mg/100 g)	1.14	17.98	2.01	1.5	15 (mg day^{-1})
Ca (mg/100 g)	57.39	40	89.50	58.0	1000 (mg day^{-1})
K (mg/100 g)	229.40	1533	139.65	366.0	3500 (mg day^{-1})
Mg (mg/100 g)	7.62	117	82.47	117.0	400 (mg day^{-1})

[a] nd: not determined by the authors.

4–11% of RDI; magnesium 82.47–117.0, 20–29% of RDI; iron 3.43–3.8, 19–21% of RDI; zinc 2–1.5, 13–10% of RDI and manganese 3.9–2.5, 195–125% of RDI. Similarly, they presented large amounts (μg/ 100 g seed) of copper (3350–1040, 52–168% RDI). The seeds are also good sources of calcium (58–89 mg/100 g, 5.8–8.9% RDI) (Carvalho *et al.*, 2011; Dias *et al.*, 2013). Therefore, these seeds are good sources of minerals, because their levels are similar to those of beans, partially or totally meeting the mineral needs established by the RDI doses shown in Table 18.1.

Antinutritive Compounds

The presence of antinutritional and/or toxic components is one of the main limiting factors for the use of the nutritional qualities of legumes (Liener, 1994). Dietary trypsin inhibitors are blamed for the poor digestibility of dietary protein and interference with the proper function of trypsin can lead to growth inhibition and pancreatic hypertrophy. Carvalho *et al.* (2011) quantified the inhibitors of trypsin of jatobá seed flour being on average 4.5 μg TI/ mg flour. Despite their antinutritional effects, trypsin inhibitors are often inactivated by appropriate heat treatment (Maia *et al.*, 2000), as shown for lupin, soybean and black beans (Trugo *et al.*, 2000).

In relation to the activity of urease, Carvalho *et al.* (2011) demonstrated that the seeds of *H. courbaril* presented values of 620 ± 28 U/ kg flour, which is low when compared with the soybean cultivars 'bays' and 'rio balsas' (Vasconcelos *et al.*, 2001).

Phytochemicals

The main secondary metabolites in the *Hymenaea* genus are sesquiterpenes, diterpenes and flavonoids. Steroids, biscoumarins, phenolic acids, proanthocyanidins and polysaccharides – xyloglucans – were also described for the genus. *H. courbaril* fruit contain sesquiterpenes, diterpenes, carotenoids, biscoumarins, flavonoids and

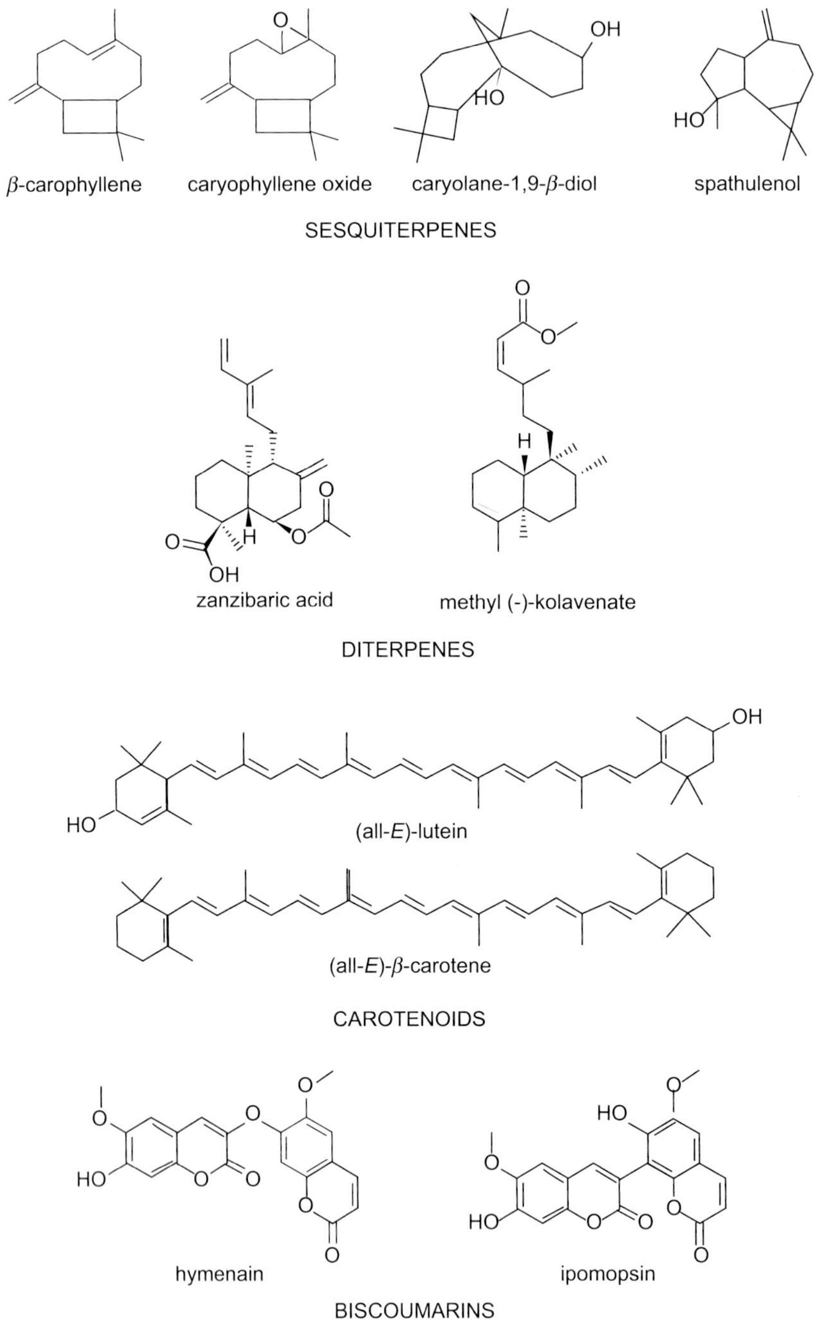

Fig. 18.2. Representative secondary metabolites from *Hymenaea courbaril* L. fruit.

xyloglucans (Fig. 18.2) (Boniface *et al.*, 2017; Oliveira *et al.*, 2018).

Phenolic compounds and carotenoids are described for the fruit pulp. The content of total phenolic compounds was determined in the fruit pulp by Folin-Ciocalteu photometric assay; the results for fruits of different regions of Brazil and Colombia ranged from 0.96 to 1.15 mg gallic acid equivalents per g of fresh fruit pulp (Contreras-Calderón *et al.*, 2011; Biazotto *et al.*, 2019). Even though Biazotto *et al.* (2019) described the analysis of the mesocarp, the same work refers to the analysed part as the edible part; thus, it may be the pulp instead of the fruit mesocarp. Figueiredo *et al.* (2016) determined the phenolic compound content for the 70% ethanol and methanol extracts of pulp and fruit peel as 3.4, 13.5, 211.3, and 187.3 mg gallic acid equivalents per g of dry extract, respectively. The phenolic compounds identified through FIA-ESI-IT-MS analysis in the pulp include genipin derivative, gardoside derivative iridoids, diosmetin-rhamnoside, cirsiliol-dihexoside, apigenin-hexoside flavones, isorhamnetin-dihexoside and quercetin-pentoside flavonols (Biazotto *et al.*, 2019).

The content of carotenoids was determined by HPLC-PDA/UV in the pulp of the fruits from three different regions of Brazil. The total carotenoid content was 11.7, 26.8 and 40.8 µg per g of fresh pulp. The major carotenoids were (all-*E*)-lutein – 5.4, 12.7 and 16.0 µg per g of fresh pulp, and (all-*E*)-*β*-carotene – 3.5, 8.0 and 13.3 µg per g of fresh pulp, (all-*E*)-neochrome, (9*Z*)-lutein, (all-*E*)-*β*-cryptoxanthin, 5,6-epoxy-*β*-carotene and (9*Z*)-*β*-carotene were the minor carotenoids (Biazotto *et al.*, 2019).

Many labdane diterpenes were isolated from the seed pods, including crotomachlin, labdanolic acid, labd-13*E*-en-8-ol-15-oic acid, methyl labd-13*E*-en-8-ol-15-oate, 13*E*-labda-7,13-dien-15-oic acid, labd-8-(17),13*E*-dien-15-oic acid, (-)-ozic acid, (-)-isoozic acid and methyl (-)-zanzibarate (Nogueira *et al.*, 2001, 2002b; Jayaprakasam *et al.*, 2007). Moreover, Nogueira *et al.* (2002b) identified methyl isoozate and methyl ozate in the seed pods and Khoo *et al.* (1973) described labdane diterpenes from the seed pod resin. The labdane diterpenes zanzibaric acid, methyl (-)-zanzibarate and isoozic acid were isolated from the fruit epicarp (Imamura *et al.*, 2004; Aguiar *et al.*, 2010). Clerodane diterpenes were also isolated from the seed pods: (5*R**,8*S**,9*S**,10*R**)-cleroda-3,13*E*-dien-15-oic acid, methyl (5*S**,8*S**,9*S**,10*R**)-cleroda-3,13*E*-dien-15-oate, (-)-kovalenic acid and methyl (-)-kovalenate (Nogueira *et al.*, 2001, 2002a).

The studies on the chemical composition of essential oils from *H. courbaril* fruit peel and seed pod resin have revealed only sesquiterpenes in their composition. The essential oils from the peel of ripe and unripe fruit were obtained by hydrodistillation, yielding 0.04 and 0.07% (w/w) on a fresh basis, respectively. GC-FID and GC-MS analysis determined the content of sesquiterpene hydrocarbons and oxygenated sesquiterpenes for ripe fruit peel (57.5 and 28.6%) and for unripe fruit peel (88.7 and 4.6%). The main essential oil constituents of the ripe fruit peel were *α*-copaene – 11.1%, spathulenol – 10.1%, *β*-selinene – 8.2%, *γ*-muurolene – 7.9%, caryophyllene oxide – 6.9%, *δ*-selinene – 5.7% and *δ*-amorphene – 5.3%. The major essential oil compounds of unripe fruit peel were germacrene D – 31.9%, *β*-caryophyllene – 27.1%, bicyclogermacrene – 6.5%, *α*-humulene – 4.2% and *α*-copaene – 4.2% (Aguiar *et al.*, 2010). In another study on the essential oil of the fruit peel, the main compounds identified by GC-MS were germacrene D – 17.6%, *β*-caryophyllene – 17.6%, caryophyllene oxide – 14.7%, *α*-copaene – 8.5%, bicyclogermacrene – 6.5%, germacrene B – 4.5%, *δ*-cadinene – 4.4% and *γ*-muurolene – 4.0% (Sales *et al.*, 2014). The sesquiterpenes caryolane-1,9-*β*-diol and spathulenol were isolated from the ripe fruit peel and from the seed pods, respectively (Jayaprakasam *et al.*, 2007; Aguiar *et al.*, 2010). The sesquiterpenes cyclosativene – 17.3%, selina-4(14),7(11)-diene – 16.2%, caryophyllene – 13.5% and *α*-himalachene – 13.5% – were identified as major components in the seed pod resin (Khoo *et al.*, 1973).

The total phenolic compound content – Folin-Ciocalteu photometric assay – in the 70% ethanol and methanol seed extracts was 464.3 and 396.7 mg gallic acid equivalents per g of dry extract, respectively (Figueiredo *et al.*, 2016). The total flavonoid content – aluminum chloride photometric assay – of the 70% ethanol seed extract was 442.2 mg rutin equivalents per g of dry extract (Spera *et al.*, 2019). Furthermore, four biscoumarins were isolated from the seeds: hymenain, ipomopsin (Simões *et al.*, 2009), hymenain-*O*-*β*-glucopyranosyl-(1'''→2'')-*O*-*α*-apiofuranosyl-(1''''→2''')-O-*α*-galactopyranoside and hymenain-7-*O*-*β*-glucopyranosyl-(1'''→2'')-*O*-*α*-apiofuranoside (Fernandes *et al.*, 2015b). The presence of coumarins in the seeds may be a concerning factor for their consumption as food and must be further investigated.

Phytochemical studies on secondary metabolites from leaves and stems were also published for the species. The total phenolic compound content – Folin-Ciocalteu photometric assay – in the 70% ethanol and methanol leaf extract was 208.2 and 236.3 mg gallic acid equivalents per g of dry extract, respectively (Figueiredo *et al.*, 2016). In addition, (-)-epicatechin and 3-*O*-rhamnosyl-3',4',5,7-tetrahydroxyflavanonol (astilbin) were isolated from the leaves (Artavia *et al.*, 1995).

The major sesquiterpene hydrocarbons identified in the leaf resin in different studies were: a) caryophyllene – 26.6%, *β*-selinene – 18.4%, *α*-selinene – 17.6%, *α*-copaene – 9.4%, and *δ*-cadinene – 6.6% (Martin *et al.*, 1972); b) α- or *β*-selinene – 45.3%, caryophyllene – 21.7%, *β*-copaene – 8.5%, *δ*-cadinene – 6.0% and γ-cadinene - 5.9% (Stubblebine *et al.*, 1975); c) *α*- and *β*-selinene – 8.4 to 65%, caryophyllene oxide – 0.1 to 30.7%, *β*-humulene – 1.7 to 12.0%, *β*-copaene – 1.0 to 8.2%, cyperene – 1.5 to 7.5% and *α*-copaene – 2.8 to 6.8% (Arrhenius and Langenheim, 1983). According to Costa *et al.* (2017), the content of essential oil in the dry leaves was 0.98% and its major constituents were caryophyllene oxide – 29.6% and *β*-caryophyllene – 18.8%. Furthermore, three labdane diterpenes were isolated from the leaves: perua-7,13-dien-15-oic acid, lab-13-en-8*β*-ol-15-oic acid and labdan-8*β*-ol-15-oic acid (Marsaioli *et al.*, 1975).

The phenolic compounds identified in the stem are B-type proanthocyanidins (Sasaki *et al.*, 2009). The flavonoids fisetin, fisetinidol, fustin, 3-*O*-methyl-fustin, taxifolin, and astilbin were identified in the trunk xylem sap (Bezerra *et al.*, 2013; Costa *et al.*, 2014). Diterpenes were also described for the stem bark: methyl (-)-copalate, methylisoozate, methylozate, methyl(-)-eperuate (labdane-type), methyl (-)-(5*R*,8*S*,9*S*,10*R*)-clerod-3-en-15-oate, methyl kovalenate, methyl (5*S**,8*S**,9*S**,10*R**)-clero-3,13-dien-15-oate and methyl (5*R**,8*S**,9*S**,10*R**)-clero-3,13-dien-15-oate (clerodane-type) (Nogueira *et al.*, 2002b). Moreover, two labdane diterpenes were isolated from the stem bark resin: copalic acid, (5*S**,9*S**,10*R**)-*ent*-labd-8(17)-en-15 ethyl acetate (Cunningham *et al.*, 1974; Bandeira *et al.*, 2015). Three *ent*-halimane diterpenes were isolated from a mixture of leaves and stems (Abdel-Kader *et al.*, 2002).

Xyloglucan is a hemicellulosic polysaccharide present in cell walls. The seeds of *H. courbaril* contain 40–45% of xyloglucan as a reserve polysaccharide. Its main chain is a cellulosic-type (1→4)-linked to *β*-D-glucan and the side chains consist of *α*-D-xylopyranosyl and *β*-D-galactopyranosyl-(1→2)-*α*-D-xylopyranosyl residues (1→6)-linked to the main chain (Lima *et al.*, 1993, 1995; Buckeridge *et al.*, 1997; Vargas-Rechia *et al.*, 1998). A fucogalactoxyloglucan was isolated from the leaves (Busato *et al.*, 2001, 2005).

Bioactive Properties

The ethnopharmacological studies conducted in traditional communities in Brazil and Nicaragua considered the species among those with the greatest versatility of indications, being useful in several diseases and different systems of the human body (Cartaxo *et al.*, 2010; Thomas, 2012; Saraiva *et al.*, 2015). Below, we report several references dealing with bioactive properties found in different parts of the plant.

Fruit

Martins *et al.* (2010) demonstrated the *in vitro* antibacterial activity of the 96% ethanol pulp extract against *S. aureus*, *E. faecalis* and *S. flexineri* by the agar diffusion method. In the broth dilution method, the value of minimum inhibitory concentration (MIC) for these bacteria was >400 μg/ml and for *E. coli* and *P. aeruginosa* was 350 μg/ml.

The *in vitro* antibacterial activity of the 96% ethanol fruit peel extract against clinical isolates of *S. aureus*, *E. coli* and *P. aeruginosa* was evaluated through the broth dilution method and the MIC value was 350 μg/ml (Martins *et al.*, 2010). The essential oil of the fruit peel presented antibacterial activity against oxacillin susceptible strains of *S. aureus*. Moreover, a synergistic modulating effect was observed when it was associated with antibiotics for clinical use (Sales *et al.*, 2014). The essential oil of the ripe fruit peel showed higher larvicidal activity against *Aedes aegypti* larvae than the essential oil of the unripe fruit – LC_{50} values of 14.8 and 28.4 μg/ml, respectively (Aguiar *et al.*, 2010).

Labdane diterpenes and spathulenol isolated from the seed pods inhibited COX-1 and -2 enzymes and lipid peroxidation. Among the tested compounds, labd-13*E*-en-8-ol-15-oic acid methyl ester and labdanolic acid inhibited COX-2 enzyme selectively – 40 and 43%, respectively – and (13*E*)-labda-7,13-dien-15-oic acid inhibited lipid peroxidation by 75% (Jayaprakasam *et al.*, 2007).

The 70% ethanol seed extract showed cytotoxic effect against B16F10 melanoma cell line, antigenotoxic effect on mice bone marrow cells and present antioxidant activity in DPPH radical assay (Spera *et al.*, 2019).

Leaves

A study including 14 Brazilian medicinal plants used in the treatment of diarrhoea, dysentery and intestinal disorders revealed that the 95% ethanol leaf extract showed the greatest effect against simian rotavirus (Cecílio *et al.*, 2012). The ethanolic and hexane extracts of *H. courbaril* leaves were the third and fourth most potent, respectively, among more than 40 extracts tested against *Leishmania amazonensis* (Ribeiro *et al.*, 2014). The essential oil of the leaves was tested for fungicidal activity against *Candida albicans*, *C. krusei*, *C. glabrata* and *C. parapsilosis* and the MIC values were 0.625 or 1.25 μl/ml; only sesquiterpenes were identified and caryophyllene oxide and β-caryophyllene were determined as the major compounds (Costa *et al.*, 2017). The leaf resin inhibited the growth *in vitro* of the fungi *Pestalotia subcuticularis*, an action which may also be related to the sesquiterpenes (Arrhenius and Langenheim, 1983).

Stem, heartwood and bark

Considering the traditional use of jatobá as an antipyretic and antimalarial, Köhler *et al.* (2002) performed a test with the petroleum ether: ethyl acetate 1:1 extract of the bark against *Plasmodium falciparum* and the IC_{50} value was 11.8 μg/ml. Similarly, the 95% ethanol bark extract showed antibacterial activity against seven different strains of *Escherichia coli* collected from pigs, with MIC values of 62.5 and 125.0 μg/ml (Fernandes *et al.*, 2015a).

The ethyl acetate extract from the stem bark strongly sequestered the DPPH radical and showed myorelaxant activity in rat trachea. Bioguided fractionation allowed the isolation of the flavonoid astilbin, whose activity indicates that it may be partially responsible for the myorelaxant action, since it reduces the contractions that depend on the influx of Ca^{2+} through voltage-dependent Ca^{2+} channels. In addition, oral treatment with the ethyl acetate extract fraction prevented hyperresponsiveness in KCl-induced contraction and decreased the total number of eosinophils and other white cells (Bezerra *et al.*, 2013).

The aqueous extract from the mixture of stem bark, leaves and heartwood of jatobá presented antifungal activity against some strains of *Cryptococcus neoformans*, evaluated through the inhibition zone from 8 to 14 mm at a concentration of 4 mg/ml (Barbosa Junior *et al.*, 2015). Moreover, the methanolic heartwood extract and its components fisetidinol and taxifolin showed antioxidant activity similar to α-tocopherol in the DPPH assay (Imai *et al.*, 2008).

The stem resin has traditionally been reported in the treatment of anaemia, prostate disorders, inflammation (Pinto and Maduro, 2003) and bronchitis, and it inhibited the 5-lipoxygenase *in vitro*, which corroborates its use as an anti-inflammatory (Braga *et al.*, 2000).

Among 16 strains of bacteria evaluated on agar diffusion method, the growth of ten strains belonging to the species *Micrococcus luteus*, *Staphylococcus aureus*, *Bacillus stearothermophylus* and *Escherichia coli* were inhibited by the 70% ethanol stem bark and and five strains of *Micrococcus luteus*, *Staphylococcus aureus* and *Streptococcus* sp were inhibited by the lyophilized stem sap. (Fernandes *et al.*, 2005). The fresh trunk xylem sap and fisetin inhibited the growth of 18 dermatophytes and *C. neoformans* with MIC values of 8–256 and 4–128 μg/ml, respectively. Fisetin showed lower toxicity against 3T3-A31 cells of Balb/c with an IC_{50} of 158 μg/ml, while the IC_{50} for the fresh xylem sap was 109 μg/ml (Costa *et al.*, 2014).

Health Benefits

In nutritional terms, it is possible to highlight the expressive content of crude fibres in jatobá pulp. These fibres are metabolized by resident colonic commensal bacteria that produce catabolic enzymes and metabolic systems for degrading dietary fibres. Interestingly, studies have shown that dietary fibre shapes microbial composition. The ability of microorganisms to metabolize specific types of fibre depends on whether their genomes encode the necessary cleavage enzymes and related proteins, such as carbohydrate-binding proteins and transporters (Thomas *et al.*, 2011). In addition, 'nutritionally specialized bacteria' – e.g. *Firmicutes*, *Actinomyces* and *Verrucomicrobia phyla* – in the initial degradation of complex substrates, such as resistant starch granules, play a crucial 'keystone' role since their degradation products can be available to other bacteria (Flint *et al.*, 2012; Ze *et al.*, 2012).

The main products of carbohydrate fermentation by colonic bacteria are the short-chain fatty acids (SCFAs): acetate, propionate and butyrate (Mortensen and Clausen, 2009). The majority of SCFAs are absorbed by colonic epithelial cells and metabolized by cells in different tissues including colon cells, and only 5–10% are excreted in the faeces. Butyrate is a main energy source for colonic cells and prevents colon carcinogenesis (Wu *et al.*, 2018). Apart from providing energy for the colon cells, SCFAs acidify the large intestine environment, thus promoting NH_4^+ taken up by the bacterial mass and cleared via the faeces, which boosts the growth of beneficial bacteria while inhibiting pathogens. They also maintain intestinal barrier function and promote the development of the intestinal immune system (Topping and Lockett, 2016). SCFAs, in particular propionate, are also capable of regulating fat synthesis and cholesterol in the liver and suppressing body weight gain by triggering glucagon secretion and increasing satiety (Maljaars *et al.*, 2008).

Phenolic compounds and carotenoids are described for the fruit pulp. The consumption of antioxidant-rich foods can enhance enzyme activity, prevent the oxidation of low-density lipoproteins (LDLs) and act on oxidation-related initiation, promotion and tumour progression processes, thus reducing the risk of chronic diseases (Vauzour *et al.*, 2010).

Another important health benefit in jatobá fruit and seed is the mineral content, because an adequate daily intake of minerals is essential for the prevention of chronic nutrition-related and degenerative diseases, including cancer, cardiovascular disease and obesity (Munoz and Dias, 2020).

Jayaprakasam *et al.* (2007) showed that labdane diterpenes and spathulenol isolated from the seed pods inhibited COX-1 and -2 enzymes and lipid peroxidation. In addition, cyclooxygenase enzymes play a critical role in inflammatory disorders. The two isozymes of cyclooxygenases – COX-1 and −2 – catalyse a rate-limiting step in prostaglandin synthesis. These enzymes have been a target for drug design to treat inflammations and related disorders.

Novel Products/Formulations

The fresh floury pulp that surrounds the seeds has a yellowish colour and it is usually not very pleasant for some people, since it has a strong smell and a slight adherence to the mouth. A Bolivian study prepared recipes for juice and cookies made with the flour of jatobá pulp which demonstrated good acceptability by the participants based on the results of the Hedonic Test (Cury *et al.*, 2011).

An innovative process for obtaining seed xyloglucan of *H. courbaril* was developed for its use as a microencapsulating agent for ascorbic acid. The objectives were to avoid the ascorbic acid oxidation and organoleptic changes, besides minimizing the formation of undesirable compounds during the cooking of fish burgers (Vicente *et al.*, 2019). Regarding its nutritional use, there is an invention of vegetable broth made with the addition of xyloglucan extracted from the endosperm of jatobá, which increases its nutritional value, makes it a source of soluble fibre and improves its physical-chemical characteristics (Moreira *et al.*, 2015).

Some patents for innovative cosmetic products were registered by Natura, a Brazilian company, using seed or leaf extracts of *H. courbaril*. As examples, the company developed a synergistic formula with several plant species that promotes modulation of collagen 1 and elastin protein, an anti-wrinkle formula and an anti-blemish serum (Reis *et al.*, 2017, 2018; Paes *et al.*, 2020). Other patents on cosmetic use include a hair tonic, a growth stimulant lotion (Fujii *et al.*, 1997) and a cream with moisturizing and skin lightening effects due to melanin inhibitory activity (Abe *et al.*, 1998).

The xyloglucan derived from jatobá also has therapeutic use. A patent registered by the Federal University of Paraná describes a curative with a bioactive membrane coated with a biopolymer (Pirich *et al.*, 2016).

Conclusions and Future Trends

H. courbaril fruit or jatobá is a PANC arising from Brazilian biodiversity which has as its main edible part the floury pulp around the seeds. This pulp is yellowish, sweet and has a characteristic odour. It is consumed fresh or used as an ingredient in traditional recipes.

The analysis of the nutritional composition of jatobá pulp highlights the fibres as the main component of the pulp. Thus, jatobá consumption may be related to several health-promoting effects such as the prevention of cardiovascular disease, diabetes, digestive tract disease, overweight and obesity. Unlike many fruits, jatobá pulp is a relevant source of protein and it can also be considered a rich source of minerals.

Jatobá seeds showed high energy values as they are rich in proteins, oil and digestible carbohydrates. They are also good sources of minerals. The seeds have the potential to be used as food or as a food ingredient, but they must be investigated further. Jatobá seeds contain trypsin inhibitors (antinutritional compounds), but with lower content than common edible legumes. The presence of coumarins in the seeds is concerning because these compounds have shown anticoagulant activity.

H. courbaril fruit present a diversity of secondary metabolites including sesquiterpenes, labdane- and clerodane-type diterpenes, carotenoids, biscoumarins and flavonoids. Phenolic compounds and carotenoids are relevant components of the fruit pulp since the consumption of these antioxidant compounds prevents degenerative diseases. The pulp also contains labdane diterpenes which inhibit COX-2 enzyme selectively and lipid peroxidation which may act in human inflammatory processes. The fruit pulp and peel showed antibacterial activity. Other parts of the plant employed in traditional medicine have presented antimicrobial, antileishmanial, antiplasmodial, myorelaxant, antioxidant and 5-lipoxygenase inhibition activities.

Besides the nutritional and health benefits of fresh jatobá pulp consumption or its use in recipes, it can be processed by the food industry to generate different products. The fruit can also be used for the development of food additives and even pharmaceutical or cosmetic ingredients.

Acknowledgements

This study was financed by the Coordenação de Aperfeiçoamento de Pessoal de Nível Superior, Brasil (CAPES), Finance Code 001 and by the Instituto Nacional de Ciência e Tecnologia em Biodiversidade e Produtos Naturais (INCT-BioNat; CNPq and FAPESP; Brasil).

References

Abdel-Kader, M., Berger, J.M., Slebodnick, C., Hoch, J., Malone, S. *et al.* (2002) Isolation and absolute configuration of ent-Halimane diterpenoids from *Hymenaea courbaril* from the Suriname rain forest. *Journal of Natural Products* 65(1), 11–15. DOI: 10.1021/np0103261.

Abe, I., Mitsunaga, T., Takagi, K., Shimomura, K. and M.P. Co Ltd (1998) Skin-lightening cosmetics containing extracts of Jatoba [JP Patent 10236943].

Aguiar, J.C.D., Santiago, G.M.P., Lavor, P.L., Veras, H.N.H., Ferreira, Y.S. *et al.* (2010) Chemical constituents and larvicidal activity of *Hymenaea courbaril* fruit peel. *Natural Product Communications* 5(12), 1977–1980. DOI: 10.1177/1934578X1000501231.

Albuquerque, U.P., Monteiro, J.M., Ramos, M.A. and de Amorim, E.L.C. (2007) Medicinal and magic plants from a public market in northeastern Brazil. *Journal of Ethnopharmacology* 110(1), 76–91. DOI: 10.1016/j.jep.2006.09.010.

Araújo, T.A.S., Alencar, N.L., de Amorim, E.L.C. and de Albuquerque, U.P. (2008) A new approach to study medicinal plants with tannins and flavonoids contents from the local knowledge. *Journal of Ethnopharmacology* 120(1), 72–80. DOI: 10.1016/j.jep.2008.07.032.

Arkcoll, B.D. (1984) Some leguminous trees providing useful fruits in the north of razil. *Pesquisa Agropecuaria Brasilia* 19, 231–234.

Arrhenius, S.P. and Langenheim, J.H. (1983) Inhibitory effects of *Hymenaea* and *Copaifera* leaf resins on the leaf fungus, *Pestalotia subcuticularis*. *Biochemical Systematics and Ecology* 11(4), 361–366. DOI: 10.1016/0305-1978(83)90037-6.

Artavia, D., Barrios, M. and Castro, O. (1995) A flavanonol rhamnoside from *Hymenaea courbaril* leaves. *Fitoterapia* 66, 91–91.

Azani, N., Babineau, M., Bailey, C.D., Banks, H., Barbosa, A.R. *et al.* (2017) A new subfamily classification of the leguminosae based on a taxonomically comprehensive phylogeny – The Legume Phylogeny Working Group (LPWG). *Taxon* 66(1), 44–77. DOI: 10.12705/661.3.

Bandeira, P.N., Santos, H.S., Albuquerque, M.R.J.R., Lemos, T.L.G. and Braz-Filho, R. (2015) A new diterpene isolated from the resin of *Hymenaea Courbaril*. *Chemistry of Natural Compounds* 51(4), 693–696. DOI: 10.1007/s10600-015-1386-5.

Barbosa Junior, A.M., Mélo, D.L.F.M.D., Almeida, F.T.C.D. and Trindade, R.D.C. (2015) Estudo comparativo da susceptibilidade de isolados clínicos de *Cryptococcus neoformans* (Sanfelice, 1895) frente a alguns antifúngicos de uso hospitalar e extratos vegetais obtidos de plantas medicinais da região semiárida sergipana. *Revista Brasileira de Plantas Medicinais* 17(1), 120–132. DOI: 10.1590/1983-084X/11_177.

Barrett, B. (1994) Medicinal plants of Nicaragua's Atlantic coast. *Economic Botany* 48(1), 8–20. DOI: 10.1007/BF02901375.

Bezerra, G.P., Góis, R.W. da S., de Brito, T.S., de Lima, F.J.B., Bandeira, M.A.M. *et al.* (2013) Phytochemical study guided by the myorelaxant activity of the crude extract, fractions and constituent from stem bark of *Hymenaea courbaril* L. *Journal of Ethnopharmacology* 149(1), 62–69. DOI: 10.1016/j.jep.2013.05.052.

Biazotto, K.R., de Souza Mesquita, L.M., Neves, B.V., Braga, A.R.C., Tangerina, M.M.P. *et al.* (2019) Brazilian biodiversity fruits: Discovering bioactive compounds from underexplored sources. *Journal of Agricultural and Food Chemistry* 67(7), 1860–1876. DOI: 10.1021/acs.jafc.8b05815.

Boniface, P.K., Baptista Ferreira, S. and Roland Kaiser, C. (2017) Current state of knowledge on the traditional uses, phytochemistry, and pharmacology of the genus *Hymenaea*. *Journal of Ethnopharmacology* 206, 193–223. DOI: 10.1016/j.jep.2017.05.024.

Botelho, S.A., Ferreira, R.A., Malavasi, M.M. and Davide, A.C. (2000) Aspectos morfológicos de frutos, sementes, plântulas e mudas de jatobá-do-Cerrado (*Hymenaea stigonocarpa* Mart. ex Hayne) – Fabaceae. *Revista Brasileira de Sementes* 22(1), 144–152. DOI: 10.17801/0101-3122/rbs.v22n1p144-152.

Braga, F.C., Wagner, H., Lombardi, J.A. and de Oliveira, A.B. (2000) Screening Brazilian plant species for in vitro inhibition of 5-lipoxygenase. *Phytomedicine: International Journal of Phytotherapy and Phytopharmacology* 6(6), 447–452. DOI: 10.1016/S0944-7113(00)80073-2.

Buckeridge, M.S., Crombie, H.J., Mendes, C.J., Reid, J.S., Gidley, M.J. *et al.* (1997) A new family of oligosaccharides from the xyloglucan of *Hymenaea courbaril* L. (Leguminosae) cotyledons. *Carbohydrate Research* 303(2), 233–237. DOI: 10.1016/s0008-6215(97)00161-4.

Busato, A.P., Vargas-Rechia, C.G. and Reicher, F. (2001) Xyloglucan from the leaves of *Hymenaea courbaril*. *Phytochemistry* 58(3), 525–531. DOI: 10.1016/s0031-9422(01)00217-5.

Busato, A.P., Vargas-Rechia, C.G., Gorin, P.A.J., Petkowicz, C.L., Tischer, C.A. *et al.* (2005) New 4-O-substituted xylosyl units in the xyloglucan from leaves of *Hymenaea courbaril*. *International Journal of Biological Macromolecules* 35(5), 277–282. DOI: 10.1016/j.ijbiomac.2005.03.007.

Cáceres, A., Girón, L.M. and Martínez, A.M. (1987) Diuretic activity of plants used for the treatment of urinary ailments in Guatemala. *Journal of Ethnopharmacology* 19(3), 233–245. DOI: 10.1016/0378-8741(87)90001-8.

Cáceres, A., López, B., Juárez, X., del Aguila, J. and García, S. (1993) Plants used in Guatemala for the treatment of dermatophytic infections. 2. Evaluation of antifungal activity of seven American plants. *Journal of Ethnopharmacology* 40(3), 207–213. DOI: 10.1016/0378-8741(93)90070-I.

Cartaxo, S.L., Souza, M.M. de A. and de Albuquerque, U.P. (2010) Medicinal plants with bioprospecting potential used in semi-arid northeastern Brazil. *Journal of Ethnopharmacology* 131(2), 326–342. DOI: 10.1016/j.jep.2010.07.003.

Carvalho, A.F.U., Farias, D.F., da Rocha-Bezerra, L.C.B., de Sousa, N.M., Cavalheiro, M.G. *et al.* (2011) Preliminary assessment of the nutritional composition of underexploited wild legumes from semi-arid Caatinga and moist forest environments of northeastern Brazil. *Journal of Food Composition and Analysis* 24(4–5), 487–493. DOI: 10.1016/j.jfca.2011.01.013.

Chaves, E.M.F. and Barros, R.F.M. (2012) Diversidade e uso de recursos medicinais do carrasco na APA da Serra da Ibiapaba, Piauí, Nordeste do Brasil. *Revista Brasileira de Plantas Medicinais* 14(3), 476–486. DOI: 10.1590/S1516-05722012000300009.

Coe, F.G. and Anderson, G.J. (1996) Screening of medicinal plants used by the Garífuna of eastern Nicaragua for bioactive compounds. *Journal of Ethnopharmacology* 53(1), 29–50. DOI: 10.1016/0378-8741(96)01424-9.

Contreras-Calderón, J., Calderón-Jaimes, L., Guerra Hernández, E. and García-Villanova, B. (2011) Antioxidant capacity, phenolic content and vitamin C in pulp, peel and seed from 24 exotic fruits from Colombia. *Food Research International* 44(7), 2047–2053. DOI: 10.1016/j.foodres.2010.11.003.

Costa, P.A. da, Ballus, C.A., Teixeira-Filho, J. and Godoy, H.T. (2010) Phytosterols and tocopherols content of pulps and nuts of Brazilian fruits. *Food Research International* 43(6), 1603–1606. DOI: 10.1016/j.foodres.2010.04.025.

Costa, C.R, de Oliveira, C.M.A., Kato, L., Fernandes, O.D.F.L., da Silva, A.L. *et al.* (2014) Antifungal and cytotoxicity activities of the fresh xylem sap of *Hymenaea courbaril* L. and its major constituent fisetin. *BMC Complementary and Alternative Medicine* 14(1), 245. DOI: 10.1186/1472-6882-14-245.

Costa, M.D.C.M.F.D, Silva, A.G. D, Silva, A.P.S. D, Lima, V.L.D.M, Bezerra-Silva, P.C. *et al.* (2017) Essential Oils from Leaves of Medicinal Plants of Brazilian Flora: Chemical Composition and Activity against Candida Species. *Medicines* 4(2), 27. DOI: 10.3390/medicines4020027.

Cunningham, A., Martin, S.S. and Langenheim, J.H. (1974) Labd-13-en-8-ol-15-oic acid in the trunk resin of amazonian *Hymenaea courbaril*. *Phytochemistry* 13(1), 294–295. DOI: 10.1016/S0031-9422(00)91323-2.

Cury, G.C., Abela Gisbert, M.D.C., Bravo, J.A., Peñarrieta, J.M. and Rendón Porcel, W.J. (2011) Estudio del fruto comestible de la especie vegetal *Hymenaea courbaril* L. *Revista Boliviana de Química* 28(2), 73–78.

Dias, L.S., Luzia, D.M.M. and Jorge, N. (2013) Physicochemical and bioactive properties of *Hymenaea courbaril* L. pulp and seed lipid fraction. *Industrial Crops and Products* 49, 610–618. DOI: 10.1016/j.indcrop.2013.06.005.

Duranti, M. (2006) Grain legume proteins and nutraceutical properties. *Fitoterapia* 77(2), 67–82. DOI: 10.1016/j.fitote.2005.11.008.

FDA (2010) Nutritional Labeling and Education Act (NLEA) Requirements (8/94-2/95). Available at: https://www.fda.gov/nutrition-labeling-and-education-act-nlea-requirements-attachment-1 (accessed 5 October 2020).

Fernandes, T.T., Santos, A.T.F.D. and Pimenta, F.C. (2005) Atividade antimicrobiana das plantas *Plathymenia reticulata*, *Hymenaea courbaril* e *Guazuma ulmifolia*. *Revista de Patologia Tropical* 34(2), 113–122. DOI: 10.5216/rpt.v34i2.1916.

Fernandes, A.W.C., Aquino, S.Á.M.C., Gouveia, G.V., Almeida, J.R.G.S. and Costa, M.M. (2015a) Atividade antimicrobiana in vitro de extratos de plantas do bioma caatinga em isolados de *Escherichia coli* de suínos. *Revista Brasileira de Plantas Medicinais* 17(4 suppl 3), 1097–1102. DOI: 10.1590/1983-084x/14_159.

Fernandes, H.P., Sinhorin, A.P., Sinhorin, V.D.G., Vieira Junior, G.M., Silva, V.C. *et al.* (2015b) New glycosylated biscoumarins from *Hymenaea coubaril* L. seeds. *Phytochemistry Letters* 13, 413–416. DOI: 10.1016/j.phytol.2015.07.016.

Figueiredo, P.A., Spera, K.D., Gomes, A. da C., Dokkedal, A.L., Saldanha, L.L. *et al.* (2016) Antioxidant activity and chemical characterization of extracts from the genus *Hymenaea*. *Research Journal of Medicinal Plants* 10(5), 330–339. DOI: 10.3923/rjmp.2016.330.339.

Flint, H.J., Scott, K.P., Duncan, S.H., Louis, P. and Forano, E. (2012) Microbial degradation of complex carbohydrates in the gut. *Gut Microbes* 3(4), 289–306. DOI: 10.4161/gmic.19897.

Fujii, W., Konishi, K., Iwazawa, R. and Suntory, L. (1997) *Hair tonic and growth stimulant compositions containing minoxidil and Hymenaea curbaril and/or Betula alnoides extracts*. JP Patent 09087148.

Imai, T., Inoue, S., Ohdaira, N., Matsushita, Y., Suzuki, R. *et al.* (2008) Heartwood extractives from the Amazonian trees *Dipteryx odorata*, *Hymenaea courbaril*, and *Astronium lecointei* and their antioxidant activities. *Journal of Wood Science* 54(6), 470–475. DOI: 10.1007/s10086-008-0975-3.

Imamura, P.M., Miranda, P.C.M.L. and Giacomini, R.A. (2004) A complete ^{1}H and ^{13}C NMR data assignment for the diterpene methyl (-)-zanzibarate by 2D spectroscopy and NOE experiments. *Magnetic Resonance in Chemistry* 42(6), 561–563. DOI: 10.1002/mrc.1379.

Jayaprakasam, B., Alexanderlindo, R., Dewitt, D. and Nair, M. (2007) Terpenoids from stinking toe (*Hymneae courbaril*) fruits with cyclooxygenase and lipid peroxidation inhibitory activities. *Food Chemistry* 105(2), 485–490. DOI: 10.1016/j.foodchem.2007.04.004.

Khoo, S.F., Oehlschlager, A.C. and Ourisson, G. (1973) Structure and stereochemistry of the diterpenes of *Hymenaea courbaril* (Caesalpinioideae) seed pod resin. *Tetrahedron* 29(21), 3379–3388. DOI: 10.1016/S0040-4020(01)93493-3.

Kinnup, V.F. and Lorenzi, H. (2014) *Plantas Alimentícias Não Convencionais (PANC) no Brasil (398-399)*. Instituto Plantarum de Estudos da Flora, São Paulo, SP/Brasil.

Köhler, I., Jenett-Siems, K., Siems, K., Hernández, M.A., Ibarra, R.A. *et al.* (2002) *In Vitro* Antiplasmodial Investigation Of Medicinal Plants From El Salvador. *Zeitschrift Für Naturforschung C* 57(3–4), 277–281. DOI: 10.1515/znc-2002-3-413.

Liener, I.E. (1994) Implications of antinutritional components in soybean foods. *Critical Reviews in Food Science and Nutrition* 34(1), 31–67. DOI: 10.1080/10408399409527649.

Lima, N.N., Reicher, F., Corrêa, J.B.C. and Ganter, J.L.M.S. (1993) Partial structure of a xyloglucan from the seeds of *Hymenaea courbaril* var. stilbocarpa (jatobá). *Ciência e Cultura* 45, 22–22.

Lima, N.N., Rechia, C.G., Ganter, J.L.M., Reicher, F. and Sierakowski, M.R. (1995) Oligosaccharides derived from the xyloglucan isolated from the seeds of *Hymenaea courbaril* var. stilbocarpa. *International Journal of Biological Macromolecules* 17(6), 413–415. DOI: 10.1016/0141-8130(96)81855-1.

Maia, F.M.M., Oliveira, J.T.A., Matos, M.R.T., Moreira, R.A. and Vasconcelos, I.M. (2000) Proximate composition, amino acid content and haemagglutinating and trypsin-inhibiting activities of some Brazilian *Vigna unguiculata* (L) Walp cultivars. *Journal of the Science of Food and Agriculture* 80(4), 453–458. DOI: 10.1002/(SICI)1097-0010(200003)80:4<453::AID-JSFA548>3.0.CO;2-X.

Maljaars, P.W.J., Peters, H.P.F., Mela, D.J. and Masclee, A.A.M. (2008) Ileal brake: A sensible food target for appetite control. A review. *Physiology & Behavior* 95(3), 271–281. DOI: 10.1016/j.physbeh.2008.07.018.

Marsaioli, A.J., De Freitas Leitão Filho, H. and De Paiva Campello, J. (1975) Diterpenes in the bark of *Hymenea coubaril*. *Phytochemistry* 14(8), 1882–1883. DOI: 10.1016/0031-9422(75)85324-6.

Martin, S.S., Langenheim, J.H. and Zavarin, E. (1972) Sesquiterpenes in leaf pocket resin of *Hymenaea courbaril*. *Phytochemistry* 11(10), 3049–3051. DOI: 10.1016/0031-9422(72)80103-1.

Mendes, F.R. and Carlini, E.A. (2007) Brazilian plants as possible adaptogens: an ethnopharmacological survey of books edited in Brazil. *Journal of Ethnopharmacology* 109(3), 493–500. DOI: 10.1016/j.jep.2006.08.024.

Martin, S., Freitas, R.A., Obayashi, E. and Sierakowski, M.R. (2003) Physico–chemical aspects of galactoxyloglucan from the seeds of *Hymenaea courbaril* and its tetraborate complex. *Carbohydrate Polymers* 54(3), 287–295. DOI: 10.1016/S0144-8617(03)00073-0.

Moreira, A.C. de O.M., Vergara, C.M.A.C., Beserra, F.J., Mendes, D., Meneses, M.A. *et al.* (2015) *Vegetable broths made with plant hemicelulose* [BR Patent 102013033232].

Mortensen, P.B. and Clausen, M.R. (2009) Short-chain fatty acids in the human colon: Relation to gastrointestinal health and disease. *Scandinavian Journal of Gastroenterology* 31(sup216), 132–148. DOI: 10.3109/00365529609094568.

Nogueira, R.T., Shepherd, G.J., Laverde, A., Marsaioli, A.J. and Imamura, P.M. (2001) Clerodane-type diterpenes from the seed pods of *Hymenaea courbaril* var. stilbocarpa. *Phytochemistry* 58(8), 1153–1157. DOI: 10.1016/s0031-9422(01)00303-x.

Nogueira, R.T., Queiroz, S.D.N. and Imamura, P.M. (2002a) Semi-preparative HPLC separation of terpenoids from the seed pods of *Hymenaea courbaril* var. stilbocarpa. *Journal of Liquid Chromatography & Related Technologies* 25(1), 59–67.

Nogueira, R.T., Giacomini, R.A., Shepherd, G.J. and Imamura, P.M. (2002b) A New ent-Clerodane Diterpene from *Hymenaea courbaril* var. *altissima*. *Journal of the Brazilian Chemical Society* 13(3), 389–391. DOI: 10.1590/S0103-50532002000300015.

Oliveira, F.G., De Souza Araújo, C., Rolim, L.A., Barbosa-Filho, J.M. and Almeida, J.R.S. (2018) The genus Hymenaea (Fabaceae): a chemical and pharmacological review. In: Rahman, A. (ed.) *Studies in Natural Products Chemistry*, Vol. 58. Elsevier, pp. 339–388.

Ordóñez Pereda, J.Á. (2005) *Tecnologia de Alimentos: Alimentos de Origem Animal*, Vol. 2. Artmed.

Paes, F., Romeu, C.C., Reis, E.A. de O., Savietto, J.P., D'annolfo, K.T. *et al.* (2020) *Composition for Modulating the Genes Responsible for General Skin Functions, Method for Modulating the Expression of Genes Responsible for General Skin Functions, and Use of a Plant Extract*. U.S. Patent 0315950.

Pinto, A.A. da C. and Maduro, C.B. (2003) Produtos e subprodutos da medicina popular comercializados na cidade de Boa Vista, Roraima. *Acta Amazonica* 33(2), 281–290. DOI: 10.1590/1809-4392200332290.

Pinto, R.B., Mansano, V.D.F. and Tozzi, A.M.G. de A. (2015) Reestablishment of *Hymenaea travassii* (Leguminosae, Caesalpinioideae), a species endemic to the Bolivian Chaco. *Phytotaxa* 219(1), 96. DOI: 10.11646/phytotaxa.219.1.8.

Pirich, C.L., Picheth, G.F., Dos Santos, L.A., Vanin, L.P., Woehl, M.A. *et al.* (2016) *Sensitive Bio-Curative Infection Compound with Bioactive Membrane Coated with Biopolymer*. BR Patent 102014026963.

Reflora - Flora do Brasil (2020) *Hymenaea courbaril* L. Available at: http://reflora.jbrj.gov.br (accessed 10 January 2021).

Reis, E.A. de O., Paes, F., Savietto, J.P., Romeu, C.C., D'annolfo, K.T. *et al.* (2017) *Anti-Wrinkle Cosmetic Compositions*. BR Patent 102016011785.

Reis, E.A. de O., Paes, F., D'annolfo, K.T., Moncayo, P.C., De Oliveira, R.A.S. *et al.* (2018) *Cosmetic Anti-Blemish Composition, Use of the Composition and Method for Anti-Blemish Treatment*. BR Patent 102016019117.

Ribeiro, T.G., Chávez-Fumagalli, M.A., Valadares, D.G., Franca, J.R., Lage, P.S. *et al.* (2014) Antileishmanial activity and cytotoxicity of Brazilian plants. *Experimental Parasitology* 143(1), 60–68. DOI: 10.1016/j.exppara.2014.05.004.

Ribeiro, R.D., Cardoso, D.B.O.S. and de Lima, H.C. (2015) A New Species of *Hymenaea* (Leguminosae: Caesalpinioideae) with A Revised Identification Key to the Genus in the Brazilian Atlantic Forest. *Systematic Botany* 40(1), 151–156. DOI: 10.1600/036364415X686440.

Rizzini, C.T. (2000) *Árvores e Madeiras Úteis Do Brasil: Manual de Dendrologia Brasileira*, 2nd edn. Edgar Blucher, São Paulo, SP/Brasil.

Sales, G.W.P., Batista, A.H.M., Rocha, L.Q. and Nogueira, N.A.P. (2014) Efeito antimicrobiano e modulador do óleo essencial extraído da casca de frutos da *Hymenaea courbaril* L. *Journal of Basic and Applied Pharmaceutical Sciences* 35(4).

Saraiva, M.E., Ulisses, A.V.R. de A., Ribeiro, D.A., de Oliveira, L.G.S., de Macêdo, D.G. *et al.* (2015) Plant species as a therapeutic resource in areas of the savanna in the state of Pernambuco, Northeast Brazil. *Journal of Ethnopharmacology* 171, 141–153. DOI: 10.1016/j.jep.2015.05.034.

Sasaki, K., Matsukura, Y., Shijima, K., Miyake, M., Fujiwara, D. *et al.* (2009) High-performance liquid chromatographic purification of oligomeric procyanidins, trimers up to nonamers, derived from the bark of Jatoba (*Hymenaea courbaril*). *Bioscience, Biotechnology, and Biochemistry* 73(6), 1274–1279. DOI: 10.1271/bbb.80747.

Schwartz, G. (2018) Jatoba-Hymenaea courbaril. In: Rodrigues, S., de Oliveira Silva, E., and Sousa de Brito, E. (eds) *Exotic Fruits*. Academic Press, London, pp. 257–261.

SiBBr (2018) Biodiversidade & Nutrição. Sistema de Informação sobre a Biodiversidade Brasileira. Available at: https://ferramentas.sibbr.gov.br/ficha/bin/view/FN/ShortName/3927_jatoba_polpa_crua (accessed 11 January 2021).

Silva, M.R., Silva, M.S., Martins, K.A. and Borges, S. (2001) Studies on the use of jatoba flour in biscuits as a source of dietary fibre containing no added simple sugars. *Ciência e Tecnologia de Alimentos* 21, 176–182.

Simões, K., Du, J., Pessoni, R.A.B., Cardoso-Lopes, E.M. and Vivanco, J.M. (2009) Ipomopsin and hymenain, two biscoumarins from seeds of *Hymenaea courbaril*. *Phytochemistry Letters* 2(2), 59–62. DOI: 10.1016/j.phytol.2008.11.003.

Spera, K.D., Figueiredo, P.A., Santos, P.C.E., Barbosa, F.C., Alves, C.P. *et al.* (2019) Genotoxicity, anti-melanoma and antioxidant activities of *Hymenaea courbaril* L. seed extract. *Anais Da Academia Brasileira de Ciencias* 91(4), e20180446. DOI: 10.1590/0001-3765201920180446.

Stubblebine, W., Langenheim, J.H. and Lincoln, D. (1975) Vegetative growth and leaf resin composition in *Hymenaea courbaril* under photoperiodic extremes. *Biochemical Systematics and Ecology* 3(4), 219–228. DOI: 10.1016/0305-1978(75)90006-X.

Thomas, A.M.E. (2012) *Plantas Útiles del Municipio de Bonanza*. Available at: http://riul.unanleon.edu.ni:8080/jspui/handle/123456789/5721 (accessed 9 January 2021).

Thomas, F., Hehemann, J.H., Rebuffet, E., Czjzek, M. and Michel, G. (2011) Environmental and gut bacteroidetes: the food connection. *Frontiers in Microbiology* 2, 93. DOI: 10.3389/fmicb.2011.00093.

Topping, D.L. and Lockett, T.J. (2016) Human Physiology and Health: Dietary Fiber, Short-Chain Fatty Acids, and Their Impact on Gut Physiology. In: *Reference Module in Food Science*. Elsevier.

Trugo, L.C., Donangelo, C.M., Trugo, N.M.F. and Bach Knudsen, K.E. (2000) Effect of heat treatment on nutritional quality of germinated legume seeds. *Journal of Agricultural and Food Chemistry* 48(6), 2082–2086. DOI: 10.1021/jf9913920.

Universidade Estadual De Campinas (UNICAMP) (2011) *Tabela Brasileira de Composição de Alimentos (TACO)*, 4th edn. NEPA -UNICAMP, Campinas/SP, Brazil.

Vargas-Rechia, C., Reicher, F., Sierakowski, M.R., Heyraud, A., Driguez, H. *et al.* (1998) Xyloglucan octasaccharide XXLGol derived from the seeds of Hymenaea courbaril acts as a signaling molecule. *Plant Physiology* 116(3), 1013–1021. DOI: 10.1104/pp.116.3.1013.

Vasconcelos, I.M., Maia, A.A.B., Siebra, E.A., Oliveira, J.T.A., Carvalho, A. d.F.F.U. *et al.* (2001) Nutritional study of two Brazilian soybean (*Glycine max*) cultivars differing in the contents of anti-nutritional and toxic proteins. *The Journal of Nutritional Biochemistry* 12(1), 55–62. DOI: 10.1016/s0955-2863(00)00148-0.

Vauzour, D., Rodriguez-Mateos, A., Corona, G., Oruna-Concha, M.J. and Spencer, J.P.E. (2010) Polyphenols and human health: prevention of disease and mechanisms of action. *Nutrients* 2(11), 1106–1131. DOI: 10.3390/nu2111106.

Vicente, A.A., Sa, D.M.A.T., Carneiro da Cunha, M.G., Farias, M.D.P. and Soares, P.A.G. (2019) Production of xyloglucan microparticle from Jatoba (*Hymenaea courbaril* var. courbaril) seed with ascorbic acid for application in fish burger. BR Patent 102017025897.

Wu, X., Wu, Y., He, L., Wu, L., Wang, X. *et al.* (2018) Effects of the intestinal microbial metabolite butyrate on the development of colorectal cancer. *Journal of Cancer* 9(14), 2510–2517. DOI: 10.7150/jca.25324.

Zappi, D.C., Filardi, F.L.R., Leitman, P., Souza, V.C., Walter, B.M.T. *et al.* (2015) Growing knowledge: an overview of seed plant diversity in Brazil. *Rodriguésia* 66(4), 1085–1113. DOI: 10.1590/2175-7860201566411.

Ze, X., Duncan, S.H., Louis, P. and Flint, H.J. (2012) *Ruminococcus bromii* is a keystone species for the degradation of resistant starch in the human colon. *The ISME Journal* 6(8), 1535–1543.

19 Ora-pro-nobis (*Pereskia aculeata* Miller)

André Gonzaga dos Santos*, Kátia Sivieri*, Bárbara Paes Miglioli da Mata, Mateus Kawata Salgaço and Luis Vitor Silva do Sacramento
School of Pharmaceutical Sciences, São Paulo State University - Unesp, Araraquara city, São Paulo state, Brazil

Abstract

The plant *Pereskia aculeata* Miller from the Cactaceae family is known in Brazil as ora-pro-nobis. Its cultivation around churches is the origin of its Latin common name. *P. aculeata* is distributed in the American continent, south and south-east Africa, and north-east and south-east Australia. In Brazil, it occurs in from the south to the north-east and central-west regions. The leaves are the main edible part of the plant, but the flowers and fruit are also eaten. The origin of its use may be related to the African slaves in Brazil. Nutritional analyses of the leaves highlight the content of proteins and essential amino acids, dietary fibre, mucilage, vitamins, iron and other minerals. The secondary metabolites identified in the leaves of *P. aculeata* are benzoic acid derivatives, phenylpropanoids, flavonoids, monoterpenes, sesquiterpenes, carotenoids, an indole alkaloid and phenethylamines. No secondary metabolites were identified in other plant parts. *P. aculeata* has shown important bioactivities, including antioxidant, anti-inflammatory, wound healing, antinociceptive, antibacterial and trypanocidal actions, as well as cytotoxicity in tumour cell lines. The phenolic compounds and carotenoids may be responsible for the antioxidant activity of its leaf extracts. The wound-healing and anti-inflammatory properties of the species have been investigated because these actions are related to uses in Brazilian traditional medicine. The wound-healing action was observed both *in vitro* and *in vivo*. The chronic and acute topical anti-inflammatory action of *P. aculeata* was demonstrated in mice. Considering these results, a pharmaceutical cream based on a hexane fraction of *P. aculeata* methanol leaf extract was developed for use in wound healing. Its leaf flour may be used as an ingredient for food enrichment and its mucilage as a thickening, gelling or emulsifying agent. Other important technological applications have been described in publications in the areas of materials and nanotechnology.

Introduction

Native plant species from South America, especially from Brazil, have been collected and exploited recently due to encouraging policies to recover the environmental balance. Brazil holds 20% of the known species of world flora due to its great biodiversity as a whole, representing over 40,000 different plant species. Non-conventional plants are being used by a portion of the population as a source of food and herbal medicines for the treatment of diseases (Oliveira *et al.*, 2012). According to Kinnup and Lorenzi (2014), an edible plant that is not part of the conventional agricultural matrix and is little-known by most people of certain regions or countries

*Corresponding author: katia.sivieri@unesp.br; andre.gonzaga@unesp.br

 Handbook of Phytonutrients in Indigenous Fruits and Vegetables (eds D. Sivakumar *et al.*)
DOI: 10.1079/9781789248067.0019

may be classified as a non-conventional edible plant, for which the Brazilian acronym is PANC.

Several FAO projects for good nutrition and health have encouraged thousands of people to work with non-conventional plant species, especially to serve low-income populations in the world. The most common species of *Pereskia* in Brazil represents one of the largest consumption fronts of PANCs (Sousa *et al.*, 2014).

Pereskia aculeata Miller (Fig. 19.1) may be highlighted in Brazilian ethnobotany due to its food and medicinal traditional uses (Pinto and Scio, 2014; Souza *et al.*, 2016; Agostini-Costa, 2020), and it is considered a PANC. The species occurs in the American continent, south and south-east Africa, and north-east and south-east Australia (Agostini-Costa, 2020). In Brazil, it occurs in the southern, south-eastern, north-eastern and central western regions (Zappi *et al.*, 2020).

The main common name of the species in Brazil is ora-pro-nobis. It has been cultivated around churches and this is the origin of its Latin common name, which means 'pray for us'. Two other edible Cactacea species are known as ora-pro-nobis: *Pereskia bleo* (Kunth) DC. and *Pereskia grandifolia* Haw. *P. aculeata* and *P. grandiflora* are sometimes confused. Another common name in Brazil is carne-de-pobre – 'meat of poor people' – which is related to the use of the leaves as a protein source (Maciel *et al.*, 2019; Agostini-Costa, 2020).

The edible parts of the plant comprise its leaves, fruit and flowers. The leaves are the most consumed part, eaten fresh in salads or cooked as an ingredient in recipes for chicken, cakes, bread and jams. Its flour is also used in recipes. The medicinal parts are the leaves and the fruit (Kinnup and Lorenzi, 2014; Santos *et al.*, 2018).

P. aculeata is a good source of proteins and amino acids, dietary fibre, vitamins, iron and other minerals. Moreover, it can promote health due to its antioxidant and other bioactive compounds and has the potential to be employed as a food additive (the mucilage and flour) and as a pharmaceutical raw material (extracts for herbal medicines indicated as wound healing and anti-inflammatory).

Botany

Pereskia aculeata Miller – Cactaceae (Angiosperm Phylogeny Group, 2009) is widely distributed in

Fig. 19.1. Leaves, flowers and fruit of *Pereskia aculeata* Miller.

the American continent. It has been recorded in North America (in restricted areas of Mexico and the USA) and most parts of Central America with emphasis on Caribbean regions. In South America, it is more common in the Brazilian territory, however, there are important occurrences in Guyana, Paraguay and Argentina (Maciel *et al.*, 2019; Agostini-Costa, 2020; Zappi *et al.*, 2020).

In Brazil, *P. aculeata* occurrence covers the eastern part of the Brazilian territory from the Maranhão and Ceará states – north-east region – to the Rio Grande do Sul state – south region, corresponding to the Pampa, Atlantic rainforest, Central Brazilian savannah and Caatinga biomes (Maciel *et al.*, 2019; Zappi *et al.*, 2020). *P. aculeata* has been introduced in other parts of the world (Egbon *et al.*, 2020) and it can be cultivated under different conditions (Lage *et al.*, 2015; Guimarães *et al.*, 2019). It has been considered an aggressive invader in South Africa and Australia (Weisser *et al.*, 1992; Paterson *et al.*, 2014a, b; Egbon *et al.*, 2020).

P. aculeata has several synonyms: *Cactus lucidus* Salisb.; *Cactus pereskia* L. and *Pereskia pereskia* (L.) H. Karst (Global Biodiversity Information Facility, 2021; Tropicos, 2021). Among its common names, the most used include ora-pro-nobis in Brazil (Rosa and Souza, 2003), barbados gooseberry (Morais *et al.*, 2019) and leafy cactus (Butterworth and Wallace, 2005).

P. aculeata is a perennial shrub with woody or semi-woody scrambled thorn branches, that can reach up to 10 m in length. The leaves are slightly curved, dark green, hairless, elliptical in shape, and mucilaginous. Thorn sets are frequent in leaf axils (Zareisedehizadeh *et al.*, 2014; Maciel *et al.*, 2019).

In the flowering, *P. aculeata* can offer an intense number of white-beige pediceled, ephemeral and scented flowers, all grouped in axillary or terminal inflorescences (panicle type). The flowers are perigynous, hermaphrodite, actinomorphic, 1-pistil with yellow-orange stamens and just one pistil with a white and lobed stigma. The ovary is considered superior: 5-carpel; 1-locule and free, inside a green-fleshy hypanthium. *P. aculeata* can produce edible fruit: when ripe these are succulent, globose, pomaceous, 'cactídio' type, with three seeds, brown or black, covered by a mucilaginous matrix. In morphological terms, the fruits are formed by hypanthium, pericarp and seeds (Rosa and Souza, 2003).

Traditional Use

As noted, *P. aculeata* is named ora-pro-nobis ('pray for us') because it has been cultivated around the churches in Minas Gerais state, Brazil (Maciel *et al.*, 2019). The African slaves in Brazil may have introduced its use as food (Brandão, 2019).

In Brazilian folk medicine, the fruits have been used as an expectorant and antisyphilitic and the leaves for emollient, anti-inflammatory and wound-healing purposes (Pio-Correa and Azeredo Pena, 1984; Almeida and Corrêa, 2012; Carvalho *et al.*, 2014; Pinto and Scio, 2014; Maciel *et al.*, 2019).

Nutritional Composition

The dark-green leaves of *P. aculeata* are widely used as a food source in underdeveloped areas (Dayrell and Vieira, 1977), as well as in the south to north-east regions of Brazil. Ora-pro-nobis leaves are used in traditional cuisine, being an ingredient of several sweet and savoury dishes. As noted, in some low-income communities, it is known as the 'meat of poor people', being the main source of protein available (Almeida *et al.*, 2014; Pinto *et al.*, 2015a). As presented in Table 19.1, several studies have showed high levels of protein in *P. aculeata* leaves – 18.5 and 28.4% w/w fresh basis – when compared to other vegetables, such as kale (1.6%), lettuce (1.3%), common maize (7.6–10.0%) and beans (18.0–20.0%) (Merce *et al.*, 2001; Girão *et al.*, 2003; Takeiti *et al.*, 2009). However, it is important to highlight that ora-pro-nobis cannot be considered as a protein source (Oliveira *et al.*, 2019a). Moreover, a comparative study of protein composition and amino acid profile in five clones of ora-pro-nobis performed by Botrel *et al.* (2019) determined the protein content in dry matter from 16.6 to 23.9%.

The levels of essential amino acids, except methionine, were considered higher than the minimum amount recommended by FAO

Table 19.1. Centesimal nutritional composition of *Pereskia aculeata* Miller leaves.

	Fresh leaves			Flour				
Nutrient	Takeiti *et al.* (2009)	Oliveira *et al.* (2019b)	Botrel *et al.* (2020)	Almeida *et al.* (2014)	Sobrinho *et al.* (2015)	Barbalho *et al.* (2016)	Cruz. *et al.* (2020)	RDI (Food and Drug Administration, 2020)
Lipid ($g.100g^{-1}$)	4.10	3.10	4.40	5.07	5.23	3.71	2.88	78 ($g.day^{-1}$)
Ash ($g.100g^{-1}$)	16.10	22.48	20.52	14.81	15.46	17.83	nd	-
Protein ($g.100g^{-1}$)	28.40	21.70	18.50	28.99	40.68	24.17	15.61	50 ($g.day^{-1}$)
Total carbohydrates ($g.100g^{-1}$)	nd	51.93	23.43	29.55	nd	49.39	nd	-
Crude fibre ($g.100g^{-1}$)	9.80	6.70	34.18	21.60	11.13	32.80	66.06	28 ($g.day^{-1}$)
Cu ($\mu g.100g^{-1}$)	1.40	8.52	2.20	1.20	nd	nd	nd	2000 ($\mu g.day^{-1}$)
Mn ($mg.100g^{-1}$)	46.40	135	64.40	43.48	nd	nd	nd	2.3 ($mg.day^{-1}$)
Fe ($mg.100g^{-1}$)	14.20	186.82	11.71	20.56	nd	nd	nd	18 ($mg.day^{-1}$)
Zn ($mg.100g^{-1}$)	26.70	27.13	2.46	7.30	nd	nd	nd	11 ($mg.day^{-1}$)
Ca ($mg.100g^{-1}$)	3420	50,317	2373	1346	nd	nd	nd	1300 ($mg.day^{-1}$)
K ($mg.100g^{-1}$)	1632	25,389	2845	3910	nd	nd	nd	3500 ($mg.day^{-1}$)
Mg ($mg.100g^{-1}$)	1900	9835	832	586	nd	nd	nd	420 ($mg.day^{-1}$)

nd: not determined by the authors.

Table 19.2. Mean values of the amino acid levels of *Pereskia aculeata* Miller leaves.

Amino acid (g.100g^{-1}) dry basis	Takeiti *et al.* (2009)	Botrel *et al.* (2019)	Silveira *et al.* (2020)	Essential amino acid requirements[a] (mg.kg^{-1} per day)
Essential				
Histidine	0.59	2.41	2.40	10
Isouleucine	1.07	3.21	3.69	20
Leucine	2.00	6.19	6.90	39
Lysine	1.43	4.26	5.34	30
Methionone	0.23	0.42	nd	Methionine + cysteine=15
Phenylalanine	1.27	4.63	nd	-
Threonine	1.00	3.75	3.00	14
Valine	1.28	4.09	4.69	26
Tryptophan	5.52	1.44	2.11	4
Non-essential				
Arginine	1.44	5.48	0.41	
Aspartic acid	1.71	6.42	0.60	-
Serine	1.00	3.94	0.20	-
Glutamic acid	2.67	9.38	0.97	-
Proline	1.11	4.78	0.58	-
Cystine	0.35	2.21	nd	-
Glycine	1.31	5.39	nd	-
Alanine	1.36	3.99	0.38	-
Tyrosine	1.21	3.19	nd	-

[a] Essential amino acids required by adults - current estimates by the FAO/WHO (2007).nd, Not determined by the authors.

as necessary for adult human consumption (Dayrell and Vieira, 1977) and the leaves also presented an excellent amino acid profile (Albuquerque *et al.*, 1991). Table 19.2 shows the content of essential and non-essential amino acids. Leucine was the major essential amino acid followed by phenylalanine and lysine. On the other hand, a study conducted by Takeiti *et al.* (2009) showed that tryptophan is the main amino acid in *P. aculeata* leaves. Regarding non-essential amino acids, glutamic acid showed the most expressive content.

Takeiti *et al.* (2009) showed 75.9% of protein digestibility to ora-pro-nobis leaves. Moreover, the same authors found the lowest content for sulfur-amino acids (methionine and cysteine), where the presence of the band at 14.4 kDa after hydrolysis could be due to a dense tertiary structure via disulfide bonds that make proteolysis difficult and, therefore, impair *in vitro* protein digestibility.

The leaves contain high amounts of fibre (6.7 to 34.2 g.100 g^{-1}) and have an especially high soluble dietary fibre content (5.2 g.100 g^{-1}) (Takeiti *et al.*, 2009). Soluble dietary fibres help to reduce the level of serum low-density lipoprotein cholesterol, decreasing postprandial blood glucose and modulating gut microbiota (Tosh and Bordenave, 2020).

Takeiti *et al.* (2009) showed that fresh ora-pro-nobis leaves contain β-carotene (4.2 mg.100 g^{-1}), vitamin A (2333 IU.100 g^{-1}), vitamin C (185.8 mg.100 g^{-1}), and folic acid (19.30 mg.100 g^{-1}). The folic acid content is higher than Fijian green leafy vegetables, such as bele or *Abelmoschus manihot* (L.) Medik. and

Chinese cabbage (0.13 and 0.065 mg.100 g^{-1}), respectively (Devi *et al.*, 2008). The content of provitamin A in the fresh leaves was determined as 121.1 to 410.8 µg retinol activity equivalent .100 g^{-1} (Agostini-Costa *et al.*, 2014). The *P. aculeata* fruit may be also a good source of provitamin A because the levels of provitamin A ranged from 79.7 to 457.9 µg retinol activity equivalent .100 g^{-1}. These amounts are higher than the values for mango and acerola: 35–215 and 35–325 µg retinol activity equivalent .100 g^{-1}, respectively (Agostini-Costa *et al.*, 2012, 2014). See more details on carotenoids composition in the Phytochemicals section of this chapter.

Microelements, also known as trace elements, are considered as essential as vitamins. Ora-pro-nobis leaves contain trace elements such as calcium, iron, zinc, magnesium and manganese (Table 19.1). EIn terms of manganese, ora-pro-nobis leaves showed a very expressive content (46.4 to 135 mg.100 g^{-1}) compared to other vegetables, such as lettuce, watercress, spinach, cabbage and arugula, which do not reach 1 mg.100 g^{-1} (Botrel *et al.*, 2020). Many of these nutrients are found in higher proportions than those observed in other plants commonly used as foods, such as cabbage and spinach (Morton, 1987). Furthermore, ora-pro-nobis leaves contain low levels of lipids, such as methyl palmitate, methyl oleate and methyl stearate (Merce *et al.*, 2001; Girão *et al.*, 2003; Takeiti *et al.*, 2009; Oliveira *et al.*, 2013).

Finally, the scientific literature has revealed the nutritional benefits of ora-pro-nobis leaves. Taking into consideration the essential amino acids, dietary fibre, vitamins and minerals, they are shown to be a good and cheap way of obtaining nutritive food.

Antinutritive Compounds

One of the main problems in exploiting a plant-based diet as a nutrient source is the presence of antinutritional factors, such as trypsin inhibitors, oxalates and tannins, among others. FAO/WHO (1991) identified several factors that contribute to lower protein digestibility of plant-based foods, such as lectins, protease inhibitors, phytic acid and tannins. Some authors have been identifying antinutritional factors in *P. aculeata* leaves, such as a protease inhibitors, lectins (Pompeu *et al.*, 2014), oxalates, and tannins (Silveira *et al.*, 2020).

Protease inhibitors have the specificity of inhibiting the proteolytic enzymes trypsin and quimotrypsin and, consequently, they reduce the protein digestion of foods (Monteiro *et al.*, 2004). Oxalate cannot be metabolized by humans and is excreted in the urine (Benevides *et al.*, 2011). Some 75% of all kidney stones are composed mainly of calcium oxalate and hyperoxaluria is one of the main factors of risk for this disease because oxalate originates from a combination of oxalate absorption from the diet (Chai and Liebman, 2005). Moreover, soluble oxalates can chelate minerals, reducing absorption, or are absorbed through the intestine and colon. Insoluble oxalates, on the other hand, are excreted in the faeces (Petroski and Minich, 2020).

Tannins may act as antioxidants by scavenging free radicals, but their ability to act as chelators inhibits the absorption of dietary minerals such as iron, copper, and zinc. In developing and low-income countries where the diet is rich in tannins, their consumption contributes to iron-deficiency anaemia in residents (Petroski and Minich, 2020). On the other hand, tannins were determined in the leaves of *P. aculeata* by low-selectivity assays (Sousa *et al.*, 2014; Silveira *et al.*, 2020) and one condensed tannin was identified (Souza *et al.*, 2021). Thus, more phytochemical studies must be developed on tannin chemistry for *P. aculeata*.

One alternative to inactivation of most or even all the activity of protease inhibitors, oxalates and tannins in ora-pro-nobis leaves is cooking at atmospheric pressure or under pressure (Lajolo and Genovese, 2002; Petroski and Minich, 2020).

Finally, it is important to highlight that *P. aculeata* leaves contain antinutritive compounds and also other compounds such as nutrients (amino acids, vitamins, minerals and others). Therefore, it remains questionable as to whether these compounds are as potentially harmful as they might seem when isolated, as they may act differently when taken within whole foods that are properly prepared. Cooking and application of heat seem to be essential for the activation of these compounds. Furthermore, their content must be further investigated in a high number of specimens.

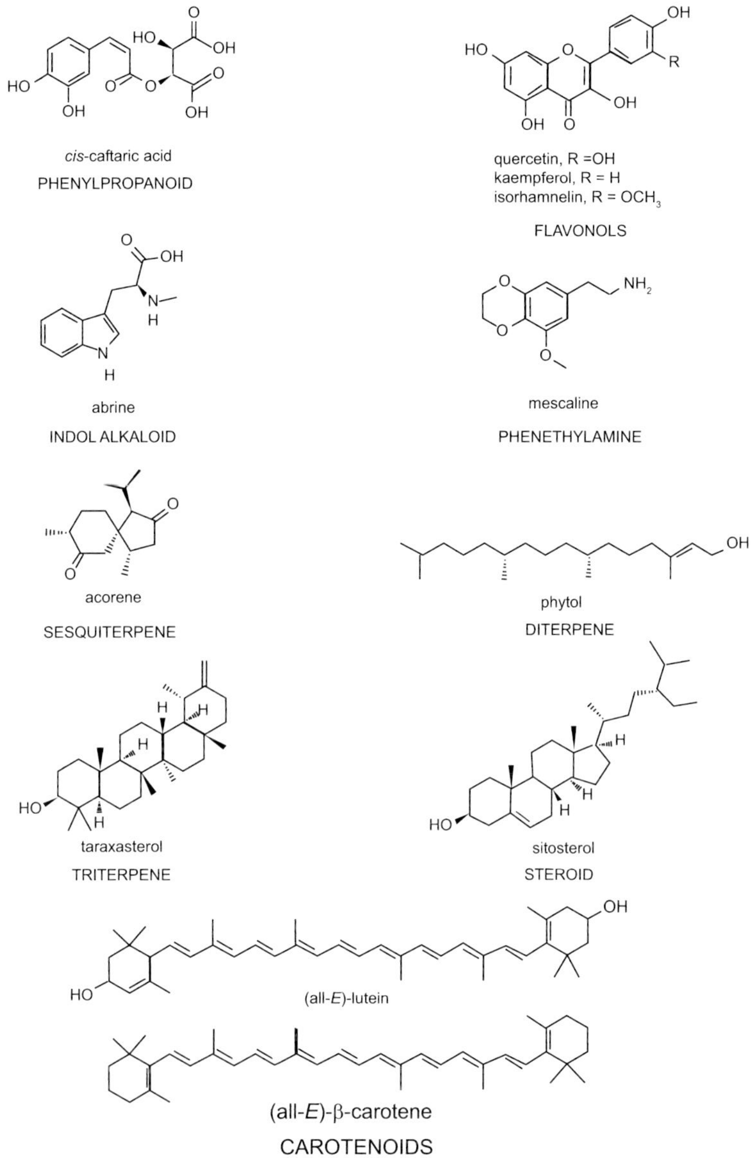

Fig. 19.2. Representative secondary metabolites from *Pereskia aculeata* Miller leaves. The flavonols are the aglycones of the main flavonoid glycosides identified in the leaves.

Phytochemicals

The phytochemical compounds identified in the leaves of *P. aculeata* (Fig. 19.2) are benzoic acid derivatives, phenylpropanoids, flavonoids, monoterpenes, sesquiterpenes, triterpenes, sterols, carotenoids, an indole alkaloid, phenethylamines and mucilage. These same compound classes were identified in other *Pereskia* species. No secondary metabolites were identified in the fruit or other parts of the plant (Pinto and Scio, 2014; Souza *et al.*, 2016; Agostini-Costa, 2020).

The content of total phenolic compounds as determined for *P. aculeata* leaves, fruit and leaf

Table 19.3. Content of total phenolic compounds in *Pereskia aculeta* Miller leaves and fruit.

Sample	Total phenolic compounds (mg/g)	Reference
Leaves		
Dry leaves	87.2[a]	Souza *et al.* (2021)
Dry leaves	0.2[b]	Almeida *et al.* (2014)
Methanol extract	17.3[c]	Pinto *et al.* (2012)
Methanol extract	15.0[a]	Souza *et al.* (2016)
Petroleum ether extract	11.8[a]	Souza *et al.* (2016)
Chloroform extract	5.2[a]	Souza *et al.* (2016)
Aqueous extract	50.1[a]	Sousa *et al.* (2014)
70% ethanol extract	95.6[a]	Sousa *et al.* (2014)
80% acetone extract	117.8[a]	Sousa *et al.* (2014)
Fruit		
Fresh fruit	0.7[a]	Agostini-Costa (2020)
Unripe fresh fruit	1.1[a]	Silva *et al.* (2018)
Intermediate fresh fruit	1.2[a]	Silva *et al.* (2018)
Ripe fresh fruit	1.2[a]	Silva *et al.* (2018)

[a] mg gallic acid equivalents per g of sample; determination by Folin Ciocalteu assay. [b] determination by Folin Dennis assay; [c] mg tannic acid equivalents per g of sample; determination by Folin Dennis assay.

extracts (Table 19.3). The content in dry leaves and fresh fruit was 0.2–87.2 mg per g of dry leaves and 0.7–1.2 mg gallic acid equivalents per g of fresh fruit. The difference observed for the content in the leaves by different authors may be due to the adopted methods, experimental errors and/or secondary metabolism variability. Among leaf extracts, 70% ethanol and 80% acetone extracts showed the higher content (95.6 and 117.8 mg gallic acid equivalents per g of extract) (Almeida *et al.*, 2014; Silva *et al.*, 2018; Agostini-Costa, 2020; Souza *et al.*, 2021).

Pinto *et al.* (2012) determined the total content of flavonoids in the methanol extract by aluminum chloride photometric assay as 13.1 mg rutin equivalents per g and the same extract showed a total phenolic compound content of 17.3 mg tannic acid equivalents per g, suggesting the predominance of flavonoids among phenolic compounds. Souza *et al.* (2021) described the flavonoid dry leaf content as 12.2 mg/g and the anthocyanin content for the leaves in dry basis as 1.2 mg cyanidin-3-glycoside equivalents per g. The flavonoid content was also determined for the unripe, intermediate and ripe fresh fruit as 0.06, 0.07 and 0.09 mg/g (Silva *et al.*, 2018). The condensed tannin content was determined by sulfuric vanillin assay for aqueous, 70% ethanol and 80% acetone leaf extracts as 2.6, 4.0 and 7.6 mg catechin equivalents per g of extract (Sousa *et al.*, 2014). The tannins were also determined through the colorimetric method of the Association of Official Analytical Chemists (1990) as 30.1 mg/g of dry leaves (Silveira *et al.*, 2020). However, just one condensed tannin was putatively identified in *P. aculeata* leaves (Souza *et al.*, 2021). Similarly, even though no saponin was identified in the species, the content of saponins in the leaves was determined by photometric assay as 0.3 mg per 100 g of dry material (Almeida *et al.*, 2014).

Three caffeic acid derivatives and seven flavonols were identified and quantified in the 70% ethanol extract of the leaves of *P. aculeata* by HPLC-PDA/UV-ESI/MS^n – w/w: *cis*-caftaric acid – 9.5%, *trans*-caftaric acid – 2.2%, caffeic acid derivative – 0.57%, quercetin-3-*O*-rutinoside – 3.6%, isorhamnetin-*O*-pentoside-*O*-rutinoside – 2.3%, quercetin-*O*-pentoside-*O*-rutinoside – 2.1%, isorhamnetin-3-*O*-rutinoside – 1.5%, kaempferol-3-*O*-rutinoside – 0.8%, quercetin-*O*-pentoside-*O*-hexoside – 0.7%, isorhamnetin-*O*-pentoside-*O*-hexoside – 0.6%. The phenylpropanoids and flavonoids were quantified based on the analytical curves of caffeic acid and quercetin-3-*O*-glucoside, respectively. Considering these data, the content in the extracts of total phenylpropanoids, total flavonoids and total phenolic compounds were determined as 12.3, 11.4 and 23.7%, respectively (Garcia *et al.*, 2019). The flavonoids petunidin, anthocyanidin and quercetin (flavonol), and three di-*tert*-butylphenol isomers were also identified through UHPLC-MS/MS (Pinto *et al.*, 2015a). Moreover, 56 compounds were putatively identified in the leaves by paper spray MS, based only on low resolution MS^1 data. The phenolic compounds as flavonoids and phenolic acids predominated with 26 identified compounds. Most of the identified flavonoids were quercetin, isorhamnetin or kaempferol glycosides (Souza *et al.*, 2021).

Pinto *et al.* (2015a) identified, by UHPLC-MS/MS, tryptamine, besides the indole alkaloid abrine and the phenethylamines mescaline and hordenine in the leaves of *P. aculeata*. Mescaline and other phenethylamines were also identified in *Pereskia bleo* (Kunth) DC. and *Pereskia grandifolia* Haw. Even though mescaline has psychedelic effects, its content in *P. aculeata* leaves must be further investigated to evaluate possible effects of plant consumption (Agostini-Costa, 2020).

The dry leaf essential oil of *P. aculeata* was obtained by hydrodistillation as a pale-yellow oil, yielding 0.02% w/w. The 24 compounds identified by GC-MS include monoterpene hydrocarbons – 0.6%, oxygenated monoterpenes – 16.0%, oxygenated sesquiterpenes – 44.9%, diterpene hydrocarbons – 5.1% and non-terpenes – 24.4%. The major compounds were acorone – 30.0%, (Z,Z)-methyl-4,6-hexadecadiene – 16.3%, 1-nonadecenol – 6.2%, (5*E*,9*E*)-farnesyl acetone – 5.7%, cyclopentadecanolide – 5.5% and phytol – 5.1% (Souza *et al.*, 2016).

Sterols, triterpenes and other minor compounds were identified by GC-MS in the hexane fraction of the methanol leaf extract of *P. aculeata* and the major compounds were sitosterol – 24.7%, taraxasterol – 11.6%, campesterol – 9.0%, taraxerol – 7.1% and stigmasterol – 6.9%; the diterpene phytol – 4.5% – was also identified (Pinto *et al.*, 2015b).

The content of carotenoids in the fruit and leaves from different accessions was determined by HPLC-PDA/UV. The main carotenoids found in the fruit were α- and β-carotene, for which the content – >20 μg/g – may be considered very high (Britton and Khachik, 2009) in some accessions (Table 19.4). The leaves presented higher total content of carotenoids than the fruit, with very high leaf content of β-carotene, lutein and zeaxanthin in some accessions (Table 19.4) (Agostini-Costa *et al.*, 2012, 2014; Oliveira *et al.*, 2019a).

Table 19.4. Carotenoid content of *Pereskia aculeta* Miller fruits and leaves (fresh basis).

Carotenoid	Fruit (μg/g)[a]	Fruit (μg/g)[b]	Leaves (μg/g)[a]	Leaves (μg/g)[c]
α-carotene	4.3–35.1	22.7	3.8–11.9	---
Trans-β-carotene	7.3–41.1	34.3	18.6–35.1	10.1
13-*cis-β*-carotene	0.3–4.8	---	0.7–2.9	---
9-*cis-β*-carotene	---	---	3.0–5.1	---
cis-β-carotene	---	2.8	---	---
Lutein	0.8–3.7	6.5	57.4–102.0	23.2
Neoxanthin	---	---	9.3–16.1	---
Violaxanthin	---	---	20.0–27.3	---
Zeaxanthin	---	---	3.3–5.3	---
α-cryptoxanthin/ zeinoxanthin	---	2.7	---	---
β-cryptoxanthin	---	2.2	---	---

[a]Agostini-Costa *et al.*, 2014; other detected carotenoids were lutein ester, α-cryptoxanthin and violaxanthin.
[b] Agostini-Costa *et al.*, 2012. [c] Oliveira *et al.*, 2019a.

The leaves and fruit of *P. aculeata* contain mucilage which is important as a dietary fibre (Silva *et al.*, 2019). The leaf mucilage was characterized as a heteropolysaccharide with a main chain consisting of β-D-galactopyranosyl-(1→4)-linked units substituted by arabinose and galactose units. Its structure includes galacturonic acid, partially esterified, and fucose units (Sierakowski *et al.*, 1987, 1990; Martin *et al.*, 2017).

Bioactive Properties

P. aculeata has shown important bioactivities, including antioxidant, anti-inflammatory, wound healing, antinociceptive, antibacterial, trypanocidal actions and cytotoxicity in tumour cell lines (Maciel *et al.*, 2019; Vieira *et al.*, 2019).

The leaves and fruit of *P. aculeata* have shown antioxidant activity in different experimental models. The antiradical activity was evaluated through DPPH assay and the EC_{50} values for aqueous, 70% ethanol and 80% acetone leaf extracts were determined as 106.1, 56.6 and 49.1 μg/ml (Sousa *et al.*, 2014). Carvalho *et al.* (2014) have determined the percentage of DPPH scavenging activity of 95% ethanol leaf extracts as 17.0–19.9%; extract concentration was 0.01% w/v. The less polar leaf extracts showed higher EC_{50} values in the DPPH assay: methanol, chloroform and petroleum ether extracts showed values of 7.1, 81.1 and 18.3 mg/ml, respectively. In another evaluation, the antiradical activity of the 70% ethanol leaf extract was higher than the activity of the trolox in DPPH and ABTS assays, showing EC_{50} values of 72.9 and 40.5 μg/ml, respectively; the EC_{50} values for the hydroxyl radical scavenging assay, the inhibition of the production of thiobarbituric acid reactive substances assay, and the oxidative haemolysis inhibition assay in 30 and 60 min were 373.5, 39.0, 57.0 and 131.0 μg/ml, respectively (Garcia *et al.*, 2019). The higher antiradical activity of more polar leaf extracts may be related to the content of phenolic compounds in the leaves (Table 19.3), which were identified mainly as flavonoids and phenolic acids. The antioxidant activity was measured through the ORAC method in the unripe, intermediate and ripe fruit and the obtained values were 1.95, 0.80 and 3.85 mmol trolox per 100 g of freeze-dried pulp, respectively (Silva *et al.*, 2018).

The 70% ethanol leaf extract of *P. aculeata* showed antibacterial activity against both Gram-positive (*Enterococcus faecalis*, *Listeria monocytogenes*) and methicillin-resistant (*Staphylococcus aureus*), as well as Gram-negative (*Escherichia coli*, *Klebsiella pneumoniae*, *Morganella morganii*, *Proteus mirabilis* and *Pseudomonas aeruginosa*) bacteria. The lower values of minimal inhibitory concentrations were observed for the Gram-positive bacteria and *K. pneumonia* (Garcia *et al.*, 2019). The petroleum ether, chloroform and methanol leaf extracts were evaluated in the agar diffusion method against Gram-positive (*Bacillus cereus* and *Staphylococcus aureus*) and Gram-negative (*Escherichia coli* and *Pseudomonas aeruginosa*) bacteria. The extracts inhibited the growth of bacteria, except chloroform and methanol extracts against *E. coli*. All these extracts showed also antifungal activity against *Penicillium expansum*, *P. citrinum*, *Aspergillus niger* and *A. versicolor*, but only the petroleum ether extract was active against all fungi (Souza *et al.*, 2016).

The aqueous leaf extract of *P. aculeata* caused 100% death of the trypomastigote form of *Trypanosoma cruzi* in 24 hours, while the dichloromethane:methanol 1:1 leaf extract caused 90% death of the parasites in 48 hours (Valente *et al.*, 2007).

The antitumoural potential of *P. aculeata* leaves has been studied through *in vitro* assays. In the cytotoxicity assay, the aqueous leaf extract – 250 μg/ml – inhibited the growth of the following cell lines: J774 murine sarcoma – 49.4%, SP2/0 murine myeloma – 49.9%, P3653 murine myeloma – 58.7% and BW murine lymphoma – 72.4%. The dichloromethane:methanol 1:1 leaf extract inhibited in 46.2% the cell proliferation of Ehrlich carcinoma. These extracts showed no cytotoxicity for erythrocytes. The *S. cerevisiae* assay showed that the cytotoxicity of these extracts is not related to the DNA damage or the inhibition of topoisomerases I and II (Valente *et al.*, 2007). The methanol leaf extract showed cytotoxicity against HL60 human promyelocytic leukaemia cell line and its fractions also inhibited the proliferation of MCF-7 human breast adenocarcinoma cell line, whereas the methanol extract was not cytotoxic against murine peritoneal macrophages (Pinto *et al.*, 2012). In another study, the treatment of

SH-SY5Y neuroblastoma cells with petroleum ether and methanol leaf extracts resulted in a low cytotoxicity - IC_{50} >2000 µg/ml and IC_{50} >1600 µg/ml, respectively, whereas the chloroform extract resulted in a stronger cytotoxicity - IC_{50} <300 µg/ml. Moreover, the same methanol extract reduced ADCY1 expression in SH-SY5Y neuroblastoma cells (Souza *et al.*, 2016).

The wound-healing and anti-inflammatory properties of the species have been investigated because these actions are related to Brazilian traditional medicine uses. The 93% ethanol leaf extract of *P. aculeata* showed wound-healing action in rats (Sartor *et al.*, 2010). Carvalho *et al.* (2014) developed morphological studies on fibroblast cell cultures submitted to the 95% ethanol leaf extract and suggested the beneficial effect of the mucilage on the fibroblasts. The methanol leaf extract of *P. aculeata* and its hexane fraction accelerated the wound healing process in mice. The effect of the hexane fraction was considered more evident as it induced the increase of blood flow rate and collagen deposition at the wound sites (Pinto *et al.*, 2016).

The anti-inflammatory effect of *P. aculeata* was demonstrated in studies on the hexane fraction of its methanol leaf extract. This fraction showed chronic and acute topical anti-inflammatory action in mice, reducing IL-6 and TNF-α levels. The authors suggested a glucocorticoid-like effect of the hexane fraction that could be related to the sterols, triterpenes and phytol which were identified in this fraction. Furthermore, no toxic effects were observed in the acute dermal irritation/corrosion test in rats (Pinto *et al.*, 2015b). The hexane fraction was employed as the active ingredient of Inflativ, a pharmaceutical cream containing 6 and 12% of the hexane fraction, which also showed chronic and acute topical anti-inflammatory activity in mice (Pinto *et al.*, 2020). The antinociceptive activity of the hydromethanolic fraction of the methanol leaf extract of *P. aculeata* was evaluated through the acetic acid-induced writhing, formalin and tail-flick tests in mice and the results suggested a peripheral and central antinociception not related to opioid receptor activation (Pinto *et al.*, 2015a).

Silva *et al.* (2017) studied the oral acute toxicity of the ethanolic leaf extract of *P. aculeata* in rats. The authors performed a single dose gavage administration of plant extract, in the concentrations of 1250, 2500 and 5000 mg of dry extract per kg of body weight. The acute toxicity investigation did not reveal any variation among the groups, neither in clinical evaluation nor in histopathological tests. The same extract did not produce cytogenotoxicity on lettuce.

Bioaccessability and Bioavailability

Studies on the evaluation of the bioaccessibility and bioavailability of *P. aculeata* are scarce. Takeiti *et al.* (2009) evaluated the *in vitro* digestibility of its leaves determining a value of 75.9%, which was considered low when compared to the results found by Almeida Filho and Cambraia (1974) for two accessions that were 79.5 and 90.2%.

Health Benefits

The leaves of *P. aculeta* are used in folk medicine as an emollient and to treat skin wounds and inflammatory processes (Duarte and Hayashi, 2005; Sartor *et al.*, 2010; Pinto and Scio, 2014). Carvalho *et al.* (2014) reported the *in vitro* wound-healing activity of its mucilage. However, there are few studies on the phytochemicals of *P. aculeta* and their therapeutic potential (Pinto *et al.*, 2012; Pinto and Scio, 2014; Agostini-Costa, 2020).

In a randomized controlled clinical study, the consumption of *P. aculeata* leaf flour was evaluated as a functional ingredient in sprinkle cookies in 24 men between 20 and 50 years old, overweight and obese (BMI between 25 and 35 kg/m²). The objective was to verify the effects on the aptitude of probiotic *Lactobacillus casei* in intestinal epithelial cells, in addition to evaluating a possible improvement in gastrointestinal symptoms, weight, body fat, glycaemia and lipid profile. After 14 weeks, it was found that the consumption of the product with *P. aculeata* flour maintained the high adherence of *L. casei* in intestinal cells, improving gastrointestinal symptoms and increasing satiety, resulting in the improvement of anthropometric and biochemical parameters of patients (Vieira *et al.*, 2019).

Besides that, the intestinal microbiota was evaluated after the consumption of biscuits prepared with *P. aculeata* flour obtained from the leaves, and no changes in the anthropometric parameters or in the nutritional status of the individuals were observed, but a reduction in flatulence and an increase in satiety were noted, suggesting the use of the species as something beneficial for the maintenance of the intestinal microbiota, with emphasis on the combination with *Lactobacillos casei* (Vieira *et al.*, 2019).

The effects of *P. aculeata* leaf juice on biochemical parameters and body composition was evaluated in male and female rats. The intake of the juice prevented the triglyceride level increasing in male rats submitted to a hypercaloric diet, decreased the levels of aspartate transaminase and alanine aminotransferase in male rats and alanine aminotransferase in females, and reduced visceral fat gain in females with a regular diet (Souza *et al.*, 2015).

Finally, consumption of *P. aculeata* leaves by the low-income population to improve nutrition reinforces their potential as a sustainable source of nutraceuticals and promising food ingredient to be used both for food enrichment and preservation.

Novel Products/Formulations

Many new products from *P. aculeta* leaves are being developed as innovations based on scientific literature data. The filed patents on the species include pharmaceutical products such as a Chinese medicine indicated for the treatment of fractures which is composed of several plants (Jiang *et al.*, 2013). A topical herbal medicine indicated as an anti-inflammatory and based on Brazilian traditional medicine has also been developed. It consists of a cream formulation containing the hexane fraction of the methanol leaf extract of *P. aculeata* and is indicated as a topical anti-inflammatory for human and veterinary uses (Fontes *et al.*, 2019).

The potential use of *P. aculeata* in the food industry has also been exploited. Some studies have suggested the use of *P. aculeata* leaf flour to enrich industrial products, as it is a good source of proteins and amino acids, dietary fibre, vitamins, iron and other minerals, besides the health benefits promoted by antioxidant compounds – a desirable feature to the industry (Takeiti *et al.*, 2009; Lima Junior *et al.*, 2013; Barbalho *et al.*, 2016).

A process for hydrocolloid extraction from *P. aculeata* leaves and branches has been developed due to its mucilage content. This hydrocolloid is odourless and may be employed by the food industry as a thickening, gelling and emulsifying agent (Resende *et al.*, 2016). As an example, hydrocolloid blends of guar gum, gum Arabic and *P. aculeata* leaf mucilage were used to produce fermented milk beverages (Amaral *et al.*, 2018). The addition of *P. aculeata* mucilage increased the protein content of the fermented milk beverages.

Bread and sausages containing the *P. aculeata* leaf flour have been produced and their sensorial and nutritional properties were positively evaluated (Martinevski *et al.*, 2013; Sobrinho *et al.*, 2015).

Lago *et al.* (2019) developed nanoemulsion systems produced with the *P. aculeata* leaf mucilage and concluded that this mucilage may be an interesting natural emulsifier for food and pharmaceutical products. Biodegradable films were produced with this mucilage plus glycerol and they may be used as biodegradable and/or edible packaging for food products (Guimarães *et al.*, 2019). Maciel *et al.* (2020) developed iron-rich chitosan-pectin microparticles laden with aqueous leaf extract of *P. aculeata* as an alternative approach to the nutritional fortification or supplementation of food products.

Conclusions and Future Trends

P. aculeata or ora-pro-nobis is a PANC and also a medicinal plant from Brazilian biodiversity. Its edible parts comprise the leaves, fruit and flowers. The leaves are the most consumed part, eaten fresh in salads or used in different recipes. In some low-income communities it is the main source of protein available. In Brazilian traditional medicine, the fruit have been used as an expectorant and antisyphilitic and the leaves as emollient, anti-inflammatory, wound-healing and auxiliary treatment for burns.

Several studies showed high levels of proteins and essential amino acids in *P. aculeata* leaves besides an excellent amino acid profile. The leaves also contain high amounts of fibre, especially soluble dietary fibre (mucilage), and they are a good source of some vitamins and minerals. The soluble dietary fibres of *P. aculeata* may promote health benefits, reducing the level of serum low-density lipoprotein cholesterol, decreasing postprandial blood glucose and modulating gut microbiota. Moreover, carotenoids and phenolic compounds identified in the leaves may be responsible for the antioxidant activity of *P. aculeata*.

Some antinutritional factors were identified in *P. aculeata* leaves – protease inhibitors, lectins, oxalates and tannins – but their content must be further investigated in a high number of specimens to determine its antinutritive potential effects.

The secondary metabolites identified in the leaves of *P. aculeata* are benzoic acid derivatives, phenylpropanoids, flavonoids, monoterpenes, sesquiterpenes, triterpenes, sterols, carotenoids, an indole alkaloid, and phenethylamines. No secondary metabolites were identified in other plant parts. In general, the chemical composition of the species must be further investigated because many compounds were just tentatively identified. Among phenolic compounds, flavonoids and phenolic acids were the main ones, which may be responsible for the antioxidant activity exhibited by the leaf extracts. The leaves contain a low content of essential oil composed mainly of oxygenated sesquiterpenes and its major compound is acorone. Mescaline was identified in the leaves and, even though it has psychedelic effects, its content in *P. aculeata* leaves must be further investigated.

P. aculeata showed important bioactivities, including antioxidant, anti-inflammatory, wound healing, antinociceptive, antibacterial, trypanocidal actions, and cytotoxicity in tumour cell lines. The topical anti-inflammatory and wound healing actions of *P. aculeata* leaf derivatives are related to Brazilian traditional medicine uses and were demonstrated through *in vivo* studies. A pharmaceutical formulation based on a hexane fraction from methanol extract of *P. aculeata* leaves showed chronic and acute topical anti-inflammatory activity in mice demonstrating the potential of the species for the development of herbal medicines.

The great potential of *P. aculeata* in the food industry includes the use of its leaf flour as an ingredient for food enrichment and its mucilage as a thickening, gelling or emulsifying agent, increasing the levels of nutrients like fibre, protein, minerals and vitamins. Other technological applications of ora-pro-nobis have been exploited in the areas of materials and nanotechnology.

Acknowledgements

This study was financed by the Coordenação de Aperfeiçoamento de Pessoal de Nível Superior, Brasil (CAPES), Finance Code 001 and the Instituto Nacional de Ciência e Tecnologia em Biodiversidade e Produtos Naturais (INCT-BioNat; CNPq and FAPESP; Brasil).

References

Agostini-Costa, T. (2020) Bioactive compounds and health benefits of Pereskioideae and Cactoideae: A review. *Food Chemistry* 327, 126961. DOI: 10.1016/j.foodchem.2020.126961.

Agostini-Costa, T.S., Wondraceck, D.C., Rocha, W.S. and Silva, D.B. (2012) Carotenoids profile and total polyphenols in fruits of *Pereskia aculeata* Miller. *Revista Brasileira de Fruticultura* 34(1), 234–238. DOI: 10.1590/S0100-29452012000100031.

Agostini-Costa, T.S., Pêssoa, G.K.A., Silva, D.B., Gomes, I.S. and Silva, J.P. (2014) Carotenoid composition of berries and leaves from a Cactaceae – *Pereskia* sp. *Journal of Functional Foods* 11, 178–184. DOI: 10.1016/j.jff.2014.09.015.

Albuquerque, M.G.P.T., Sabaa-Srur, A.U.O. and Freiman, L.O. (1991) Composição centesimal e escore de amino-ácidos em três espécies de "ora-pro-nobis". *Boletim Sociedade Brasileira de Ciência e Tecnologia de Alimentos* 25(1), 7–12.

Almeida Filho, J.D. and Cambraia, J. (1974) Estudo do valor nutritivo do "ora-pro-nobis"(*Pereskia aculeata* Mill). *Revista Ceres* 21(114), 105–111.

Almeida, M.E.F. and Corrêa, A.D. (2012) Utilização de cactáceas do gênero *Pereskia* na alimentação humana em um município de Minas Gerais. *Ciência Rural* 42(4), 751–756. DOI: 10.1590/S0103-84782012000400029.

Almeida, M.E.F., Junqueira, A.M.B., Simão, A.A. and Corrêa, A.D. (2014) Caracterização química das hortaliças não-convencionais conhecidas como ora-pro-nobis. *Bioscience Journal* 30(3).

Amaral, T.N., Junqueira, L.A., Prado, M.E.T., Cirillo, M.A., de Abreu, L.R. *et al.* (2018) Blends of *Pereskia aculeata* Miller mucilage, guar gum, and gum Arabic added to fermented milk beverages. *Food Hydrocolloids* 79, 331–342. DOI: 10.1016/j.foodhyd.2018.01.009.

Angiosperm Phylogeny Group (2009) An update of the Angiosperm Phylogeny Group classification for the orders and families of flowering plants: APG III. *Botanical Journal of the Linnean Society* 161(2), 105–121. DOI: 10.1111/j.1095-8339.2009.00996.x.

Association of Official Analytical Chemists (1990) *Official Methods of Analysis of the Association of the Analytical Chemists*, 12th edn. Association of Official Analytical Chemists, Washington, DC.

Barbalho, S.M., Guiguer, É.L., Marinelli, P.S., do Santos Bueno, P.C., Pescinini-Salzedas, L.M. *et al.* (2016) *Pereskia aculeata* Miller flour: Metabolic effects and composition. *Journal of Medicinal Food* 19(9), 890–894. DOI: 10.1089/jmf.2016.0052.

Benevides, C.M.D.J., Souza, M.V., Souza, R.D.B. and Lopes, M.V. (2011) Fatores antinutricionais em alimentos: revisão. *Segurança Alimentar e Nutricional* 18(2), 67–79. DOI: 10.20396/san.v18i2.8634679.

Botrel, N., Godoy, R.D.O., Madeira, N., Amaro, G. and Melo, R.D.C. (2019) Estudo comparativo da composição proteica E do perfil de aminoácidos em cinco clones de ora-pro-nóbis. Embrapa Hortaliças-Boletim de Pesquisa E Desenvolvimento (INFOTECA-E).

Botrel, N., Freitas, S., Fonseca, M.J. de O., Melo, R.A. de C. e and Madeira, N. (2020) Valor nutricional de hortaliças folhosas não convencionais cultivadas no Bioma Cerrado. *Brazilian Journal of Food Technology* 23, e2018174. DOI: 10.1590/1981-6723.17418.

Brandão, M. das G.L. (2019) *Plantas Úteis e Medicinais Na Obra de Frei Vellozo*. MG, Brasil.

Britton, G. and Khachik, F. (2009) Carotenoids in food. In: Britton, G., Pfander, H. and Liaaen-Jensen, S. (eds) *Carotenoids Volume 5: Nutrition and Health*. Birkhäuser Basel, pp. 45–66.

Butterworth, C.A. and Wallace, R.S. (2005) Molecular phylogenetics of the leafy cactus genus Pereskia (Cactaceae). *Systematic Botany* 30(4), 800–808.

Carvalho, E.G., Soares, C.P., Blau, L., Menegon, R.F. and Joaquim, W.M. (2014) Wound healing properties and mucilage content of *Pereskia aculeata* from different substrates. *Revista Brasileira de Farmacognosia* 24(6), 677–682. DOI: 10.1016/j.bjp.2014.11.008.

Chai, W. and Liebman, M. (2005) Effect of different cooking methods on vegetable oxalate content. *Journal of Agricultural and Food Chemistry* 53(8), 3027–3030. DOI: 10.1021/jf048128d.

Cruz., A.F., Savicki., A., Frentzel., A.E., Adam., I.P., Prado., L.O. *et al.* (2020) Non-conventional edible plants: the use of "ora-pro-nobis" (*Pereskia aculeata* Mill., Cactaceae) leaves in human consumption. *Visão Acadêmica* 21(3), 19–33. DOI: 10.5380/acd.v21i3.76001.

Dayrell, M.D.S. and Vieira, E.C. (1977) Leaf protein concentrate of the cactacea *Pereskia aculeata* Mill. I. Extraction and composition. *Nutrition Reports International* 15(5), 539–545.

Devi, R., Arcot, J., Sotheeswaran, S. and Ali, S. (2008) Folate contents of some selected Fijian foods using tri-enzyme extraction method. *Food Chemistry* 106(3), 1100–1104. DOI: 10.1016/j.foodchem.2007.07.037.

Duarte, M.R. and Hayashi, S.S. (2005) Estudo anatômico de folha e caule de *Pereskia aculeata* Mill. (Cactaceae). *Revista Brasileira de Farmacognosia* 15(2), 103–109. DOI: 10.1590/S0102-695X2005000200006.

Egbon, I.N., Paterson, I.D., Compton, S. and Hill, M. (2020) Evolution of growth traits in invasive *Pereskia aculeata* (Cactaceae): testing the EICA hypothesis using its specialist herbivore, *Catorhintha schaffneri* (Coreidae). *Pest Management Science* 76(12), 4046–4056. DOI: 10.1002/ps.5959.

Fontes, E.S., Mello, D., Silva, J., Pinto, N.C.C. and Universidade Federal de Juiz de Fora (2019) *Phytotherapeutic pharmaceutical formulation containing hexane partition of methanol extract of Pereskia aculeata used in treatment of dermatitis*. BR Patent 102015017575.

Food and Agriculture Organization/World Health Organization (1991) *Evaluation of Protein Quality*. FAO, Food Nutrition. Report of the Joint FAO/WHO expert consultation on protein quality evaluation, Rome.

Food and Agriculture Organization/World Health Organization (2007) *Protein and Amino Acid Requirements in Human Nutrition*. Report of a Joint WHO/FAO/UNU Expert Consultation, WHO Tech Rep Ser No. 935, Geneva.

Food and Drug Administration (2020) Daily value and percent daily value: changes on the new nutrition and supplement facts labels. Available at: https://www.fda.gov/food/food-labeling-nutrition/changes-nutrition-facts-label (accessed 6 April 2022).

Garcia, J.A.A., Corrêa, R.C.G., Barros, L., Pereira, C., Abreu, R.M.V. *et al.* (2019) Phytochemical profile and biological activities of "Ora-pro-nobis" leaves (*Pereskia aculeata* Miller), an underexploited superfood from the Brazilian Atlantic Forest. *Food Chemistry* 294, 302–308. DOI: 10.1016/j.foodchem.2019.05.074.

Girão, L.V.C., Silva Filho, J.C., Pinto, J.E.B.P. and Bertolucci, S.K.V. (2003) Avaliação da composição bromatológica de ora-pronobis. *Anais Congresso Brasileiro de Olericultura* 43.

Global Biodiversity Information Facility - GBIF (2021) *Pereskia aculeata* Mill. Available at: https://www.gbif.org/species/5384215 (accessed 2 March 2021).

Guimarães, J.R.A., Gomes, J.A.O., Teixeira, D.A., Bonfim, F.P.G. and Evangelista, R.M. (2019) Agronomic performance and protein content of *Pereskia aculeata* Mill. using organic chicken. *Australian Journal of Crop Science* 13(2), 179–184. DOI: 10.21475/ajcs.19.13.02.p868.

Jiang, M., Jiang, W., Jiang, Q. and Nanning Duoling Biotechnology Co. Ltd (2013) *Drug for Treating Fractures*. CN Patent 103223022.

Kinnup, V.F. and Lorenzi, H. (2014) *Plantas Alimentícias Não Convencionais (PANC) no Brasil*. Instituto Plantarum de Estudos da Flora, São Paulo, SP/Brasil, pp. 272–277.

Lage, D.A., Tirado, M. da. S.`, Vanicore, S.R., Sabino, K.C.C. and Albarello, N. (2015) Production of betalains from callus and cell suspension cultures of *Pereskia aculeata* Miller, an unconventional leafy vegetable. *Plant Cell, Tissue and Organ Culture (PCTOC)* 122(2), 341–350. DOI: 10.1007/s11240-015-0771-x.

Lago, A.M.T., Neves, I.C.O., Oliveira, N.L., Botrel, D.A., Minim, L.A. *et al.* (2019) Ultrasound-assisted oil-in-water nanoemulsion produced from *Pereskia aculeata* Miller mucilage. *Ultrasonics Sonochemistry* 50, 339–353. DOI: 10.1016/j.ultsonch.2018.09.036.

Lajolo, F.M. and Genovese, M.I. (2002) Nutritional significance of lectins and enzyme inhibitors from legumes. *Journal of Agricultural and Food Chemistry* 50(22), 6592–6598. DOI: 10.1021/jf020191k.

Lima Junior, F.A., Conceição, M.C., Vilela de Resende, J., Junqueira, L.A., Pereira, C.G. *et al.* (2013) Response surface methodology for optimization of the mucilage extraction process from *Pereskia aculeata* Miller. *Food Hydrocolloids* 33(1), 38–47. DOI: 10.1016/j.foodhyd.2013.02.012.

Maciel, V.B.V., Yoshida, C.M.P. and Goycoolea, F.M. (2019) Agronomic cultivation, chemical composition, functional activities and applications of Pereskia species - a mini review. *Current Medicinal Chemistry* 26(24), 4573–4584. DOI: 10.2174/0929867325666180926151615.

Maciel, V.B.V., Yoshida, C.M.P., Boesch, C., Goycoolea, F.M. and Carvalho, R.A. (2020) Iron-rich chitosan-pectin colloidal microparticles laden with ora-pro-nobis (*Pereskia aculeata* Miller) extract. *Food Hydrocolloids* 98, 105313. DOI: 10.1016/j.foodhyd.2019.105313.

Martin, A.A., de Freitas, R.A., Sassaki, G.L., Evangelista, P.H.L. and Sierakowski, M.R. (2017) Chemical structure and physical-chemical properties of mucilage from the leaves of *Pereskia aculeata*. *Food Hydrocolloids* 70, 20–28. DOI: 10.1016/j.foodhyd.2017.03.020.

Martinevski, C.S., Oliveira, V.D., Rios, A.D.O., Flores, S.H. and Venzke, J.G. (2013) Utilização de bertalha (Anredera cordifolia (TEN.) Steenis) e ora-pro-nobis (*Pereskia aculeata* Mill.) na elaboração de pães. *Alimentos e Nutrição* 24(3), 272.

Merce, A.L., Landaluze, J.S., Mangrich, A.S., Szpoganicz, B. and Sierakowski, M.R. (2001) Complexes of arabinogalactan of *Pereskia aculeata* and Co2+, Cu2+, Mn2+, and Ni2+. *Bioresource Technology* 76(1), 29–37. DOI: 10.1016/s0960-8524(00)00078-x.

Monteiro, M.R.P., Costa, N.M.B., Oliveira, M.G.D.A., Pires, C.V. and Moreira, M.A. (2004) Qualidade protéica de linhagens de soja com ausência do Inibidor de Tripsina Kunitz e das isoenzimas Lipoxigenases. *Revista de Nutrição* 17(2), 195–205. DOI: 10.1590/S1415-52732004000200006.

Morais, L.C., Neves, I.C.O., Batista, G.A., Silva, M.L.M., Valentim, T.T. *et al.* (2019) Protein recovery from Barbados gooseberry (*Pereskia aculeata* Miller) leaves by salting out and isoeletric precipitation. *Revista Mexicana de Ingeniería Química* 18(2), 419–430. DOI: 10.24275/uam/izt/dcbi/revmexingquim/2019v18n2/Morais.

Morton, J.F. (1987) Barbados gooseberry. In: Morton, J.F. (ed.) *Fruits of Warm Climates*. Florida Flair Books, Miami, pp. 349–351.

Oliveira, V.B., Yamada, L.T., Fagg, C.W. and Brandão, M.G.L. (2012) Native foods from Brazilian biodiversity as a source of bioactive compounds. *Food Research International* 48(1), 170–179. DOI: 10.1016/j.foodres.2012.03.011.

Oliveira, D.C.S., Wobeto, C., Zanuzo, M.R. and Severgnini, C. (2013) Composição mineral e teor de ácido ascórbico nas folhas de quatro espécies olerícolas não-convencionais. *Horticultura Brasileira* 31(3), 472–475. DOI: 10.1590/S0102-05362013000300021.

Oliveira, H.A.B. de, Anunciação, P.C., Silva, B.P. da, Souza, Â.M.N. de, Pinheiro, S.S. *et al.* (2019a) Nutritional value of non-conventional vegetables prepared by family farmers in rural communities. *Ciência Rural* 49(8). DOI: 10.1590/0103-8478cr20180918.

Oliveira, N.L., Rodrigues, A.A., Oliveira Neves, I.C., Teixeira Lago, A.M., Borges, S.V. *et al.* (2019b) Development and characterization of biodegradable films based on *Pereskia aculeata* Miller mucilage. *Industrial Crops and Products* 130, 499–510. DOI: 10.1016/j.indcrop.2019.01.014.

Paterson, I.D., Vitorino, M.D., de Cristo, S.C., Martin, G.D. and Hill, M.P. (2014a) Prioritisation of potential agents for the biological control of the invasive alien weed, *Pereskia aculeata* (Cactaceae), in South Africa. *Biocontrol Science and Technology* 24(4), 407–425. DOI: 10.1080/09583157.2013.864382.

Paterson, I.D., Mdodana, L.A., Mpekula, O., Mabunda, B.D.X. and Hill, M.P. (2014b) A promising biological control agent for the invasive alien plant, *Pereskia aculeata* Miller (Cactaceae), in South Africa. *Biocontrol Science and Technology* 24(10), 1083–1095. DOI: 10.1080/09583157.2014.919439.

Petroski, W. and Minich, D.M. (2020) Is there such A thing as "Anti-Nutrients"? A narrative review of perceived problematic plant compounds. *Nutrients* 12(10), E2929. DOI: 10.3390/nu12102929.

Pinto, N.D.C.C. and Scio, E. (2014) The biological activities and chemical composition of *Pereskia* species (Cactaceae)--a review. *Plant Foods for Human Nutrition* 69(3), 189–195. DOI: 10.1007/s11130-014-0423-z.

Pinto, N.D.C.C., Santos, R., Machado, D.C., Florêncio, J.R., Antinarelli, L.M.R. *et al.* (2012) Cytotoxic and antioxidant activity of *Pereskia* aculeata Miller. *Pharmacology Online* 3, 63–69.

Pinto, N.D.C.C., Duque, A.P. do N., Pacheco, N.R., Mendes, R. de F., Motta, E.V. da S. *et al.* (2015a) *Pereskia aculeata*: A plant food with antinociceptive activity. *Pharmaceutical Biology* 53(12), 1780–1785. DOI: 10.3109/13880209.2015.1008144.

Pinto, N.D.C.C., Machado, D.C., da Silva, J.M., Conegundes, J.L.M., Gualberto, A.C.M. *et al.* (2015b) *Pereskia aculeata* Miller leaves present in vivo topical anti-inflammatory activity in models of acute and chronic dermatitis. *Journal of Ethnopharmacology* 173, 330–337. DOI: 10.1016/j.jep.2015.07.032.

Pinto, N.D.C.C., Cassini-Vieira, P., Souza-Fagundes, E.M. de, Barcelos, L.S., Castañon, M.C.M.N. *et al.* (2016) Pereskia aculeata Miller leaves accelerate excisional wound healing in mice. *Journal of Ethnopharmacology* 194, 131–136. DOI: 10.1016/j.jep.2016.09.005.

Pinto, N. de C.C., Maciel, M. de S.F., Rezende, N.D.S., Duque, A.P. do N., Mendes, R. de F. *et al.* (2020) Preclinical studies indicate that INFLATIV, an herbal medicine cream containing *Pereskia aculeata*, presents potential to be marketed as a topical anti-inflammatory agent and as adjuvant in psoriasis therapy. *The Journal of Pharmacy and Pharmacology* 72(12), 1933–1945. DOI: 10.1111/jphp.13357.

Pio-Correa, M. and Azeredo Pena, D. (1984) *Dicionário das plantas úteis do Brasil e das exóticas cultivadas*, 1st edn. Instituto Brasileiro de Desenvolvimento Florestal, Rio de Janeiro.

Pompeu, D.G., Carvalho, A.D.S., Costa, O.F.D., Galdino, A.S., Gonçalves, D.B. *et al.* (2014) Anti-nutritional factors and "in vitro" digestibility of leaves of *Pereskia aculeata* Miller. *Biochemistry and Biotechnology Reports* 3(1), 1–9.

Resende, J.V., Conceição, M.C., Lima Junior, F.A., Vilela, M.B., Junqueira, L.A. *et al.* (2016) *Processo de extração de hidrocolóides da Pereskia aculeata Miller*. BRPI1102426A2.

Rosa, S. da and Souza, L. de (2003) Morfo-anatomia do fruto (hipanto, pericarpo e semente) em desenvolvimento de *Pereskia aculeata* Miller (Cactaceae). *Acta Scientiarum. Biological Sciences* 25(2), 415–428. DOI: 10.4025/actascibiolsci.v25i2.2046.

Santos, A.Q., Santos, R.X. and Marisco, G. (2018) Atividades biológicas, toxicológicas e parâmetros nutricionaisda *Pereskia aculeata* Miller: uma revisão bibliográfica. *Scientia Amazonia* 7(2).

Sartor, C.F.P., do Amaral, V., Guimarães, H.E.T., de Barros, K.N., Felipe, D.F. *et al.* (2010) Estudo da Ação Cicatrizante das Folhas de Pereskia aculeata. *Saúde e Pesquisa* 3(2).

Sierakowski, M.-R., Gorin, P.A., Reicher, F. and Corres, J.B.C. (1987) Some structural features of a heteropolysaccharide from the leaves of the cactus *Pereskia aculeata*. *Phytochemistry* 26(6), 1709–1713. DOI: 10.1016/S0031-9422(00)82273-6.

Sierakowski, M.-R., Gorin, P.A.J., Reicher, F. and Corrêa, J.B.C. (1990) Location of O-acetyl groups in the heteropolysaccharide of the cactus *Pereskia aculeata*. *Carbohydrate Research* 201(2), 277–284. DOI: 10.1016/0008-6215(90)84243-N.

Silva, D.O., Seifert, M., Nora, F.R., Bobrowski, V.L., Freitag, R.A. *et al.* (2017) Acute toxicity and cytotoxicity of *Pereskia aculeata*, a highly nutritious cactaceae plant. *Journal of Medicinal Food* 20(4), 403–409. DOI: 10.1089/jmf.2016.0133.

Silva, A.P.G. da, Spricigo, P.C., Freitas, T.P. de, Acioly, T.M. da S., Alencar, S.M. de *et al.* (2018) Ripe Ora-pro-nobis (*Pereskia aculeata* Miller) fruits express high contents of bioactive compounds and antioxidant capacity. *Revista Brasileira de Fruticultura* 40(3). DOI: 10.1590/0100-29452018749.

Silva, S.H., Neves, I.C.O., Oliveira, N.L., de Oliveira, A.C.F., Lago, A.M.T. *et al.* (2019) Extraction processes and characterization of the mucilage obtained from green fruits of *Pereskia aculeata* Miller. *Industrial Crops and Products* 140, 111716. DOI: 10.1016/j.indcrop.2019.111716.

Silveira, M.G., Picinin, C.T.R., Cirillo, M.Â., Freire, J.M. and Barcelos, M.D.F.P. (2020) Nutritional assay *Pereskia* spp.: unconventional vegetable. *Anais Da Academia Brasileira de Ciencias* 92 Suppl 1, e20180757. DOI: 10.1590/0001-3765202020180757.

Sobrinho, S.S., Costa, L.L., Gonçalves, C.A.A. and Campagnol, P.C.B. (2015) Emulsified cooked sausages enriched with flour from ora-pro-nobis leaves (*Pereskia aculeata* miller). *International Food Research Journal* 22(1).

Sousa, R.M., Lira, C.S., Rodrigues, A.O., Morais, S.A., Queiroz, C.R. *et al.* (2014) Atividade antioxidante de extratos de folhas de ora-pro-nóbis (*Pereskia aculeata* Mill.) usando métodos espectrofotométricos e voltamétricos in vitro. *Bioscience Journal* 30(3).

Souza, M. da S.S., Barbalho, S.M., Guiguer, E.L., Araújo, A.C., dos Santos Bueno, P.C. *et al.* (2015) Effects of *Pereskia aculeata* Miller on the biochemical profiles and body composition of Wistar rats. *Journal of Biosciences and Medicines* 03(7), 82–89. DOI: 10.4236/jbm.2015.37010.

Souza, L.F., Caputo, L., Inchausti De Barros, I.B., Fratianni, F., Nazzaro, F. *et al.* (2016) *Pereskia aculeata* Miller (Cactaceae) leaves: chemical composition and biological activities. *International Journal of Molecular Sciences* 17(9), E1478. DOI: 10.3390/ijms17091478.

Souza, A.H. de, Silva, B.M., Silva, E.C. da, Augusti, R., Melo, J.O.F. *et al.* (2021) Influence of thermal processing on the characteristics and chemical profile of ora-pro-nobis by PS/MS paper spray. *Research, Society and Development* 10(2), e12110212119. DOI: 10.33448/rsd-v10i2.12119.

Takeiti, C.Y., Antonio, G.C., Motta, E.M.P., Collares-Queiroz, F.P. and Park, K.J. (2009) Nutritive evaluation of a non-conventional leafy vegetable (*Pereskia aculeata* Miller). *International Journal of Food Sciences and Nutrition* 60(sup1), 148–160. DOI: 10.1080/09637480802534509.

Tosh, S.M. and Bordenave, N. (2020) Emerging science on benefits of whole grain oat and barley and their soluble dietary fibers for heart health, glycemic response, and gut microbiota. *Nutrition Reviews* 78(Suppl 1), 13–20. DOI: 10.1093/nutrit/nuz085.

Tropicos (2021) Missouri Botanical Garden. Available at: http://www.tropicos.org (accessed 2 March 2021).

Valente, L.M.M., Scheinvar, L.A., Silva, G.C., Antunes, A.P., Santos, F.A.L. *et al.* (2007) Evaluation of the antitumor and trypanocidal activities and alkaloid profile in species of Brazilian Cactacea. *Pharmacognosy Magazine* 11(3), 167.

Vieira, C.R., da Silva, B.P., do Carmo, M.A.V., Azevedo, L., Nogueira, D.A. *et al.* (2019) Effect of *Pereskia aculeata* Mill. in vitro and in overweight humans: A randomized controlled trial. *Journal of Food Biochemistry* 43(7), e12903. DOI: 10.1111/jfbc.12903.

Weisser, P.J., Smith, E.C.A., Backer, A.P. and Van Eeden, S. (1992) Flora and vegetation of the Mbonambi Beach Arcuate Scar on the Zululand dune barrier, Natal, South Africa. *Bothalia* 22(2), 289–294. DOI: 10.4102/abc.v22i2.852.

Zappi, D., Taylor, N., Santos, M.R. and Larocca, J. (2020) *Pereskia* in Lista de Espécies da Flora do Brasil. Jardim Botânico do Rio de Janeiro. Available at: http://floradobrasil.jbrj.gov.br/reflora/listaBrasil (accessed 2 March 2021).

Zareisedehizadeh, S., Tan, C.-H. and Koh, H.-L. (2014) A review of botanical characteristics, traditional usage, chemical components, pharmacological activities, and safety of *Pereskia* bleo (Kunth) DC. *Evidence-Based Complementary and Alternative Medicine* 2014, 326107. DOI: 10.1155/2014/326107.

20 Bacupari Fruit (*Garcinia brasiliensis* Mart.)

André Gonzaga dos Santos*, Kátia Sivieri, Bárbara Paes Miglioli da Mata, Isabela Jacob Moro, Mateus Kawata Salgaço and Luis Vitor Silva do Sacramento
School of Pharmaceutical Sciences, São Paulo State University - Unesp, Araraquara city, São Paulo state, Brazil

Abstract

Garcinia brasiliensis Mart. (Clusiaceae) is a tree species, the fruit of which are called bacupari in Brazil, a word that originated from the Tupi language. Its use by Indigenous people was first reported in the 16th century. The species occurs in all five Brazilian geographical regions, being common in the Amazon. The yellow fruit contains a white, mucilaginous, sweet and acid edible pulp which is consumed fresh or as an ingredient. The pulp is a source of some minerals. The secondary metabolites identified in the fruit are polyprenylated benzophenones, xanthones, biflavonoids and sesquiterpenes of the essential oil of the epicarp. Other parts of the plant also contain, as main components, the mentioned non-volatile compounds. *G. brasiliensis* has shown important bioactivities, including antibacterial, antiproteolytic, leishmanicidal, schistosomicidal, antioxidant, antitumour, antinociceptive and anti-inflammatory. The extracts or compounds have the potential to be employed in the development of drugs, for example for neglected diseases and cancer. The ethanolic extract of bacupari bark was able to reduce obesity, biometrics, lipogenesis and NAFLD in rats fed a high-fat diet. The antioxidant compounds of *G. brasiliensis* can influence metabolic homeostasis and the composition of the intestinal microbiota – important mechanisms related to obesity. Bacupari is an interesting source for the development of new products including functional foods, antioxidants for food preservation and cosmetics for skin photoprotection. The chemistry of the plant and its biological actions must be further investigated and exploited by researchers and industry.

Introduction

Edible plants which are not part of the conventional agricultural matrix and are little-known by most people of certain regions or countries have been classified as non-conventional edible plants (PANCs) in Brazil, according to Kinupp and Lorenzi (2014). The PANCs have been highlighted as promising plants to be incorporated into the human food matrix of a high number of people. An important example of a Brazilian PANC is *Garcinia brasiliensis* Mart., a tree species that belongs to the Clusiaceae family, formerly Guttiferae. The indigenous use of the plant was first reported in the 16th century (Santa-Cecília *et al.*, 2013; DATAPLAMT, 2020).

The pulp of its yellow fruit is the main edible part, and is white, mucilaginous, sweet and acid. In Brazil, the fruit is called bacupari and in some regions the plant is cultivated in people's backyards for fruit consumption. It is eaten fresh or employed as an ingredient in porridge, bread,

*Corresponding author: andre.gonzaga@unesp.br

 Handbook of Phytonutrients in Indigenous Fruits and Vegetables (eds D. Sivakumar *et al.*)
DOI: 10.1079/9781789248067.0020

ice cream, regional dishes and beverages (Santa-Cecília *et al.*, 2013; Kinupp and Lorenzi, 2014; Espirito Santo *et al.*, 2020).

Studies on the nutritional composition of bacupari pulp are scarce, but the published data highlights the mineral content. Phenolic compounds were the major secondary metabolites described for the fruit, which also contain essential oil. Bacupari extracts and their isolated compounds have shown many bioactivities which are very interesting for the food, cosmetic and pharmaceutical industries.

Botany

Garcinia L. (Clusiaceae) is common in lowland tropical forests and comprises more than 250 species. This genus has several striking morphological characteristics for its identification, present in shrubs or medium-sized trees (Sweeney, 2008).

Garcinia brasiliensis Mart. – synonym *Rheedia brasiliensis* (Mart.) Planch. & Triana – is a leafy native species reported in all five Brazilian geographical regions, but not endemic, being common in the Amazon region (Pio-Corrêa, 1984; Santa-Cecília *et al.*, 2013; Lorenzi, 2014; Muniz, 2020).

Reaching up to 13 m in height, it has a rounded canopy and cylinder trunk, covered by thin, brown and rough skin. It blooms between March and August and the fruit ripen from September to January (Lorenzi, 2014). In general, the leaves are coriaceous when mature, dark green above and paler below. The younger stems are green, glabrous, releasing yellow latex when broken (Harley, 1977). Flexible leaves can occur in some specimens (Santa-Cecília *et al.*, 2013).

The inflorescences occur in leaf axil fascicles. The numerous long-pedicellate flowers – masculine 10–15: 20–30 linear and translucent stamens with coloured-ferruginous anthers with longitudinal opening; hermaphrodite 3–5: superior ovary with ovarian disc and calipitriform stigma – are white-green and have a soft aroma, showing half-orbiculate fragile sepals (two to three) and obovate petals (four to five). The smooth fruit are indehiscent, of an ellipsis berry yellow type, 2–2.5 cm long and 2–4 cm in largest diameter, with a rounded or slightly pronounced apex with remaining stigma, and a fleshy acid pulp involving one to four brown smooth seeds (Lorenzi, 2014; Muniz, 2020) (Fig. 20.1).

Traditional Use

The use of *G. brasiliensis* by Brazilian indigenous people has been reported since the 16th century (Hoehne, 1946; Sousa, 1987). The fruit of the plant is known in Brazil as bacupari, bacupari-miúdo, bacupari-mirim, bacopari, bacopari-do-rio, bacuripari, bacuri, bacurizinho, bacuri-liso, pacuri, sacopari, limão-do-mato and poroco; in Bolivia it is called guapomo. The names bacupari or bacopari come from the Tupi word yuakuparí.

Fig. 20.1. Bacupari fruit (*Garcinia brasiliensis* Mart.).

Table 20.1. Nutritional composition of *Garcinia brasiliensis* Mart. pulp, peel and seed from Brazil compared to recommended daily intake (RDI) standards based on an intake of 2000 kcal or 8368 kJ (FDA, 2010; Souza *et al.*, 2016).

Nutrient	Pulp	Peel	Seed	RDI
Moisture (g/100 g)	86.10	82.37	41.22	-
Ash (g/100 g)	0.15	0.45	1.16	-
Lipid (g/100 g)	0.76	1.23	4.65	65 (g day^{-1})
Protein (g/100 g)	0.69	0.76	2.34	50 (g day^{-1})
Cu (µg/100 g)	1.46	1.73	7.11	2000 (µg day^{-1})
Mn (mg/100 g)	4.98	9.61	19.61	2.0 (mg day^{-1})
Fe (mg/100 g)	1.94	3.29	5.36	18 (mg day^{-1})
Zn (mg/100 g)	2.28	2.32	5.19	15 (mg day^{-1})
Mg (mg/100 g)	55.77	90.08	257.83	400 (mg day^{-1})

Besides its consumption as food, the plant has been employed in construction, woodwork, carpentry, tannery and for ornamental purposes (Pio-Corrêa, 1984; Pott and Pott, 1994; Dicio, 2009; Kinupp and Lorenzi, 2014; DATAPLAMT, 2019; Espirito Santo *et al.*, 2020; Del Melo *et al.*, 2021).

In Brazilian folk medicine, its leaves have been indicated for inflammation of the urinary tract, arthritis, pain, tumours, peptic ulcers and wound healing (Pio-Corrêa, 1984; Ferreira *et al.*, 2012; Santa-Cecília *et al.*, 2013); the bark for urinary diseases; and the seeds as wound-healing agents and for bruise treatment (Pott and Pott, 1994).

Nutritional Composition

The nutritional composition of bacupari pulp, peel and seed is shown in Table 20.1. Literature data is scarce for this fruit. The protein content in the pulp was determined as 0.69%, a value that is slightly lower when compared to other fruit such as açaí (*Euterpe oleracea* Mart.) and abiu (*Pouteria caimito* (Ruiz & Pav.) Radlk.), with protein content of 1.35 and 0.84%, respectively. Proteins of animal origin have higher biological value compared to proteins of plant origin (Kinupp and Barros, 2006), therefore the determination of the protein content in plant species is important in the search for plant protein sources (Aletor and Adeogun, 1995). Bacupari pulp presented a low content of lipids at 0.76%.

The pulp showed good levels of some minerals. The content of magnesium was 55.77 mg/100 g, 14% of RDI; iron 1.94 mg/100 g, 11% of RDI; zinc 2.28 mg/100 g, 15% of RDI; and manganese 4.98 mg/100 g, 250% of RDI.

Phytochemicals

The *Garcinia* genus is chemically characterized by the occurrence of flavonoids as biflavonoids, xanthones, polyprenylated benzophenones, triterpenes and sesquiterpenes. The secondary metabolites described for *G. brasiliensis* fruit comprise polyprenylated benzophenones, xanthones, biflavonoids and sesquiterpenes (Fig. 20.2) (Ferreira *et al.*, 2012; Kumar *et al.*, 2013; Espirito Santo *et al.*, 2020).

The content of total phenolic compounds of the ethanol extract of the fruit epicarp was determined by Folin-Ciocalteu photometric assay as 69.8 mg gallic acid equivalents per g of extract and the total flavonoid content was determined by aluminium chloride photometric assay as 3.4 and 22.5 mg quercetin equivalents per g of extract (Figueiredo *et al.*, 2014; Moreira *et al.*, 2017, 2018). In a comparative study of different plant parts, the ethanol and aqueous extracts of the seeds, leaves and bark showed total phenolic compound content ranging from 82.8 to 344.0 mg gallic acid equivalents per g of extract; and the total flavonoid contents ranging from 6.5 to 47.1 mg quercetin equivalents per g

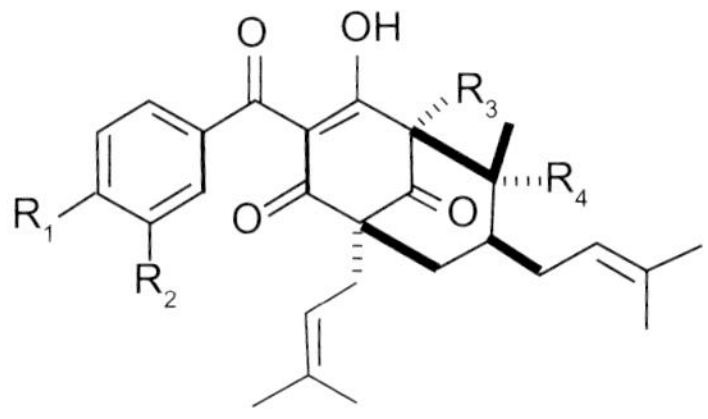

garciniaphenone: $R_1 = R_2 = R_3 = H$; $R_4 = CH_3$

guttiferone A: $R_1 = R_2 = OH$; $R_3 = R_4 =$ 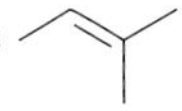

7-*epi*-clusianone: $R_1 = R_2 = H$; $R_3 =$ 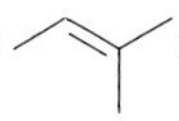; $R_4 = CH_3$

BENZOPHENONES

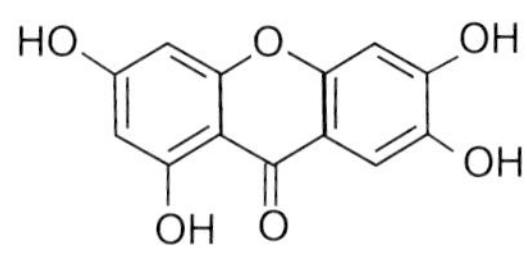

norathyriol

XANTHONE

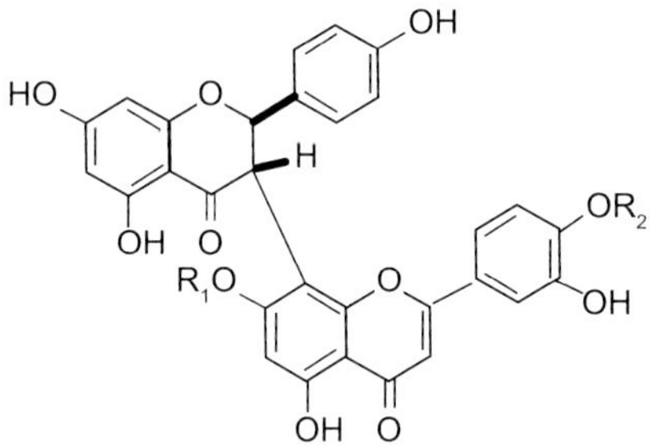

morelloflavone: $R_1 = R_2 = H$

morelloflavone-4'''-*O*-β-D-glucoside:R_1 =H; R_2= glycose

morelloflavone-7''-*O*-β-D-glucoside:R_1 = glycose; R_2=H

BIFLAVONOIDS

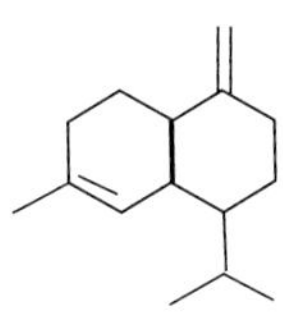

γ-muurolene

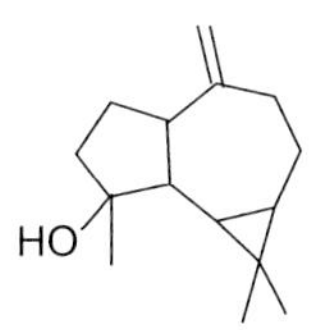

spathulenol

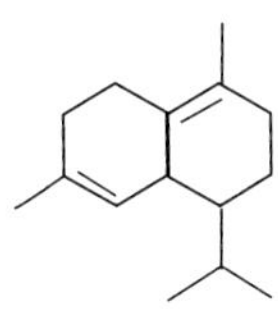

δ-cadinene

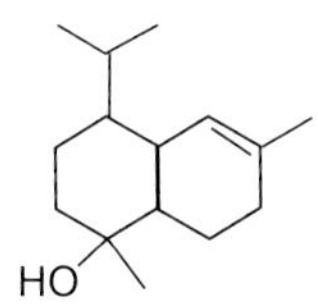

torreyol

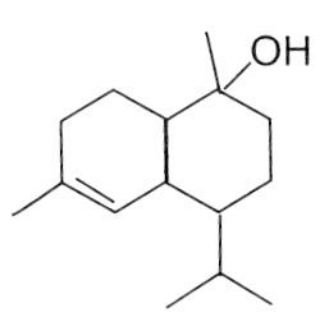

α-cadinol

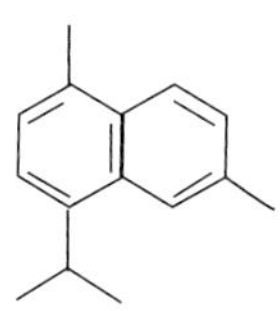

cadalene

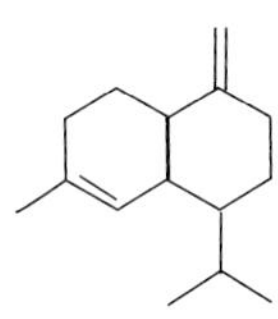

γ-cadinene

SESQUITERPENES

Fig. 20.2. Representative secondary metabolites from *Garcinia brasiliensis* Mart. fruit.

of extract. The extracts with the highest content of phenolic compounds and flavonoids were the leaf aqueous extract and the seed ethanol extract, respectively (Naves *et al.*, 2019).

The benzophenones are major secondary metabolites in the different parts of the fruit. Three polyprenylated benzophenones were isolated from the fruit pericarp of *G. brasiliensis*: garciniaphenone, 7-*epi*-clusianone and guttiferone A (Derogis *et al.*, 2008; Martins *et al.*, 2008a, 2009; Murata *et al.*, 2010; Pereira *et al.*, 2010; Carvalho-Silva *et al.*, 2012). High performance liquid chromatography (HPLC)HPLC-UV analysis of the hexane extract of the fruit pericarp has determined 7-*epi*-clusianone as its major secondary metabolite with the contents of 20.2 and 26.3% w/w (Pereira *et al.*, 2010; Martins *et al.*, 2011); the content of garciniaphenone and guttiferone A were determined as 6.3 and 2.1% w/w, respectively (Pereira *et al.*, 2010). Studies on the fruit epicarp showed the presence of the benzophenones guttiferone A and 7-*epi*-clusianone (Figueiredo *et al.*, 2014; Castro *et al.*, 2015, 2018; Ionta *et al.*, 2015; Silva *et al.*, 2020) and the major component in the ethanol epicarp extract was 7-*epi*-clusianone 14.0% w/w, HPLC-UV (Moreira *et al.*, 2017, 2018). Guttiferone A and 7-*epi*-clusianone were isolated from the seeds (Martins *et al.*, 2007, 2009) and their content was determined by HPLC-UV in hexane seed extract and 95% ethanol seed extract as 10.1 and 17.0% w/w for guttiferone A and 7.0 and 3.8% w/w for 7-*epi*-clusianone, respectively (Martins *et al.*, 2011).

Among the xanthones, compounds biosynthetically related to the benzophenones and which are distributed in other plant organs, only norathyriol or 1,3,6,7-tetrahydroxyxanthone was isolated from the fruit epicarp of *G. brasiliensis* (Gontijo *et al.*, 2012b; Silva *et al.*, 2020).

The fruit epicarp contains the biflavones morelloflavone or fukugetin, morelloflavone-4‴-*O*-β-D-glucoside and morelloflavone-7″-*O*-*β*-D-glucoside or fukugiside (Gontijo *et al.*, 2012a, b; Assis *et al.*, 2013; Castro *et al.*, 2015; Silva *et al.*, 2020). The content of morelloflavone in the ethanol extract of the fruit epicarp was determined by HPLC-UV as 3.6% w/w (Moreira *et al.*, 2017, 2018).

The yield of the volatile oil from the fresh fruit epicarp obtained by hydrodistillation was estimated as 0.4% w/w. The 38 identified compounds by GC-MS include 82.5% of sesquiterpenes: 38.7% of sesquiterpene hydrocarbons and 43.8% of oxygenated sesquiterpenes. The major components were γ-muurolene (10.3%), spathulenol (8.7%), δ-cadinene (8.3%), torreyol (8.0%), α-cadinol (7.0%), cadalene (6.3%) and γ-cadinene (5.3%) (Martins *et al.*, 2008b).

Other parts of the plant also contain mainly benzophenones, biflavonoids and/or xanthones. The secondary metabolites isolated from *G. brasiliensis* leaves include: 7-*epi*-clusianone and garcinol (prenylated benzophenones); morelloflavone, amentoflavone and podocarpusflavone (biflavonoids); procyanidin (condensed tannin); gallic acid; and lupeol (pentacyclic triterpene) (Corrêa *et al.*, 2009; Santa-Cecília *et al.*, 2011, 2012; Arwa *et al.*, 2015). Moreover, biflavonoids and other flavonoids were identified in the leaves by employing hyphenated techniques (Zan *et al.*, 2018; Naves *et al.*, 2019).

Polyprenylated xanthones were isolated from the roots: from the roots: macluraxanthone; rheediaxanthones A, B, and C; rheediachromenoxanthone; isorheediaxanthone B; 4″,5″-dihydro-1,5-dihydroxy-6',6'-dimethylpyran(2',3':6,7)-4″,4″,5″-trimethylfuran(2″,3″:3,4)xanthone; 1,3,5-trihydroxy-6',6' dimethylpyran (2',3':6,7)-4-(1,2-dimethylprop-2-enyl) xanthone; 1,3,5,6-tetrahydroxy-4-(1,1-dimethylprop-2-enyl)-7-(3-methylbut-2-enyl)xanthone; 8-deoxygartanin; pyranojacareubin; and 7-prenyljacareubin (Monache *et al.*, 1984a, b). The following biflavonoids such as volkensiflavone, fukugetin, and I3-naringenin-II8-eriodictyol were also isolated from the roots: volkensiflavone, fukugetin, and I3-naringenin-II8-eriodictyol (Botta *et al.*, 1984).

Bioactive Properties

G. brasiliensis is a plant employed by people in Brazil and other countries due to its different nutritional and medicinal properties and the resulting health benefits. Many papers on the species describe the bioactivities of pharmacological interest of its secondary metabolites, highlighting the biflavonoids and 7-*epi*-clusianone.

The 1,3,6,7-tetrahydroxyxanthone and the biflavonoids morelloflavone, morelloflavone-7"-*O*-β-D-glucoside and morelloflavone-4'"-*O*-β-D-glucoside, isolated from the fruit epicarp, showed radical scavenging activity in the DPPH assay with EC_{50} values similar or even lower than the positive controls, morelloflavone being the most active compound (Gontijo *et al.*, 2012a, b). The ethyl acetate fraction of the methanol leaf extract obtained by LLE was tested in the DPPH, reducing power measured by ferricyanide Prussian blue, β-carotene bleaching inhibition and lipid peroxidation inhibition by thiobarbituric acid reactive substances assays. The EC_{50} values for the fraction were similar or lower than EC_{50} values for the positive control. Five biflavonoids, four flavones, two flavonols and one flavan-3-ol were identified by HPLC-PDA/UV-ESI/MS^n and the main components were morelloflavone-7"-*O*-glucoside and GB2a-glucoside (Zan *et al.*, 2018). Sunny and Krishnakumar (2017) evaluated the antioxidant activity of the ethanol leaf extract fractions, also obtained by LLE, in the hydrogen peroxide scavenging and ferric reducing power assays, the ethyl acetate fraction being the most active and with the higher content of total phenolic compounds. The ethanolic and aqueous extracts of the leaves, bark and seeds showed radical scavenging activity on DPPH, presenting EC_{50} values from 23.1 to 135.6μg/ml (Naves *et al.*, 2019).

The prenylated benzophenones from seeds and fruit, guttiferone A, 7-*epi*-clusianone, garciniaphenone and two non-prenylated benzophenone semi-synthetic derivatives inhibited the enzymes papain, trypsin and cathepsins B and G, highlighting the inhibitory activity of guttiferone A which can be used as an antiproteolytic drug for the treatment of diseases like cancer (Martins *et al.*, 2009). The biflavonoid morelloflavone inhibited human tissue kallikreins, which are related to some diseases such as cancer (Santos *et al.*, 2016). Moreover, morelloflavone also inhibited the papain and more interestingly inhibited cruzain, the major cysteine protease of *Trypanosoma cruzi*, the infectious agent of Chagas disease (Assis *et al.*, 2013).

Pereira *et al.* (2011) reported the protease inhibiting activity of bacupari fruit extracts. These extracts showed *in vitro* inhibition of papain and trypsin. The proteases were particularly inhibited by the ethyl acetate extract which was very active in commercial proteases papain and trypsin. These results inspired the evaluation of the activity of the extracts on *Leishmania* proteases to find a new natural product with activity against leishmaniasis. The ethyl acetate extract and its biflavonoid morelloflavone showed inhibition of the *Leishmania* proteases, but they were inactive on *Leishmania* cells *in vitro*. Moreover, they showed no cytotoxic effect on mammalian cells. In another study on antileishmanial activity, the biflavonoids morelloflavone, morelloflavone-4'"-*O*-β-D-glucoside, morelloflavone-7"-*O*-β-D-glucoside and semi-synthetic derivatives of morelloflavone showed leishmanicidal activity against promastigote and amastigote forms of *L.* (*L.*) *amazonensis*. The compounds with a lower number of hydroxyls, substituted by acyloxy or methoxy groups, showed greater leishmanicidal activity as observed for morelloflavone semi-synthetic derivatives. These compounds also showed antiproteolytic activity on the *L.* (*L.*) *mexicana* cysteine protease r-CPB2.8 and its isoform r-CPB3, the non-glycosylated compounds being more active. In addition, the compounds showed low cytotoxicity on mammalian cells (Gontijo *et al.*, 2012a). The hexane extract of *G. brasiliensis* pericarp and its prenylated benzophenones 7-*epi*-clusianone, garciniaphenone and guttiferone A were active on promastigote and amastigote forms of *Leishmania* (*L.*) *amazonensis* and showed low toxicity in mammalian cells. The structure-activity relationship study for benzophenones, considering the IC_{50} values, showed that the activity on *L.* (*L.*) *amazonensis* is dependent on the prenyl and phenolic groups, and inversely proportional to the lipophilicity (Pereira *et al.*, 2010).

In another evaluation of the potential of *G. brasiliensis* for the treatment of neglected tropical diseases, the ethanol extract of its fruit epicarp and its main components were tested *in vitro* on *Schistosoma mansoni*. The benzophenone 7-*epi*-clusianone was considered an effective compound against *S. mansoni* adult worms, cercariae and schistosomula, with extensive effects on the mortality rate, reproductive fitness, parasite tegument morphology and adult *S. mansoni* worm excretory (Castro *et al.*, 2015).

The LLE fractions of the leaf methanol extract showed cytotoxic action in different tumour cell lines (Zan *et al.*, 2018). Sales *et al.* (2015) demonstrated that the bezophenone 7-*epi*-clusianone has anticancer activity in glioblastoma cells, showing effects on cell growth, cell cycle dynamics, apoptosis and colony formation. In another study with different tumour cell lines, 7-*epi*-clusianone showed a dose-dependent cytotoxic action and garciniaphenone showed cytostatic activity. The antiproliferative action of garciniaphenone and 7-*epi*-clusianone was related to cathepsin inhibition (Murata *et al.*, 2010).

Some studies have evaluated the antimicrobial activity of *G. brasiliensis* chemical constituents. The hexane extracts of the peel and seeds and their component 7-*epi*-clusianone showed antibacterial activity against *Streptococcus mutans* in the MIC/MBC assays and reduced the biofilm viability. Thus, this benzophenone demonstrated the potential to control cariogenic microorganisms and to prevent dental caries (Almeida *et al.*, 2008). The hexane extract of pericarp and the ethanol extract of seeds, besides their prenylated benzophenones 7-*epi*-clusianone and guttiferone A, were active against other Gram-positive bacteria, *Staphylococcus aureus* and *Bacillus cereus*, but were not active against *Escherichia coli*, a Gram-negative bacterium, and the fungus *Candida albicans* (Naldoni *et al.*, 2009). The ethyl acetate fraction of the methanol leaf extract, constituted by biflavonoids and other flavonoids, showed significant antimicrobial activity against *Candida albicans*, Gram-positive and Gram-negative bacteria (Zan *et al.*, 2018). The biflavonoids volkensiflavone, fukugetin and fukugiside, isolated from the hydroalcoholic leaf extract of *Garcinia gardneriana* (Planch. & Triana) Zappi, were active against *B. cereus*, *E. coli*, *Pseudomonas aeruginosa* and *S. aureus*, with lower MIC values for the Gram-positive bacteria (Verdi *et al.*, 2004).

The ethanol extract of the *G. brasiliensis* peel presented antiobesity activity which was related to the down-regulation of lipogenic gene FAS and TNF-α protein and up-regulation of IL10 protein, decreasing adipogenese, adipocyte size and body weight. The main constituents of the extract were 7-*epi*-clusianone and morelloflavone (Moreira *et al.*, 2017).

The photoprotective and photochemopreventive actions of the ethanol extract of *G. brasiliensis* fruit epicarp were demonstrated through *in vitro* and *in vivo* assays. The polyprenylated benzophenones 7-*epi*-clusianone and guttiferone A were determined as the main extract constituents and the 7-*epi*-clusianone was at least partially responsible for the photoprotection activity (Figueiredo *et al.*, 2014).

The 7-*epi*-clusianone produces an endothelium-dependent vasodilator effect in rat aortic rings at low concentrations. A vasoconstractile effect was observed at higher concentrations with the inhibition of NO synthase (Cruz *et al.*, 2006). This compound also can act as an antispasmodic agent with potential applications for the treatment of airflow limitation triggered by allergens in humans (Coelho *et al.*, 2008). The antianaphylatic action of 7-*epi*-clusianone isolated from the fruit pericarp of *G. brasiliensis* was observed in guinea pigs. The prenylated benzophenone prevented the allergen-induced tissue histamine release and smooth muscle spasm produced by allergen, histamine, acetylcholine and high K^+ depolarization (Neves *et al.*, 2007).

The prenylated benzophenone 7-*epi*-clusianone exhibited anti-inflammatory and antinociceptive activities in the experimental models of carrageenan-induced paw oedema and peritonitis induced by lipopolysaccharide in rats; and the acetic acid-induced writhing, formalin and hot-plate tests in mice (Santa-Cecília *et al.*, 2011). Interestingly, another prenylated benzophenone isolated also from *Garcinia achachairu* Rusby (guttiferone A) presented gastroprotective action in the HCl/ethanol-induced ulcer model in mice (Niero *et al.*, 2012). Moreover, the *in vitro* anti-inflammatory activity of the hexane, dichloromethane and ethyl acetate fractions of the leaf methanol extract was evaluated through NO inhibition assay and the dichloromethane fraction was the most effective (Zan *et al.*, 2018).

The *in vivo* anti-inflammatory action of the biflavonoids morelloflavone and I3-naringenin-II8-eriodictyol isolated from *G. gardneriana* was demonstrated by Castardo *et al.* (2008) and Otuki *et al.* (2011). The compounds reduced mouse paw oedema, ear oedema, myeloperoxidase activity, leucocyte migration and reactive oxygen species production in the skin. The analgesic effects of four biflavonoids also isolated from *G. gardneriana* (morelloflavone, morelloflavone-7"-*O*-*β*-D-glucoside, volkensiflavone, and I3-naringenin-II8-eriodictyol) were evaluated in mice through the formalin-induced pain model. The compounds showed significant analgesic activity in the second phase of the test – inflammatory pain (Luzzi *et al.*, 1997).

Health Benefits

Different parts of bacupari, including the seeds, fruit, leaves and bark have been related to health benefits. Studies have shown that compounds extracted from the plant have relevant biological properties, such as antimicrobial (Almeida *et al.*, 2008), antinociceptive, anti-inflammatory (Santa-Cecília *et al.*, 2011), antiproteolytic, leishmanicidal, antioxidant (Gontijo *et al.*, 2012a, b) and antitumour (Sales *et al.*, 2015).

Several studies have shown that eating foods or natural extracts rich in antioxidants can influence metabolic homeostasis, as well as the composition of the intestinal microbiota. These are important mechanisms for the prevention and treatment of obesity (Moraes *et al.*, 2014; Nair *et al.*, 2014). The use of bacupari in Brazilian folk medicine may be related to the phenolic compounds in the plant. Some of these compounds like benzophenones, flavonoids and xanthones present in these species exert biological activities, such as antifungal, anti-inflammatory, antitumor and antioxidant effects (Martins *et al.*, 2008b; Santa-Cecília *et al.*, 2011; García-Conesa, 2015).

Interestingly, Araújo *et al.* (2019) showed that intervention with 300 mg/kg of the ethanolic extract of bacupari bark per body weight for eight weeks was able to reduce obesity, biometrics, lipogenesis and NAFLD in rats fed a high-fat diet. Gontijo *et al.* (2012b) observed antioxidant effects of the biflavonoid morelloflavone present in the fruit epicarp, which can act together with dietary fibre, probably producing radical elimination. The synergistic effects of these food components with endogenous antioxidant systems can promote cell rebalancing in obese animals.

The *in vivo* study developed by Araújo *et al.* (2019) showed the enhancement of microbiota homeostasis with an decrease of phylum *Firmicutes* and *Proteobacteria* and an increase of *Bacteroidetes* after the consumption of bacupari. As described before, bacupari is rich in phenolic compounds and many studies have revealed that the specific groups of intestinal bacteria can transform phenolic compounds into bioactive metabolites, contributing to the maintenance of intestinal homeostasis, stimulating the growth of beneficial bacteria (*Lactobaciliaceae* and *Bifidobacterium* spp.) and inhibiting pathogenic bacteria, thus exerting a prebiotic effect on gut microbiota (Parkar *et al.*, 2013; Zhu *et al.*, 2018).

Novel Products/Formulations

A bibliometric review that analysed papers published from 2007 to 2018 demonstrated the trend of research with *G. brasiliensis*: the main interest is related to its bioactive compounds and respective beneficial health effects (Del Melo *et al.*, 2021). However, to date, no patents have been registered for products and/or formulations based on *G. brasiliensis*. Thus, we can consider bacupari as an interesting source for new products or formulations, due to the bioactive properties found in its different parts. For example, due to its antioxidant and photochemopreventive activity, the peel of the fruit can be used in the production of cosmetics for skin photoprotection (Figueiredo *et al.*, 2014; Castro *et al.*, 2019; Del Melo *et al.*, 2021).

Polyprenylated benzophenones, such as 7-epiclusianone, may be important in the discovery and development of drugs, such as for allergic disorders, neglected diseases such as leishmaniasis, treatment of airflow limitation and cardiovascular diseases, due to their vasodilator effect (Cruz *et al.*, 2006; Neves *et al.*, 2007; Coelho *et al.*, 2008; Gontijo *et al.*, 2012a). They may also generate interest in future studies as an alternative in the control of diseases associated with Gram-positive bacteria (Naldoni *et al.*, 2009).

Regarding the area of food sciences and technology, more research is needed to investigate the use of bacupari for the development of functional foods in industry, mainly studies with the pulp – the edible part of this fruit. The literature describes the use of seeds from other fruit to obtain starch, demonstrating that there is a tendency to reuse agri-food by-products for introduction into the food industry. The extraction of starch from the seeds or bark of bacupari has been suggested by some authors (Del Melo *et al.*, 2021).

Conclusions and Future Trends

G. brasiliensis is a native plant of Brazil called bacupari which is distributed in all five Brazilian geographical regions. It may be considered a PANC and a medicinal plant. Its indigenous use has been reported since the 16th century and it has been employed in construction, carpentry, tannery and as an ornamental plant.

The edible part is the pulp fruit that is eaten fresh or used as an ingredient in different traditional recipes. The nutritional composition of the fruit has been underinvestigated. The literature data shows that the pulp is a relevant source of some minerals like magnesium, zinc and manganese. The protein content in the pulp was determined as 0.69% and the lipid content as 0.76%.

The main secondary metabolites described for *G. brasiliensis* fruit are phenolic compounds (polyprenylated benzophenones, xanthones and biflavonoids) which are responsible for the bioactivities exhibited by its extracts. The fruit epicarp also contains essential oil composed of sesquiterpenes, the three major components being γ-muurolene, spathulenol and δ-cadinene. The main secondary metabolites identified in the leaves were flavonoids and prenylated benzophenones and in the roots were prenylated xanthones. The chemistry of the plant must be further investigated both in terms of secondary metabolites and nutrients.

The bioactive compounds of *G. brasiliensis* and their respective beneficial health effects are the main investigation subject on the species described in the scientific literature. Different parts of *G. brasiliensis* and their secondary metabolites have shown pharmacological activities, such as antimicrobial, antinociceptive, anti-inflammatory, antiproteolytic, leishmanicidal, antioxidant and antitumour. The use of bacupari in Brazilian folk medicine may be related to the phenolic compounds in the plant. Some of these compounds present in these species exert biological activity, such as antifungal, anti-inflammatory, antitumour and antioxidant effects.

The antioxidant compounds of *G. brasiliensis*, as morelloflavone, can influence metabolic homeostasis and the composition of the intestinal microbiota; important mechanisms for the prevention and treatment of obesity. The enhancement of microbiota homeostasis with the decrease of phylum *Firmicutes* and *Proteobacteria* and the increase of *Bacteroidetes* was observed in animals after the consumption of bacupari, thus exerting a prebiotic effect on gut microbiota. Moreover, the ethanolic extract of the bacupari bark was able to reduce obesity, biometrics, lipogenesis and NAFLD in rats fed a high-fat diet.

Bacupari is an interesting source for the development of new products, even though no patents have been registered for products based on this plant species to date. The use of bacupari for the development of functional foods in industry or the use of its extracts as antioxidants for food preservation are interesting alternatives. The potential use of the seeds by the agri-food industry to produce flour can be also investigated. Furthermore, the peel of the fruit may be used in the development of cosmetics for skin photoprotection.

Acknowledgements

This study was financed by the Coordenação de Aperfeiçoamento de Pessoal de Nível Superior, Brasil (CAPES), Finance Code 001 and by the Instituto Nacional de Ciência e Tecnologia em Biodiversidade e Produtos Naturais (INCT-BioNat; CNPq and FAPESP; Brasil).

References

Aletor, V.A. and Adeogun, O.A. (1995) Nutrient and anti-nutrient components of some tropical leafy vegetables. *Food Chemistry* 53(4), 375–379. DOI: 10.1016/0308-8146(95)99830-S.

Almeida, L.S.B., Murata, R.M., Yatsuda, R., Dos Santos, M.H., Nagem, T.J. *et al.* (2008) Antimicrobial activity of *Rheedia brasiliensis* and 7-epiclusianone against *Streptococcus* mutans. *Phytomedicine : International Journal of Phytotherapy and Phytopharmacology* 15(10), 886–891. DOI: 10.1016/j.phymed.2007.12.003.

Araújo, F. de O., Moreira, M.E. de C., Lima, C.F., Toledo, R.C.L., de Sousa, A.R. *et al.* (2019) Bacupari (*Garcinia brasiliensis*) extract modulates intestinal microbiota and reduces oxidative stress and inflammation in obese rats. *Food Research International* 122, 199–208. DOI: 10.1016/j.foodres.2019.04.012.

Arwa, P.S, Zeraik, M.L., Ximenes, V.F., da Fonseca, L.M., Bolzani, V. da S. *et al.* (2015) Redox-active biflavonoids from *Garcinia brasiliensis* as inhibitors of neutrophil oxidative burst and human erythrocyte membrane damage. *Journal of Ethnopharmacology* 174, 410–418. DOI: 10.1016/j.jep.2015.08.041.

Assis, D.M., Gontijo, V.S., de Oliveira Pereira, I., Santos, J.A.N., Camps, I. *et al.* (2013) Inhibition of cysteine proteases by a natural biflavone: behavioral evaluation of fukugetin as papain and cruzain inhibitor. *Journal of Enzyme Inhibition and Medicinal Chemistry* 28(4), 661–670. DOI: 10.3109/14756366.2012.668539.

Botta, B., Mac-Quhae, M.M., Monache, D.J., Monache, F.D. and De Mello, J.F. (1984) Chemical investigation of the genus *Rheedia*, v. biflavonoids and xanthochymol. *Journal of Natural Products* 47(6), 1053–1053. DOI: 10.1021/np50036a033.

Carvalho-Silva, L.B., Oliveira, M. do V., Gontijo, V.S., Oliveira, W.F., Derogis, P.B.M.C. *et al.* (2012) Antioxidant, cytotoxic and antimutagenic activities of 7-epi-clusianone obtained from pericarp of *Garcinia brasiliensis*. *Food Research International* 48(1), 180–186. DOI: 10.1016/j.foodres.2012.03.003.

Castardo, J.C., Prudente, A.S., Ferreira, J., Guimarães, C.L., Monache, F.D. *et al.* (2008) Anti-inflammatory effects of hydroalcoholic extract and two biflavonoids from *Garcinia gardneriana* leaves in mouse paw oedema. *Journal of Ethnopharmacology* 118(3), 405–411. DOI: 10.1016/j.jep.2008.05.002.

Castro, A.P., de Mattos, A.C.A., Pereira, N.A., Anchieta, N.F., Silva, M.S. *et al.* (2015) Potent schistosomicidal constituents from Garcinia brasiliensis. *Planta Medica* 81(9), 733–741. DOI: 10.1055/s-0035-1545927.

Castro, A.P., Kawano, T., Spelta, L.E.W., de Castro, A.T., Pereira, N.A. *et al.* (2018) *In vivo* schistosomicidal activity of 7-epiclusianone and its quantification in the plasma of healthy and *Schistosoma mansoni* infected mice using UPLC-MS/MS. *Phytomedicine : International Journal of Phytotherapy and Phytopharmacology* 38, 66–73. DOI: 10.1016/j.phymed.2017.11.001.

Castro, A.P., Morais Ruela, A.L., Amaral, J.G., dos Santos, M.H., Pereira, G.R. *et al.* (2019) A stability-indicating method by LC-UV for analysis of 7-epi-clusianone extracted from Garcinia brasiliensis fruits and a novel prenylated benzophenone from the oxidation of this molecular marker. *Analytical Methods* 11(23), 3023–3032. DOI: 10.1039/C9AY00318E.

Coelho, L.P., Serra, M.F., Pires, A.L. de A., Cordeiro, R.S.B., Rodrigues e Silva, P.M. *et al.* (2008) 7-Epiclusianone, a tetraprenylated benzophenone, relaxes airway smooth muscle through activation of the nitric oxide-cGMP pathway. *The Journal of Pharmacology and Experimental Therapeutics* 327(1), 206–214. DOI: 10.1124/jpet.108.138032.

Corrêa, R.S., Coelho, C.P., Santos, M.H., Ellena, J. and Doriguetto, A.C. (2009) Lupeol. *Acta Crystallographica Section C* 65(3), 97–99. DOI: 10.1107/S0108270109004910.

Cruz, A.J., Lemos, V.S., dos Santos, M.H., Nagem, T.J. and Cortes, S.F. (2006) Vascular effects of 7-epiclusianone, a prenylated benzophenone from *Rheedia gardneriana*, on the rat aorta. *Phytomedicine: International Journal of Phytotherapy and Phytopharmacology* 13(6), 442–445. DOI: 10.1016/j.phymed.2005.01.014.

DATAPLAMT (2019) Banco de dados e amostras de plantas aromáticas, medicinais e tóxicas / UFMGI nformation about bacupari. Available at: http://www.dataplamt.org.br/v3-novaversao-block/#/planta/?idPlanta=2005 (accessed 22 March 2021).

DATAPLAMT (2020) Banco de dados e amostras de plantas aromáticas, medicinais e tóxicas / UFMG Information about bacupari. Available at: http://www.dataplamt.org.br/v3-novaversao-block/#/planta/?idPlanta=2005

Del Melo, A.M., Almeida, F.L.C., Cavalcante, A.M. de M., Ikeda, M., Barbi, R.C.T. *et al.* (2021) *Garcinia brasiliensis* fruits and its by-products: Antioxidant activity, health effects and future food industry trends – A bibliometric review. *Trends in Food Science & Technology* 112, 325–335. DOI: 10.1016/j.tifs.2021.04.005.

Derogis, P.B.M.C., Martins, F.T., de Souza, T.C., de C Moreira, M.E., Souza Filho, J.D. *et al.* (2008) Complete assignment of the 1H and 13C NMR spectra of garciniaphenone and keto-enol equilibrium statements for prenylated benzophenones. *Magnetic Resonance in Chemistry* 46(3), 278–282. DOI: 10.1002/mrc.2166.

Dicio - Dicionário online de português (2009) Definition of bacupari. Available at: https://www.dicio.com.br/bacupari/#:~:text=Bacupari%2Dmi%C3%BAdo%3A%20a)%20%C3%A1rvore,Do%20tupi%20yuakupar%C3%AD (accessed 22 March 2021).

Espirito Santo, B.L.S. do, Santana, L.F., Kato Junior, W.H., de Araújo, F. de O., Bogo, D. *et al.* (2020) Medicinal Potential of *Garcinia* Species and Their Compounds. *Molecules* 25(19), 4513. DOI: 10.3390/molecules25194513.

FDA (2010) Nutritional Labeling and Education Act (NLEA) Requirements (8/94-2/95). Available at: https://www.fda.gov/nutrition-labeling-and-education-act-nlea-requirements-attachment-1 (accessed 5 October 2020).

Ferreira, R.O., Carvalho, M.G. de and da Silva, T.M.S. (2012) Ocorrência de biflavonoides em Clusiaceae: aspectos químicos e farmacológicos. *Química Nova* 35(11), 2271–2277. DOI: 10.1590/S0100-40422012001100035.

Figueiredo, S.A., Vilela, F.M.P., da Silva, C.A., Cunha, T.M., Dos Santos, M.H. *et al.* (2014) *In vitro* and *in vivo* photoprotective/photochemopreventive potential of *Garcinia brasiliensis* epicarp extract. *Journal of Photochemistry and Photobiology. B, Biology* 131, 65–73. DOI: 10.1016/j.jphotobiol.2014.01.004.

García-Conesa, M.T. (2015) Dietary polyphenols against metabolic disorders: How far have we progressed in the understanding of the molecular mechanisms of action of these compounds? *Critical Reviews in Food Science and Nutrition* 57(9), 1769–1786. DOI: 10.1080/10408398.2014.980499.

Gontijo, V.S., Judice, W.A.S., Codonho, B., Pereira, I.O., Assis, D.M. *et al.* (2012a) Leishmanicidal, antiproteolytic and antioxidant evaluation of natural biflavonoids isolated from *Garcinia brasiliensis* and their semisynthetic derivatives. *European Journal of Medicinal Chemistry* 58, 613–623. DOI: 10.1016/j.ejmech.2012.06.021.

Gontijo, V.S., de Souza, T.C., Rosa, I.A., Soares, M.G., da Silva, M.A. *et al.* (2012b) Isolation and evaluation of the antioxidant activity of phenolic constituents of the *Garcinia brasiliensis* epicarp. *Food Chemistry* 132(3), 1230–1235. DOI: 10.1016/j.foodchem.2011.10.110.

Harley, R.M. (1977) Missouri Botanical Garden. Available at: http://www.tropicos.org/Specimen/100607552 (accessed 20 April 2021).

Hoehne, F.C. (1946) *Frutas Indígenas*. Secretaria da Agricultura, Indústria e Comércio/Instituto de Botânica, São Paulo, SP/Brasil.

Ionta, M., Ferreira-Silva, G.A., Niero, E.L., Costa, É.D., Martens, A.A. *et al.* (2015) 7-Epiclusianone, a Benzophenone Extracted from *Garcinia brasiliensis* (Clusiaceae), Induces Cell Cycle Arrest in G1/S Transition in A549 Cells. *Molecules* 20(7), 12804–12816. DOI: 10.3390/molecules200712804.

Kinupp, V.F. and Barros, I.B.I. (2006) Protein and mineral contents of native species, potential vegetable and fruits. *Ciência e Tecnologia de Alimentos* 28, 846–857.

Kinupp, V.F. and Lorenzi, H. (2014) *Plantas Alimentícias Não Convencionais (PANC) no Brasil*. Instituto Plantarum de Estudos da Flora, São Paulo, SP/Brasil.

Kumar, S., Sharma, S. and Chattopadhyay, S.K. (2013) The potential health benefit of polyisoprenylated benzophenones from *Garcinia* and related genera: ethnobotanical and therapeutic importance. *Fitoterapia* 89, 86–125. DOI: 10.1016/j.fitote.2013.05.010.

Lorenzi, H. (2014) *Árvores brasileiras: manual de identificação e cultivo de plantas arbóreas do Brasil*, v1, 6th edn. Instituto Plantarum de Estudos da Flora, São Paulo, SP/Brasil.

Luzzi, R., Guimarães, C.L., Verdi, L.G., Simionatto, E.L., Delle Monache, F. *et al.* (1997) Isolation of biflavonoids with analgesic activity from *Rheedia gardneriana* leaves. *Phytomedicine : International Journal of Phytotherapy and Phytopharmacology* 4(2), 141–144. DOI: 10.1016/S0944-7113(97)80060-8.

Martins, F.T., Cruz Jr., J.W., Derogis, P.B.M.C., Santos, M.H. dos, Veloso, M.P. *et al.* (2007) Natural polyprenylated benzophenones: keto-enol tautomerism and stereochemistry. *Journal of the Brazilian Chemical Society* 18(8), 1515–1523. DOI: 10.1590/S0103-50532007000800011.

Martins, F.T., Camps, I., Doriguetto, A.C., dos Santos, M.H., Ellena, J. *et al.* (2008a) Crystal structure of garciniaphenone and evidences on the relationship between Keto-Enol tautomerism and configuration. *Helvetica Chimica Acta* 91(7), 1313–1325. DOI: 10.1002/hlca.200890143.

Martins, F.T., Doriguetto, A.C., de Souza, T.C., de Souza, K.R.D., Dos Santos, M.H. *et al.* (2008b) Composition, and anti-inflammatory and antioxidant activities of the volatile oil from the fruit peel of *Garcinia brasiliensis*. *Chemistry & Biodiversity* 5(2), 251–258. DOI: 10.1002/cbdv.200890022.

Martins, F.T., Assis, D.M., Dos Santos, M.H., Camps, I., Veloso, M.P. *et al.* (2009) Natural polyprenylated benzophenones inhibiting cysteine and serine proteases. *European Journal of Medicinal Chemistry* 44(3), 1230–1239. DOI: 10.1016/j.ejmech.2008.09.018.

Martins, F.T., dos Santos, M.H., Coelho, C.P., Barbosa, L.C.A., Dias, G.C. *et al.* (2011) A powder X-ray diffraction method for detection of polyprenylated benzophenones in plant extracts associated with HPLC for quantitative analysis. *Journal of Pharmaceutical and Biomedical Analysis* 54(3), 451–457. DOI: 10.1016/j.jpba.2010.09.010.

Monache, G.D., Monache, F.D., Waterman, P.G., Crichton, E.G. and De Limas, R.A. (1984a) Minor xanthones from *Rheedia gardneriana*. *Phytochemistry* 23(8), 1757–1759. DOI: 10.1016/S0031-9422(00)83485-8.

Monache, G.D., Botta, B., de Mello, J.F., de Barros Coelho, J.S. and Menichini, F. (1984b) Chemical investigation of the Genus *Rheedia*, IV. Three new Xanthones from *Rheedia brasiliensis*. *Journal of Natural Products* 47(4), 620–625. DOI: 10.1021/np50034a009.

Moraes, A.C.F., Silva, I.T., Almeida-Pititto, B. and Ferreira, S.R.G. (2014) Intestinal microbiota and cardiometabolic risk: mechanisms and diet modulation. *Arquivos Brasileiros de Endocrinologia e Metabologia* 58(4), 317–327. DOI: 10.1590/0004-2730000002940.

Moreira, M.E. de C., Natal, D.I.G., Toledo, R.C.L., Ramirez, N.M., Ribeiro, S.M.R. *et al.* (2017) Bacupari peel extracts (*Garcinia brasiliensis*) reduce high-fat diet-induced obesity in rats. *Journal of Functional Foods* 29, 143–153. DOI: 10.1016/j.jff.2016.11.001.

Moreira, M.E. de C., de Oliveira Araújo, F., de Sousa, A.R., Toledo, R.C.L., Dos Anjos Benjamin, L. *et al.* (2018) Bacupari peel extracts (*Garcinia brasiliensis*) reduces the biometry, lipogenesis and hepatic steatosis in obese rats. *Food Research International* 114, 169–177. DOI: 10.1016/j.foodres.2018.08.004.

Muniz, F.H. (2020) Garcinia in Flora do Brasil 2020. Jardim Botânico do Rio de Janeiro. Available at: http://floradobrasil.jbrj.gov.br/reflora/floradobrasil/FB6851 (accessed 22 March 2021).

Murata, R.M., Yatsuda, R., dos Santos, M.H., Kohn, L.K., Martins, F.T. *et al.* (2010) Antiproliferative effect of benzophenones and their influence on cathepsin activity. *Phytotherapy Research* 24(3), 379–383. DOI: 10.1002/ptr.2954.

Nair, A.R., Elks, C.M., Vila, J., Del Piero, F., Paulsen, D.B. *et al.* (2014) A blueberry-enriched diet improves renal function and reduces oxidative stress in metabolic syndrome animals: potential mechanism of TLR4-MAPK signaling pathway. *PloS One* 9(11), 1–12. DOI: 10.1371/journal.pone.0111976.

Naldoni, F.J., Claudino, A.L.R., Cruz, J.W., Chavasco, J.K., Faria e Silva, P.M. *et al.* (2009) Antimicrobial activity of benzophenones and extracts from the fruits of *Garcinia brasiliensis*. *Journal of Medicinal Food* 12(2), 403–407. DOI: 10.1089/jmf.2007.0622.

Naves, V.M.L., dos Santos, M.H., Ribeiro, I.S., da Silva, C.A., Silva, N.C. *et al.* (2019) Antimicrobial and antioxidant activity of *Garcinia brasiliensis* extracts. *South African Journal of Botany* 124, 244–250. DOI: 10.1016/j.sajb.2019.05.021.

Neves, J.S., Coelho, L.P., Cordeiro, R.S.B., Rodrigues e Silva, P.M. *et al.* (2007) Antianaphylactic properties of 7-epiclusianone, a tetraprenylated benzophenone isolated from *Garcinia brasiliensis*. *Planta Medica* 73(7), 644–649. DOI: 10.1055/s-2007-981534.

Niero, R., Dal Molin, M.M., Silva, S., Damian, N.S., Maia, L.O. *et al.* (2012) Gastroprotective effects of extracts and guttiferone A isolated from *Garcinia achachairu* Rusby (Clusiaceae) against experimentally induced gastric lesions in mice. *Naunyn-Schmiedeberg's Archives of Pharmacology* 385(11), 1103–1109. DOI: 10.1007/s00210-012-0788-1.

Otuki, M.F., Bernardi, C.A., Prudente, A.S., Laskoski, K., Gomig, F. *et al.* (2011) *Garcinia gardneriana* (Planchon & Triana) Zappi. (Clusiaceae) as a topical anti-inflammatory alternative for cutaneous inflammation. *Basic & Clinical Pharmacology & Toxicology* 109(1), 56–62. DOI: 10.1111/j.1742-7843.2011.00689.x.

Parkar, S.G., Trower, T.M. and Stevenson, D.E. (2013) Fecal microbial metabolism of polyphenols and its effects on human gut microbiota. *Anaerobe* 23, 12–19. DOI: 10.1016/j.anaerobe.2013.07.009.

Pereira, I.O., Marques, M.J., Pavan, A.L.R., Codonho, B.S., Barbiéri, C.L. *et al.* (2010) Leishmanicidal activity of benzophenones and extracts from *Garcinia brasiliensis* Mart. fruits. *Phytomedicine : International Journal of Phytotherapy and Phytopharmacology* 17(5), 339–345. DOI: 10.1016/j.phymed.2009.07.020.

Pereira, I.O., Assis, D.M., Juliano, M.A., Cunha, R.L.O.R., Barbieri, C.L. *et al.* (2011) Natural products from *Garcinia brasiliensis* as Leishmania protease inhibitors. *Journal of Medicinal Food* 14(6), 557–562. DOI: 10.1089/jmf.2010.0018.

Pio-Corrêa, M. (1984) *Dicionário das plantas úteis do Brasil e das exóticas cultivadas*, 1st edn. Ministério da Agricultura,Instituto Brasileiro de Desenvolvimento Florestal, Rio de Janeiro.

Pott, A. and Pott, V.J. (1994) *Plantas do pantanal*. EMBRAPA/ SPI/CPAP, Corumbá, MS/Brasil.

Sales, L., Pezuk, J.A., Borges, K.S., Brassesco, M.S., Scrideli, C.A. *et al.* (2015) Anticancer activity of 7-epiclusianone, a benzophenone from *Garcinia brasiliensis*, in glioblastoma. *BMC Complementary and Alternative Medicine* 15, 393. DOI: 10.1186/s12906-015-0911-1.

Santa-Cecília, F.V., Freitas, L.A.S., Vilela, F.C., Veloso, C. de C., da Rocha, C.Q. *et al.* (2011) Antinociceptive and anti-inflammatory properties of 7-epiclusianone, a prenylated benzophenone from *Garcinia brasiliensis*. *European Journal of Pharmacology* 670(1), 280–285. DOI: 10.1016/j.ejphar.2011.08.032.

Santa-Cecília, F.V., Santos, G.B., Fuzissaki, C.N., Derogis, P.B.M.C., Freitas, L.A.S. *et al.* (2012) 7-epiclusianone, the natural prenylated benzophenone, inhibits superoxide anions in the neutrophil respiratory burst. *Journal of Medicinal Food* 15(2), 200–205. DOI: 10.1089/jmf.2011.0043.

Santa-Cecília, F.V., Abreu, F.A., da Silva, M.A., de Castro, E.M. and dos Santos, M.H. (2013) Estudo farmacobotânico das folhas de *Garcinia brasiliensis* Mart. (Clusiaceae). *Revista Brasileira de Plantas Medicinais* 15(3), 397–404. DOI: 10.1590/S1516-05722013000300013.

Santos, J.A.N., Kondo, M.Y., Freitas, R.F., dos Santos, M.H., Ramalho, T.C. *et al.* (2016) The natural flavone fukugetin as a mixed-type inhibitor for human tissue kallikreins. *Bioorganic & Medicinal Chemistry Letters* 26(5), 1485–1489. DOI: 10.1016/j.bmcl.2016.01.039.

Silva, C.A., Jauhiainen, A., Betts, J. and Da Silva, C.A. (2020) Evaluating four modes of extraction to analyse bioactive compounds in *Garcinia brasiliensis* (bacupari) by high-performance liquid chromatography diode-array detection (HPLC-DAD). *Natural Product Research* 1–5.

Sousa, G.S. (1987) *Tratado descritivo do Brasil em 1587*. Companhia Editora Nacional, São Paulo, SP/ Brasil.

Souza, P.M., Araucária, F., Dos Santos, V.J., Rotta, E.M. and Visentainer, J. (2016) *Composição Nutricional e Atividade Antioxidante Da Fruta Nativa Bacupari (Garcinia Brasiliensis)*. Universidade Estadual de Maringá, Paraná, Brazil.

Sunny, S. and Krishnakumar, K. (2017) Evaluation of in vitro antioxidant properties of *Garcinia brasiliensis* leaf extract. *International Journal of Pharmaceutical Science and Research* 8, 763.

Sweeney, P.W. (2008) Phylogeny and floral diversity in the genus *Garcinia* (Clusiaceae) and relatives. *International Journal of Plant Sciences* 169(9), 1288–1303. DOI: 10.1086/591990.

Verdi, L.G., Pizzolatti, M.G., Montanher, A.B.P., Brighente, I.M., Júnior, A.S. *et al.* (2004) Antibacterial and brine shrimp lethality tests of biflavonoids and derivatives of Rheedia gardneriana. *Fitoterapia* 75(3–4), 360–363. DOI: 10.1016/j.fitote.2003.12.023.

Zan, R.A., Fernandes, Â., Jedoz, S., Oludemi, T., Calhelha, R.C. *et al.* (2018) Bioactive properties and phytochemical assessment of Bacupari-anão (*Garcinia brasiliensis* Mart.) leaves native to Rondônia, Brazil. *Food & Function* 9(11), 5621–5628. DOI: 10.1039/c8fo01474d.

Zhu, W., Winter, M.G., Byndloss, M.X., Spiga, L., Duerkop, B.A. *et al.* (2018) Precision editing of the gut microbiota ameliorates colitis. *Nature* 553(7687), 208–211. DOI: 10.1038/nature25172.

21 Monkey Jack (*Artocarpus lakoocha*)

Md Tariqul Islam Shajib[1]*, Manzur-E-Mohsina Ferdous[1], Ambreen Naz Shah[1], Ummai Habiba[1] and Sheikh Nazrul Islam[2]

[1]Section of Scientific Research, Care to Association, Slangerup, Denmark; [2]Institute of Nutrition and Food Science, University of Dhaka, Dhaka, Bangladesh

Abstract

Monkey jack is a tropical minor fruit getting more scientific attention in South and South-East Asian countries after the unveiling of its phytonutrient or phytochemical importance. In this chapter, the fruit's importance in Bangladesh is discussed in relation to its nutritional, medicinal, therapeutic and edible values. Specifically, nutritional composition data from four studies, and phytonutrient data from eight studies has been gathered and is here discussed. Also addressed is the fact that due to its ability to endure adverse climatic conditions and a wide range of soils, this fruit can play an important role in feeding the increasing population by extending the growing season and area of propagation.

Introduction

In Bangladesh, there are many varieties of local fruit with good nutritional, phytochemical and medicinal qualities. These fruit trees can be seen growing in homesteads, forest areas and on roadsides without any cultivation or care (Shajib *et al.*, 2013). The fruit are regarded as minor/underutilized, because there is less awareness of their nutritional value. Minor fruit play a very significant role in human nutrition and some of them have received much attention in Bangladesh (Shajib *et al.*, 2013). Nutritional composition, phytochemical constituents and bioactive components of minor fruit have shown a direct connection with human wellbeing in south Asian countries including Bangladesh (Shajib *et al.*, 2013; Gupta *et al.*, 2020). *Artocarpus lakoocha* Roxb. (English name, monkey jack) is among the most nutritious of the minor fruit in Asia. Scientists have now discovered the health benefits of monkey jack, including nutritional, phytonutritional and pharmacological components (Shajib *et al.*, 2013) and have utilized it in the food and pharmacological fields as part of new product development aimed at improving health (Khalid *et al.*, 2017). Ashitha *et al.* (2020) reported that the remarkable phytonutrients of monkey jack are gaining popularity in both traditional and modern medical science uses in South Asian countries. Phytonutrients or phytochemicals are natural chemicals produced by plants for keeping them healthy and protecting them from insects and against UV rays. They are also impart potential benefits for human health. In this chapter, the nutritional, phytonutritional and pharmacological composition of monkey jack fruit are discussed.

*Corresponding author: agt.tariqbd@yahoo.com

DOI: 10.1079/9781789248067.0021

Minor Fruit

Bangladesh is bestowed with both tropical and subtropical climatic conditions. There are many species that produce minor edible fruit that form genetic resources and play a dynamic role as food, nutrition and medicine. They contribute approximately 8.38% production to the total fruit yield in Bangladesh (BBS, 2011). Roy (2007) reported 120 species of minor fruit in Bangladesh, cultivated and wild. Due to new discoveries of and uses for wild fruit, their use and research of them is increasing. There are at least 30 species in Bangladesh used for their therapeutic properties (Rahman and Rahman, 2014). Some are antidiabetic and rich in antioxidants. Furthermore, many minor fruit can be used in jams, jellies, sauces, pickles, juices and as food additives (Pasha and Uddin, 2019).

Diversification for food security: the role of minor fruit

Food security is of basic importance. Crop diversification measures the sustainability of food security, and is also an efficient tool for mitigating the adverse effects of climate change (Ijaz *et al.*, 2019). It is essential to observe the role of minor fruit in food security. They are grown in domestic gardens, small orchards and in natural forests, thus a large area is not required to justify their economy of production. They possess both food and market value and a few of them are available during the whole year. Increased production of minor and underutilized fruit by extending their growing season offers a major opportunity in the heavily populated countries for nutritional security (Reddy *et al.*, 2012). Minor indigenous fruit are rich in vitamins and minerals. Increased production of minor fruit and mass awareness of their nutritional values can help to prevent many diseases caused by nutritional deficiency. Moreover, large-scale cultivation will enrich the food composition table in Bangladesh which is essential for health, nutrition and food policy programme planning, as well as maintaining biodiversity while protecting these plants from extinction (Shajib *et al.*, 2013).

Bangladeshi Monkey Jack

Artocarpus lakoocha Roxb. (commonly known as monkey jack fruit or dewa in Bangladesh, lakuchi in India) is a highly nutritious tropical fruit (Hossain *et al.*, 2016) available in Bangladesh, India, Thailand and South-East Asia (Saha, 2019). In Bangladesh, this medium-/large-sized deciduous fruit tree can be found throughout the country (Ghani, 2003). The freshly eaten fruit can help overcome malnutrition and create protection against some diseases (Shajib *et al.*, 2013; Hossain *et al.*, 2016). Curries, pickles and sauces are made using the unripe fruit and the male flower spike. The tree is an important source of firewood, timber and fresh fodder for lactating animals. In addition, monkey jack is used to make juices and certain items for food (Sarala and Krishnamurthy, 2014; Gupta *et al.*, 2020a).

Fig. 21.1. *Artocarpus lakoocha*: (a) fruit on a tree; (b) ripe fruit.

Botanical Aspects

Morphological characterization

Monkey jack is a deciduous tree of medium to large size. Trees are 10–15 m tall, shaded a brownish-grey colour with a milky latex. The simple alternate, petiolate leaves are 16–25 cm long and 6–15 cm elliptic, ovate, oblong, cuspidate-acuminate apex, dark green, midrib more prominent beneath. Flowers are unisexual in the axillary, globose head. This monoecious plant bears orange-yellow male flowers and greenish female flowers. The irregular round-shaped compound fruit are derived from the swollen flower heads. The male flowers are mostly 3.5 cm long and 3.1 cm wide, whereas the female flowers are mostly 1.8 cm long and 1.3 cm wide. The fruit are syncarp, globular, smooth velvety, 5–8 cm in diameter, and are considered as a typical wild fruit developed with the spine in the skin (Fig. 21.1). The orange-yellow pulp has a sweet juicy taste with a medium pleasant smell and soft texture. Flowering is seen during February–April and fruit sets in May–July. The seedcoat is white and thin; the seed contains white sticky latex and is recalcitrant (Islam *et al.*, 2016).

Classification of monkey jack (Table 21.1)

Table 21.1. Taxonomic position of monkey jack (Cronquist, 1988).

Kingdom	Plantae
Division	Magnoliophyte
Class	Magnoliopsida
Order	Rosales
Family	Moraceae
Genus	*Artocarpus*
Species	*A. lakoocha* Roxb.

Geographical distribution and conservation

Artocarpus lakoocha is widely distributed in India, Nepal, Bhutan, Bangladesh, Srilanka, Vietnam, Thailand, Myanmar and Malaysia, in areas with an annual rainfall of 700–2000 mm and a dry season (up to three months). Deep permeable soils with good moisture are required for its cultivation. It grows mainly in tropical and subtropical regions of South and South-East Asia; however, the populations are adversely affected because of poor seed viability despite widespread utilization in many places (Jøker and Adhikari, 2003; Vanajakshi *et al.*, 2016). Not enough attempts have been made to develop this species, despite its ability to survive unfavourable climatic conditions and grow in various types of soil. More characterization and evaluation studies of this important minor fruit are needed and steps should be taken to establish advance cultivars with selected traits (Islam *et al.*, 2016). Monkey jack is mainly cross-pollinated, and so seed conservation is the only way to secure the gene pool. A high level of genetic variability in this species is at risk of extinction because of large-scale urbanization, lack of developmental activities and agricultural land in marginal, tribal and coastal regions (Islam *et al.*, 2016). Reduction in genetic variability of the plant species plays the key role to enter into the extinction vortex (Govindaraj *et al.*, 2015).

Germplasm management and cultivation

A single fruit's weight ranges between 250 and 300 g and contains 10–30 seeds (Orwa *et al.*, 2009). Seeds are collected from the mature fruit; seeds extracted from unripe fruit have low viability. The traditional procedure is that the seeds are taken out of the fruit just before sowing; however, seeds stay viable for a week in this way. The moisture content of seeds is 50–55% when extracted from mature fruit. Seeds having 30% moisture content lose viability in 50% of cases as they are intoleraant to drying and low moisture content. Seeds with a moisture content of 50% can be stored at 5°C in cold storage, but 70% moisture gave a better result as the seeds still maintained 55% germination after four weeks to almost 0% after eight weeks (Jøker and Adhikari, 2003). The tree is typically propagated by seeds, and vegetative propagation has also been tried using softwood or hard root stem cutting and grafting (Islam *et al.*, 2016). For seedling preparation, seeds are sown in pots (two seeds in 10 × 18 cm) or seedbeds (1 kg seeds per m2) with 20–25% compost. A spacing of 1 m × 1 m is recommended for better yields of fodder production (Orwa *et al.*, 2009).

Rich Sources of Nutrients

A. lakoocha fruit contains a unique combination of various health-promoting elements including fibre, iron, minerals, phytochemicals, vitamin B complex and a good source of antioxidants, such as vitamin C and beta-carotene (Shajib *et al.*, 2013; Hossain *et al.*, 2016) (Table 21.2). It contains a large amount of water, which can be dried to develop a powder (Sarala and Krishnamurthy, 2014).

In Bangladesh, another fruit of the Moraceae family, jackfruit (*Artocarpus heterophyllus*) is the national fruit. Jackfruit is ranked second in production among the fruit of Bangladesh and third in cultivation (Banglapedia, 2021). Monkey jack is a close comparative, and contains higher amounts of carotenoid (4609 µg/100 g) than jackfruit (710 µg/100 g) (FAO-NFPCSP, 2010; Shajib *et al.*, 2013) indicating that this fruit is also an exceptionally good source of pro-vitamin A to prevent night blindness among children. Monkey jack is an excellent source of ascorbic acid of – 171.07 mg/100 g; higher than 35.32 mg/100 g in jackfruit – whereas the minimum daily requirement is recommended at 10 mg or a little less for prevention of vitamin

Table 21.2. Comparison of nutritional composition between fresh monkey jack and jackfruit (Jahan *et al.*, 2011; Shajib *et al.*, 2013; Goswami and Chacrabati, 2016; Gupta *et al.*, 2020).

Content in edible portion	Monkey jack	Jackfruit
Water (%)	84.25	86.21
Carbohydrate (g)	9.47	9.11
Protein (g)	1.03	1.91
Fat (g)	0.71	0.18
Crude fibre (g)	2.41	1.27
Vitamins		
Vitamin A (IU)	423	175–540
Vitamin C (mg/100 g)	171.07	35.32
Vitamin B-1 (thiamine)	0.02 (µg %)	0.03–0.09 (mg/100 g)
Vitamin B-2 (riboflavin) (µg %)	0.15 (µg %)	0.05–0.4 (mg/100 g)
Vitamin B-3 (niacin) (µg %)	0.3 (µg %)	-
Beta-carotene (µg /100 g)	3718.16	4401.82
Total carotenoid (µg /100 g)	4609	710
Macrominerals		
Sodium (mg/100 g)	46	66
Potassium (mg/100 g)	785	623
Magnesium (mg/100 g)	55	35
Phosphorus (mg/100 g)	66.06	50.72
Calcium (mg/100 g)	47	29
Microminerals		
Iron (mg/100 g)	15.09	5.08
Zinc (mg/100 g)	1.68	1.13
Manganese (mg/100 g)	3.76	3.07
Copper (mg/100 g)	1.31	0.87

C deficiency/scurvy for adults (Jukes, 1974; Food Agriculture Organization, 1998). Monkey jack has a relatively higher amount of carbohydrate (9.47 mg/100 g), fat (0.71 mg/100 g), and energy (48.39 kcal) compared to jackfruit which contains 9.11 mg/100 g, 0.18 mg/100g and 45.70 kcal of carbohydrate, fat and energy respectively (Table 21.2). It also has superior macro- and micromineral content, such as three times more iron (15.09 mg/100 g), compared to jackfruit (5.08 mg/100 g). This minor fruit has higher amounts of various mineral nutritional compositions, presented in Table 21.2. Monkey jack can fulfil the daily needs of microminerals to a certain extent (Shajib *et al.*, 2013).

Phytonutrients in monkey jack

The natural compounds produced by plants are called phytonutrients, which function with other essential nutrients to improve health by preventing and even treating diseases. Phytonutrients/phytochemicals include the secondary metabolites present in medicinal plants (Krishnaiah *et al.*, 2009). Monkey jack exhibits a significant quantity of phytonutrients, including flavonoids, phenols, tannins, saponins, steroids, glycosides and lignin. Krishnamurthy and Sarala (2013) screened secondary metabolites of monkey jack in different organic solvents such as petroleum ether, chloroform, ethanol and aqueous extracts for qualitative phytochemicals. They observed that the level of alkaloids was high, followed by lignins, phenols, flavonoids and tannins.. There was a preliminary study by Pandey and Bhatnagar (2009) on phytochemicals from bark extracts which showed the presence of triterpenoids, flavonoids, tannins and glycosides. Monkey jack contains alkaloid compounds which are responsible for therapeutic values, and are confirmed to have pharmacological properties such as hypotensive, anticonvulsant, antiprotozoal, antimicrobial and antimalarial (Table 21.3). Phenols are composed of a large and ubiquitous group of plant metabolites that exhibit inhibition of angiogenesis and cell proliferation, which also works for atherosclerosis and cardiovascular protection. Monkey jack contains the naturally occurring compounds, flavonoids, which are well known for their antioxidant and antiradical properties. Tannins are water-soluble polyphenols that have anticarcinogenic, antimutagenic and antioxidative properties and give protection from oxidative damage (Krishnamurthy and Sarala, 2013).

Table 21.3. Phytochemicals in *A. lakoocha*.

Phytonutrient	Plant part	Health benefits	References
Oxyresveratrol	Leaves	Antiviral cytotoxic, anti-HSV, anti-HIV	Chuanasa *et al.* (2008)
Artocarpin	Seed	Coagulant, mitogenic, antibacterial	Chowdhury and Chandra (1987)
α-amyrin acetate	Fruit pericarp	Anti-inflammatory, hypersensitivity oedema, radical scavenging, antihyperlipidemic	Okoye *et al.* (2014)
β-amyrin acetate	Fruit pericarp	-	-
Alkaloids	Fruit	Anti-inflamatory, analgesic, hypnotic	Krishnamurthy and Sarala (2013)
Phenols	Fruit	Antioxidant, antiviral, anticancer, anti-inflammatory	Krishnamurthy and Sarala (2013)
Flavonoids	Fruit	Antioxidant	Krishnamurthy and Sarala (2013)
Glycosides	Fruit	Antioxidant activity, anticancer, antitumour activity	Krishnamurthy and Sarala (2013)
Saponins		Lower cancer risks, lower blood glucose	Shi *et al.* (2004); Krishnamurthy and Sarala (2013)

The methanol fraction of *A. lakoocha* leaves showed a higher number of phytochemicals (i.e. total phenolic content) at 3175.21±290.43mg GAE/g dry extract, and flavonoids (1173.15±47.52mg GAE/g dry extract). The amount significantly differed except for tannin, where content was 811.76±44.50, 905.88±130.28 and 923.53±95.21 mg TAE/g dry extract in ethyl acetate, acetone and methanol fractions, respectively.

The highest antioxidant activity was also seen in DPPH and ABTS radical scavenging assays with IC_{50} of (111.98±34.20) µg/ml and (138.26±0.66) µg/ml, respectively. All the fractions showed a significant correlation between the concentrations and antioxidant activity. Between 5 and 6 mg/ml concentration of the extract effectively secured Lambda-DNA damage and showed bactericidal and fungicidal activities (Bhattacharya *et al.*, 2019).

Artocarpus species are full of phenolic compounds and reveal plentiful properties for the development of nutraceuticals and value-added food-based products from edible fruit (Gupta *et al.*, 2020). Borah *et al.* (2017) explained the antioxidant profile of resveratrol (a type of phenol) of monkey fruit as they observed the significant antioxidant activity with IC_{50} of 53.24µg/ml. *A. lakoocha* is rich in resveratrol (0.09 g w/w), and the amount may differ in various plant parts and soil fertilities. Beside flavonoids, triterpenoids, steroids and stilbenoids, the extract has a high content of oxyresveratrol (trans-2,4,3',5'-tetrahydroxystilbene) which is a derivative of resveratrol. It contained 13.09% of oxyresveratrol DW from heartwood extract solvent by 70% ethanol (Jiratanakittiwat *et al.*, 2020).

The edible fruit of monkey jack have nutraceutical values (Kumar *et al.*, 2010), which include numerous medicinal properties, such as antidiarrheal, antiglycation, antiageing, anti-inflammatory, analgesic, antiviral, anticancerous, antibacterial, cytotoxic, insecticidal and pancreatic lipase inhibitory activity (Gautam and Patel, 2014). Some studies have showed that the fruit contain phytooxyresvertrol exhibited antioxidant and antiglycation activities. The fruit has compounds such as amyrin acetate and lupeol acetate, obtaining antihyperglycemic and hypolipidemic effects, which are essential in the production of diabetes and atherosclerosis medicines (Gupta *et al.*, 2020a). Methanol extract of fruit pericarp showed tannins and alkaloids, which can be used as a treatment for bacterial infection (Kumar *et al.*, 2010). The methanol extract (100mg/ml) of fruit pericarp showed an anthelmintic effect, which can control roundworms (e.g. ascariasis and hookworm's infections). Carbohydrates, saponins, phytosterols, flavonoids, tannins and phenols have mosquito larvicidal activity (Khanna and Kannabiran, 2007). The larvicidal effect was dose-dependent and could be used to prevent dengue and chikungunya infections (Kumar *et al.*, 2010). The methanol extract of fruit showed antioxidative properties, free radical scavenging activities and inhibition of enzymes involved in fatty acid synthesis and cholesterol (Tarbiat, 2018). The extract also showed anticoagulant activity and inhibited oxidation of HDL and LDL *in vitro*. Significant antimicrobial activity of fruit extract has been observed against bacterial strains *Escherichia coli*, *Bacillus subtilis*, *Staphylococcus aureus*, *Salmonella typhi*, *Bacillus cereus*, *Salmonella paratyphi* and two fungal strains, *Candida albicans* and *Candida krusei* (Saha, 2019). The hydroalcoloholic extract of the fruits showed an extraordinary estrogenic activity on female Sprague-Dawley rats without ovaries in an experiment performed by (Gautam *et al.*, 2016). The fruit extract worked against *candida* infection and secondary dermatitis in a research using agar well diffusion method due to its strong anti-biofilm property (Senapong *et al.*, 2014).

The bark of monkey jack possesses strong antioxidant activity by producing methanolic extracts (flavonoids) and phenolic content (Pandey and Bhatnagar, 2009). It contains 8.5% tannin, which is used for skin lesion and wound treatment (Tomar *et al.*, 2015). Phytooxyresveratrol (25µg/ml) extracted from the bark has effective antiageing activity and prevents DNA damage and cellular ageing due to its antioxidant and antiglycation defences (Suwannalert *et al.*, 2012).

There are two stilbene derivatives, lakoochins A (1) and B (2) in the roots, which have antimycobacterial activity. Lakoochins A (1) showed cytotoxicity against the breast cancer (BC) cell line but expressed inactivity toward nasopharyngeal carcinoma (KB) cells; however compound B (2) was cytotoxic against both the BC and KB cell lines (Puntumchai *et al.*, 2004).

A. lakoocha's seed contains oxyresveratrol, which inhibits the development of skin lesions significantly, having more potentiality in comparison to resveratrol (Shin *et al.*, 1998). A study on high cholesterol diet (HCD)-fed rat showed that oxyresveratrol helps in reducing body and liver weight (Hwang *et al.*, 2015). The combination of oxyresveratrol with microemulsion and absorbic acid is a potential anti-browning agent for fresh-cut vegetables (He *et al.*, 2017). The seed demonstrated the highest haemagglutination activity (14,400 units/g fwt) and the leaves had the lowest (5 units/g fwt) according to a study performed by Wongkham *et al.* (1995) in various organs of the plant except for fruit flesh. The activity was found not to be organ-specific which is exhibited because of the presence of artocarpins (ALA I and ALA II). The leaves and flowers are an excellent source of bioactive compounds; thus, these are used as fresh fodder for lactating animals (Gupta *et al.*, 2020a). The flowers contain many secondary metabolites which revealed high antioxidant and antimicrobial activities. However, the experiment was performed only on male flowers to evaluate nutritional, mineral, antioxidant and antimicrobial potential; female flowers showed variation in compounds and biological activities. The concentration of 200 g/ml from the flower extract acted as a strong antibacterial agent against *Salmonella typhimurium* (Gupta et *al.*, 2020a).

A. lakoocha extract contains oxyresveratrol, which exhibited antibacterial effects against both Gram-negative bacteria (*Aggregatibacter actinomycetemcomitans, Porphyromonas gingivalis*) and Gram-positive bacteria (*Streptococcus mutans, Streptococcus sobrinus*) (Teanpaisan *et al.*, 2014). The antimicrobial and antibiofilm activities were found to be effective against dental cavities and oral infection-causing pathogens in a study conducted on artificial biofilm *in vitro*. Both ethanolic and methanolic extract of *Artocarpus lakoocha* bark showed antibacterial activity against five oral inhabiting bacteria (*Escherichia coli, Streptococcus mutans, Streptococcus sanguinis, Porphyromonas gingivalis* and *Fusobacterium nucleatum)* by using agar well diffusion method (Nath and Boruah, 2019).

Seven of 39 patients experienced immediate vomiting while being treated with 5 g of crude aqueous extract from monkey jack wood to eliminate tapeworm (Charoenlarp *et al.*, 1981). On the other hand, 3 g of that extract gave a better result along with fewer side effects to treat taeniasis (Charoenlarp *et al.*, 1989). The different concentrations of *A. lakoocha*'s crude extract reduced the adult parasite's (*Fasciola gigantica*) motility and killed them after one or two days of incubation, which also inhibited larval movement (Saowakon *et al.*, 2009). The crude extracts showed anthelmintic activity against the rumen fluke parasite (*Paramphistomum cervi*) and gave a better result than albendazole (anthelmintic medication) in an SEM-based study (Preyavichyapugdee *et al.*, 2017).

Antioxidant Properties

Antioxidant phytochemicals significantly contribute to the maintenance of good health, as well as the prevention and treatment of chronic diseases caused by oxidative stress such as cardiovascular disease and cancer. Many studies have been performed to learn the protective roles of phytochemicals along with their proper utilization. Imbalance between antioxidant defence and excessive production of free radicals leads to many diseases and speeds up ageing (Goze *et al.*, 2009).

The plant (*A. lakoocha*) contains phenolic compounds, flavonoids and phenolic acids (Jasprica *et al.*, 2007) which are strong antioxidants and play a key role in the absorption and neutralization of some free radicals, reducing singlet or triplet oxygen or peroxide decomposition (Javanmardi *et al.*, 2003). Presence of phenols (325.63 mg gallic acid equivalent), flavonoids (521.98 mg quercetin equivalent) and tannins (124.03 g tannin equivalent) in 1 g dry weight (DW) of the heartwood extract (Singhatong *et al.*, 2010) ensures the strong antioxidant, anti-inflammatory and neuroprotective properties of this plant (Maneechai *et al.*, 2009). A recent study suggested anti-inflammatory activities of the crude extraction from heartwoods, enriched with oxyresveratrol, inhibits the LPS-stimulated inflammatory response through the blocking of the Akt, MAPK, and NF-_B signalling pathways (Hankittichai *et al.*, 2020). Anti-inflammatory activities of the plant have been known for its high content of stilbenoids such as pinostilbene, desoxyrhapontigenin, pterostilbene, resveratrol, and oxyresveratrol (Maneechai *et al.*, 2012).

The plant produces phenolic compounds (lavonoids and tannins), inhibited lipid peroxidation, active oxygen and chelate iron ions, free radical scavengers and inactivated lipoxygenase (Singhatong *et al.*, 2010). Tannin possesses chemopreventive action against carcinogens, mutagens and free radical scavengers, preventing damaging reactions (Sudha and Gnanamani, 2008). Plant extracts were found to be preventive against the development of various diseases by scavenging free radicals (e.g. diabetes, rheumatoid arthritis, Alzheimer's) (Jasprica *et al.*, 2007). Oxyresveratrol of ethanolic extraction of monkey jack plant supported strong tyrosinase-inhibitory activity (Maneechai *et al.*, 2009), and is used as a skin-whitening agent (Veerawat *et al.*, 2013).

Ethanolic extraction of *A. lakoocha* contained antioxidants and polyphenolic compounds, especially tannins and flavonoids. However, the high concentration may be toxic as the unspent tannins can act as pro-oxidants, which induce oxidative stress (Singhatong *et al.*, 2010).

Health Benefits

More than 80% of the world's population depends on traditional herbal medicines for their basic health treatment (Akerele, 1993). Two-thirds of medicinal plants among 50,000 different species are collected from the wild for healing ailments (Canter *et al.*, 2005; Vijayan *et al.*, 2007). Monkey jack has various traditional medicinal applications in South Asia (Vanajakshi *et al.*, 2016). Extracts of the different plant parts are used traditionally for their many nutraceutical values, such as for curing liver-related diseases, diarrhoea, diabetes, tapeworm infections, skin problems, malaria, hypertension, taeniasis and other health-related issues (Charoenlarp *et al.*, 1989; Gautam and Patel, 2014; Gupta *et al.*, 2020). The plant is also used as an anti-skin ageing agent and anti-inflammatory in traditional Thai medicine (Povichit *et al.*, 2010). The saturated extract is used as a poultice for skin irritation and the bark as a headache remedy (Orwa *et al.*, 2009). Powdered tree bark is applied to cure wounds, skin lesions and to clear out purulent matter. The bark is also useful as an astringent and for stomach and liver diseases (Tomar *et al.*, 2015). Moreover, antioxidants from the plant play a significant role in cardiovascular disease prevention (Pandey and Bhatnagar, 2009).

The medicinal values of monkey jack fruit are accepted as conventional remedies, as well as in modern medical science (Ashitha *et al.*, 2020). The fruit is significantly used in the treatment of stomach and liver disease (Joshee *et al.*, 2002) as the sour-sweet fruit pulp acts as a liver refresher. It can also treat arthritic swelling, dysentery and skin infections, and be used for wound cleaning (Vanajakshi *et al.*, 2016). This fruit contains a variety of inhibitory factors (e.g. antibacterial, antiviral, antifungal, antitubercular, antiarthritic, antiplatelet, tyrosinase inhibitory, cytotoxicity and H2O2 scavenging activity) against herpes simplex virus (HSV) and human immunodeficiency virus (HIV) (Gupta *et al.*, 2020). Furthermore, it helps in treating inflammatory arthritis, in preventing skin diseases, in cleaning out wounds and in the treatment of dysentery, if absorbed with goat milk and herbs (Vanajakshi *et al.*, 2016).

The milky latex and seeds are good purgatives; the seeds are also used for stomach and liver diseases (Tomar *et al.*, 2015). The roots are used as a refresher and the leaves to treat dropsy (Perry, 1980).

Conclusion

Monkey jack is a combined source of food and energy. This fruit and its tree have been extensively studied and yet practical development of their potential is still lacking – for example, the establishment of trees to increase local farmers' income, the marketing and selling of fruit, awareness in local communities, inhibitory factors, treatment effects and so on. Extending the production area will enable not only Bangladesh but also other South and South-East Asian countries to overcome food insecurity to a great extent.

References

Akerele, O. (1993) Nature's medicinal bounty: don't throw it away. *World Health Forum* 14(4), 390–395.

Ashitha, G.N., Sunny, A.C. and Nisha, R. (2020) Effect of Pre-harvest and Post-harvest Hexanal Treatments on Fruits and Vegetables: A Review. *Agricultural Reviews* 41(OF), 124–131. DOI: 10.18805/ag.R-1928.

Banglapedia (2021). Available at: https://en.banglapedia.org/index.php/Jackfruit

BBS (2011) *Statistical Yearbook of Bangladesh*. Bangladesh Bureau of Statistics, Ministry of Planning, Govt. of Bangladesh, Dhaka, pp. 1–554

Bhattacharya, E., Dutta, R., Chakraborty, S. and Mandal Biswas, S. (2019) Phytochemical profiling of *Artocarpus lakoocha* Roxb. leaf methanol extract and its antioxidant, antimicrobial and antioxidative activities. *Asian Pacific Journal of Tropical Biomedicine* 9(11), 484–492. DOI: 10.4103/2221-1691.270984.

Borah, H.J., Singhal, R. and Hazarika, S. (2017) Artocarpus lakoocha Roxb.: An untapped bioresource of resveratrol from North East India, its extractive separation and antioxidant activity. *Industrial Crops and Products* 95, 75–82. DOI: 10.1016/j.indcrop.2016.10.015.

Canter, P.H., Thomas, H. and Ernst, E. (2005) Bringing medicinal plants into cultivation: opportunities and challenges for biotechnology. *Trends in Biotechnology* 23(4), 180–185. DOI: 10.1016/j.tibtech.2005.02.002.

Charoenlarp, P., Radomyos, P. and Harinasuta, T. (1981) Treatment of taeniasis with Puag-Haad: a crude extract of *Artocarpus lakoocha* wood. *The Southeast Asian Journal of Tropical Medicine and Public Health* 12(4), 568–570

Charoenlarp, P., Radomyos, P. and Bunnag, D. (1989) The optimum dose of Puag-Haad in the treatment of taeniasis. *Journal of the Medical Association of Thailand = Chotmaihet Thangphaet* 72(2), 71–73

Chowdhury, B.A. and Chandra, R.K. (1987) Biological and health implications of toxic heavy metal and essential trace element interactions. *Progress in Food & Nutrition Science* 11(1), 55–113

Chuanasa, T., Phromjai, J., Lipipun, V., Likhitwitayawuid, K., Suzuki, M. *et al.* (2008) Anti-herpes simplex virus (HSV-1) activity of oxyresveratrol derived from Thai medicinal plant: mechanism of action and therapeutic efficacy on cutaneous HSV-1 infection in mice. *Antiviral Research* 80(1), 62–70. DOI: 10.1016/j.antiviral.2008.05.002.

Cronquist, S.E. (1988) *The Evolution and Classification of Flowering Plants*. New York Botanical Garden, Bronx

Gautam, P. and Patel, R. (2014) Artocarpus lakoocha Roxb: an overview. *European Journal of Complementary and Alternative Medicine* 1(1), 10–14.

FAO-NFPCSP (2010). Available at: http://www.nfpcsp.org/Sk_Nazrul_Islam-PR11-08.pdf

Food Agriculture Organization (1998) "*Vitamin and mineral requirements in human nutrition: Report of a joint FAO/WHO expert consultation, 2nd edition.*"

Gautam, P., Jawaid, T. and Patel, R. (2016) Estrogenic activity of fruit extract of Artocarpus Monkey fruit in female Sprague-Dawley rats. *Pharma Chemica* 8(15), 27–33.

Ghani, A. (2003) *Medicinal plants of Bangladesh: chemical constituents and uses*, 2nd edn. Asiatic Society of Bangladesh, Dhaka

Goswami, C. and Chacrabati, R. (2016) Jackfruit (*Artocarpus heterophylus*). In: Simmonds, M. and Preedy, V.R. (eds) *Nutritional Composition of Fruit Cultivars*. Academic Press, London, pp. 317–335

Govindaraj, M., Vetriventhan, M. and Srinivasan, M. (2015) Importance of genetic diversity assessment in crop plants and its recent advances: an overview of its analytical perspectives. *Genetics Research International* 2015, 431487. DOI: 10.1155/2015/431487.

Goze, I., Alim, A., Tepe, A.S., Sokmen, M., Sevgi, K. *et al.* (2009) Screening of the antioxidant activity of essential oil and various extracts of Origanum rotundifolium Boiss. from Turkey. *Journal of Medicinal Plants Research* 3(4), 246–254.

Gupta, A.K., Pathak, U., Medhi, M., Mastinu, A., Sikarwar, M.S. *et al.* (2020) Botanical, Chemical and Pharmacological Properties of *Artocarpus lakoocha* (Monkey Fruit): A Review. *Agricultural Reviews* 42(Of), 1–12. DOI: 10.18805/ag.R-1989.

Gupta, A.K., Rather, M.A., Kumar Jha, A., Shashank, A., Singhal, S. *et al.* (2020a) *Artocarpus lakoocha* Roxb. and *Artocarpus heterophyllus* Lam. Flowers: New Sources of Bioactive Compounds. *Plants* 9(10), 1329. DOI: 10.3390/plants9101329.

Hankittichai, P., Buacheen, P., Pitchakarn, P., Na Takuathung, M., Wikan, N. *et al.* (2020) *Artocarpus lakoocha* extract inhibits LPS-induced inflammatory response in RAW 264.7 macrophage cells. *International Journal of Molecular Sciences* 21(4), 1355. DOI: 10.3390/ijms21041355.

He, J., Zhu, Q., Dong, X., Pan, H., Chen, J. *et al.* (2017) Oxyresveratrol and ascorbic acid O/W microemulsion: Preparation, characterization, anti-isomerization and potential application as antibrowning agent on fresh-cut lotus root slices. *Food Chemistry* 214, 269–276. DOI: 10.1016/j.foodchem.2016.07.095.

Hossain, M.F., Islam, M.A., Akhtar, S. and Numan, S.M. (2016) Nutritional value and medicinal uses of monkey fruit (*Artocarpus lakoocha*). *International Research Journal of Biological Sciences* 5(1), 2278–3202

Hwang, D., Jo, S.-P., Lee, J., Kim, J.-K., Kim, K.-H. *et al.* (2015) Antihyperlipidaemic effects of oxyresveratrol-containing *Ramulus mori* ethanol extract in rats fed a high-cholesterol diet. *Journal of Functional Foods* 19, 353–362. DOI: 10.1016/j.jff.2015.09.039.

Ijaz, M., Nawz, A., Sami, U., Muhammad, R., Ullah, A. *et al.* (2019) Crop Diversification and Food Security. *Agronomic Crops* 607–621. DOI: 10.1007/978-981-32-9151-5_26.

Islam, M.S., Hasan, M.F., Ali, M., Robbani, M. and Hossain, M.T. (2016) Socioeconomic potential of monkey jack: a promising underutilized fruit in Bangladesh. *International Journal of Innovative Research* 1(1), 40–44.

Jahan, S., Gosh, T., Begum, M. and Saha, B.K. (2011) Nutritional profile of some tropical fruits in Bangladesh: specially anti-oxidant vitamins and minerals. *Bangladesh Journal of Medical Science* 10(2), 95–103. DOI: 10.3329/bjms.v10i2.7804.

Jasprica, I., Bojic, M., Mornar, A., Besic, E., Bucan, K. *et al.* (2007) Evaluation of antioxidative activity of Croatian propolis samples using DPPH* and ABTS*+ stable free radical assays. *Molecules (Basel, Switzerland)* 12(5), 1006–1021. DOI: 10.3390/12051006.

Javanmardi, J., Stushnoff, C., Locke, E. and Vivanco, J.M. (2003) Antioxidant activity and total phenolic content of Iranian *Ocimum* accessions. *Food Chemistry* 83(4), 547–550. DOI: 10.1016/S0308-8146(03)00151-1.

Jiratanakittiwat, K., Satirapipathkul, C. and Charnvanich, D. (2020) The influences of extraction on the quantity of Oxyresveratrol from *Artocarpus lakoocha* Roxb. *International Journal of Bioscience, Biochemistry and Bioinformatics* 10(2), 110–116. DOI: 10.17706/ijbbb.2020.10.2.110-116.

Joshee, N., Bastola, D.R., Agrawal, V.P. and Yadav, A.K. (2002) Monkey fruit: a multipurpose tree of warm climate. In: Janick, J. and Whipkey, A. (eds) *Trends in New Crops and New Uses*. ASHS Press, Alexandria, Virginia, pp. 405–406

Jøker, D. and Adhikari, B. (2003) Seed leaflet: *Artocarpus lakoocha* Roxb. Danida Forest Seed Centre. Denmark. Available at: https://sl.ku.dk/rapporter/seed-leaflets/filer/artocarpus-lakoocha-73-int.pdf (accessed 6 April 2022)

Jukes, T.H. (1974) Are Recommended Daily Allowances for Vitamin C Adequate? (ascorbic acid/diet). *Proc. Nat. Acad. Sci. USA* 71(5), 1949–1951.

Khalid, S., Khalid, N., Khan, R.S., Ahmed, H. and Ahmad, A. (2017) A review on chemistry and pharmacology of Ajwa date fruit and pit. *Trends in Food Science and Technology* 63, 60–69. DOI: 10.1016/j.tifs.2017.02.009.

Khanna, V.G. and Kannabiran, K. (2007) Larvicidal effect of *Hemidesmus indicus*, *Gymnema sylvestre*, and *Eclipta prostrata* against *Culex qinquifaciatus* mosquito larvae. *African Journal of Biotechnology* 6(3), 307–311

Krishnaiah, D., Devi, T., Bono, A.S. and Sarbatly, R. (2009) Studies on phytochemical constituents of six Malaysian medicinal plants. *Journal of Medicinal Plants Research* 3(2), 067–072

Krishnamurthy, S.R. and Sarala, P. (2013) Phytochemical studies of *Artocarpus gomezianus* Wall. Ex Trecul. Var. lakoocha Roxb. fruits collected from various altitudes of central western Ghats. Indian Journal of Natural Products and Resources. *Indian Journal of Natural Products and Resources* 4(4), 398–411

Kumar, M.B.S., Kumar, M.C.R., Bharath, A.C., Kumar, H.R.V., Prashith Kekuda, T.R. *et al.* (2010) Screening of selected biological activities of Artocarpus lakoocha Roxb (Moraceae) fruit pericarp. *Journal of Basic and Clinical Pharmacy* 1(4), 239–245.

Maneechai, S., Likhitwitayawuid, K., Sritularak, B., Palanuvej, C., Ruangrungsi, N. *et al.* (2009) Quantitative analysis of oxyresveratrol content in *Artocarpus lakoocha* and "Puag-Haad." *Medical Principles and Practice* 18(3), 223–227. DOI: 10.1159/000204354.

Maneechai, S., De-Eknamkul, W., Umehara, K., Noguchi, H. and Likhitwitayawuid, K. (2012) Flavonoid and stilbenoid production in callus cultures of *Artocarpus lakoocha*. *Phytochemistry* 81, 42–49. DOI: 10.1016/j.phytochem.2012.05.031.

Nath, P.C. and Boruah, S. (2019) Antimicrobial activity of ethanolic and methanolic extract of *Artocarpus lakoocha* Wall. Ex Roxb. (Moraceae) against five different oral bacterial strains. *International Journal of Current Microbiology and Applied Sciences* 8(3), 1321–1325. DOI: 10.20546/ijcmas.2019.803.156.

Okoye, I.S., Coomes, S.M., Pelly, V.S., Czieso, S., Papayannopoulos, V. *et al*. (2014) MicroRNA-containing T-regulatory-cell-derived exosomes suppress pathogenic T helper 1 cells. *Immunity* 41(1), 89–103. DOI: 10.1016/j.immuni.2014.05.019.

Orwa, C., Mutua, A., Kindt, R., Jamnadass, R. and Anthony, S. (2009) Agroforestry Database: A tree reference and selection guide version 4.0. Kenya: World Agroforestry Centre. Available at: https://www.worldagroforestry.org (accessed 6 April 2022)

Pandey, A.S.P. and Bhatnagar, S.P. (2009) Antioxidant and phenolic content of the bark of *Artocarpus lakoocha*. *The Pharma Review* 5, 173–186

Pasha, M.K. and Uddin, S.B. (2019) Minor edible fruits of Bangladesh. *Bangladesh Journal of Plant Taxonomy* 26(2), 299–313. DOI: 10.3329/bjpt.v26i2.44596.

Perry, L.M. (1980) *Medicinal Plants of East and Southeast Asia: Attributed Properties and Uses*. MIT Press, Cambridge, pp. 149–150

Povichit, N., Phrutivorapongkul, A., Suttajit, M. and Leelapornpisid, P. (2010) Antiglycation and antioxidant activities of oxyresveratrol extracted from the heartwood of *Artocarpus lakoocha* Roxb. *Maejo International Journal of Science and Technology* 4(3), 454–461

Preyavichyapugdee, M., Sangfuang, M., Preyavichyapugdee, N., Pootaeng-on, Y. and Sophon, P. (2017) The *in vitro* effect of a crude extract from *Artocarpus lakoocha* Roxb on Paramphistomum cervi. *Science, Engineering and Health Studies* 36–46

Puntumchai, A., Kittakoop, P., Rajviroongit, S., Vimuttipong, S., Likhitwitayawuid, K. *et al*. (2004) Lakoochins A and B, new antimycobacterial stilbene derivatives from *Artocarpus lakoocha*. *Journal of Natural Products* 67(3), 485–486. DOI: 10.1021/np030429e.

Rahman, M. and Rahman, J. (2014) Medicinal Value and Nutrient Status of Indigenous Fruits in Bangladesh. *Nova Journal of Medical and Biological Sciences* 03(04), 1–10. DOI: 10.20286/nova-jmbs-030423.

Reddy, V.K., Pavithra, C.B. and Geeta, B. (2012) Minor fruits for overall livelihood security and economic growth. In: *International Conference on Agricultural & Horticultural Sciences*, Hyderabad International Convention Centre, India, September 14–12, 2012

Roy, M. (2007) *Bangladesher Phol*. Dibyaprakash, Dhaka, pp. 1–315

Saha, S. (2019) *In vitro* Antimicrobial Activity of Bangladeshi *Artocarpus lakoocha* Roxb Fruits. *International Journal of Current Microbiology and Applied Sciences* 8(2), 438–444. DOI: 10.20546/ijcmas.2019.802.049.

Saowakon, N., Tansatit, T., Wanichanon, C., Chanakul, W., Reutrakul, V. *et al*. (2009) *Fasciola gigantica*: anthelmintic effect of the aqueous extract of *Artocarpus lakoocha*. *Experimental Parasitology* 122(4), 289–298. DOI: 10.1016/j.exppara.2009.04.011.

Sarala, P. and Krishnamurthy, S.R. (2014) Monkey jack: underutilized, edible medicinal plant, nutritional attributes and traditional foods of western ghats, karnataka, india. *Indian Journal of Traditional Knowledge* 13(3), 508–518.

Senapong, S., Puripattanavong, J. and Teanpaisan, R. (2014) Anticandidal and antibiofilm activity of *Artocarpus lakoocha* extract. *Songklanakarin Journal of Science and Technology* 36(4), 451–457

Shajib, M.T.I, Kawser, M., Nuruddin Miah, M., Begum, P., Bhattacharjee, L. *et al*. (2013) Nutritional composition of minor indigenous fruits: cheapest nutritional source for the rural people of Bangladesh. *Food Chemistry* 140(3), 466–470. DOI: 10.1016/j.foodchem.2012.11.035.

Shi, J., Arunasalam, K., Yeung, D., Kakuda, Y., Mittal, G. *et al*. (2004) Saponins from edible legumes: chemistry, processing, and health benefits. *Journal of Medicinal Food* 7(1), 67–78. DOI: 10.1089/109662004322984734.

Shin, N.H., Ryu, S.Y., Choi, E.J., Kang, S.H., Chang, I.M. *et al*. (1998) Oxyresveratrol as the potent inhibitor on dopa oxidase activity of mushroom tyrosinase. *Biochemical and Biophysical Research Communications* 243(3), 801–803. DOI: 10.1006/bbrc.1998.8169.

Sudha, M. and Gnanamani, A. (2008) In vitro studies on evaluation of pro and antioxidant properties of unspent tannins. *Current Science* 95, 87–89.

Singhatong, S., Leelarungrayub, D. and Chaiyasut, C. (2010) Antioxidant and toxicity activities of *Artocarpus lakoocha* Roxb, Heartwood extract. *Journal of Medicinal Plant Research* 4(10), 947–953

Suwannalert, P., Povichit, N., Puchadapirom, P. and Junking, M. (2012) Anti-aging activity and non-toxic dose of phytooxyresveratrol from *Artocarpus lakoocha* Roxb. *Tropical Journal of Pharmaceutical Research* 11(1), 69–74. DOI: 10.4314/tjpr.v11i1.9.

Tarbiat, S. (2018) *In vitro* antioxidant, anti-inflammatory and lipid lowering activities of *Artocarpus lakoocha* fruit extract and its implication on treatment of dyslipidemia. *International Journal of Pharmaceutical Sciences Review and Research* 52(2), 75–82

Teanpaisan, R., Senapong, S. and Puripattanavong, J. (2014) *In vitro* antimicrobial and antibiofilm activity of *Artocarpus lakoocha* (Moraceae) extract against some oral pathogens. *Tropical Journal of Pharmaceutical Research* 13(7), 1149–1155

Tomar, A., Srivastava, A. and Dubey, K. (2015) *Need of Conservation Approach of Underutilized Fruits: A Potential of Local Resource. International Day for Biological Diversity - Biodiversity for Sustainable Development*. Uttar Pradesh Biodiversity Board, Uttar Pradesh, India

Vanajakshi, M., Virupaksha, J.H. and Maria, S. (2016) A review on pharmacological actions of *Artocarpus lakoocha* Roxb. *Journal of Pharmacology and Pharmacodynamics* 8(4), 181–184

Veerawat, T., Nithitanakool, S., Junhunkit, T., Ponpanich, L., Nopporn, N. *et al.* (2013) Liposomes: A novel carrier system for *Artocarpus lakoocha* extracts to improve skin whitening. *Journal of Asian Association of Schools of Pharmacy* 2, 243–253.

Vijayan, A., Liju, V.B., Reena, J.V., Parthipan, B. and Renuka, C. (2007) Traditional remedies of Kani Tribes of Kottor reserve forest Agasthavanam, Thiruvanthapuram, Kerela. *Indian Journal of Traditional Knowledge* 6(4), 589–594

Wongkham, S., Wongkham, C., Boonsiri, P., Simasathiansophon, S., Trisonthi, C. *et al.* (1995) Isolectins from seeds of *Artocarpus lakoocha*. *Phytochemistry* 40(5), 1331–1334. DOI: 10.1016/0031-9422(95)00535-f.

22 Black Plum (*Vitex doniana*)

Faith Seke*, Vimbainashe E. Manhivi and Dharini Sivakumar
Phytochemical Food Network Research Group, Department of Crop Sciences, Tshwane University of Technology, Pretoria West, South Africa

Abstract

Vitex doniana (black plum) is an indigenous fruit of the Lamiaceae family that grows in many regions in Africa. Black plum fruit pulp is a good source of vitamin C, phenolic compounds including flavonoids, and minerals such as magnesium, potassium and calcium. Due to the presence of these components, particularly the phenolic compounds, the fruit has received some attention and its consumption has increased. These properties have been linked to a variety of health benefits, making the fruit a possible ingredient in the development of functional foods. This chapter therefore summarizes studies on the nutritional and phytochemical content, antioxidant and antidiabetic capacity of *Vitex doniana* fruit, stressing its importance.

Introduction

The relationship between diet and health has increased consumer demand for more knowledge about nutraceutical-rich diets, including indigenous fruit consumption. Indigenous fruit have been reported to play a crucial role in improving diet quality and food and nutrition security (Mbhenyane, 2017). Globally, shifting from traditional foods to more refined western diets has been linked to an increase in the prevalence of chronic non-communicable diseases. Bioactive constituents of various indigenous fruit have beneficial effects on health and help to avoid these chronic diseases (Liu, 2013). Indigenous fruit have been shown to exhibit antioxidant activity in *in vitro* and *in vivo* studies, as well as in humans, due to the presence of phytochemicals. It is proposed that indigenous fruit antioxidants strengthen the antioxidant defence mechanism (Wang *et al.*, 2010). The phytochemicals in some indigenous fruit may account for increased overall serum antioxidant capacity, as other studies have shown the existence of anthocyanins in human serum, in addition to the vitamins present in the fruit (Mazza *et al.*, 2002; Harasym and Oledzki, 2014).

Vitex doniana (black plum) is a common multifunctional species in Africa, and of considerable socioeconomic significance. Dialect names include Uchakiri (Igbo) Dinyar (Hausa), Galbihi (Fulani), Ori-nla (Yoruba) and black plum (English), from West Africa (Senegal) to East Africa (Somalia) and southern Africa. The fruits are eaten (Kranjac-Berisavljevic and Gandaa, 2013) or used for juices, sweeteners and jams (Ajenifujah-Solebo and Aina, 2011), and sold at local and national markets (Mapongmetsem *et al.*, 2012). In fact, there is a high variation in the organoleptic and morphological features of the fruit. It has many uses with promising economic potential for eradicating poverty in rural and periurban

*Corresponding author: fayeesk@gmail.com

DOI: 10.1079/9781789248067.0022

areas in Africa. Interestingly, *Vitex doniana* fruit is a good source of antioxidant phenolics. Phenolic compounds are a major group of phytochemicals and *Vitex doniana* fruit has been reported to contain flavonoids, stilbenes, tannins and phenolic acids (Aiwonegbe *et al.*, 2018). The phenolics contribute to protection against degenerative diseases, and their effects on health have been mostly attributed to their antioxidant properties. Improving our knowledge on their bioavailability is an important step to understand the possible mechanism of their health effects and to characterize the impact of *Vitex doniana* fruit in human healthy life.

This review summarizes data related to the nutritional and phytochemical composition, antioxidant and antidiabetic capacity of *Vitex doniana* fruit, emphasizing the importance and biological activities of phenolics in *Vitex doniana* for promoting human health.

Taxonomy and Structure

Vitex doniana Sweet belongs to *Vitex* l. genus. The species *Vitex* has been considered to be in the Verbenaceae family by different authors but in recent works it has been transferred into Lamiaceae, based on different evidence. *Vitex doniana* Sweet (syn. *Vitex cuneata* Thonn.) is the most widespread *Vitex* species in Africa (Rani and Sharma, 2013). *Vitex doniana* is a diploid species, with basic chromosome number of ($2n = 32$). *Vitex doniana* is a medium-sized deciduous tree, 8–18 m high, occasionally 25 m, with a heavy rounded crown and a clear bole up to 5 m. Its bark is rough, pale brown or greyish-white, rather smooth with narrow vertical fissures. The bases of old trees have oblong scales.

Leaves

Leaves are opposite, glabrous and are 14–34 cm long. They usually have five leaflets on stalks of 6–14 cm long. Leaflets are distinctly stalked, ovate, obovate-elliptic or oblong, entire, are 8–22 cm long, and 2–9 cm wide.

Flowers

Vitex doniana exhibits hermaphroditism, where both functional male and female organs are in the same flower. The flower petals are white except on the largest lobe, where they are purple, in dense opposite and axillary cymes (Dadjo, 2018). Flowers are numerous and small, blue or violet, 3–12 cm in diameter, only a few being open at a time. Calyx and pedicels are densely hairy.

Fruit

The fruit are oblong, about 3 cm long and green when young, turning purplish black on ripening and with a starchy black pulp (Fig. 22.1). Each fruit contains one hard, conical seed, of 1.5–2 cm long, and 1–1.2 cm wide. The tree fruits between March and April, and its fruit are broadly ellipsoid.

Ecology

The wide geographical distribution of *Vitex doniana* reflects its wide ecological tolerance. *Vitex doniana*

Fig. 22.1. *Vitex doniana* (black plum).

Table 22.1. Nutritional, minerals, sugars and vitamins composition of *Vitex doniana* fruit pulp (Agbede and Ibitoye, 2007; Vunchi *et al.*, 2011).

Proximate composition g/100g (DB)		Minerals and vitamin C mg/100g (DB)		Sugars mg/kg (DB)	
Moisture content	16.66	Ca	30.27	Reducing sugars	474
Ash content	11.50	K	15.70	Dextrose	460
Crude protein	8.24	P	16.50	Fructose	89.3
Crude fibre	0.58	Mn	0.0052	Maltose	750.9
Crude fat	34.62	Fe	5.20	Lactose	632.5
Vitamin A	0.27	Cu	2.70	Sucrose	40,000
Vitamin B1	18.33				
Vitamin B2	4.80	Na	10.40		
Vitamin B6	20.45				
Vitamin C	35.58	Mg	20.10		

DB: dry basis.

exists in a variety of habitats, from forest to savannah, often in tropical conditions and along rivers, at an altitude of up to 2000m. In southern Africa they are more abundant in Zimbabwe, Malawi, Zambia, Mozambique and South Africa (Limpopo). This is a deciduous forest tree found in coastal forest, river and lowland forests and deciduous forests, as high as highland grasslands (Hounkpevi *et al.*, 2018). It occurs in environments with average annual rainfall of 750–2000mm and a high-water table is required. Optimal level growth is achieved in an average temperature between 10°C and 3°C. When it grows in open areas, a very thick bark develops, acting as fire protection. It has a much cleaner, thinner bark in swamps. It occurs predominantly on alluvial soils in a variety of soils of varying origins.

Nutritional Value

The nutritional composition of *Vitex doniana* fruit comprises amino acids, fatty acids, vitamins and minerals (Tables 22.1 and 22.2) (Vunchi *et al.*, 2011; Osum *et al.*, 2013). Reports on the mineral content of *Vitex doniana* have showed that iron, phosphate, zinc, calcium and sodium are present (Vunchi *et al.*, 2011; Amah and Okogeri, 2019). The fruit also have a high value of vitamin C, carbohydrate and lipid. It is suggested that the species could be promoted as a carbohydrate and lipid supplement for cereal-based diets in rural areas, while its modest calcium value could be used for oesteomalacia management. Agbede and Ibitoye (2007) researched both the sugar content and the antinutritional factors in the fruit. Qualitative research on the crude extract has detected the presence of moderate amounts of carbohydrate and small amounts of protein (Nwachukwu and Uzoeto, 2010). The total lipid composition has been reported to be 150mg/g and palmatic acid, mystric acid, stearic acid and arachidic acid were reported in small amounts in the fruit (Glew *et al.*, 1997). These results correspond with those obtained from studies on the fatty acid composition of blackcurrants, strawberries, cranberries and bilberries (Helbig *et al.*, 2008). All essential and non-essential amino acids were reported in *Vitex doniana* fruit,

Table 22.2. Amino acid composition of *Vitex doniana* fruit pulp (mg/g DW) (Glew *et al.*, 1997).

Alanin	1.36	Histidine	0.46
Aspartic acid	2.64	Isoleucine	1.11
Glutamic acid	2.93	Leucine	1.71
Serine	1.45	Lysine	0.38
Arginine	1.53	Methionine	0.17
Cysteine	0.74	Phenylalanine	1.03
Glycine	1.32	Threonine	1.07
Proline	1.65	Valine	1.26
Tyrosine	0.82		

and the results were comparable to the amino acids found in marula and bush plums.

The blackish pulp of the fruit is sweet and edible. It is eaten raw and tastes like prunes (Dadjo *et al.*, 2012). It is also used for juice, syrup, wine, liquor and jam production (Adejumo *et al.*, 2013). Jam formulated from the fruit indicated no significant difference in taste, colour and overall acceptability from commercial plum jam (Ajenifujah-Solebo and Aina, 2011). Black plum jam is also preferred for consistency and spreadability. The pulp syrup can be used as a nutritious sweetener instead of other syrups.

Table 22.3. Phytochemical composition of *Vitex doniana*.

Antinutrient	Fruit	Reference
Tannin (mg/g)	1.44	Agbafor and Nwachukwu (2011)
Phytate (mg/g)	0.32	Agbafor and Nwachukwu (2011)
Saponin (g/100 g)	6.48	Agbafor and Nwachukwu (2011)
Alkaloids (g/100 g)	0.90	Agbafor and Nwachukwu (2011)
Phenols (mg/g)	96.4	Agbafor and Nwachukwu (2011)
Total phenolic content (mg GA/100 g)	94.67	Ampofo-Asiama *et al.* (2020)
Flavonoids (mg/g)	20.8	Agbafor and Nwachukwu (2011)
Total flavonoid content (μg RE/g)	18.03	Ampofo-Asiama *et al.* (2020)

DB, dry basis.

Phytochemical Composition and Properties

The non-nutrient plant chemical compounds or the bioactive components are normally referred to as phytochemicals (*phyto*, from the Greek meaning plant) (Saxena *et al.*, 2013). Phytochemicals prevent infections from disease and damage caused by other plants, herbivores or insects, or add to the fragrance, taste and colour of plants (Saxena *et al.*, 2013). Phytochemicals can be classified into two parts based on their roles and characteristics, the primary and the secondary metabolites. Primary metabolites play a key role in cell metabolism and reproduction and are necessary for plant survival. These metabolites include carbohydrates, lipids, proteins, nucleotides, fatty acids and steroids (Canarini *et al.*, 2019). The medicinal value of a plant depends on its bioactive phytochemical constituents, which produce a definite physiological action in the body. Several researchers have reviewed the phytochemical composition of *Vitex doniana*. The following phytochemicals have been reported to be present in the *Vitex doniana*: fruit alkaloids, flavonoids, saponins, tannins steroids, anthraquinones and terpenes (Table 22.3) (Agbafor and Nwachukwu, 2011). Suleiman and Yusuf (2008) also reported the presence of anthraquinones, flavonoids, saponins, sterols and tannins in the *Vitex doniana* fruit. The medicinal and health-promoting properties of the fruit may be related to their phytochemical constituents. For instance, tannins are known to be useful in the treatment of inflamed or ulcerated tissues and they have been shown to have remarkable activity in cancer prevention and treatment (Aiyegoro and Okoh, 2010). Saponins are known to produce an inhibitory effect on inflammation and have hypocholesteremic and antidiabetic effects. Cardiac glycosides are naturally cardioactive drugs used in the treatment of congestive heart failure and cardiac arrhythmia. GC–MS profiling of *Vitex doniana* fruit pulp extract revealed the presence of flavanol (galangin), flavone (apigenin and chrysin), prenylated flavonoid (abyssinone), phenol (catechol), polyphenol (ellagitannin), saponin (kalopanaxsaponin A, platycodin D and saikosaponin), phytosterol (β-sitosterol, campsterol and ecdysterol), loliolid, lignin (4-keto pinoresinol) and 6-O-aprofuramosyl. Anthraquinones have been confirmed to have astringent, purgative, anti-inflammatory, mild antitumour and bactericidal effects (Agbafor and Nwachukwu, 2011). Alkaloids, the most important among all phytochemicals, are known to be pharmacologically active, and their activity is felt throughout the autonomic nervous system, the blood vessels, diuretic advancement, the respiratory system, the gastrointestinal tract, the uterus and in malignant diseases, infections and malaria. Terpenes or terpenoids have antihepatotoxic characteristics, thus helping to reduce damage to the liver. In addition, they also have antimicrobial or antiseptic properties (Edeoga *et al.*, 2005). It should be noted that steroidal compounds are of importance and interest in pharmacy due to their relationship with sex hormones (Okwu, 2001). Phenolic compounds have also been found to act as antioxidants and possess a variety of medicinal properties, such as anticancer, anti-inflammatory and antidiabetes. It is documented in the literature that flavonoids demonstrate significant antioxidant action on human health and fitness and act either by scavenging or chelating processes. It has been proposed that phenolics are useful for diarrhoea treatment because of their antibacterial activity and effect in the reduction of the gastrointestinal transit time (Ríos and Recio, 2005; Palombo, 2006), supporting the observations registered for the *Vitex* spp. fruits and their use as antidiarrhoeic agents. The aqueous extract of *Vitex doniana* was reported to have inhibited the pig intestine contractions induced by treatment with acetylcholine; in addition, it showed a dose-response effect in the reduction of the gastrointestinal transit time and in the inhibition of diarrhoea induced by castor oil (Suleiman and Yusuf, 2008).

A flavonoid-rich diet has been reported to mitigate neurodegenerative diseases and to avoid natural or irregular loss of cognitive ability (Solanki *et al.*, 2015). As a result, the methanol extract of *Vitex doniana* fruit may be a suitable option for the treatment of neurodegenerative diseases (Tshikalange *et al.*, 2017). The high presence of flavonoids and phenols in the methanolic extract of *Vitex doniana* fruit makes them an essential source of antioxidants which can be used as a food supplement. These fruit extracts may also serve as a significant source of bioactive compounds in the prevention and management of oxidative stress disorders.

Most flavonoids are bound to sugars (glycosides) in plants and they are present sometimes as aglycones. Most flavonoids, except for certain flavan-3-ols and proanthocyanidins, are not totally absorbed and enter the circulatory system (Clifford *et al.*, 2013). The mechanism underlying the flavonoid-mediated reduction of ROS in plants is still unclear. Antioxidant flavonoids are present in the chloroplast, which indicates the role of singlet oxygen scavengers and chloroplast outer membrane stabilizers (Hernández *et al.*, 2009). They have attracted much attention in relation to prevention of degenerative diseases such as atherosclerosis. Antioxidant flavonoids may effectively control key steps of cell growth and differentiation, thus acting regulating the development of the whole plant and individual organs. Flavonoids are usually oxidized by radicals, leading to a more stable, less-reactive radical (Patel, 2008). In this reaction flavonoids stabilize the ROS by reacting to become a radical flavonoid. This is achieved due to high reactive hydroxyl group of the flavonoids as shown below.

$$\text{Flavonoid OH} + R\bullet \rightarrow \text{flavonoid O}\bullet + RH$$

where R• is a free radical and O• is an oxygen free radical.

Vitex doniana fruit is rich in isoflavones which are widely appreciated and are presently the subject of intensive research and discussion, because they help protect against hormone-related abnormalities such as breast and prostate cancer, osteosarocoma, lung carcinoma and ovarian cancer (Reiter *et al.*, 2009).

Anthocyanins are a class of flavonoids with antioxidant potential. They are distinct from other flavonoids due to their ability to form flavylium cations (Mazza *et al.*, 2002). One of these is peonidin, responsible for the colour found in *Vitex doniana* fruit (purplish blue, also influenced by the abundant supply of the hydroxyl group). Anthocyanins may have a significant chemopreventive effect due to their antioxidant properties in scavenging reactive oxygen and reactive nitrogen species or by chelating trace metals implicated in free radical formation (Kampa *et al.*, 2007). As reported by Klotz and Sies (2003) some anthocyanins scavenge superoxides directly, as well as peroxynitrite. Certain flavonoids can serve as antioxidants by inhibiting free radical enzyme-generating activity such as xanthine oxidase and nitric oxide synthase. Quercetin, rutin and silibin have been shown to inhibit xanthine oxidase activity whilst silibin was reported to prevent dependent dose of nitric oxide (Panche *et al.*, 2016). Quercetin is a flavonol, known to protect DNA from oxidative damage resulting from the attack of $^{\bullet}OH$, H_2O_2, and $O_2^{\bullet}$ on DNA oligonucleotides. However, quercetin is reported to be a carcinogenic agent at high concentrations of cupric ion, by improving DNA disruption via ROS (Nimse and Pal, 2015). Hence, when determining the protective or degenerative effects of quercetin and other bioflavonoids, it is very important to note the concentration of chelating metal ions such as copper or iron.

Effects of Postharvest Technology on Phytochemicals

Vitex doniana fruit is seasonal and highly perishable due to its moisture content, hence the need for postharvest techniques that can help in preserving the fruit. Also, because *Vitex doniana* fruit is rich in antioxidants, appropriate methods of postharvest treatment are necessary for maintaining these bioactive phytochemicals. Drying methods, storage temperatures, storage duration and packaging methods have been reported to affect the content of some bioactive compounds (Tseng and Zhao, 2012). Drying is a basic method of removing excess moisture from a natural or industrial product to achieve the required specification of the moisture content. This process is energy intensive, reduces fruit moisture content to a certain level and hence inhibits the activity of enzymes, bacteria, yeasts and moulds. Without spoilage, fruits can thus be processed and preserved for a long time. It has been reported that different drying methods have different impacts on the nutritional and antinutritional properties of dried fruit (Bonazzi and Dumoulin, 2011). In some research it has been suggested that sun-drying reduces the loss of anthocyanins compared to hot-air drying (Norkaew *et al.*, 2017). In a study by Wojdyło *et al.* (2009) the effect of different drying methods on the bioactive compounds in strawberries suggested that drying destroyed anthocyanins, flavanols and ascorbic acid, and there was a significant decrease in antioxidant activity. Also, conventional and vacuum-drying decreased antioxidant activity, however vacuum-microwave drying showed the most favourable results. In another study, freeze-drying improved retention of phytochemicals in organic berries during processing and in some cases it even increased the concentration of phytochemicals (Sablani *et al.*, 2011). In the same study it was also observed that blanching treatment prior to air-drying increased the retention of phytochemicals in dried berries.

Antioxidant Properties

It is well established that plants develop several secondary metabolites to interact with the surrounding environment. The main class of secondary metabolites is phenol alkaloids. Phenols are usually categorized either as simple low molecular weight phenols or as high molecular weight polyphenols (Hagerman *et al.*, 1998). The quantity and type of polyphenols produced by plants varies considerably between species (El Gharras, 2009). In addition, polyphenols provide plant tissue regeneration by oxidative polymerization of enzymes after mechanical damage. These compounds play an important role in plant growth and development, providing effective defence against pathogens and pests, as well as contributing to fruit and vegetable colour, sensory characteristics and nutritional properties (Naczk and Shahidi, 2006). These compounds play a critical role in plant growth and reproduction, providing trenchant protection against pathogens and pests, as well as contributing to fruit and vegetable colour, sensory characteristics and nutritional properties (Škerget *et al.*, 2005).

Hamzah *et al.* (2013) DPPH scavenging activities of the methanolic extract of fruits of *Vitex doniana* revealed that the fruit is a DPPH radical scavenger in a dose-dependent manner. The DPPH scavenging activity of *Vitex doniana* fruit extract is higher than the DPPH scavenging activity of most commonly consumed fruit, as shown in Table 22.4. This significant scavenging ability in *Vitex doniana* fruit extract could be attributed to its high content of phenols and flavonoids. Flavonoids and phenolic acids are the most important groups of secondary metabolites and bioactive compounds in plants (Ghasemzadeh and Ghasemzadeh, 2011). Flavonoids play significant role in protecting biological systems against the harmful effects of oxidative processes on macromolecules, such as carbohydrates, proteins, lipids and DNA. Reducing power is a novel antioxidant defence

Table 22.4. DPPH scavenging activity of methanolic extracts of *Vitex doniana* fruit and other commonly consumed fruit.

Fruit	DPPH scavenging activity %	Reference
Vitex doniana fruit	45	Hamzah *et al.*, 2013
Red grapefruit	34.3	Park *et al.*, 2015
Peaches	11.9	Park *et al.*, 2015
Bananas	33.1	Park *et al.*, 2015
Lemons	35.1	Park *et al.*, 2015
Oranges	34.9	Park *et al.*, 2015

mechanism in which electron transfer and hydrogen abstraction are the two mechanisms available to affect this property (Gülçin *et al.*, 2005). The reductive powers of the methanolic extracts of fruit of *Vitex doniana* were assessed based on their ability to reduce Fe (III) to Fe (II). The results revealed that the methanolic extracts of the fruit of *Blighia sapida* had a higher reducing power than methanolic extract of *Vitex doniana* at a concentration of 100 μg/ml. Since the reducing power of a compound is related to its electron transfer ability and hence may serve as an indicator of its potential activity, the antioxidant activity of this extract might partially be due to its reducing ability. The Frap values of ripe fruit of *Vitex doniana* (545 $\mu MFe^{2+}/100\,g$) were shown to be significantly higher than ascorbic acid and gallic acid, 417.71 and 442.80 μMFe^{2+} / 100 g respectively. These observations were attributed to the high phenolic and flavonoid composition present in *Vitex doniana*.

In a study by Ajiboye, 2015 the possible hepatoprotective potential of *Vitex doniana* fruit methanolic extract was examined with the aim to provide possible insights to its potential in the management of diseases associated with oxidative stress. Data from the study showed that methanol extract of *Vitex doniana* fruit stalled paracetamol mediated protein oxidation, lipid peroxidation and DNA fragmentation. This may be by acting as a free radical and reactive oxygen species scavenger or by antioxidant enzymes that enhance the activity of inherent phytochemical components such as β-sitosterol, platycodin D, apigenin, saikosaponin, chrysin and ellagitanin in the extract. As a result, the fruit extract exhibited excellent potential to be used as a dietary supplement/functional food because of its prophylactic role.

Vitamin C (ascorbic acid)

Ascorbic acid is a water-soluble antioxidant. It also serves as a chain breaker to stop the chain reaction to lipid peroxidation. In this reaction, to become radical ascorbate it donates an electron to the lipid radical (LOO•). Two ascorbate radical molecules can react quickly to generate an ascorbate molecule and a dehydroascorbate molecule which has no scavenging activity. Dehydroascorbate can be repurposed to ascorbate by adding two oxidoreductase catalysed electrons. In addition, ascorbate in cell membranes can react with GSH to regenerate vitamin E (Duarte and Lunec, 2005). The vitamin C content of *Vitex doniana* fruit (35 mg/100 g) when compared with fruit like mangos (22 mg/100 g), peaches (8 mg/100 g), bananas (17 mg/100 g), red plums (4 mg/100 g) and persimmons (12 mg/100 g), pineapple, lime and apricot (25 mg/100 g), and apples (6 mg/100 g) (Hiwilepo-van Hal *et al.*, 2012; Park *et al.*, 2015) suggests that it has a higher vitamin C.

Oxidative Stress and Human Diseases

Oxidative stress is the uncontrolled accumulation of reactive oxygen species (ROS) in cells and tissues, and therefore cannot be neutralized by the antioxidant method (Rahman *et al.*, 2012). An imbalance in this protective mechanism may cause cellular molecules such as DNA, proteins and lipids to suffer damage. ROS are usually released in limited amounts within the body, and are significant compounds involved in controlling processes that involve maintaining cell homeostasis and functions such as signal transduction, gene expression and receptor activation (Valko *et al.*, 2016). Proteins, lipids and DNA/RNA are the main targets of oxidative stress and changes in these molecules can increase the chances of mutagenesis. Overproduction of ROS/RNS, especially over an extended period, may cause damage to the cellular structure and functions, and may trigger somatic mutations and preneoplastic and neoplastic transformations (Khansari *et al.*, 2009). Excessive development of ROS in cells and tissues may be detrimental if not promptly removed. Indeed, excessive production of ROS/RNS can cause permanent damage to cells that results in necrotic and apoptotic processes leading to cell death. Most illnesses, including atherosclerosis (Linke *et al.*, 2005), cancer (Husain *et al.*, 2001), neurodegenerative disorders such as Alzheimer's disease (Selley *et al.*, 2002), Parkinson's disease and Huntington's disease (Chaitanya *et al.*, 2010) are linked to oxidative stress. Oxidative stress also contributes to hyperoxia and irradiation disruption to the tissues. Studies suggest that oxidative stress plays a substantial role in the pathogenesis and progression of metabolic syndrome-related illnesses such as obesity, hypertension, diabetes, dyslipidemia and cardiovascular-related diseases. Cancer is another ROS-related illness, as it was proposed that ROS induce oncogenes such as Jun and Fos whose overexpression is directly linked with lung cancer (Szabo *et al.*, 2001). In lung cancers, ROS can genetically manipulate p53 thus losing its apoptosis role and acting as an oncogene. Previously, the occurrence of gastric cancer was thought to be related to higher production of ROS and RNS from human stomach infection induced by *Helicobacter pylori* (Vasavidevi *et al.*, 2006). Excess ROS contributes to urolithiasis in human kidneys. ROS has also been reported to cause damage to cell components in cartilage leading to osteoarthritis and has been shown to be involved in damaging the pancreatic islet cells. Moreover, hyperglycemia triggers the development of ROS in human kidney tubular and mesangial cells, causing functional and structural changes in glomeruli which cause diabetic nephropathy (Verzola *et al.*, 2004).

Antidiabetic Properties

Type 2 diabetes is associated with oxidative stress-mediated complications and hyperglycemia, that are known to be significant underlying factors for disease pathogenesis (Brownlee, 2001). As a result, agents which can scavenge free radicals and have strong antioxidants as well as inhibitory properties of α-glucosidase and amylase play a significant role in the treatment and prevention of T2D and related complications. It has been strongly suggested that the regular consumption of plants with high anthocyanin content is beneficial in the prevention and management of diabetes (Fig. 22.2). Postprandial excursions in blood glucose levels reach a maximum within the first 60 minutes after a carbohydrate-rich meal. Therefore, to prevent the negative side effects of postprandial hyperglycemia, a fast-acting postprandial blood glucose lowering agent is desirable. Interestingly, in a study by (Ibrahim *et al.*, 2019) the fruit extracts of *Vitex doniana* at low and high doses had fast inhibitory actions against postprandial hyperglycemia induced by maltose and sucrose within the first 30 minutes. The *Vitex doniana* extract induced a stronger tolerance of maltose, indicating the existence of different inhibitors of maltase. This concept was further supported by the finding that a greater increase in the tolerance of maltose was obtained at a lower dose. This observation was attributed to the fact that at higher dose, the extract contains indigenous absorbable sugars that may overwhelm the maltase inhibitory activity. The fruits of *Vitex doniana* have been reported to be rich in sugars, tannins, phytin and other phytochemicals (Ganapaty and Vidyadhar, 2005; Vunchi *et al.*, 2011). These phytochemicals, and other probably uncharacterized fruit constituents, may be individually or synergistically responsible for the inhibiting activity of maltase observed. It was also concluded that the use of *Vitex doniana* the development of a standardized antihyperglycemic supplements is possible.

Conclusion

This review shows that *Vitex doniana* fruit can supplement the diet of many rural residents by supplying vital micronutrients and health benefits, as well as being an alternative to income, especially during times of famine. Nevertheless, very little research has been done on the biological activities of *Vitex doniana* fruit, much of the work being centred on the bark, leaves

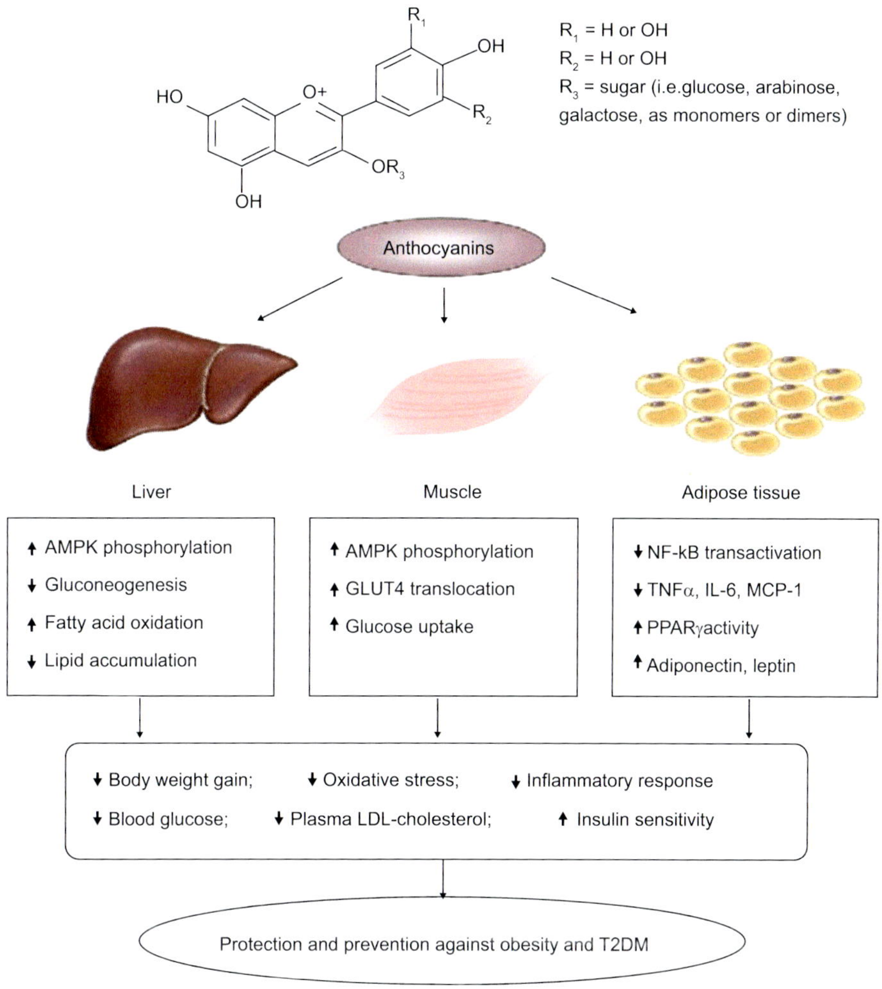

Fig. 22.2. Presumed biological mechanisms underlying the effect of anthocyanins on obesity and type 2 diabetes (Guo and Ling, 2015).

and roots. It is also necessary to determine the specific role of each phenolic compound and to conduct detailed biological intervention studies. Further studies are needed to determine the *in vivo* potential of these fruits for the management of human diseases caused by oxidative stress. Studies on bioaccessibility and bioavailability and *in vitro* digestibility of bioactive compounds in *Vitex doniana* fruit are needed. Also, there is a need to pay attention to the added value of processing *Vitex doniana* fruit. Significantly reducing nutrient losses, antioxidants, organoleptic properties and the content of potential toxic alkaloids during processing is important to produce a nutritious fruit product, since processing affects the overall quality of fruit products. Thus, optimizing state of the art processing methods and assays of nutritional and sensory quality of *Vitex doniana* fruit products is crucial for the implementation of preservation procedures and for the consequent promotion of *Vitex doniana* fruit consumption. Therefore, the improvement of food production processes by optimizing conservation techniques as a sustainable solution to malnutrition in rural transition areas needs to be investigated.

References

Kranjac-Berisavljevic, G. and Gandaa, B.Z. (2013) Importance of bush grape (Vitex spp..) as food and medicinal plant to the Dagarti speaking people of the Upper West Region of Northern Ghana. *Acta Horticulturae* 979, 669–673. DOI: 10.17660/ActaHortic.2013.979.74.

Adejumo, A.A., Alaye, S.A., Ajagbe, R.O., Abi, E.A. and Adedokun, F.T. (2013) Nutritional and anti-nutritional composition of black plum (*Vitex doniana*). *Journal of Natural Sciences Research* 3(12), 144–148.

Agbafor, K.N. and Nwachukwu, N. (2011) Phytochemical analysis and antioxidant property of leaf extracts of *Vitex doniana* and *Mucuna pruriens*. *Biochemistry Research International* 2011, 459839. DOI: 10.1155/2011/459839.

Agbede, J.O. and Ibitoye, A.A. (2007) Chemical composition of black plum (*Vitex doniana*): an under-utilized fruit. *Journal of Food Agriculture and Environment* 5(2), 95.

Aiwonegbe, A.E., Iyasele, J.U. and Izevbuwa, N.O. (2018) Proximate composition, phytochemical and antimicrobial screening of the methanol and acetone extracts of *Vitex doniana* fruit pulp. *Ife Journal of Science* 20(2), 317. DOI: 10.4314/ijs.v20i2.12.

Aiyegoro, O.A. and Okoh, A.I. (2010) Preliminary phytochemical screening and *in vitro* antioxidant activities of the aqueous extract of *Helichrysum longifolium* DC. *BMC Complementary and Alternative Medicine* 10(1), 1–8. DOI: 10.1186/1472-6882-10-21.

Ajenifujah-Solebo, S.O. and Aina, J.O. (2011) Physico-chemical properties and sensory evaluation of jam made from black-plum fruit (*Vitex doniana*). *African Journal of Food, Agriculture, Nutrition and Development* 11(3). DOI: 10.4314/ajfand.v11i3.66629.

Ajiboye, T.O. (2015) Standardized extract of Vitex doniana sweet stalls protein oxidation, lipid peroxidation and DNA fragmention in acetaminophen-induced hepatotoxicity. *Journal of Ethnopharmacology* 164, 273–282. DOI: 10.1016/j.jep.2015.01.026.

Amah, U.J. and Okogeri, O. (2019) Nutritional and phytochemical properties of wild black plum (*Vitex doniana*) seed from Ebonyi State. *International Journal of Horticulture, Agriculture and Food Science* 3(1), 32–36. DOI: 10.22161/ijhaf.3.1.5.

Ampofo-Asiama, J., Debrah, D.T., Quaye, B., Abakah, B., Owusu, E.T. *et al.* (2020) Phytochemical and physicochemical properties of some underutilized tropical fruits. *African Journal of Agriculture and Food Science* 3(1), 1–8.

Bonazzi, C. and Dumoulin, E. (2011) Quality changes in food materials as influenced by drying processes. *Modern Drying Technology* 3, 1–20.

Brownlee, M. (2001) Biochemistry and molecular cell biology of diabetic complications. *Nature* 414(6865), 813–820. DOI: 10.1038/414813a.

Canarini, A., Kaiser, C., Merchant, A., Richter, A. and Wanek, W. (2019) Root exudation of primary metabolites: mechanisms and their roles in plant responses to environmental stimuli. *Frontiers in Plant Science* 10, 157. DOI: 10.3389/fpls.2019.00157.

Chaitanya, G.V., Steven, A.J. and Babu, P.P. (2010) PARP-1 cleavage fragments: signatures of cell-death proteases in neurodegeneration. *Cell Communication and Signaling* 8(1), 31. DOI: 10.1186/1478-811X-8-31.

Clifford, M.N., van der Hooft, J.J.J. and Crozier, A. (2013) Human studies on the absorption, distribution, metabolism, and excretion of tea polyphenols. *The American Journal of Clinical Nutrition* 98(6 Suppl), 1619S-1630S. DOI: 10.3945/ajcn.113.058958.

Dadjo, C. (2018) *Assessment of nutritional variation and establishment of an efficient tissue culture propagation protocol for* Vitex doniana *Sweet (lamiaceae) from Benin (Western Africa)*. Doctoral dissertation, JKUAT-PAUSTI.

Dadjo, C., Assogbadjo, A.E., Fandohan, B., Glèlè Kakaï, R., Chakeredza, S. *et al.* (2012) Uses and management of black plum (Vitex doniana Sweet) in Southern Benin. *Fruits* 67(4), 239–248. DOI: 10.1051/fruits/2012017.

Duarte, T.L. and Lunec, J. (2005) Review: When is an antioxidant not an antioxidant? A review of novel actions and reactions of vitamin C. *Free Radical Research* 39(7), 671–686. DOI: 10.1080/10715760500104025.

Edeoga, H.O., Okwu, D.E. and Mbaebie, B.O. (2005) Phytochemical constituents of some Nigerian medicinal plants. *African Journal of Biotechnology* 4(7), 685–688. DOI: 10.5897/AJB2005.000-3127.

El Gharras, H. (2009) Polyphenols: food sources, properties and applications - a review. *International Journal of Food Science & Technology* 44(12), 2512–2518. DOI: 10.1111/j.1365-2621.2009.02077.x.

Ganapaty, S. and Vidyadhar, K.N. (2005) Phytoconstituents and biological activities of Vitex - a review. *Journal of Natural Remedies* 5(2), 75–95.

Ghasemzadeh, A. and Ghasemzadeh, N. (2011) Flavonoids and phenolic acids: Role and biochemical activity in plants and human. *Journal of Medicinal Plants Research* 5(31), 6697–6703. DOI: 10.5897/JMPR11.1404.

Glew, R.H., Vanderjagt, D.J., Lockett, C., Grivetti, L.E., Smith, G.C. *et al.* (1997) Amino acid, fatty acid, and mineral composition of 24 indigenous plants of Burkina Faso. *Journal of Food Composition and Analysis* 10(3), 205–217. DOI: 10.1006/jfca.1997.0539.

Gülçin, I., Berashvili, D. and Gepdiremen, A. (2005) Antiradical and antioxidant activity of total anthocyanins from *Perilla pankinensis* Decne. *Journal of Ethnopharmacology* 101(1–3), 287–293. DOI: 10.1016/j.jep.2005.05.006.

Guo, H. and Ling, W. (2015) The update of anthocyanins on obesity and type 2 diabetes: experimental evidence and clinical perspectives. *Reviews in Endocrine Metabolic Disorders* 16(1), 1–13. DOI: 10.1007/s11154-014-9302-z.

Hagerman, A.E., Riedl, K.M., Jones, G.A., Sovik, K.N., Ritchard, N.T. *et al.* (1998) High molecular weight plant polyphenolics (tannins) as biological antioxidants. *Journal of Agricultural and Food Chemistry* 46(5), 1887–1892. DOI: 10.1021/jf970975b.

Hamzah, R.U., Egwim, E.C., Kabiru, A.Y. and Muazu, M.B. (2013) Phytochemical and in vitro antioxidant properties of the methanolic extract of fruits of Blighia Sapida, Vitellaria Paradoxa and Vitex Doniana. *Oxidants Antioxidants in Medical Science* 2(3), 215–221. DOI: 10.5455/oams.090513.

Harasym, J. and Oledzki, R. (2014) Effect of fruit and vegetable antioxidants on total antioxidant capacity of blood plasma. *Nutrition*30(5), 511–517. DOI: 10.1016/j.nut.2013.08.019.

Helbig, D., Böhm, V., Wagner, A., Schubert, R. and Jahreis, G. (2008) Berry seed press residues and their valuable ingredients with special regard to black currant seed press residues. *Food Chemistry* 111(4), 1043–1049. DOI: 10.1016/j.foodchem.2008.05.017.

Hernández, I., Alegre, L., Van Breusegem, F. and Munné-Bosch, S. (2009) How relevant are flavonoids as antioxidants in plants? *Trends in Plant Science* 14(3), 125–132. DOI: 10.1016/j.tplants.2008.12.003.

Hiwilepo-van Hal, P., Bosschaart, C., van Twisk, C., Verkerk, R. and Dekker, M. (2012) Kinetics of thermal degradation of vitamin C in marula fruit (*Sclerocarya birrea* subsp. caffra) as compared to other selected tropical fruits. *LWT - Food Science and Technology* 49(2), 188–191. DOI: 10.1016/j.lwt.2011.12.038.

Hounkpevi, A., Kouassi, E.K. and Kakai, R.G. (2018) Effects of climatic variability and local environment patterns on the ecology and population structure of the multipurpose plant species, *Vitex doniana* Sweet (Lamiaceae) in Benin. *Tropical Ecology* 59(1), 129–143.

Husain, S.S., Szabo, I.L., Pai, R., Soreghan, B., Jones, M.K. *et al.* (2001) MAPK (ERK2) kinase—a key target for NSAIDs-induced inhibition of gastric cancer cell proliferation and growth. *Life Sciences* 69(25–26), 3045–3054. DOI: 10.1016/S0024-3205(01)01411-4.

Ibrahim, M.A., Isah, M.B., Yunusa, I., Hamza, S.A., Isah, S. *et al.* (2019) Phytochemical composition of *Vitex doniana*. *Analytical Chemistry Letters* 9(6), 863–875. DOI: 10.1080/22297928.2020.1722221.

Kampa, M., Nifli, A.P., Notas, G. and Castanas, E. (2007) Polyphenols and cancer cell growth. *Reviews of Physiology, Biochemistry and Pharmacology* 159, 79–113. DOI: 10.1007/112_2006_0702.

Khansari, N., Shakiba, Y. and Mahmoudi, M. (2009) Chronic inflammation and oxidative stress as a major cause of age-related diseases and cancer. *Recent Patents on Inflammation & Allergy Drug Discovery* 3(1), 73–80. DOI: 10.2174/187221309787158371.

Klotz, L.O. and Sies, H. (2003) Defenses against peroxynitrite: selenocompounds and flavonoids. *Toxicology Letters* 140–141, 125–132. DOI: 10.1016/s0378-4274(02)00511-8.

Linke, A., Adams, V., Schulze, P.C., Erbs, S., Gielen, S. *et al.* (2005) Antioxidative effects of exercise training in patients with chronic heart failure: increase in radical scavenger enzyme activity in skeletal muscle. *Circulation* 111(14), 1763–1770. DOI: 10.1161/01.CIR.0000165503.08661.E5.

Liu, R.H. (2013) Health-promoting components of fruits and vegetables in the diet. *Advances in Nutrition* 4(3), 384S-392S. DOI: 10.3945/an.112.003517.

Mapongmetsem, P.M., Djoumessi, M.C., Yemele, M.T., Fawa, G., Doumara, G. *et al.* (2012) Domestication of *Vitex doniana* (Verbenaceae): influence of substrate type, hormonal stimulation of leaf area and position of the node on the stem cuttings. *Journal of Agriculture and Environment for International Development* 106(1), 23–45.

Mazza, G., Kay, C.D., Cottrell, T. and Holub, B.J. (2002) Absorption of anthocyanins from blueberries and serum antioxidant status in human subjects. *Journal of Agricultural and Food Chemistry* 50(26), 7731–7737. DOI: 10.1021/jf020690l.

Mbhenyane, X.G. (2017) Indigenous foods and their contribution to nutrient requirements. *South African Journal of Clinical Nutrition* 30(4), 5–7.

Naczk, M. and Shahidi, F. (2006) Phenolics in cereals, fruits and vegetables: occurrence, extraction and analysis. *Journal of Pharmaceutical and Biomedical Analysis* 41(5), 1523–1542. DOI: 10.1016/j.jpba.2006.04.002.

Nimse, S.B. and Pal, D. (2015) Free radicals, natural antioxidants, and their reaction mechanisms. *Research Advances* 5(35), 27986–28006. DOI: 10.1039/C4RA13315C.

Norkaew, O., Boontakham, P., Dumri, K., Noenplab, A.N.L., Sookwong, P. *et al.* (2017) Effect of post-harvest treatment on bioactive phytochemicals of Thai black rice. *Food Chemistry* 217, 98–105. DOI: 10.1016/j.foodchem.2016.08.084.

Nwachukwu, E. and Uzoeto, H.O. (2010) Antimicrobial activities of leaf of *Vitex doniana* and *Cajanus cajan* on some bacteria. *Researcher* 2(3), 37–47.

Okwu, D.E. (2001) Evaluation of the chemical composition of indigenous spices of southeastern Nigeria. *Global Journal of Pure Applied Science* 7, 455–459.

Osum, F.I., Okonkwo, T.M. and Okafor, G.I. (2013) Effect of processing methods on the chemical composition of *Vitex doniana* leaf and leaf products. *Food Science & Nutrition* 1(3), 241–245. DOI: 10.1002/fsn3.31.

Palombo, E.A. (2006) Phytochemicals from traditional medicinal plants used in the treatment of diarrhoea: modes of action and effects on intestinal function. *Phytotherapy Research* 20(9), 717–724. DOI: 10.1002/ptr.1907.

Panche, A.N., Diwan, A.D. and Chandra, S.R. (2016) Flavonoids: an overview. *Journal of Nutritional Science* 5, e47. DOI: 10.1017/jns.2016.41.

Park, Y.-S., Im, M.H., Ham, K.-S., Kang, S.-G., Park, Y.-K. *et al.* (2015) Quantitative assessment of the main antioxidant compounds, antioxidant activities and FTIR spectra from commonly consumed fruits, compared to standard kiwi fruit. *LWT - Food Science and Technology* 63(1), 346–352. DOI: 10.1016/j.lwt.2015.03.057.

Patel, J.M. (2008) A review of potential health benefits of flavonoids. *Lethbridge Undergraduate Research Journal (LURJ)* 3(2).

Rahman, T., Hosen, I., Islam, M.M.T. and Shekhar, H.U. (2012) Oxidative stress and human health. *Advances in Bioscience and Biotechnology* 03(7), 997–1019. DOI: 10.4236/abb.2012.327123.

Rani, A. and Sharma, A. (2013) The genus *Vitex*: A review. *Pharmacognosy Reviews* 7(14), 188–198. DOI: 10.4103/0973-7847.120522.

Reiter, E., Reiter, E., Beck, V., Medjakovic, S. and Jungbauer, A. (2009) Isoflavones are safe compounds for therapeutical applications - evaluation of *in vitro* data. *Gynecological Endocrinology* 25(9), 554–580. DOI: 10.1080/09513590802596461.

Ríos, J.L. and Recio, M.C. (2005) Medicinal plants and antimicrobial activity. *Journal of Ethnopharmacology* 100(1–2), 80–84. DOI: 10.1016/j.jep.2005.04.025.

Sablani, S.S., Andrews, P.K., Davies, N.M., Walters, T., Saez, H. *et al.* (2011) Effects of air and freeze drying on phytochemical content of conventional and organic berries. *Drying Technology* 29(2), 205–216. DOI: 10.1080/07373937.2010.483047.

Saxena, M., Saxena, J., Nema, R., Singh, D. and Gupta, A. (2013) Phytochemistry of medicinal plants. *Journal of Pharmacognosy and Phytochemistry* 1(6).

Selley, M.L., Close, D.R. and Stern, S.E. (2002) The effect of increased concentrations of homocysteine on the concentration of (E)-4-hydroxy-2-nonenal in the plasma and cerebrospinal fluid of patients with Alzheimer's disease. *Neurobiology of Aging* 23(3), 383–388. DOI: 10.1016/s0197-4580(01)00327-x.

Škerget, M., Kotnik, P., Hadolin, M., Hraš, A.R., Simonič, M. *et al.* (2005) Phenols, proanthocyanidins, flavones and flavonols in some plant materials and their antioxidant activities. *Food Chemistry* 89(2), 191–198. DOI: 10.1016/j.foodchem.2004.02.025.

Solanki, I., Parihar, P., Mansuri, M.L. and Parihar, M.S. (2015) Flavonoid-based therapies in the early management of neurodegenerative diseases. *Advances in Nutrition* 6(1), 64–72. DOI: 10.3945/an.114.007500.

Suleiman, M.M. and Yusuf, S. (2008) Antidiarrheal activity of the fruits of Vitex doniana in laboratory animals. *Pharmaceutical Biology* 46(6), 387–392. DOI: 10.1080/13880200802055826.

Tseng, A. and Zhao, Y. (2012) Effect of different drying methods and storage time on the retention of bioactive compounds and antibacterial activity of wine grape pomace (Pinot Noir and Merlot). *Journal of Food Science* 77(9), H192-201. DOI: 10.1111/j.1750-3841.2012.02840.x.

Tshikalange, T.E., Modishane, D.C. and Tabit, F.T. (2017) Antimicrobial, antioxidant, and cytotoxicity properties of selected wild edible fruits of traditional medicinal plants. *Journal of Herbs, Spices & Medicinal Plants* 23(1), 68–76. DOI: 10.1080/10496475.2016.1261387.

Valko, M., Jomova, K., Rhodes, C.J., Kuča, K. and Musílek, K. (2016) Redox-and non-redox-metal-induced formation of free radicals and their role in human disease. *Archives of Toxicology* 90(1), 1–37. DOI: 10.1007/s00204-015-1579-5.

Vasavidevi, V. B, Suryakar, A.N., Ankush, R.D., Katkam, R.V. and Deshpande, K.H. (2006) Depleted nitrite and enhanced oxidative stress in urolithiasis. *Indian Journal of Clinical Biochemistry* 21(2), 177–180. DOI: 10.1007/BF02912938.

Verzola, D., Bertolotto, M.B., Villaggio, B., Ottonello, L., Dallegri, F. *et al.* (2004) Oxidative stress mediates apoptotic changes induced by hyperglycemia in human tubular kidney cells. *Journal of the American Society of Nephrology* 15(1 suppl), S85–7. DOI: 10.1097/01.asn.0000093370.20008.bc.

Vunchi, M.A., Umar, A.N., King, M.A., Liman, A.A., Jeremiah, G. *et al.* (2011) Proximate, vitamins and mineral composition of *Vitex doniana* (black plum) fruit pulp. *Nigerian Journal of Basic and Applied Sciences* 19(1). DOI: 10.4314/njbas.v19i1.69352.

Wang, J., Zhao, J., Liu, H., Zhou, L., Liu, Z. *et al.* (2010) Chemical analysis and biological activity of the essential oils of two valerianaceous species from China: Nardostachys chinensis and *Valeriana officinalis*. *Molecules* 15(9), 6411–6422. DOI: 10.3390/molecules15096411.

Wojdyło, A., Figiel, A. and Oszmiański, J. (2009) Effect of drying methods with the application of vacuum microwaves on the bioactive compounds, color, and antioxidant activity of strawberry fruits. *Journal of Agricultural and Food Chemistry* 57(4), 1337–1343. DOI: 10.1021/jf802507j.

23 Natal Plum (*Carissa macrocarpa*)

Faith Seke[1]*, Vimbainashe E. Manhivi[2] and Dharini Sivakumar[2]
[1]Department of Horticulture, Tshwane University of Technology, Pretoria West, South Africa; [2]Phytochemical Food Network Group, Department of Crop Sciences, Tshwane University of Technology, Pretoria West, South Africa

Abstract
The fruit of *Carissa macrocarpa* (Natal plum) is one of the indigenous berry fruit commonly found in KwaZulu-Natal in South Africa. It is recognized for its high concentration of phenolic compounds, especially anthocyanins, which have a phenolic structure that contributes to their antioxidant activity. The presence of these bioactive compounds has led to the recognition and increase of consumption of the fruit due to the health benefits that are linked to it. These characteristics make this berry a potential ingredient in the development of functional foods. This chapter therefore discusses the fruit in terms of its physiology, uses, nutritional and phytochemical composition, and the changes that occur during ripening, storage and processing. Furthermore, it reports on the findings of its biological activity, as well as research efforts into its intake factors and human health benefits.

Introduction

For many years, indigenous fruit have contributed to food security, nutrition, health and income in rural communities throughout sub-Saharan Africa. It has been reported that incorporating indigenous fruit into the diet can help with a variety of nutrition-related issues (Bvenura and Sivakumar, 2017). *Carissa macrocarpa* (Natal plum) is a common multifunctional indigenous fruit (Moodley *et al.*, 2011), mainly sold regionally and grown by residents. The fruit is usually found in summer, but if the seasons are favourable, it can also be produced at any time of year. Over the years Natal plum fruit has been consumed raw, added to deserts or made into a jam (Abbas *et al.*, 2014). Despite its diverse uses and potential as a greater crop, the Natal plum is still not produced commercially. Natal plum is reported to contain high levels of vitamin C and all age groups, including children, love it. Natal plum has the potential to contribute to food security in rural areas (Moodley *et al.*, 2012).

Carissa macrocarpa can be an essential component of the diet due to the presence of bioactive compounds. Phenolic compounds are among the chemical components that have been identified in the fruit, namely flavonoids (anthocyanins, flavonols, flavones, flavanols, flavanones and isoflavonoids), tannins, ascorbic acid (vitamin C) and phenolic acids (Ndou *et al.*, 2019). The phenolic compounds contribute to protection against degenerative diseases, and their effects on health have been commonly attributed to their antioxidant properties (Seeram, 2008). Improvement in knowledge on their bioavailability is an important step to understand the possible mechanisms of action and their impact on human health (Szajdek and Borowska, 2008).

*Corresponding author: fayeesk@gmail.com

 Handbook of Phytonutrients in Indigenous Fruits and Vegetables (eds D. Sivakumar *et al.*)
DOI: 10.1079/9781789248067.0023

Due to the positive relationship between diet and health, the market demand for more knowledge on healthier foods, including indigenous fruit, has been enhanced (Paredes-López *et al.*, 2010). The consumption of indigenous fruit has been related to the prevention of chronic diseases due to the bioactive compounds they contain. Metabolic behaviour of bioactive compounds is manifested primarily by their tendency to scavenge reactive oxygen species (ROS) (Szajdek and Borowska, 2008). Popular bioactive compounds found in fruit include ascorbic acid (vitamin C), phenolic compounds and β-carotene. Consumers favour natural plant-source antioxidants because of their efficacy. In addition, indigenous fruits also provide customers with essential nutrients such as vitamins, minerals, dietary fibres, proteins and sugars. The Food and Agricultural Organization (FAO) advocates dietary diversity through the restoration and sustainable use of biodiversity for nutrition and agriculture (Achaglinkame *et al.*, 2019). It can be concluded that Natal plum fruit can be part of a daily basic diet as a source of nutritional compounds with high bioactive capacity.

This chapter provides a comprehensive literature summary of studies that have been done on the uses, nutritional and phytochemical composition, bioavailability and bioaccessibility of Natal plum fruit.

Table 23.1. Botanical classification of Natal plum (Patel, 2013).

Kingdom	Plantae
Subdivision	Angiosperms
Class	Asterids
Order	Gentianales
Family	Apocynaceae
Genus	*Carissa*
Species	*Carissa macrocarpa*

Fig. 23.1. *Carissa macrocarpa*: evergreen, small, twiggy shrub with star-shaped white flowers.

Botany

The *Carissa macrocarpa* tree is a shrub which is native to Natal, South Africa. The fruit are known as Natal plum/Num num or Amantangula (Sibiya *et al.*, 2021). The Natal plum is part of the dogbane family, which is also known by its botanical name Apocynaceae. The botanical details of Natal plum are shown in Table 23.1. The tree grows to heights of up to 3 m. It is an evergreen, small, twiggy shrub with star-shaped white flowers, as shown in Fig. 23.1. The fruit ripens in summer and is red to crimson in colour when ripe. *Carissa macrocarpa* is a densely branched shrub with stems that contain long, strong, bifurcated spines (Moodley *et al.*, 2011). The spines carry clusters of white fragrant flowers that develop into red, plum-like, fleshy fruit, the inside of which contains a red flesh and approximately 12 small, brown, flat seeds (Kirsten, 2013). All parts of the plant contain a milky latex, including the fruit. Some *Carissa* species are used in traditional medicine systems in various parts of the world. Among these, *Carissa edulis* is used in traditional African medicine (Ibrahim *et al.*, 2007), *Carissa carandas* in Ayurvedic systems of medicine (Hegde *et al.*, 2009) and *Carissa opaca* by the hakims in Pakistan (Abbasi *et al.*, 2010). Except for *Carissa macrocarpa*, these other species of *Carissa* have drawn extensive scientific interest.

Natal plum fruit is green in colour and turns red to dark red as it ripens due to the presence of anthocyanins (Fig. 23.2) (Moodley *et al.*, 2012). In a study by Ndou *et al.* (2019), the

Fig. 23.2. Different maturity stages of Natal plum from immature green stage to mature red stage.

fruit longitudinal and transverse diameter and the mass were reported to be lower at the early stages of maturity when the fruit is still green, but significantly increased after the break stage when the fruit is pink. In the same study it was also observed that the fruit firmness decreased significantly with the advancement of maturity stage from green to deep-red. This decrease in fruit firmness has been attributed to softening of the fruit because of physiological changes in the cell wall structure caused by pectolytic enzyme action and solubilization of peptic polysaccharides (Atkinson and MacRae, 2007).

Traditional Uses

Carissa macrocarpa has been used to treat coughs and venereal diseases in South Africa. The leaves are also used to treat livestock diarrhoea and the fruit have antiviral properties against HIV and hepatitis (Moodley *et al.*, 2011; Souilem *et al.*, 2019). The fruit can be enjoyed raw when fully ripe. It can also be added to salads. Boiling the fruit makes a sauce which can be used as pie or tart filling. Furthermore, Natal plum can be preserved by making jam, juice, sherbet and jelly (Abbas *et al.*, 2014). Commercialization of Natal plum products is still rare. The fruit is still underutilized. Recently, the fruit was mixed with mango to make a leather (dried fruit pulp) (Mphaphuli *et al.*, 2020). However, much more can be done with this promising fruit. Natal plum is also a popular landscaping plant. The tree is grown as a hedge or fence owing to its thorns.

Nutritional Value

Proximate composition

The proximate composition content of *Carissa macrocarpa* fruit is presented in Table 23.2. Natal plum contains a wide variety of macronutrients which include carbohydrates, fats, proteins, fibre and ash. The moisture content of fresh Natal plum is 78% (Souilem *et al.*, 2019). Carbohydrates are the major macronutrient in the fruit while protein is in very low quantities but comparable to avocados (Wehmeyer, 1966; Souilem *et al.*, 2019; Sibiya *et al.*, 2020). The relatively high carbohydrate content of the fruit has seen it recommended as an aid in the prevention of malnutrition caused by lack of protein and energy (Moodley *et al.*, 2012). The recommended dietary allowance (RDA) of protein for children, adult males, adult females, pregnant women and lactating mothers is 28, 63, 50, 60 and 65 g, respectively; 100 g

Table 23.2. The nutritional composition of Natal plum fruit and other indigenous commercial fruit.

	Total protein g/100 g	Dietary fibre g/100 g	Total fat g/100 g	Carbohydrates g/100 g	Reference
Natal plum (*Carissa macrocarpa*)	1.96	4.68	0.87	21.57	Sibiya *et al.* (2021)
Indigenous fruit					
Mobola plum (*Parinari curatellifolia*)	2.61	8.88	0.25	21.18	Sibiya *et al.* (2021)
Wild mango *(Cordyla Africana*)	0.004	3.06	3.55	19.94	Sibiya *et al.* (2021)
Black plum (*Vitex doniana*)	8.24	0.58	34.62	28.40,	Vunchi *et al.* (2011)
Monkey orange (*Strychnos innocua* Del.)	3.97%	Trace	20.0	75.53	Hassan *et al.* (2014)
Ma-Mao (*Antidesma velutinosum Blum*)	1.6	15.6	1.1	22.5	Judprasong *et al.* (2013)
Hog plum or Ma-Kok-Pa (*Spondias pinnata* (L.f.) Kurz)	0.4	7.5	0.1	16.5	Judprasong *et al.* (2013)
Camu-camu (*Myrciaria dubia*)	0.4	0.1	0.2	0.44	Akter *et al.* (2011)
Commercial fruit					
Strawberry	0.67	2	0.3	7.68	Hidalgo and Almajano (2017)
Mango	0.8	1.6	0.4	15	USDA (2019)
Papaya	0.47	1.7	0.26	10.8	USDA (2019)
Banana	1.09	2.6	70.33	22.8	USDA (2019)
Avocado	1.96	6.80	15.40	8.64	Dreher and Davenport (2013)

of fresh Natal plum contributes between 1.8 to 7% of the daily protein requirements for children. When dried or concentrated into products, the protein contribution can be increased. Dietary fibre content in Natal plum fruit is generally higher than the sub-Saharan African indigenous and other commonly consumed fruit save for mobola plum (Table 23.2). The RDA of fibre are 19–25 g, 21–38 g, 28 g and 29 g for children, adults, pregnant and breast-feeding mothers, respectively – 100 g of fresh Natal plum therefore contributes 18.8% of the fibre requirements (based on an average RDA of 25 g). This is a very significant contribution. Total fat content was highest in Natal plum fruit when compared with all sub-Saharan African indigenous and commercial fruit listed in Table 23.2. An analysis of the fatty acid content and the possible uses of the fat may be necessary for value addition.

Vitamins, organic acids and soluble sugars

Natal plum contains a variety of vitamins (Table 23.3). Vitamin C functions as a redox buffer which can reduce reactive oxygen and radical species thereby making Natal plum fruit a good antioxidant. The presence of vitamin E in *Carissa macrocarpa* fruit was reported by Souilem *et al.* (2019). This vitamin is of the upmost importance for human nutrition and health through the prevention of many diseases including hypertension and cardiovascular disease. Vitamin E (α-tocopherol) presents a relatively high biological activity and has been reported to directly quench oxidizing radicals, thus preventing the propagation of several degenerative diseases (Lobo *et al.*, 2010). Considering the higher total fat content of the fruit, vitamin E (fat soluble) may be found in higher amounts compared to other similar fruits. Regarding the free sugar composition, fructose was the main soluble sugar found in the fruit, followed by glucose (Table 23.3). The total glucose and fructose content are above 9% and hence may render a sweet taste to the fruit. The fruit also contains organic acids such as oxalic acid, shikimic acid and citric acid. Natal plum contains 1.54 g/100 g fresh weight (FW) of citric acid and this contributes to the the sour/tart taste of the fruit. Quinic (1.6 g/100 g FW) and shikimic (2.1 mg/100 g FW) acids are cyclohexane carboxylic acids and important metabolites. Quinic acid is converted to shikimic acid at the end of the shikimate/phenylpropanoid pathway (Ghosh *et al.*, 2012).

Table 23.3. Vitamin, sugar and organic acid content of Natal plum based on FW (Ndou *et al.*, 2019; Souilem *et al.*, 2019).

Component	Content
Fructose g/100 g	5.6–9.4
Glucose g/100 g	3.7–4.3
Oxalic acid mg/100 g	20
Quinic acid g/100 g	1.6
Shikimic mg/100 g	2.1
Vitamin C mg/100 g	10
Citric acid g/100 g	1.54
α-tocopherol mg/100 g	0.91
γ-tocopherol mg/100 g	0.56

Minerals

The fruit is reportedly abundant in minerals, in order from highest to lowest: calcium, magnesium, iron, zinc, managanese and lead (1012, 768, 46.6, 16.1, 14.6, 8.72 and 8.50 μg/g respectively) (Fig. 23.3). The mineral composition is superior to some of the commonly consumed fruits, namely avocados, mangos and blackberries (Moodley *et al.*, 2012; Dreher and Davenport, 2013; Guedes *et al.*, 2013; Maldonado-Celis *et al.*, 2019). The maximum permissible Pb content of fruit in South Africa is 200 μg/kg (Department of Health, 2016). The reported Pb content (8.5 μg/g) is too high and research is necessary to show how planting and postharvest processing may reduce this heavy metal. Daily consumption of Natal plum fruit significantly contributes to the RDA for essential minerals (Moodley *et al.*, 2011). The recommended daily allowances for calcium and magnesium are 1000 mg and 310 mg, respectively. Most adults get around 1.6 and 3% of their RDA for calcium and magnesium, respectively, from eating 10.0 g of Natal plum fruit. Cu and Zn in foodstuffs are restricted to 30 mg/kg and 40 mg/kg, respectively, by the South African Department of Health (Department of Health, 2016). Natal plum however exhibits safe levels of Cu and Zn. Natal plum is therefore a potential mineral source even in poor communities.

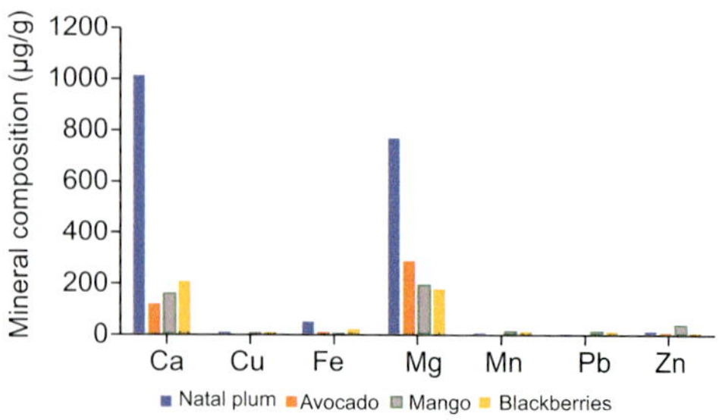

Fig. 23.3. Mineral composition of Natal plum, avocados, mango and blackberries (Moodley *et al.*, 2012; Dreher and Davenport, 2013; Guedes *et al.*, 2013; Maldonado-Celis *et al.*, 2019).

Fatty acids

The fruit has been reported to be rich in monounsaturated and essential fatty acids with a ratio of linoleic acid to α-linolenic acid (Table 23.4) in line with the recommended cardiac health range (Moodley *et al.*, 2012). These results were comparable to the observations in a study by Souilem *et al.* (2019) where monosaturated fatty acids were observed as a major group of compounds present (61.6%), mainly caused by the presence of oleic acids. Fatty acid content is also high (23.0%) and is influenced by the high content of palmitic acid (C16:0) and polyunsaturated fatty acid (15.46%) caused by the predominance of linoleic acid (C18:2n6). Compared to the commonly consumed avocado fruit Natal plum fruit has a lower fatty acid content though some of the fatty acids such as capric acid and stearic acid are comparable (Table 23.4). However, Natal plum fruit has a very low oleic acid content (0.16 mg/g) compared to avocados (55.38 mg/g) and it is reported to be the main fatty acid in avocados. *Carissa macrocarpa* fruit are highly recommended for their anticariogenic properties due to their richness in oleanolic acid and B-amyrin that can be used as a natural alternative to aspirin. It also presents hepatoprotective, hypoglycemic, antilipidemic and antiproliferative effects (Moodley *et al.*, 2011).

Table 23.4. Fatty acid profile of Natal plum fruit (mg/g dry weight – DW) (Moodley, 2012; Carvalho *et al.*, 2015; Souilem *et al.*, 2019).

Fatty acid	Natal plum (mg/g)	Avocado (mg/g)
Caprylic acid (C8:0)	0.045	-
Capric acid (C10:0)	0.079	0.2
Lauric acid (C12:0)	0.128	1.8
Myristic acid (C14:0)	1.89	0.04
Penta-decanoic acid (C15:0	0.132	-
Palmitic acid (C16:0)	0.89	18.14
Palmitoleic acid (C16:1)	7.65	10.84
Heptadecanoic acid (C17:0)	0.306	-
Stearic acid (C18:0)	0.76	1.52
Oleic acid (C18:1n9)	0.16	55.38
Linoleic acid (C18:2n6)	0.35	13.80
α-linolenic acid (C18:3n3)	0.18	1.16
Arachidic acid(C20:0)	0.02	-
Eicosanoic acid (C20:1)	0.5	0.15
Behenic acid (C22:0)	0.10	-
Tricosanoic acid (C23:0)	0.155	-
Lignoceric acid (C24:0)	0.242–0.95	-

Amino acids

Amino acids are needed by the human body to carry out biological processes that give cells structure, as well as transport and store nutrients. The body cannot function without 20 amino acids, which are divided into essential and non-essential amino acids (Ha and Zemel, 2003). Since essential amino acids cannot be synthesized by the human body and must be derived from food, they are considered indispensable. All essential and non-essential amino acids were reported in Natal plum fruit, though the results were comparatively lower than some of the amino acids reported in wild mango (*Cordyla africanna*), kei apple (*Dovyalis longispina*), water berry (*Syzygium cordatum*), milk berry (*Manilkara mochisia*) and white olives (*Halleria lucida*) (Table 23.5). The amino acids that have higher concentrations in the fruit include aspartic acid (0.43 g/100 g), glutamic acid (0.43 g/100 g), H-O proline (0.40 g/100 g), arginine (0.35 g/100 g) and lysine (0.30 g/100 g). The amino acid composition of these wild edible indigenous fruit of southern Africa are comparable and even higher than the World Health Organization (WHO) standard protein recommended daily requirements for adults (WHO, 2007). The requirement for lysine has received most attention given its nutritional significance as the likely limiting amino acid in staple foods, such as wheat. Natal plum meets the lysine required daily intake (3 mg/kg) recommended by WHO for adults. The histidine concentration in Natal plum fruit (0.14 g/100 g) is above the WHO standard protein recommended daily requirement (0.10 g/100 g). Although the concentration of the other essential amino acids is lower than the recommended daily requirements, the consumption of Natal plum fruit can have a contribution to healthy growth

Table 23.5. Amino acid composition of Natal plum fruit and some of the commonly consumed indigenous fruit in southern Africa (g/100g DW) (Sibiya *et al.*, 2021).

	Natal plum (*Carissa macrocarpa*)	Wild mango (*Cordyla africanna*)	Kei apple (*Dovyalis longispina*)	Water berry (*Syzygium cordatum*)	Milk berry (*Manilkara mochisia*)	White olives (*Halleria lucida*)
Arginine	0.35	0.29	0.39	0.52	0.30	0.55
Serine	0.20	0.28	0.24	0.23	0.21	0.64
Aspartic acid	0.43	0.47	0.45	0.55	0.34	0.62
Glutamic acid	0.48	0.64	0.57	0.58	0.45	0.85
Glycine	0.17	0.22	0.19	0.22	0.20	0.46
Threonine	0.17	0.22	0.20	0.19	0.17	0.30
Alanine	0.21	0.34	0.28	0.31	0.23	0.40
Tyrosine	0.05	0.14	0.04	0.14	0.11	0.27
Proline	0.15	0.24	0.19	0.19	0.19	0.27
H-O proline	0.40	0.19	0.07	0.11	0.06	0.07
Methionine	0.07	0.05	0.04	0.06	0.05	0.08
Valine	0.17	0.29	0.22	0.25	0.23	0.39
Phenylalanine	0.14	0.21	0.18	0.20	0.15	0.31
Isoleucine	0.14	0.23	0.16	0.19	0.17	0.30
Leucine	0.23	0.31	0.27	0.32	0.28	0.47
Histidine	0.14	0.26	0.08	0.12	0.23	1.56
Lysine	0.30	0.53	0.27	0.49	0.37	0.58

and development as well as the maintenance of healthy immune function.

Phenolics in Natal plum

Total phenolics

The total phenolic content of Natal plum fruit is generally determined using the Folin Ciocalteu assay. A study on the effect of ripening on Natal plum fruit showed that the fruit had relatively high total phenolics ranging from 575.4 to 1134 mg GAE/100 g dry weight (DW) (Table 23.5) (Ndou *et al.*, 2019). The study showed clear seasonal variations in the total phenolic content, which also reduced with ripening. The total phenolic content reported in other indigenous fruits like *Vitex doniana* (601.4 mg GAE/100 g), *Vitex kiniesis* (719.83 mg GAE/100 g), *Vaccinium myrtilus* (670.9 mg GAE/100 g), *Rubus coesin* (355.3 mg GAE/100 g), *Prunus cerasus vulgaris* (429.5 mg GAE/100 g) and *Prunus domestica* (303.9 mg GAE/100 g) has not reached the high levels of Natal plum (Marinova *et al.*, 2005; Ochieng and Nandwa, 2010). More research is necessary to fully utilize this promising fruit. It is necessary to understand any seasonal variations that might have induced an increase in the total phenolic content and then mimic them when planting this shrub. According to a study on *Carissa macrocarpa*, phytochemicals are affected by fruit maturity stage, with M4 (red) and M5 (deep-red) showing high levels of antioxidant and phenolic compounds. Natal plum fruit had the highest total phenolic concentration during the mature dark green (M1) and light green (M2) stage and the ascorbic acid, FRAP, and DPPH levels increased with maturity (Ndou *et al.*, 2019).

Hydroxycinnamic acids

The major hydroxycinnamic acids in Natal plum are caffeic, ferulic, coumaric, dicaffeoyl tartaric acid and chlorogenic acid (Table 23.5). These were shown to consistently decrease with fruit ripening in a study carried out over two seasons. This may have been due to a shift from production of hydroxycinnamic acids to anthocyanins during ripening (Ndou *et al.*, 2019). Similar results have been reported in four highbush blueberry cultivars (*Vaccinium corymbosum* L.) and hydroxycinnamic acids were identified as the most predominant group at all stages of ripening (Castrejón *et al.*, 2008). However, as the blueberries ripened a decrease in hydroxycinnamic acids was observed. It was insinuated that during the ripening there is a shift in the pool of total phenolics toward anthocyanin synthesis, and an overall decline in the content of other phenolic components. In another study, several hydroxycinnamic acids were detected in different *Ribes* species namely *p*-coumaric acid derivatives, ferulic acid derivatives, caffeic acid and 3-*O*-caffeoylquinic acid. A decrease in the hydroxycinnamic acid content was also observed among the species and this can be attributed to the cellular activity being remarkably high during ripening and most of the biomolecules being metabolized (Payasi *et al.*, 2009).

Flavonoids

Generally, Natal plum fruit contains higher amounts of flavonoids. Quercetin 3-*O*-rhamnosyl glucoside was by far the most abundant flavonol in the fruit (Table 23.6). Epicatechin, luteolin, quercetin 3-*O*-rhamnosyl glucoside, apigenin, eriodictyol 7-*O*-glucoside and catechin in Natal plum fruit have been found to decrease during ripening (Ndou *et al.*, 2019). A similar trend has been reported on the polyphenol content during the ripening of some fruit species like *Prunus persica* (Belhadj *et al.*, 2016), blueberries (*Vaccinium corymbosum*) (Castrejón *et al.*, 2008) and blackberries (Siriwoharn *et al.*, 2004) where polyphenolic content was found to decrease towards ripening. However, Naringenin 4-*O*-glucoside increased with fruit ripening. A higher content of polyphenols in unripened fruit justifies the phenomenon of fruit protection from various fruit-borne diseases during the pre-maturation stage (Lattanzio *et al.*, 2008). Radioactive labelling of carbon atoms may however help determine exactly what happens to these phenolics, which may contribute to formation of other phenolic compounds during ripening.

Table 23.6. Changes in phenolic compounds with ripening of Natal plum (Ndou *et al.*, 2019).

Phenolic component	Content (mg/kg DW)	Change during ripening
Catechin	79–136	Decrease with ripening
Epicatechin	21–71	Decrease with ripening
Quercertin 3-*O*-rhamnosyl glucoside	146.9–564.4	Decrease with ripening
Luteolin	2.97–4.92	Decrease with ripening
Apigenin	3.33–11.08	Decrease with ripening
Eriodictyol 7-*O*-glucoside	14.45–40.56	Decrease with ripening
Naringenin 4-*O*-glucoside	29.85–102.43	Increase with ripening
Caffeic acid	0.87–3.67	Decrease with ripening
Ferulic acid	1.04–5.91	Decrease with ripening
p-coumaric acid	0.87–1.74	Decrease with ripening
Dicaffeoyl tartaric acid	1.25–3.92	Decrease with ripening
Chlorogenic acid	1.04–2.64	Decrease with ripening
Cyanidin 3-*O*-pyranoside	ND- 80.74	Increase with ripening
Cyanidin 3-*O*-sambubioside	ND-75.82	Increase with ripening
Cyanidin 3-*O*-glucoside	1.10–98.4	Increase with ripening

Anthocyanins

The anthocyanins present in the extracts of Natal plum fruit are cyanidin derivatives including cyanidin 3-*O*-pyranoside, cyanidin 3-*O*-β-sambubioside and cyanidin 3-*O*-glucoside (Table 23.6). These compounds steadily increased with maturation to a deep-red level (Ndou *et al.*, 2019). A similar shift in the concentration of polyphenols more toward anthocyanin biosynthesis was recorded in *Myrica esculenta*, *Morus alba* and *Berberis asiatica* (Belwal *et al.*, 2019). The accumulation of anthocyanin pigments during fruit maturity can also be attributed to the upregulation of phenylpropanoid pathway (PAL) and chalcone synthase enzyme, which are involved in anthocyanin biosynthesis (Whale and Singh, 2007). Cyanidin derivatives are the major pigments in Natal plum fruit, which are not detected in the green fruit. Natal plum extracts from the deep-red stage may be used as food colourants but research is necessary to ensure their stability during product shelf life.

Effect of Postharvest Techniques on Phytochemicals and Antioxidant Capacity

Fresh fruit, with a moisture content of more than 80%, are labelled as highly perishable produce, and their preservation by sun-drying, which goes back a few decades, has always resulted in low quality and substance contamination (Sagar and Suresh Kumar, 2010). In most cases, the most traditional drying methods yield lower quality products and need longer drying periods (Moses *et al.*, 2014; Mahiuddin *et al.*, 2018; Rahman *et al.*, 2018). Therefore, there is a significant need to establish innovative drying technologies, considering operational ability, process management, time requirements, cost economy, product quality, protection and environmental aspects. According to the review by Kamiloglu *et al.* (2016), freeze-drying is superior to other drying methods in terms of the retention of antioxidant compounds. Recent studies, which examined the effects of different drying methods on the phytochemical content and antioxidant function of various fruit and vegetable products, showed a substantial increase and decrease in phytochemical content (Martínez-Las Heras *et al.*, 2014; Youssef and Mokhtar, 2014). However, phytochemicals and antioxidant activity tend to rise at 60°C because, at higher temperatures, the cell wall breaks down and releases attached phenols and flavonoids, allowing an increase in the antioxidant properties of the dried substance, resulting in reduced oxidation reactions (Khan and Karim, 2017; Khan *et al.*, 2018). The study on the effect of freeze-drying on the antioxidant activity of Natal plum fruit revealed that 2,2-diphenylpicrylhydrazyl (DPPH) radical scavenging capacity was not affected significantly by freeze-drying (Seke *et al.*, 2021). However, the 2, 2'-Azino-Bis-3-Ethylbenzothiazoline-6-Sulfonic Acid (ABTS) radical scavenging capacity of Natal plum decreased significantly with freeze-drying and a similar trend was also observed with the ferric reducing antioxidant power (FRAP) activity of Natal plum.

Fruit quality gradually deteriorates over time, particularly during cold storage. The storage temperature and relative humidity, the atmospheric composition and previous background of the fruit are among the factors that influence the rate of decay and hence the shelf life. Polyphenols are considered to be very unstable at room temperature during storage (Howard *et al.*, 2012). Storage of Natal plum at different temperatures for different times showed that low temperature storage reduced anthocyanin degradation compared to a higher storage temperature (Mphaphuli *et al.*, 2020). Storage at 2°C resulted in chilling injury and loss of marketable fruit weight but increased the anthocyanin content. Storage at 4°C reduced chilling injury and losses in antioxidant capacity. Mphaphuli *et al.* (2020) observed that anthocyanins in Natal plum accumulated due to cold stress. Low temperatures induced the biosynthesis of anthocyanins in blood Moro and blond Pera oranges in response to cold temperature stress, with higher up-regulation of the enzymes involved in the phenylpropanoid pathway (Carmona *et al.*, 2017). Optimizing the production and storage of functional foods is a critical step in reducing phytochemical depletion. Fruit leather (dried fruit pulp) made from mango and Natal plum fruit has also proven to be an effective way to preserve bioactive compounds (Mphaphuli *et al.*, 2020). Mango and Natal plum (2:1) fruit leathers had the highest levels of cyanidin-3-*O*-glucoside chloride, cyanidin-3-*O*-sambubioside, epicatechin, apigenin, kaempferol, luteolin, quercetin-3-*O*-rhamnosyl glucoside, catechin, quinic and chlorogenic acids, as well as *in vitro* antioxidant capacity. More postharvest research is needed to determine the optimal storage conditions such as lighting, duration, temperature and atmospheric gas composition for retaining phenolic compounds in Natal plum fruit.

Potential Health Benefits

Various studies have focused on the identification, utilization and beneficial properties of bioactive compounds in functional foods based on clinical trials over the years (Paredes-López *et al.*, 2010). According to the WHO, bioactive compounds, especially those found in small colourful fruits, are important in preventing health issues such as cardiovascular disease, diabetes, cancer and obesity (WHO, 2002). Berries, such as Natal plum fruit, may play a greater part in formulating healthy diets because they contain a wide variety of phenolic compounds. Scavenging or detoxifying radicals, preventing the formation of radicals, influencing the cell cycle, tumour suppression, signal transduction control, apoptosis, detoxifying enzymes and metabolism are all biologically essential modes of action for phenolic compounds (Huang *et al.*, 2009). Some of the main phenolic compounds that have been identified in Natal plum fruit are anthocyanin derivatives (cyanidin-3-*O*-pyranoside, cyanidin 3-*O*-β-sambubioside (Cy-3-Sa), cyanidin-3-*O*-glucoside (Cy-3-G), naringenin 4′-*O*-glucoside, rutin, quercetin 3-*O*-rhamnosyl galactoside, quercetin 3-*O*-rhamnosyl glucoside and ascorbic acid (Ndou *et al.*, 2019), reportedly responsible for antioxidant capacity (Table 23.6). Cyanidin 3-*O*-glucoside has been reported to have decreased proliferation of pre-adipocytes and mature adipocytes, decrease in adipocyte differentiation and lipolysis (Kim *et al.*, 2012). Cyanidin 3-*O*-Sambubioside from *Hibiscuss sabdariffa* also inhibited angiotensin converting enzyme, showing high blood pressure reduction potential (Ojeda *et al.*, 2010). Black mulberry extract-enriched pasta also showed a reduced glycaemic index and inhibition of α-amylases and α-glucosidases (Yazdankhah *et al.*, 2019). Rutin has been extensively studied for its anticancer properties, and it is reported to actively inhibit cancer cell growth through cell cycle arrest and/or apoptosis, as well as proliferation, angiogenesis, and/or metastasis in colorectal cell lines (Ben Sghaier *et al.*, 2016). Human leukemia HL-60 cells showed antileukemic potential as evidenced by the reduction of the tumour size when rutin was added (Lin *et al.*, 2012). Rutin analogue, quercetin, also showed a dose-related inhibition of the growth of ovarian cancer cell line OVCA 433 (Jayameena *et al.*, 2018). Furthermore, protochatechuic acid possesses antioxidant, antibacterial, anticancer, antiulcer, antidiabetic and antiageing activity (Kakkar and Bais, 2014). Immune-boosting supplements are expensive and mostly out of reach for poor and rural populations, but readily available indigenous fruit like Natal plum help to provide daily nutrients and minerals. Moodley *et al.* (2012) confirmed that *Carissa macrocarpa* also has enthnomedicinal value and great health benefits. Souilem *et al.* (2019) reported that the fruit did not show any sign of toxicity against non-tumour cells, and it can be concluded that *Carissa macrocarpa* fruit could be included in a daily basic diet as a source of nutritional compounds with high bioactive capacity. These findings show the potential of Natal plum in the development of anticancer therapies as the fruit possess a considerable amount of bioactive compounds.

Effect of *in vitro* digestion on bioactive compounds and antioxidant activity

Physicochemical factors such as temperature, pH and ionic strength, as well as biochemical factors such as the presence and action of enzymes, all influence the stability of dietary polyphenols and their interaction-affected release from the food matrix during digestion. As a result, the post-digestion content of these phytochemicals may differ significantly from the ingested native amount in food (Alminger *et al.*, 2014).

The *in vitro* digestion of Natal plum fruit showed that the total phenolic content decreased during digestion as show in Table 23.7. Undigested Natal plum fruit exhibited the highest amount of total phenolic content (2265.77 mg/100 g), and the gastric fraction showed 1786.13 mg/100. The intestinal fraction had the lowest content at 1399.18 mg/100 g (Seke *et al.*, 2021). Phenolic glycosides, which are made up of anthocyanins and quercetin glycosides, were found to be the most abundant compounds (66.67%). Secondary fragments corresponded to phyloglucinaldehyde, which is a degradation product of cyanidin-based anthocyanins produced after subsequent deglycosylation followed by cleavage of the aglycone (Sinela *et al.*, 2017). When compared to the undigested Natal plum, the gastric fraction had slightly lower concentrations of flavonoid aglycones like kaempferol, quercetin and naringenin,

Table 23.7. Changes in phenolic acid components in Natal plum fruit during different processing methods on a DW basis (Seke *et al.*, 2021).

	Freeze-dried Natal plum (undigested)	Gastric digested Natal plum (DW)	Intestinal digested Natal plum (DW)
TPC (mg GAE/100 g)	2265.77^a ±1.23*	1786.13^b± 0.08	1399.18^c ± 0.16
Phenolic components (mg/kg)			
Anthocyanins			
Cyanidin 3-*O*-sambubioside	1631.4^a ± 1.2	1193.2^b ± 1.5	525.3^c ± 3.1
Cyanidin 3-*O*-glucoside	955.2^a ± 1.9	347.6^b ± 1.5	155.7^c ± 1.8
Hydrobenzoic acids			
Ellagic acid	42.4^b±0.1	49.6^a±0.5	11.3^c±0.2
Gallic acid	5.5^b± 0.1	2.1^c±0.1	8.8^a±0.1
Protocatechuic acid	167.6^a±0.9	59.6^c±0.2	88.6^b±0.5
Hydroxycinnamic acids			
p-coumaric acid	140.3^a±1.6	55.4^c±0.5	85.7^b±0.2
Ferulic acid	165.1^a±0.3	49.6^c±0.3	73.5^b±0.3
Caffeic acid	168.4^a±0.1	53.8^b±0.5	40.8^c±0.4
Chlorogenic acid	28.1 ± 0.7	ND	ND
Syringic acid	62.2^a±0.2	47.5^b±0.6	26.5^c±0.2
Flavonols			
Catechin	491.8^a±2.3	354.1^a±2.8	210.4^c±1.4
Epicatechin	19.3^b±0.2	12.2^b±0.2	6.7^c±0.2
Kaempferol	333.5^a±1.2	70.9^c±3.5	259.9^b±3.1
Quercetin	279.7^a±1.4	88.6^c±5.9	112.1^b±0.9
Flavanone			
Naringenin	48.3^a±0.2	17.2^c±0.1	22.3^b±0.2
Flavones			
Apigenin	9.7^a±0.6	3.8^c±0.1	5.9^b±0.1
Luteolin	4.6^a±0.1	1.3^b±0.1	1.7^b±0.1
Flavonoid glycoside			
Quercitrin (quercetin *O*-glycoside)	5.3^a ± 1.9	1.0^b ±0.1	0.9^c ± 0.1
Quercetin 3-galactoside 7-rhamoside	7.8^a ± 0.6	0.9^c ± 0.1	3.8^b ± 0.1
Rutin (quercetin-3-*O*-rutinoside)	34.6^a ± 1.8	0.9^c ± 0.1	18.5^b ± 1.4
Antioxidant activity			
DPPH (IC_{50} for mg/mL)	0.88^b ± 0.03	0.94^a± 0.18	0.85^b±0.08
ABTS (IC_{50} mg/mL)	0.699^c ± 0.12	0.717^b± 0.45	0.835^a± 0.16
FRAP (mM VCEAC/g)	493.54^a ± 1.40	482.54^a ± 1.11	399.42^b± 0.88

while the intestinal fraction had higher bioaccessibility. According to Sinela *et al.* (2017) quercetin glycosides were affected by acidic pH in the gastric fraction, and the amount of quercetin glycosides that were bioavailable in the intestinal fraction differed depending on the form of quercetin glycoside. Strawberry (Ariza *et al.*, 2017), blackberry, blueberry and raspberry (Marhuenda *et al.*, 2016) all showed a decreasing trend in total phenols in the intestinal fraction compared to the gastric phase, which was likely due to interactions with macronutrients (Alminger *et al.*, 2014) or the production of other phenolic derivatives as a result of polymerization and oxidation. Furthermore, when compared to the undigested Natal plums, the concentrations of naringenin, apigenin and luteolin in the gastric fraction were affected, while naringenin is highly bioavailable in the intestinal stage. The differences in pH between the gastric and intestinal phases had no effect on luteolin. To fully understand this phenomenon, researchers will need to investigate the structure of Natal plum cells as well as the position of phenolics. According to Mosele *et al.* (2016), the higher catechin concentration (354.1 mg/kg) in the intestinal phase could be due to partial hydrolysis of proanthocyanidins at intestinal pH. However, FRAP activity was comparable in the gastric and undigested fractions but decreased in the intestinal fraction. The DPPH, ABTS and FRAP activities were decreased in this study due to the lower concentration of phenolic compounds in the intestinal fraction relative to the original methanolic fraction of the Natal plum fruit. The positive correlation with FRAP activity ($r = 0.87$) confirmed this. Furthermore, pH changes between the gastric and intestinal phases have been reported to affect antioxidant capacity by Bouayed *et al.* (2011). Tagliazucchi *et al.* (2010) found that grape extracts rich in phenolics (e.g. quercetin) provided more protection in the intestine than the stomach phase and had higher antioxidant capacity when measured in ABTS.

Inhibitory activity of α-glucosidase, digested and undigested

The carbohydrate digestion enzyme α-glucosidase has been identified as a therapeutic target for the regulation of postprandial hyperglycemia, which is the first metabolic deformity to appear in type 2 diabetes mellitus (Bermano *et al.*, 2020). The study on the effect of *in vitro* digestion on Natal plum fruit revealed that the inhibition capacity of freeze-dried Natal plum powder, on the other hand, was higher in the undigested freeze-dried sample than in the gastric and intestinal phases. The α-glucosidase IC_{50} had a strong negative correlation with syringic acid, Cy-3-Sa and Cy-3-G (R2 values 0.76, 0.99 and 0.85, respectively), indicating that as these phenolics increased, so did the inhibition. Tadera *et al.* (2006) suggested that the presence of an unsaturated C-ring, hydroxylation on the C-ring at the 3-position and an increase in the number of hydroxyl groups on the B-ring all improved α-glucosidase inhibitory activity. Contrary to these findings the IC_{50} values of different red cabbage varieties were lower than the activity observed with freeze-dried Natal plum powder.

Antimicrobial activity

The antimicrobial activity of different extracts from Natal plum fruit has been reported previously (Moodley *et al.*, 2011). The dichloromethane (DCM) extract from *Carissa macrocarpa* fruit contained three compounds (β-amyrin, methyl oleanolate and 3β-hydroxyolean-11-en-28,1β-olide) while the methanol extract contained oleanolic acid. The pentacyclic triterpenes β-amyrin, methyl oleanolate, oleanolic acid and ursolic acid showed moderate antibacterial activity (MICs in the range 0.12–1.0 mg ml-1) while the oleanolic acid derivative, 3β-hydroxyolean-11-en-28,13β-olide has good antibacterial activity (MICs in the range 0.06–0.12 mg ml-1) for all bacterial strains (*E. coli* (ATCC 25922), *P. aeruginosa* (ATCC 35032), *K. pneumoniae* (ATCC 700603), *E. faecium* (ATCC 19434), *S. aureus* (ATCC 25923), *S. aureus* (ATCC 43300) and *S. saprophyticus* (ATCC 35552)). Pentacyclic triterpenoids, oleanolic acid, ursolic acid and their derivatives have a range of pharmacological benefits without substantial toxicity (Cargnin and Gnoatto, 2017). There was little difference in the degree of action between Gram-positive and Gram-negative bacteria, suggesting that antibacterial activity of these pentacyclic triterpenes is not influenced by the cell wall differences in the bacterial species (Moodley, 2012). Therefore, Natal plum fruit are promising in the development of novel therapeutic agents.

Furthemore, incorporation of Natal plum in food products may render some preservative or bacterial load reduction action.

Conclusion

Natal plum is a source of a wide range of nutritional and bioactive compounds. It contains most of the minerals, vitamin C, vitamin E, phenolic compounds and also essential fatty acids. The consumption of Natal plum as part of a diverse diet may improve nutritional status. Furthermore, Natal plum consumption may reduce the occurrence of lifestyle and degenerative diseases. Commercialization of Natal plum may boost income generation, while producing functional foods. Natal plum is a promising fruit for the functional food sector.

References

Abbas, M., Rasool, N., Riaz, M., Zubair, M., Abbas, M. *et al.* (2014) GC-MS profiling, antioxidant, and antimicrobial studies of various parts of *Carissa grandiflora*. *Bulgarian Chemical Communications* 46, 831–839.

Abbasi, A.M., Khan, M.A., Ahmed, M. and Zafar, M. (2010) Herbal medicines used to cure various ailments by the inhabitants of Abbottabad district, North West Frontier Province, Pakistan. *Indian Journal of Traditional Knowledge* 09(1), 175–183.

Achaglinkame, M.A., Aderibigbe, R.O., Hensel, O., Sturm, B. and Korese, J.K. (2019) Nutritional characteristics of four underutilized edible wild fruits of dietary interest in Ghana. *Foods (Basel, Switzerland)* 8(3), 104. DOI: 10.3390/foods8030104.

Akter, M.S., Oh, S., Eun, J.B. and Ahmed, M. (2011) Nutritional compositions and health promoting phytochemicals of camu-camu (*Myrciaria dubia*) fruit: A review. *Food Research International* 44(7), 1728–1732. DOI: 10.1016/j.foodres.2011.03.045.

Alminger, M., Aura, A.-M., Bohn, T., Dufour, C., El, S.N. *et al.* (2014) *In vitro* models for studying secondary plant metabolite digestion and bioaccessibility. *Comprehensive Reviews in Food Science and Food Safety* 13(4), 413–436. DOI: 10.1111/1541-4337.12081.

Ariza, M.T., Forbes-Hernandez, T.Y., Giampieri, F., Gasparrini, M., Soria, C. *et al.* (2017) Effects of in vitro gastrointestinal digestion on strawberry polyphenols stability. *Acta Horticulturae* (1156), 389–396. DOI: 10.17660/ActaHortic.2017.1156.58.

Atkinson, R.G. and MacRae, E.A. (2007) I. 13 kiwifruit. *Biotechnology in Agriculture and Forestry* 60, 329–346.

Belhadj, F., Somrani, I., Aissaoui, N., Messaoud, C., Boussaid, M. *et al.* (2016) Bioactive compounds contents, antioxidant and antimicrobial activities during ripening of *Prunus persica* L. varieties from the North West of Tunisia. *Food Chemistry* 204, 29–36. DOI: 10.1016/j.foodchem.2016.02.111.

Belwal, T., Pandey, A., Bhatt, I.D., Rawal, R.S. and Luo, Z. (2019) Trends of polyphenolics and anthocyanins accumulation along ripening stages of wild edible fruits of Indian Himalayan region. *Scientific Reports* 9(1), 1–11. DOI: 10.1038/s41598-019-42270-2.

Ben Sghaier, M., Pagano, A., Mousslim, M., Ammari, Y., Kovacic, H. *et al.* (2016) Rutin inhibits proliferation, attenuates superoxide production and decreases adhesion and migration of human cancerous cells. *Biomedicine & Pharmacotherapy = Biomedecine & Pharmacotherapie* 84, 1972–1978. DOI: 10.1016/j.biopha.2016.11.001.

Bermano, G., Stoyanova, T., Hennequart, F. and Wainwright, C.L. (2020) Seaweed-derived bioactives as potential energy regulators in obesity and type 2 diabetes. *Advances in Pharmacology* 87, 205–256. DOI: 10.1016/bs.apha.2019.10.002.

Bouayed, J., Hoffmann, L. and Bohn, T. (2011) Total phenolics, flavonoids, anthocyanins and antioxidant activity following simulated gastro-intestinal digestion and dialysis of apple varieties: Bioaccessibility and potential uptake. *Food Chemistry* 128(1), 14–21. DOI: 10.1016/j.foodchem.2011.02.052.

Bvenura, C. and Sivakumar, D. (2017) The role of wild fruits and vegetables in delivering a balanced and healthy diet. *Food Research International* 99(Pt 1), 15–30. DOI: 10.1016/j.foodres.2017.06.046.

Cargnin, S.T. and Gnoatto, S.B. (2017) Ursolic acid from apple pomace and traditional plants: A valuable triterpenoid with functional properties. *Food Chemistry* 220, 477–489. DOI: 10.1016/j.foodchem.2016.10.029.

Carmona, L., Alquézar, B., Marques, V.V. and Peña, L. (2017) Anthocyanin biosynthesis and accumulation in blood oranges during postharvest storage at different low temperatures. *Food Chemistry* 237, 7–14. DOI: 10.1016/j.foodchem.2017.05.076.

Carvalho, C.P., Bernal, E.J., Velásquez, M.A. and Cartagena, V.J.R. (2015) Fatty acid content of avocados (*Persea americana* Mill. cv. Hass) in relation to orchard altitude and fruit maturity stage. *Agronomía Colombiana* 33(2), 220–227. DOI: 10.15446/agron.colomb.v33n2.49902.

Castrejón, A.D.R., Eichholz, I., Rohn, S., Kroh, L.W. and Huyskens-Keil, S. (2008) Phenolic profile and antioxidant activity of highbush blueberry (*Vaccinium corymbosum* L.) during fruit maturation and ripening. *Food Chemistry* 109(3), 564–572. DOI: 10.1016/j.foodchem.2008.01.007.

Department of Health (2016) *Government Notices, Foodstuffs, Cosmetics and Disinfectants Act (Act No.54 of 1972)*. Government Gazette, South Africa.

Dreher, M.L. and Davenport, A.J. (2013) Hass avocado composition and potential health effects. *Critical Reviews in Food Science and Nutrition* 53(7), 738–750. DOI: 10.1080/10408398.2011.556759.

Ghosh, S., Chisti, Y. and Banerjee, U.C. (2012) Production of shikimic acid. *Biotechnology Advances* 30(6), 1425–1431. DOI: 10.1016/j.biotechadv.2012.03.001.

Guedes, M.N.S., Abreu, C.M.P. de, Maro, L.A.C., Pio, R., Abreu, J.R. de *et al*. (2013) Chemical characterization and mineral levels in the fruits of blackberry cultivars grown in a tropical climate at an elevation. *Acta Scientiarum. Agronomy* 35(2), 191–196. DOI: 10.4025/actasciagron.v35i2.16630.

Ha, E. and Zemel, M.B. (2003) Functional properties of whey, whey components, and essential amino acids: mechanisms underlying health benefits for active people (review). *The Journal of Nutritional Biochemistry* 14(5), 251–258. DOI: 10.1016/s0955-2863(03)00030-5.

Hassan, L.G., Abdulmumin, U., Umar, K.J., Ikeh, O.P. and Aliero, A.A. (2014) Nutritional and anti-nutritional composition of *Strychnosinnocua* Del. (Monkey Orange) fruit pulp grown in Zuru, Nigeria. *Nigerian Journal of Basic and Applied Sciences* 22(1–2), 33. DOI: 10.4314/njbas.v22i1.6.

Hegde, K., Thakker, S.P., Joshi, A.B., Shastry, C.S. and Chandrashekhar, K.S. (2009) Anticonvulsant activity of *Carissa carandas* Linn. root extract in experimental mice. *Tropical Journal of Pharmaceutical Research* 8(2), 117–125. DOI: 10.4314/tjpr.v8i2.44519.

Hidalgo, G.I. and Almajano, M.P. (2017) Red fruits: extraction of antioxidants, phenolic content, and radical scavenging determination: A review. *Antioxidants* 6(1), 7. DOI: 10.3390/antiox6010007.

Howard, L.R., Prior, R.L., Liyanage, R. and Lay, J.O. (2012) Processing and storage effect on berry polyphenols: challenges and implications for bioactive properties. *Journal of Agricultural and Food Chemistry* 60(27), 6678–6693. DOI: 10.1021/jf2046575.

Huang, W.Y., Cai, Y.Z. and Zhang, Y. (2009) Natural phenolic compounds from medicinal herbs and dietary plants: potential use for cancer prevention. *Nutrition and Cancer* 62(1), 1–20. DOI: 10.1080/01635580903191585.

Ibrahim, H., Abdulrahman, E.M., Shok, M., Ilyas, N., Musa, K.Y. *et al*. (2007) Comparative analgesic activity of the root bark, stem bark, leaves, fruits and seeds of *Carissa edulis* VAHL (Apocynaceae). *African Journal of Biotechnology* 6(10), 1233–1235.

Jayameena, P., Sivakumari, K., Ashok, K. and Rajesh, S. (2018) Rutin: A potential anticancer drug against human colon cancer (HCT116) cells. *International Journal of Biology, Pharmacy and Allied Sciences* 7(9), 731–1745. DOI: 10.31032/IJBPAS/2018/7.9.4532.

Judprasong, K., Charoenkiatkul, S., Thiyajai, P. and Sukprasansap, M. (2013) Nutrients and bioactive compounds of Thai indigenous fruits. *Food Chemistry* 140(3), 507–512. DOI: 10.1016/j.foodchem.2013.01.057.

Kakkar, S. and Bais, S. (2014) A review on protocatechuic acid and its pharmacological potential. *International Scholarly Research Notices* 2014, 952943–952943. DOI: 10.1155/2014/952943.

Kamiloglu, S., Toydemir, G., Boyacioglu, D., Beekwilder, J., Hall, R.D. *et al*. (2016) A review on the effect of drying on antioxidant potential of fruits and vegetables. *Critical Reviews in Food Science and Nutrition* 56(Suppl 1), S110–S129. DOI: 10.1080/10408398.2015.1045969.

Khan, M.I.H. and Karim, M.A. (2017) Cellular water distribution, transport, and its investigation methods for plant-based food material. *Food Research International* 99(Pt 1), 1–14. DOI: 10.1016/j.foodres.2017.06.037.

Khan, M.I.H., Nagy, S.A. and Karim, M.A. (2018) Transport of cellular water during drying: An understanding of cell rupturing mechanism in apple tissue. *Food Research International* 105, 772–781. DOI: 10.1016/j.foodres.2017.12.010.

Kim, H.-K., Kim, J.N., Han, S.N., Nam, J.-H., Na, H.-N. *et al.* (2012) Black soybean anthocyanins inhibit adipocyte differentiation in 3T3-L1 cells. *Nutrition Research* 32(10), 770–777. DOI: 10.1016/j.nutres.2012.06.008.

Kirsten, K. (2013) *Garden Plant and Flowers*. Struik Publishers, Cape Town.

Lattanzio, V., Kroon, P.A., Quideau, S. and Treutter, D. (2008) Plant phenolics—secondary metabolites with diverse functions. *Recent Advances in Polyphenol Research* 1, 1–35. DOI: 10.1002/9781444302400.

Lin, J.-P., Yang, J.-S., Lin, J.-J., Lai, K.-C., Lu, H.-F. *et al.* (2012) Rutin inhibits human leukemia tumor growth in a murine xenograft model in vivo. *Environmental Toxicology* 27(8), 480–484. DOI: 10.1002/tox.20662.

Lobo, V., Patil, A., Phatak, A. and Chandra, N. (2010) Free radicals, antioxidants and functional foods: Impact on human health. *Pharmacognosy Reviews* 4(8), 118–126. DOI: 10.4103/0973-7847.70902.

Mahiuddin, Md., Khan, Md.I.H., Duc Pham, N. and Karim, M.A. (2018) Development of fractional viscoelastic model for characterizing viscoelastic properties of food material during drying. *Food Bioscience* 23, 45–53. DOI: 10.1016/j.fbio.2018.03.002.

Maldonado-Celis, M.E., Yahia, E.M., Bedoya, R., Landázuri, P., Loango, N. *et al.* (2019) Chemical composition of mango (*Mangifera indica* L.) fruit: Nutritional and phytochemical compounds. *Frontiers in Plant Science* 10, 1073. DOI: 10.3389/fpls.2019.01073.

Marhuenda, J., Alemán, M.D., Gironés-Vilaplana, A., Pérez, A., Caravaca, G. *et al.* (2016) Phenolic composition, antioxidant activity, and in vitro availability of four different berries. *Journal of Chemistry* 2016, 1–7. DOI: 10.1155/2016/5194901.

Marinova, D., Ribarova, F. and Atanassova, M. (2005) Total phenolics and total flavonoids in Bulgarian fruits and vegetables. *Journal of the University of Chemical Technology and Metallurgy* 40, 255–260.

Martínez-Las Heras, R., Heredia, A., Castelló, M.L. and Andrés, A. (2014) Influence of drying method and extraction variables on the antioxidant properties of persimmon leaves. *Food Bioscience* 6, 1–8. DOI: 10.1016/j.fbio.2014.01.002.

Moodley, R. (2012) Phytochemical and analytical studies on two indigenous medicinal plants found in KwaZulu-Natal: *Carissa macrocarpa* and *Harpephyllum caffrum* (Doctoral dissertation).

Moodley, R., Chenia, H., Jonnalagadda, S.B. and Koorbanally, N. (2011) Antibacterial and anti-adhesion activity of the pentacyclic triterpenoids isolated from the leaves and edible fruits of *Carissa macrocarpa*. *Journal of Medicinal Plants Research* 5, 4851–4858.

Moodley, R., Koorbanally, N. and Jonnalagadda, S.B. (2012) Elemental composition and fatty acid profile of the edible fruits of Amatungula (*Carissa macrocarpa*) and impact of soil quality on chemical characteristics. *Analytica Chimica Acta* 730, 33–41. DOI: 10.1016/j.aca.2011.11.066.

Mosele, J.I., Macià, A., Romero, M.P. and Motilva, M.J. (2016) Stability and metabolism of *Arbutus unedo* bioactive compounds (phenolics and antioxidants) under in vitro digestion and colonic fermentation. *Food Chemistry* 201, 120–130. DOI: 10.1016/j.foodchem.2016.01.076.

Moses, J.A., Norton, T., Alagusundaram, K. and Tiwari, B.K. (2014) Novel drying techniques for the food industry. *Food Engineering Reviews* 6(3), 43–55. DOI: 10.1007/s12393-014-9078-7.

Mphaphuli, T., Slabbert, R.M. and Sivakumar, D. (2020) Storage temperature and time changes of phenolic compounds and antioxidant properties of Natal plum (*Carissa macrocarpa*). *Food Bioscience* 38, 100772. DOI: 10.1016/j.fbio.2020.100772.

Ndou, A., Tinyani, P.P., Slabbert, R.M., Sultanbawa, Y. and Sivakumar, D. (2019) An integrated approach for harvesting Natal plum (*Carissa macrocarpa*) for quality and functional compounds related to maturity stages. *Food Chemistry* 293, 499–510. DOI: 10.1016/j.foodchem.2019.04.102.

Ochieng, C.O. and Nandwa, B.O. (2010) Proximate composition, phenolic content and antioxidant activities of three black plum (*Vitex* sp.) fruits: preliminary results. *Journal of Food Technology* 8(3), 118–125. DOI: 10.3923/jftech.2010.118.125.

Ojeda, D., Jiménez-Ferrer, E., Zamilpa, A., Herrera-Arellano, A., Tortoriello, J. *et al.* (2010) Inhibition of angiotensin convertin enzyme (ACE) activity by the anthocyanins delphinidin- and cyanidin-3-O-sambubiosides from *Hibiscus sabdariffa*. *Journal of Ethnopharmacology* 127(1), 7–10. DOI: 10.1016/j.jep.2009.09.059.

Paredes-López, O., Cervantes-Ceja, M.L., Vigna-Pérez, M. and Hernández-Pérez, T. (2010) Berries: Improving human health and healthy aging, and promoting quality life—A review. *Plant Foods for Human Nutrition* 65(3), 299–308. DOI: 10.1007/s11130-010-0177-1.

Patel, S. (2013) Food, pharmaceutical and industrial potential of *Carissa* genus: an overview. *Reviews in Environmental Science and Bio/Technology* 12(3), 201–208. DOI: 10.1007/s11157-012-9306-7.

Payasi, A., Mishra, N.N., Chaves, A.L.S. and Singh, R. (2009) Biochemistry of fruit softening: an overview. *Physiology and Molecular Biology of Plants* 15(2), 103–113. DOI: 10.1007/s12298-009-0012-z.

Rahman, M.M., Joardder, M.U.H. and Karim, A. (2018) Non-destructive investigation of cellular level moisture distribution and morphological changes during drying of a plant-based food material. *Biosystems Engineering* 169, 126–138. DOI: 10.1016/j.biosystemseng.2018.02.007.

Sagar, V.R. and Suresh Kumar, P. (2010) Recent advances in drying and dehydration of fruits and vegetables: a review. *Journal of Food Science and Technology* 47(1), 15–26. DOI: 10.1007/s13197-010-0010-8.

Seeram, N.P. (2008) Berry fruits: compositional elements, biochemical activities, and the impact of their intake on human health, performance, and disease. *Journal of Agricultural and Food Chemistry* 56(3), 627–629. DOI: 10.1021/jf071988k.

Seke, F., Manhivi, V.E., Shoko, T., Slabbert, R.M., Sultanbawa, Y. *et al.* (2021) Effect of freeze drying and simulated gastrointestinal digestion on phenolic metabolites and antioxidant property of the Natal Plum (*Carissa macrocarpa*). *Foods* 10(6), 1420. DOI: 10.3390/foods10061420.

Sibiya, N.P., Kayitesi, E. and Moteetee, A. (2020) Mineral composition of selected indigenous wild southern African fruits. *South African Journal of Botany* 132, 87–94. DOI: 10.1016/j.sajb.2020.04.014.

Sibiya, N.P., Kayitesi, E. and Moteetee, A.N. (2021) Proximate analyses and amino acid composition of selected wild indigenous fruits of Southern Africa. *Plants* 10(4), 721. DOI: 10.3390/plants10040721.

Sinela, A., Rawat, N., Mertz, C., Achir, N., Fulcrand, H. *et al.* (2017) Anthocyanins degradation during storage of *Hibiscus sabdariffa* extract and evolution of its degradation products. *Food Chemistry* 214, 234–241. DOI: 10.1016/j.foodchem.2016.07.071.

Siriwoharn, T., Wrolstad, R.E., Finn, C.E. and Pereira, C.B. (2004) Influence of cultivar, maturity, and sampling on blackberry (*Rubus* L. Hybrids) anthocyanins, polyphenolics, and antioxidant properties. *Journal of Agricultural and Food Chemistry* 52(26), 8021–8030. DOI: 10.1021/jf048619y.

Souilem, F., Dias, M.I., Barros, L., Calhelha, R.C., Alves, M.J. *et al.* (2019) Amantagula fruit (*Carissa macrocarpa* (Eckl.) A.DC.): Nutritional and phytochemical characterization. *Plant Foods for Human Nutrition* 74(1), 76–82. DOI: 10.1007/s11130-018-0703-0.

Szajdek, A. and Borowska, E.J. (2008) Bioactive compounds and health-promoting properties of berry fruits: a review. *Plant Foods for Human Nutrition (Dordrecht, Netherlands)* 63(4), 147–156. DOI: 10.1007/s11130-008-0097-5.

Tadera, K., Minami, Y., Takamatsu, K. and Matsuoka, T. (2006) Inhibition of α-glucosidase. and α-amylase. by flavonoids. *Journal of Nutritional Science and Vitaminology* 52(2), 149–153. DOI: 10.3177/jnsv.52.149.

Tagliazucchi, D., Verzelloni, E., Bertolini, D. and Conte, A. (2010) In vitro bio-accessibility and antioxidant activity of grape polyphenols. *Food Chemistry* 120(2), 599–606. DOI: 10.1016/j.foodchem.2009.10.030.

US Department of Agriculture (2019) *Agricultural Research Service*. FoodData Central.

Vunchi, M.A., Umar, A.N., King, M.A., Liman, A.A., Jeremiah, G. *et al.* (2011) Proximate, vitamins and mineral composition of *Vitex doniana* (black plum) fruit pulp. *Nigerian Journal of Basic and Applied Sciences* 19(1). DOI: 10.4314/njbas.v19i1.69352.

Wehmeyer, A.S. (1966) The nutrient composition of some edible wild fruits found in the Transvaal. *South African Medical Journal = Suid-Afrikaanse Tydskrif Vir Geneeskunde* 40(45), 1102–1104.

Whale, S.K. and Singh, Z. (2007) Endogenous ethylene and color development in the skin of 'Pink Lady'apple. *Journal of the American Society for Horticultural Science* 132(1), 20–28. DOI: 10.21273/JASHS.132.1.20.

World Health Organization (2002) *The World Health Report 2002: Reducing Risks, Promoting Healthy Life.* World Health Organization, Geneva, Switzerland.

World Health Organization (2007) *Protein and Amino Acid Requirements in Human Nutrition. Report of a Joint FAO/WHO/UNU Expert Consultation (WHO Technical Report Series 935).* World Health Organization, Geneva, Switzerland.

Yazdankhah, S., Hojjati, M. and Azizi, M.H. (2019) The antidiabetic potential of black mulberry extract-enriched pasta through inhibition of enzymes and glycemic index. *Plant Foods for Human Nutrition* 74(1), 149–155. DOI: 10.1007/s11130-018-0711-0.

Youssef, K.M. and Mokhtar, S.M. (2014) Effect of drying methods on the antioxidant capacity, color and phytochemicals of *Portulaca oleracea* L. leaves. *Journal of Nutrition & Food Sciences* 04(6), 1. DOI: 10.4172/2155-9600.1000322.

24 Brazilian Cherry (*Eugenia uniflora* L.)

Adaucto Bellarmino Pereira-Netto*
Department of Botany, Universidade Federal do Paraná, Curitiba, Brazil

Abstract

Brazilian cherry (*Eugenia uniflora* L.), popularly known as 'pitangueira' or Surinam cherry, is highly appreciated in various Central and South America countries, like Brazil, due to its reddish and sour-sweet fruit. Brazilian cherry fruit also present highly desirable human health-promoting features. The serving size of raw Brazilian cherry (173 g) provides 50.56% of vitamin C and 18.57% of vitamin A human daily needs, besides being a low-calorie fruit (33 calories per 100g of fruit). However, probably the most relevant feature of Brazilian cherry fruit is the fact that they are rich in antioxidants, carotenoids and anthocyanins. In addition, essential oil from the leaves and fruit of Brazilian cherry has been used in folk medicine to treat various diseases for hundreds of years. In this chapter, we describe several bioactive properties of the extracts and essential oils obtained from Brazilian cherry.

Introduction

Eugenia uniflora L. (Myrtaceae), commonly known as Brazilian cherry, Surinam cherry or pitangueira, is a species native to Brazil, Uruguay, French Guiana and Suriname (van der Merwe *et al.*, 2004; Wilson, 2009) although it can also be found in Argentina, Australia, the Caribbean Islands, China, Egypt, India, Nigeria, Paraguay, Spain and southern USA (Consolini and Sarubbio, 2002; Lima *et al.*, 2002; Oliveira *et al.*, 2009). Similar to many exotic fruit, Brazilian cherries have traditionally been consumed where they are produced. However, such fruit are becoming increasingly popular worldwide due to their nutritional value and exotic, appealing flavour. In addition, Brazilian cherry contains compounds considered to present health-promoting properties, including therapeutic properties against important human diseases. Brazilian cherry has been used in traditional South American native medicine for over 500 years (Alonso, 1998).

Brazilian cherry fruit are wild harvested and also harvested from commercial plantations. Accumulation of anthocyanins, carotenoids and total soluble solid content, along with reduction in firmness, are directly related to progress in ripening. Brazilian cherry is harvested once, when the fruit have an orange, red or purple colour. Respiration rates are inversely related to progress in ripening, at least until the fruit begin to present signs of deterioration, when respiration rates increase. These findings suggest that Brazilian cherry fruit do not present climacteric behaviour (dos Santos *et al.*, 2002; Sanches *et al.*, 2017). Loss of fruit firmness is known to rapidly reduce shelf life in Brazilian cherry (Sanches *et al.*, 2017). However, postharvest treatment with L-cystein or calcium chlorate can reduce this loss of fruit firmness (Sanches *et al.*, 2017). Since Brazilian cherry fruit are highly perishable, their short shelf life limits marketability and potential expansion. Attempts to enhance Brazilian cherry shelf life using low storage temperatures showed that refrigeration (7.2–10.0°C) could keep the fruit

*apereira@ufpr.br

 Handbook of Phytonutrients in Indigenous Fruits and Vegetables (eds D. Sivakumar *et al.*)
DOI: 10.1079/9781789248067.0024

Brazilian cherry – purple variety

Brazilian cherry – red variety

Fig. 24.1. Ripened fruit of the purple and red varieties of Brazilian cherry (*Eugenia uniflora*). The ovate, glossy leaves, without stipules, and held in opposite pairs, are also seen.

fresh without damaging marketability for up to five days. This allows growers to consolidate their harvest during the week into a single product (Griffis, Jr. *et al.*, 2013).

The aim of this chapter is to summarize the information available on the phytonutrient constituents of Brazilian cherry, along with its nutritional and functional properties, and human health benefits.

Botany

Brazilian cherry are trees or shrubs, 3 to 10 m in height, with edible fruit. When immature, the fruit present a green-coloured epicarp. This epicarp turns orange, red or purple, depending on the cultivar, with maturity (Fig. 24.1). In orange fruit, carotenoids are the major colourant, while in red fruit a combination of carotenoids and anthocyanins are responsible for the colour, and in purple fruit, anthocyanins are the major colourant. Fruit from the Brazilian cherry look like a small pumpkin, presenting seven to ten longitudinal grooves ranging from 1.5 to 5 cm in diameter, and a persistent calyx. Brazilian cherry fruit are berries that present a succulent endocarp (pulp), 3 to 5 mm in thickness, with a unique and pleasant taste, sweet and sour at the same time. This endocarp is similar in colour to the skin. The pulp tends to be lighter in colour when compared to the skin (Fig. 24.2). Brazilian cherry fruit bear one fairly large, round seed (Fig. 24.2) or two or three smaller seeds, each with a flattened side, more or less attached to the flesh by a few slender fibres. Brazilian cherry has racemose inflorescences with the presence of inserted pedicellate flowers in the leaf axils. These long-stalked, hermaphrodite flowers have four delicate, recurved, white petals and a conspicuous central cluster of prominent white stamens ending in pale-yellow anthers. Leaves are ovate, without stipules, glossy and held in opposite pairs. Young leaves are typically pink in colour, becoming dark glossy green with ageing, reaching up to 4 cm in length.

Traditional Use

Brazilian cherry fruit is consumed either fresh or processed as candy, fruit compotes, ice cream, jelly, juice, liquor and wine. The leaves are rich in essential oil used in the cosmetic industry for the production of body oil, bath and face soaps, shampoos, hair conditioners and perfumes because of its astringent properties and pleasant smell.

(a) (b)

Fig. 24.2. (a) Whole ripe fruit; (b) transversal section of a fruit of the orange cultivar of Brazilian cherry (*Eugenia uniflora*). The epicarp, along with the endocarp and single seed, are shown.

The leaves of Brazilian cherry are used for treating diarrhoea, fever, hypertension and inflammation, besides uses as a diuretic and to lower blood glucose levels (Gomes *et al.*, 2020). Pharmacological uses for Brazilian cherry oil, extracted from the leaves include antimicrobial, antifungal and antidepressive activities, antinociceptive and hypothermic effects and anti-leishmania action. Alcoholic extracts of Brazilian cherry leaf present larvicidal activity against *Aedes aegyptii*, leishmanicidal activity and cytotoxic and antioxidant effects (Figueiredo *et al.*, 2019). Brazilian cherry leaf and bark tea is used by local communities to cure colic, diarrhoea, headache, rheumatism, sore throat and also in the control of hypertension (Rovedder *et al.*, 2016).

Nutritional Composition

Brazilian cherry fruit are rich in calcium, magnesium, phosphorus and potassium (Bender, 2006), vitamin C (ascorbic acid) (Freyre *et al.*, 2000; Lima *et al.*, 2002), vitamin E, caffeine, γ-sitosterol (a phytosterol known to reduce blood cholesterol levels) (Oliveira *et al.*, 2009), and also present notable amounts of pro-vitamin A (β-carotene) (Bender, 2006), lycopene (Porcu and Rodriguez-Amaya, 2008) (Table 24.1) and anthocyanins (Lima *et al.*, 2002) (Tables 24.1 and 24.2).

The amount of β-carotene found in Brazilian cherry is about 23-fold higher than the amount of

Table 24.1. Organic compounds well known for their relevance for human health, abundantly found and quantified in mature Brazilian cherry fruit.

Organic compound	Amount (mg/100 g fresh fruit)	References
β-carotene	11.98	Porcu and Rodriguez-Amaya, 2008
Lycopene (all-E)	71.10	Bender, 2006
Total anthocyanins	26.00	Lima *et al.*, 2002

Table 24.2. Amount of cyanidin 3-*O*-glucoside and myricetin 3-*O*-rhamnoside (mg aglycone equivalents/100 g dry weight – DW) in the pulp extracted from fruit of red and purple varieties of Brazilian cherry at different developmental stages (Celli *et al.*, 2011).

Fruit	Developmental stage	Cyanidin 3-*O*-glucoside (mg aglycone/100 g DW)	Myricetin 3-*O*-rhamnoside (mg aglycone/100 g DW)
Red variety	Green	Not detected	141.6±0.7
	Yellow	11.8±0.1	179.9±0.7
	Orange	18.8±0.1	213.7±0.9
	Red	31.0±1.2	226.1±4.0
Purple variety	Green	7.4±0.0	319.8±1.8
	Yellow	59.1±0.8	354.6±2.8
	Red	345.0±2.3	282.1±1.5
	Purple	1,689.6±36,6	279.6±1.3

β-carotene found in fruit known to present high amounts of β carotene, like tomato (Nasir *et al.*, 2015), which presents 0.51 mg of β-carotene per 100 g of fresh weight (FW) (Pinela *et al.*, 2012), or papaya (~0.3 mg per 100 g of FW, Schweiggert *et al.*, 2014). The amount of β-carotene found in Brazilian cherry exceeds 370% of the daily requirement of an adult's body.

Similarly, the amount of lycopene found in Brazilian cherry (Table 24.1) is significantly higher than the amount of lycopene found in other fruit known to present high amounts of lycopene. For example, the amount of lycopene found in Brazilian cherry is 147% higher than the amount of lycopene found in tomato, which presents 2.9 mg of lycopene per 100 g of fresh fruit (https://fdc.nal.usda.gov/fdc-app.html#/food-details/1999634/nutrients). The amount of lycopene found in Brazilian cherry is higher even when compared to fruit like watermelon, which presents 4.8 mg of lycopene per 100 g of FW (Edwards *et al.*, 2003). Although there is currently no recommended daily intake for lycopene, comparison between the lycopene content of Brazilian cherry and other fruit rich in lycopene indicates that Brazilian cherry could enrich human diet as one of the richest known dietary sources of lycopene.

Phytochemicals

A broad variety of phytochemicals has been identified in leaf extracts from Brazilian cherry, including flavonols, terpenoids, tannins, saponins (Meira *et al.*, 2020), steroids, essential oils (Gomes *et al.*, 2020), anthraquinones (Sobeh *et al.*, 2020), mono- and triterpenic compounds and cineol (reviewed in Consolini and Sarubbio, 2002). Analysis by high performance liquid chromatography coupled with mass-spectrometry (LC-MS) of leaf ethyl acetate fraction revealed myricetin-3-*O*-rhamnoside, by far, as the major polyphenolic identified. Myricetin rhamnoside and myricetin galloyl-rhamnoside followed myricetin-3-*O*-rhamnoside (Table 24.2) in abundance (Sobeh *et al.*, 2020). These three myricetin derivatives, together, made up 42.75% of abundance, out of 22 polyphenolics identified in Brazilian cherry leaf ethyl acetate fraction. Gas chromatography-mass spectrometry (GC-MS) analysis revealed that selina-1,3,7(11)-trien-8-one, oleic acid, phytol and γ-sitosterol were the most abundant compounds identified in supercritical CO_2-extraction of Brazilian cherry leaf. Other compounds identified in these extracts included α-tocopherol (a type of vitamin E), palmitic acid, vitamin C and friedelin (Canabarro *et al.*, 2020). Importantly, α-tocopherol is known for its anticarcinogenic activity and to prevent Alzheimer's disease; phytol is broadly employed in cosmetics, detergents, household cleaners, shampoos and soap, due to its remarkable aroma; γ-sitosterol presents antidiabetic activity; and friedelin shows analgesic, anti-inflammatory and antipyretic activity. Although compounds of pharmacological interest like α-tocopherol, γ-sitosterol and friedelin have been undoubtedly identified in Brazilian cherry extracts through techniques like GC-MS and nuclear magnetic resonance (NMR), unfortunately, reports on non-comparative (absolute) quantification of these compounds are rare or even non-existent.

Analysis of flavonols, a class of polyphenolic secondary metabolites, by LC-MS from fruit from the red and purple varieties of Brazilian cherry throughout the fruit development stages revealed that myricetin 3-*O*-rhamnoside was the major flavonoid glycoside identified in red fruit at all developmental stages (Table 24.2). This anthocyanin, along with β-carotene and lycopene are responsible, at least partially, for the red colour of the red variety of Brazilian cherry. Another anthocyanin, cyaniding 3-*O*-glycoside was found to occur in fruit of both the red and purple varieties of Brazilian cherry (Celli *et al.*, 2011). A remarkable accumulation of cyanidin 3-*O*-glycoside was found for fruit of the purple variety at the two latest developmental stages. At the purple stage, the last developmental stage, cyanidin glycoside made up 59% of the total flavonoid glycosides. It is noteworthy that in fruit of the red variety, at the final developmental stage, red state, cyanidin glycoside accounted for only 4% of the total flavonoid glycosides (Celli *et al.*, 2011). Cyanidin 3-*O*-glycoside was later confirmed as by far the most abundant anthocyanin in purple fruit of Brazilian cherry, accounting, in this study, for 82.7% of the total anthocyanin (Tambara *et al.*, 2018). Because of the characteristic red-purple colour of cyanidin 3-*O*-glucoside, this anthocyanin is likely a major pigment contributing to the dark red-purple colour of mature fruit of the purple Brazilian cherry. Fruit of the purple variety at the purple stage showed a content of cyanidin 3-*O*-glucoside 4.9-fold higher when compared to fruit at the red stage (Celli *et al.*, 2011). The highest amount of cyanidin 3-*O*-glucoside found in the purple variety of Brazilian cherry is several times higher when compared to fruit known to present significant amounts of this anthocyanin, like strawberry and high-bush blueberry (Fig. 24.3), and similar when compared to fruit known to present the highest known amounts of cyanidin 3-*O*-glucoside, like blackberry (Fig. 24.3). In Brazilian cherry, most of the anthocyanins are found in the peel, 0.42% of FW, while in the pulp, anthocyanins are found only at 0.03% of the pulp FW. However, the amount of anthocyanins in the pulp are not neglectable once the pulp holds almost 80% of the whole fruit FW (Bicas *et al.*, 2011) and pulp is the major product of Brazilian cherry.

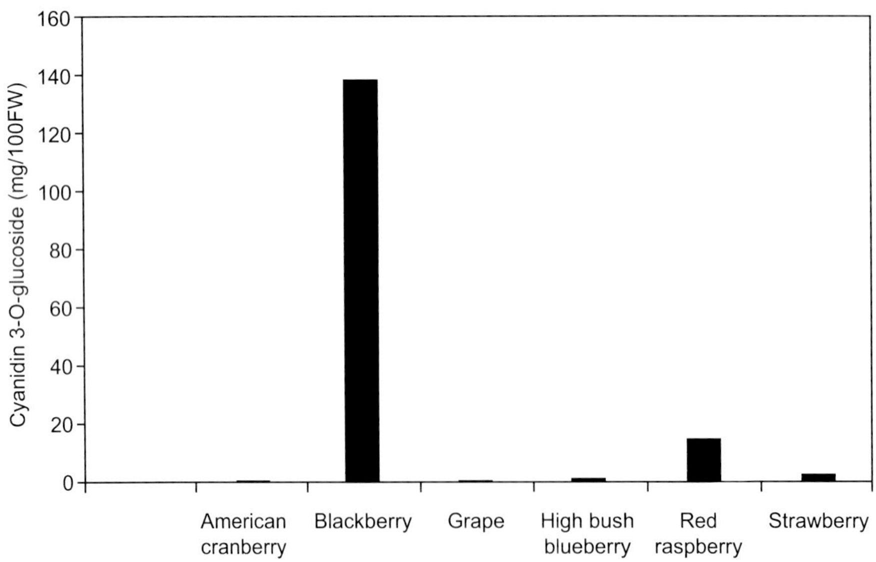

Fig. 24.3. Amount of cyanidin 3-*O*-glucoside in the pulp of various berries.

GC/MS analysis along with determination of Kovatz indexes indicated atractylone and curzerene as the major constituents of Brazilian cherry seed oil (Lago *et al.*, 2011). Curzerene is a rearrangement product of the heat-sensitive furanodiene. Curzerene, a terpenoid also found as a major component of myrrh oil, along with furanodiene, yields the unique scent that characterizes the Brazilian cherry (Figueiredo *et al.*, 2019).

Antinutritive Compounds

Reports on harmful compounds in Brazilian cherry that interfere with the absorption of biomolecules and hamper their bioavailability to humans are not common. However, oxalates and other compounds like cyanogenic glycosides are present in Brazilian cherry fruit – tough cyanogenic glycosides are found in very low concentrations. Because of this, Brazilian cherry fruit are considered to be safe, regarding cyanide poisoning, due to the fact that cyanogenic glycosides found in this fruit are far below the harmful level of 4.6 mg/kg (Onwudiwe *et al.*, 2018). Oxalate is the antinutritional compound found in the highest concentration in Brazilian cherry fruit. Oxalate is known to reduce calcium absorption and to contribute to the formation of kidney stones. However, since oxalate is found at 34% of the total antinutritional content of the pulp, this content is even considered to have a positive impact on the health of consumers due to a potential enhancement of the bioavailability of essential dietary minerals of the pulp of Brazilian cherry fruit. In fact, the level of all of the antinutritional compounds found in Brazilian cherry fruit is considered to be safe for human consumption (Onwudiwe *et al.*, 2018).

Implications of Food Processing on Nutrients and Phytochemicals

Ripe Brazilian cherry fruit rapidly deteriorate, which results in short shelf life. However, pulp freezing has been used commercially to overcome reduced shelf life. Nevertheless, pulp freezing for 90 days, for example, results in 14% loss in total carotenoids (Lopes *et al.*, 2005). Lyophilization and spray-drying have been used, more recently, for the preservation of Brazilian cherry pulp. These processes allowed a reduction of over 80% on the humidity of the pulp (Borges, 2015). However, losses in anthocyanins and carotenoids observed after processing demonstrate that these preservation efforts require improvement. Please see more information on the implications of processing on Brazilian cherry health promoting compounds below in the bioaccessability/bioavailability section.

Bioactive Properties

Health benefits

Antinociceptive (pain killer) activity

Essential oil extracted from the leaves of Brazilian cherry by hydrodistillation, its pentane fraction and an isolated mixture of sesquiterpenes, all provided orally presented peripheric (acetic acid-induced abdominal constrictions) and central (hot plate test) antinociceptive (pain killer) activity, besides presenting hypothermic activity in albino Swiss mice (Amorim *et al.*, 2009).

Antisepsis activity

A flavonoid-rich fraction obtained from the leaves of Brazilian cherry was shown to reduce sepsis, the leading cause of mortality in intensive care units in various countries, in a mice-based model that mimics sepsis in humans.

Lifespan enhancement

Caenorhabditis elegans mutants that show increased sensitivity to oxidative stress and reduced lifespan, treated with purple Brazilian cherry extracts, presented enhanced lifespan (Tambara *et al.*, 2018).

Antimicrobial activity

The essential oil from seeds of Brazilian cherry presented activity against two Gram-positive bacteria, *Streptococcus equi* and *Staphylococcus epidermis*. Furthermore, this oil also presented activity against opportunistic pathogenic yeasts, i.e., six species of the genus *Candida* (*Candida albicans*, *C. dubliniensis*, *C. glabrata*, *C.krusei*, *C. parapsilosis* and *C. tropicalis*) and three species of the genus *Cryptococcus* (*Cryptococcus grubii*, *Cryptococcus gattii* and *C. neoformans*) (Lago *et al.*, 2011). Victoria *et al.* (2012) reported that the essential oil from Brazilian cherry presented antimicrobial activity against *Candida guilliermondii* and *Candida lipolytica*, *Listeria monocytogenes* and *Staphylococcus aureus*.

Antiparasite activity

Diseases of parasitic origin are very common in developing countries. Treatment of such diseases, leishmaniasis and trypanosomiasis included, is often challenging. Brazilian cherry leaf extract presented trypanocidal and leishmanicidal activity (de Souza *et al.*, 2017; Dos Santos *et al.*, 2020).

Hepatoprotective activity

An ethyl acetate fraction of leaves from the red variety of Brazilian cherry showed hepatoprotective effects in rats presenting carbon tetrachloride (CCl_4)-induced acute liver injury (Sobeh *et al.*, 2020).

Blood pressure control activity

Intraperitoneal administration of leaf crude aqueous extract of Brazilian cherry was demonstrated to present a significant hypotensive activity in normotensive anaesthetized rats (Meira *et al.*, 2020). These results scientifically support the empirical use of Brazilian cherry in folk medicine for the treatment of hypertension.

Protection against acute kidney injury

Acute kidney injury is a condition characterized by a fast and potentially reversible reduction in kidney function, associated with increased risk of death and long hospitalization periods in intensive care units, which leads to high costs and major economic impacts in health systems. Treatment of Wistar rats with Brazilian cherry leaf ethanolic extract prevented acute kidney injury-induced reduction in glomerular filtration rate and renal blood flow, and promoted a rise in renal vascular resistance as well. In addition, the ethanolic extract also prevented enhanced oxidative stress and apoptosis, demonstrating its kidney protective effect against acute kidney injury development (Meira *et al.*, 2020).

Antitumoural activity

Leaf oil from Brazilian cherry presented cytotoxic activity against prostate (PC-3), liver (Hep-G2) and breast (Hs578T) cancer cell lines (Ogunwande *et al.*, 2005). A research panel in which a dichloromethane extract from leaves of Brazilian cherry was tested for antiproliferative activity against nine human cancer cell lines demonstrated that the use of this dichloromethane extract resulted in growth inhibition of the all nine tumour cell lines tested (Gomes *et al.*, 2020). Using the 3-(4,5-dimethyl-2-thiazol)-2,5-diphenyl-2H-tetra zolium bromide cell-proliferation (MTT) assay, Figueiredo *et al.* (2019) demonstrated that an essential oil extracted from Brazilian cherry displayed cytotoxic activity against human cancer cell lines HCT-116 (colon), AGP-01 (gastric), SKMEL-19 (melanoma) and MRC-5 (normal human fibroblast).

Bioaccessability and Bioavailability

Brazilian cherry fruit are very rich in carotenoids (especially lycopene and β-carotene) and anthocyanins (Table 24.1). However, bioaccessibility of such compounds in food is very low because they are susceptible to structural changes and degradation in the gut. In order to overcome this problem, various strategies have been combined, such as structure design approaches and/or delivery systems engineering, aiming at enhancing solubility, and preserving, protecting and increasing bioaccessibility and bioactivity of health promoting compounds. Among these strategies, the most widely employed is encapsulation in emulsion-based systems. Microemulsions have been produced through the application of direct processing (high-speed homogenization and ultrasound) of the whole pulp together with surfactants and maize oil. Microemulsions have been shown to be able to increase the amount of carotenoids and anthocyanins released from Brazilian cherry pulp. Final recovery of carotenoids, for example, after passing by a dynamic gastrointestinal system that simulates the human gastrointestinal system, is higher for microemulsions compared to unprocessed pulps, demonstrating that microemulsions are capable of increasing β-carotene and lycopene bioaccessibility (Berni *et al.*, 2020).

The bioavailability of both lycopene and β-carotene from tomato products has been shown to increase with heat and/or homogenization, processes that break down plant cell walls, allowing release of carotenoids. In addition, heat processing of tomato into various products such as ketchup and sauces induces isomerization of lycopene from *trans* to *cis* configuration, which increases lycopene bioavailability. In fact, plasma concentrations of lycopene have been shown to be enhanced after ingestion of canned tomato juice (Edwards *et al.*, 2003). Although such processing sounds promising as a tool to enhance bioaccessability/bioavailability of health promoting compounds like carotenoids and anthocyanins from Brazilian cherry, this is yet to be tested.

Novel Products/Formulations

The high number of compounds like anthocyanins and carotenoids, along with essential oils found in the seeds, leaves and fruit of the Brazilian cherry, especially in the purple variety, justify a broader use of this species by the food industry for the development of new naturally coloured, antioxidant enriched, health-promoting products. These promising products might be effectively used in therapies where antitumoural, antinociceptive (pain killer), antisepsis, antimicrobial, antiparasite and hepatoprotective activities, blood pressure control and protection against acute kidney injury are needed. After further research, Brazilian cherry products might help to prolong lifespan in other organisms besides *Caenorhabditis elegans*. The low toxicity shown by extracts and essential oils from Brazilian cherry has demonstrated the potential of these extracts and oils to be used as an alternative or adjuvant to the currently used high toxicity, synthetic drugs-based therapies.

Development of new chemotherapeutic drugs, based in quercetin derivatives, (major flavonoids found in Brazilian cherry extracts known to present selective antiproliferative activity, including induction of cell apoptosis, and suppression of migratory and invasive properties in several cancer cell lines, melanoma cells included), might be potential candidates for use in anticancer therapy. In addition, since β-sitosterol has been shown to be effective against breast cancer cells (Chai *et al.*, 2008), a new Brazilian cherry-derived β-sitosterol containing chemotherapeutic drugs might be useful in treatment against breast cancer.

Since the essential oil from Brazilian cherry was shown to present cytotoxic activity against various cancer cell lines, it is a potential candidate for anticancer therapy, as a preventive or adjuvant treatment, especially in metastatic tumours, in which therapy is still ineffective.

Conclusions and Future Trends

Since safety tests like the first-tier *Allium cepa* test have demonstrated that aqueous extracts of Brazilian cherry present neither significant genotoxic or clastogenic effects (Yajía *et al.*, 1997), or other safety concerns, when used in doses demonstrated to be biologically effective against various diseases, the use of these extracts might be considered as an alternative or adjuvant in therapies in which Brazilian cherry extracts have been demonstrated to be effective.

Development of modern technologies like microencapsulation, lyophilization and spray-drying for the preservation of compounds of interest extracted from Brazilian cherries are expected to stimulate the use of these compounds in various applications in industries like the cosmetic industry and especially in human-health promoting initiatives.

Acknowledgement

The author thanks CNPq-Brazil (grant 311946/2014-3) for financial support.

References

Alonso, J.R. (1998) *Tratado de Fitomedicina: Bases Clínicas y Farmacológicas*. Isis Ediciones, Buenos Aires, Argentina.

Amorim, A.C.L., Lima, C.K.F., Hovell, A.M.C., Miranda, A.L.P. and Rezende, C.M. (2009) Antinociceptive and hypothermic evaluation of the leaf essential oil and isolated terpenoids from *Eugenia uniflora* L. (Brazilian Pitanga). *Phytomedicine: International Journal of Phytotherapy and Phytopharmacology* 16(10), 923–928. DOI: 10.1016/j.phymed.2009.03.009.

Bender, D.A. (2006) *Benders' Dictionary of Nutrition and Food Technology*, 8th edn. Woodhead Publishing, Cambridge, UK.

Berni, P., Pinheiro, A.C., Bourbon, A.I., Guimarães, M., Canniatti-Brazaca, S.G. *et al.* (2020) Characterization of the behavior of carotenoids from pitanga (*Eugenia uniflora*) and buriti (*Mauritia flexuosa*) during microemulsion production and in a dynamic gastrointestinal system. *Journal of Food Science and Technology* 57(2), 650–662. DOI: 10.1007/s13197-019-04097-7.

Bicas, J.L., Molina, G., Dionísio, A.P., Barros, F.F.C., Wagner, R. *et al.* (2011) Volatile constituents of exotic fruits from Brazil. *Food Research International* 44(7), 1843–1855. DOI: 10.1016/j.foodres.2011.01.012.

Borges, K.C. (2015) Pitanga (*Eugenia uniflora*) desidratada por atomização e liofilização: Características físico-químicas, compostos bioativos e efeito sobre longevidade, estresse oxidativo e neurotoxicidade induzida em modelos in vivo *Caenorhabditis elegans*. Ph.D. thesis, Universidade Federal do Rio Grande do Norte, Natal, Brazil.

Canabarro, N.I., Veggi, P.C., Vardanega, R., Mazutti, M.A. and Ferreira, M. do C. (2020) Techno-economic evaluation and mathematical modeling of supercritical CO2 extraction from *Eugenia uniflora* L. leaves. *Journal of Applied Research on Medicinal and Aromatic Plants* 18, 100261. DOI: 10.1016/j.jarmap.2020.100261.

Celli, G.B., Pereira-Netto, A.B. and Beta, T. (2011) Comparative analysis of total phenolic content, antioxidant activity, and flavonoids profile of fruits from two varieties of Brazilian cherry (*Eugenia uniflora* L.) throughout the fruit developmental stages. *Food Research International* 44(8), 2442–2451. DOI: 10.1016/j.foodres.2010.12.036.

Chai, J.W., Kuppusamy, U.R. and Kanthimathi, M.S. (2008) Beta-sitosterol induces apoptosis in MCF-7 cells. *Malaysian Journal of Biochemistry and Molecular Biology* 16, 28–30.

Consolini, A.E. and Sarubbio, M.G. (2002) Pharmacological effects of *Eugenia uniflora* (Myrtaceae) aqueous crude extract on rat's heart. *Journal of Ethnopharmacology* 81(1), 57–63. DOI: 10.1016/s0378-8741(02)00039-9.

de Souza, C.E.S, Pereira da Silva, A.R., Rocha, J.E., Vega Gomez, M.C., Rolóm, M. *et al.* (2017) LC–MS characterization, anti-kinetoplastide and cytotoxic activities of natural products from *Eugenia jambolana* Lam. and *Eugenia uniflora*. *Asian Pacific Journal of Tropical Biomedicine* 7(9), 836–841. DOI: 10.1016/j.apjtb.2017.08.007.

dos Santos, A.F., Silva, S.M. and Mendonça, R.M.N. (2002) Alterações fisiológicas durante a maturação de pitanga (*Eugenia uniflora* L.). In: *Proceedings of the Interamerican Society for Tropical Horticulture*, Vol. 1. FAO, Rome, Italy, pp. 52–54.

Dos Santos, C., Mizobucchi, A.L., Escaramboni, B., Lopes, B.P., Angolini, C.F.F. *et al.* (2020) Optimization of *Eugenia punicifolia* (Kunth) D. C. leaf extraction using a simplex centroid design focused on extracting phenolics with antioxidant and antiproliferative activities. *BMC Chemistry* 14(1), 34. DOI: 10.1186/s13065-020-00686-2.

Edwards, A.J., Vinyard, B.T., Wiley, E.R., Brown, E.D., Collins, J.K. *et al.* (2003) Consumption of watermelon juice increases plasma concentrations of lycopene and β-carotene in humans. *The Journal of Nutrition* 133(4), 1043–1050. DOI: 10.1093/jn/133.4.1043.

Figueiredo, P.L.B., Pinto, L.C., da Costa, J.S., da Silva, A.R.C., Mourão, R.H.V. *et al.* (2019) Composition, antioxidant capacity and cytotoxic activity of *Eugenia uniflora* L. chemotype-oils from the Amazon. *Journal of Ethnopharmacology* 232, 30–38. DOI: 10.1016/j.jep.2018.12.011.

Freyre, M., Baigorria, C., Rozycki, V., Bernardi, C.M. and Charpentier, M. (2000) Vegetales silvestres sub explotados del chaco argentino y su potencial como recurso alimentício. *Archivos Latino Americanos de Nutrición* 50, 394–399.

Gomes, D.B., Zanchet, B., Serpa, P.Z., Locateli, G., Miorando, D. *et al.* (2020) Chemical Analysis and in Vitro Antiproliferative Potential of *Eugenia uniflora* L. (Myrtaceae). *European Journal of Medicinal Plants* 31, 34–44. DOI: 10.9734/ejmp/2020/v31i630246.

Griffis, Jr., J.L., McDonald, T.G., Manners, M.M. and Bingham, J.P. (2013) Some effects of refrigerated storage on postharvest quality of ripe fruits of the tropical, purple-fruited pitanga (Eugenia uniflora L.). *Acta Horticulturae* 1088(1088), 113–118. DOI: 10.17660/ActaHortic.2015.1088.14.

Lago, J.H.G., Souza, E.D., Mariane, B., Pascon, R., Vallim, M.A. *et al.* (2011) Chemical and biological evaluation of essential oils from two species of myrtaceae - *Eugenia uniflora* L. and *Pliniatrunci flora* (O. berg) kausel. *Molecules* 16(12), 9827–9837. DOI: 10.3390/molecules16129827.

Lima, V.L.A.G. de, Mélo, E. de A. and Lima, D.E. da S. (2002) Fenólicos e carotenóides totais em pitanga. *Scientia Agricola* 59(3), 447–450. DOI: 10.1590/S0103-90162002000300006.

Lopes, A.S., Mattietto, R. de A. and Menezes, H.C. de (2005) Estabilidade da polpa de pitanga sob congelamento. *Food Science and Technology* 25(3), 553–559. DOI: 10.1590/S0101-20612005000300026.

Meira, E.F., Oliveira, N.D., Mariani, N.P., Porto, M.L., Severi, J.A. *et al.* (2020) *Eugenia uniflora* (pitanga) leaf extract prevents the progression of experimental acute kidney injury. *Journal of Functional Foods* 66, 103818. DOI: 10.1016/j.jff.2020.103818.

Nasir, M.U., Hussain, S. and Jabbar, S. (2015) Tomato processing, lycopene and health benefits: a review. *Science Letters* 3, 1–5.

Ogunwande, I., Olawore, N., Ekundayo, O., Walker, T., Schmidt, J. *et al.* (2005) Studies on the essential oils composition, antibacterial and cytotoxicity of *Eugenia uniflora* L. *International Journal of Aromatherapy* 15(3), 147–152. DOI: 10.1016/j.ijat.2005.07.004.

Oliveira, A.L., Kamimura, E.S. and Rabi, J.A. (2009) Response surface analysis of extract yield and flavour intensity of Brazilian cherry (*Eugenia uniflora* L.) obtained by supercritical carbon dioxide extraction. *Innovative Food Science and Emerging Technologies* 10(2), 189–194. DOI: 10.1016/j.ifset.2008.11.002.

Onwudiwe, N.N., Njoku, O.U., Nwodo, O.F.C., Parker, J.E. and Ogugofor, M.O. (2018) Evaluation of the nutritional compositions and analgesic effects of the flavonoid fraction of *Eugenia uniflora* ripe fruit pulp. *American Journal of Ethnomedicine* 05(1), 7. DOI: 10.21767/2348-9502.100007.

Pinela, J., Barros, L., Carvalho, A.M. and Ferreira, I.C.F.R. (2012) Nutritional composition and antioxidant activity of four tomato (*Lycopersicon esculentum* L.) farmer' varieties in Northeastern Portugal homegardens. *Food and Chemical Toxicology* 50(3–4), 829–834. DOI: 10.1016/j.fct.2011.11.045.

Porcu, O.M. and Rodriguez-Amaya, D.B. (2008) Variation in the carotenoid composition of the lycopene-rich Brazilian fruit *Eugenia uniflora* L. *Plant Foods for Human Nutrition* 63(4), 195–199. DOI: 10.1007/s11130-008-0085-9.

Rovedder, A.P.M., Piazza, E.M., Thomas, P.A., Felker, R.M., Hummel, R.B. *et al.* (2016) Potential medicinal use of forest species of the Deciduous Seasonal Forest from Atlantic Forest Biome, South Brazil. *Brazilian Archives of Biology and Technology* 59, 1–11. DOI: 10.1590/1678-4324-2016150329.

Sanches, A.G., Costa, J.M., Silva, M.B., Moreira, E.G.S. and Cosme, S.S. (2017) Tratamentos químicos na manutenção da qualidade pós-colheita em frutos de pitanga (*Eugenia uniflora* L.). *Nativa* 5(4), 257–262. DOI: 10.5935/2318-7670.v05n04a05.

Schweiggert, R.M., Kopec, R.E., Villalobos-Gutierrez, M.G., Högel, J., Quesada, S. *et al.* (2014) Carotenoids are more bioavailable from papaya than from tomato and carrot in humans: a randomised cross-over study. *The British Journal of Nutrition* 111(3), 490–498. DOI: 10.1017/S0007114513002596.

Sobeh, M., Hamza, M.S., Ashour, M.L., Elkhatieb, M., El Raey, M.A. *et al.* (2020) A polyphenol-rich fraction from *Eugenia uniflora* exhibits antioxidant and hepatoprotective activities *in vivo*. *Pharmaceuticals* 13(5), 84. DOI: 10.3390/ph13050084.

Tambara, A.L., de Los Santos Moraes, L., Dal Forno, A.H., Boldori, J.R., Gonçalves Soares, A.T. *et al.* (2018) Purple pitanga fruit (*Eugenia uniflora* L.) protects against oxidative stress and increase the lifespan in *Caenorhabditis elegans* via the DAF-16/FOXO pathway. *Food and Chemical Toxicology* 120, 639–650. DOI: 10.1016/j.fct.2018.07.057.

van der Merwe, M.M., Wyk, A.E. van and Botha, A.M. (2004) Molecular phylogenetic analysis of *Eugenia* L. (Myrtaceae), with emphasis on southern African taxa. *Plant Systematics and Evolution* 251(1), 21–34. DOI: 10.1007/s00606-004-0160-0.

Victoria, F.N., Lenardão, E.J., Savegnago, L., Perin, G., Jacob, R.G. *et al.* (2012) Essential oil of the leaves of *Eugenia uniflora* L.: antioxidant and antimicrobial properties. *Food and Chemical Toxicology* 50(8), 2668–2674. DOI: 10.1016/j.fct.2012.05.002.

Wilson, P.G. (2009) Conspectus of the genus *Eugenia* (Myrtaceae) in the Philippines. *Garaden's Bulletin* 60, 399–410.

Yajía, M.E., González, C.F. and Lorca, C.L. (1997) Toxicidad general y genotoxicidad de los extractivos de *Eugenia uniflora* L. (Myrtaceae). *Anales de La Sociedad Argentina Para La Investigación de Productos Aromáticos* 15, 229–237.

25 Monkey Orange (*Strychnos spinosa* Lam.)

Adeyemi Oladapo Aremu[1,2,3]*, Abiodun Olusola Omotayo[2] and Mack Moyo[4]
[1]Indigenous Knowledge Systems Centre, Faculty of Natural and Agricultural Sciences, North-West University, Mmabatho, North West Province, South Africa; [2]Food Security and Safety Niche Area, Faculty of Natural and Agricultural Sciences, North-West University, Mmabatho, North West Province, South Africa; [3]School of Life Sciences, University of KwaZulu-Natal Pietermaritzburg, Scottsville, South Africa; [4]Department of Horticulture, Faculty of Applied Sciences, Durban University of Technology, Durban, South Africa

Abstract

Spiny monkey orange (*Strychnos spinosa* Lam.) is one of the neglected indigenous fruit trees occurring in tropical and subtropical Africa. This chapter provides a critical appraisal of the nutritional composition, phytochemical content and health-promoting attributes of *Strychnos spinosa*. Spiny monkey orange remains a commonly consumed edible fruit for nutrition and energy to meet the needs of humans and animals. Nutritional analyses have revealed the fruit as a rich source of carbohydrates, crude protein, fats, fibre and minerals. The fruit is known for diverse health-promoting benefits including the management of non-communicable diseases and sexually transmitted infections among different ethnic groups in Africa. Total phenol, flavonol and proanthocyanidin content in *Strychnos spinosa* fruit were approximately 40 mg/100 g FW gallic acid equivalents (GAE), 55 mg/100 g FW catechin equivalents and 0.407% leucocyanidin equivalents, respectively. *Strychnos spinosa* fruit extract and juice have demonstrated varying degrees of biological activity including antioxidant, antimicrobial and anti-tick effects. However, the majority of the existing evidence on these aforementioned therapeutic properties has been based on *in vitro* models, thereby limiting their clinical significance.

Introduction

Strychnos spinosa Lam. (commonly known as spiny monkey orange) is one of over 200 members of the largest genus of the family Loganiaceae that has remained of great interest due to its diverse nutritional, phytochemical and pharmacological properties (Ohiri *et al.*, 1983; Rasoanaivo *et al.*, 2002; Sitrit *et al.*, 2003; Tittikpina *et al.*, 2020). Globally, members of the genus *Strychnos*, especially *Strychnos spinosa*, are receiving more attention as potential new crops and food sources due to their inherent ability to tolerate severe environmental conditions (Sitrit *et al.*, 2003; Ngadze *et al.*, 2017; Nitcheu Ngemakwe *et al.*, 2017; van Rayne *et al.*, 2020). *Strychnos spinosa* fruit has commercial potential for the production of jam, jelly and other related processed products (Mizrahi *et al.*, 2002; Van Wyk, 2011; Rodrigues *et al.*, 2018). However, the current extensive dependence on the wild

*Corresponding author: Oladapo.Aremu@nwu.ac.za

 Handbook of Phytonutrients in Indigenous Fruits and Vegetables (eds D. Sivakumar *et al.*)
DOI: 10.1079/9781789248067.0025

population of *Strychnos spinosa* remains a major challenge for its successful commercialization (Sitrit *et al.*, 2003; Ngadze *et al.*, 2017; Rodrigues *et al.*, 2018). As a result, the potential of *Strychnos spinosa* for intensive cultivation was explored at the Besor Experimental Station in the western Negev Desert, Israel (Sitrit *et al.*, 2003). Preliminary evidence suggested a high degree of variability in terms of their growth, yield, fruit size, ripening season and taste (Mizrahi *et al.*, 2002).

Strychnos spinosa is considered as a small fruit tree or shrub which is indigenous to tropical and subtropical Africa (Van Wyk and Gericke, 2017). Morphologically, the fruits are large, similar in size and shape to apples, pears and oranges, which makes them easy to handle (National Research Council, 2008). The *Strychnos* species is widely represented in Africa, America and Asia (Ngadze, 2018; Rasoanaivo *et al.*, 2002). *Strychnos spinosa* has several traditional uses, and is known to be a rich source of nutrition and phytochemicals, thereby suggesting its potential health benefits (Nitcheu Ngemakwe *et al.*, 2017; Ngadze, 2018). In addition to the increasing importance of food and nutritional security, the ecological advantages of *Strychnos spinosa* make the plant an ideal candidate for commercialization (Sitrit *et al.*, 2003; Thiombiano *et al.*, 2013; Rodrigues *et al.*, 2018; Avakoudjo *et al.*, 2020; Tittikpina *et al.*, 2020). This chapter provides a critical appraisal of its nutritional composition, phytochemical content and health-promoting benefits. It is anticipated that consolidated information on *Strychnos spinosa* will be essential in any attempt to unravel and explore its functional food potential and therapeutic properties.

Botany

Strychnos spinosa is an accepted name for the plant but approximately 30 synonyms currently exist (The Plant List, 2013). The plant has several common names among different ethnic groups in Africa, especially in South Africa where it has a name among the majority of the official languages (Table 25.1). The common name of 'spiny monkey orange' is associated with the presence of spines on the tree and the frequent consumption of the fruit by monkeys (Rodrigues *et al.*, 2018).

Table 25.1. Overview of the taxonomy and general features of *Strychnos spinosa* (spiny monkey orange).

Feature	Details	References
Botanical name	*Strychnos spinosa* Lam.	The Plant List, 2013
Family	Loganiaceae	
Synonyms	*Brehmia spinosa* Harv. *Strychnos euryphylla* Gilg & Busse *Strychnos mueghe* Chiov. *Strychnos sansibariensis* Gilg *Strychnos spinosa* subsp. *spinosa* *Strychnos volkensii* Gilg	
Common/local names in South Africa		
Afrikaans	Groenklapper, doringklapper	Van Wyk and Gericke, 2017; Shai *et al.*, 2020; SANBI, 2021
English	Spiny monkey orange/green monkey orange	
Northern Sotho	Morapa	
Sepulana	Mashala	
Swati	umKwakwa	
Tswana	Nsala	
Vhavenda	Muramba	
Zulu	Umhala	

Fig. 25.1. Morphology of *Strychnos spinosa* (spiny monkey orange) indicating: (a) the fruiting monkey orange tree; (b) mature green fruit; (c) ripe fruit.

Strychnos spinosa ranges considerably from 1 to 7 m in height, and bears edible, round-shaped fruit (Fig. 25.1). The plant is characterized by the presence of simple, oval and glabrous leaves which are arranged in opposite directions with axillary spines. The unripe fruit has a bright-green woody peel of about 3–4 mm thick and turns yellow-brown upon ripening (Van Wyk and Gericke, 2017). The fruit size varies (10–15 cm in diameter) and the fruit contain flat seeds and have a pleasantly sourish taste, rather like apricots (Skead, 2009). The fruit has a pale brown, edible, juicy, sweet-sour pulp and contains hard-brown (1–3 cm) seeds, which may range from 53 to 69 seeds/fruit (Sitrit *et al.*, 2003).

Traditional Use

Strychnos spinosa remains a popular fruit tree in several African countries and is often considered a source of income for rural households (Thiombiano *et al.*, 2013; Bruschi *et al.*, 2014; Magwede *et al.*, 2019; Avakoudjo *et al.*, 2020; Dejene *et al.*, 2020; Shai *et al.*, 2020). In many countries, the edible fruit remains a source of energy and nutrition for humans and several animals (e.g. rodents, elephants and monkeys). In Namibia, the fruit is used for preparing local beer/beverages (Cheikhyoussef and Embashu, 2013).

The therapeutic value of the fruit has been well highlighted in folk medicine (Fig. 25.2). The use of *Strychnos spinosa* for managing health conditions such as inflammation and pain remains a common practice in West Africa (Avakoudjo *et al.*, 2020). Evidence of the use of *Strychnos spinosa* fruit as fish poison for fishing has been recorded in Mozambique (Bruschi *et al.*, 2014) and Zambia (Neuwinger, 2004).

Nutritional Composition

The fruit of spiny monkey is rich in carbohydrates, crude protein, fats, fibre and energy (Table 25.2). Furthermore, the presence of vitamin C (Table 25.3), as well as micro- and macro-minerals in *Strychnos spinosa* fruit has been documented (Table 25.4). Indigenous fruits, which are often endowed with high levels of vitamin C, play an important role in the food and nutrition security of local communities due to availability. Amarteifio and Mosase (2006) and Nhukarume *et al.* (2010) have reported considerable levels of ascorbic acid in ripe *Strychnos spinosa* fruit. The vitamin C content

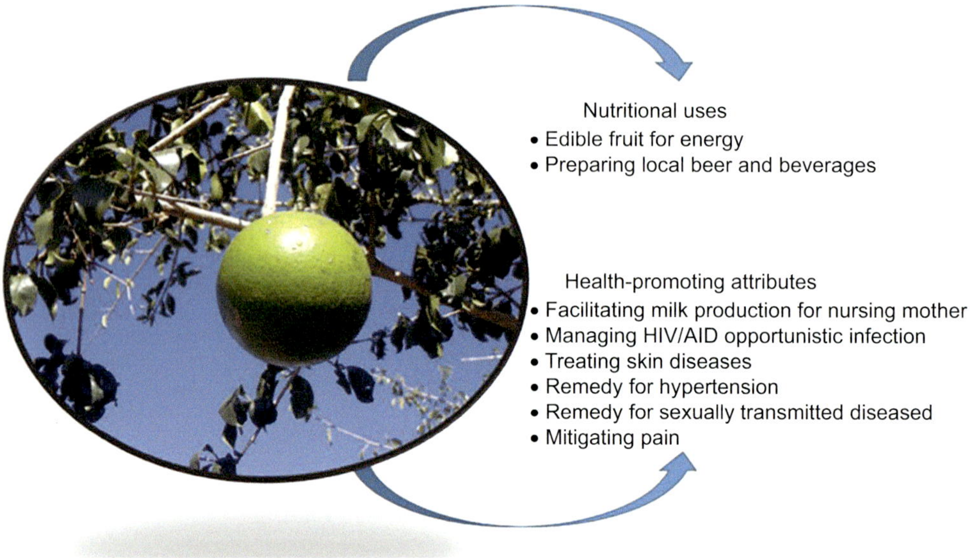

Fig. 25.2. Traditional uses of *Strychnos spinosa* (spiny monkey orange) fruit among local communities in Africa. Nutritional uses (Cheikhyoussef and Embashu, 2013; Thiombiano *et al.*, 2013; Bruschi *et al.*, 2019; Magwede *et al.*, 2019; Dejene *et al.*, 2020; Shai *et al.*, 2020); and health-promoting attributes (Mabogo, 1990; Ndubani and Höjer, 1999; Chigora *et al.*, 2007; Doka and Yagi, 2009; De Wet *et al.*, 2013; Chinsembu *et al.*, 2019; Avakoudjo *et al.*, 2020).

Table 25.2. Proximate composition of edible portion of *Strychnos spinosa* fruit.

Component	Amarteifio and Mosase, 2006	Lockett *et al.*, 2000	Saka and Msonthi, 1994	Sitrit *et al.*, 2003
Dry matter (%)	19.7	10.77	22.1	nr
Ash (%)	4.6	2.58	4.1	nr
Crude protein (%)	3.3	11.7	5.4	nr
Fat (%)	nr	1.94	31.2	nr
Fibre (%)	nr	nr	17.6	nr
Acid detergent lignin (%)	4.4	nr	nr	nr
Acid detergent fibre (%)	6.1	23.96	nr	nr
Neutral detergent fibre (%)	6.2	nr	nr	nr
Total carbohydrate (%)	nr	59.82	42.1	nr
Energy value (kJ/100 g)	nr	nr	1923	nr
Total soluble sugar (%)	nr	nr	nr	16–22
Total sugar (mg/g FW)	nr	nr	nr	111–150
Total acidity (μmol H^+/g FW)	nr	nr	nr	204–242

nr: not reported.

Table 25.3. Comparison of vitamin C content (mg/100 g fresh weight – FW) in edible portion of *Strychnos spinosa* with commercial and indigenous fruit.

Fruit	Quantity	Technique	References
Spiny monkey orange (*Strychnos spinosa*)	88	AOAC, 1996	Amarteifio and Mosase, 2006
Marula (*Sclerocarya birrea*)	133	DCPIP	Dlamini and Dube, 2008
	267	HPLC	Borochov-Neori *et al*., 2008
Baobab (*Adansonia digitata*)	300	AOAC, 1970	Nour *et al*., 1980
Blackcurrant (*Ribes nigrum*)	181	HPLC	Atkinson *et al*., 2013
Kiwi (*Actinidia chinensis*)	82	HPLC	Kvesitadze *et al*., 2001
Orange, navel (*Citrus sinensis*)	43.96	HPLC	Najwa and Azrina, 2017
	57.9	HPLC	Phillips *et al*., 2018
	59.1	-	Mieszczakowska-Frąc *et al*., 2021
Mango (*Mangifera indica*)	18.2	UV–Vis spectrophotometry	Septembre-Malaterre *et al*., 2016
	15.97	Titration	Sogi *et al*., 2012
	36.4	-	Mieszczakowska-Frąc *et al*., 2021
Apple (*Malus domestica*)	12.98	HPLC	Średnicka-Tober *et al*., 2020
	4.6	-	Sogi *et al*., 2012

DCPIP, 2,6-dichlorophenolindophenol; HPLC, high performance liquid chromatography.

of spiny monkey orange fruit is comparable to kiwi fruit, and greater than the level in commercial orange, mango and apple (Table 25.3). However, marula, baobab and blackcurrant fruit have higher vitamin C content than spiny monkey orange (Table 25.3). The recommended dietary allowance (RDA) for vitamin C in adults is 45 mg/day, an amount that is easily obtained from a single navel orange (WHO, 2004). The European Food Safety Authority (EFSA) recommends 90 mg/day and 80 mg/day of vitamin C for men and women, respectively (EFSA, 2013). The RDA for infants and children ranges from 25 to 35 mg/day (WHO, 2004). The water-soluble vitamin contributes to the non-enzymatic antioxidant activity of fruit and vegetables. Vitamin C exhibits chain-breaking scavenger ability against reactive oxygen species (ROS) and reactive nitrogen species (RNS), which can damage low-density lipoproteins in mammalian tissues (Baiano and Del Nobile, 2016). Consumption of indigenous fruit such as *Strychnos spinosa* provides a source of ascorbate and alleviates nutrition insecurity for local communities.

Strychnos spinosa fruit are rich sources of minerals (Table 25.4), which are important for maintaining human health. The main macro-minerals found in spiny monkey orange include phosphorus, potassium, calcium and magnesium. Several studies have reported variable mineral content levels (Table 25.4), which makes comparison of values between different studies difficult. The variation can be attributed to phenotype diversity, climatic factors and soil chemistry in different geographic regions as well as the analytical methods used (Ngadze *et al*., 2017). The phosphorus content varies from 12 mg/100 g fresh weight (FW) (Sitrit *et al*., 2003) to 168 mg/100 g FW (Lockett *et al*., 2000). Likewise reported values for potassium vary from 421 mg/100 g FW (Sitrit *et al*., 2003) to 1968.3 mg/100 g FW (Saka and Msonthi, 1994) compared to 962 mg/100 g for navel oranges (Liu *et al*., 2020). The WHO

Table 25.4. Mineral composition (mg/100g FW) of edible part of *Strychnos spinosa* fruit relative to commercial and indigenous fruit.

	Spiny monkey orange				Marula	Baobab	Orange, navel
Minerals	Amarteifio and Mosase, 2006	Lockett *et al.*, 2000	Saka and Msonthi, 1994	Sitrit *et al.*, 2003	Hiwilepo-van Hal *et al.*, 2014	Chadare *et al.*, 2008	Liu *et al.*, 2020
Phosphorus	66	168	108	12–13	262	106	86
Calcium	56	130	14.9	20.1–25	481	302	232
Magnesium	49	141	43.0	23.6–38.3	310	195	152
Potassium	1370	nr	1968.3	421–582	212	1794	962
Sodium	21.7	nr	25.3	6–16.5	1.5	14.8	nr
Zinc	0.22	1.08	nr	0.18–0.29	0.17	1.7	0.52
Copper	nr	0.24	nr	0.040–0.054	0.45	0.9	0.35
Manganese	nr	2.74	nr	nr	nr	0.7	0.81
Iron	0.11	4.39	13.6	0.26–0.42	2.5	4.3	2.43

nr- not reported.

recommended dietary intake for potassium is 3510 mg/day (WHO, 2012). Based on potassium content obtained by Amarteifio and Mosase (2006), the daily intake requirement can be achieved from 256 g of fruit. Magnesium content varied in spiny monkey orange and ranged from 43 mg/100 g (Saka and Msonthi, 1994) to 141 mg/100 g (Lockett *et al.*, 2000) compared to the recommended intake level of 220 and 260 mg/day for adult women and men, respectively (WHO, 2004). At a concentration of 141 mg/100 g FW (Lockett *et al.*, 2000), adult men would require about two 100 g portions of spiny monkey orange fruit pulp. Different studies have obtained varied calcium and magnesium concentrations in spiny monkey orange, which are lower than marula, baobab and navel orange (Table 25.4). Lockett *et al.* (2000) have reported a concentration of 130 mg/100 g of fruit. The recommended calcium intake in the USA and Canada is 1200 mg/day (Rodríguez-Rodríguez *et al.*, 2010), which translates to about nine 100 g portions of spiny monkey orange. *Strychnos spinosa* fruit used in the study by Sitrit *et al.* (2003) were cultivated in Israel, which may illustrate the impact of climate, soil chemistry and agro-technical conditions like fertilization and irrigation, on uptake of ions by the plants. The observed variation in values between different studies could also be attributed to stage of ripening and analysed component, that is, flesh, pulp or juice. Notwithstanding, the information gives vital insights on the nutritional composition of spiny monkey orange and its potential as a source of valuable nutrients.

Strychnos spinosa fruit contain important micro-minerals, namely iron, zinc, copper and manganese (Table 25.4). Deficiency of micro-minerals impairs growth and increases susceptibility to infections and risk of mortality especially in children (Shajib *et al.*, 2013). To some extent, intake of spiny monkey fruit can contribute to the daily requirement of zinc, iron, copper and manganese. However, there is no consistency in the presence of some minerals – for example, a manganese concentration of 2.74 mg/100 g FW was reported by Lockett *et al.* (2000). On the other hand, presence of iron was reported by multiple studies, confirming its ubiquitous occurrence in *Strychnos spinosa* fruit (Table 25.4). However, there is a notable 123-fold variation in iron concentration between Saka and Msonthi (1994) and Amarteifio and Mosase (2006) with values of 13.6 mg/100 g FW and 0.11 mg/100 g FW, respectively. Table 25.4 shows higher zinc and copper concentrations in orange compared to spiny monkey orange. However, spiny monkey orange has greater manganese content than the commercial navel orange, marula and baobab fruit (Table 25.4). The reported extreme variations in mineral content of spiny monkey orange warrant further investigation using advanced analytical techniques such as inductively coupled plasma-optical emission spectrophotometry (ICP-OES). Clearly, there remains a paucity of knowledge on nutritional composition of the wide diversity of *Strychnos spinosa* genotypes in sub-Saharan Africa. Hence, it is essential to comprehensively explore the nutritional composition and sensory properties of *Strychnos spinosa* from different geographic regions.

Sensory Properties and Volatile Compounds

Table 25.5 provides an overview on the sensory properties and chemicals in *Strychnos spinosa*. The fruits have yellow juicy, fleshy and sweet-sour pulp that is highly aromatic due to a complex mixture of volatile compounds (Ngadze *et al.*, 2017). The main volatiles in ripe spiny monkey orange fruit that contribute to this delicate complex aroma profile include trans-isoeugenol, eugenol, chavicol, *p*-trans-anol [4-(1-propenyl)-phenol], dihydroeugenol, trans-β-ocimene, 2,6-dimethoxyphenol and vanillin (Sitrit *et al.*, 2003). Eugenol and trans-isoeugenol impart the distinctive pungent clove aroma of *Strychnos spinosa* fruit. The complex aroma is characterized as a mixture of multiple flavours, notably clove, pineapple, citrus, apricot and melon. Flavour (the multidimensional set of aroma and taste attributes) is one of the fundamental determinants of fruit quality that influences consumer preferences and liking (Farcuh *et al.*, 2020). A range of chemicals contribute to the flavour of fruit including those that interact with taste receptors and the volatile compounds detected by olfactory receptors (Klee and Tieman, 2018). Obviously, a wide range of products can be obtained from *Strychnos spinosa* fruit, including novel flavours (Van Wyk, 2011).

Table 25.5. Sensory properties and phytochemicals in edible portion of *Strychnos spinosa*.

Characteristics	Description	References
Taste	Tart/fermented acid-sweet	Mizrahi *et al*., 2002
Aroma volatiles	Major compound (>75%): trans-isoeugenol – 4.762 mg/g FW Other compounds: eugenol – 307 µg/g FW; chavicol – 172 µg/g FW; ρ-trans-anol – 647.5 µg/g FW; dihydroeugenol – 123.5 µg/g FW	Sitrit *et al*., 2003
Aroma	Clove	Mizrahi *et al*., 2002
Colour	Yellow (when ripe)	Sitrit *et al*., 2003; Ngadze, 2018
Acidity	0.77%	Amarteifio and Mosase, 2006
pH	2.6–3.33 3.96	Sitrit *et al*., 2003 Amarteifio and Mosase, 2006

Chemicals that influence taste are sugars (mainly sucrose, glucose and fructose), acids (mainly citrate, malate and ascorbate) and, in some fruit, bitter chemicals (Klee and Tieman, 2018; Farcuh *et al*., 2020). Ripe spiny monkey orange fruit exhibit a tart acidic taste, which can be attributed to the presence of citric acid, succinic acid, ascorbic acid and malic acid, and low pH (Sitrit *et al*., 2003; Amarteifio and Mosase, 2006). The high acidity levels affect ratings for sweet, sour and other flavour properties and can significantly prolong the storage life of fresh *Strychnos spinosa* fruit (Marsh *et al*., 2004; Ngadze *et al*., 2017). According to Malundo *et al*. (2001), acids are a major determinant for the sour flavour in fruit. The pulp sweetness of *Strychnos spinosa* fruit can be attributed to its soluble sugar content, which is mainly composed of sucrose, a disaccharide, as well as the monosaccharides glucose and fructose (Sitrit *et al*., 2003). Soluble sugars are known to enhance most flavour attributes in fresh fruit except for sour taste (Malundo *et al*., 2001).

Phytochemicals

The phytochemical content of *Strychnos spinosa* fruit is important to explain their nutritional and health promoting benefits. Based on spectrophotometric measurement, the total phenolic, flavonol and proanthocyanidin content of *Strychnos spinosa* fruit samples was approximately 40 mg/100 g FW gallic acid equivalents (GAE), 55 mg/100 g FW catechin equivalents and 0.407% leucocyanidin equivalents, respectively (Nhukarume *et al*., 2010). Among the nine volatile compounds (phenylpropanoids) present in the fruit peel of *Strychnos spinosa* at the mature yellow stage, the major compound, *trans*-isoeugenol, accounted for >75% of the total volatiles (Sitrit *et al*., 2003).

Health Benefits and Bioactive Properties

The therapeutic potential of *Strychnos spinosa* has been demonstrated using different biological assays including antioxidant, antimicrobial and anti-tick effects (Fig. 25.3). Evidence of the antioxidant potential of the fruit extract and juice have been demonstrated using both free radical scavenging power and β-carotene inhibition models (Nhukarume *et al*., 2010; Kabine *et al*., 2015). However, β-carotene inhibition of the fruit juice was moderate and significantly lower that butylated hydroxyl anisole, the positive control (Nhukarume *et al*., 2010).

Noteworthy broad-spectrum antibacterial and antifungal activity of *Strychnos spinosa* (fruit) organic extracts has been reported (Fig. 25.3). As indicated by Nciki *et al*. (2016), the organic extract of *Strychnos spinosa* fruit had a minimum inhibitory concentration (MIC) value of 500 and 1500 µg/ml against *Brevibacterium agri* and *Pseudomonas aeruginosa*, respectively. Likewise, the fruit extract had MIC value of 380 µg/ml against *Trichophyton mentagrophytes*, a fungal strain associated with skin diseases in

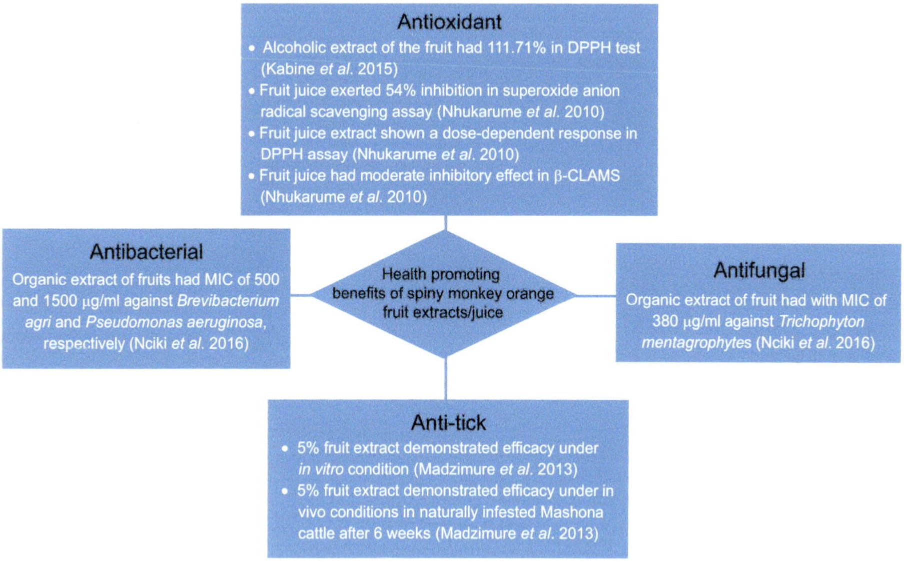

Fig. 25.3. Overview of the potential health benefits of *Strychnos spinosa* fruit extract and juice (Nhukarume *et al.*, 2010; Madzimure *et al.*, 2013; Kabine *et al.*, 2015; Nciki *et al.*, 2016).

humans. In an attempt to explore the anti-tick potential, high efficacy ratio was observed for the 5% fruit extract against *Rhipicephalus (Boophilus) decoloratus* tick larvae over 24- and 48-hour durations (Madzimure *et al.*, 2013). On the basis of the promising *in vitro* response, the potential anti-tick effect of the fruit extract at different dilutions (5, 10, and 20%, w/v) was evaluated. After six weeks, the efficacy ratio was significantly higher in the 5% *Strychnos spinosa*-treated cattle than in the untreated control but lower ($p<0.05$) than that for the positive control (Madzimure *et al.*, 2013).

Conclusions and Future Perspectives

The importance of *Strychnos spinosa* across many ethnic groups in Africa is well exemplified by its popularity as an edible fruit and source of nutrition. The high nutritional content of *Strychnos spinosa* fruit suggests its role as source of vital nutrients, vitamins and minerals for the promotion of good health and general wellbeing. The fruit can supplement dietary intake of carbohydrates, proteins and minerals such as calcium, phosphorus, potassium, iron, copper, manganese and zinc. Furthermore, its fruit pulp can be sun-dried, thereby extending shelf life and availability. Phytochemical profiling revealed the presence of important compounds with therapeutic value. In view of the increasing importance and potential of *Strychnos spinosa* as a novel functional health product/formulation for maintaining general wellbeing and mitigating different disease conditions, several aspects require further research. For instance, there is a need for studies focusing on plant morphological and genetic diversity as well pollination biology of *Strychnos spinosa*. Identification of antinutritive compound(s) is essential in order to achieve complete chemical characterization of the fruit so as to enhance its acceptability in functional novel products/formulations. Further exploration of the therapeutic effects of *Strychnos spinosa* is essential, as this aspect is currently incomplete given that the majority of the existing evidence is *in vitro*-based, which is known to have limited clinical significance. The preharvest and postharvest biologies (including storage) as well as the implication of food processing on the nutrients and phytochemicals need to be fully elucidated as an important aspect for commercialization.

Acknowledgements

AOO is a recipient of Postdoctoral fellowship from the North-West University, Mmabatho, South Africa. In addition, the National Research Foundation of South Africa supported this work under Grant No. 129179, Incentive Funding for Rated Researchers (MM) and Grant UID: 109508, Incentive Funding for Rated Researchers (AOA). We appreciate Ms Moloko Mojapelo (Department of Agriculture, Forestry and Fisheries Directorate Plant Production: Indigenous Food Crops Division) for supplying the photographs of spiny monkey orange.

Declaration of Competing Interest

We hereby declare no conflict of interest with regards to this chapter. The National Research Foundation (NRF) had no role in the design of the study; in the collection, analyses, or interpretation of data; in the writing of the chapter, or in the decision to publish.

References

Amarteifio, J.O. and Mosase, M.O. (2006) The chemical composition of selected indigenous fruits of Botswana. *Journal of Applied Sciences and Environmental Management* 10(2), 43–47. DOI: 10.4314/jasem.v10i2.43659.

AOAC (1970) *Official Methods of Analysis*, 11th edn. Association of Official Analytical Chemists, Washington DC, USA.

AOAC (1996) *Official Methods of Analysis*, 16th edn. Association of Official Analytical Chemists, Washington DC, USA.

Atkinson, C.J., Davies, M.J., Taylor, J.M. and Longbottom, H. (2013) Linking ascorbic acid production in *Ribes nigrum* with fruit development and changes in sources and sinks. *Annals of Botany* 111(4), 703–712. DOI: 10.1093/aob/mct026.

Avakoudjo, H.G.G., Hounkpèvi, A., Idohou, R., Koné, M.W. and Assogbadjo, A.E. (2020) Local, knowledge, uses, and factors determining the use of *Strychnos spinosa* organs in Benin (West Africa). *Economic Botany* 74(1), 15–31. DOI: 10.1007/s12231-019-09481-0.

Baiano, A. and Del Nobile, M.A. (2016) Antioxidant compounds from undefined: Biosynthesis, occurrence, and extraction systems. *Critical Reviews in Food Science and Nutrition* 56(12), 2053–2068. DOI: 10.1080/10408398.2013.812059.

Borochov-Neori, H., Judeinstein, S., Greenberg, A., Fuhrman, B., Attias, J. *et al.* (2008) Phenolic antioxidants and antiatherogenic effects of marula (*Sclerocarrya birrea* subsp. *caffra*) fruit juice in healthy humans. *Journal of Agricultural and Food Chemistry* 56(21), 9884–9891. DOI: 10.1021/jf801467m.

Bruschi, P., Mancini, M., Mattioli, E., Morganti, M. and Signorini, M.A. (2014) Traditional uses of plants in a rural community of Mozambique and possible links with Miombo degradation and harvesting sustainability. *Journal of Ethnobiology and Ethnomedicine* 10, 59. DOI: 10.1186/1746-4269-10-59.

Bruschi, P., Sugni, M., Moretti, A., Signorini, M.A. and Fico, G. (2019) Children's versus adult's knowledge of medicinal plants: an ethnobotanical study in Tremezzina (Como, Lombardy, Italy). *Revista Brasileira de Farmacognosia* 29(5), 644–655. DOI: 10.1016/j.bjp.2019.04.009.

Chadare, F.J., Linnemann, A.R., Hounhouigan, J.D., Nout, M.J.R. and Van Boekel, M.A.J.S. (2008) Baobab food products: a review on their composition and nutritional value. *Critical Reviews in Food Science and Nutrition* 49(3), 254–274. DOI: 10.1080/10408390701856330.

Cheikhyoussef, A. and Embashu, W. (2013) Ethnobotanical knowledge on indigenous fruits in Ohangwena and Oshikoto regions in Northern Namibia. *Journal of Ethnobiology and Ethnomedicine* 9, 34. DOI: 10.1186/1746-4269-9-34.

Chigora, P., Masocha, R. and Mutenheri, F. (2007) The role of indigenous medicinal knowledge (IMK) in the treatment of ailments in rural Zimbabwe: the case of Mutirikwi communal lands. *Journal of Sustainable Development in Africa* 9, 26–43.

Chinsembu, K.C., Syakalima, M. and Semenya, S.S. (2019) Ethnomedicinal plants used by traditional healers in the management of HIV/AIDS opportunistic diseases in Lusaka, Zambia. *South African Journal of Botany* 122, 369–384. DOI: 10.1016/j.sajb.2018.09.007.

De Wet, H., Nciki, S. and van Vuuren, S.F. (2013) Medicinal plants used for the treatment of various skin disorders by a rural community in northern Maputaland, South Africa. *Journal of Ethnobiology and Ethnomedicine* 9, 51. DOI: 10.1186/1746-4269-9-51.

Dejene, T., Agamy, M.S., Agúndez, D. and Martín-Pinto, P. (2020) Ethnobotanical survey of wild edible fruit tree species in lowland areas of Ethiopia. *Forests* 11(2), 177. DOI: 10.3390/f11020177.

Dlamini, N.R. and Dube, S. (2008) Studies on the physico-chemical, nutritional and microbiological changes during the traditional preparation of marula wine in Gwanda, Zimbabwe. *Nutrition & Food Science* 38(1), 61–69. DOI: 10.1108/00346650810848025.

Doka, I.G. and Yagi, S.M. (2009) Ethnobotanical survey of medicinal plants in West Kordofan (Western Sudan). Ethnobotanical Leaflets, 13, 1409-1416.

EFSA Panel on Dietetic Products, Nutrition and Allergies (NDA) (2013) Scientific opinion on dietary reference values for vitamin c. *EFSA Journal* 11(11), 3418. DOI: 10.2903/j.efsa.2013.3418.

Farcuh, M., Copes, B., Le-Navenec, G., Marroquin, J., Cantu, D. *et al.* (2020) Sensory, physicochemical and volatile compound analysis of short and long shelf-life melon (*Cucumis melo* L.) genotypes at harvest and after postharvest storage. *Food Chemistry* 8, 100107. DOI: 10.1016/j.fochx.2020.100107.

Hiwilepo-van Hal, P., Bille, P.G., Verkerk, R., van Boekel, M.A.J.S. and Dekker, M. (2014) A review of the proximate composition and nutritional value of Marula (*Sclerocarya birrea* subsp. *caffra*). *Phytochemistry Reviews* 13(4), 881–892. DOI: 10.1007/s11101-014-9352-6.

Kabine, O., Samba, B.M., Fatoumata, B., Namagan, K., Luopou, H.N, *et al.* (2015) Anti-oxidative activity of fruit extracts of some medicinal plants used against chronic diseases (diabetes, hypertension) in Kankan, Guinea. *Journal of Plant Sciences* 3, 1–5.

Klee, H.J. and Tieman, D.M. (2018) The genetics of fruit flavour preferences. *Nature Reviews Genetics* 19(6), 347–356. DOI: 10.1038/s41576-018-0002-5.

Kvesitadze, G.I., Kalandiya, A.G., Papunidze, S.G. and Vanidze, M.R. (2001) Identification and quantification of ascorbic acid in kiwi fruit by high-performance liquid chromatography. *Applied Biochemistry and Microbiology* 37(2), 215–218. DOI: 10.1023/A:1002848302873.

Liu, G., Chen, Y., He, X., Yao, F., Guan, G. *et al.* (2020) Seasonal changes of mineral nutrients in the fruit of navel orange plants grafted on trifoliate orange and citrange. *Scientia Horticulturae* 264, 109156. DOI: 10.1016/j.scienta.2019.109156.

Lockett, C.T., Calvert, C.C. and Grivetti, L.E. (2000) Energy and micronutrient composition of dietary and medicinal wild plants consumed during drought. Study of rural Fulani, Northeastern Nigeria. *International Journal of Food Sciences and Nutrition* 51(3), 195–208. DOI: 10.1080/09637480050029700.

Mabogo, D.E.N. (1990) *The Ethnobotany of the Vhavenda*. MSc thesis, University of Pretoria, South Africa.

Madzimure, J., Nyahangare, E.T., Hamudikuwanda, H., Hove, T., Belmain, S.R. *et al.* (2013) Efficacy of *Strychnos spinosa* (Lam.) and *Solanum incanum* L. aqueous fruit extracts against cattle ticks. *Tropical Animal Health and Production* 45(6), 1341–1347. DOI: 10.1007/s11250-013-0367-6.

Magwede, K., van Wyk, B.-E. and van Wyk, A.E. (2019) An inventory of Vhavenḓa useful plants. *South African Journal of Botany* 122, 57–89. DOI: 10.1016/j.sajb.2017.12.013.

Malundo, T.M.M., Shewfelt, R.L., Ware, G.O. and Baldwin, E.A. (2001) Sugars and acids influence flavor properties of Mango (*Mangifera indica*). *Journal of the American Society for Horticultural Science* 126(1), 115–121. DOI: 10.21273/JASHS.126.1.115.

Marsh, K., Attanayake, S., Walker, S., Gunson, A., Boldingh, H. *et al.* (2004) Acidity and taste in kiwifruit. *Postharvest Biology and Technology* 32(2), 159–168. DOI: 10.1016/j.postharvbio.2003.11.001.

Mieszczakowska-Frąc, M., Celejewska, K. and Płocharski, W. (2021) Impact of innovative technologies on the content of vitamin C and its bioavailability from processed fruit and vegetable products. *Antioxidants* 10(1), 54. DOI: 10.3390/antiox10010054.

Mizrahi, Y., Nerd, A. and Sitrit, Y. (2002) New fruits for arid climates. In: Janick, J. and Whipkey, A. (eds) *Trends in New Crops and New Uses*. ASHS Press, Alexandria, Virginia, USA, pp. 378–384.

Najwa, F.R. and Azrina, A. (2017) Comparison of vitamin C content in citrus fruits by titration and high performance liquid chromatography (HPLC) methods. *International Food Research Journal* 24, 726–733.

National Research Council (2008) *Lost Crops of Africa: Vol. III Fruits*. National Academies Press, Washington, DC, USA.

Nciki, S., Vuuren, S., van Eyk, A. and de Wet, H. (2016) Plants used to treat skin diseases in northern Maputaland, South Africa: antimicrobial activity and *in vitro* permeability studies. *Pharmaceutical Biology* 54(11), 2420–2436. DOI: 10.3109/13880209.2016.1158287.

Ndubani, P. and Höjer, B. (1999) Traditional healers and the treatment of sexually transmitted illnesses in rural Zambia. *Journal of Ethnopharmacology* 67(1), 15–25. DOI: 10.1016/s0378-8741(99)00075-6.

Neuwinger, H.D. (2004) Plants used for poison fishing in tropical Africa. *Toxicon: Official Journal of the International Society on Toxinology* 44(4), 417–430. DOI: 10.1016/j.toxicon.2004.05.014.

Ngadze, R.T. (2018) *Value addition of Southern African monkey orange (Strychnos spp.): composition, utilization and quality*. PhD thesis, Wageningen University, The Netherlands.

Ngadze, R.T., Linnemann, A.R., Nyanga, L.K., Fogliano, V. and Verkerk, R. (2017) Local processing and nutritional composition of indigenous fruits: The case of monkey orange (*Strychnos* spp.) from Southern Africa. *Food Reviews International* 33(2), 123–142. DOI: 10.1080/87559129.2016.1149862.

Nhukarume, L., Chikwambi, Z., Muchuweti, M. and Chipurura, B. (2010) Phenolic content and antioxidant capacities of *Parinari curatelifolia*, *Strychnos spinosa* and *Adansonia digitata*. *Journal of Food Biochemistry* 34, 207–221. DOI: 10.1111/j.1745-4514.2009.00325.x.

Nitcheu Ngemakwe, P.H., Remize, F., Thaoge, M.L. and Sivakumar, D. (2017) Phytochemical and nutritional properties of underutilised fruits in the southern African region. *South African Journal of Botany* 113, 137–149. DOI: 10.1016/j.sajb.2017.08.006.

Nour, A., Magboul, B. and Kheiri, N. (1980) Chemical composition of baobab fruit (*Adansonia digitata* L.). *Tropical Science* 22, 383–388.

Ohiri, F.C., Verpoorte, R. and Baerheim Svendsen, A. (1983) The African *Strychnos* species and their alkaloids: A review. *Journal of Ethnopharmacology* 9(2–3), 167–223. DOI: 10.1016/0378-8741(83)90032-6.

Phillips, K.M., Tarrago-Trani, M.T., McGinty, R.C., Rasor, A.S., Haytowitz, D.B. *et al.* (2018) Seasonal variability of the vitamin C content of fresh fruits and vegetables in a local retail market. *Journal of the Science of Food and Agriculture* 98(11), 4191–4204. DOI: 10.1002/jsfa.8941.

Rasoanaivo, P., Martin, M.T., Guittet, E. and Frappier, F. (2002) New contributions to the structure elucidation and pharmacology of *Strychnos* alkaloids. In: Atta ur, R. (ed.) *Studies in Natural Products Chemistry*. Elsevier, Amsterdam, The Netherlands, pp. 1029–1072.

Rodrigues, S., de Brito, E.S. and de Oliveira Silva, E. (2018) Maboque/Monkey Orange - *Strychnos spinosa*. In: Rodrigues, S., de Oliveira Silva, E., and de Brito, E.S. (eds) *Exotic Fruits*. Academic Press, London, pp. 293–296.

Rodríguez-Rodríguez, E., Navia Lombán, B., López-Sobaler, A.M. and Ortega Anta, R.M. (2010) Review and future perspectives on recommended calcium intake. *Nutricion Hospitalaria* 25(3), 366–374.

Saka, J.D.K. and Msonthi, J.D. (1994) Nutritional value of edible fruits of indigenous wild trees in Malawi. *Forest Ecology and Management* 64(2–3), 245–248. DOI: 10.1016/0378-1127(94)90298-4.

SANBI (2021) South African National Biodiversity Institute's plantzafrica.com. Available at: http://pza.sanbi.org/strychnos-spinosa (accessed 30 April 2021).

Septembre-Malaterre, A., Stanislas, G., Douraguia, E. and Gonthier, M.-P. (2016) Evaluation of nutritional and antioxidant properties of the tropical fruits banana, litchi, mango, papaya, passion fruit and pineapple cultivated in Réunion French Island. *Food Chemistry* 212, 225–233. DOI: 10.1016/j.foodchem.2016.05.147.

Shai, K.N., Ncama, K., Ndhlovu, P.T., Struwig, M. and Aremu, A.O. (2020) An Exploratory Study on the Diverse Uses and Benefits of Locally-Sourced Fruit Species in Three Villages of Mpumalanga Province, South Africa. *Foods* 9(11), 1581. DOI: 10.3390/foods9111581.

Shajib, M.T.I, Nuruddin Miah, M., Begum, P., Bhattacharjee, L., Hossain, A. *et al.* (2013) Nutritional composition of minor indigenous fruits: cheapest nutritional source for the rural people of Bangladesh. *Food Chemistry* 140(3), 466–470. DOI: 10.1016/j.foodchem.2012.11.035.

Sitrit, Y., Loison, S., Ninio, R., Dishon, E., Bar, E. *et al.* (2003) Characterization of monkey orange (*Strychnos spinosa* Lam.), a potential new crop for arid regions. *Journal of Agricultural and Food Chemistry* 51(21), 6256–6260. DOI: 10.1021/jf030289e.

Skead, C.J. (2009) *Historical Plant Incidence in Southern Africa: A Collection of Early Travel Records in Southern Africa*. South African National Biodiversity Institute, Pretoria, South Africa.

Sogi, D.S., Siddiq, M., Roidoung, S. and Dolan, K.D. (2012) Total phenolics, carotenoids, ascorbic acid, and antioxidant properties of fresh-cut mango (*Mangifera indica* L., cv. Tommy Atkin) as affected by infrared heat treatment. *Journal of Food Science* 77(11), C1197-202. DOI: 10.1111/j.1750-3841.2012.02933.x.

Średnicka-Tober, D., Barański, M., Kazimierczak, R., Ponder, A., Kopczyńska, K. *et al.* (2020) Selected antioxidants in organic vs. conventionally grown apple fruits. *Applied Sciences* 10(9), 2997. DOI: 10.3390/app10092997.

The Plant List (2013) The Plant List: A working list of all species. Available at: http://www.theplantlist.org/tpl1.1/record/kew-2598152 (accessed 30 April 2021).

Thiombiano, D.N.E., Lamien, N., Castro-Euler, A.M., Vinceti, B., Agundez, D. *et al.* (2013) Local communities demand for food tree species and the potentialities of their landscapes in two ecological zones of Burkina Faso. *Open Journal of Forestry* 03(3), 79–87. DOI: 10.4236/ojf.2013.33014.

Tittikpina, N.K., Atakpama, W., Hoekou, Y., Diop, Y.M., Batawila, K. *et al.* (2020) *Strychnos spinosa* Lam: Comprehensive review on its medicinal and nutritional uses. *African Journal of Traditional, Complementary and Alternative Medicines* 17(2), 8–28. DOI: 10.21010/ajtcam.v17i2.2.

van Rayne, K.K., Adebo, O.A. and Ngobese, N.Z. (2020) Nutritional and Physicochemical Characterization of *Strychnos madagascariensis* Poir (Black Monkey Orange) Seeds as a Potential Food Source. *Foods* 9(8), 1060. DOI: 10.3390/foods9081060.

Van Wyk, B.E. (2011) The potential of South African plants in the development of new food and beverage products. *South African Journal of Botany* 77(4), 857–868. DOI: 10.1016/j.sajb.2011.08.003.

Van Wyk, B.-E. and Gericke, N. (2017) *People's Plants - a Guide to Useful Plants of Southern Africa*, 2nd edn. Briza Publications, Pretoria, South Africa.

WHO (2004) *Vitamin and Mineral Requirements in Human Nutrition*, 2nd edn. World Health Organization. Available at: https://www.who.int/nutrition/publications/micronutrients/9241546123/en/ (accessed 30 April 2021).

WHO (2012) *Guideline: Potassium Intake for Adults and Children*. World Health Organization, Geneva, Switzerland. Available at: http://www.who.int/nutrition/publications/guidelines/potassium_intake/en

26 African Star Apple (*Chrysophyllum albidum*)

Olubukola Motunrayo Odeyemi[1] and Olaniyi Amos Fawole[2]*
[1]Department of Horticulture, Federal University of Agriculture, Abeokuta, Ogun State, Nigeria; [2]Postharvest Research Laboratory, Department of Botany and Plant Biotechnology, University of Johannesburg, Johannesburg, South Africa

Abstract
African star apple (*Chrysophyllum albidum*) is a wild plant native to many parts of tropical Africa. The sweet, pleasantly acid, fleshy fruit pulp is relished as a snack, and sometimes the fruit peel is chewed to turn it into gum. The fruit is rich in macronutrients, micronutrients and dietary phytochemicals. There is great potential for the fruit to improve the human health, nutrition, food security and income of the communities engaged in its production. It also has broad medicinal applications in ethnomedicine. The fruit are produced in considerably large quantities on the tree, and are mostly under-utilized except for consumption in the fresh form. Therefore, substantial postharvest losses are experienced during the production season due to inadequate utilization, storage and processing facilities. The economic importance of the fruit can be enhanced if the technology for extending shelf life is developed. Processing this fruit will curb wastage during its season and make it available in different forms for both consumption and industrial use.

Introduction

African star apple (*Chrysophyllum albidum*) is a tropical forest fruit tree found native to many parts of Africa (Edem *et al.*, 1984). It belongs to the group of unexploited indigenous fruits that grow wild (Opeke, 2005). The ideal habitat is the lowland tropical rainforest zones (Essien *et al.*, 1995). Its natural occurrences have been reported in diverse ecozones in Nigeria, Uganda, Niger Republic, Cameroon and Cote d'Ivoire (Bada, 1997). The plant often grows to a height of 36.5 m though it may be smaller (Bada, 1997), producing wild fruits which are good sources of phytonutrients, antioxidants such as vitamin C, carotenoids and some minerals. The plant features prominently in the compound agroforestry system for fruit, food, cash income and other auxiliary uses, including environmental uses (Kang, 1992).

The African star apple plays a tremendous role in socioeconomic development and has great economic value in tropical Africa due to its diverse industrial, medicinal and food uses (Aiyelaagbe *et al.*, 1999; Adebayo *et al.*, 2012; Dadegnon *et al.*, 2015). The fruit is famous and has gained popularity in urban and rural areas where it is widely distributed in the lowland rainforest zones and frequently found in villages (Onwe *et al.*, 2020). The fruit is highly used and appreciated in southern Benin, where it is called Azongogwe or Azonbobwe in the local language (Dah-Dovonon, 2002). In Nigeria, it is called Agbalumo or Osan in the south-west, Udala in the south-east and

*Corresponding author: olaniyif@uj.ac.za

 Handbook of Phytonutrients in Indigenous Fruits and Vegetables (eds D. Sivakumar *et al.*)
DOI: 10.1079/9781789248067.0026

Agwaluma in the northern region in the local languages (Adepoju and Adeniji, 2012). While in Ghana, it is fondly called Alasa (Adekanmi and Olowofoyeku, 2020).

African star apple fruit are produced in large quantities on the tree. They are mostly under-utilized except for consumption in the fresh form. The fruit is made up of a sweet, pleasantly acid fleshy pulp, eaten as snacks by both children and adults; the peel is sometimes chewed to form stable chewing gum (Olubode *et al.*, 2019). According to Abiodun and Oladapo (2011), African star apple fruit are an excellent source of vitamins and iron, add flavour to diets and are used as raw materials in some manufacturing industries. This contributes to the nutrition, food security and income of local communities.

The fruit, despite the considerable production on the tree, are seasonal. The ripe fruit are available from December through March, sometimes extending to April in different production areas. Availability of fruit during the dry season provides much-needed sustainable nutrition for the populace. Traditionally, the fruit are not harvested from the trees but left to drop naturally on the forest floor where they are picked (Amusa *et al.*, 2003). However, fruit can also be harvested either by shaking or dislodging with forked sticks at the peak of ripening when they are most susceptible to mould, bacteria and insects (Wills *et al.*, 1998). The fruit are usually kept under ambient conditions with an estimated shelf life of three to five days, after which they deteriorate. Sometimes maggots can be found within the pulp due to poor postharvest handling. The fruit are handled in traditional woven baskets during storage and marketing. These traditional baskets can initiate mechanical damage to the fruit, resulting in huge postharvest losses. These losses are experienced during the seasonal glut due to inadequate utilization, storage and processing facilities. Therefore, preserving the fruit by any economical technique is important in the region of production (Falade and Aworh, 2004).

Botany

Chrysophyllum is derived from the Greek word for 'golden leaf', which refers to the colour of the leaves of some of its species (Adekanmi and Olowofoyeku, 2020). The tree is in the family Sapotaceae of the order Ericales comprising 70 genera and 800 species. Twenty-three genera and over 300 species are found in West Africa, including *Chrysophyllum*, *Mimusops*, *Breviea*, *Kanton*, *Mnanikara* and *Delipdora* (Gill, 1988). Each of these species produces a milky latex (Gill, 1988).

The African star apple is mostly grown as small or solitary plantings in village and home gardens, raised from seed rather than vegetative propagation and produced more by tradition than horticultural technology (Edem *et al.*, 1984; National Research Council, 2008). The tree is evergreen and can grow up to 25–40 m in height with a mature girth varying between 1.5 to 2 m (Edem *et al.*, 1984; Opeke, 2005). The stem is a straight and long fluted bole with a small buttress at the base (Edem *et al.*, 1984). The bark is thin and light brown and, when incised, exudes a gummy latex (Jayeoba *et al.*, 2007). The leaves are elliptic to oblong (Opeke, 2005).

The tree flowers between April and June and sets fruit in July. The fruit are referred to as berry and are green in colour at the unripe stage (Fig. 26.1a). The fruit ripen from December to March/April. They are large, ovoid to subglobose at the ripe stage, pointed at the apex, up to 6 cm long and 5 cm in diameter, shaped like an orange or apple (Fig. 26.1b) (Adepoju and Adeniji, 2012). According to Olubode *et al.* (2019), the fruit peel turns yellowish orange or orange to golden yellow when ripe but is rusty orange when overripe.

The pulp is fleshy, and the colour can vary from orange to pinkish or light yellow when the fruit is ripe (Fig. 26.2a) (Onyegbule *et al.*, 2020). A milky juice is secreted within the pulp. The pulp can turn into gum when chewed over a considerable period (Jayeoba *et al.*, 2010). About three to five seeds are arranged in a star-shaped pattern within the pulp from where the fruit derives its name. The seeds are dark brown or blackish, obliquely ellipsoid to obovoid, up to 2.8 cm long and 1.2 cm wide (Fig. 26.2b) (Adebayo *et al.*, 2012). The seedcoat is hard, bony, shiny and dark brown and, when broken, reveals white-coloured cotyledons (Adebayo *et al.*, 2012; Onyegbule *et al.*, 2020).

(a) (b)

Fig. 26.1. African star apple fruit (*Chrysophyllum albidum*): (a) unripe; (b) ripe.

(a) (b)

Fig. 26.2. African star apple fruit: (a) pulp; (b) seeds.

Traditional Uses

African star apple is used in traditional medicine in the treatment of numerous sicknesses, diseases and disorders in many countries. Herbal remedies are either prepared from dry plants or freshly collected plant parts. The bark is commonly used, followed by the leaves, roots, seeds and fruit in order of preference in ethnomedicine (Houessou *et al.*, 2012). The stem, bark and leaves are reportedly either used alone or in combination with other plant parts to treat malaria in Lagos, south-west Nigeria (Ishola *et al.*, 2014). The bark and roots are widely used for treating wounds, sprains and bruises in southern Nigeria (Ajayi *et al.*, 2020b), while the leaves are used in pyrexia treatment (Onyegbule *et al.*, 2020). The bark is also used as a remedy for yellow fever and malaria. In contrast, the leaves are used as emollients and for the treatment of skin eruptions, diarrhoea and stomach ache which are as a result of infections and inflammatory reactions (Idowu et al., 2006). Houessou *et al.* (2012) reported the use of the plant in Benin: the leaves are frequently used to treat malaria, blood pressure and anaemia; the roots in the treatment of sterility, sexual asthenia and asthma; the seeds mostly to treat intestinal worms and hemorrhoids; the bark for treating cough, jaundice and yellow fever; and the fruit against avitaminosis (pellagra) and dental decay.

Furthermore, in Ijebu, south-west Nigeria, the dried seeds of African star apple are ground into powder and mixed with honey for the treatment of cancer (Segun *et al.*, 2018). The leaves and cotyledons from the seed are used as ointments in the treatment of vaginal and dermatological infections in western Nigeria (Okoli and Okere, 2010). The fruit pulp is used to manage diabetes in folklore (Ajayi *et al.*, 2020a), while the seeds and root extracts are used to arrest bleeding from fresh wounds, and to enhance the wound-healing process (Okoli and Okere, 2010).

Nutritional Composition

The moisture content of fruit parts has been reported to range from 53.50% to 86.30% within the peel, pulp, juice and pericarp (Abiodun and Oladapo, 2011). The juice part of the fruit had the highest amount of moisture. The different parts of the fruit's moisture content ranged between 58.9 and 78.3% (Table 26.1). Edem *et al.* (1984) reported a moisture content of 58.9 and 67.5% for the peel and pulp, respectively (Table 26.1). The difference in values obtained may be due to variety, stage of maturity, location and other environmental factors. In comparison to African star apple fruit, the moisture content obtained from pineapple (87.63%) (Odeyemi and Ojetayo, 2018) and pawpaw fruit (90.5%) juice (Odeyemi *et al.*, 2014) was higher. The relatively high moisture content of African star apple fruit could suggest fast perishability and short shelf life. High moisture content also encourages microbial growth and increases enzymatic reaction rates, and hence deterioration. The crude protein of the pulp and peel was low (2–9%) compared with values for protein-rich seeds such as cowpea, soybean, melon, canola and groundnut. However, it was richer than the protein content of tubers, like cassava and yam, and some fruits such as sweet orange, pineapple and shea butter (Edem *et al.*, 1984; Adepoju and Adeniji, 2012; Onyekwelu *et al.*, 2015; Odeyemi and Ojetayo, 2018; Odeyemi *et al.*, 2019). Generally, the fruit are poor sources of crude protein. The protein content of African star apple fruit is low but when consumed in resource-poor communities they can make a significant contribution by ameliorating protein deficiency.

The pulp had higher crude protein, fat and crude fibre values than the seed kernel. In contrast, the carbohydrate value was higher in the seed kernel (Onyekwelu *et al.*, 2015). The amount of crude fibre may influence the digestibility of the fruit. It may also help maintain the normal internal distention of the intestinal tract and aid peristaltic movements (Edem *et al.*, 1984). The ash content value ranged between 1.66% and 4.6% within the fruit parts.

Table 26.1. Nutritional composition of African star apple fruit.

	Edem *et al.*, 1984		Onyekwelu *et al.*, 2015		Adepoju and Adeniji, 2012
Parameter	Peel	Pulp	Seed kernel	Pulp	Pulp
Moisture (%)	58.9	67.5	78.3	75.93	74.7
Ash (%)	4.6	3.4	1.66	1.48	2.2
Crude protein (%)	6.1	8.8	2.68	3.45	4.0
Fat (%)	-	-	0.71	1.61	5.5
Crude fibre (%)	14.5	4.0	1.94	16.43	4.3
Carbohydrates (%)	62.4	68.7	14.71	1.10	9.6
Gross energy (kcal/100 g)	385.5	447.0	-	-	119.8

Fruit peel recorded 4.6% ash which may be considered high for fruit and may indicate the amount of high molecular weight elements in fruit peel (Edem *et al.*, 1984). The fruit pulp, with low caloric value, may be advantageous for its suitability for consumption as a snack by the obese. Carbohydrate content reported by Edem *et al.* (1984) was high, especially within the pulp (68.7 %) compared with values obtained in pineapple (10.91%) and sweet orange (0.38%) (Odeyemi and Ojetayo, 2018; Odeyemi *et al.*, 2019). However, Onyekwelu *et al.* (2015) and Adepoju and Adeniji (2012) observed low carbohydrate content in the pulp and concluded that this may be responsible for the fruit's lack of sugary taste. Carbohydrates are essential components of human nutrition and enable the body to produce the energy required for normal function (Onyekwelu *et al.*, 2015). Plants are the major source of carbohydrates in human diets.

Ajayi *et al.* (2020a) also studied the free soluble starch content of African star apple fruit parts (pulp, cotyledon, seedcoat and pulp coat). The author reported free soluble starch of 3.13 g/100 g in pulp and 4.42 g/100 g in cotyledon, while the free soluble sugar ranged from 21.48 g/100 g (cotyledon) to 28.63 g/100 g (pulp). This could be the reason for the low starch and high sugar content of the African star apple fruit parts (Ajayi *et al.*, 2020). The amylose content of the fruit ranged from 2.01 g/100 g (pulp coat) to 2.31 g/100 g (cotyledon). In contrast, the amylopectin content of the pulp coat is reduced in comparison to other fruit parts. The amylose/amylopectin ratio of the fruit parts ranges from 0.37 to 0.44. The fruit parts had estimated glycaemic indices ranging from 13.63 (pulp coat) to 29.10 (cotyledon) (Ajayi *et al.*, 2020).

The fruit are high in micronutrients such as potassium (682.58 mg/100 g), calcium (391 mg/100 g), magnesium (181.17 mg/100 g), phosphorus (251.83 mg/100 g), zinc (8.32 mg/100 g), copper (5.71 mg/100 g) and manganese (4.98 mg/100 g) but low in sodium (43.33 mg/100 g) and iron (2.25 mg/100 g) (Adepoju and Adeniji, 2012) . The micronutrient content of the fruit pulp had index of nutritional quality (INQ) values above 1.0, hence the fruit pulp can be considered to be very nutritious (Table 26.2).

However, Edem *et al.* (1984) observed that the peel had higher calcium and potassium value than the pulp; there was no trace of chromium and cobalt in the fruit pulp or the peel. The fruit can be a good source of minerals that are essential for proper growth and development in the human body when consumed. The low sodium content of the fruit makes it suitable for consumption for everyone, especially the hypertensive.

Chemical, Flavour and Aroma Properties

Chemical

The chemical properties of the African star apple differ by fruit part. Abiodun and Oladapo (2011) observed the pH value range between 4 and 5.5 within the fruit peel, pulp, juice and pericarp, while Aina and Adesina (1991) reported a low pH value of 2.80 (Table 26.3). The pH value and titratable acidity (1.20 %) of the peel were the lowest, suggesting the fruit's acidic nature.

Table 26.2. Index of Nutritional Quality (INQ) of African star apple fruit pulp (Adepoju and Adeniji, 2012).

Nutrient	Nutrient content of fresh pulp/100 g	US RDA	Contribution to US RDA (%)	INQ (%)
Potassium	682.58	2000	34.1	6.6
Calcium	391.00	1000	39.1	7.5
Magnesium	181.17	420	43.1	8.3
Phosphorus	251.83	1000	25.2	4.8
Iron	2.25	18	12.5	2.4
Zinc	8.322	15	55.5	10.7

Table 26.3. Chemical properties of African Star apple fruit.

	Abiodun and Oladapo, 2011				Edem *et al.*, 1984		Adepoju and Adeniji, 2012
Parameter	Peel	Pulp	Juice	Pericarp	Peel	Pulp	Pulp
Total solids (%)	46.50	31.30	13.40	35.70	-	-	-
pH (5% solution)	4.01	4.20	4.45	5.03	-	-	-
Titratable acidity (%)	1.20	3.80	4.20	2.60	-	-	-
Total sugar (% Brix)	0.90	3.20	5.40	3.60	-	-	-
Ascorbic acid mg/100 g	106.00	125.00	128.50	112.00	239.1	446.1	91.62

Low pH has been reported to inhibit the growth of undesirable microorganisms (Abiodun and Oladapo, 2011). This may be responsible for the non-appearance of infection, sometimes on the peel, when the pulp is diseased at harvest. The total solids content ranged between 9.80 and 12.90% with the different fruit parts, while the sugar content value ranged from 0.90 to 5.40% Brix within the fruit part. The juice had the highest sugar content (5.4% Brix), while the peel had the least sugar content (0.9% Brix). Fawole and Opara (2013) reported higher sugar content in pomegranate fruit at full ripe stage than the value of sugar content in African star apple.

The ascorbic acid of African star apple fruit was found to range from 91.62 mg/100 g to 446.1 mg/100 g within the fruit parts (Table 26.3). This is higher than figures typically reported for some more popular tropical fruit like mango, pawpaw and pineapple (Aina and Adesina, 1991). However, it is comparable to the values obtained in kiwi fruit (91.20 mg/100 g) (Dumbrava *et al.*, 2016). This wide range in ascorbic acid content could be attributed to differences in growing location, variety, fruit maturity stage at harvest and method of analysis. The value of ascorbic acid (446.1 mg/100 g) in the pulp is higher than the recommended dietary allowance (RDA) (Edem *et al.*, 1984). Therefore, the fruit is an excellent source of ascorbic acid, and this may also contribute to its acidic taste. The carotene content of the fruit varied with the fruit part, but the highest values were obtained from the peel (15.0 mg/100 g). The carotene content in African star apple fruit (1.380 mg/100 g) was higher than that obtained in hog plum (0. 381 mg/100 g), bush mango (1.267 mg/100 g) and monkey cola (0.172 mg/100 g) (Table 26.4). The high carotene may account for the yellow colour of the ripe fruit. Thus, the fruit is a good source of carotene when consumed.

Flavour and aroma

The flavour and aroma are both important evaluating signals that determine the acceptability of food by consumers. Lasekan *et al.* (2013) identified 59 odour-active compounds in African star apple fruit using a headspace solid-phase microextraction coupled with gas chromatography-mass spectrometry. Application of aroma extract dilution analysis (AEDA) to the aroma distillates from the fruit further revealed 45 odour-active compounds in the flavour dilution (FD) factor range of 4–128,

Table 26.4. Carotene content of African star apple fruit compared with other fruit (Olayiwola *et al.*, 2013; Odeyemi *et al.*, 2014, 2019).

Fruit	Carotene content mg/100 g
African star apple	1.380
Hog plum	0.381
Bush mango	1.267
Monkey cola	0.172
Papaya	0.028
Plantain	4.810

while ten of the odour-active compounds were identified as key odourants. The ten key odourants identified were ethyl acetate, acetyl methyl carbinol, methylhexanoate, sabinene, p-cymere, methylbenzoate, geraniol, cis-x-bergomotene, acetophenone and ethyl dodecanoate (Lasekan *et al.*, 2013). There is a positive correlation between the very sweet/sweet varieties of African star apple and key odourants identified in the study. Further analysis revealed that ethyl hexanoate, limonene, geraniol, ethyl-(E)-2-butenoate and ethyl dodecanoate contributed intensely to the aroma of the very sweet variety while the bitter variety exhibited a strong positive correlation with acetyl methyl carbinol, ethyl-(E)-2-butenoate, methylhexanoate, hexadecanoic and octadecanoic acids respectively (Lasekan *et al.*, 2013). These volatile compounds play an important role in the organoleptic characteristics, while the aroma constituent determines the varietal character in fruit (Lasekan *et al.*, 2013).

In another study, nine fruit types were also identified based on ripe fruit size, colour and taste which varied from being sweet to sour (Oyebade *et al.*, 2011). The sour-tasting African star apple fruit were reported to have higher nutritive value than the sweet-tasting fruit (Ureigho and Ekeke, 2010). The variation of the fruit peel colour is said to be correlated with the pulp taste. Therefore, the fruit's taste depends on the stage of maturity, which makes it very sour or sweet. However, on most occasions, the pulp is tart in taste and chewing the pulp results in a significant quantity of gum residue (Abiodun and Oladapo, 2011).

Phytochemical Composition

African star apple fruit is documented to produce abundant health-promoting phytochemicals. Antioxidant qualities of plant foods have been associated with the presence of an array of important phenolics and flavonoids. The total phenolic content of African star apple fruit reported by Ajayi *et al.* (2020) in the pulp was 0.141 mg/100 g and 0.230 mg/100 g in the pulp coat, while the total flavonoid content was 0.053 mg/ 100 g in the pulp and 0.579 mg/100 g in the pulp coat. Ibrahim *et al.* (2017), in their preliminary phytochemical investigation, also revealed the presence of alkaloids, tannins, saponins, flavonoids and terpenoids in the edible parts of African star apple. However, cardiac glycosides were absent in the peel, while phlobatannin was absent in both the seed shell pericarp and the peel of the fruit (Ibrahim *et al.*, 2017). The seed shell pericarp had the highest content of alkaloids (25.80%) and tannins (10.19%). The peel had the highest content of saponins (0.41%) and flavonoids (17.23%). The total phenolics of African star fruit (398.23 mg/100 g) was significantly higher than monkey cola (121.39 mg/100 g) but comparable with bush mango (382.20 mg/100 g) (Olayiwola *et al.*, 2013) and pomegranate (289–450 mg/100 g) (Fawole *et al.*, 2012). In contrast, total anthocyanins in African star apple (52.25 mg/100 g) were higher than values obtained in bush mango (27.28 mg/100 g) (Olayiwola *et al.*, 2013). Akaloids, flavonoids, terpenoids, saponins, steroids, anthraquinones, cardiac glycosides, phenolics, tannins and amino acids were absent in the seed gum of African star apple fruit. In addition, compounds found in the seed cotyledon include three alkaloids: eleagnine (1, 2, 3, 4 tetrahydro-1-methyl-β-carboline); tetrahydro-2-methylharman (1, 2, 3, 4-tetrahydro-1, 2 dimethyl-β-carboline) and skatole (3-methylindole) (Idowu *et al.*, 2016).

Muhammad and Abubakar (2016) revealed from their study that the aqueous extract of African star apple seed kernel contained alkaloids (3.56 g/100 g), tannins (2.184 g/100 g), cardiac glycosides (1.88 g/100 g), saponins (0.24 g/100 g) and flavonoids (0.15 g/100 g). The presence of these phytochemicals suggests the potential of the kernel of African star apple for use as a nutraceutical (Muhammad and Abubakar, 2016). Akaneme (2008) also reported tannins, flavonoids, terpenoids, proteins, carbohydrates and resins as the phytochemicals found in African star apple fruit.

Classification of the extracts with high-performance liquid chromatography revealed that the major polyphenolic constituents of African star apple fruit parts (pulp, cotyledon, seedcoat and pulp coat) are beta-amyrin acetate, eleagine, epicatechin, epigallocatechin, skatole, stigmasterol and tetrahydro-2-methyl harman (Table 26.5). Gentisic acid and myricetin-3-rhamnoside were not present in the pulp

Table 26.5. Phenolic components of aqueous extract of African star apple fruit parts (Ajayi *et al.*, 2020a).

Compound	Pulp mg/g	Cotyledon	Seedcoat	Pulp coat
Beta-amyrin acetate	2.42	4.24	2.83	2.66
Eleagnine	0.86	0.34	1.46	1.31
Epicatechin	0.98	1.39	1.27	1.24
Epigallocatechin	0.78	0.34	0.97	0.87
Gentisic acid	-	0.67	0.44	0.21
Myricetin rhamnoside	0.50	-	0.78	0.64
Procyanidin B5	-	-	0.54	0.54
Skatole	0.84	0.32	1.59	1.49
Stigmasterol	3.74	2.47	5.82	5.25
Tetrahydro-2-methylharman	1.05	0.35	1.66	1.59

and cotyledon, respectively. However, they were present in the seedcoat and pulp coat. Procyanidin-B5 was also contained in all the fruit parts except from the pulp and cotyledon (Table 26.5). African star apple fruit peel extract has also been shown to be rich in phenolic compounds, including eleagnine, myricetin and rhamnoside, and fatty acids, with potential antinociceptive and anti-inflammatory properties (Ibrahim *et al.*, 2017).

mould count did not exceed limits specified by the US Pharmacopoeia. Objectionable organisms such as *Salmonella* spp., *Shigella* spp., *Escherichia coli*, *Klebsiella* spp. and *Pseudomonas aeroginosa* were absent in the seed gum. This implies the stability of this biomaterial which could be due to its low and acceptable moisture content that would not encourage microbial contamination and spoilage.

Functional Properties

The ability of the fruit parts to scavenge ABTS radicals showed that the pulp had the least value (0.0370 mmol. TEAC/g). In comparison, the cotyledon had the highest value (0.0548 mmol. TEAC/g). The reducing power of the fruit parts can be ranked in the order: pulp coat (15.82 mg AAE/g) > pulp (10.23 mg AAE/g) > seedcoat (9.75 mg AAE/g) > cotyledon (5.52 mg AAE/g) (Ajayi *et al.*, 2020a). Thus, African star apple fruit parts (pulp, cotyledon, seedcoat and pulp coat) exhibited significant antioxidant activities as revealed by their ability to scavenge free radicals and their effective ferric reducing power. In terms of biological activity, they concluded that the seed gum had an acceptable microbial load and safe toxicity profile. According to the authors, the total aerobic microbial counts and total combined yeast and

Health Benefits and Clinical Studies

The seed cotyledons of African star apple possess very high alkaloids and tannins: the latter show antiallergic, anti-inflammatory, antimicrobial and anticancer activity (Muhammad and Abubakar, 2016). The presence of high alkaloids (which may be a toxic chemical element) in the seed cotyledon has an anti-inflammatory effect that helps control all indications of gastritis, oesophagitis, enteritis and irritating bowel disorders (Muhammad and Abubakar, 2016). The seed cotyledon is also used as a remedy for fever, as a treatment of vaginal infection and as an external application for skin eruption diseases. Furthermore, the presence of tannins is responsible for antidiarrhoeal activity, while saponins are used as a dietary supplement, expectorant and anti-inflammatory agent (Muhammad and Abubakar, 2016).

Evaluating the potential of African star apple fruit in wound care showed that the cotyledon extract exhibited haemostatic, antimicrobial and wound-healing activities (Muhammad and Abubakar, 2016). The presence of phytochemicals in the fruit may contribute to wound-healing activity by suppressing inflammatory reactions in injured tissues (Muhammad and Abubakar, 2016). Idowu *et al.* (2016) reported that antibacterial and antioxidant activities of epicatechin, epigallocatechin and procyanidin B5 isolated from African star apple stembark substantiate the folkloric uses. More so, eleagnine isolated from the seed cotyledon is responsible for the ethnomedical uses as an antimicrobial agent (Idowu et al., 2006).

Ajayi *et al.* (2020a) concluded that the low glycaemic indices and the inhibitory effects of the African fruit parts (pulp, cotyledon, seedcoat and pulp coat) on carbohydrate hydrolysing enzymes (α-amylase and α-glucosidase) due to their phytochemical constituents showed that the fruit parts could be of great benefit as a functional dietary supplement to manage or control diabetes-related disorders. African star apple fruit can be a good source of potassium, calcium and phosphorus, needed for electrolyte balance, neurotransmission and strong bones and teeth (Roth and Townsend, 2003). The amount of crude fibre may influence the digestibility of the fruit and may also help to maintain normal internal distention of the intestinal tract and thus aid peristaltic movements (Edem *et al.*, 1984).

Furthermore, low calorie, sugar and high vitamin content qualify African star apple fruit as suitable for the obese; very low sodium content qualifies it as a good fruit for the hypertensive; while its low carbohydrate content underscores its low value of simple sugar – hence it can be consumed by diabetic patients (Adepoju and Adeniji, 2012). Ologunagba *et al.* (2018), in their study, showed that at a concentration of 10% w/w, African star apple seed gum had comparable tablet properties with 10% w/w acacia gum that could be employed as a binder in the formulation of paracetamol tablets using the wet granulation method. Ihekwereme *et al.* (2018) reported that the methanol extract of the fruit pulp of African star apple possesses prophylactic and curative properties against murine models of malaria.

Antinutritive Compounds and Toxicological Properties

The antinutrient profile showed African star apple fruit to contain phylate, oxalates, tannins, saponins, alkaloids, phytin phosphorus, trypsin inhibitors, hydrocyanic acid (HCN) and nicotinic acid (Table 26.6). The tannin level ranged from 0.59×100^{-4} to 627 mg/100 g in the fruit parts (pulp, peel and seed kernel). This was higher than the value reported for bitter kola pulp (0.05×10^{-4} mg/100 g) (Onyekwelu *et al.*, 2015) but comparable with values reported for pummelo juice (36.99 mg/100 g) (Ani and Abel, 2018). Tannins are known to bind irreversibly to proteins and render them indigestible by the intestinal enzyme. The effect of such a high tannin content on consumers of a large amount of the fruit should be considered (Edem *et al.*, 1984). The oxalate level reported by Edem *et al.* (1984) in African star apple pulp (167 mg/100 g) was high when compared with values obtained in bitter kola pulp (0.31 mg/100 g) and pummelo juice (63.0 mg/100 g) (Onyekwelu *et al.*, 2015; Ani and Abel, 2018) but lower than values reported in gardenia fruit (1100 mg/100 g) (Bello *et al.*, 2008). The HCN level ranged from 5.4 to 6.8 mg/100 g dry sample. A dose of 35 mg of HCN is considered lethal to man; hence a huge amount of the fruit would have to be consumed to attain the lethal dose (Edem *et al.*, 1984). Phytate level ranged from 0.047 to 1.6 in the fruit pulp and was higher than the values reported in bitter kola (0.016 mg/100 g) but lower than values reported in pummelo juice (86.0 mg/100 g) (Onyekwelu *et al.*, 2015; Ani and Abel, 2018). Phytic acid is known to interfere with the absorption and utilization of calcium, magnesium and iron. It may also influence the availability of phosphorus. Thus, phytic acid can influence bone calcification. Though these toxicants occur at non-toxic levels, their interfering effect on the utilization of essential nutrients should be considered (Edem *et al.*, 1984). Adepoju and Adeniji (2012) also revealed that the fruit pulp was very low in antinutritional factors, and the possibility of intoxication or malabsorption of nutrients from other sources when a large quantity of the fruit is consumed is very unlikely, while trypsin inhibitors had the highest value among the

Table 26.6. Antinutrient composition of fresh African star apple evaluated by different authors.

	Onyekwelu *et al.*, 2015		Edem *et al.*, 1984		Adepoju and Adeniji, 2012	Ureigho and Ekeke, 2010
Parameter	Seed kernel	Pulp	Peel	Pulp	Pulp	Pulp
Phytates (mg/100 g)	2.29	0.65	0.8	1.6	0.047	-
Oxalates (mg/100 g)	1.01	0.89	211	167	0.534	-
Tannins (mg/100 g)	0.59×100^{-4}	0.6×100^{-4}	264	627	1.442	-
Saponins (mg/g)	0.06	0.09	-	-	-	-
Alkaloids (mg/g)	0.11×100^{-2}	0.05×100^{-2}	-	-	-	-
Phytin phosphorus (mg/g)	0.65	0.64	-	-	-	-
Trypsin inhibitors (mg/100 g)	-	-	-	-	3.838	-
Hydrocyanic acid (mg/100 g)	-	-	5.4	6.8	-	-
Nicotinic acid (mg/100 g)						0.68

Table 26.7. Comparative evaluation of jams made from African star apple (*Chrysophyllum albidum*) and African mango (*Irvingia gabonensis* var. *gabonensis*) with commercial jam made from conventional mango (*Mangifera indica*) (Aworh, 2014).

Type of jam	Appearance	Colour	Flavour	Consistency	Overall acceptability
African star apple jam	7.8^a	7.9^a	7.7^a	7.1^a	7.6^a
African mango jam	6.4^b	7.1^a	5.1^b	5.5^b	5.7^b
Commercial mango jam	7.3^{ab}	7.5^a	7.2^a	7.4^a	7.5^a

Rated on a scale of 1 = dislike extremely, 9 = like extremely. Values in a column with different superscripts are significantly different ($p<0.05$) according to Tukey's test.

antinutrients studied. Ihekwereme *et al.* (2018) revealed that repeated dose toxicity extract is not harmful to experimental animals. The pulp extract's acute toxic dose is beyond 5000 mg/kg per oral. All the animals survived even at 5000 mg/kg, which shows that the extract might be considered relatively safe at very high doses (Ihekwereme *et al.*, 2018), in turn suggesting that large quantities of the fruit pulp could be eaten by consumers without any restraint or harm.

The very sweet African star apple fruit was observed to contain nicotinic acid, a toxic material according to nutrient classification (Ureigho and Ekeke, 2010). Also, Shobo *et al.* (2019) reported butanolic extract from the seed cotyledon of African star apple to cause mild to moderate toxicity at higher portions. There is the need to protect against aimless utilization and apply caution in using the seed cotyledon.

Food Applications

Pulp

Fruit pulp is an excellent source of raw material for industries. It is used to formulate diet soft drinks in some beverage industries, as well as jam, stewed fruit, marmalade, syrup and jelly. Aworh (2014), in a comparative study of jams made from African star apple, African mango (*Irvingia gabonensis*) and conventional mango (*Magifera indica*) pulp, found that African star apple jam was most preferred in terms of appearance by the assessors (Table 26.7).

The colour, flavour and consistency of African star apple jam were also preferred over African mango jam but comparable with conventional mango jam. This shows that African star apple jam has a good prospect for commercial jam production. Fruit juices can also be produced from African star apple using simple procedures suitable for home preparation or small-scale commercial production. However, even though the extracted juice has an attractive deep-red colour it also has a rather harsh flavour (Okoro, 2001 cited in Aworh, 2015). Jam made from African star apple by the open-kettle process has an attractive deep-red colour, and that made from hog plum has a golden yellow colour (Aworh, 2015).

Sensory evaluation conducted by Aina and Adesina (1991) also indicated that African star apple jams were preferred to plum jams and storage trials showed that the fruit jams maintained good stability as reflected by the small decreases in acid and sugar levels. African star apple jam stored at ambient temperature for 16 weeks maintained good stability (Aina and Adesina, 1991).

Furthermore, laboratory studies on the production of fruit leathers (dried fruit pulps)/candies from African star apple revealed that the best products, in terms of consumer acceptance, cost and energy savings, were obtained when ripe African star apple fruit pulp was immersed in sugar syrup (60°B; 1:4 ratio (w/w) of fruit pulp to sugar syrup) at ambient temperature (27°C) for eight hours, followed by hot-air drying at 60°C (Falade and Aworh, 2004). The gum extracted from the fruit has also been characterized and used as a binder and suspending agent (Bakre and Sholuade, 2018).

Seed/kernel

Oil could be derived from the seed of the African star apple. In parts of West Africa, the seeds are gathered together to produce essential oil and extract oil for soap-making or cooking purposes (National Research Council, 2008). Adebayo *et al.* (2012) conclusively stated that African star apple seed oil could be used as domestic and industrial oil. Although the oil yield is low, the physicochemical properties indicated that it is non-drying and may be used to produce soap, lubricating oil and candles. However, it may not be suitable for oil paint, varnishes and surface coatings due to its non-drying attribute.

Conclusion and Future Prospects

African star apple fruit is rich in macronutrients, micronutrients and dietary phytochemicals. The fruit has great potential for improving human health, nutrition, food security and the income of rural populations. This underutilized fruit has broad medicinal purposes in ethnomedicine, thanks to the presence of health-giving compounds in the different parts of the fruit. As demonstrated in this review, there is a huge potential for exploring African star apple fruit for health benefits, primarily through agro-processing. The fruit is produced in considerably large quantities, but it is mostly under-utilized except for consumption in the fresh form. Inadequate utilization, storage and processing facilities are the primary cause of the substantial postharvest losses experienced during the production season of the fruit. Therefore, improving the postharvest systems of African star apple fruit through the optimization of preservation techniques and development of value-added functional products will enhance the fruit's economic importance, making it available in different forms for both consumption and industrial use.

Acknowledgements

OMO acknowledge the support of the Department of Horticulture, Federal University of Agriculture, Abeokuta (FUNAAB) and pictorial contributions of Mr Alpheus Simon, research assistant, FUNAAB. OAF acknowledges the financial support of the University of Johannesburg and the National Research Foundation, South Africa.

References

Abiodun, O.A. and Oladapo, A.S. (2011) Physico-chemical properties of African star apple (*Chrysophylum albidum*) components. *Nutrition & Food Science* 41(1), 8–11. DOI: 10.1108/00346651111102847.

Adebayo, S.E., Orhevba, B.A., Adeoye, P.A., Musa, J.J. and Fasae, O.J. (2012) Solvent extraction and characterization of oil from African star apple (*Chrysophyllum albidum*) seeds. *Academic Research International* 3(2), 178–183.

Adekanmi, D.G. and Olowofoyeku, A.E. (2020) African star apple: potentials and application of some indigenous species in Nigeria. *Journal of Applied Sciences and Environmental Management* 24(8), 1307–1314. DOI: 10.4314/jasem.v24i8.1.

Adepoju, O.T. and Adeniji, P.O. (2012) Nutrient composition and micronutrient potential of three wildly grown varieties of African star apple (*Chrysophyllum albidum*) from Nigeria. *African Journal of Food Science* 6(12), 344–351. DOI: 10.5897/AJFS12.045.

Aina, J.O. and Adesina, A.A. (1991) Quality attributes of jams from low-usage tropical fruits. *Food Chemistry* 40(3), 345–351. DOI: 10.1016/0308-8146(91)90118-8.

Aiyelaagbe, I.O.O., Adeola, A.O., Labode, P., Ladipo, D.O. and Obisesan, K.O. (1999) *Chrysophyllum albidum* in the farming systems of Nigeria: Its prevalence, farmers preference and agroforestry potential. In: Denton, O.A., Ladipo, D.O., Adejoro, M.A., and Sarumi, M.B. (eds) *Proceedings of a National Workshop on Potentials of the Star Apple in Nigeria*. CENRAD, Ibadan, Nigeria.

Ajayi, O.B., Oyetayo, F.L. and Akomolafe, S.F. (2020a) Starch composition, glycemic indices, antioxidant properties and carbohydrate hydrolyzing enzymes activities of African star apple fruit parts. *BMC Complementary Medicine and Therapies* 20(1), 260. DOI: 10.1186/s12906-020-03053-9.

Ajayi, A.M., Badaki, V.B., Ariyo, O.O., Ben-Azu, B., Asejeje, F.O. *et al.* (2020b) *Chrysophyllum albidum* fruit peel attenuates nociceptive pain and inflammatory response in rodents by inhibition of pro-inflammatory cytokines and COX-2 expression through suppression of NF-κB activation. *Nutrition Research* 77, 73–84. DOI: 10.1016/j.nutres.2020.03.004.

Akaneme, F.I. (2008) Identification and preliminary phytochemical analysis of herbs that can arrest threatened miscarriage in Orba and Nsukka towns of Enugu state. *African Journal of Biotechnology* 7, 6–11.

Amusa, N.A., Ashaye, O.A. and Oladapo, M.O. (2003) Biodeterioration of the African star apple (*Chrysophylum albidum*) in storage and the effect on its food value. *African Journal of Biotechnology* 2(3), 56–59. DOI: 10.5897/AJB2003.000-1011.

Ani, P.N. and Abel, H.C. (2018) Nutrient, phytochemical, and antinutrient composition of *Citrus maxima* fruit juice and peel extract. *Food Science and Nutrition* 6(3), 653–658. DOI: 10.1002/fsn3.604.

Aworh, O.C. (2014) *Lesser-Known Nigerian Fruits and Vegetables: Postharvest Handling, Utilization and Nutritional Value*. Ibadan University Press, Ibadan, Nigeria.

Aworh, O.C. (2015) Promoting food security and enhancing Nigeria's small farmers' income through value-added processing of lesser-known and under-utilized indigenous fruits and vegetables. *Food Research International* 76, 986–991. DOI: 10.1016/j.foodres.2015.06.003.

Bada, S.O. (1997) Preliminary information on the ecology of *Chrysophillum albidum* G. Don, in West and Central Africa. In: Denton, O.A., Ladipo, D.O., Adetoro, M.A., and Sarumi, M.B. (eds) *Proceedings of a National Workshop on the Potentials of the Star Apple in Nigeria*. pp. 16–25.

Bakre, L. and Sholuade, T. (2018) Formulation and in vitro evaluation of Ciprofloxacin:matrix tablets: Effect of drug-polymer ratio. *Tropical Journal of Natural Product Research* 2(5), 250–253. DOI: 10.26538/tjnpr/v2i5.9.

Bello, M.O., Falade, O.S., Adewusi, S.R.A. and Olawore, N.O. (2008) Studies on the chemical compositions and antinutrients of some lesser known Nigeria fruits. *African Journal of Biotechnology* 7(21), 3972–3979.

Dadegnon, S., Gbemavo, C., Ouinsavi, C. and Sokpon, N. (2015) Morphological variation and ecological structure of *Chrysophyllum albidum* G.Don. *International Journal of Plant and Soil Science* 5(1), 25–39. DOI: 10.9734/IJPSS/2015/11385.

Dah-Dovonon, J.Z. (2002) *Rapport du Bénin. In Réseau Espèces Ligneuses Alimentaires. Compte rendu de la première Réunion du Réseau: 11–13 Décembre 2000*. Ouagadougou, Burkina Faso, IPGRI.

Dumbrava, D.G., Moldovan, C., Raba, D.N., Popa, M.V. and Druga, M. (2016) Evaluation of antioxidant activity, polyphenols and vitamin C content of some exotic fruits. *Journal of Agroalimentary Processes and Technology* 22(1), 13–16.

Edem, D.O., Eka, O.U. and Ifon, E.T. (1984) Chemical evaluation of nutritive value of the fruit of African Star apple (*Chrysophyllum albidum*). *Food Chemistry* 14(4), 303–311. DOI: 10.1016/0308-8146(84)90085-2.

Essien, E.U., Esenowo, G.J. and Akpanabiatu, M.I. (1995) Lipid composition of lesser known tropical seeds. *Plant Foods for Human Nutrition* 48(2), 135–140. DOI: 10.1007/BF01088309.

Falade, K.O. and Aworh, O.C. (2004) Adsorption isotherms of osmo-oven dried African star apple (*Chrysophyllum albidum*) and African mango (*Irvingia gabonensis*) slices. *European Food Research and Technology* 218(3), 278–283. DOI: 10.1007/s00217-003-0843-8.

Fawole, O.A. and Opara, U.L. (2013) Changes in physical properties, chemical and elemental composition and antioxidant capacity of pomegranate (cv. Ruby) fruit at five maturity stages. *Scientia Horticulturae* 150, 37–46. DOI: 10.1016/j.scienta.2012.10.026.

Fawole, O.A., Opara, U.L. and Theron, K.I. (2012) Chemical and phytochemical properties and antioxidant activities of three pomegranate cultivars grown in South Africa. *Food and Bioprocess Technology* 5(7), 2934–2940. DOI: 10.1007/s11947-011-0533-7.

Gill, L.S. (1988) *Taxonomy of Flowering Plants*. Africana-Feb Publisher Ltd, Ibadan, Zaria.

Houessou, L.G., Lougbegnon, T.O., Gbesso, F.G.H., Anagonou, L.E.S. and Sinsin, B. (2012) Ethnobotanical study of the African star apple (*Chrysophyllum albidum* G. Don) in the Southern Benin (West Africa). *Journal of Ethnobiology and Ethnomedicine* 8, 40. DOI: 10.1186/1746-4269-8-40.

Ibrahim, H.O., Osilesi, O., Adebawo, O.O., Onajobi, F.D., Karigidi, K.O. *et al.* (2017) Nutrients compositions and phytochemical contents of edible parts of *Chrysophyllum albidum* fruit. *Journal of Nutrition & Food Sciences* 7, 2.

Idowu, T.O., Ilwalewa, E.O., Aderogba, M., Akinpelu, B.A. and Ogundaini, A. (2006) Antinociceptive, anti-inflammatory and antioxidant activities of eleagnine: an alkaloid isolated from *Chrysophyllum albidum* seed cotyledons. *Journal of Biological Sciences* 6(6), 1029–1034. DOI: 10.3923/jbs.2006.1029.1034.

Idowu, T.O., Ogundaini, A.O., Adesanya, S.A., Onawunmi, G.O., Osungunna, M.O. *et al.* (2016) Isolation and characterization of chemical constituents from *Chrysophyllum albidum* G. Don-Holl. stem-bark extracts and their antioxidant and antibacterial properties. *African Journal of Traditional, Complementary, and Alternative Medicines* 13(5), 182–189. DOI: 10.21010/ajtcam.v13i5.24.

Ihekwereme, C., Melidem, C., Maduka, I. and Okoyeh, J. (2018) In Vivo Antiplasmodial and toxicological effects of extracts of fruit pulp of *Chrysophyllum albidium* G. Don (Sapotaceae). *Tropical Journal of Natural Product Research* 2(3), 126–131. DOI: 10.26538/tjnpr/v2i3.5.

Ishola, I.O., Oreagba, I.A., Adeneye, A.A., Adirije, C., Oshikoya, K.A. *et al.* (2014) Ethnopharmacological survey of herbal treatment of malaria in Lagos, Southwest Nigeria. *Journal of Herbal Medicine* 4(4), 224–234. DOI: 10.1016/j.hermed.2014.08.001.

Jayeoba, O.J., Ige, M.M., Omolaiye, J.A., Gbadamosi, S.O., Ogunbanjo, R.O. *et al.* (2007) Chemical composition and physical properties of African star apple (Chrysophyllum albidum). *ASSET* 7(1), 37–42.

Jayeoba, O.J., Ige, M.M., Omolaiye, J.A., Gbadamosi, S.O., Ogunbanjo, R.O. *et al.* (2010) Chemical composition and physical properties of African star apple(*Chrysophyllum albidum*). *ASSET (Series A)* 7, 37–42.

Kang, B.T. (1992) *Multipurpose Tree Screening and Evaluation for Agroforestry in the Humid Lowland of West Africa*. IITA, Ibadan, Nigeria, p. 13.

Lasekan, O., Khatib, A., Juhari, H., Patiram, P. and Lasekan, S. (2013) Headspace solid-phase microextraction gas chromatography -mass spectrometry determination of volatile compounds in different varieties of African star apple fruit (*Chrysophyllum albidum*). *Food Chemistry* 141(3), 2089–2097. DOI: 10.1016/j.foodchem.2013.05.081.

Muhammad, S.A. and Abubakar, S.M. (2016) Qualitative and quantitative determination of phytochemicals in aqueous extract of *Chrysophyllum albidum* seed kernel. *Biosciences, Biotechnology Research Asia* 13(2), 1201–1206. DOI: 10.13005/bbra/2153.

National Research Council (2008) *Lost Crops of Africa: Volume III - Fruit*. The National Academies Press, Washington, DC.

Odeyemi, O.M. and Ojetayo, A.E. (2018) Physico-chemical properties, proximate composition and sensory evaluation of Pineapple fruits (*Ananas comosus*) stored in different media. In: *Proceedings of the 36th Annual Conference of the Horticultural Society of Nigeria*, Nasarawa State University, Keffi, Nasarawa State, Nigeria, pp. 699–703.

Odeyemi, O.M., Aiyelaagbe, I.O.O. and Olubode, O.O. (2019) Quality changes in sweet orange (Citrus sinensis. cv Agege 1) with fruit maturity in tropical humid climate. *African Journal of Organic Agriculture and Ecology* 2, 45–53.

Odeyemi, O.M., Aiyelaagbe, I.O.O., Bodunde, J.G. and Babalola, S.O. (2014) Influence of variety and production system on postharvest quality of pawpaw (*Carica papaya*. L) fruits stored in different media. *Nigerian Journal of Horticultural Science* 19, 53–60.

Okoli, B.J. and Okere, O.S. (2010) Antimicrobial activity of the phytochemical constituents of *Chrysophyllum albidum* G. Don-Holl. *Journal of Research of the National Institute of Standards and Technology* 8, 1.

Olayiwola, I.O., Akinfenwa, V.O., Oguntona, C.O., Sanni, S.A., Onabanjo, O.O. *et al*. (2013) Phytonutrients, antioxidant and mineral composition of some wild fruits in South west Nigeria. *Nigerian Food Journal* 31(2), 33–40. DOI: 10.1016/S0189-7241(15)30074-6.

Ologunagba, M., Azubuike, C., Sadiku, O. and Silva, B. (2018) Evaluation of the binding potential of *Chrysophyllum albidum* seed Linn (Family: Sapotaceae) gum in paracetamol tablet formulation. *Tropical Journal of Natural Product Research* 2(3), 136–139. DOI: 10.26538/tjnpr/v2i3.7.

Olubode, O.O., Yisau, O.P., Olubobe, A.A. and Oyegoke, C.O. (2019) Growth medium and soil amendment influence on seedling growth responses of African star apple (*Chrysophyllum albidum*). *Journal of Botanical Research* 1(3).

Onwe, D.N., Umani, K.C., Olosunde, W.A. and Ossom, I.S. (2020) Comparative analysis of moisture-dependent physical and mechanical properties of two varieties of African star apple (*Chrysophyllum albidum*) seeds relevant in engineering design. *Scientific African* 8, e00303. DOI: 10.1016/j.sciaf.2020.e00303.

Onyegbule, F., Ezenwa, C., Bruce, S. and Umeokoli, B.O. (2020) Standardization, chemical composition and antipyretic evaluation of the methanol leaf extract and fractions of *Chrysophyllum albidum* (Sapotaceae). *Tropical Journal of Natural Product Research* 4(6), 216–222. DOI: 10.26538/tjnpr/v4i6.1.

Onyekwelu, J.C., Oyewale, O., Stimm, B. and Mosandl, R. (2015) Antioxidant, nutritional and antinutritional composition of *Garcinia kola* and *Chrysophyllum albidum* from rainforest ecosystem of Ondo State. *Journal of Forestry Research* 26(2), 417–424. DOI: 10.1007/s11676-015-0068-2.

Opeke, L.K. (2005) *Tropical Tree Crops*. Spectrum Book Ltd, Ibadan, Nigeria.

Oyebade, B.A., Ekeke, B.A. and Adeyemo, F.C. (2011) Fruits categorization and diagnostic analysis of *Chrysophyllum albidum* in Nigeria. *Advances in Applied Science Research* 2, 7–15.

Roth, A.R. and Townsend, C.E. (2003) *Nutrition and Diet Therapy*, 8th edn. Delmar Learning, Thomson Learning Inc, Canada, p. 132.

Segun, P.A., Ogbole, O.O. and Ajaiyeoba, E.O. (2018) Medicinal plants used in the management of cancer among the Ijebus of Southwestern Nigeria. *Journal of Herbal Medicine* 14, 68–75. DOI: 10.1016/j.hermed.2018.04.002.

Shobo, A.A., Daniyan, M.O. and Olayiwola, G. (2019) Toxicity evaluation of the extract and fraction of *Chrysophyllum albidum* seed cotyledons in rats. *SOJ Pharmacy & Pharmaceutical Sciences* 6(1), 1–12.

Ureigho, U.N. and Ekeke, B.A. (2010) Nutrient values of *Chrysophyllum albidum* Linn African star apple as a domestic income plantation species. *African Research Review* 4(2), 50–56. DOI: 10.4314/afrrev.v4i2.58288.

Wills, R., McGlasson, B., Graham, D. and Joyce, D. (1998) *Post-Harvest. An Introduction to the Physiology and Handling of Fruits, Vegetables and Ornamentals*. CAB International, Wallingford, UK.

27 Phalsa (*Grewia asiatica* L.)

Manjeshwar Shrinath Baliga[1*], Giriyapura Srikantachar Pavankumar[2], Karkala Sreedhara Ranganath Pai[3], Harish Kumar Bastimal[3], Beguvalli Subash Inchara[4] and Karkala Usha Sreedhar Pai[4]

[1]Mangalore Institute of Oncology, Mangalore, Karnataka, India; [2]Sri Lakshmi Group of Institution, Bengaluru, Karnataka, India; [3]Department of Pharmacology, Manipal College of Pharmaceutical Sciences, Manipal Academy of Higher Education (MAHE), Karnataka, India; [4]Sahyadri Science College, Kuvempu University, Shimoga, Karnataka, India

Abstract

Grewia asiatica (L.), commonly known as Indian Phalsa, is a medium sized evergreen tree originally indigenous to the Himalayan region of India and today found growing throughout South-East Asia. The fruit are the most desired plant part and are produced annually. They have wide dietary and medicinal use. The plant has been used extensively in various traditional and folk systems of medicine to treat a range of ailments. Phytochemical studies have confirmed that the plant contains triterpenoids, steroids, glycosides, flavones, lignanes, phenolics, alkaloids, lactones and organic acids. Pharmacological studies carried out in accordance with modern medicine guidelines with various preclinical models of experiment have shown that the plant possesses free radical scavenging, antioxidant, anti-inflammatory, analgesic, antimicrobial, anticancer, radioprotective, immunomodulatory and antidiabetic activities, thereby validating many of its ethnotherapeutical uses. This review addresses its ethnomedicinal uses, chemical constituents and validated pharmacological observations.

Introduction

India is recognized to be one of the biodiversity hotspots of the world and the flora and fauna of the geographical area is unique (Nayar *et al.*, 2009; Bertzky *et al.*, 2013). Some indigenous Indian plants have over time been cultivated in other parts of the world (FAO, 1982; Yadav, 1999; Orwa *et al.*, 2009; Akram *et al.*, 2019; Krishna *et al.*, 2019). One such originally indigenous Indian fruiting plant is *Grewia asiatica*, which belongs to the family Tiliaceae and the order Malvales (Table 27.1).

The genus contains nearly 40 species and the plants *G. tenax*, *G. hirsuta*, *G. damine*, *G. Lasiodiscus*, *G. optiva*, *G. biloba*, *G. bicolor*, *G. tiliaefolia*, *G. flavescens* are found growing wild in various terrains, although with a preference for slightly alkaline soil. The plant is commonly referred to as Indian Phalsa in English, Parushaka in Sanskrit and Phalash in Hindi (Akram *et al.*, 2019; Krishna *et al.*, 2019). Other vernacular names are listed in Table 27.2. *Grewia asiatica* belongs to the genus Grewia which is exemplified for containing shrubs and trees that are principally seen in

*Corresponding author: msbaliga@gmail.com

 Handbook of Phytonutrients in Indigenous Fruits and Vegetables (eds D. Sivakumar *et al.*)
DOI: 10.1079/9781789248067.0027

Table 27.1. Taxonomic details of *Grewia asiatica*.

Taxonomic tree	
Domain	Eukaryota
Kingdom	Plantae
Phylum	Spermatophyta
Subphylum	Angiospermae
Class	Dicotyledonae
Order	Malvales
Family	Tiliaceae
Genus	*Grewia*
Species	*Grewia asiatica*

the warmer tropical regions of the world (FAO, 1982; Yadav, 1999; Orwa *et al.*, 2009).

In India, the plants are reported to have been indigenous to the Himalayan region and are found growing in almost all parts of Uttar Pradesh, Bihar, Rajasthan, Haryana, Punjab, Gujarat, Maharashtra, Andhra Pradesh and Madhya Pradesh, and also in Pakistan, Bangladesh, Sri Lanka and Nepal. Phalsa has been a neglected plant for many years and is now regarded as a minor fruit crop that grows in limited amounts in the villages and forest regions of India (Samson, 1986; Ali *et al.*, 2001; Akram *et al.*, 2019; Krishna *et al.*, 2019).

The plants are a source of income, in particular for poor families. The fruit (Fig. 27.1)

Table 27.2. Names for *Grewia asciatica* L .in different languages.

Language	Names
English	Indian phalsa, falsa
Sanskrit	Mriduphal, Alpasthi, Dhanvanacchada, Dhanvanah, Dhanvanchhada, Dharmana, Giripilu, Mriduphala, Nagadalapam, Nilacharma, Nilamandala, Parapara, Paravata, Parusa, Parusaka, Parusakah, Parusakam, Parusam, Parusha, Parushaka, Porusha, Purusha, Roshana
Hindi	Bhyunl, Dhaman, Dhamani, Dhamin, Karra, Parusha, Phalasa, Phalsa, Pharsa, Pharsula, Pharvah, Phulsa, Shukri
Arabic	Phalasah
Persian	Falseh, Palasah
Urdu	Phalasah, Phalsa Taza
Tibetan	Barusaka, Parusaka
Assamese	Man-Bijal
Bengali	Phalsa and Shukri
Gujarati	Shukri
Kannada	Phulsha
Konkani	Phalsi
Malayalam	Chadicha
Marathi	Phalsa
Odiya	Pharosakoli
Punjabi	Dhaman
Sindhi	Pharaho
Tamil	Unnu, Palica, Palicam, Palisa, Tadachi, Tadachit
Telugu	Phutiki, Jana, Nallajana, Peddajana, Pethajana, Phalsa
Laos	Nhap
Philippines	Bariuan-gulod
Thailand	Malai, potaohai, ya-khithut

Fig. 27.1. *Grewia asiatica* L. plant with raw and ripe berries (Wikipedia).

are widely eaten alone or with some salt and are also used to make jam, squash, pickle, pies and chutneys (Goyal, 2012; Zia-Ul-Haq *et al.*, 2013; Akram *et al.*, 2019; Krishna *et al.*, 2019). The fruit have a short shelf life and a fully ripe one perishes in 24 hours until and unless it is stored at refrigeration temperature (Akram *et al.*, 2019). Because of this the fruit are not available in the cities. However, more recently, the juice of the fruit, prepared and marketed, has gained public attention. The stems and the bark can be used to make rope, baskets and as a soap substitute, while the leaves are a good cattle fodder (Zia-Ul-Haq *et al.*, 2012a).

Botany

In India, two morphotypes of *Grewia asiatica* plants are found, the tall and short varieties, and they grow very well in areas that have distinct summer and winter seasons (Orwa *et al.*, 2009; Akram *et al.*, 2019). However, irrespective of the morphotype, the plants are very hardy, drought resistant and rarely infested with pests. The plants are propagated by stem cuttings and through seeds. They do not need any deliberate care, like watering or adding of manure. The plants grow extremely well in alluvial soil while also preferring clay or sandy soil in the arid regions of India (Orwa *et al.*, 2009).

From a morphological perspective, the plants are classified as a shrub or small tree and grow to a maximum of 3.5 m (12 ft) high. The bark is greyish-white or greyish-brown in colour and the leaves vary from broadly heart-shaped to obliquely ovate with serrated margins (Morton, 1987; Orwa *et al.*, 2009; Goyal, 2012; Zia-Ul-Haq *et al.*, 2013; Akram *et al.*, 2019; Krishna *et al.*, 2019). The flowers are densely crowded axillary cymes and each flower is yellow in colour and about 2 cm in length (Orwa *et al.*, 2009; Goyal, 2012; Zia-Ul-Haq *et al.*, 2013; Akram *et al.*, 2019; Krishna *et al.*, 2019).

The fruit are small, round in shape, light green in colour when raw and cherry red or purplish red, to dark purple/dark blue or black when fully ripe. The fruit ripens in about two months. Each drupe has a rather large seed and small pulp (Goyal, 2012; Zia-Ul-Haq *et al.*, 2013Akram *et al.*, 2019; Krishna *et al.*, 2019). In India, the fruit are formed from late summer through to the following spring (i.e. from March to November). The ripe fruit are soft and taste sweetish with a tinge of acid; they are highly perishable (Akram *et al.*, 2019; Krishna *et al.*, 2019). They have a pleasant flavour, similar to that of grapes, and are rich in carbohydrates, proteins, minerals and vitamins A and C (Akram *et al.*, 2019; Krishna *et al.*, 2019). In one season, a well mature plant produces around 9–11 kg fruit yield (Morton (1987); Akram *et al.*, 2019).

Table 27.3. Nutritional composition of phalsa fruit (Yadav, 1999).

Nutrient content	Nutrient in 100 g
Protein (g)	1.57
Total lipid (fat) (g)	<0.1
Carbohydrate (g)	21.1
Moisture content	76.3
Ash (g)	1.1
Dietary fibre (g)	5.53
Calories (kcal)	90.5
Calcium (mg)	136
Iron (mg)	1.08
Phosphorus (mg)	24.2
Potassium (mg)	372
Sodium (mg)	17.3
Vitamin B1, thiamine (mg)	0.02
Vitamin B2 (mg)	0.264
Vitamin, B3, niacin (mg)	0.825
Vitamin C, ascorbic acid (g)	4.385
Vitamin A, retinol (μg)	16.11

Proximate Composition and Phytochemistry

Phytochemical studies have shown that the *Grewia asiatica* plant contains triterpenoids, steroids, glycosides, flavones, lignanes, phenolics, alkaloids, lactones and organic acids (Akram *et al.*, 2019). The fruit have been the most investigated plant part and reports suggest that they are very low in calorific value and fat, rich in carbohydrate, proteins, minerals and vitamins A and C (Akram *et al.*, 2019; Krishna *et al.*, 2019) (Table 27.3). The fruit contain amino acids such as proline, glutaric acid, lysine and phenyl alanine, and carbohydrates such as glucose and aldopentose (Akram *et al.*, 2019) (Table 27.3). The are rich in micronutrients (Khan *et al.*, 2006; Akram *et al.*, 2019). Details of their composition in wet and dry conditions are shown in Table 27.4. Reports also suggest that composition varies with the long and short varieties of the plant (Akram *et al.*, 2019) (Table 27.5). On a comparative note, when compared with the commonly consumed apple, banana, orange and mango, phalsa has less carbohydrates than banana and mango and more iron and phosphorus than the other fruits (Gopalan *et al.*, 2004) (Table 27.6).

Table 27.4. Mineral content of the fruit based on dry or wet conditions (Khan *et al.*, 2006).

Nutrient content	μg/100 g dry weight	mg/100 g fresh weight
Cobalt	33	0.99
Chromium	36	1.08
Copper	16	0.48
Nickel	87	2.61
Zinc	48	144
Iron	1695	140.8

The phytochemical constituents and their ratios change with the maturing status of the fruit. Phenols, tannins, steroids and flavonoids like kaempferol, quercetin and their glycoside mixtures like naringenin-7-*O*-β-D-glucoside, 3,5-diglucoside, 3-*O*-β-D-glucoside, quercetin, catechins, tannins and cyanidin-3-glucoside (Ali *et al.*, 1982; Gupta *et al.*, 2006; Akram *et al.*, 2019) are all found. With regard to the other plant parts, chemical studies have shown that

Table 27.5. Characteristics of tall and dwarf types of *Grewia asiatica* (Phalsa). (Dhawan *et al.*, 1993).

Content (%)	Tall	Dwarf
Edible portion	91.30	90.79
Seed	8.70	9.21
Juice yield	67.50	65.90
Pomace	32.50	34.10
Moisture	76.80	74.83
Total sugars	5.73	7.95
Reducing sugars	1.24	0.99
Non-reducing sugars	4.49	6.96
Titrable acidity	1.48	1.12
Fruit protein	3.13	1.89
Seed protein	8.75	7.00
Pulp protein	1.40	7.00

the leaves contain diterpenes, acetal derivatives, fats, glycosides, triterpenoids, sterols, polyphenols, aglycones and tannins (Chattopadhyay and Pakrashi, 1975; Ali *et al.*, 1982; Akram *et al.*, 2019) while the flowers contain quercetin 3-O-β-D-glucoside, quercetin, β-sitosterol, naringenin 7-O-β-D-glucoside, naringenin and lactone 3,21,24-trimethyl-5,7-dihydroxy hentriacontanoic acid (Lakshmi *et al.*, 1976; Akram *et al.*, 2019). The stem and bark contain lupeol, betulin, friedelin, lupenone, β-amyrin and β-sitosterol (Akram *et al.*, 2019). Some of the important phytochemicals are represented in Fig. 27.2.

Traditional Uses

Since antiquity, in the various traditional and folk systems of medicine practised in the Indian subcontinent and other parts of South-East Asia, phalsa has had a mention. It is an integral part of many pharmacopeias aimed at maintenance of general health as well as mitigating ailments (Morton, 1987). Phalsa has been used in Ayurveda since time immemorial and finds mention in the work of Charaka and Sushruta (Subhose *et al.*, 2005). The Santal tribes in India are known to use a decoction of the root bark for rheumatism. In the Sind region of Pakistan, folk medicine experts prepare infusions of the bark as a demulcent (Ullah *et al.*, 2012; Zia-Ul-Haq *et al.*, 2013). In villages, the root bark is used to treat rheumatism and urinary tract problems (Sisodia and Singh, 2009). The leaves are found to be useful as a skincare agent – reports suggest that a paste made by grinding the leaves is topically applied for pustular eruptions, to heal wounds and cuts and to mitigate irritation and pain caused by rashes (Zia-Ul-Haq *et al.*, 2012a).

Table 27.6. Comparison of nutritional value of *Grewia asiatica* with apple, banana, orange and mango (Gopalan *et al.*, 2004).

	Species				
Nutrient	Apple (*Malus domestica*)	Phalsa (*Grewia asiatica*)	Banana (*Musa paradisiaca*)	Orange (*Citrus* X *sinensis*)	Mango (*Mangifera indica*)
Energy (kcal)	59	72	116	48	74
Moisture (g)	85	81	70	88	81
Protein (g)	0	1	1	1	1
Fat (g)	0	1	0	0	0
Minerals (g)	0	1	1	0	0
Fibre (g)	1	1	0	0	1
Carbohydrates (g)	13	15	27	11	17
Calcium (mg)	10	129	17	26	14
Phosphorus (mg)	14	39	36	20	16
Iron (mg)	1	3	0	0	1

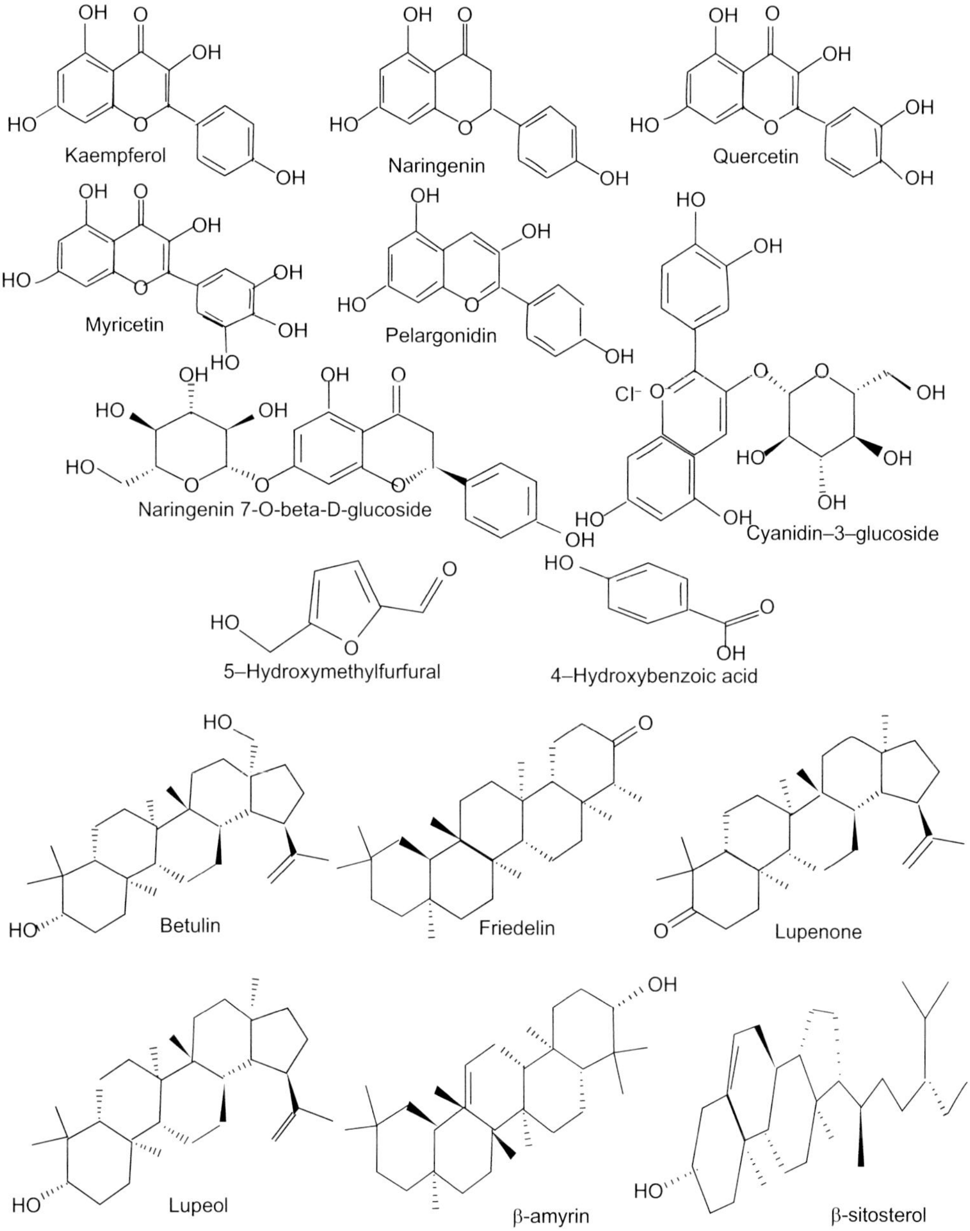

Fig. 27.2. Important phytochemicals present in *Grewia asiatica* L.

The fruit are the most commonly used medicinal part, employed as a coolant and to treat burning sensations and fever. The fruit are supposed to possess astringent, cooling and stomachic properties. A spirit distilled from the fruit to make sherbet is reported to possess cardioprotective effects. The fruit are also reported to be useful for blood and liver disorders, anorexia, indigestion, thirst, toxaemia, stomatitis, hiccough, asthma, spermatorrhoea, fevers and diarrhoea, and in treating ailments affecting the throat, tuberculosis and rectifying sexual debility (Sharma and Sisodia, 2009).

Scientifically Validated Pharmacological Properties

In the very recent past, the scientific fraternity has taken an active interest in validating the medical uses attributed to the plant and fruit of *Grewia asiatica*. Experiments have been conducted using the accepted preclinical pharmacological protocols followed in the modern system of medicine using both *in vitro* and animal models of study. In the subsequent sections, the findings are addressed in detail.

Free radical scavenging and antioxidant activity

The electophilic reactive oxygen species (ROS) and reactive nitrogen species (RNS), when produced in surplus, inflict damage to DNA, lipids, proteins and other biomolecules. Accordingly, antioxidants are needed to prevent the formation/nullify the deleterious effects of ROS and RNS (Leopoldini *et al.*, 2010). With regard to phalsa, in one of the earliest reports Gupta and co-workers observed that the leaf extract was also effective in 1,1-diphenyl-2-picrylhydrazyl (DPPH) and NO scavenging assay and the results comparable to those of ascorbic acid and quercetin, used as standards (Gupta *et al.*, 2007). Subsequently Asghar *et al.* (2008) studied the antioxidant activity of fresh and stored samples of seed, peel and fruit pulp and observed that the fruit and its various parts possessed good antioxidant activity and that the extracts prepared when fruit were fresh were better than when frozen in the 2,2'-azino-bis(3-ethylbenzothiazoline-6-sulfonic acid (ABTS) and modified N,N-dimethyl-p-phenylenediamine (DMPD) decolorization assays (Asghar *et al.*, 2008).

The fruit peel has been reported to possess the best antioxidant value, followed by pulp and seeds (Asghar *et al.*, 2008), and polyphenolics were present in higher concentration in the peel (Asghar *et al.*, 2008). The aqueous, methanolic and hydroalcoholic extracts of the roots have also been studied for their scavenging effects in ferric reducing antioxidant power (FRAP), DPPH, ABTS and NO assays and observed to be effective (Sharma *et al.*, 2016). The aqueous (both using hot and cold extraction methods), acetone, ethyl acetate, methanol and petroleum ether extracts of the leaves have been investigated for the nitric oxide radical scavenging assay, reducing power ability and total antioxidant properties, and the results suggest that hot aqueous extract had potent nitric oxide radical scavenging, while the hot aqueous extract had superior reducing power ability and total antioxidant properties (Goswami *et al.*, 2018).

Anti-inflammatory activity

Chronic or recurrent inflammation in the tissues results in oxidative stress and this may play a contributory role in the initiation and/or progression and promotion of cancer cell growth (Philip *et al.*, 2004; Surh and Na, 2008). Regular administration of non-steroidal anti-inflammatory drugs (NSAIDs) decreases the risks of certain cancers, thereby asserting the fact that modulation of cellular signalling involving chronic inflammatory responses by anti-inflammatory agents offers an important strategy in molecular target-based chemoprevention and cytoprotection (Surh and Na, 2008). However, long-term administration of NSAIDs can cause ulcerations of the gastrointestinal tract which at times can be a major clinical problem. Preclinical studies with rats have shown that methanolic and aqueous extract of the bark at 200 and 400 mg/kg was effective in mitigating the carrageenan-induced oedema in laboratory rats in a concentration-dependent manner. The methanolic extract was observed to be better in anti-inflammatory activity than the aqueous extract but inferior to the standard indomethacin used as a positive control (Paviaya *et al.*, 2013).

Analgesic activity

Pain is defined as an unpleasant sensory and emotional experience associated with actual or potential tissue damage, or described in terms of such damage (de Sousa, 2011). It is a sensation described as a multidimensional experience, in which several aspects are involved: motivational, emotional, sensory-discriminative, affective and cognitive (de Sousa, 2011). Analgesic drugs currently in use are either narcotics or non-narcotics, which have proven side and toxic effects (de Sousa, 2011). This has necessitated the need for non-toxic agents that are safe and preferably of dietary origin (de Sousa, 2011).

Paviaya *et al.* (2004) evaluated the analgesic potential of the methanolic and aqueous extract of the bark at 200 and 400 mg/kg in an acetic acid-induced writhing test model used to assess peripheral analgesic effect and Eddy's hot plate test to affirm the central analgesic effect (Paviaya *et al.*, 2013). The results indicated that both extracts were effective and caused concentration-dependent analgesic effects comparable to those of the standard drugs indomethacin and pentazocine used for the acetic acid-induced writhing test model and Eddy's hot plate test, respectively (Paviaya *et al.*, 2013).

Antiemetic activity

Emesis, which is a scientific term for the colloquially used vomiting, a forcible emptying of the stomach, is a distressing physiological event associated with severe morbidity. In extreme conditions emesis requires hospitalization and may even lead to death (Singh *et al.*, 2016). Animal studies by Yaqeen *et al.* (2008) showed that the crude alcoholic extract of the fruit were effective as an antiemetic agent against the standard apomorphine-induced emesis in the dog model of study and that a dose of 120 mg/kg was effective as an antiemetic agent and comparable to the standard drugs maxolon (metocloroamide) and chlorpromazine, used as control (Yaqeen *et al.*, 2008). Subsequent studies have shown that the methanolic extract of leaves (50 mg/kg and 100 mg/kg dose) exhibited antiemetic action against the anhydrous copper sulfate induced emesis in five-day old chickens (Zia-Ul-Haq *et al.*, 2012b). However, the antiemetic effects were not comparable to the mediation demonstrated by the positive control drug metoclopramide (50 mg/kg) (Zia-Ul-Haq *et al.*, 2012b). Together, both these observations suggest the moderate usefulness of the plant as an antiemetic agent, and further studies may be beneficial.

Neuroprotective effects

Ageing is a natural phenomenon and recent data suggest that at a global level the number of people above the age of 70 is increasing. However, on the down side, a significant percentage of elderly people suffer from memory loss and amnesia and this negates the benefit of long life (Wimo *et al.*, 2013). From a mechanistic point of view, the hippocampus and cortex are the principal regions of the brain responsible for maintaining memory and the levels of the neurotransmitter acetylcholine (ACh), released via cholinergic projections from the basal forebrain to these regions, are derailed in old age, compromising cognitive function, causing memory loss (Hasselmo, 2006). Conventionally, acetylcholinesterase (AChE) inhibitors have been widely used to reduce cognitive decline but are associated with adverse effects like hepatotoxicity and nausea (Rountree *et al.*, 2009), and this has necessitated the need for alternative medicines that are effective and devoid of side effects in treating amnesia.

In Ayurveda, the fruit of phalsa have been used as chemical drug to reduce cognitive decline and improve neurostimulation (Paul *et al.*, 2019). Seminal studies by Paul and co-workers have shown that the methanol extract of the fruit was effective in mitigating scopolamine-induced memory loss in rats as observed in the behavioural assays, preventing neuronal cell damage and maintaining the cytoarchitecture, increasing the neurotransmitter acetylcholine and reducing oxidative stress (Paul *et al.*, 2019). Cumulatively all these aspects affirm the ethnomedicinal use of the fruit in Ayurveda as a neuroprotective agent and a promising nootropic candidate for further drug development.

Antidiabetic effects

Diabetes mellitus (DM), a disease as old as mankind, is today a leading cause of chronic metabolic disorder worldwide. Recent reports from the International Diabetes Federation suggest that in 2017, 451 million adults were living with diabetes and that predictions are that in 2045 the number will be 693 million cases worldwide (International Diabetes Federation, 2017; Cho *et al.*, 2018). The ailment is characterized by deficiency in insulin. An abnormal increase in blood sugar level, altered carbohydrate, protein and lipid metabolism affects the vascular system and compromises the general health of the affected individual (Cho *et al.*, 2018).

The extract of the leaves was studied for its antihyperglycaemic effects in alloxan-induced diabetes in mice (Bhangale *et al.*, 2013) and rats (Patil *et al.*, 2011) and was observed to be effective. Additionally, the ethanolic extract of the bark was also effective as an antidiabetic, antihyperlipidemic and found to possess antioxidant properties in alloxan-induced diabetic rats (Khatune *et al.*, 2016). When compared to diabetic control, the oral administration of the extract to diabetic rats caused a decrease in the levels of serum glutamate-oxaloacetate transaminase (SGOT) and creatinine kinase-MB levels, restoring the levels of liver glycogen in a concentration-dependent manner. The histopathological analysis of the pancreas showed a marked improvement in normal cellular population size in the islets of Langerhans (Khatune *et al.*, 2016).

Subsequent studies have also shown that the fruit extract possesses antihyperglycaemic effects against streptozotocin-induced hyperglycaemia in rats and mediates these effects by triggering anti-inflammatory and antioxidant mechanisms (Khattab *et al.*, 2015). Clinical studies with healthy volunteers have also shown that the fruit has a low glycaemic index of 5.34 and a modest hypoglycaemic activity. Additionally, reports also suggest that the chloroform, hexane and ethanol extracts of the fruit are effective in mitigating the D-glucose induced genesis of ROS and oxidative burst activity in whole blood of the volunteers, indicating clinical usefulness (Mesaik *et al.*, 2013).

Antimicrobial activity

Globally, drug resistance in bacteria and fungi is emerging as a major health issue and the need for effective, novel antimicrobial agents is an impending quest. Laboratory experiments have shown that plant foliage has appreciable antibacterial action on *Proteus mirabilis*, *Citrobacter* sp., *Pseudomonas aeruginosa*, *Escherichia coli*, *Salmonella typhi*, *Micrococcus luteus*, *Staphylococcus aureus* and *Bacillus subtilis* (Zia-Ul-Haq *et al.*, 2011). The extract showed moderate activity on the fungi *Aspergillus effusus*, *A. parasiticus*, *A. niger*, *Saccharomyces cerevisiae*, *Candida albicans*, *Yersinia aldovae*, *Fusarium solani*, *Macrophomina phaseolina* and *Trichophyton rubrum* (Zia-Ul-Haq *et al.*, 2011).

The ethanol extract of the bark and fruit has been shown to be potent against *S. aureus*, *E. coli* and *P. vulgaris*, while similar observations were unseen for *Bacillus subtilis*, *S. epidermidis*, *S. pneumoniae P. mirabilis*, *Salmonella typhi* para A, *S. typhi* para B and *Shigella dysenteriae* (Israr *et al.*, 2012). The aqueous, methanolic and hydroalcoholic extracts of the roots have also been studied for their antimicrobial effects on *Bacillus subtilis*, *Escherichia coli*, *Klebsiella pneumonia*, *Pseudomonas aeruginosa*, *Staphylococcus aureus*, *Candida albicans* and *Aspergillus fumigates*, and the results suggest that the hydroalcoholic extract possesses better activity than the other two extracts (Sharma *et al.*, 2016).

With regard to the fruit, fractions rich in flavanols and polyphenols (except those containing anthocyanins) isolated from the crude methanol extracts of pulp and peel were effective on *Salmonella typhi*, while *Bacillus subtilis* and *E. coli* were resistant (Siddiqi *et al.*, 2011). The phenolic acid fraction was also effective against the fungal pathogens *Penicillium notatum*, *Aspergillus niger*,

A. flavus, *Microsporum gypseum*, *Trichophyton mentagrophytes* and *T. rubrum* (Siddiqi *et al.*, 2011).

The leaf extract was effective against *Candida albicans* while being ineffective against *Aspergillus niger* (Siddiqi *et al.*, 2011). Studies have also shown that ethanolic extract of the fruit was effective against a range of bacteria like *Bacillus subtilis*, *B. cereus*, *Staphylococcus aureus*, *Enterococcus faecalis*, *Escherichia coli*, *Listeria monocytogeneses*, *Salmonella typhimurium*, *Shigella flexneri* and *Pseudomonas aerugenosa*, with superior activity observed in the Gram-positive bacteria (Gupta *et al.*, 2012).

Antimalarial activity

Globally, malaria is one of the world's oldest known parasitic ailments and remains a major health problem, especially in the tropical and developing countries of South-East Asia and sub-Saharan Africa (World Health Organization, 2015). In humans, the infection is caused by four different species of *Plasmodium* (*Plasmodium falciparum*, *Plasmodium vivax*, *Plasmodium ovale and Plasmodium malariae*), and can be lethal to children, pregnant women, the geriatic population and people with comorbidities (Rich *et al.*, 2009). Of the four species, malaria caused by *Plasmodium falciparum* is the most life threatening (Rich *et al.*, 2009) and has become resistant to current treatments that include standard drugs like chloroquine, sulfadoxine/pyrimethamine and artemisinin (O'Brien *et al.*, 2011; Tschan *et al.*, 2012), necessitating the need for new agents.

A key malarial enzyme, enoyl-acyl carrier protein (enoyl-ACP) reductase, is being investigated as a possible drug target, principally because it does not have any human homologues and has a vital role in the membrane construction and energy production of the parasite. Cell-free assays with enoyl-ACP reductase have shown that the methanolic extract of the leaves of *Grewia asiatica* was effective and caused 69% inhibition, indicating its usefulness (Zia-Ul-Haq *et al.*, 2012b). Detailed drug development studies are warranted, especially with preclinical animal models to understand its clinical utility.

Anticancer activity

Cancer is characterized by uncontrolled growth of abnormal cells with potential to metastasize, and is one of the most common causes of death throughout the world (Ferlay *et al.*, 2019). Recent reports from the International Agency for Research on Cancer (IARC) suggest that in 2018 about 9.6 million deaths and around 18.1 million new cases were registered worldwide (Ferlay *et al.*, 2019). In the light of these observations, the need for novel anticancer agents is vital.

Phalsa *et al.* (2011) investigated the cytotoxic effects of aqueous extract of fruit and leaves in cultured human neoplastic cell lines HEK-293 (epidermal kidney cancer cell line), NCI-H522 (lung cancer cell line), HELA (cervical cancer cell line), Hep2 (laryngeal cancer cell line) and MCF-7 (breast cancer cell line) *in vitro* (Marya *et al.*, 2011). The investigators observed that the extract of the fruit was effective on lung (IC_{50} = 59.03 μg/ml) and breast (IC_{50} = 58.65 μg/ml); while the leaf extract was potent on breast (IC_{50} = 50.37 μg/ml) and Hep-2 (IC_{50} = 61.23 μg/ml) (Marya *et al.*, 2011). Subsequent studies have also showed that the methanol extract of the leaves is effective on human cancer cell lines HL-60 (acute myeloblastic leukemia), K-562 (chronic myelogenic leukemia), MCF-7 and hela (cervical epithelial carcinoma), and IC_{50} values of 53.70 μg/ml (HL-60), 54.90 μg/ml (K-562), 199.5 μg/ml (MCF-7) and 177.8 μg/ml (HeLa) were observed (Kakoti *et al.*, 2011).

The methanolic extract of the fruit was also observed to be effective against HeLa, MCF-7 and HepG-2 cells, and an IC_{50} of >100, 68.91 and >250 μg/ml were respectively observed for the investigated cell lines (Kour *et al.*, 2016). The methanolic extract of the fruit when evaluated at a concentration of 50 μg/ml has been reported to inhibit growth of A-498 (renal), MDA-MB-435 (melanoma), A-549 (lung) and HCT-116 (colon) cancer cells by 81%, 80%, 78%–75% and 72%, respectively (Kour *et al.*, 2016). In addition to this, studies with male Swiss albino mice injected with transplantable Ehrlich's ascites carcinoma (EAC) cells showed that the intraperitoneal administration of 250 and 500 mg/kg of methanolic extract of leaves was effective in increasing the lifespan of the tumour-bearing mice by 41.22 and 61.06%, respectively (Kakoti *et al.*, 2011). Together, all these observations suggest that the cytotoxic effect on the cell differed with the plant part and the phytochemicals present.

Immunomodulatory effect

In humans the immune system plays a vital role in protecting against various infectious and potentially harmful microorganisms and eliminates any abnormal growth in the body (Yatim and Lakkis, 2015). A weak immune system makes the individual susceptible to infection and increases the chances of constant dependence on medical care (Yatim and Lakkis, 2015). In the traditional and folk systems of medicine, plant and plant-based products are used to modulate the immune system and many plants and formulations are known to enhance the activity of the immune system (Kumar *et al.*, 2012). These agents are known as immunostimulants and play a key role in improving the innate immunity of the individual (Kumar *et al.*, 2012). Scientific studies have shown that the ethanolic extract of leaves at 200 and 400 mg/kg caused immunostimulant effects and negated the cyclophosphamide-induced decrease in total WBC, in neutrophil counts and haemoglobin levels, and increased the phagocytic index in carbon clearance assay (Kumar *et al.*, 2012). Together all these observations indicate the usefulness of the leaves as a potential immunostimulant for clinical development.

Radioprotective effect

Ionizing radiation, which is an important treatment modality, has its set of side effects which when severe increase the morbidity of the patient (Baskar *et al.*, 2012). Under these conditions, chemicals that can selectively protect the normal tissues against radiation injury are very handy, allowing the continued use of radiation to obtain better cancer control and cure (Arora *et al.*, 2005; Baskar *et al.*, 2012). However, the synthetic compounds investigated have been shown to be inadequate for clinical application due to the systemic toxicity they induce at their optimal protective doses (Arora *et al.*, 2005).

Experiments have shown that the methanolic extract of the fruit when administered for 15 days before exposure to a supra-lethal dose of radiation was effective and that 700 mg/kg was very effective in reducing radiation sickness and mortality (Sisodia and Singh, 2009). The extract was also observed to mitigate radiation-induced damage to the hematopoietic system (Gupta *et al.*, 2007) and the liver (Sharma and Sisodia, 2010), and to improve learning and memory in animals exposed to radiation (Sisodia and Singh, 2009).

Mechanistic studies have indicated that the extract prevents radiation-induced damage by preventing loss of protein, cholesterol, nucleic acids and the cell's predominant antioxidant, glutathione, by reducing the levels of lipid peroxidation and concomitantly increasing the activities of the antioxidant enzymes catalase and superoxide dismutase (Sisodia *et al.*, 2008a, b; Sisodia and Singh, 2009; Sharma and Sisodia, 2010), and by scavenging free radicals (Sharma and Sisodia, 2009).

Conclusion

Studies clearly suggest that phalsa possesses diverse health benefits such as free radical scavenging, antimicrobial, anti-inflammatory, analgesic, anticancer, radioprotective, immunomodulatory and antidiabetic activity effects. Future studies should consider isolated phytochemicals as this will improve our understanding of the mechanism of action responsible for the various beneficial effects. More advanced approaches such as genomics, transcriptomics and metabolomics applied to studies of this plant may be very advantageous to increase our understanding of genes and their expression for the biosynthesis of bioactive metabolites. Moreover, it facilitates the genetic manipulation to improve the biosynthetic pathways to enhance the nutritional quality and the shelf life of the fruit.

References

Akram, Z., Perveen, R., Noreen, A., Inayat, M., Hussain, M. *et al.* (2019) Morphological, phytochemical, and pharmacological studies of *Grewia asiatica*: a Review. *RADS Journal of Pharmacy and Pharmaceutical Sciences* 7, 53–62.

Ali, S.I., Khan, N.A. and Husain, I. (1982) Flavonoid constituents of *Grewia asiatica*. *Journal of Scientific Research* 4, 55–56.

Ali, H.S., Ghaffoor, A., Waseem, K. and Nadeem, M.A. (2001) Growth and yield response of phalsa (*Grewia asiatica* L.) to various pruning intensities and dates. *Journal of Biological Sciences* 1(7), 548–550. DOI: 10.3923/jbs.2001.548.550.

Arora, R., Gupta, D., Chawla, R., Sagar, R., Sharma, A. *et al.* (2005) Radioprotection by plant products: present status and future prospects. *Phytotherapy Research* 19(1), 1–22. DOI: 10.1002/ptr.1605.

Asghar, M.N., Khan, I.U., Sherin, L. and Ashfaq, M. (2008) Evaulation of antioxidant activity of *Grewia asiatica* berry using 2, 2'-azinobis-(3-ethylbenzothiazoline-6-sulphonic acid) and N, N-dimethyl p-phenylenediamine radical cations decolourization assays. *Asian Journal of Chemistry* 20(7), 5123.

Baskar, R., Lee, K.A., Yeo, R. and Yeoh, K.W. (2012) Cancer and radiation therapy: current advances and future directions. *International Journal of Medical Sciences* 9(3), 193–199. DOI: 10.7150/ijms.3635.

Bertzky, B., Shi, Y., Hughes, A., Engels, B., Ali, M.K. *et al.* (2013) *Terrestrial Biodiversity and the World Heritage List: Identifying Broad Gaps and Potential Candidate Sites for Inclusion in the Natural World Heritage Network*. IUCN, Gland, Switzerland and UNEP-WCMC, Cambridge, UK.

Bhangale, J., Acharya, S. and Deshmukh, T. (2013) Antihyperglycaemic activity of ethanolic extract of *Grewia asiatica* (L.) leaves in alloxan induced diabetic mice. *World Journal of Pharmaceutical Research* 2(5), 1486–1500.

Chattopadhyay, S. and Pakrashi, S.C. (1975) Studies on Indian medicinal plants. XXXIV. triterpenes from *Grewia asiatica*. *The Journal*.

Cho, N.H., Shaw, J.E., Karuranga, S., Huang, Y., da Rocha Fernandes, J.D. *et al.* (2018) IDF Diabetes Atlas: Global estimates of diabetes prevalence for 2017 and projections for 2045. *Diabetes Research and Clinical Practice* 138(3), 271–281. DOI: 10.1016/j.diabres.2018.02.023.

de Sousa, D.P. (2011) Analgesic-like activity of essential oils constituents. *Molecules* 16(3), 2233–2252. DOI: 10.3390/molecules16032233.

Dhawan, K., Malhotra, S., Dhawan, S.S., Singh, D. and Dhindsa, K.S. (1993) Nutrient composition and electrophoretic pattern of protein in two distinct types of phalsa (*Grewia subinequalis* DC). *Plant Foods for Human Nutrition (Dordrecht, Netherlands)* 44(3), 255–260. DOI: 10.1007/BF01088320.

FAO (1982) Food and agricultural organization of United Nations. Fruit-bearing forest trees: technical notes. *FAO-Forestry* 34–177.

Ferlay, J., Colombet, M., Soerjomataram, I., Mathers, C., Parkin, D.M. *et al.* (2019) Estimating the global cancer incidence and mortality in 2018: GLOBOCAN sources and methods. *International Journal of Cancer* 144(8), 1941–1953. DOI: 10.1002/ijc.31937.

Gopalan, C., Rama Sastri, B.V. and Balasubramanian, S.C. (2004) *Nutritive Value of Indian Foods*. National Institute of Nutrition, ICMR, Hyderabad, India.

Goswami, S., Jain, R. and Masih, H. (2018) Comparative evaluation of *in vitro* antioxidant analysis of various leaves extracts for selected medicinal plants. *Journal of Pharmacognosy and Phytochemistry* 7, 1477–1481.

Goyal, P.K. (2012) Phytochemical and pharmacological properties of the genus Grewia: a review. *International Journal of Pharmacy and Pharmaceutical Sciences* 4, 72–78.

Gupta, M.K., Sharma, P.K., Ansari, S.H. and Lagarkha, R.E.K.H.A. (2006) Pharmacognostical evaluation of *Grewia asiatica* fruits. *International Journal of Plant Sciences* 1, 249–251.

Gupta, M.K., Lagarkha, R., Sharma, D.K., Sharma, P.K., Singh, R. *et al.* (2007) Antioxidant activity of the successive extracts *Grewia asiatica* leaves. *Asian Journal of Chemistry* 19, 3417.

Gupta, P., Sharma, A. and Verma, A.K. (2012) GC/MS profiling and antimicrobial effect of six indian tropical fruit residues against clinically pathogenic bacterial strain. *International Journal of Advances in Pharmaceutical Research* 3, 1229–1235.

Hasselmo, M.E. (2006) The role of acetylcholine in learning and memory. *Current Opinion in Neurobiology* 16(6), 710–715. DOI: 10.1016/j.conb.2006.09.002.

International Diabetes Federation (2017) IDF Diabetes Atlas, 8th Edn. Available at: https://diabetesatlas.org/upload/resources/previous/files/8/IDF_DA_8e-EN-final.pdf (accessed 12 April 2022).

Kakoti, B.B., Selvan, V.T., Manikandan, L., Gupta, M., Mazumder, U.K. *et al.* (2011) Antitumor and *in vitro* cytoxicity activity of the methanolic extract of *Grewia asiatica* against Ehrilch's ascites carcinoma cell lines. *Pharmacologyonline* 3, 956–960.

Khan, A.S., Hussain, A. and Khan, F. (2006) Nutritional importance of micronutrients in some edible wild and unconventional fruits. *Journal of the Chemical Society of Pakistan* 28, 576–582.

Khattab, H.A.H., El-Shitany, N.A., Abdallah, I.Z.A., Yousef, F.M. and Alkreathy, H.M. (2015) Antihyperglycemic potential of *Grewia asiatica* fruit extract against streptozotocin-induced hyperglycemia in rats: anti-inflammatory and antioxidant mechanisms. *Oxidative Medicine and Cellular Longevity* 2015, 549743. DOI: 10.1155/2015/549743.

Khatune, N.A., Rahman, B.M., Barman, R.K. and Wahed, M.I.I. (2016) Antidiabetic, antihyperlipidemic and antioxidant properties of ethanol extract of *Grewia asiatica* Linn. bark in alloxan-induced diabetic rats. *BMC Complementary and Alternative Medicine* 16, 295. DOI: 10.1186/s12906-016-1276-9.

Kour, N., Sharma, V. and Singh, S.K. (2016) Phalsa possesses in vitro cytotoxic efficacy against human cancer cells. In: *Natural Resource Management: Ecological Perspectives. Indian Ecological Society: International Conference*, p. 454.

Krishna, H., Saroj, P.L., Maheshwari, S.K., Singh, R.S., Meena, R.K. *et al.* (2019) Underutilized fruits of arid & semi-arid regions for nutritional and livelihood security. *International Journal of Minor Fruits, Medicinal and Aromatic Plants* 5(2), 1–14.

Kumar, D., Arya, V., Kaur, R., Bhat, Z.A., Gupta, V.K. *et al.* (2012) A review of immunomodulators in the Indian traditional health care system. *Journal of Microbiology, Immunology, and Infection* 45(3), 165–184. DOI: 10.1016/j.jmii.2011.09.030.

Lakshmi, V., Agarwal, S.K. and Chauhan, J.S. (1976) A new δ-lactone from the flowers of *Grewia asiatica*. *Phytochemistry* 15(9), 1397–1399. DOI: 10.1016/S0031-9422(00)97125-5.

Leopoldini, M., Rondinelli, F., Russo, N. and Toscano, M. (2010) Pyranoanthocyanins: a theoretical investigation on their antioxidant activity. *Journal of Agricultural and Food Chemistry* 58(15), 8862–8871. DOI: 10.1021/jf101693k.

Marya, B., Dattani, K.H., Patel, D.D., Patel, P.D., Patel, D. *et al.* (2011) *In vitro* cytotoxicity evaluation of aqueous fruit and leaf extracts of *Grewia asiatica* using MTT assay. *Journal for Medicinal Chemistry, Pharmaceutical Chemistry, Pharmaceutical Sciences and Computational Chemistry* 3, 282–287.

Mesaik, M.A., Ahmed, A., Khalid, A.S., Jan, S., Siddiqui, A.A. *et al.* (2013) Effect of *Grewia asiatica* fruit on glycemic index and phagocytosis tested in healthy human subjects. *Pakistan Journal of Pharmaceutical Sciences* 26(1), 85–89.

Morton, J.F. (1987) Phalsa. In: *Fruits of Warm Climates*. ISBN: 0-9610184-1-0; Distributed by Creative Resource Systems, Inc, Winterville, N.C. 28590, USA, pp. 276–277.

Nayar, M.P., Singh, A.K. and Nair, K.N. (2009) *Agrobiodiversity Hotspots in India: Conservation and Benefit Sharing*, Vol. 2. Chandu Press, Shakarpur, Delhi.

O'Brien, C., Henrich, P.P., Passi, N. and Fidock, D.A. (2011) Recent clinical and molecular insights into emerging artemisinin resistance in *Plasmodium falciparum*. *Current Opinion in Infectious Diseases* 24(6), 570–577. DOI: 10.1097/QCO.0b013e32834cd3ed.

Orwa, C., Mutua, A., Kindt, R., Jamnadass, R. and Anthony, S. (2009) Agroforestree database: a tree reference and selection guide version 4.0. Kenya: World Agroforestry Centre. Available at: https://www.worldagroforestry.org/publication/agroforestree-database-tree-reference-and-selection-guide-version-40 (accessed 12 April 2022).

Patil, P., Patel, M.M. and Bhavsar, C.J. (2011) Preliminary phytochemical and hypoglycemic activity of leaves of *Grewia asiatica* L. *Research Journal of Pharmaceutical, Biological and Chemical Sciences* 2, 516–520.

Paul, S., Sharma, S., Paliwal, S.K. and Kasture, S.B. (2019) Protective action of *Grewia asiatica* (phalsa) berries against scopolamine-induced deficit in learning and memory using behavior paradigms in rats. *Advances in Traditional Medicine* 20(2), 243–253. DOI: 10.1007/s13596-019-00376-y.

Paviaya, U.S., Kumar, P., Wanjari, M.M., Thenmozhi, S. and Balakrishnan, B.R. (2013) Analgesic and anti-inflammatory activity of root bark of *Grewia asiatica* Linn. in rodents. *Ancient Science of Life* 32(3), 150–155. DOI: 10.4103/0257-7941.122998.

Philip, M., Rowley, D.A. and Schreiber, H. (2004) Inflammation as a tumor promoter in cancer induction. *Seminars in Cancer Biology* 14(6), 433–439. DOI: 10.1016/j.semcancer.2004.06.006.

Rich, S.M., Leendertz, F.H., Xu, G., LeBreton, M., Djoko, C.F. *et al.* (2009) The origin of malignant malaria. *Proceedings of the National Academy of Sciences of the United States of America* 106(35), 14902–14907. DOI: 10.1073/pnas.0907740106.

Rountree, S.D., Chan, W., Pavlik, V.N., Darby, E.J., Siddiqui, S. *et al.* (2009) Persistent treatment with cholinesterase inhibitors and/or memantine slows clinical progression of Alzheimer disease. *Alzheimer's Research & Therapy* 1(2), 7. DOI: 10.1186/alzrt7.

Samson, J.A. (1986) *The Minor Tropical Fruits. Tropical Fruits*. Longman Inc, New York.

Sharma, K.V. and Sisodia, R. (2009) Evaluation of the free radical scavenging activity and radioprotective efficacy of *Grewia asiatica* fruit. *Journal of Radiological Protection* 29(3), 429–443. DOI: 10.1088/0952-4746/29/3/007.

Sharma, C., Malgaonkar, M., Sangvikar, S.G., Murthy, S.N. and Pawar, S.D. (2016) *In vitro* evaluation of antimicrobial and antioxidant profile of *Grewia* L. root extracts. *Journal of Applied Life Sciences International* 7(1), 1–9. DOI: 10.9734/JALSI/2016/26748.

Sharma, K.V. and Sisodia, R. (2010) Hepatoprotective efficacy of Grewia asiatica fruit against oxidative stress in Swiss albino mice. *Iranian Journal of Radiation Research* 8, 75–85.

Siddiqi, R., Naz, S., Ahmad, S. and Sayeed, S.A. (2011) Antimicrobial activity of the polyphenolic fractions derived from *Grewia asiatica*, *Eugenia jambolana* and *Carissa carandas*. *International Journal of Food Science & Technology* 46(2), 250–256. DOI: 10.1111/j.1365-2621.2010.02480.x.

Singh, P., Yoon, S.S. and Kuo, B. (2016) Nausea: a review of pathophysiology and therapeutics. *Therapeutic Advances in Gastroenterology* 9(1), 98–112. DOI: 10.1177/1756283X15618131.

Sisodia, R. and Singh, S. (2009) Biochemical, behavioural and quantitative alterations in cerebellum of Swiss albino mice following irradiation and its modulation by *Grewia asiatica*. *International Journal of Radiation Biology* 85(9), 787–795. DOI: 10.1080/09553000903009555.

Sisodia, R., Ahaskar, M., Sharma, K.V. and Singh, S. (2008a) Modulation of radiation-induced biochemical changes in cerebrum of Swiss albino mice by *Grewia asiatica*. *Acta Neurobiologiae Experimentalis* 68, 32–38.

Sisodia, R., Singh, S., Sharma, K.V. and Ahaskar, M. (2008b) Post treatment effect of *Grewia asiatica* against radiation-induced biochemical alterations in Swiss albino mice. *Journal of Environmental Pathology, Toxicology and Oncology* 27(2), 113–121. DOI: 10.1615/jenvironpatholtoxicoloncol.v27.i2.40.

Subhose, V., Srinivas, P. and Narayana, A. (2005) Basic principles of pharmaceutical science in Ayurvĕda. *Bulletin of the Indian Institute of History of Medicine (Hyderabad)* 35(2), 83–92.

Surh, Y.J. and Na, H.K. (2008) NF-κB and Nrf2 as prime molecular targets for chemoprevention and cytoprotection with anti-inflammatory and antioxidant phytochemicals. *Genes & Nutrition* 2(4), 313–317. DOI: 10.1007/s12263-007-0063-0.

Tschan, S., Kremsner, P.G. and Mordmüller, B. (2012) Emerging drugs for malaria. *Expert Opinion on Emerging Drugs* 17(3), 319–333. DOI: 10.1517/14728214.2012.702754.

Ullah, W., Uddin, G. and Siddiqui, B.S. (2012) Ethnic uses, pharmacological and phytochemical profile of genus Grewia. *Journal of Asian Natural Products Research* 14(2), 186–195. DOI: 10.1080/10286020.2011.639764.

Wimo, A., Jönsson, L., Bond, J., Prince, M., Winblad, B. *et al.* (2013) The worldwide economic impact of dementia 2010. *Alzheimer's & Dementia* 9(1), 1–11. DOI: 10.1016/j.jalz.2012.11.006.

World Health Organization (2015) World malaria report 2015. World Health Organization. Available at: http s://apps.who.int/iris/handle/10665/200018

Yadav, A.K. (1999) Phalsa: a potential new small fruit for Georgia. In: Janick, J. (ed.) *Perspectives on New Crops and New Uses*. ASHS Press, Alexandria, VA, pp. 348–352.

Yaqeen, Z., Sohail, T., Saleem, M. and Rehman, Z.U. (2008) Evaluation of antiemetic activities of alcoholic extract of *Grewia asiatica* in experimental model dog. *Biological Sciences-PJSIR* 51, 212–215.

Yatim, K.M. and Lakkis, F.G. (2015) A brief journey through the immune system. *Clinical Journal of the American Society of Nephrology* 10(7), 1274–1281. DOI: 10.2215/CJN.10031014.

Zia-Ul-Haq, M., Ahmad, M., Mehjabeen, M., Jehan, N., Ahmad, S. *et al.* (2011) Antimicrobial screening of selected flora of Pakistan. *Archives of Biological Sciences* 63(3), 691–695. DOI: 10.2298/ABS1103691Z.

Zia-Ul-Haq, M., Raza Shah, M., Qayum, M. and Ercisli, S. (2012a) Biological screening of selected flora of Pakistan. *Biological Research* 45(4), 375–379. DOI: 10.4067/S0716-97602012000400008.

Zia-Ul-Haq M., Shahid, S.A. and Muhammed, S. (2012b) Antimalarial, antiemetic and antidiabetic potential of *Grewia asiatica* L. leaves. *Journal of Medicinal Plants Research* 6(16), 3087–3092. DOI: 10.5897/JMPR12.028.

Zia-Ul-Haq, M., Stanković, M.S., Rizwan, K. and Feo, V.D. (2013) *Grewia asiatica* L., a food plant with multiple uses. *Molecules* 18(3), 2663–2682. DOI: 10.3390/molecules18032663.

28 Mexican Plum (*Spondias purpurea* L.)

Miriam Monserrat Ferrer*, María del Rocío Ruenes-Morales, Patricia Irene Montañez-Escalante and Héctor Estrada-Medina
Departamento de Manejo y Conservación de Recursos Naturales Tropicales, Universidad Autónoma de Yucatán, Mérida, Yucatán, México

Abstract

This chapter examines the nutritional content and phytochemical production of *Spondias purpurea* fruit and the potential positive effects on human health resulting from the consumption of the fresh fruit, or its use as a raw material to prepare other food products. Information on the centesimal content of nutrient types, vitamins and minerals of at least one traditional variant/ecotype of the species was compiled from seven different countries. The content of chlorophylls, carotenoids, anthocyanins and phenolic compounds from the species was studied, due to the potential antioxidant capacity, in at least one traditional variant/ecotype from six different countries where it is grown. Further studies are required to identify the traditional variants/ecotypes with higher nutritional value and the factors that influence nutritional and bioactive compound variations. The fruit of the species should be considered a nutraceutical food due to the high content of fibre, potassium, iron, zinc, vitamin C, carotenoids, phenols and anthocyanins, which give it a good antioxidant capacity. Due to the larger content of bioactive compounds in the epicarp, consumption of the whole fruit is recommended. Further studies are needed to investigate acceptance, safety and integral management of the fruit in the development of new products in bakery, beverage and food preservation, and the cosmetic industries.

Introduction

Spondias purpurea L. (Anacardiaceae) is a multipurpose tree native to Mesoamerica where it is found as a floristic element in the low-height deciduous forests and medium-height sub-evergreen forests. It became naturalized in the West Indies and the arid stripe of north-east South America. Some varieties are cultivated in Florida, the West Indies, Bahamas, Philippines, Nigeria and Senegal (Miller, 2011; Mitchell and Daly, 2015). The importance of *S. purpurea* is associated with its contribution to the food sovereignty of Mesoamerican populations. The fruit is collected from secondary vegetation and forests or is grown in traditional home gardens, hedges and boundaries, or on pastures in the countries to which it is native (Moz-Preza *et al.*, 2003; Ruenes-Morales *et al.*, 2010; Miller, 2011) or in which it has been naturalized (Macía and Barfod, 2000; Metzel and Montagnini, 2014; Buriticá Céspedes and Cartagena Valenzuela, 2015; Vael, 2015).

Spondias purpurea fruit is consumed fresh, but it is also cooked in dishes that form part of the traditional food systems of Mexico's native population (Zizumbo-Villarreal *et al.*, 2012, 2014; Salazar *et al.*, 2016) and of the

*Corresponding author: mferrer@correo.uady.mx

DOI: 10.1079/9781789248067.0028

mestizo population (Macía and Barfod, 2000; Vélez-Coria and Cruz-León, 2012; Medeiros and Albuquerque, 2014). Cultivation of the species as a fruit tree is promising because of its adaptation to tolerate drought (Borchert, 1994). Its phenotypic plasticity that allows it to grow in different types of soil, temperatures and altitudes facilitates its establishment in different environmental conditions (Álvarez-Vargas *et al.*, 2017; Vargas-Simón, 2018). The main challenges for the marketing/production of the crop are the susceptibility of the fruit to the attack of the fruit fly, *Anastrepha* spp. and *Ceratitis capitata* (Wied.) and the postharvest management of the fruit (Maldonado-Astudillo *et al.*, 2014; Mohammed *et al.*, 2019).

Botany

The species is a deciduous bush or tree 10 to 15 m high with a diameter between 40 and 80 cm in wild populations. It is a dioecious species and it is assumed that under cultivation the individuals are hermaphrodite, but some populations have been reported to be parthenocarpic (Avitia-García, 1997). The most updated botanic description of the species is presented in the monograph of the genus (Mitchell and Daly, 2015). It can be identified by its grey bark, compound leaves of oblong or elliptic shape with 5 to 12 pairs of oblong or elliptic petioles that may be puberulent with smooth or serrated margins. The flowers present five pink fused sepals and red to purple petals (up to 3 mm) that aggregate in racemose panicles (up to 1 cm in length and 8 cm in width). In wild individuals, the fruit is an oblong to cylindrical drupe (1.5–3.5 cm long and 1.0–2.5 cm wide) and its colour is orange to intense red. Different cultivated plants have yellow, orange, red or purple fruit (2.5 to 7 cm long and 2 to 3.7 cm wide) (Fig. 28.1). The endocarp (called bone, caroço or seed) is yellow and fibrous, with one or more seeds – or none in parthenocarpic variants. The tree loses its leaves in the dry season of the year and begins its flowering and fruit production in the same period. Certain cultivated variants fructify at the beginning and the end of the summer rains (Ruenes-Morales *et al.*, 2010; Miller, 2011). The species is known by more than 180 common names (Mitchell and Daly, 2015). The most used in Spanish are ciruela, ciruelo, ciruela mexicana, jobo, jocote, ovo; in Portuguese ciriguela, seriguela, siriguela; in English red mombin, purple mombin, hogplum; and in French prune d'Espagne, prune de Chile, prune rouge.

Traditional Use

The most common use of *S. purpurea* is as food, including the fruit, young leaves and mature leaves. *Spondias purpurea* fruit is edible when the epidermis is still green and brilliant (sipon) and when it is ripe and has a reddish colour. Sauces, broths and gruels (atoles) of pre-Hispanic origin are still prepared with the fruit of *S. purpurea*, plus other typical Mesoamerican elements from the cornfields (milpa) and/or home gardens such as maize dough, flat maize cakes (tortillas), ground pumpkin seeds and

Fig. 28.1. Mexican plum (*Spondias purpurea* L.) fruit.

chili peppers (Zizumbo-Villarreal *et al.*, 2012, 2014; Salazar *et al.*, 2016). The shoots are also consumed. In the 17th century, the ground leaves of the tree were used as a spice in Brazil (Tomchinsky and Ming, 2019). In colonial times, the fruit began to be used to produce vinegar, pastries (Ortíz-Gil, 2017), wine and liquors (Godoy *et al.*, 2003). These uses expanded to incorporate salted or sweet preserves and refreshing beverages that are occasionally frozen to make popsicles or ice cream (Macía and Barfod, 2000; Moz-Preza *et al.*, 2003; Vélez-Coria and Cruz-León, 2012).

The use of the bark, leaves and occasionally the fruit of *S. pupurea* is documented in the traditional medicine of Mexico's native population to treat skin conditions, digestive upsets, fungal infections and eye diseases (Ruenes-Morales *et al.*, 2010), and is part of the pharmacopeia in those countries where the species was naturalized (Heckel, 1897; Meléndez, 1982; Godínez Caraballo and Volpato, 2008; Bieski *et al.*, 2012; Bravo *et al.*, 2017; Tantengco *et al.*, 2018). Because of its multipurpose nature, the foliage of this Mesoamerican species is harnessed as forage to feed farm animals and is thus found in agroforestry systems such as pastures, hedges and boundaries (Macía and Barfod, 2000; Chamorro Hernández, 2014; Metzel and Montagnini, 2014). *Spondias purpurea* stakes are planted at house boundaries and in agricultural fields and are used to build hedgerows (Moz-Preza *et al.*, 2003; Vargas-Simón, 2018). Sticks from the tree are reported to be used as firewood (Macía and Barfod, 2000; Chamorro Hernández, 2014; Medeiros and Albuquerque, 2014; Metzel and Montagnini, 2014).

Nutritional Composition

The centesimal composition evaluation of 37 *S. purpurea* traditional variants/ecotypes from six countries in the Central American region is reported below (Table 28.1). *S. purpurea* fruit content: ash (0.2–1%), lipids (0.1–1.5%), ether extract (0–0.4%), crude protein (0.1–1.4%), total carbohydrates (2.9–28.9%), total fibre (0.2–18.1%), soluble dietary fibre (0.1–7.4%), insoluble dietary fibre (2.8–10.2%). Observed values fall in the range of 31 out of 38 Mesoamerican fruit for all nutrients excepting dietary fibre. Compared with more commercial fruit such as guava, avocado and zucchini, *S. purpurea* have ten times lower lipid and half the crude protein, and two times more carbohydrates and fibre (Gómez-Maqueo *et al.*, 2020). Mineral macro- and micronutrients were obtained for 38 traditional varieties/ecotypes from seven countries in the Central American region (Table 28.1). A portion of 100 g of *S. purpurea* fruit can have 0.004 to 34 mg of calcium, 0.0272 to 53 mg of phosphorus, 0.0007 to 4.73 mg of iron, 0.1 to 9.5 mg of sodium and 0.13 to 260 mg of potassium. Micronutrient content range in 100 g of fruit is: copper 0.12–0.97 mg, zinc traces 1.77 mg, Mn 0.530 mg; there were only traces of selenium, chromium and lithium (Fernandes, 2012). These range values for the minerals are within the range of values for most of the 38 Mesoamerican fruit, except for phosphorus and iron whose maximum concentrations are higher than most fruit and comparable to those from avocado for phosphorus and from chilaca pepper and yellow mombin (*S. mombin* L.) for iron (Gómez-Maqueo *et al.*, 2020). Pulp and endocarp nutritional content of ash, lipids, protein, fibre, calcium, iron, sodium and zinc are lower than in the epicarp, and potassium is lower still (Lima, 2009; Bramont *et al.*, 2018).

The content of vitamins has been recorded for 18 ecotypes/varieties (Table 28.1); the activity of vitamin A was estimated for Brazil as 42 UI (Núcleo de Estudos e Pesquisas em Alimentação – NEPA Universidade Estadual de Campinas – UNICAMP, 2011) and for Venezuela as 33 UI (Instituto Nacional de Nutrición, 2001). The concentration of carotenoids in the five variants/ecotypes in which the extraction was carried out with ethanol, acetone and petroleum ether was 0.05–0.76 mg/100 g (Cravioto *et al.*, 1952; Munsell *et al.*, 1953; Kozioł and Macía, 1998; Zielinski *et al.*, 2014), consistently lower than in the other 61 variants/ecotypes (0.50–9.02 mg/100 g FW – fresh weight) in which the extraction was carried out with ethanol:acetone:hexane (Table 28.2). In addition to those shown in Table 28.1, ascorbic acid concentrations in 15 accessions varied between 0.6 mg and 33.3 mg per 100 g of the fruit (Almeida *et al.*, 2011; Moo-Huchin *et al.*, 2014; Poveda Núñez and Robalino Martínez, 2014; Solorzano-Morán *et al.*, 2015; Mendonça and

Table 28.1. Centesimal composition of nutrients (expressed in %) of minerals and vitamins A and C macronutrients and micronutrients (expressed in mg/100 g) in the fruit of *Spondias purpurea* from countries where it is cultivated. Data of one variety/ecotype are presented for each region. When the study reports more than one region it is noted in parentheses.

Country/ region	Ashe	Lipids	Ether extract	Crude protein	Carbohydrate	Fibre	Soluble dietary fibre	Insoluble dietary fibre	Ca	P	Fe	Na	K	Mg	Cu	Zn	Mn	Vitamin B1	Vitamin B2	Vitamin B3	Vitamin C	Reference
Brazil	0.21	0.24		0.47.	2.91	2.82	0.10	2.78														(Berni *et al.*, 2019)
Brazil	0.4–0.9	0.76–1.25	0.47–0.73	19.12–30.97					0.35–6.78			2.41–2.45	8.45									, (Bramont *et al.*, 2018)[a]
Brazil									14.4	1.9	1.5	0.1	185.6	6.2	0.1	0.310	0.001					(Fernandes, 2012)
Brazil	0.91–1.11	1.34–1.97		0.86–1.59	19.25–21.54	0.36–2.74			13.17–18.63		6.77–9.22		366.49–469.53			0.57–1.10					0.00–32.15	(Lima, 2009)[a]
Brazil	0.70	0.40		1.40	18.90	3.90			27.0	49.0	0.4	2.0	248.0	18.0	0.1	0.500	0.060	0.4	tr	tr	27	(Núcleo de Estudos e Pesquisas em Alimentação – NEPA Universidade Estadual de Campinas – UNICAMP, 2011)
Centroamerica	0.80	0.10		0.90	22.00				22.0	40.0	0.6											(Menchú *et al.*, 2007)
Colombia									20.0	30	0.3							0.08	0.02	0.5	30	(FAO/ LATINFOODS, 2010)
Ecuador					13.77				12.0	24.0	(260)	(2799)	260.00	12.00	0.25	0.160		0.07	0.03	1	45	(Ortiz Rodea, 2013) (salted sample)
Ecuador	0.70	0.20		0.70	19.10	0.50			17.0	42	0.72	6	250.00			0.002		0.084	0.04	1	49	(Kozioł and Macía, 1998)
Ecuador			0.03–0.09	0.10–0.18		79.1–86.8	0.2	78.9–86.6	6.8–9.6	33.2–42.7	0.09–0.52							0.035–0.106	0.014–0.023	0.61–1.38	26.4–31.5	(Munsell *et al.*, 1953) (2 varieties/ ecotypes)

Continued

Table 28.1. Continued

Country/ region	Ashe	Lipids	Ether extract	Crude protein	Carbohydrate	Fibre	Soluble dietary fibre	Insoluble dietary fibre	Ca	P	Fe	Na	K	Mg	Cu	Zn	Mn	Vitamin B1	Vitamin B2	Vitamin B3	Vitamin C	Reference
El Salvador	0.34–073	0.22–1.47	0.12–0.35	0.85–1.35	12.16–28.86				0.03–0.08	0.03–0.05	0.001–0.003		0.13–0.55	0.03–0.05		0.001						(Moz-Preza *et al.*, 2003) (7 varieties/ ecotypes)
Mexico	2.28		0.27	2.92		8.12																(Carrillo Martínez, 2017)
Mexico	0.20–0.60		0.00–0.28	0.62–1.12	8.17–26.33	0.39–2.78			16.00–34.00	19–38	1.31–3.68							0.04–0.12	0.03–0.07	0.32–0.70	22.5–62.8	(Cravioto *et al.*, 1952) (8 varieties/ ecotypes)
Mexico				0.07–0.58					2.50–5.10	1.30–2.80	1.57–4.76	4.00–9.50	6.20–24.00	2.60–4.60	0.40–0.97	1.03–1.77	0.330					(Ramírez Hernández *et al.*, 2008) (12 varieties/ ecotypes)
Mexico						17.54–18.11	7.07–7.39	10.15–11.04														(Moo-Huchin *et al.*, 2014) (2 varieties/ ecotypes)
Peru	1.00	0.20			21.50				20.00	53.00	0.90							0.05	0.06	1.44	36.8	(FAO/ LATINFOODS, 2010)
Venezuela	0.50	0.80		0.90	17.90				25.00	39.00	1.10							0.11	0.08	0.60	53.0	(Instituto Nacional de Nutrición, 2001)
Venezuela						8.1–43.9	4.00–31.1	4.10–12.70														(Pire Sierra *et al.*, 2010) [a]

[a] Comparison among structures.

Table 28.2. Carotenoid, chlorophyll, anthocyanin content in *Spondias purpurea* fruit from the different countries where it is cultivated.

Country	Carotenoids (mg/100 g fresh weight)	β-carotene (mg/100 g fresh weight)	β-cryptoxanthin (mg/100 g fresh weight)	Lutein (mg/100 g fresh weight)	Total chlorophyll (mg/100 g fresh weight)	Anthocyanins (mg/100 g fresh weight)	Reference
Brazil						1.35	Almeida *et al.*, 2011
Brazil						0.66–1.87	Bramont *et al.*, 2018[a]
Brazil		28.9	59.4	16.3			Berni *et al.*, 2019
Brazil		107.80 (DW)	92.60 (DW)	139.20 (DW)			Costa, 2018
Brazil						0.47 (DW)	Gregoris *et al.*, 2013
Brazil					0.30–9.81	0.14–0.82	Figueiredo, 2018 +
Brazil		0–3.4	3.6–54.1	0.2–30.6			Lima, 2009[a]
Brazil	2.75				228.6	0.00	Mendonça and Vieites, 2019
Brazil						0.30	Stafussa *et al.*, 2018
Brazil	0.19					3.18	Zielinski *et al.*, 2014
Costa Rica	8.90	103.80	24.3				Furtado *et al.*, 2004
Costa Rica		73.30	42.9				Monge-Rojas and Campos Nunes, 2013
Ecuador	0.12						Kozioł and Macía, 1998
Ecuador	0.76–0.78						Munsell *et al.*, 1953 (2 varieties/ecotypes)
Ecuador	2.26–2.39						Poveda Núñez and Robalino Martínez, 2014 (2 varieties/ecotypes)

Continued

Table 28.2. Continued

Country	Carotenoids (mg/100 g fresh weight)	β-carotene (mg/100 g fresh weight)	β-cryptoxanthin (mg/100 g fresh weight)	Lutein (mg/100 g fresh weight)	Total chlorophyll (mg/100 g fresh weight)	Anthocyanins (mg/100 g fresh weight)	Reference
Mexico	1.17–9.02						Álvarez-Vargas *et al.*, 2017 (36 varieties/ ecotypes)
Mexico	0.05–0.27						Cravioto *et al.*, 1952 (8 varieties/ecotypes)
Mexico	1.25					0.06 (DW)	Guillen-Poot, 2019
Mexico		17.16–25.07				1.67–1.77	Moo-Huchin *et al.*, 2014 (2 varieties/ ecotypes)
Mexico	0.50–32.70						Solorzano-Morán *et al.*, 2015[a] (10 varieties/ecotypes)
Mexico	0.37–8.54						Suárez-Vargas *et al.*, 2017
Mexico		0.0001–0.0017 (DW)		0.001–0.064 (DW			Villa-Hernández *et al.*, 2017 (7 varieties/ecotypes)
Panama	0.18						Murillo *et al.*, 2010
Venezuela	1.98						Instituto Nacional de Nutrición, 2001

[a] Comparison among structures; + comparison among structures and ripening stages (DW, concentration expressed as dry weight)

Vieites, 2019). Vitamin B1 (thiamine) values ranged from 0.023 mg to 1.2 mg; vitamin B2 (riboflavin) from 0.014 mg to 0.08 mg; niacin from 0.06 mg to 1.44 mg per 100 g of fruit. Vitamin A and C concentration values were in the range for those recorded in most of the 38 Mesoamerican fruit, whereas the maximum values recorded for vitamin B complex were higher, and comparable to the ones for poblano pepper and mamey (Gómez-Maqueo *et al.*, 2020). The fruit's concentration of γ-tocopherol was estimated as 0.13 and α-tocopherol as 1.32 (components of vitamin E) for the species cultivated in Costa Rica (Monge-Rojas and Campos Nunes, 2013).

Phytochemicals

The content of chlorophylls, carotenoids and anthocyanins – pigments that give colour to *S. purpurea* fruit – has been recorded. These pigments accumulate in the epicarp and are associated with the green, orange and purple colours of the fruit, respectively. Pigment variants are associated with the ripening stage (Maldonado-Astudillo *et al.*, 2014). Chlorophyll concentration reports range from 0.3 to 228.6 mg/100 g FW in three variants/ecotypes, and anthocyanin content in ten variants/ecotypes (Table 28.2). When the fruit ripened, chlorophyll concentration decreased from 9.81 to 0.3 mg/100 g and anthocyanins increased from 0.14 to 0.82 mg/100 g (Figueiredo, 2018) with higher concentrations in the epicarp than in the pulp (0.19 and 3.18, respectively). Up to nine types of carotenoid (auroxanthin, lutein, β-cryptoxanthin, zeinoxanthin, luteoxanthin myristate, -luteoxanthin palmitate, ζ-carotene, α-carotene, b-carotene) have been quantified with high-performance liquid chromatography (HPLC) and HPLC-DAD-MS/MS (Furtado *et al.*, 2004; Monge-Rojas and Campos Nunes, 2013; Costa, 2018; Berni *et al.*, 2019) of which β-carotenes, β-cryptoxanthin and lutein stand out, with lower content in the pulp and greater in the epicarp (Lima, 2009; Villa-Hernández *et al.*, 2017), with large variation associated with the management of the crop, storage management, extraction methods and ripening stage.

Variation in the content of phenolic compounds is high, ranging from 0.005 to 3900 mg gallic acid equivalents (GAE)/100 g, both when expressed as DW (0.0287–2586 mg GAE/100 g) in 15 varieties/ecotypes of plums from Brazil and Mexico (Omena *et al.*, 2012; Gregoris *et al.*, 2013; Reis *et al.*, 2015; Silva *et al.*, 2016; Villa-Hernández *et al.*, 2017; Silva, 2018c; Guillen-Poot, 2019) and FW (0.005–3900 mg GAE/100 g) in 105 variants/ecotypes from Mexico, Brazil, and the Philippines (da Silva *et al.*, 2012a; Moo-Huchin *et al.*, 2014; Zielinski *et al.*, 2014; Quintão, 2015; Solorzano-Morán *et al.*, 2015; Recuenco *et al.*, 2016; Reis, 2016; Álvarez-Vargas *et al.*, 2017; Suárez-Vargas *et al.*, 2017; Bramont *et al.*, 2018; Figueiredo, 2018; Hernández-Ruiz *et al.*, 2018; Stafussa *et al.*, 2018; Mendonça and Vieites, 2019; Villareal Fuente *et al.*, 2019). This variation is associated with differences in pre-extraction treatment (Reis *et al.*, 2015; Silva, 2018c), the solvent used in the extraction (Quintão, 2015; Hernández-Ruiz *et al.*, 2018; Mendonça and Vieites, 2019), the analysed structure (epicarp, pulp mesocarp, or endocarp) (Omena *et al.*, 2012; Solorzano-Morán *et al.*, 2015; Álvarez-Vargas *et al.*, 2017; Suárez-Vargas *et al.*, 2017; Villa-Hernández *et al.*, 2017; Bramont *et al.*, 2018), the ripening stage of the fruit (Suárez-Vargas *et al.*, 2017; Figueiredo, 2018), the environmental conditions in which it grows (Álvarez-Vargas *et al.*, 2017) and genotype differences (da Silva *et al.*, 2012b; Moo-Huchin *et al.*, 2014). Phenol concentration in the epicarp can be two to seven times greater than in the pulp (Solorzano-Morán *et al.*, 2015; Álvarez-Vargas *et al.*, 2017; Villa-Hernández *et al.*, 2017; Bramont *et al.*, 2018), while the endocarp has twice as many phenols as the epicarp and 20 times more than the pulp (Omena *et al.*, 2012). The identification of phenols for the species of the genus was revised in the work of Sameh *et al.* (2018). Ripe fruit hold two times more phenols as compared with unripe fruit, both in the endocarp and the pulp (Suárez-Vargas *et al.*, 2017) with a reduction in phenols in the first two ripening stages, followed by an increase until it reaches a concentration 2.5 times greater than that of the third stage (Suárez-Vargas *et al.*, 2017; Figueiredo, 2018). The concentrations of flavonoids ranged from 0.0090 to 907.5 mg of quercetin equivalents in 16 variants/ecotypes from Mexico, Brazil and the Philippines, which depends on the pretreatment of the samples, the solvent used for the extraction (Quintão, 2015; Reis *et al.*, 2015; Hernández-Ruiz *et al.*, 2018), the analysed structure – seven to ten times more flavonoids in the endocarp than in the pulp (Suárez-Vargas *et al.*, 2017; Villa-Hernández *et al.*, 2017), the ripening stage of the fruit – 0.8 and 1.5 times more flavonoids in the pulp and epicarp of the ripe fruit (Suárez-Vargas *et al.*, 2017) and genotype differences (Moo-Huchin *et al.*, 2014).

Implications of Food Processing on Nutrients and Phytochemicals

In a review on the postharvest physiology of the fruit of *S. purpurea*, Maldonado-Astudillo *et al.* (2014) reported that ascorbic acid reduces its concentration during storage, that chlorophylls are not metabolized when CO_2 increases in airtight containers and that darkness prevents the formation of anthocyanins. Under these conditions, the fruit do not develop the orange colours that indicate the presence of anthocyanins. Storage of the fruit reduces the bioactive compounds (vitamin C, carotenoids and phenolic compounds), so a decrease in antioxidative activity is expected (Maldonado-Astudillo *et al.*, 2014). A study evaluating the joint effect of storage and compression of yellow and Cuernavaqueña varieties on the content of carotenoids and phenols found an increase in the concentration of carotenoids in pulp and epicarp in the unripe fruit in one of the two varieties analysed (from the fourth day of storage on), particularly when they were subjected to biocedence compression levels (Salazar-Díaz, 2019). *Spondias purpurea* stored fruit cultivated in Brazil were exposed to 2 kJ/m^2 UV-C light, which extended their life up to 12 days (three more days than controls), reducing the decrease in vitamin C, phenolic compounds and antioxidant activity in around 60% compared with those that were not irradiated (Sanches *et al.*, 2018).

Bioaccessibility and Bioavailability

Bioaccessibility – the release of carotenes from the nutritional matrix to be accessible for absorption – was tested with digestion experiments *in vitro* with *S. pupurea* fruit cultivated in Brazil, finding that 9-cis-β-carotene and (all-E)-zeaxanthin were not bioaccessible, while (all-E)-β-cryptoxanthin, (all-E)-β-carotene, (all-E)-lutein, (all-E)-α-Carotene), 13-cis-β-carotene, and 15-cis-β-carotene bioaccessibility varied between 29.07 and 21.72%, higher values than those that have been reported for papaya and lower than those reported for mango (Berni *et al.*, 2019). The potential bioavailability – the proportion of minerals that are absorbed and utilized by the body (calcium, iron and zinc) was considered high for the fruit of *S. purpurea* cultivated in Brazil, based on phytic acid and mineral ratios, that were lower than the thresholds that are considered of low bioavailability (0.19 vs. 1.56 for Ca; 1.79 vs. 14; and 4.39 vs. 10 (Marin *et al.*, 2009).

Health Benefits

The benefits of consuming the fruit of *S. prupurea* stem mostly from the antioxidant effects of the fruit's extracts. Like many of the native Mesoamerican fruit, plum is a good source of vitamins and minerals (Gómez-Maqueo *et al.*, 2020). Compared to the recommendations of the American Dietetic Association (US Department of Agriculture, 2021) ingestion of 100 g of *S. purpurea* fruit provides low quantities of proteins and lipids, and moderate to good quantities of fibre for adult requirements (greater than that recommended for children between 1 and 3 years old); less than 1% of sodium, copper and selenium; between 2.5 and 12% of calcium, phosphorus and potassium; and between 22.5 and 68% of iron, magnesium, zinc and manganese (Table 28.3). Consumption of the fruit enhances the development of intestinal flora and the speed of intestinal transit. It may reduce blood cholesterol and glucose concentrations because of its high content of soluble fibre, while its content of insoluble fibre has a laxative effect (Pire Sierra *et al.*, 2010). The potential high bioavailability of calcium, iron and zinc make *S. purpurea* fruit an important nutraceutical food that can help in the prevention of nutritional and chronic-degenerative diseases, while the concentration of vitamins. A, C and E, and mineral micronutrients such

Table 28.3. Comparison of the nutritional content of *S. purpurea* fruit with the values suggested by the American Dietetics Association for children and adults (maximum values were selected for women and men between 18 and 51 years old) (U.S. Department of Agriculture and U.S. Department of Health and Human Services, December 2020).

Data	Recorded value		Suggested intake		Maximum contribution to diet %		Minimum contribution to diet %	
	Max	Min	Adult	Child	Adult	Child	Adult	Child
Lipids	1.50	0.100	7	17	8.60	21.00	0.01	0.01
Crude protein	1.40	0.100	13	56	2.50	10.80	0.00	0.01
Carbohydrate	28.90	2.900	130	130	22.20	22.20	0.02	0.02
Fibre	18.10	0.200	14	33	54.90	129.40	0.01	0.01
Ca	34.00	0.000	700	1300	2.62	4.86	0.00	0.00
P	53.00	0.027	460	1250	4.24	11.52	0.00	0.01
Fe	4.73	0.001	7	18	26.28	67.57	0.00	0.01
Na	9.50	0.100	1500	2300	0.41	0.63	0.00	0.01
K	260.00	0.130	3000	4700	5.53	8.67	0.00	0.00
Mg	18.00	0.023	80	420	4.29	22.50	0.01	0.03
Cu	0.97	0.100	340	900	0.11	0.29	0.01	0.03
Zn	1.77	0.000	3	11	16.09	59.00	0.00	0.01
Mn	0.53	0.001	1.2	2.3	23.04	44.17	0.04	0.08
Se	0.00	0.001	20	55	0.00	0.01	0.00	0.01
Vitamin B1	0.12	0.023	0.5	1.2	10.00	24.00	1.92	4.60
Vitamin B2	0.08	0.014	0.5	1.3	6.15	16.00	1.08	2.80
Vitamin B3	1.44	0.060	6	16	9.00	24.00	0.38	1.00
Vitamin C	62.80	0.600	15	90	69.78	418.67	0.67	4.00

as zinc and selenium are important for skin and hair health (Marin *et al.*, 2009).

The antioxidant potential of the pulp of the fruit of the species has been evaluated by *in vitro* assays with the scavenging of 2,2-diphenyl-1-picrylhydrazyl (DPPH) (Vasco *et al.*, 2008; Ruenes-Morales *et al.*, 2010; Almeida *et al.*, 2011; Omena *et al.*, 2012; Chamorro Hernández, 2014; Moo-Huchin *et al.*, 2014; Reis, 2016; Silva *et al.*, 2016; Álvarez-Vargas *et al.*, 2017; Suárez-Vargas *et al.*, 2017; Hernández-Ruiz *et al.*, 2018; Stafussa *et al.*, 2018; Mendonça and Vieites, 2019; Villareal Fuente *et al.*, 2019); 2,2' azino-bis(3-ethylbenzothiazoline-6-sulfonic acid) (ABTS) (Almeida *et al.*, 2011; Omena *et al.*, 2012; da Silva *et al.*, 2012b; Gregoris *et al.*, 2013; Moo-Huchin *et al.*, 2014; Quintão, 2015; Solorzano-Morán *et al.*, 2015; Reis, 2016; Álvarez-Vargas *et al.*, 2017; Suárez-Vargas *et al.*, 2017; Villa-Hernández *et al.*, 2017; Hernández-Ruiz *et al.*, 2018; Silva, 2018c; Stafussa *et al.*, 2018; Villareal Fuente *et al.*, 2019); 2,2'-Azobis(2-amidinopropane) dihydrochloride, (AAPH) (Poveda Núñez and Robalino Martínez, 2014); the ferric reducing ability of plasma (FRAP) (Omena *et al.*, 2012; Silva *et al.*, 2016; Álvarez-Vargas *et al.*, 2017; Suárez-Vargas *et al.*, 2017; Silva, 2018c; Guillen-Poot, 2019; Villareal Fuente *et al.*, 2019); the Cupric reducing antioxidant capacity (CUPRAC) (Omena *et al.*, 2012) and β-carotene bleaching (Guillen-Poot, 2019). All the assays found that the fruit of *S. purpurea* has antioxidant potential, with greater activity in the epicarp (Solorzano-Morán *et al.*, 2015; Álvarez-Vargas *et al.*, 2017; Suárez-Vargas *et al.*, 2017; Villa-Hernández *et al.*, 2017) and the maximum in the endocarp (Omena *et al.*, 2012); greater antioxidant potential in ripe fruit (Suárez-Vargas *et al.*, 2017) and with the best methods of extraction of the active compounds

(Quintão, 2015; Hernández-Ruiz *et al.*, 2018; Silva, 2018c; Mendonça and Vieites, 2019), which is related to a higher content of carotenoids, phenolic compounds or anthocyanins. Because of the high content of carotenoids and the high bioavailability, the species is considered a good source of pro-vitamin A. In particular, the bioaccessibility of lutein and the presence of zeaxhantin in the fruit may help to prevent diseases of the ocular macula (Murillo *et al.*, 2010; Costa, 2018; Berni *et al.*, 2019), and β-carotene protects the epidermis against the harmful effects of UV and blue radiation; β-cryptoxanthin, because of its bioaccesibility and bioavailability, plays an important role in the prevention of cancer, diabetes, liver and inflammatory diseases (Jiao *et al.*, 2019). For fruit extracts, *in vitro* activity of α-glucosidase was found (IC50 287.63 mg/ml), but there was no activity of α-amylase, neither growth inhibition of Vero, SiHa, HeLa, MDA, and MCF-7 cellular lines (Guillen-Poot, 2019) nor destruction of blood cells (Hernández-Ruiz *et al.*, 2018).

Novel Products/Formulations

In the regions in which the species is cultivated, the fruit is prepared in salt or sweet preserves (Macía and Barfod, 2000; Moz-Preza *et al.*, 2003; Ruenes-Morales *et al.*, 2010; Chamorro Hernández, 2014; Guillen-Poot, 2019). The incorporation of the juice and preserves of the fruit in the baking (Quisingo-Chiza, 2018; Chastinet Braga *et al.*, 2019; Quinzo-Gómez, 2019) and alcohol beverage industries (Ramalho *et al.*, 2014; Corrêa, 2017; Gómez Solís and Morga Ayala, 2017; Silva, 2018a), the transformation of the epicarp and endocarp in flours (Nascimento, 2014; Albuquerque *et al.*, 2016; Menezes Filho *et al.*, 2019; de Abreu *et al.*, 2021; Gutiérrez, 2021) and as a preserver of food products from animal origin (Silva, 2018b; Vieira *et al.*, 2020) have recently been evaluated. It is considered that in these food products, many of the nutrient and bioactive components will be maintained, so they could be considered nutraceutical. However, only a few studies have proved it. In the most complete study, Paixão (2016) found that treatment with cold plasma can increase β-carotenes, lycopene-precursors of vitamin A, niacin and piridoxine (vitamin B3 and B6) concentrations. Aditionally ascorbic acid and vitamin C were at levels similar to those of the untreated juice, while the phenolic compounds concentration was slightly increased. The antioxidant potential, measured as (DPPH) was increased, while measured as (ABTS) was at levels similar to those of untreated juice (Paixão, 2016). Carrillo Martínez (2017) proved that in a beverage prepared with the pulp of the fruit: the antioxidative activity (ABTS) is sustained — although decreased by 57% — when pasteurized and is higher than that reported for apple juice and beer; it has high dietary fibre content, greater than that of commercial orange juice, beer and wine; it presents a delay in glucose release *in vitro*; and it has an average glycaemic index estimated *in vitro*. An increase in antioxidant activity (DPPH and FRAP) of artisan beer with plum pulp and epicarp added was found, with higher concentrations of phenols and similar calories to those of wheat beer (Silva, 2018a). It is considered that the pulverized pulp or its aqueous extract can be used as a meat preserver for meatballs (Silva, 2018b) or fish paté (Vieira *et al.*, 2020); or even be introduced into the cosmetic industry as a sun blocker (Silva *et al.*, 2016).

Conclusions and Future Trends

The ease of cultivating and maintaining *S. purpurea* in home gardens make it a key element in the promotion of healthy food in Mesoamerica and South America. Consumption of the fresh fruit of *S. purpurea* or in different preparations should be promoted because of its high fibre, potassium, iron, zinc, vitamin C, carotenoid, phenol and anthocyanin content, which give it a good antioxidant capacity. Further studies are required to identify the traditional variants/ecotypes with higher nutritional value and the factors that influence nutritional and bioactive compound variations. Consumption of the pulp has been promoted for marketing, but since the epicarp has more bioactive compounds, consumption of the whole fruit should be promoted. Integration of the species in the preserves, bakery, beverage, food preservation and cosmetic industries could be beneficial considering the perishable nature of the fruit. Further studies are needed to analyse acceptance, safety and integral management of the fruit in the development of new products.

References

Albuquerque, J.G.D., Duarte, A.M., Conceição, M.L.D. and Aquino, J.D.S. (2016) Integral utilization of seriguela fruit (*Spondias purpurea* L.) in the production of cookies. *Revista Brasileira de Fruticultura* 38(3), e-229. DOI: 10.1590/0100-29452016229.

Almeida, M.M.B., de Sousa, P.H.M., Arriaga, Â.M.C., do Prado, G.M., Magalhães, C.E. de C. *et al.* (2011) Bioactive compounds and antioxidant activity of fresh exotic fruits from northeastern Brazil. *Food Research International* 44(7), 2155–2159. DOI: 10.1016/j.foodres.2011.03.051.

Álvarez-Vargas, J.E., Alia-Tejacal, I., Chavez-Franco, S.H., Colinas-León, M.T., Nieto-Ángel, D. *et al.* (2017) Ciruelas mexicanas (*Spondias purpurea* L) de clima húmedo y seco: calidad, metabolitos funcionales y actividad antioxidante. *Interciencia* 42, 653–660.

Avitia-García, E. (1997) Estructura floral y anatomia del aborto de ovulos y semillas en ciruela mexicana (*Spondias purpurea* L.). *Horticultura Mexicana* 5, 282–288.

Berni, P., Campoli, S.S., Negri, T.C., de Toledo, N.M.V. and Canniatti-Brazaca, S.G. (2019) Non-conventional tropical fruits: characterization, antioxidant potential and carotenoid bioaccessibility. *Plant Foods for Human Nutrition* 74(1), 141–148. DOI: 10.1007/s11130-018-0710-1.

Bieski, I.G.C., Rios Santos, F., de Oliveira, R.M., Espinosa, M.M., Macedo, M. *et al.* (2012) Ethnopharmacology of medicinal plants of the Pantanal region (Mato Grosso, Brazil). *Evidence-Based Complementary and Alternative Medicine* 2012, 272749. DOI: 10.1155/2012/272749.

Borchert, R. (1994) Soil and stem water storage determine phenology and distribution of tropical dry forest trees. *Ecology* 75(5), 1437–1449. DOI: 10.2307/1937467.

Bramont, W.B., Leal, I.L., Umsza-Guez, M.A., Guedes, A.S., Alves, S.C.O. *et al.* (2018) Comparação da composição centesimal, mineral e fitoquímica de polpas e cascas de dez diferentes frutas. *Revista Virtual de Química* 10, 811–823. DOI: 10.21577/1984-6835.20180059.

Bravo, M., Arteaga, M.I. and Herrera, F.F. (2017) Bioinventario de especies subutilizadas comestibles y medicinales en el norte de Venezuela. *Boletín Latinoamericano y Del Caribe de Plantas Medicinales y Aromáticas* 16, 347–360.

Buriticá Céspedes, P. and Cartagena Valenzuela, J.R. (2015) Neotropical and introduced fruits of special taste and consistency consumed in Colombia. *Revista Facultad Nacional de Agronomía Medellín* 68(2), 7589–7617. DOI: 10.15446/rfnam.v68n2.50948.

Carrillo Martínez, G.G. (2017) *Formulacion de Una Bebida Elaborada a Base de Ciruela (Spondias Purpurea L.) y Evaluación de Su Actividad Nutracéutica*. Maestría En Ciencias En Desarrollo de Productos Bióticos, Instituto Politécnico Nacional.

Chamorro Hernández, L.P. (2014) *Caracterización Físico-Química Del Ovo (Spondias Purpurea L) de Ambuquí. Universidad Pública de Navarra*. Máster Universitario en Tecnología y Calidad en las Industrias Agroalimentarias por la Universidad Pública de Navarra. Nekazaritzako Elikagaien Industrietako Teknologiako eta Kalitateko Unibertsitate Masterra Nafarroako Unibertsitate Universidad Pública de Navarra/Teknologiako eta Kalitateko Unibertsitate Masterra Nafarroako Unibertsitate.

Chastinet Braga, R., Lima Monteiro, L.L. de, Bezerra do Nascimento, K.K., Rabelo e Silva, F.M. and Fernandes Lima, A. (2019) Elaboração e Caracterização de Mousse de Siriguela (*Spondias purpurea*) Adicionado de Pólen Apícola. *Conexõ:Es - Ciência e Tecnologia* 13(5), 84. DOI: 10.21439/conexoes.v13i5.1817.

Corrêa, A. (2017) *Avaliação do processo fermentativo durante a produção de vinho de ciriguela (Spondias purpurea L.) por três isolados de Saccharomyces cerevisae*. Bacharelado em Ciências Biológicas, Universidade Federal da Grande Dourados.

Costa, A.G. (2018) *Carotenoides e compostos fenólicos de Spondias do nordeste brasileiro: Composição e bioacessibilidade*, Doutor em Ciência de Alimentos, Universidade Estadual de Campinas, São Paulo, Brazil.

Cravioto, R.O., Massieu, G., Guzmán García, J. and Calvo de laTorre, J. (1952) Valor nutritivo de plantas alimenticias de yucatán. *Boletín de La Oficina Sanitaria Panamericana (OSP)* 32, 328–339.

da Silva, Q.J., Moreira, A.C.C.G., de Almeida Melo, E. and de Lima, V.L.A.G. (2012a) Compostos fenólicos e atividade antioxidante de genótipos de ciriguelas (*Spondia purpurea* L). *Brazilian Journal of Food & Nutrition/Alimentos e Nutrição* 23, 73–80.

da Silva, Q.J., Moreira, A.C.C.G., Melo, E. d. A. and de Lima, V.L.A.G. (2012b) Phenolic compounds and antioxidant activity of red mombin genotypes (*Spondias purpurea* L.)/Compostos fenolicos e atividade antioxidante de genotipos de ciriguelas (*Spondias purpurea* L.). *Brazilian Journal of Food and Nutrition* 23, 73.

de Abreu, D.J.M., de Moraes, I.A., Asquieri, E.R. and Damiani, C. (2021) Red mombin (*Spondias purpurea* L.) seed flour as a functional component in chocolate brownies. *Journal of Food Science and Technology* 58(2), 612–620. DOI: 10.1007/s13197-020-04574-4.

FAO/LATINFOODS (2010) Tabla de Composición de Alimentos de América Latina. Available at: http://latinfoods.org/tablas-de-composicion-de-alimentos-por-pais (accessed 27 December 2020).

Fernandes, L.G. (2012) *Avaliação mineralógica e dos teores de vitamina C das polpas in natura purpurea de Spondias tuberosa Arruda Câmara (umbu), Spondias, L.(ciriguela) e Spondias sp.(cajarana do sertão)*. Mestrado em Ciências Florestais, Universidade Federal de Campina Grande, Paraíba, Brazil.

Figueiredo, V. (2018) *Potencial Funcional de Frutos Da Cirigueleira de Plantio Comercial Do Sertão Paraibano Durante a Maturação*. Engenheira Agrônoma. Universidade Federal da Paraíba, Paraíba, Brazil.

Furtado, J.D., Siles, X. and Campos, H. (2004) Carotenoid concentrations in vegetables and fruits common to the Costa Rican diet. *International Journal of Food Sciences and Nutrition* 55(2), 101–113. DOI: 10.1080/09637480410001666522.

Godoy, A., Herrera, T. and Ulloa, M. (2003) *Más allá del pulque y el tepache: las bebidas alcohólicas no destiladas indígenas de México*, Universidad Nacional Autónoma de México.

Godínez Caraballo, D. and Volpato, G. (2008) Plantas medicinales que se venden en el mercado El Río, Camagüey, Cuba. *Revista Mexicana de Biodiversidad* 79(1), 217–241. DOI: 10.22201/ib.20078706e.2008.001.529.

Gómez Solís, J.D. and Morga Ayala, C. (2017) *Elaboración de licor de jocote (Spondias purpurea L.)*, Licenciatura en Gastronomía, Universidad de Ciencias y Artes de Chiapas.

Gómez-Maqueo, A., Escobedo-Avellaneda, Z. and Welti-Chanes, J. (2020) Phenolic compounds in Mesoamerican fruits-characterization, health potential and processing with innovative technologies. *International Journal of Molecular Sciences* 21(21), E8357. DOI: 10.3390/ijms21218357.

Gregoris, E., Pereira Lima, G.P., Fabris, S., Bertelle, M., Sicari, M. *et al.* (2013) Antioxidant properties of Brazilian tropical fruits by correlation between different assays. *BioMed Research International* 2013, 132759. DOI: 10.1155/2013/132759.

Guillen-Poot, M.A. (2019) *Las plantas y frutas de la cocina tradicional yucateca como alimentos funcionales en la dieta diaria, Maestro en Ciencias*, Centro de Investigación Científica de Yucatán, Mérida, Yucatán, México.

Gutiérrez, T.J. (2021) In vitro and in vivo digestibility from bionanocomposite edible films based on native pumpkin flour/plum flour. *Food Hydrocolloids* 112, 106272. DOI: 10.1016/j.foodhyd.2020.106272.

Heckel, É. (1897) *Les Plantes Médicinales et Toxiques de La Guyane Française'(Catalogue Raisonné et Alphabétique)'p Le Dr Édouard Heckel*. Protat Freres.

Hernández-Ruiz, K.L., Ruiz-Cruz, S., Cira-Chávez, L.A., Gassos-Ortega, L.E., Ornelas-Paz, J. de J. *et al.* (2018) Evaluation of antioxidant capacity, protective effect on human erythrocytes and phenolic compound identification in two varieties of plum fruit (*Spondias* spp.) by UPLC-MS. *Molecules (Basel, Switzerland)* 23(12), E3200. DOI: 10.3390/molecules23123200.

Instituto Nacional de Nutrición (2001) *Tabla de composición de alimentos para uso práctico*. Caracas Venezuela, Instituto Nacional de Nutrición - Ministerio de Salud y Bienestar Social.

Jiao, Y., Reuss, L. and Wang, Y. (2019) β-Cryptoxanthin: chemistry, occurrence, and potential health benefits. *Current Pharmacology Reports* 5(1), 20–34. DOI: 10.1007/s40495-019-00168-7.

Kozioł, M.J. and Macía, M.J. (1998) Chemical composition, nutritional evaluation, and economic prospects of *Spondias purpurea* (Anacardiaceae). *Economic Botany* 52(4), 373–380. DOI: 10.1007/BF02862067.

Lima, I. d. C.G.S. (2009) *Seriguela (Spondias Purpurea L.): Propriedades Físico-Químicas e Desenvolvimento de Geléia de Doce de Corte e Aceitabilidade Desses Produtos*. Dissertação,Mestrado em Ciência e Tecnologia de Alimentos, Universidade Federal Rural do Rio de Janeiro.

Macía, M.J. and Barfod, A.S. (2000) Economic botany of *Spondias purpurea* (Anacardiaceae) in Ecuador. *Economic Botany* 54(4), 449–458. DOI: 10.1007/BF02866544.

Maldonado-Astudillo, Y.I., Alia-Tejacal, I., Núñez-Colín, C.A., Jiménez-Hernández, J., Pelayo-Zaldívar, C. *et al.* (2014) Postharvest physiology and technology of *Spondias purpurea* L. and *S. mombin* L. *Scientia Horticulturae* 174, 193–206. DOI: 10.1016/j.scienta.2014.05.016.

Marin, A.M.F., Siqueira, E.M.A. and Arruda, S.F. (2009) Minerals, phytic acid and tannin contents of 18 fruits from the Brazilian savanna. *International Journal of Food Sciences and Nutrition* 60(Suppl 7), 180–190. DOI: 10.1080/09637480902789342.

Medeiros, M.F.T. and Albuquerque, U.P. (2014) Food flora in 17th century northeast region of Brazil in Historia Naturalis Brasiliae. *Journal of Ethnobiology and Ethnomedicine* 10, 50. DOI: 10.1186/1746-4269-10-50.

Meléndez, E.N. (1982) *Plantas medicinales de Puerto Rico: folklore y fundamentos científicos*. La Editorial, UPR.

Menchú, M.T., Méndez, H., Barrera, M.A. and Ortega, L. (2007) *Tabla de Composición de Alimentos de Centroamérica*. Guatemala, Organización Panamericana de la Salud.

Mendonça, V.Z. d. and Vieites, R.L. (2019) Propiedades físicoquímicas de frutas exóticas nativas de Brasil. *Acta Agronómica* 68, 175–181.

Menezes Filho, A.C.P. de, Oliveira Filho, J., Deminski, G., Jesus, A., Andrade, M. *et al.* (2019) Avaliação colorimétrica e caracterização morfológica por microscopia óptica de alta resolução das farinhas dos frutos do jatobá, jambolão e siriguela. *Multi-Science Journal* 2(1), 16–22. DOI: 10.33837/msj.v2i1.544.

Metzel, R. and Montagnini, F. (2014) From farm to forest: factors associated with protecting and planting trees in a Panamanian agricultural landscape. *Bois & Forets Des Tropiques* 322(322), 3. DOI: 10.19182/bft2014.322.a31225.

Miller, A. (2011) *Spondias*. In: Kole, C. (ed.) *Wild Crop Relatives: Genomic and Breeding Resources*. Springer, Berlin, Heidelberg, pp. 203–212. DOI: 10.1007/978-3-642-21250-5.

Mitchell, J.D. and Daly, D.C. (2015) A revision of *Spondias* L. (Anacardiaceae) in the neotropics. *PhytoKeys* 1–92.

Mohammed, M., Bridgemohan, P., Graham, O., Wickham, L., Bridgemohan, R.S.H. *et al.* (2019) Postharvest physiology, biochemistry and quality management of chili plum (*Spondias purpurea* var. *lutea*): a review. *Journal of Food Research* 8(3), 1. DOI: 10.5539/jfr.v8n3p1.

Monge-Rojas, R. and Campos Nunes, H. (2013) *Tabla de composicion de alimentos de Costa Rica: carotenoides y tocoferoles, Tres Ríos*. Universidad de Harvard,INCIENSA, Tres Ríos, Costa Rica.

Moo-Huchin, V.M., Estrada-Mota, I., Estrada-León, R., Cuevas-Glory, L., Ortiz-Vázquez, E. *et al.* (2014) Determination of some physicochemical characteristics, bioactive compounds and antioxidant activity of tropical fruits from Yucatan, Mexico. *Food Chemistry* 152, 508–515. DOI: 10.1016/j.foodchem.2013.12.013.

Moz-Preza, J., Navarrete-Jiménez, A. and Núñez-López, M. (2003) *Caracterización Morfológica de Cinco Materiales de Jocote de Verano Spondias Spp. Con Valor Comercial e Investigativo En La Zona Occidental de El Salvador*. Tesis de Ingeniería agronómica,Universidad de El Salvador. Available at: http://ri.ues.edu.sv/id/eprint/1631/1/13100966.pdf (accessed 12 April 2022).

Munsell, H.E., Castillo, R., Zurita, C. and Portilla, J.M. (1953) Production, uses, and composition of foods of plant origin from Ecuador. *Journal of Food Science* 18(1–6), 319–342. DOI: 10.1111/j.1365-2621.1953.tb17724.x.

Murillo, E., Meléndez-Martínez, A.J. and Portugal, F. (2010) Screening of vegetables and fruits from Panama for rich sources of lutein and zeaxanthin. *Food Chemistry* 122(1), 167–172. DOI: 10.1016/j.foodchem.2010.02.034.

Nascimento, M.R.F. (2014) *Caracterização e aproveitamento das farinhas dos caroços de abacate (Persea gratíssima Gaertner f.), jaca (Artocarpus heterophyllus L.) e seriguela (Spondias purpurea L.) para elaboração de biscoitos tipo cookies*. Doutorado em Ciência e Tecnologia de Alimentos, Universidade Federal Rural do Rio de Janeiro, Rio de Janeiro.

Núcleo de Estudos e Pesquisas em Alimentação – NEPA Universidade Estadual de Campinas – UNICAMP (2011) *Tabela Brasileira de Composição de Alimentos TBCA-USP: 2001-2004*. Núcleo de Estudos e Pesquisas em Alimentação – NEPA Universidade Estadual de Campinas – UNICAMP.

Omena, C.M.B., Valentim, I.B., Guedes, G. d. S., Rabelo, L.A., Mano, C.M. *et al.* (2012) Antioxidant, anti-acetylcholinesterase and cytotoxic activities of ethanol extracts of peel, pulp and seeds of exotic Brazilian fruits: Antioxidant, anti-acetylcholinesterase and cytotoxic activities in fruits. *Food Research International* 49, 334–344.

Ortiz Rodea, A. (2013) *Evaluacion del sistema de produccion bovino de Zacazonapan desde un enfoque silvopastoril*. Tesis de Doctorado, Universidad Autónoma del Estado de México.

Ortíz-Gil, S. (2017) *El uso de plantas alimenticias en el Valle de México y su periferia: un estudio de cambios, persistencias y pervivencias prehispánicas durante la Época Colonial*. Estudios para la protección del patrimonio arqueológico e histórico, Maestra en Estudios Arquológicos, El Colegio de Michoacán, A. C, Michoacán, Mexico.

Paixão, L.M.N. (2016) *Efeito do processamento por plasma frio sobre as características de qualidade de suco de seriguela (Spondias purpurea, L)*. Mestrado em Ciência e Tecnologia de Alimentos, Universidade Federal do Ceará, Fortaleza.

Pire Sierra, M.C., Garrido, E., González, H. and Pérez, H. (2010) Estudio comparativo del aporte de fibra alimentaria en cuatro tipos de frutas de consumo común en Venezuela. *Interciencia* 35, 939–944.

Poveda Núñez, M.C. and Robalino Martínez, D. del R. (2014) *Determinación de la influencia de las zonas de producción sobre el contenido de componentes bioactivos y la capacidad antioxidante de cinco frutas andinas*, Universidad Técnica de Ambato. Facultad de Ciencia e Ingeniería en Alimentos. Carrera de Ingeniería Bioquímica.

Quintão, W. d. S.C. (2015) *Maturação, compostos fenólicos e atividade antioxidante de seriguelas (Spondias purpurea L.) cultivadas no cerrado*, Universidade De Brasília Faculdade De Ceilândia.

Quinzo-Gómez, K.P. (2019) *Desarrollo de Una Fórmula Para Elaborar Yogur Artesanal de Dos Sabores: Aguacate (Persea Americana Mill) y Ciruela (Spondias Purpurea L)*. Ingeniería Agroindustrial, Universidad Católica de Santiago de Guayaquil.

Quisingo-Chiza, A.R. (2018) *Aplicación gastronómica del hobo en la repostería clásica*. Tecnológo en Alimentos y Bebidas, Universidad de las Américas.

Ramalho, S.A., Gualberto, N.C., Neta, M.T.S.L., Batista, R.A., Araújo, S.M. *et al*. (2014) Catechin and epicatechin contents in wines obtained from Brazilian exotic tropical fruits. *Food and Nutrition Sciences* 05(5), 449–457. DOI: 10.4236/fns.2014.55053.

Ramírez Hernández, B.C., Pimienta Barrios, E., Castellanos Ramos, J., Múñoz Urias, A., Palomino Hasbach, G. *et al*. (2008) Sistemas de producción de *Spondias purpurea* (Anacardiaceae) en el centro-occidente de México. *Revista de Biología Tropical* 56, 675–687. DOI: 10.15517/rbt.v56i2.5616.

Recuenco, M.C., Lacsamana, M.S., Hurtada, W.A. and Sabularse, V.C. (2016) Total phenolic and total flavonoid contents of selected fruits in the Philippines. *Philippine Journal of Science* 145, 275–281.

Reis, B.C. (2016) *Frutas exóticas cultivadas no Cerrado do Distrito Federal: maturação, composição fenólica, atividade antioxidante, aplicação e estudo de estabilidade em micropartículas*. Mestrado em Ciências e Tecnologias em Saúde, Universidade de Brasília, Brasília.

Reis, L.C.B., Carneiro, L.M., Branco, C.R.C. and Branco, A. (2015) Comparison of conventional microwave and focused microwave-assisted extraction to enhance the efficiency of the extraction of antioxidant flavonols from jocote pomace (*Spondias purpurea* L.). *Plant Foods for Human Nutrition* 70(2), 160–169. DOI: 10.1007/s11130-015-0473-x.

Ruenes-Morales, M. d. R., Casas, A., Jiménez-Osornio, J.J. and Caballero, J. (2010) Etnobotánica de *Spondias purpurea* L. (Anacardiaceae) en la península de Yucatán. *Interciencia* 35, 247–254.

Salazar-Díaz, J.S. (2019) *Daño por compresión en frutos de ciruela mexicana (Spondias purpurea L.)*. Maestría en Ciencias Agropecuarias y Desarrollo Rural Universidad Autónoma del Estado de Morelos.

Salazar, C., Zizumbo-Villarreal, D., Colunga-GarcíaMarín, P. and Brush, S. (2016) Contemporary Maya food system in the lowlands of northern Yucatan. In: Lira, R., Casas, A., and Blancas, J. (eds) *Ethnobotany of Mexico*. Springer, New York, NY, pp. 133–150.

Sameh, S., Al-Sayed, E., Labib, R.M. and Singab, A.N. (2018) Genus *Spondias*: A phytochemical and pharmacological review. *Evidence-Based Complementary and Alternative Medicine* 2018, 1–13. DOI: 10.1155/2018/5382904.

Sanches, A.G., da Silva, M.B., Moreira, E.G.S. and dos Santos, E.X. (2018) Atraso na maturação e qualidade pós-colheita de seriguela exposta a radiação ultravioleta-C. *Revista Nativa, Sinop* 6(3), 225. DOI: 10.31413/nativa.v6i3.4498.

Silva, A.A. d (2018a) *Produção, caracterização físico-química e análise sensorial de cerveja artesanal de trigo adicionada de polpa e casca de seriguela (Spondias purpurea L.) e casca de laranja (Citrus sinensis L.)*. DTA - Dissertações defendidas na UFC(Mestrado em Ciência e Tecnologia de Alimentos), Universidade Federal do Ceará, FORTALEZA,Brazil. Available at: https://repositorio.ufc.br/bitstream/riufc/42605/3/2018_dis_aasilva.pdf (accessed 12 April 2022).

Silva, I.S. d. S. (2018b) *Elaboração, caracterização física, físico-química e avaliação da estabilidade lipídica de almôdega adicionada de extrato de seriguela (Spondias purpurea L.)*. Bacharelado em Nutrição, Universidade Federal de Campina Grande, Cuité-PB. Available at: http://dspace.sti.ufcg.edu.br › bitstream › riufcg.

Silva, M.C.C. d (2018c) *Caracterização in vitro do potencial antioxidante de extrato obtido a partir da seriguela*. Bacharelado em Nutrição, Universidade Federal de Campina Grande, Cuité, Paraíba, Brasil.

Silva, R.V., Costa, S.C.C., Branco, C.R.C. and Branco, A. (2016) In vitro photoprotective activity of the *Spondias purpurea* L. peel crude extract and its incorporation in a pharmaceutical formulation. *Industrial Crops and Products* 83, 509–514. DOI: 10.1016/j.indcrop.2015.12.077.

Solorzano-Morán, S., Alia-Tejacal, I., Rivera-Cabrera, F., López-Martínez, V., Pérez-Flores, L.J. *et al.* (2015) Quality attributes and functional compounds of Mexican plum (*Spondias purpurea* L.) fruit ecotypes. *Fruits* 70(5), 261–270. DOI: 10.1051/fruits/2015027.

Stafussa, A.P., Maciel, G.M., Rampazzo, V., Bona, E., Makara, C.N. *et al.* (2018) Bioactive compounds of 44 traditional and exotic Brazilian fruit pulps: phenolic compounds and antioxidant activity. *International Journal of Food Properties* 21(1), 106–118. DOI: 10.1080/10942912.2017.1409761.

Suárez-Vargas, A., Juárez-lópez, P., López-martínez, V., Flores, L.J.P., Sánchez, D.G. *et al.* (2017) Botânica e fisiologia/ botany and physiology antioxidant activity and physicochemical parameters in cuernavaqueña mexican plum (*Spondias purpurea* L.) at different ripening stages. *Revista Brasileira de Fruticultura* 39(4), e-787. DOI: 10.1590/0100-29452017787.

Tantengco, O.A.G., C. Condes, M.L., T. Estadilla, H.H. and M. Ragragio, E. (2018) Ethnobotanical survey of medicinal plants used by Ayta communities in Dinalupihan, Bataan, Philippines. *Pharmacognosy Journal* 10(5), 859–870. DOI: 10.5530/pj.2018.5.145.

Tomchinsky, B. and Ming, L.C. (2019) As plantas comestíveis no Brasil dos séculos XVI e XVII segundo relatos de época. *Rodriguésia* 70, e03792017. DOI: 10.1590/2175-7860201970040.

US Department of Agriculture (2021) FoodData Central. Available at: https://ndb.nal.usda.gov/ (accessed 3 January 2021).

U.S. Department of Agriculture and U.S. Department of Health and Human Services (December 2020) DietaryGuidelines.gov. Available at: (accessed 10 September 2022).

Vael, L. (2015) *Ethnobotanical study of the plant use in the natural landscape of two mestizo communities in the Ucayali region of the Peruvian Amazon*. Universiteit Ghent, Ghent.

Vargas-Simón, G. (2018) Ciruela/Mexican Plum—*Spondias purpurea* L. In: Rodrigues, S., de Oliveira Silva, E., and de Brito, E.S. (eds) *Exotic Fruits*. Academic Press, London, pp. 141–152.

Vasco, C., Ruales, J. and Kamal-Eldin, A. (2008) Total phenolic compounds and antioxidant capacities of major fruits from Ecuador. *Food Chemistry* 111(4), 816–823. DOI: 10.1016/j.foodchem.2008.04.054.

Vélez-Coria, O. and Cruz-León, A. (2012) Evidencias del proceso de trasformacion de los frutos de ciruela: Las recetas. In: Cruz-León, A., Pita-Duque, Á., and Rodríguez-Haros, B. (eds) *Jocotes, Jobos, Abales o Ciruelas Mexicana*. Universidad Autónoma Chapingo, Chapingo, México, pp. 111–123.

Vieira, R.M., Farias, M.D.P., Ambrósio, L.S.A., Sá, D.M.A.T., Andrade, F.J.E.T. *et al.* (2020) Avaliação microbiológica e sensorial de patê de peixe adicionado de pó de siriguela (*Spondias purpurea* L.)/ microbiological and sensorial evaluation of fish paté added from siriguela powder (*Spondias purpurea* L.). *Brazilian Journal of Development* 6, 12933–12942. DOI: 10.34117/bjdv6n3-235.

Villa-Hernández, J.M., Mendoza-Cardoso, G., Mendoza-Espinoza, J.A., Vela-Hinojosa, C., Díaz de León-Sánchez, F. *et al.* (2017) Antioxidant capacity in vitro and in vivo of various ecotypes of Mexican plum (*Spondias purpurea* L.). *Journal of Food Science* 82(11), 2576–2582. DOI: 10.1111/1750-3841.13862.

Villareal Fuente, J.M., Alia Tejacal, I., Xóchitl Doraneli Pérez-Pérez, X.D., Espinoza-Zaragoza, S., Marroquín-Agreda, F.J. *et al.* (2019) Caracterización fisicoquímica de frutos de ciruela mexicana (*Spondias purpurea* L.) en el Soconusco, Chiapas. *Ecosistemas y Recursos Agropecuarios* 6(17), 219–229. DOI: 10.19136/era.a6n17.1949.

Zielinski, A.A.F., Ávila, S., Ito, V., Nogueira, A., Wosiacki, G. *et al.* (2014) The association between chromaticity, phenolics, carotenoids, and in vitro antioxidant activity of frozen fruit pulp in Brazil: an application of chemometrics. *Journal of Food Science* 79(4), C510-6. DOI: 10.1111/1750-3841.12389.

Zizumbo-Villarreal, D., Flores-Silva, A. and Colunga-García Marín, P. (2012) The archaic diet in Mesoamerica: Incentive for milpa development and species domestication. *Economic Botany* 66(4), 328–343. DOI: 10.1007/s12231-012-9212-5.

Zizumbo-Villarreal, D., Flores-Silva, A. and Colunga-GarcíaMarín, P. (2014) The food system during the formative period in West Mesoamerica. *Economic Botany* 68(1), 67–84. DOI: 10.1007/s12231-014-9262-y.

29 Soursop (*Annona muricata* L.)

Annegowda Hardur Venkatappa[1] and Rajeev Bhat[2]*
[1]Sri Adichunchanagiri College of Pharmacy, Adichunchanagiri University, Nagamangala, Mandya, Karnataka, India; [2]ERA-Chair for Food (By-) Products Valorization Technologies, Estonian University of Life Sciences, Tartu, Estonia

Abstract

Soursop (*Annona muricata* L.) is one of the widely consumed and marketed fruit of the Annonaceae family. It possesses high nutraceutical and biopharmaceutical value (nutritional and therapeutic value). Soursop fruit are sweet with an exotic flavour, and loaded with essential vitamins, minerals and antioxidant-rich polyphenolic compounds. The fruit are established for their bioactivity such as antioxidant, anticancer, antimicrobial, larvicidal and insecticidal properties. However, soursop fruit have been linked with induced toxicity, owing to some of the compounds present. The waste portions, mainly peel and seed, have been recognized to encompass ample amounts of bioactive components. Soursop fruit are climacteric and perish rapidly, thus necessitating the application of various postharvest techniques to preserve them for an extend period. In this chapter, information on botanical aspects, cultivation, traditional uses, presence of bioactive components, nutraceutical value and postharvest preservation along with potential applications is discussed.

Introduction

Ever since the beginning of human civilization, fruit and vegetables have been considered as two important components of a healthy diet. Regular consumption of fruit and vegetables not only provides nutrition and energy, but also helps in the prevention of various degenerative diseases related to cardiovascular systems, inflammation, cancer and others. Among a wide range of popular fruit available on the international market, fruit belonging to Annonaceae family remain the most popular and are widely admired for their smooth and creamy flesh, exotic flavour and sweet taste (Márquez Cardozo *et al.*, 2013). Soursop fruit is native to Central and South America, but often encountered in the majority of tropical and subtropical regions, such as Australia, Bermuda, Brazil, China, Colombia, Cuba, India, Malaysia, New Zealand, Panama and Vietnam. The fruit contributes partially to regional economic growth, being sold as fresh fruit or exotic processed products (Shashirekha *et al.*, 2008; Rabelo *et al.*, 2016). The top three producers of soursop are Mexico, Venezuela and Brazil (Pinto *et al.*, 2005; Márquez Cardozo *et al.*, 2013).

This fruit has more importance in comparison with other species of Annonaceae family members, mainly due to commercial importance, nutritional and medicinal values (Pinto *et al.*, 2005). Thus, it can be considered as a fruit with nutraceutical or biopharmaceutical value. However, the seeds of the fruit are toxic (1–2 cm in length and 0.3–0.6 g in weight) and occur from very few to up to 200 in number (Moghadamtousi *et al.*, 2015). Soursop fruit

*Corresponding author: rajeev.bhat@emu.ee

DOI: 10.1079/9781789248067.0029

have a tinge of papaya, apple, pineapple, mango and strawberry flavour with the edible portion being 67% of the total weight. The fruit have an attractive aroma and exotic taste of the flesh part (pulp) and can be used either fresh or frozen. The fruit are used to prepare juice (sherbets), smoothies, ice cream, milkshake and as an ingredient to make desserts. In South America, this plant is also grown as an ornamental/garden plant.

Though many research works are available on this fruit, information remains scattered on the composition, nutraceutical value and postharvest technologies employed for its preservation for an extended time, emphasizing the importance of recording this informatio in a single document, as done in this book chapter, wherein information regarding the nutritional and therapeutic values, food applications and postharvest technologies are covered.

Botanical Description

Annona muricata L. belongs to the family Annonaceae and is popularly known as soursop, graviola, guyabano, durian belanda and mamphal. The plants are small, slender, erect evergreen trees reaching a height of up to 5–8 m at full maturity. Leaves are oblong to ovate to cylindrical, 14–16 cm in length and 5–7 cm in width. Soursop trees are highly sensitive to frost and temperatures below 5°C (Mishra *et al.*, 2013). The fruit are large, oval, heart-shaped, dark green coloured when unripe, and slightly light green-coloured on ripening. It is one of the larger fruit of the genus *Annona* weighing up to 10 kg (average weight being 3–4 kg) (Tovar-Gómez *et al.*, 2011). The rind or skin portion of the fruit has short, soft and pointed spines. The pulp is white, cottony, fibrous and juicy, surrounded by a central soft pith/core (Fig. 29.1).

Traditional Uses

Since time immemorial, soursop fruit has been traditionally used as a natural medicine to cure various ailments. Immature fruit are used to treat skin diseases, skin rash, cankers, fever, malaria, peptic ulcer, dysentery and colic diseases. In addition, the peel of the immature fruit is used in treating atonic dyspepsia (Khan *et al.*, 1998). The acidic pulp of the immature fruit is used to treat liver diseases and to cure foot parasite (Pinto *et al.*, 2005). In Jamaica and the West Indies, the fruit has been used to reduce fever, dysentery and gastric ulcers (Badrie and Schauss, 2010), to treat malaria and as an antiviral and abortificient agent (Osorio *et al.*, 2007). In the Philippines and Uganda, it is used to treat hypertension, cancer and diabetes (Ong and Kim, 2014), while in Brazil, the fruit is used as a galactogogue. In Bolivia, soursop fruit are used to treat kidney disorders, and in the Dominican Republic to treat asthmatic symptoms, arthritis pain, neuralgia and rheumatism (Badrie and Schauss, 2010; Vandebroek *et al.*, 2010; Hajdu and Hohmann, 2012; Moghadamtousi *et al.*, 2015). In South America, the fruit are claimed to have a tranquillizing effect (Hasrat *et al.*, 1997). In Malaysia, the fruit juice is used to treat stomach pain and hypertension. In the Philippines the fruit is used in the management of diabetes, while in Peru it is used to treat obesity, gastritis, dyspepsia and inflammation (Daddiouaissa and Amid, 2018).

Nutritional Composition

The fruit harvest season can influence the biochemical composition of soursop fruit. Those harvested during the dry season are reported to have higher values of sugar, acidity and ascorbic acid than fruit harvested during the wet season. However, the volume of alcoholic components and protein were much higher when harvested during the wet season (Omoifo, 2004). The quantity of fruit pulp accounts for 75% of the total weight followed by peel (20%) and seed (5%). The edible pulp portion contains ~81% of water, 17% carbohydrates, 1% protein, 0.5% lipids and 3.3% fibre. The amount of crude protein and crude fibre was detected to be higher in the peel, whereas the amount of crude fat, moisture and carbohydrate was greater in the pulp (Akomolafe and Ajayi, 2015). The fruit has 1% acidity and mainly contains citric acid and slight amounts of ascorbic, malic and

(a) (b) (c) (d)

Fig. 29.1. Soursop fruit (*Annona muricata* L.): (a) whole fruit on plant; (b) freshly harvested fruit; (c) cut-open fruit showing white cottony pulp and seeds; (d) dried seeds (photo: authors).

isocitric acid. The acidity of the fruit pulp was recorded to increase corresponding with the increased production of ascorbic acid, and this was maximum after harvesting, which was attributed to increased metabolic activity (Pareek *et al.*, 2011). The total amount of sugars in soursop generally ranges between 16–17%. The sugar level (total soluble solids 1.02 ± 0.28%; sucrose 2.97 ± 0.24%; glucose 3.60 ± 0.27%) has been observed to increase during the ripening process (owing to hydrolysis of starch) (Badrie and Schauss, 2010; Márquez Cardozo *et al.*, 2012). The total amount of sugars reaches concentration approximately five days after harvesting. The increased soluble solids/sugars during ripening is attributed to the breakdown of polysaccharides by enzymes such as amylase, cellulase and polygalacturonase (Paull *et al.*, 1983). The glycaemic index and glycaemic load is reported to be 32 and 3, respectively, which is very low and thus requires no dietary restrictions (Passos *et al.*, 2015). In addition, 100 g of fruit contains 2 IU vitamin A (retinol), 0.07 mg vitamin B1 (thiamine), 0.05 mg vitamin B2 (riboflavin), 0.08 mg vitamin B12, 0.2 mg vitamin B3, 1.2 mg vitamin B5, 22.6 mg of vitamin C and 0.08 mg of vitamin E (Márquez Cardozo *et al.*, 2012). The vital minerals present in 100 g of fruit pulp are potassium (278 mg), that can help in the prevention of hypertension, calcium (14 mg) and phosphorus (27 mg), both helpful in bone and teeth formation. Further, phosphorus, magnesium, sodium, potassium and zinc are reported to be higher in the peel compared to the pulp (Akomolafe and Ajayi, 2015). Magnesium (0.04 mg/100 g pulp) functions like a co-factor of many enzymes involved in various metabolic reactions and protein synthesis along with 85 mg of tannins (Coria-Téllez *et al.*, 2017, 2018). Various enzymes like pectinase, catalase and peroxidase are found in the pulp of soursop fruit (Arbaisah *et al.*, 1997). Pectin-esterase is a heat-resistant enzyme that can be used for gelatinization and precipitation of pectin in purées and juice, thus reducing the cloudiness. The amylase formed during ripening helps in converting polysaccharides into simple sugars. Enzymes like polygalactouronase, cellulase and polyphenol oxidase were also found in the fruit, responsible for cell wall degradation and fruit colour change to brown (Bezerra *et al.*, 2003). In Table 29.1, the nutritional value and bioactive components in an edible portion (100 g) of soursop fruit are compared with other fruit in the Annonaceae family.

Table 29.1. Nutritional and other bioactive components in edible portion (100 g) of soursop fruit compared with other fruit of the Annonaceae family (Pinto *et al.*, 2005; Badrie and Schauss, 2010; Márquez Cardozo *et al.*, 2012; Moo-Huchin *et al.*, 2014; Padmanabhan and Paliyath, 2016).

	Fruit type (Annonaceae family)			
Component	Soursop (*Annona muricata* L.)	Cherimoya (*Annona cherimola* L.)	Custard apple (*Annona reticulata* L.)	Sugar apple (*Annona squamosa* L.)
Moisture content	~81%	~79%	~73–77%	~71%
Protein (g)	~1	1.57	1.7	2.06
Fat (g)	~0.5	0.68	0.60	0.3
Carbohydrates (g)	16.84	17.71	25.20	23.64
Fibre	3.3	3.0	2.4	4.4
Calcium (mg)	14	10	30	24
Potassium (mg)	278	287	382	247
Phosphorus (mg)	27	26	21	32
Iron (mg)	0.64	0.27	0.71	0.60
Magnesium (mg)	21	17	18	21
Vitamin A (IU)	2	5	33	6
Vitamin B5 (mg)	1.2	0.8	0.7	0.9
Thiamine (mg)	0.07	0.101	0.080	0.110
Vitamin B12 (mg)	0.08	0.12	0.12	0.13
Riboflavin (mg)	0.05	0.131	0.100	0.113
Niacin (mg)	0.900	0.644	0.500	0.883
Ascorbic acid (mg)	22.6	11.5	30	37.4
Vitamin E (mg)	0.08	0.27	-	-
Titratable acidity	1.02	-	0.66	-
pH	3.70	-	-	-
Total soluble solids (°Brix)	~11.0	-	17.75	-
Energy (kilocalories)	65	75	101	94
Anthocyanin (mg/100 g)	6.44	-	1.55	0.73
Flavonoids (mg of quercetin/100 g)	9.32	-	418.24	-

During ripening, the green colour (due to chlorophyll pigment) of the fruit peel gets diminished leaving behind a slight yellowish colour (owing to carotenoid pigment). Further, ripening of fruit leads to change in the colour from green to brown due to the oxidation and polymerization of phenols by the enzyme polyphenol oxidase. During this time, the fruit skin becomes smooth and soft. Simultaneously, there is a significant decrease in the pH of the pulp from 5.5 to 3.7 within three to four days of ripening (oweing to an increase in ascorbic acid, malic acid and citric acid content, respectively) (Paull *et al.*, 1983). Furthermore, the flavour of the fruit varies and is related to respiration and ethylene production (with the ethylene peak being 250–350 ml/kg/h) accompanied by decreased weight. It was also observed that on the first day of harvest, the fruit start producing CO_2 (100 ml/kg/h), reaching a maximum (350 ml/kg/h) on the fourth day (at 25–30°C). After this, there occurs a bland flavour with loss of astringency, owing to overripening of fruit. During this stage, fermentation of sugars and a decreased level of organic acids and phenols occurs (Badrie and Schauss, 2010; Pareek *et al.*, 2011; Márquez Cardozo *et al.*, 2012).

Chemical Constituents

Systematic phytochemical research has been carried out on the pulp of soursop fruit, which revealed the presence of various types of major phytoconstituents like alkaloids (1.9 mg), tannins (65.98 mg), flavonoids (9.32 mg), saponins (0.17 mg) and anthocyanins (6.44 mg) in 100 g of fruit pulp (Pinto *et al.*, 2005; Onyechi *et al.*, 2012). Each fruit is reported to contain 14–226 μmol annonaceous acetogenins in 100 g of fruit pulp (Bonneau *et al.*, 2017). These are responsible for sporadic atypical Parkinsonism or dementia syndromes (Lannuzel *et al.*, 2007). Studies have revealed the neurodegenerative effects of annonacin (Yamada *et al.*, 2014; Höllerhage *et al.*, 2015). Acetogenins have been isolated from the fruit pulp (by column chromatography and by HPLC), and among them annonacin was the major acetogenin identified (Ragasa *et al.*, 2012; Sun *et al.*, 2014). Acetogenin, the main bioactive compound in soursop, is a lipophilic secondary metabolite derived from long-chain fatty acids, which are responsible for pharmacological activity (Alali *et al.*, 1999).

The by-products of soursop are reported to contain phenolic acids like cinnamic acid, coumaric acid, syringic acid, procateic acid, gallic acid, caffeic acid, 4-hydroxybenzoic acid, chlorogenic acid and neochlorogenic acid, that serve as a source of antioxidants (Adefegha *et al.*, 2015; Aguilar-Hernández *et al.*, 2019). In Table 29.2, phytoconstituents/bioactive compounds in various parts of soursop fruit and their bioactivities are shown.

Aroma compounds

Gas chromatography/mass spectroscopic analysis (GC-MS) of the fruit pulp during various stages of ripening revealed the presence of esters, aldehydes, alcohols, terpenes and ketones. Further, esters of aliphatic acids were detected to be the dominant odour compound (~51%) followed by 2-hexenoic acid ethyl ester (8.6%), 2-octenoic acid methyl ester (5.4%) and 2-butenoic acid methyl ester (2.4%), which were present in the essential oils extracted from fruit pulp (Jirovetz *et al.*, 1998). However, half-ripe and unripe fruit contained ~70% and 60% of esters, respectively. The solid-phase microextraction (SPME) and GC-MS study of fruit pulp revealed the presence of 21 compounds among which 12 compounds were identified to be esters like methyl hexanoate, methyl 2-hexenoate, ethyl caproate, ethyl 2-hexenoate, methyl caprylate, methyl crotonate, ethyl butyrate, butyl acetate, ethyl crotonate, methyl caproate, etc. Pulp also contained mono- and sesquiterpenes like β-caryophyllene, 1,8-cineole, linalool, limonene, α-terpineol, linalyl propionate, α-ocimene, β-myrcene, δ-limonene and calarene. Moreover, alcohol, such as 1-pentanol, 2-hexanol, 2-heptanol, 1-octanol and isopulegol, was detected in the essential oil of the fruit (Augusto *et al.*, 2000; Pino *et al.*, 2001). The compound methyl butanoate is responsible for the pineapple-like, fruity and sweet flavour; compounds like methyl hexenoate, methyl pentenoate and methyl 2-butanoate are responsible for fruity and sweet flavour. Ethyl hexanoate is

Table 29.2. Phytoconstituents/bioactive compounds in various parts of soursop fruit and their bioactivities.

Fruit portion	Bioactive compounds	Compounds identified	Bioactivity	Reference
Whole fruit	Flavonoids	Myricetin, fisetin, morin, kaempferol, isorhamnetin. quercetin 3-*O*-glucoside, and quercetin	Antioxidant activity	Lako *et al*. (2007) Rubio-Melgarejo *et al*. (2020)
Whole fruit	Phenolic acids	Ferulic acid and *p*-coumaric acid	Antioxidant activity	Rubio-Melgarejo *et al*. (2020)
Whole fruit	Acetogenins	Bullatacin	Potent against L1210 murine leukemia	Badrie and Schauss (2010)
Whole fruit	Acetogenins	Muricin (M and N), muricenin	Muricin M & N shown potent antiproliferative activity against human prostate cancer (PC-3) cells	Sun *et al*. (2016)
Whole fruit	Acetogenins	Annonamuricin (A, B and C)	Demonstrated potent antiproliferative activity against human prostate cancer (PC-3) cells	Sun *et al*. (2017)
Pulp	Acetogenins	Cis-annoreticuin, sabadelin	Cytotoxic against human hepatoma carcinoma cell line (Hep G2)	Ragasa *et al*. (2012)
Pulp	Phenolics	Gallic acid, chlorogenic acid, caffeic acid	Antioxidants	Blancas-Benítez *et al*. (2019)
Pulp	Phenolics	Gallic acid, cinnamic acid, coumaric acid, 4-hydroxybenzoic acid, chlorogenic acid	Antioxidants	Aguilar-Hernández *et al*. (2019)
Seeds	Phenolic acids	Gallic acid, chlorogenic acid, caffeic acid, ferulic acid, *p*-coumaric acid, *m*-coumaric acid, *0*-coumaric acid	Antioxidant activity	Menezes *et al*. (2019)
Seeds	Flavonoids	Rutin, quercetin, catechin	Antioxidant activity	Menezes *et al*. (2019)
Seeds	Fatty acids	Methyl palmitate, methyl oleate, methyl stearate, 10-nonadecanol	Larvicidal against *Aedes aegypti*	Komansilan *et al*. (2012)
Seed	Acetogenins	Annoreticuin-9-one	Cytotoxic to human pancreatic tumour cell line (PACA-2), human prostate adenocarcinoma (PC-3) and human lung carcinoma cells (A-549)	Ragasa *et al*. (2012)
Seed	Phenolics	Cinnamic acid, syringic acid, procateic acid, gallic acid, chlorogenic acid, caffeic acid, 4-hydroxybenzoic acid, neochlorogenic acid	Antioxidants	Aguilar-Hernández *et al*. (2019)

responsible for fruity flavour; methyl heptenoate and methyl octanoate are responsible for orange flavour; methyl cinnamate for strawberry flavour; β-myrcene for balsamic and spice flavour; δ-limonene for citrus and mint flavour; and linalool for lavender flavour (Santana *et al.*, 2017). Such essential oils are considered to be of use to enhance the flavour of processed food products.

Pharmacological and bioactivity

A wide range of pharmacological and bioactivity has been reported for soursop fruit, thus supporting its use in traditional medicine. Some of the important activities include antimalarial, antiparasitic and anticancer potency (Zafra-Polo *et al.*, 1998; Chih *et al.*, 2001). Acetogenins like annonacin, gigantetrocin, isoannonacin, isoannonain-10-one and goniothalamicin obtained from the defatted seeds of the fruit are reported to exhibit cytotoxic effects against A-549 lung carcinoma, MCF-7 breast carcinoma and HT-29 colon adenocarcinoma human tumour cell lines (Rieser *et al.*, 1993). The study conducted by Aguilar-Hernández *et al.* (2020a) revealed the peel and seeds to be a potential source of acetogenins, and these researchers identified sonication to be an ideal method for extraction.

The freeze-dried fruit pulp extract exhibited analgesic and anti-inflammatory activities in rodents (Ishola *et al.*, 2014). Methanolic extracts (obtained by Soxhlet extraction) of seeds and pulp of *A. muricata* possess potential cytotoxic effect against cell lines of prostate (PC3) and cervical cancer (HeLa). Besides, moderate activity against breast cancer cell lines (MCF-7) were recorded (Raybaudi-Massilia *et al.*, 2015). Acetogenins like muricins M and N, and muricenin, have been reported to exhibit cytotoxic activity against human prostate cancer cells (PC-3) (Sun *et al.*, 2016). Aqueous extract of the fruit inhibited proliferation of breast cancer cells (T47D) with least toxicity in comparison with tamoxifen (Fidianingsih and Handayani, 2014). The bioactive acetogenin was identified as bullatacin, which was found to be 10^4 to 10^5 times more effective than doxorubicin against A549 and MCF-7 cell lines (Hopp *et al.*, 1997). Bullatacin was also found to be 300 times more active as taxol in L1210 murine leukemia-bearing mice (Ahammadsahib *et al.*, 1993). Acetogenins of the fruit extract induced cytotoxicity by inhibiting the mitochondrial complex I (NADH: ubiquinone oxidoreductase) of the electron transport chain which decreases ATP production causing apoptosis (Miyoshi *et al.*, 1998). In addition, acetogenin nano-suspension (144.4 nm particles) prepared using hydroxypropyl-beta-cyclodextrin and soybean lecithin significantly increased the cytotoxicity against both HeLa and HepG2 cancer cell lines compared to normal acetogenins. The cytotoxic effect of acetogenin nanosuspension in H22 tumour-bearing mice was 1/10th in comparison with normal acetogenins. Additionally, intravenous administration of the acetogenin nanoparticle shown good efficacy compared to the oral administration of the same (Hong *et al.*, 2016). Fruit extracts also selectively inhibited the human breast-cancer cell lines via down-regulation of EGFR (Dai *et al.*, 2011). Annonacin obtained from fruit extract led to growth arrest and apoptosis of MCF-7 cell lines and attenuated MCF-7 xenograft tumour growth in nude mice (Ko *et al.*, 2011). A beverage prepared using the pulp of soursop and blackberry with yogurt inactivated the breast tumour cells (MCF-7) and prostate tumour cells (PC3) along with exhibiting good antioxidant activity (Zambrano *et al.*, 2018). The methanolic and aqueous extracts of fruit pulp induced apoptosis as well as G_0/G_1 cell cycle arrest in MCF-7 cell lines. They also showed good antioxidant and anti-inflammatory activities (Prasad *et al.*, 2020). Methanolic extracts of fruit pulp showed potential anticancer activity against human hepatic cancer cells (HepG02) (Hemalatha *et al.*, 2020).

The fruit pulp extracted with methanol-acetone has been reported to exhibit potent antioxidant activity (González *et al.*, 2017). Fruit peel possessed significant chelating ability, hydroxyl radical scavenging activity and ferric reducing antioxidant properties compared to pulp (Akomolafe and Ajayi, 2015). The fruit pulp inhibited α-amylase and α-glucosidase and reduced the assimilation of glucose into blood in diabetic patients (Agu *et al.*, 2019). Administration of fruit juice to experimental rabbits revealed dose-dependent hypocholesterolemic activity (Jimoh *et al.*, 2018). The fruit extract also exhibited a neuroprotective effect by maintaining the transmission of various neurotransmitters in scopolamine-induced amnesia with the enhancement of cellular glutathione antioxidant enzymes in an animal model (Al-Brakati *et al.*, 2019).

The fruit pulp showed significant antibacterial activity against *Enterobacter aerogenes* and *Salmonella typhimurium*, and antifungal activity against *Colletotrichum gloeosporioides* and *Rhizopus stolonifer* (León-Fernández *et al.*, 2019). The fruit fermented for over a week exhibited higher antimicrobial activity in comparison with unfermented fruit (Otto *et al.*, 2015). In one of the studies, the pericarp of the soursop fruit significantly inhibited α-amylase and α-glucosidase enzymes, and showed antidiabetic activity and inhibition of angiotensin-converting enzymes under *in vitro* conditions. The result of this study also showed the free radical scavenging and ferrous chelating activities to be strongly correlated with polyphenolic compound and flavonoid content in the pericarp.

Alkaloids present in the fruit (e.g. annonaine, nor-nuciferine, asimilobine) are reported to possess antidepressant activity (proved in animal models), and this was attributed to their moderate affinity towards 5-HT_{1A} receptor and to the inhibitory effect on dopamine reuptake (Hasrat *et al.*, 1997). Freeze-dried unripe fruit extract did not show any mortality even at a dose of 4000 mg/kg in acute toxicity studies. The fruit extract with a dose of 200 mg/kg possessed peripheral and central analgesic activity through acting on opioid systems and exhibiting anti-inflammatory activity via inhibiting cyclooxygenase activity and nitric oxide generation (Ishola *et al.*, 2014). Soursop fruit extract mitigated caffeine-induced toxicity on weight of testes, epididymes, sperm motility, count and sperm head abnormalities. This effect was correlated to the presence of rich amounts of vitamin C and the antioxidant potency of the fruit (Ekaluo *et al.*, 2013).

Bioassay guided isolation of bioactives from soursop fruit seeds against parasitic *Leishmania* sp. like *donovani* and *Mexicana*, which led to the isolation of acetogenins like annonacinone and corossolone (Vila-Nova *et al.*, 2011). Annonacin isolated from pulp of soursop fruit is reported to induce nigral and striatal neurodegeneration *in vivo* (Champy *et al.*, 2004) and was also found to cause neuronal cell death in neuron culture (Escobar-Khondiker *et al.*, 2007). Acetogenin, a neurotoxic found in soursop fruit (annonacin), is the cause of 'Guadeloupean' Parkinsonism This is mainly because of neurodegeneration potency of the annonacin as it depletes the ATP supply to mitochondrial cells of rat striatal neurons (Escobar-Khondiker *et al.*, 2007; Bonneau *et al.*, 2015). It is also suggested that the amount annonacin accumulated by the daily consumption of soursop fruit for over a year (as recorded after administering intravenously in rats) can lead to brain lesions (Champy *et al.*, 2004).

Ethanol extracts of soursop fruit seeds in combination with *Piper nigrum* showed five times higher synergistic larvicidal activity (Grzybowski *et al.*, 2013). In addition, seed extracts demonstrated good larvicidal and insecticidal effects against *Aedes albopictus* and *Culex quinquefasciatus* (Ravaomanarivo *et al.*, 2014).

Implications of Food Processing and Preservation on Nutrients, Phytochemicals and Overall Fruit Quality

After attaining market maturity, the fruit are handpicked and washed with fresh or chlorinated water to remove adhering dust, soil particles and pesticides (if any) present on the surface. Later the fruit are peeled by hand followed by extraction using a blender without mixing the peel and without breaking the seeds as they contain toxic chemicals. The soursop fruit are usually processed into juices, ice creams, nectar (sweetened pulp) or sherbet (Pinto *et al.*, 2005). A refreshing drink is prepared in Cuba and Brazil using fruit pulp, milk and sugar, served as 'champol' (Badrie and Schauss, 2010). In these regions, the fruit pulp is used in preparation of ice cream, juice blends, sherbets, nectars, syrups, milkshakes, jams, jellies and yoghurts. The fruit powder is also used in the preparation of fruit bars and fruit flakes (Umme *et al.*, 1999, 2001; Gratão *et al.*, 2007).

The fruit pulp is processed and preserved either by freezing or by pasteurization for commercial purposes. If the pulp and nectar are pasteurized at 85°C and 90.6°C, respectively, they can be preserved for over a year. The quality of the processed products mainly depends on the amount of total sugars, vitamin C retained, pectinesterase activity, acidity and viscosity. Higher temperatures (around 93°C) decreased the quality of processed (frozen) fruit and this was opined to be altered by varying the vitamin C content and pectinesterase activity. During the preparation of juice, jam and marmalade, nearly 18% of the pulp is mixed with 11% of sugar and 0.02% of sodium benzoate and sodium metabisulfide. This is mixed with water and heated (at 100°C for 15 minutes). The soursop nectar (of 10 and 15%), on incorporation with yoghurt, showed good acceptance levels and had appreciable amounts of zinc, calcium and phosphorus (Lutchmedial *et al.*, 2004).

Pasteurization of soursop purée (at 78.8°C for a period of 69 s at a pH of 3.7) retained its sensory qualities along with improvement in the colour and appearance compared to non-pasteurized purée (Umme *et al.*, 1999). Soursop puree stored at –20°C showed higher stability in comparison with the puree stored at 4 and 15°C (Umme *et al.*, 1999). Addition of ascorbic acid to the pasteurized purée retained the flavour of the nectar; an antioxidant inhibited the polyphenol oxidase mediated pulp darkening of the fruit (Umme *et al.*, 2001).

For long-term preservation, the juice was subjected to pasteurization at 85°C followed by storing at various temperatures (4, 10 and 25°C). The pasteurized juice retained its consistency and quality without any visual changes for 12 weeks when preserved at temperatures of 4 and 10°C, whereas in the case of juice stored at 25°C after three to five weeks there was a decrease in the pH, total soluble solids and increased titratable acidity along with spoilage (Ampofo-Asiama and Quaye, 2019). In the majority of cases, soursop fruit juice viscosity is high and turbid. Addition of α-amylase to fruit juice reduced the viscosity along with a reduction in turbidity. This was attributed to the liquefaction of starch found in the fruit by α-amylase enzyme (Atolagbe *et al.*, 2016).

The amount of soursop fruit harvested globally is meeting only a small proportion of global demand, and this necessitates sustainable production and increased cultivation of this species. Around 30% of soursop fruit loss occurs during the postharvest period. After harvesting, the fruit need to be processed and preserved for longer, but various constraints like short storage life, the fragile nature of the peel, loss occurring in flavour due to thermo-sensitiveness, inactivation of enzymes (cell wall degrading and browning enzymes) and uneven ripening of fruit render these steps very tedious. When stored at 15°C, nearly nine days are required for ripening, while at 21°C ripening takes seven days and at 22–23°C it takes six days (Lima *et al.*, 2004).

Postharvest management technologies can help to overcome the problems associated with the marketing of the fruit only in the regions close to the cultivating areas and this helps to provide consumers with excellent quality of the product. Some of the techniques include selection of the fruit, disinfection, pre-cooling, drying of residual moisture, waxing, storage and transport (Jiménez-Zurita *et al.*, 2017). Understanding the postharvest physiology of soursop fruit is necessary for the establishment of handling procedures and for recognition of ideal packaging conditions. Fruit can be preserved for a long time at room temperature (25 ± 1°C) if subjected to postharvest preservation technologies like refrigeration, coating and modified atmosphere. In addition, accurate harvest time of the fruit is also important to find out the shelf life of the fruit. Soursop, being a climacteric fruit, have different periods of maturity and ripen at different intervals. Therefore, immature fruit are often harvested and subjected to forced ripening in storage. In Colombia, the fruit is verified for its maturity by pressing the fruit between the thumbs. Great care must be taken to ensure that the fruit is physiologically mature otherwise it can exhibit irregular maturation with off-taste. If fruit are harvested when they are ripened then they have reduced shelf life. It is not recommended to leave the fruit on the tree until maturation as it may fall and not meet the requirements for successful sale (Pareek *et al.*, 2011).

The shelf life of soursop fruit can be prolonged either by covering with waxes (carnauba wax, candelilla wax, polyethylene wax, etc.) or by refrigeration. In fruit covered by candelilla wax, maturity is delayed and the fruit weight loss during ripening reduced. Similarly, postharvest shelf life of fruit was extended up to eight days when fruit were stored at 15–22°C (Berumen-Varela *et al.*, 2019). Another study revealed a seven-day delay in reaching maturity after treating fruit with 1-methylcyclopropene (1-MCP) followed by storing at 16°C without chilling injury (Espinosa *et al.*, 2013). Soursop fruit, when treated with 1-MCP and polyethylene wax and stored at 15.4°C, retained total soluble solids, titratable acidity, pH, soluble sugars and reducing sugars during storage up to 15 days (Lima *et al.*, 2004). The emulsion prepared using 1-MCP with carnauba wax, candelilla wax and silicone oils when applied on soursop fruit delayed the ripening up to 15 days of storage compared to a control batch reaching ripening stage within six to seven days (Tovar-Gómez *et al.*, 2011). The flash-pasteurized nectar could be stored for a period of a year at 30°C without any noticeable loss in the quality of the nectar (Benero *et al.*, 1974). Soursop purée if maintained at −23°C, is reported to be preserved for nearly 400 days (Pinto *et al.*, 2005).

Soursop fruit when coated with the emulsion of 1-methylcyclopropene (1-MCP) and beeswax (15:85 v/v) were preserved for a period of 14–15 days at 16°C when compared to the preservation at 25°C for only a period of six days (Moreno-Hernández *et al.*, 2014). Application of 1.0 and 1.5% of chitosan on harvested soursop fruit prevented fungal infection and weight loss after three, six and nine days. Additionally, the pH, total soluble solids, firmness and titratable acidity was not altered by chitosan application (Ramos-Guerrero *et al.*, 2020).

Browning of soursop is very common as it is a climacteric fruit and browning of fruit is mainly catalysed by the enzyme polyphenol oxidase. Browning of fruit along with the preservation of its nutrients and flavour can be ensured by adapting advanced green technologies like microwaves and ultra-sonication. Exposure to microwaves reduced polyphenol oxidase enzyme activity in the fruit by 57% at 70 W in 30 s and ultrasound treatment reduced the activity of polyphenol oxidase by 43% at 120 W in 220 s (Palma-Orozco *et al.*, 2019). So, soursop fruit pulp can be preserved for an extended time by converting it into dry powder. Fruit pulp dried by spray-drying method maintained antioxidant potency and nutritional value with minimal changes in aroma and volatile components (Neta *et al.*, 2019).

Conclusion and Outlook

Soursop is a recognized nutraceutically valued fruit imparting rich health benefits. The edible pulp encompasses appreciable amounts of vital minerals, vitamins and antioxidant-rich polyphenolic compounds. The fruit imparts a rich exotic flavour that makes it popular. Since the fruit is widely consumed, development of quantitative assays to identify bioactives is needed. Although soursop is a good nutritional source, the presence of various toxic acetogenins affects dopaminergic and other neurons that are responsible/contribute towards developing atypical Parkinsonism when this fruit is consumed in large quantities. The neurotoxic effect is restricted only to the isolated acetogenins, not the pulp of the fruit, and hence future study is required to ensure the safety of the fruit. However, these acetogenins are also reported to possess cytotoxic potency against various human cancer cell lines that necessitates studies to decide the amount of fruit to be consumed to obtain the required therapeutic effect with minimal side effects. When it comes to overall safety, the amount of soursop fruit consumption mainly depends on the individual's age and associated comorbid diseases, if any. Consumption of this fruit needs to be strictly avoided among pregnant and breastfeeding mothers. There is great scope to explore the waste and by-products of soursop fruit with a view to obtaining value-added bioactive compounds. Overall, to conclude, soursop fruit can be a power-packed nutraceutically valued source if consumed with certain precautions.

References

Adefegha, S.A., Oyeleye, S.I. and Oboh, G. (2015) Distribution of phenolic contents, antidiabetic potentials, antihypertensive properties, and antioxidative effects of soursop (*Annona muricata* L.) fruit parts in vitro. *Biochemistry Research International* 2015, 347673. DOI: 10.1155/2015/347673.

Agu, K.C., Eluehike, N., Ofeimun, R.O., Abile, D., Ideho, G. *et al.* (2019) Possible anti-diabetic potentials of *Annona muricata* (soursop): inhibition of α-amylase and α-glucosidase activities. *Clinical Phytoscience* 5(1), 1–13. DOI: 10.1186/s40816-019-0116-0.

Aguilar-Hernández, G., García-Magaña, M. de L., Vivar-Vera, M. de L.Á., Sáyago-Ayerdi, S.G., Sánchez-Burgos, J.A. *et al.* (2019) Optimization of ultrasound-assisted extraction of phenolic compounds from *Annona muricata* by-products and pulp. *Molecules* 24(5), 904. DOI: 10.3390/molecules24050904.

Aguilar-Hernández, G., Vivar-Vera, M. de L.Á., García-Magaña, M. de L., González-Silva, N., Pérez-Larios, A. *et al.* (2020a) Ultrasound-assisted extraction of total acetogenins from the soursop fruit by Response Surface Methodology. *Molecules* 25(5), 1139. DOI: 10.3390/molecules25051139.

Ahammadsahib, K.I., Hollingworth, R.M., McGovren, J.P., Hui, Y.H. and McLaughlin, J.L. (1993) Mode of action of bullatacin: a potent antitumor and pesticidal annonaceous acetogenin. *Life Sciences* 53(14), 1113–1120. DOI: 10.1016/0024-3205(93)90547-g.

Akomolafe, S.F. and Ajayi, O.B. (2015) A comparative study on antioxidant properties, proximate and mineral compositions of the peel and pulp of ripe *Annona muricata* (L.) fruit. *International Food Research Journal* 22(6), 2381–2388.

Al-Brakati, A.Y., Al Omairi,, Kassab, R.B., Lokman, M.S., Elmahallawy, E.K. *et al.* (2019) Soursop fruit extract mitigates scopolamine-induced amnesia and oxidative stress via activating cholinergic and Nrf2/HO-1 pathways. *Metabolic Brain Disease* 34(3), 853–864. DOI: 10.1007/s11011-019-00407-2.

Alali, F.Q., Liu, X.X. and McLaughlin, J.L. (1999) Annonaceous acetogenins: recent progress. *Journal of Natural Products* 62(3), 504–540. DOI: 10.1021/np980406d.

Ampofo-Asiama, J. and Quaye, B. (2019) Effect of storage temperature on the physicochemical, nutritional and microbiological quality of pasteurised soursop (*Annona muricata* L.) Juice. *African Journal of Food Science* 13, 38–47.

Arbaisah, S.M., Asbi, B.A., Junainah, A.H. and Jamilah, B. (1997) Purification and properties of pectinesterase from soursop (*Anona muricata*) pulp. *Food Chemistry* 59(1), 33–40. DOI: 10.1016/S0308-8146(96)00043-X.

Atolagbe, O.M., Ajayi, A.A. and Edegbo, O. (2016) Characterization of α-amylase from soursop (*Annona muricata* Linn.) fruits degraded by *Rhizopus stolonifer*. *Pakistan Journal of Biological Sciences* 19(2), 77–81. DOI: 10.3923/pjbs.2016.77.81.

Augusto, F., Valente, A.L., dos Santos Tada, E. and Rivellino, S.R. (2000) Screening of Brazilian fruit aromas using solid-phase microextraction-gas chromatography-mass spectrometry. *Journal of Chromatography. A* 873(1), 117–127. DOI: 10.1016/s0021-9673(99)01282-0.

Badrie, N. and Schauss, A.G. (2010) Soursop (*Annona muricata* L.): Composition, nutritional value, medicinal uses, and toxicology. In: Watson, R. and Preedy, V. (eds) *Bioactive Foods in Promoting Health*. Academic Press, Cambridge, Massachusetts, USA, pp. 621–643.

Benero, J.R., Collazo de Rivera, A.L. and De George, L.M.I. (1974) Studies on the preparation and shelf-life of soursop, tamarind, and blended soursop-tamarind soft drinks. *The Journal of Agriculture of the University of Puerto Rico* 58(1), 99–104. DOI: 10.46429/jaupr.v58i1.10709.

Berumen-Varela, G., Hernández-Oñate, M.A. and Tiznado-Hernández, M.E. (2019) Utilization of biotechnological tools in soursop (*Annona muricata* L.). *Scientia Horticulturae* 245, 269–273. DOI: 10.1016/j.scienta.2018.10.028.

Bezerra, V.S., de Lima Filho, J.L., Montenegro, M.C.B.S.M., Araújo, A.N. and da Silva, V.L. (2003) Flow-injection amperometric determination of dopamine in pharmaceuticals using a polyphenol oxidase biosensor obtained from soursop pulp. *Journal of Pharmaceutical and Biomedical Analysis* 33(5), 1025–1031. DOI: 10.1016/s0731-7085(03)00412-6.

Blancas-Benítez, F.J., Montalvo-González, E., González-Aguilar, G.A. and Sáyago-Ayerdi, S.G. (2019) In vitro bioaccessibility and release kinetics of phenolic compounds from guava (*Psidium guajava* L.) and soursop (*Annona muricata* L.) pulp. *TIP Revista Especializada En Ciencias Químico-Biológicas* 22, 1–7.

Bonneau, N., Cynober, T., Jullian, J.C. and Champy, P. (2017) ^{1}H qNMR quantification of annonaceous acetogenins in crude extracts of Annona muricata L. fruit pulp. *Phytochemical Analysis* 28(4), 251–256. DOI: 10.1002/pca.2668.

Bonneau, N., Schmitz-Afonso, I., Brunelle, A., Touboul, D. and Champy, P. (2015) Method development for quantification of the environmental neurotoxin annonacin in Rat plasma by UPLC-MS/MS and application to a pharmacokinetic study. *Journal of Chromatography. B, Analytical Technologies in the Biomedical and Life Sciences* 1004, 46–52. DOI: 10.1016/j.jchromb.2015.09.039.

Champy, P., Höglinger, G.U., Féger, J., Gleye, C., Hocquemiller, R. *et al.* (2004) Annonacin, a lipophilic inhibitor of mitochondrial complex I, induces nigral and striatal neurodegeneration in rats: possible relevance for atypical parkinsonism in Guadeloupe. *Journal of Neurochemistry* 88(1), 63–69. DOI: 10.1046/j.1471-4159.2003.02138.x.

Chih, H.W., Chiu, H.F., Tang, K.S., Chang, F.R. and Wu, Y.C. (2001) Bullatacin, a potent antitumor annonaceous acetogenin, inhibits proliferation of human hepatocarcinoma cell line 2.2.15 by apoptosis induction. *Life Sciences* 69(11), 1321–1331. DOI: 10.1016/s0024-3205(01)01209-7.

Coria-Téllez, A.V., Montalvo-González, E. and Obledo-Vázquez, E.N. (2017) Soursop (*Annona Muricata*). In: Yahia, E.M. (ed.) *Fruit and Vegetable Phytochemicals: Chemistry and Human Health*, 2nd edn. Wiley-Blackwell, Chichester, UK, pp. 1243–1252.

Coria-Téllez, A.V., Montalvo-Gónzalez, E., Yahia, E.M. and Obledo-Vázquez, E.N. (2018) *Annona muricata*: A comprehensive review on its traditional medicinal uses, phytochemicals, pharmacological activities, mechanisms of action and toxicity. *Arabian Journal of Chemistry* 11(5), 662–691. DOI: 10.1016/j.arabjc.2016.01.004.

Daddiouaissa, D. and Amid, A. (2018) Anticancer activity of acetogenins from *Annona Muricata* fruit. *IIUM Medical Journal Malaysia* 17(3). DOI: 10.31436/imjm.v17i3.236.

Dai, Y., Hogan, S., Schmelz, E.M., Ju, Y.H., Canning, C. *et al.* (2011) Selective growth inhibition of human breast cancer cells by graviola fruit extract *in vitro* and *in vivo* involving downregulation of EGFR expression. *Nutrition and Cancer* 63(5), 795–801. DOI: 10.1080/01635581.2011.563027.

Ekaluo, U.B., Ikpeme, E.V., Ibiang, Y.B. and Omordia, F.O. (2013) Effect of soursop (*Annona muricata* L.) fruit extract on sperm toxicity Induced by caffeine in albino rats. *Journal of Medical Sciences* 13(1), 67–71. DOI: 10.3923/jms.2013.67.71.

Escobar-Khondiker, M., Höllerhage, M., Muriel, M.-P., Champy, P., Bach, A. *et al.* (2007) Annonacin, a natural mitochondrial complex I inhibitor, causes tau pathology in cultured neurons. *The Journal of Neuroscience* 27(29), 7827–7837. DOI: 10.1523/JNEUROSCI.1644-07.2007.

Espinosa, I., Ortiz, R.I., Tovar, B., Mata, M. and Montalvo, E. (2013) Physiological and physicochemical behavior of soursop fruits refrigerated with 1-methylcyclopropene. *Journal of Food Quality* 36(1), 10–20. DOI: 10.1111/jfq.12011.

Fidianingsih, I. and Handayani, E.S. (2014) *Annona muricata* aqueous extract suppresses T47D breast cancer cell proliferation. *Universa Medicina* 33(1), 19–26.

González, E.M., Fernández, A.E.L., Sáyago-Ayerdi, S.G., Estrada, R.M.V. and Vallejo, L.G.Z. (2017) In vitro antioxidant capacity of crude extracts and acetogenin fraction of soursop fruit pulp. *Pharmaceutica Analytica Acta* 8, 1–7.

Gratão, A.C.A., Silveira, V. and Telis-Romero, J. (2007) Laminar flow of soursop juice through concentric annuli: Friction factors and rheology. *Journal of Food Engineering* 78(4), 1343–1354. DOI: 10.1016/j.jfoodeng.2006.01.006.

Grzybowski, A., Tiboni, M., Silva, M.A.N., Chitolina, R.F., Passos, M. *et al.* (2013) Synergistic larvicidal effect and morphological alterations induced by ethanolic extracts of *Annona muricata* and *Piper nigrum* against the dengue fever vector *Aedes aegypti*. *Pest Management Science* 69(5), 589–601. DOI: 10.1002/ps.3409.

Hajdu, Z. and Hohmann, J. (2012) An ethnopharmacological survey of the traditional medicine utilized in the community of Porvenir, Bajo Paraguá Indian Reservation, Bolivia. *Journal of Ethnopharmacology* 139(3), 838–857. DOI: 10.1016/j.jep.2011.12.029.

Hasrat, J.A., De Bruyne, T., De Backer, J.P., Vauquelin, G. and Vlietinck, A.J. (1997) Isoquinoline derivatives isolated from the fruit of *Annona muricata* as 5-HTergic 5-HT1A receptor agonists in rats: unexploited antidepressive (lead) products. *The Journal of Pharmacy and Pharmacology* 49(11), 1145–1149. DOI: 10.1111/j.2042-7158.1997.tb06058.x.

Hemalatha, G., Sivakumari, K., Rajesh, S. and Shyamala Devi, K. (2020) Phytochemical profiling, anticancer and apoptotic activity of Graviola (*Annona muricata*) fruit extract against human hepatocellular carcinoma (HepG-2) cells. *International Journal of Zoology and Applied Biosciences* 5, 32–47.

Höllerhage, M., Rösler, T.W., Berjas, M., Luo, R., Tran, K. *et al.* (2015) Neurotoxicity of dietary supplements from Annonaceae species. *International Journal of Toxicology* 34(6), 543–550. DOI: 10.1177/1091581815602252.

Hong, J., Li, Y., Xiao, Y., Li, Y., Guo, Y. *et al.* (2016) Annonaceous acetogenins (ACGs) nanosuspensions based on a self-assembly stabilizer and the significantly improved anti-tumor efficacy. *Colloids and Surfaces. B, Biointerfaces* 145, 319–327. DOI: 10.1016/j.colsurfb.2016.05.012.

Hopp, D.C., Zeng, L., Gu, Z.M., Kozlowski, J.F. and McLaughlin, J.L. (1997) Novel mono-tetrahydrofuran ring acetogenins, from the bark of *Annona squamosa*, showing cytotoxic selectivities for the human pancreatic carcinoma cell line, PACA-2. *Journal of Natural Products* 60(6), 581–586. DOI: 10.1021/np9701283.

Ishola, I.O., Awodele, O., Olusayero, A.M. and Ochieng, C.O. (2014) Mechanisms of analgesic and anti-inflammatory properties of *Annona muricata* Linn. (Annonaceae) fruit extract in rodents. *Journal of Medicinal Food* 17(12), 1375–1382. DOI: 10.1089/jmf.2013.0088.

Jiménez-Zurita, J.O., Balois-Morales, R., Alia-Tejacal, I., Juárez-López, P., Jiménez-Ruíz, E.I. *et al.* (2017) Topics of postharvest handling of soursop fruit (*Annona muricata* L.). *Revista Mexicana de Ciencias Agrícolas* 8, 1155–1167. DOI: 10.29312/remexca.v8i5.115.

Jimoh, O.A., Ayedun, E.S., Oyelade, W.A., Oloruntola, O.D., Daramola, O.T. *et al.* (2018) Protective effect of soursop (*Annona muricata* linn.) juice on oxidative stress in heat stressed rabbits. *Journal of Animal Science and Technology* 60, 28. DOI: 10.1186/s40781-018-0186-4.

Jirovetz, L., Buchbauer, G. and Ngassoum, M.B. (1998) Essential oil compounds of the *Annona muricata* fresh fruit pulp from Cameroon. *Journal of Agricultural and Food Chemistry* 46(9), 3719–3720. DOI: 10.1021/jf980204n.

Khan, M.R., Kornine, K. and Omoloso, A.D. (1998) Antibacterial activity of some Annonaceae. *Part1. Fitoterapia (Milano)* (69), 367–369.

Komansilan, A., Abadi, A.L., Yanuwiadi, B. and Kaligis, D.A. (2012) Isolation and identification of biolarvicide from soursop (*Annona muricata* linn) seeds to mosquito (*Aedes aegypti*) larvae. *International Journal of Engineering & Technology* 12, 28–32.

Ko, Y.M., Wu, T.Y., Wu, Y.C., Chang, F.R., Guh, J.Y. *et al.* (2011) Annonacin induces cell cycle-dependent growth arrest and apoptosis in estrogen receptor-α-related pathways in MCF-7 cells. *Journal of Ethnopharmacology* 137(3), 1283–1290. DOI: 10.1016/j.jep.2011.07.056.

Lako, J., Trenerry, V.C., Wahlqvist, M., Wattanapenpaiboon, N., Sotheeswaran, S. *et al.* (2007) Phytochemical flavonols, carotenoids and the antioxidant properties of a wide selection of Fijian fruit, vegetables and other readily available foods. *Food Chemistry* 101(4), 1727–1741. DOI: 10.1016/j.foodchem.2006.01.031.

Lannuzel, A., Höglinger, G.U., Verhaeghe, S., Gire, L., Belson, S. *et al.* (2007) Atypical parkinsonism in Guadeloupe: A common risk factor for two closely related phenotypes? *Brain : A Journal of Neurology* 130(Pt 3), 816–827. DOI: 10.1093/brain/awl347.

León-Fernández, A.E., Martínez-Cárdenas, L., Zepeda-Vallejo, L.G., Arteaga-Garibay, R.I., Gutiérrez-Martínez, P. *et al.* (2019) Antibacterial, antifungal, antioxidant and toxic effect of fractioned extracts from soursop pulp. *Revista Bio Ciencias* (6), 1–17.

Lima, M.A.C.D., Alves, R.E., Filgueiras, H.A.C. and Lima, J.R.G. (2004) The use of wax and 1-methylcyclopropene on refrigerated storage of soursop fruit (*Annona muricata* L.). *Revista Brasileira de Fruticultura* 26, 433–437.

Lutchmedial, M., Ramlal, R., Badrie, N. and Chang-Yen, I. (2004) Nutritional and sensory quality of stirred soursop (*Annona muricata* L.) yoghurt. *International Journal of Food Sciences and Nutrition* 55(5), 407–414. DOI: 10.1080/09637480400002800.

Márquez Cardozo, C.J., Villacorta Lozano, V., Yepes Betancur, D.P., Ciro Velásquez, H.J. and Cartagena Valenzuela, J.R. (2012) Physiological and physico-chemical characterization of the soursop fruit (*Annona muricata* L. cv Elita). *Revista Facultad Nacional de Agronomía Medellín* 65, 6477–6486.

Márquez Cardozo, C.J., Cartagena Valenzuela, J.R. and Correa Londoño, G.A. (2013) Determination of soursop (*Annona muricata* L. cv. Elita) fruit volatiles during ripening by electronic nose and gas chromatography coupled to mass spectroscopy. *Revista Facultad Nacional de Agronomía Medellín* 66, 7117–7128.

Menezes, E.G.T., Oliveira, É.R., Carvalho, G.R., Guimarães, I.C. and Queiroz, F. (2019) Assessment of chemical, nutritional and bioactive properties of *Annona crassiflora* and *Annona muricata* wastes. *Food Science and Technology* 39(suppl 2), 662–672. DOI: 10.1590/fst.22918.

Mishra, S., Ahmad, S., Kumar, N. and Sharma, B.K. (2013) *Annona muricata* (the cancer killer): A review. *The Global Journal of Pharmaceutical Research* 2, 1613–1618.

Miyoshi, H., Ohshima, M., Shimada, H., Akagi, T., Iwamura, H. *et al.* (1998) Essential structural factors of annonaceous acetogenins as potent inhibitors of mitochondrial complex I. *Biochimica et Biophysica Acta* 1365(3), 443–452. DOI: 10.1016/s0005-2728(98)00097-8.

Moghadamtousi, S.Z., Fadaeinasab, M., Nikzad, S., Mohan, G., Ali, H.M. *et al.* (2015) *Annona muricata* (Annonaceae): A review of its traditional uses, isolated acetogenins and biological activities. *International Journal of Molecular Sciences* 16(7), 15625–15658. DOI: 10.3390/ijms160715625.

Moo-Huchin, V.M., Estrada-Mota, I., Estrada-León, R., Cuevas-Glory, L., Ortiz-Vázquez, E. *et al.* (2014) Determination of some physicochemical characteristics, bioactive compounds and antioxidant activity of tropical fruits from Yucatan, Mexico. *Food Chemistry* 152, 508–515. DOI: 10.1016/j.foodchem.2013.12.013.

Moreno-Hernández, C.L., Sáyago-Ayerdi, S.G., García-Galindo, H.S., Mata-Montes De Oca, M. and Montalvo-González, E. (2014) Effect of the application of 1-methylcyclopropene and wax emulsions on proximate analysis and some antioxidants of soursop (*Annona muricata* L.). *The Scientific World Journal* 2014, 1–7. DOI: 10.1155/2014/896853.

Neta, M.T.S.L, de Jesus, M.S., da Silva, J.L.A., Araujo, H.C.S., Sandes, R.D.D. *et al.* (2019) Effect of spray drying on bioactive and volatile compounds in soursop (*Annona muricata*) fruit pulp. *Food Research International* 124, 70–77. DOI: 10.1016/j.foodres.2018.09.039.

Omoifo, C.O. (2004) Biochemical composition of soursop fruit, *Annona muricata* L., as affected by two harvest seasons. *Tropical Agricultural Research and Extension* 7, 1–8.

Ong, H.G. and Kim, Y.-D. (2014) Quantitative ethnobotanical study of the medicinal plants used by the Ati Negrito indigenous group in Guimaras island, Philippines. *Journal of Ethnopharmacology* 157, 228–242. DOI: 10.1016/j.jep.2014.09.015.

Onyechi, U., Ibeanu, U., Nkiruka, V., Eme, E.P. and Madubike, K. (2012) Nutrient phytochemical composition and sensory evaluation of soursop (*Annona muricata*) pulp and drink in South Eastern Nigeria. *International Journal of Basic and Applied Sciences* 12, 53–57.

Osorio, E., Arango, G.J., Jiménez, N., Alzate, F., Ruiz, G. *et al.* (2007) Antiprotozoal and cytotoxic activities in vitro of Colombian Annonaceae. *Journal of Ethnopharmacology* 111(3), 630–635. DOI: 10.1016/j.jep.2007.01.015.

Otto, R.B., Nankwanga, M. and Sesaazi, D. (2015) Comparison of antibacterial activities of fermented with those of unfermented *Annona muricata* (L) fruit extracts. *International Journal of Current Microbiology and Applied Sciences* 4, 696–707.

Padmanabhan, P. and Paliyath, G. (2016) Annonaceous Fruits. In: Caballero, B., Finglas, P.M., and Toldrá, F. (eds) *Encyclopedia of Food and Health*. Academic Press, Cambridge, Massachusetts, USA, pp. 169–173.

Palma-Orozco, G., Marrufo-Hernández, N.A., Tobías, I. and Nájera, H. (2019) Purification and biochemical characterization of polyphenol oxidase from soursop (*Annona muricata* L.) and its inactivation by microwave and ultrasound treatments. *Journal of Food Biochemistry* 43(3), e12770. DOI: 10.1111/jfbc.12770.

Pareek, S., Yahia, E.M., Pareek, O.P. and Kaushik, R.A. (2011) Postharvest physiology and technology of Annona fruits. *Food Research International* 44(7), 1741–1751. DOI: 10.1016/j.foodres.2011.02.016.

Passos, T.U., Sampaio, H.A. de C., Sabry, M.O.D., Melo, M.L.P. de, Coelho, M.A.M. *et al.* (2015) Glycemic index and glycemic load of tropical fruits and the potential risk for chronic diseases. *Food Science and Technology (Campinas)* 35(1), 66–73. DOI: 10.1590/1678-457X.6449.

Paull, R.E., Deputy, J. and Chen, N.J. (1983) Changes in organic acids, sugars, and headspace volatiles during fruit ripening of soursop (*Annona muricata* L.). *Journal of the American Society for Horticultural Science* 108, 931–934.

Pino, J.A., Agüero, J. and Marbot, R. (2001) Volatile components of soursop (*Annona muricata* L.). *Journal of Essential Oil Research* 13, 140–141.

Pinto, A.C. de Q., Cordeiro, M.C.R., de Andrade, S.R.M., Ferreira, F.R., de C. Filgueiras, H.A. *et al.* (2005) Annona species. In: Williams, J.T., Smith, R.W., Hughes, A., Haq, N., and Clement, C.R. (eds) *Annona Sp. Monograph*. International Center for Underutilized Crops, Southampton, UK, pp. 1–123.

Prasad, S.K., Veeresh, P.M., Ramesh, P.S., Natraj, S.M., Madhunapantula, S.V. *et al.* (2020) Phytochemical fractions from *Annona muricata* seeds and fruit pulp inhibited the growth of breast cancer cells through cell cycle arrest at G0/G1 phase. *Journal of Cancer Research and Therapeutics* 16, 1235.

Rabelo, S.V., Quintans, J.D.S.S., Costa, E.V., da Silva Almeida, J.R.G., Júnior, L.J.Q. *et al.* (2016) *Annona* species (Annonaceae) oils. In: Prredy, V. (ed.) *Essential Oils in Food Preservation, Flavor and Safety*. Academic Press, Cambridge, Massachusetts, USA, pp. 221–229.

Ragasa, C.Y., Soriano, G., Torres, O.B., Don, M.J. and Shen, C.C. (2012) Acetogenins from *Annona muricata*. *Pharmacognosy Journal* 4(32), 32–37. DOI: 10.5530/pj.2012.32.7.

Ramos-Guerrero, A., González-Estrada, R.R., Romanazzi, G., Landi, L., Gutiérrez-Martínez, P. *et al.* (2020) Effects of chitosan in the control of postharvest anthracnose of soursop (*Annona muricata*) fruit. *Revista Mexicana de Ingeniería Química* 19(1), 99–108. DOI: 10.24275/rmiq/Bio527.

Ravaomanarivo, L.H.R., Razafindraleva, H.A., Raharimalala, F.N., Rasoahantaveloniaina, B., Ravelonandro, P.H. *et al.* (2014) Efficacy of seed extracts of *Annona squamosa* and *Annona muricata* (Annonaceae) for the control of *Aedes albopictus* and *Culex quinquefasciatus* (Culicidae). *Asian Pacific Journal of Tropical Biomedicine* 4(10), 798–806. DOI: 10.12980/APJTB.4.2014C1264.

Raybaudi-Massilia, R., Suárez, A., Arvelo, F., Sojo, F., Mosqueda-Melgar, J. *et al.* (2015) An analysis in-vitro of the cytotoxic, antioxidant and antimicrobial activity of aqueous and alcoholic extracts of *Annona muricata* L. seed and pulp. *British Journal of Applied Science & Technology* 5(4), 333–341. DOI: 10.9734/BJAST/2015/13587.

Rieser, M.J., Fang, X.P., Rupprecht, J.K., Hui, Y.H., Smith, D.L. *et al.* (1993) Bioactive single-ring acetogenins from seed extracts of *Annona muricata*. *Planta Medica* 59(1), 91–92. DOI: 10.1055/s-2006-959614.

Rubio-Melgarejo, A., Balois-Morales, R., Palomino-Hermosillo, Y.A., López-Guzmán, G.G., Ramírez-Ramírez, J.C. *et al.* (2020) Phytochemical and antioxidant dynamics of the soursop fruit (*Annona muricata* L.) in response to *colletotrichum* spp. *Journal of Food Quality* 2020. DOI: 10.1155/2020/3180634.

Santana, K.L. de, Galvão, M. de S., Jesus, M.S. de, Nogueira, J.P. and Narain, N. (2017) HS-SPME optimization and extraction of volatile compounds from soursop (*Annona muricata* L.) pulp with emphasis on their characteristic impact compounds. *Food Science and Technology* 37(2), 250–260. DOI: 10.1590/1678-457x.20916.

Shashirekha, M.N., Baskaran, R., Jaganmohan Rao, L., Vijayalakshmi, M.R. and Rajarathnam, S. (2008) Influence of processing conditions on flavour compounds of custard apple (*Annona squamosa* L.). *LWT - Food Science and Technology* 41(2), 236–243. DOI: 10.1016/j.lwt.2007.03.005.

Sun, S., Liu, J., Kadouh, H., Sun, X. and Zhou, K. (2014) Three new anti-proliferative Annonaceous acetogenins with mono-tetrahydrofuran ring from graviola fruit (*Annona muricata*). *Bioorganic & Medicinal Chemistry Letters* 24(12), 2773–2776. DOI: 10.1016/j.bmcl.2014.03.099.

Sun, S., Liu, J., Zhou, N., Zhu, W., Dou, Q.P. *et al.* (2016) Isolation of three new annonaceous acetogenins from Graviola fruit (*Annona muricata*) and their anti-proliferation on human prostate cancer cell PC-3. *Bioorganic & Medicinal Chemistry Letters* 26(17), 4382–4385. DOI: 10.1016/j.bmcl.2015.06.038.

Sun, S., Liu, J., Sun, X., Zhu, W., Yang, F. *et al.* (2017) Novel Annonaceous acetogenins from Graviola (*Annona muricata*) fruits with strong anti-proliferative activity. *Tetrahedron Letters* 58(19), 1895–1899. DOI: 10.1016/j.tetlet.2017.04.016.

Tovar-Gómez, B., Mata-Montes de Oca, M., García-Galindo, H.S. and Montalvo-González, E. (2011) Efecto de emulsiones de cera y 1-metilciclopropeno en la conservación poscosecha de guanabana. *Revista Chapingo. Serie Horticultura* 17(SPE1), 53–61.

Umme, A., Salmah, Y., Jamilah, B. and Asbi, B.A. (1999) Microbial and enzymatic changes in natural soursop puree during storage. *Food Chemistry* 65(3), 315–322. DOI: 10.1016/S0308-8146(98)00215-5.

Umme, A., Bambang, S.S., Salmah, Y. and Jamilah, B. (2001) Effect of pasteurisation on sensory quality of natural soursop puree under different storage conditions. *Food Chemistry* 75(3), 293–301. DOI: 10.1016/S0308-8146(01)00151-0.

Vandebroek, I., Balick, M.J., Ososki, A., Kronenberg, F., Yukes, J. *et al.* (2010) The importance of botellas and other plant mixtures in Dominican traditional medicine. *Journal of Ethnopharmacology* 128(1), 20–41. DOI: 10.1016/j.jep.2009.12.013.

Vila-Nova, N.S., Morais, S.M., Falcão, M.J.C., Machado, L.K.A., Beviláqua, C.M.L. *et al.* (2011) Leishmanicidal activity and cytotoxicity of compounds from two *Annonacea* species cultivated in Northeastern Brazil. *Revista Da Sociedade Brasileira de Medicina Tropical* 44(5), 567–571. DOI: 10.1590/s0037-86822011000500007.

Yamada, E.S., Respondek, G., Müssner, S., de Andrade, A., Höllerhage, M. *et al.* (2014) Annonacin, a natural lipophilic mitochondrial complex I inhibitor, increases phosphorylation of tau in the brain of FTDP-17 transgenic mice. *Experimental Neurology* 253, 113–125. DOI: 10.1016/j.expneurol.2013.12.017.

Zafra-Polo, M.C., Figadère, B., Gallardo, T., Tormo, JoséR. and Cortes, D. (1998) Natural acetogenins from annonaceae, synthesis and mechanisms of action. *Phytochemistry* 48(7), 1087–1117. DOI: 10.1016/S0031-9422(97)00917-5.

Zambrano, A., Raybaudi-Massilia, R., Arvelo, F. and Sojo, F. (2018) Cytotoxic and antioxidant properties in vitro of functional beverages based on blackberry (*Rubus glaucus* Benth) and soursop (*Annona muricata* L) pulps. *Functional Foods in Health and Disease* 8(11), 531. DOI: 10.31989/ffhd.v8i11.541.

30 Dabai (*Canarium odontophyllum* Miq.)

Lee-Hoon Ho[1], Thuan-Chew Tan[2] and Rajeev Bhat[3*]
[1]*Faculty of Bioresources and Food Industry, Universiti Sultan Zainal Abidin, Malaysia;* [2]*Food Technology Division, School of Industrial Technology, Universiti Sains Malaysia, Malaysia;* [3]*ERA-Chair for Food (By-) Products Valorization Technologies, Estonian University of Life Sciences, Tartu, Estonia*

Abstract
The present chapter is designed to provide information on the potential uses of dabai, an indigenous fruit of Borneo (Sarawak region of Malaysia). This fruit holds high nutraceutical value, is a vital source of food and possesses potential therapeutic properties. Apart from describing the nutritional composition, in this chapter we have focused on the bioactive compounds, antioxidant activities and health benefits of dabai fruit. Further, this chapter discusses the progress made on the potential uses of dabai fruit as a functional ingredient in the development of novel food products. The implications of postharvest storage and processing on the physicochemical qualities of dabai fruit are also addressed and details are provided on the by-products of dabai (post-oil production), including defatted pulp, peel and seed portions that can be investigated with the aim of obtaining bioactive polyphenolic compounds. Maximal utilization of dabai fruit with minimal waste generation can support the concept of a circular economy as well as ensure food security in a region. We hope the information provided in this chapter may popularize this fruit and contribute to its commercial exploitation in the development of functional food products and other applications in the pharmaceutical and cosmeceutical industries.

Introduction

According to the Food and Agriculture Organization of the United Nations (FAO), to ensure global food security, it is vital to recognize the contributions of natural resources, intensify the consumption of nutritious foods and healthy diets, improve living standards among rural communities and enhance the resilience of those communities (FAO, 2019). Accordingly, sustainable use of natural resources possessing food and therapeutic value is vital to support global wellbeing.

In Malaysia, mainly in the Borneo/Sarawak region (a state located in the eastern part of Malaysia) a wide diversity of natural flora offers a range of indigenous edible fruit and vegetables to the local/rural communities. Some of these indigenous fruit and vegetables are commercially cultivated and are a part of the Sarawakian diet. However, many indigenous fruit with nutraceutical potential remains underexploited or underutilized. There are several indigenous popular fruit of the Sarawak region, such as asam embang, belimbing merah (wild starfruit), dabai, engkala, keranji, mata kucing and salak. These fruit are a good source of essential

*Corresponding author: rajeev.bhat@emu.ee

DOI: 10.1079/9781789248067.0030

macro- and microminerals, as well as being rich in vitamins (Chua *et al.*, 2015; Khoo *et al.*, 2016; Tan and Azlan, 2016). Recently, some researchers have reviewed selected indigenous fruit of the region, such as bambangan and durian nyekak, and have reported them to contain an abundance of nutrients and phytochemicals, mainly those of polyphenolic compounds. Among the indigenous fruit, dabai (*Canarium odontophyllum* Miq.) is one of the most popular and has great potential to be developed as a 'super fruit' of the region. Currently, oweing to its nutritional quality (Pin and Daniel, 2017), the Agriculture Department of Sarawak has promoted dabai fruit as a 'specialty fruit'. Due to several findings and reports on the health benefits of dabai fruit to humans, studies on the use of dabai as a functional ingredient while developing novel food products have gained much popularity. Dabai fruit have been used as a functional ingredient in bakery-based products such as in the preparation of cakes, cookies, jams, etc. Incorporation of the fruit extract has not only enhanced the taste/flavour of the products but also increased the nutritional value. In Sarawak, the edible part of dabai is also incorporated into local cuisine, such as fried rice and omelette. The fruit pulp has been developed to produce sauces and pastes to be used as ingredients in cooking (Khoo *et al.*, 2016). Traditionally, dabai fruit are soaked in warm water (~50°C; 15 to 20 minutes) to soften them and then consumed with salt or soy sauce, as part of a meal or as a savoury snack (Ding and Tee, 2010; Pin and Daniel, 2017; Hassan *et al.*, 2020).

In this chapter, we discuss the nutritional composition and health-promoting phytochemicals of dabai fruit. The information is envisaged to be of use in popularizing this fruit as well as for highlighting its suitability for commercial exploitation in the development of functional food products and use in the pharmaceutical and cosmeceutical industries.

Botanical Description and Marketability

Dabai (scientific name: *Canarium odontophyllum*; family: Burseraceae; order: Sapindales; class: Eudicotyledoneae) grows from seed, and the tree (Fig. 30.1a) is upright (growing up to 20–25 m), medium-sized, with large pinnate leaves. The tree starts to flower and sets fruit (Fig. 30.1b) at between six and eight years. The tree has a lifespan of ~40 years. A mature dabai tree can yield up to 300 kg of fruit per season (during May–June and December–January), depending on the weather pattern (Pin and Daniel, 2017). Each tree can bring in a gross return of about US$2500 per annum. There are five main stages of maturation in dabai fruit. During the first stage (lasting up to six days) flowers open progressively until anthesis of the inflorescence is completed; in the second stage (seven to ten days) fruitlets emerge; in the third stage (eight to ten weeks) the immature white-coloured fruit are formed (Fig. 30.1c); in the fourth stage (four to five weeks) the semi-ripe fruit begin to change in colour from white to pinkish-red and then turn dark purple or black. Finally, during the fifth stage (approximately five weeks) fully matured fruit develop (Fig. 30.1d) (Yuon and Brooke, 2013). The recommended harvest time is ~120 days after flowering. At the maturity stage, the peel is dark purple in colour and has low chromaticity (~52.83°), while the pulp contains ~1.61% total soluble solids,~0.2% citric acid and has a pH of 5.88 (Ding and Tee, 2010, 2011). Harvesting of ripe dabai fruit in the natural habitat is often done by cutting down the fruit bunches using a knife (with the assistance of ladder). The fruit are then made to fall off in a net, placed below the canopy of the tree. Harvesting weather needs to be dry and early morning is generally preferred (Ding, 2011). There are reports that indicate dabai fruit to be widely cultivated in orchards (Khoo *et al.*, 2016).

Dabai is a seasonal indigenous fruit of Borneo Island in South-East Asia – Malaysia (Sabah and Sarawak) and Indonesia (Kalimantan and Sumatra). It is also native to the Philippines (Palawan) (Ding and Tee, 2010; Khoo *et al.*, 2016; Ting *et al.*, 2017). In Malaysia, dabai trees grows naturally in the wild/natural forests of Sarawak, especially in the divisions of Sibu, Sarikei, Limbang and Kapit (Ding and Tee, 2010; Chua *et al.*, 2015; Pin and Daniel, 2017). Locally, dabai fruit is called Dabei, Sibu olive, tropical olive, Sarawak olive or Zaitun tropika, because the fruit resembles olives in terms of external appearance, texture and flavour (Ding

Fig. 30.1. (a) Dabai tree (*Canarium odontophyllum* Miq.); (b) tree with immature fruit; (c) immature fruit; (d) mature fruit before harvest; (e) mature fruit after harvest; (f) cross-section of fruit; (g) dabai seeds; (h) seed kernel (phtographs: Mr. Lau Cheng Yuon).

and Tee, 2010). However, these fruit are not related to olives or the olive family. Until now, the classification of dabai fruit as a climacteric or non-climacteric fruit also remains unclear (Ding and Diana, 2013).

As a fruit, dabai are oblong drupe berries (size ranging from 3 to 4.0 cm long by 2.0 to 2.5 cm wide) with outer thin and edible peel (epidermis), white coloured when unripe and turning to dark purple/black when ripe (Fig. 30.1e). The pulp (mesocarp), also called aril, which is an edible part, is either cream or yellow in colour, 0.4 to 0.7 cm thick (Fig. 30.1f), and has a unique taste and oily texture like avocado. The hard, thick seed (endocarp) (Fig. 30.1g) (size ranging from 2.5 to 3.5 cm long and 1.6 to 2.0 cm in diameter) is triangle-shaped and contains the edible kernel (Fig. 30.1h), present in single form within the edible pulp. Dabai fruit contain ~62% of pulp,~37% of stony seed and 6% of peel (Hassan *et al.*, 2020). According to Pin and Daniel (2017), high-quality dabai fruit are characterized by their large size (~18g per fruit) with thick, yellow coloured pulp, possessing a nutty aroma with a fine and creamy texture. The taste is neutral and sometimes slightly sour.

Dabai fruit is sold at varying prices depending on the season, overall quality and seasonal demands. It is sold with high prices ranging between US$4.00 to 9.00/kg at the early harvesting period. However, the price of dabai fruit can drop to US$1.50/kg, owing to surplus harvest during peak season (Cangao, 2014). Dabai fruit is extremely perishable with a short shelf life, owing to a high respiration rate when stored at ambient/room temperature. The fruit can last for about two to three days after harvest and the shiny peel will wrinkle/shrink leading to quality deterioration. Hence, adopting appropriate postharvest handling practices is very important to minimize postharvest losses during transportation and marketing. Numerous studies have been conducted to investigate the right packaging material and storage conditions for retaining the overall quality of dabai fruit, as well as to delay maturation after harvesting (Ding and Tee, 2010; Ding and Diana, 2013; Tan and Azlan, 2016). Before consumption, dabai fruit pulp needs to be separated from the

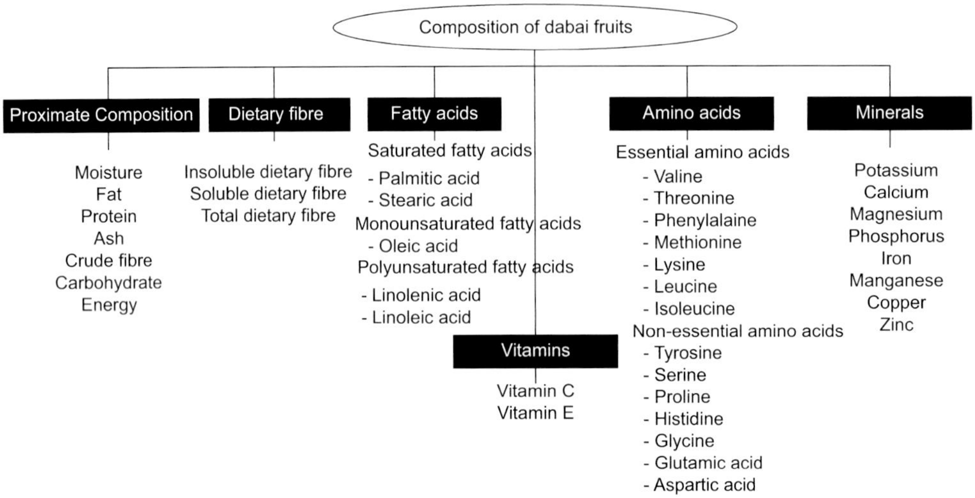

Fig. 30.2. Composition of dabai fruit (authors).

seed and this can be comparable to that of olives. However, in the majority of instances the peel of dabai is consumed along with the pulp.

Nutritional Composition

Dabai fruit is considered a rich source of nutraceutically valued components. The fruit are high in nutrition (Chua *et al.*, 2015; Pin and Daniel, 2017) (Fig. 30.2), with the pulp containing ample amounts of fat (26.2 g/100 g), carbohydrates (22.1 g/100 g), fibre (4.3 g/100 g), protein (3.8 g/100 g) and energy (339 kcal) (Tan and Azlan, 2016) (Table 30.1). The amount of energy in dabai fruit pulp is comparable with the calorie value of several commonly consumed tropical fruit such as mango (~321 kcal), banana (~338 kcal) and papaya (~356 kcal) (Bala and Bashar, 2017). The high-energy value of dabai pulp indicates that it has great potential to be explored as a new fruit source of energy. This fruit being highly nutritious can contribute significantly towards meeting the daily dietary needs of local populations, as well as supporting regional food security.

The fat content is reported to be high in dabai fruit (between 21–26 g/100 g) (Chew *et al.*, 2011). Interestingly, the amount of fat in dabai pulp is comparatively higher than other fat-rich fruit, such as olive (23 g/100 g) or avocado (29 g/100 g). In terms of fatty acids, dabai fruit is reported to contain nearly 12 times more monounsaturated fatty acids (e.g. oleic acid) than avocado (Chua *et al.*, 2015; Morais *et al.*, 2017) (Table 30.2). According to Chua *et al.* (2015), in dabai fruit, the major saturated fatty acids are stearic and palmitic, while oleic, linolenic and linoleic acids are the major unsaturated fatty acids (Table 30.1). According to Pin and Daniel (2017), the fatty acid of dabai fruit has a high ratio of polyunsaturated fatty acids (PUFA) to saturated fatty acids (SFA) (i.e. PUFA:SFA > 1.0), especially in the genotype of 'Bulat' (3.45) (Chua *et al.*, 2015). This ratio is desirable especially for prevention of cardiovascular disease. On the other hand, Ding and Tee (2011) have reported the fruit pulp to be devoid of any starch. By comparison with commonly consumed fruit (e.g. mango, banana, papaya), dabai fruit pulp has much higher amounts of saturated, monounsaturated and polyunsaturated fatty acids (linolenic acid and linoleic acid) (Chew *et al.*, 2011; Morais *et al.*, 2017) (Table 30.2). Further, the oil extracted from dabai fruit contains nearly 70% of the recommended dietary allowance (RDA) of linolenic acid (n-3 PUFA), the RDA being between 1100–1600 mg/day for adults (aged between 19 to 70 years). In addition, the

Table 30.1. Composition of dabai fruit pulp (on fresh weight – FW – basis) (Chew *et al.*, 2011; Chua *et al.*, 2015; Tan and Azlan, 2016).

Composition	Amount
Proximate composition (g/100 g)	
Moisture	41.3
Fat	26.2
Protein	3.8
Ash	2.3
Crude fibre	4.3
Carbohydrate	22.1
Energy (kcal)	339
Dietary fibre (g/100 g)	
Insoluble dietary fibre	9.3–23.0
Soluble dietary fibre	0.0–0.2
Total dietary fibre	9.5–19.1
Fatty acids (g/100 ml of oil extracted from dabai pulp)	
Saturated fatty acids	10.1–54.9
Palmitic acid (C16:0)	10.0–46.5
Stearic acid (C18:0)	4.9–8.6
Monounsaturated fatty acid	32.3–54.7
Oleic acid (C18:1)	6.03–45.2
Polyunsaturated fatty acids	10.5–35.2
Linolenic acid (C18:3)	5.14
Linoleic acid (C18:2)	2.6–29.2
Amino acids (% of total amino acids)	
Essential	
Valine	3.1–3.7
Threonine	3.2–3.8
Phenylalaine	4.6–5.2
Methionine	0.9–1.1
Lysine	5.4–5.9
Leucine	7.2–8.2
Isoleucine	2.4–2.8
Non-essential	
Tyrosine	2.4–2.9
Serine	2.4–2.9
Proline	4.3–4.7
Histidine	2.0–2.4
Glycine	4.0–4.4
Glutamic acid	18.3–19.6
Aspartic acid	25.3–30.9
Cystine	0.3–0.8
Arginine	2.5–3.2
Alanine	4.8–5.1
Minerals (mg/100 g)	
Potassium	810.0
Calcium	200.0
Magnesium	106.0
Phosphorus	65.0
Iron	1.3
Manganese (mg/kg)	8.0
Copper (mg/kg)	7.0
Zinc (mg/kg)	4.7
Vitamins (mg/100 g)	
Vitamin C	2.6
Vitamin E	25.7

fruit contains 40% higher linoleic acid (n-6 PUFA) content than the recommended level of RDA which is between 11,000–17,000 mg/day for adults (aged between 19 to 70 years) (USA NIH (National Institute of Health), 2021). According to Morais *et al.* (2017), the presence of unsaturated fatty acids, specifically those of essentials omega-3 (n-3) and omega-6 (n-6), is vital for human health. Hence, this fruit holds high potential as a new alternative heathy fat source when developing food products (e.g. as a replacement of cooking oil and fat spread).

Dietary fibre, which is considered as a non-nutritional component, due to its characteristic non-digestible nature and inability to be absorbed in the gastrointestinal tract is now much preferred by consumers. Today, dietary fibre is well established to impart a wide range of positive health effects in humans such as helping in regular bowel movement, hemorrhoids, cancer, etc. (Rowland *et al.*, 2018). The positive effects of dietary fibre intake have been linked to its ability to lower the glycaemic index by reducing food digestion rates in the body, thereby slowing down glucose absorption in

Table 30.2. Fat (g/100 g) and fatty acid (mg/100 g) composition of dabai fruit compared with avocado, mango, banana and papaya (Chew *et al.*, 2011; Chua *et al.*, 2015; Morais *et al.*, 2017).

Composition	Dabai fruit	Avocado	Mango	Banana	Papaya
Fat	26.20	28.60	0.38	1.70	1.40
Fatty acids					
Palmitic acid (C16:0)	46,500.00	2751.20	193.34	242.60	202.50
Stearic acid (C18:0)	8600	91.40	7.56	16.90	10.8
Oleic acid (C18:1)	45,200.00	3667.70	75.90	26.50	70.10
Linolenic acid (C18:3)	5140.00	189.20	84.04	186.40	146.30
Linoleic acid (C18:2)	29,200.00	1963.1	8.36	145.2	24.7

the gastrointestinal tract along with contributing towards improved insulin sensitivity (Wee and Henry, 2020). Therefore, intake of foods high in dietary fibre may reduce the risk of diabetes mellitus. Intake of dietary fibre is also associated with overcoming constipation and gut motility (Barber *et al.*, 2020). The fibre in dabai fruit is mainly insoluble dietary fibre, and ranges between 9.3 to 23.0 g/100 g, whereas soluble dietary fibre is present in lower/negligible amounts (<0.2 g/100 g) (Table 30.1) (Chua *et al.*, 2015).

With regard to amino acids, almost 17 types of amino acid were detected in dabai fruit, among which seven were essential amino acids (valine, threonine, phenylalanine, methionine, lysine, leucine and isoleucine) and ten were non-essential amino acids (tyrosine, serine, proline, histidine, glycine, glutamic acid, aspartic acid, cystine, arginine and alanine) (Chew *et al.*, 2011). Glutamic acids (18.3–19.6% of total amino acids) and aspartic acids (25.3–30.9% of total amino acids) (non-essential amino acid) are the major amino acids of dabai fruit, while methionine (essential amino acid) (0.90–1.10% of total amino acids) was found to be present at low concentrations (Chew *et al.*, 2011) (Table 30.1). Since both aspartic and glutamic acid (which are responsible for palatable taste) accounted for 45–49% of the total amino acids, dabai fruit can be considered as a natural flavour enhancer, and a potential monosodium glutamate (MSG) substitute.

Dabai fruit is also reported to be a rich source of macro- and microminerals, such as potassium (810 mg/100 g), calcium (200 mg/100 g), magnesium (106 mg/100 g) and phosphorus (65.0 mg/100 g), which are essential for maintaining the physiology of the body. Remarkably, calcium concentration is higher than that of avocado (54.9 mg/100 g), banana (25.4 mg/100 g), papaya (184.7 mg/100 g) (Morais *et al.*, 2017) and mango (7–21 mg/100 g) (Maldonado-Celis *et al.*, 2019). Dabai fruit is also reported to contain higher levels of magnesium than avocado (82.1 mg/100 g), banana (71.7 mg/100 g) (Morais *et al.*, 2017) and mango (8–38 mg/100 g) (Maldonado-Celis *et al.*, 2019). However, the fruit has lower potassium content than that of avocado (1,195.2 mg/100 g), banana (1,004.8 mg/100 g) and papaya (1,562.5 mg/100 g) (Morais *et al.*, 2017), but is comparable with mango (120–617 mg/100 g) (Maldonado-Celis *et al.*, 2019). Dabai fruit also has appreciable amounts of iron (1.3 mg/kg), manganese (8.0 mg/kg), copper (7.0 mg/kg) and zinc (4.7 mg/kg) (Tan and Azlan, 2016). In a study by Basri and Bakar (2014), dabai fruit was reported to have high concentrations of the beneficial trace elements such as molybdenum and iron. Chew *et al.* (2011) concluded that the macro- and micronutrients of dabai fruit can be affected by the growing conditions such as the climate, year of production, growing region/place of production, cultivar, ripening stage and planting practices. Dabai fruit is considered as an excellent natural source of calcium. This indicates that dabai fruit pulp (which is ~600 g) has high potential to be developed into a mineral supplement, perhaps in capsule form, which could provide adequate daily intake of calcium for adults (RDA for calcium is 1000–1200 mg/day) (USA NIH, 2021).

With regard to vitamins, vitamin C and E content in dabai fruit has been investigated. Fresh dabai fruit pulp contains 2.6mg/100g, while the peel has 5.9mg/100g of vitamin C (Pin and Daniel, 2017). Hence, it has been suggested that consumption of the whole fruit (peel and pulp) is required to obtain optimal intake of vitamin C. Consumption of dabai fruit along with other vitamin C-rich fruit such as mango (13–93mg/100g) (Maldonado-Celis *et al.*, 2019), kiwi fruit (93–161mg/100g) and papaya (60mg/100g) (Pinnamaneni, 2017; Richardson *et al.*, 2018) can meet RDA levels of vitamin C (the recommendation for adults aged between 19 to 70 years is 75–90mg per day) (USA NIH, 2021).

Regarding vitamin E, dabai pulp contains 25.7mg/100g (Tan and Azlan, 2016). The same researchers recorded high amounts of γ-tocopherol (11.0mg/100g) and α-tocopherol (1.0mg/100g) in the oil extracted from the fruit kernel, while fruit pulp oil was devoid of vitamin E. The vitamin E content of dabai kernel oil was significantly higher than olive oil (α-tocopherol: 1.0mg/100g; γ-tocopherol: non-detectable).

Phytochemical Properties

Reports have demonstrated high antioxidant activity in dabai fruit (Nagendra Prasad *et al.*, 2010; Shakirin *et al.*, 2010; Chew *et al.*, 2011; Ali-Hassan and Abu-Bakar, 2013) (Fig. 30.3). The high antioxidant activity of dabai fruit pulp and peel is due to the presence of naturally occurring bioactive compounds/phytochemicals (Table 30.3) (Shakirin *et al.*, 2010; Chew *et al.*, 2011). Some of these include catechins, anthocyanins and proanthocyanidins (Chew *et al.*, 2012). In a study conducted by Chew *et al.* (2011) and Nagendra Prasad *et al.* (2010), total phenolics and flavonoids showed a positive correlation with the exhibited antioxidant activities (Table 30.3). When comparing dabai fruit with some popular fruit rich in phytochemicals, such as berries and currants, the total anthocyanins were found to be comparable. However, overall total flavonoid content of dabai fruit was much higher (Chew *et al.*, 2012). When compared with some of the indigenous fruit of Malaysia, dabai fruit on average have higher polyphenolic content than belimbing buluh (*Averrhoa bilimbi*), bacang (*Mangifera foetida*), jambu mawar (*Syzygium jambos*) and durian nyekak (*Durio kutejensis*), but lower than that of rambai (*Baccaurea motleyana*), katak puru (*Cynometra cauliflora*), beruas (*Garcinia hombroniana*) and Melaka (*Phyllanthus emblica*) (Khoo *et al.*, 2016).

Traditionally, dabai fruit are consumed whole (pulp and peel), without any further processing. These fruit are recommended for oil production, and the resulting oil is claimed

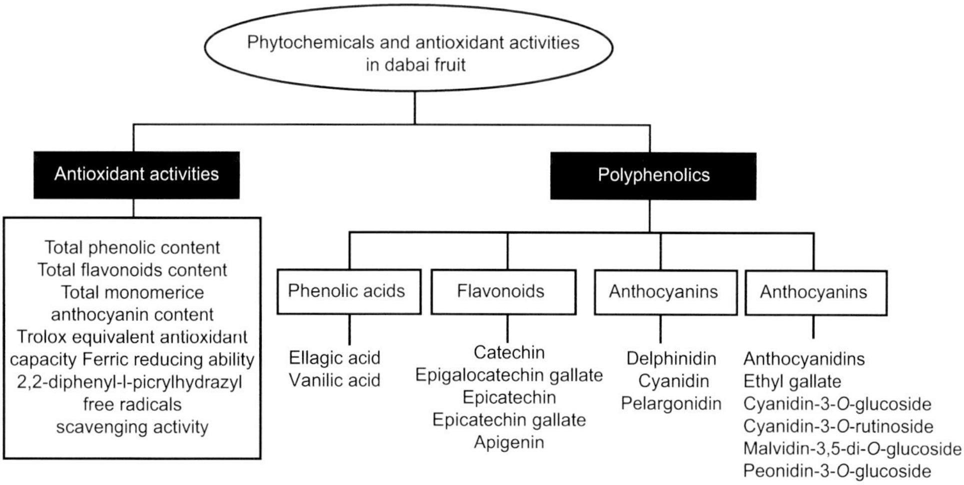

Fig. 30.3. Phytochemical and antioxidant activities in dabai fruit (authors).

Table 30.3. Antioxidant activity and phytochemical content (mg/g DW) in purple and red dabai fruit from different districts in Sarawak, Malaysia (Chew *et al.*, 2011, 2012).

Phytochemical	Purple dabai			Red dabai,
	Kanowit	Kapit	Song	Sarikei
Phenolic acids				
Ellagic acid	0.09 ± 0.02	0.21 ± 0.11	0.16 ± 0.07	0.10 ± 0.01
Vanilic acid	0.01 ± 0.00	0.02 ± 0.02	0.01 ± 0.00	0.01 ± 0.00
Flavonoids				
Catechin	3.01 ± 0.06	4.00 ± 0.58	3.22 ± 0.29	0.33 ± 0.02
Epigalocatechin gallate	0.28 ± 0.01	0.25 ± 0.06	0.24 ± 0.04	0.16 ± 0.01
Epicatechin	0.09 ± 0.01	0.10 ± 0.04	0.08 ± 0.01	0.07 ± 0.00
Epicatechin gallate	0.04 ± 0.01	0.05 ± 0.01	0.03 ± 0.01	0.03 ± 0.00
Apigenin	0.09 ± 0.01	0.12 ± 0.02	0.08 ± 0.02	0.08 ± 0.00
Anthocyanidins				
Delphinidin	0.02 ± 0.00	0.11 ± 0.03	0.04 ± 0.03	nd
Cyanidin	0.06 ± 0.01	0.07 ± 0.01	0.06 ± 0.01	0.03 ± 0.00
Pelargonidin	tr	tr	tr	tr
Anthocyanins				
Ethyl gallate	0.02 ± 0.00	0.03 ± 0.01	0.01 ± 0.00	0.01 ± 0.00
Cyanidin-3-*O*-glucoside	0.34 ± 0.02	0.39 ± 0.07	0.32 ± 0.07	0.03 ± 0.00
Cyanidin-3-*O*-rutinoside	1.16 ± 0.03	1.85 ± 0.34	1.41 ± 0.13	0.07 ± 0.00
Malvidin-3,5-di-*O*-glucoside	0.07 ± 0.02	0.20 ± 0.07	0.09 ± 0.04	nd
Peonidin-3-*O*-glucoside	tr	tr	tr	nd
Antioxidant activities				
TPC	19.44 ± 2.90	33.21 ± 6.11	22.81 ± 5.91	9.50 ± 0.42
TFC	55.46 ± 9.11	103.92 ± 24.60	53.31 ± 12.15	20.46 ± 5.62
TMAC	2.06 ± 0.53	2.49 ± 0.60	2.05 ± 0.30	0.08 ± 0.00
TEAC	0.45 ± 0.05	0.68 ± 0.09	0.38 ± 0.06	0.20 ± 0.00
FRAP	0.93 ± 1.35	1.74 ± 0.32	1.08 ± 0.31	0.27 ± 0.02
DPPH	89.45 ± 0.50	88.14 ± 1.21	88.12 ± 1.09	90.96 ± 0.22

Results expressed in means ± standard deviation
TPC represents total phenolic content (mg gallic acid equivalent/g dried weight);
TFC represents total flavonoid content (mg rutin equivalent/g DW);
TMAC represents total monomeric anthocyanin content (mg monomeric anthocyanin pigment/g DW);
TEAC represents trolox equivalent antioxidant capacity (mmol trolox equivalent/g DW);
FRAP represents ferric reducing ability (mmol of Fe^{2+}/g DW);
DPPH represents 2,2-diphenyl-1-picrylhydrazyl free radicals scavenging activity (% inhibition);
tr represents trace; nd represents not detected (<10 ng/injection at 20 μl).

to be comparable to olive and palm oils (Azlan *et al.*, 2010; Jelani *et al.*, 2017). A study by Jelani *et al.* (2017) showed that supercritical carbon dioxide (sCO_2) and hexane extractions produced dabai oil with high antioxidant properties. sCO_2-extracted dabai oil had higher polyphenolic content, while hexane-extracted dabai oil possessed higher (up to four times higher) amounts of total flavonoids.

By-products of dabai (post-oil production) include the defatted pulp, peel and seed portion, which can be explored to obtain bioactive polyphenolic compounds that possess health-promoting properties (Khoo *et al.*, 2012, 2013; Ali-Hassan and Abu-Bakar, 2013; Abdul Kadir *et al.*, 2020). Different types of solvent extraction (methanol, ethanol, ethyl acetate, acetone and water) were evaluated by Khoo *et al.* (2012) who reported methanol to exhibit higher efficiency in extracting polyphenolic compounds (mainly catechin and epigallocatechin). Among the various by-products studied, the peel portion consisted of maximal amounts of polyphenolic compounds, which subsequently exhibited higher antioxidant activities (Azrina *et al.*, 2010; Khoo *et al.*, 2012).

Pharmacological Properties

During the season, dabai fruit are generally consumed as part of the daily diet. However, reports on the usage of dabai fruit in traditional medicine are scarce (Khoo *et al.*, 2016). Nevertheless, numerous studies (both *in vivo* and *in vitro*) have been conducted during the last decade to prove the pharmacological efficacy of dabai fruit (Table 30.4). Findings from some of these reports have shown the fruit (a couple of the studies even tested leaf extract from the dabai plant) to possess high hypolipidemic (Shakirin *et al.*, 2012; Khoo *et al.*, 2013; Mokiran *et al.*, 2014; Faridah Hanim *et al.*, 2015; Abdul Kadir *et al.*, 2018, 2020), anti-atherosclerotic (Shakirin *et al.*, 2012), anticancer (Basri *et al.*, 2015), hypoglycaemic, antidiabetic (Saari *et al.*, 2017) and antimicrobial (Basri *et al.*, 2014a, Basri *et al.*, 2014b) activities, along with exhibiting anticholinesterase effects (Ali-Hassan and Abu-Bakar, 2013) (Fig. 30.4). It was hypothesized that the natural phytochemicals present in dabai fruit are responsible for these observed effects (Table 30.3). Hence, further studies are required to identify the key bioactive compounds having pharmacological effects in dabai fruit. Undeniably, dabai fruit have high potential to be used as a health-promoting ingredient in disease treatment (such as hypercholesterolemia and yeast infection), and health maintenance. However, human trials and clinical studies are required to enable a better understanding of the mechanism of action, toxicity (both acute and chronic) and efficacy of these extracts on humans prior to the commercialization as a phytotherapeutic agent.

Potential for Developing Food Products

Dabai fruit is a highly perishable commodity when fresh or after harvesting and this subsequently leads to significant postharvest loss during transportation and storage (Ding and Diana, 2013). During the peak season, enormous volumes of dabai fruit remain discarded and go to waste or are sold at a very low price. Hence, it is a necessity that new products are developed to increase the use and consumption of this fruit. Research findings on nutritional values and physicochemical properties have supported the potential to explore dabai fruit and develop new food products. To completely exploit dabai fruit during food product development, the Semongok Agricultural Research Centre of Sarawak (ARC), Malaysia, has an industrial collaboration with other research institutions, including the Malaysian Research and Development Institute (MARDI) and the Malaysian Palm Oil Board (MPOB) with the aim of developing dabai-based food products (Khoo *et al.*, 2016). Several value-added products from dabai fruit have been successfully developed, including pickles, seasoning pastes, dipping sauces, desserts, salad dressings, frozen pulp and dried seaweed rolls (maki, Japanese food), along with other functional foods (Liew and Lau, 2012; Hassan *et al.*, 2019). According to Hassan *et al.* (2020), dabai paste is popular and can be purchased from local markets/retail shops in Sarawak, Malaysia. Dabai fried rice is a popular hawkers' dish as well. Development of these new food products will help to diversify the use of dabai, while ensuring continuity of supply during the off-season. It will thus help to develop the processing of local fruit (Pin and Daniel, 2017) and ensure food security.

Rahman and Thajudin (2015) utilized this fruit as an ingredient in preparing fruit sauce. The fruit sauce was prepared using dabai fruit purée (45%), sugar (30%), water (18%), salt (0.9%), modified cornstarch (1.53%), xanthan gum (0.1%), citric acid (0.4%), margarine (4%) and sodium benzoate (0.07%). The developed

Table 30.4. Pharmacological properties of dabai fruit and leaves.

Reference	Part of the fruit/plant	Findings
Shakirin *et al*. (2012)	Defatted dabai pulp	Defatted dabai pulp showed cholesterol-lowering and antiatherosclerotic effects. It reduced plasma low-density lipoprotein cholesterol, total cholesterol, thiobarbiturate reactive substance levels and atherosclerotic plaque formation in hypercholesterolemic rabbits fed with 50 mg/kg of the defatted dabai pulp for 56 days
Ali-Hassan and Abu-Bakar (2013)	Dabai pulp and seed extracts	Methanolic extracts of dabai pulp and seed showed anticholinesterase effect, wherein at 100 µg/ml, the pulp and seed extracts inhibited acetylcholinesterase activity (22.4 and 18.6%, respectively) No anticholinesterase effect was observed for aqueous extracts
Khoo *et al*. (2013)	Defatted dabai peel extract	Cyanidin-3-glucoside-rich extract of the defatted dabai peel showed cardioprotective properties, wherein inhibition in the production of malondialdehyde and ability to maintain the physiological regulation between the oxidants and the antioxidants under *in vivo* conditions were found in hypercholesterolemic rabbits fed with 2 g/day of defatted dabai peel extract for 56 days
Basri *et al*. (2014a)	Dabai seed coat extract	Acetone extract of dabai seed coat showed promising antibacterial effect against *Acinetobacter baumannii* and *Proteus mirabilis*. Terpenoids and tannins in the extract accounted for this effect
Basri *et al*. (2014b)	Dabai pulp extract	Dabai pulp extract did not show any antimicrobial activity. However, both hexane and acetone extracts of dabai pulp exhibited antifungal effects against *Candida glabrata*, while no antifungal effects were observed for methanolic and aqueous extracts
Mokiran *et al*. (2014)	Defatted dabai extract	Defatted dabai extract showed cholesterol lowering effect, wherein they reduced the plasma cholesterol and low-density lipoprotein levels and increased high-density lipoprotein in induced obese–diabetic rats fed with 300 mg extract/kg body weight of dabai extract for 28 days It had a short-term glucose-lowering effect, wherein an increased sensitivity of insulin and reduced insulin resistance were found in induced obese–diabetic rats
Basri *et al*. (2015)	Dabai leaves extract	Both aqueous and solvent extracts of dabai leaves showed cytotoxic effect against human colorectal cancer cell HCT 116, with IC_{50} values of 0.08 ± 0.003, 0.10 ± 0.011, and 0.40 ± 0.162 mg/ml for acetone, methanol, and aqueous extracts, respectively

Continued

Table 30.4. Continued

Reference	Part of the fruit/plant	Findings
Faridah Hanim *et al.* (2015)	Dabai (pulp and kernel) oil	Dabai oil showed protective effect against oxidative stress, wherein the oil increased plasma total antioxidant status and reduced catalase activity in hypercholesterolemic rabbits fed with 2% of dabai oil for 28 days
Saari *et al.* (2017)	Dabai leaves aqueous extract	Dabai leaves aqueous extract showed hypoglycaemic effect, and reduced blood glucose levels in the streptozotocin-induced diabetic rats fed with 300 mg/kg for 28 days
Abdul Kadir *et al.* (2018)	Supercritical carbon dioxide extracted dabai pulp oil	Supercritical carbon dioxide extracted dabai pulp oil showed cholesterol lowering effects, wherein diminishing levels of total cholesterol, triglyceride and low-density lipoprotein levels were found in hypercholesterolemic rats fed with 2% of the dabai pulp oil for 28 days
, Basri *et al.*, 2018	Dabai leaves aqueous extract	Dabai leaves aqueous extract (3–15 mg/ml) showed vasorelaxant effect in rat thoracic aorta It was hypothesized that the terpenoid constituents that are contained in the extract may contribute to vasorelaxant effect via modulation of K^+ channel
Abdul Kadir *et al.* (2020)	Supercritical carbon dioxide extracted dabai pulp extract	Supercritical carbon dioxide extracted dabai pulp extract showed cholesterol-lowering effects, whereby a 30 days diet treatment with 2% of the dabai pulp extract significantly reduced the level of total cholesterol, triglyceride, low-density lipoprotein and inflammatory markers, and significantly increased the level of antioxidant profile in hypercholesterolemic rats

fruit sauce contained 60% moisture, 36% carbohydrates, 1% crude protein, 0.9% crude fibre, 0.75% crude fat and 0.50% ash. The developed fruit sauce had a characteristic lightness (L^*) value of 39.75, redness ($+a^*$) value 6.12, yellowness ($+b^*$) 12.42 along with viscosity 24.42 cP, pH 3.15, and 34 °Brix. Further, these researchers, by referring to the Malaysia Food Regulation Act (1985) for sauce products (Regulation 342 for chilli sauce and Regulation 343 for tomato sauce or tomato ketchup or tomato catsup) indicated dabai fruit sauce to be safe for human consumption with no signs of fermentation or quality loss even after 56 days of storage.

Along with food product development, dabai fruit has also shown high potential as an alternative ingredient (plant-based fat) to improve the nutritional value of food products. Hassan *et al.* (2019) evaluated the nutritional quality, sensory properties and shelf life of cake formulated with dabai fruit (pulp with peel). From their results, the nutritional value, especially the protein, was shown to increase proportionally to the increase in the substitution levels of dabai fruit (pulp with peel) for butter in cake-making. The amount of fat decreased as the amount of dabai fruit increased. For mineral content (i.e. sodium, potassium, calcium, magnesium, iron and zinc), an increase in proportion with dabai fruit (pulp with peel) content was observed, mainly in the cake formulation (Hassan *et al.*, 2019). Apart from this, the cake prepared using dabai fruit (replacing butter at 10%) received the highest score from sensory panelists. In addition, the cake could be stored up to 35 days under chilled conditions (4–8°C) and showed no signs of spoilage.

Recently, Hassan *et al.* (2020) formulated dabai fruit (pulp with peel) into cookies and analysed the physicochemical and sensory attributes

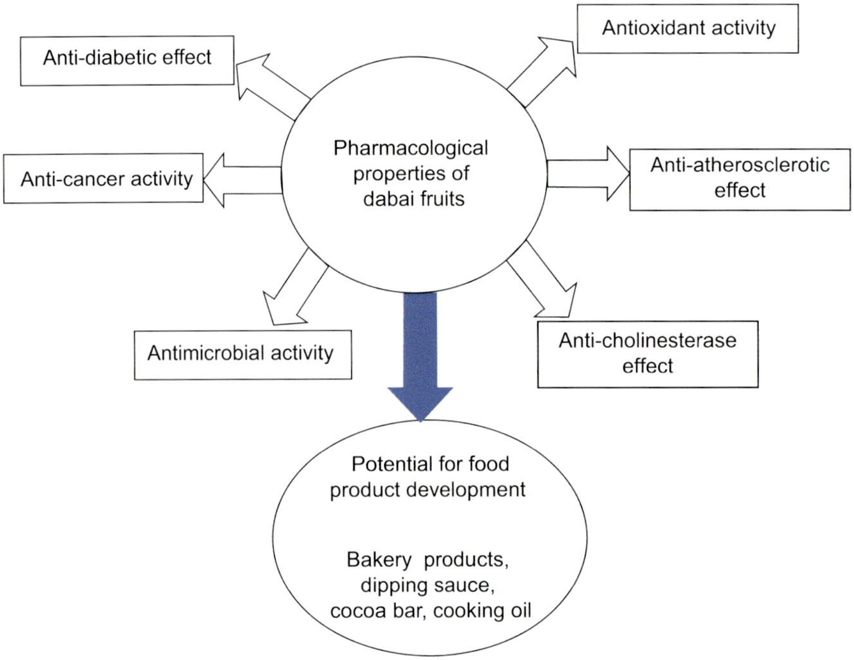

Fig. 30.4. Pharmacological properties of dabai fruit and their potential for food product development (authors).

of the product. From their results, the amount of crude protein and fibre increased corresponding to the increase in the substitution level of dabai fruit (for butter) during cookie preparation. However, the amount of crude fat significantly decreased in cookies containing dabai fruit as compared with control cookies. These researchers reported that amounts of dabai fruit added to the cookie formulation significantly influenced the colour, wherein cookies containing dabai fruit exhibited lower L^* (darker colour) and b^* (toward yellowness) values, but higher a^* (toward redness) value than the controls. They also observed that cookies made using butter substituted with 15% dabai fruit had the highest score for overall acceptability. The researchers concluded by indicating the feasibility of using dabai fruit as a future food ingredient during development of new food products, especially cookies.

In accordance with dabai pulp and kernel being rich sources of saturated fat (Chua *et al.*, 2015), Azlan *et al.* (2020) developed cocoa bars by using different parts of dabai, such as kernel, oil and oleoresin. They also formulated dark cocoa bars by partially substituting dabai kernel for cocoa butter at 5, 10, 15, 20, 25 and 35% levels. These researchers reported the cocoa bars with oleoresin (40%) to have received the lowest score from the sensory panelists. However, dark cocoa bars with 5–35%dabai kernel showed improved formulation: dark cocoa bars with 20% dabai kernel presented a higher overall acceptability score than dark cocoa bars with 5% dabai kernel. Apart from this, cocoa bar with oleoresin (20%)+oil (20%) of dabai exhibited higher antioxidant activity than with oleoresin (40%) or with kernel (5%)+oil (5%)+cocoa butter (30%). However, all the cocoa butter containing different parts of dabai showed lower antioxidant activity than dark cocoa bar and butylated hydroxytoluene (BHT) (standard) as measured by thiobarbituric acid assay. In addition, among the dark cocoa bars formulated by partially substituting dabai kernels at different percentages, dark cocoa bars with 20% dabai kernel demonstrated the lowest lipid peroxidation value. Thus, dark cocoa bars formulated with 20% dabai kernel could be suggested to have a protective effect toward inflammatory diseases. In addition, dark cocoa bars with 5% dabai kernel had the highest polyphenolic content and antioxidant activity (2,2 Diphenyl-1-picrylhydrazyl (DPPH) radical scavenging ability). Azlan *et al.* (2020) concluded that further research should focus on improving the sensory attributes (especially, crispiness and taste) of dark cocoa bars formulated with dabai kernel.

The fatty acid composition of dabai fruit is comparable to palm oil (Azrina *et al.*, 2007). Palm oil is reported to contain 42% palmitic acid, 35% linoleic acid and 11% linolenic acid (Azlan *et al.*, 2010). It is notable that fatty acid compositions of dabai pulp oil were proved to be the same with palm oil, while the fatty acid composition of oil extracted from the dabai kernel was comparable with cocoa butter (Jelani *et al.*, 2017). Hence, dabai fruit (pulp and kernel) could be processed and used as cooking oil and/or cocoa butter in the near future. In addition, Tan and Azlan (2016) have suggested that the oil extracted from pulp or kernel has lower viscosity compared to saturated fats, thus confirming suitability for use in designing new formulations for butter and margarine to improve the spreadability of the end product. Apart from this, Azrina *et al.* (2013) reported defatted dabai pulp to exhibit a higher ability to scavenge free radicals than the simvastatin drug. Thus, the defatted dabai pulp is recommended for development as a new drug for hypolipidemia (Azrina *et al.*, 2013).

Implications of Postharvest Handling and Processing on Quality

Dabai is an extremely perishable fruit with a short shelf life, attributed to the high respiration rate (1,411.6 ml/kg/h) at ambient temperature (~20°C). In this condition, dabai fruit can only last for about two or three days after harvest and thereafter the fruit wrinkle, shrink and dry (Ding and Tee, 2011; Ding and Diana, 2013). This limits the marketing and distribution potential. Currently, dabai fruit is limited to nearby towns in Sabah, Sarawak and Brunei. As surplus dabai fruit production during peak crop seasons can lead to a reduction in fruit prices (Pin and Daniel, 2017), good postharvest handling and storage strategies/practices (Fig. 30.5) are required to prolong the shelf life, while retaining the quality of the fruit as well as minimizing wastage. Dabai fruit can penetrate the international market if a breakthrough in the techniques for postharvest handling practices and storage is made.

Lau and Fatimah (2007) studied the quality of dabai fruit during cold storage under freezing conditions (−18°C). Fruit were packed using different packaging methods: bulk storage in polyethylene plastic bag (12.5 kg of fruit); polyvinyl chloride (PVC) container (250 g of fruit); polyethylene plastic bag (250 g of fruit); and vacuum-packed plastic bag (250 g of fruit) prior to cold storage for different storage durations (i.e. zero or fresh fruit, one, three and six months). Then, the frozen fruit was steeped in boiling water (for five minutes), followed by sensory evaluation. The frozen dabai fruit received a slightly lower score (3) on creamy taste as compared with that of fresh fruit (3.5) when stored by adopting different packaging methods for the entire study period. All the frozen fruit tasted less sour (0.5–1.0) after storing for a month, but increased (1.0–2.0) to about the same level as fresh fruit (2.0) after six months of storage.

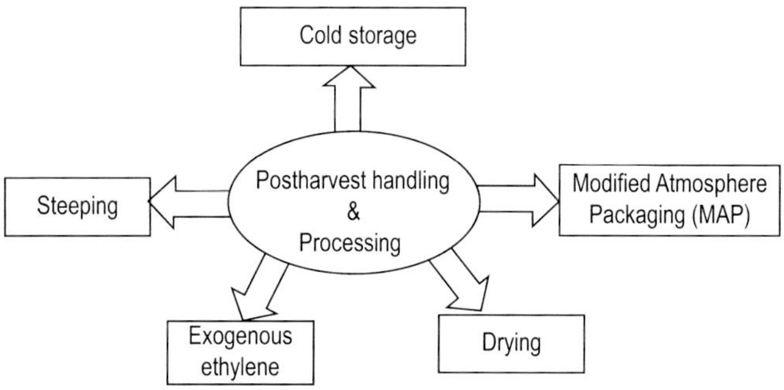

Fig. 30.5. Postharvest handling practices and processing techniques of dabai fruit (authors).

Comparing the packaging methods, fruit packed in PVC containers received high scores (1.42) in sourness taste after three months storage and the highest score (2.42) at six months of cold storage. This indicates that packaging methods have a strong influence on the decreasing and/or increasing trend of sourness levels. Cold storage duration with different packaging methods did not significantly affect the sensory acceptability of pulp colour: the colour remained pale yellow even after six months of cold storage. However, the frozen steeped fruit were found to receive lower scoring (3.5–4.5) in physical appearance than fresh fruit (5.0), but their quality was still acceptable. The researchers concluded that all the different packaging methods adopted maintained fruit quality satisfactorily under freezing conditions (Lau and Fatimah, 2007).

In the frozen state, the fruit supply can be made available during the off-season and can be marketed to a wider domestic market (Lau and Fatimah, 2007). Dabai fruit packed using different packaging methods (e.g. bulk storage in polyethylene plastic bags with ~12.5 kg of fruit; PVC container with 250 g of fruit; polyethylene plastic bag with 250 g of fruit and vacuum packed plastic bag with 250 g of fruit), followed by storing under freezing conditions (Lau and Fatimah, 2007) showed an extended shelf life of up to six months. In addition, fruit packed in polyethylene plastic bags and stored at 14°C had an extended shelf life of up to eight days (Dominic, 2006).

In some instances, traditionally, dabai fruit are eaten fresh, soon after steeping the fruit in warm water. However, nutrients are often leached out from the fruit during warm-water steeping. Ding and Tee (2010) applied exogenous ethylene instead of warm water, as an alternative method to soften the fruit. In this study, dabai fruit was exposed to a range of exogenous ethylene (i.e. 0, 10, 50, and 100 ml/ l) for different exposure durations (i.e. 0, 24, 48, and 75 hours) at 27°C. The peel lightness (L^*) values of the dabai fruit were observed to increase significantly (from ~38 to 48, authors presented data in figure) after treating with 0 and 10 ml/l ethylene, ethylene exposure duration up to 24 hours. Inversely, the lightness of dabai pulp decreased as ethylene exposure duration increased regardless of ethylene level. Apart from this, the fruit treated with ethylene showed a decrease in the firmness value to palatable level after treating with 10 ml/l ethylene for 24 hours. Total soluble solids, titratable acidity and pH were also determined by Ding and Tee (2010). In general, higher ethylene concentration and longer ethylene exposure periods led to an increase in total soluble solids (from ~2.0 to ~9.0%, authors presented data in figure) and total titratable acidity values (from ~0.19 to ~0.37% citric acid, authors have presented data in figure) of the fruit and a decreased pH value (from ~5.2 to ~4.0, authors presented data in figure). The researchers also suggested that exposure of dabai fruit to 10 ml/l concentration ethylene exposure for 24 hours at 27°C was sufficient to soften dabai fruit, thereafter senescence could occur (Ding and Tee, 2010).

Further, Ding and Tee (2011) evaluated the postharvest quality (e.g. biochemical properties and physical characteristics) of dabai fruit collected from Papar, Sabah, Malaysia. Results on biochemical composition revealed an increase in total titratable acidity, mainly contributed by citric (from 67.48 to 99.72 mg/g), malic (from 61.69 to 71.75 mg/g) and succinic acids (from 23.79 to 47.12 mg/g), and total soluble solids (from 1.61 to 10.95%). The pH value (from 5.88 to 5.04) and pulp hardness decreased tremendously by 92% after treatment by steeping the fruit at 60°C for 15 minutes. Besides, the authors also found dabai fruit to produce high concentrations of ethylene (from 4.56 to 89.17 μL/kg/h) after undergoing the steeping process – therefore, it is not recommended to mix dabai fruit with other fresh fruit crops during postharvest handling (Ding and Tee, 2011).

Drying is one of the postharvest processing techniques frequently applied to reduce moisture content. Hot-air drying is most often used to dry agrifood commodities to preserve the quality of the final product. Basri *et al.* (2012) studied the drying characteristics of dabai fruit via a mathematical model. They found that drying at 55°C under a relative humidity of 10%, and air velocity of 1 m/s was the optimum condition for dabai fruit, with the appropriate equations using the Page model (i.e. drying equation: $MR = \exp(-1.3279\, t^{1.3308})$) that produced 93.5% accuracy.

In another study, Ding and Diana (2013) applied postharvest technology such as modified atmosphere packaging (MAP) to maintain the quality of dabai fruit while prolonging their shelf life. Dabai fruit were packed in four different packaging materials (low-density polyethylene plastic bag with vacuum, low-density polyethylene plastic bag lined with a layer of kitchen towel, low-density polyethylene plastic bag and control, unwrapped) and then stored under cold conditions (10°C) for eight days. These packaging methods did not affect the colour (hue value) of dabai peel (283.02–286.05°) and pulp (74.33–88.34°) when compared with the control set (fresh dabai); 282.80° (peel) and 82.39° (pulp). However, during the storage period, L^* values of dabai peel (from 31.51 to 29.98) significantly decreased. MAP was found to be the best packaging method. It slowed down the respiration rate with lower CO_2 production. Considering the storage days, respiration rate decreased as storage days progressed. Thus, the fruit can be stored for a longer period with the aid of MAP. On the other hand, packaging methods were found to be incapable of affecting the production of ethylene in dabai fruit, however this was significantly (i.e. increasing trend) influenced by storage days. Regarding chemical composition, packaging methods did not affect soluble solid concentration, titratable acidity and pH values. Dabai fruit packed in a low-density polyethylene plastic bag with vacuum had a low firmness value, but the texture was observed to be 'soggy' and the fruit had a 'foul, sour' odour. Based on the findings, dabai fruit were recommended to be packed in a low-density polyethylene plastic bag as compared to other packaging materials during storage.

Conclusions and Future Trends

Based on the extensive studies undertaken on dabai fruit, it is evident that they are nutritious and possess many therapeutic effects. The fruit can be used as a functional food ingredient while developing various conventional food products (e.g. pastes, pickles, juices, jams, sauces, etc.) and value-added products (e.g. dabai oil). As limited research has been undertaken on dabai fruit, a large gap exists in identifying the presence of toxic antinutrient components (if any). It is understandable that some indigenous, underutilized wild fruit may contain antinutritional factors, such as hydrogen cyanide, phytates, alkaloids and oxalates, however, the presence of antinutrients in dabai might be remote, as this fruit has been consumed regularly over a very long period without any reports of negative health impacts.

To ensure dabai does not remain an indigenous fruit consumed only by local popualtions in the Borneo region (Malaysia or in Brunei), initiatives from the local government to promote this fruit's nutraceutical value and resulting fruit products to the mainstream international market assume much importance. This will not only ensure that local farming communities gain additional income and a sustainable living, but will also provide local food industries with the opportunity to explore the potential of developing more dabai-based functional foods.

As the therapeutic values/pharmacological properties remain well recognized for dabai fruit (established traditionally or via *in vivo* studies in animal models), it would be interesting to undertake human trials for a better understanding of the pharmacological properties. Providing some basic information on toxicity can be of use while exploiting the potential of the non-edible seeds and kernels of dabai fruit, especially when consideration/focus is on utilizing them in food or pharmaceutical or cosmeceutical applications. Furthermore, the by-products obtained post-oil production can also be explored as a potential source of bioactive compounds. Finally, maximal utilization of dabai fruit with minimal waste generation is expected to support the circular economy concept as well as ensure food security of the region.

Acknowledgements

The authors would like to thank Mr Lau Cheng Yuon and the Agricultural Research Centre of Sarawak (ARC), Semongok, Malaysia for providing photos of dabai fruit. RB acknowledges an ongoing project (VALORTECH at EMU), which has received funding from the European Union's Horizon 2020 research and innovation programme under grant agreement No. 810,630.

References

Abdul Kadir, N.A.A., Azlan, A., Abas, F. and Ismail, I.S. (2018) Beneficial effect of supercritical carbon dioxide extracted (SC-CO_2) dabai (Canarium odontophyllum) pulp oil in hypercholesterolemia-induced SPF Sprague-Dawley rats. *Natural Product Communications* 13(12), 1934578X1801301. DOI: 10.1177/1934578X1801301205.

Abdul Kadir, N.A.A., Azlan, A., Abas, F. and Ismail, I.S. (2020) Effect of defatted dabai pulp extract in urine metabolomics of hypercholesterolemic rats. *Nutrients* 12(11), E3511. DOI: 10.3390/nu12113511.

Ali-Hassan, S.-H. and Abu-Bakar, M.F. (2013) Antioxidative phytochemicals and anti-cholinesterase activity of native kembayau (*Canarium Odontophyllum*) fruit of Sabah, Malaysian Borneo. *Journal of Nutrition & Food Sciences* 04(1), 249. DOI: 10.4172/2155-9600.1000249.

Azlan, A., Prasad, K.N., Khoo, H.E., Abdul-Aziz, N., Mohamad, A. *et al.* (2010) Comparison of fatty acids, vitamin E and physicochemical properties of *Canarium odontophyllum* Miq. (dabai), olive and palm oils. *Journal of Food Composition and Analysis* 23(8), 772–776. DOI: 10.1016/j.jfca.2010.03.026.

Azlan, A., Khoo, H.E., Shapie, W.K.W., Abd Kadir, N.A. and Sultana, S. (2020) Nutritional quality and sensory evaluation of dabai-fortified cocoa bar. *International Journal of Food Properties* 23(1), 1324–1336. DOI: 10.1080/10942912.2020.1800031.

Azrina, A., Amin, I. and Muhammad, I. (2007) Proximate composition and characterization of fat in *Canarium odontophyllum*. In: *Proceedings of the 10th ASEAN Food Conference 2007: Food for Mankind – Contribution of Science and Technology*, Kuala Lumpur, Malaysia.

Azrina, A., Nurul Nadiah, M.N. and Amin, I. (2010) Antioxidant properties of methanolic extract of *Canarium odontophyllum* fruit. *International Food Research Journal* 17, 319–326.

Azrina, A., Amin, I., Amom, Z. and Shakirin, F.H. (2013) Effect of defatted dabai (*Canarium odontophyllum* Miq.) pulp ingestion on lipid peroxidation and antioxidant status of hypercholesterolemic-induced rabbits. In: *4th International Conference on Biomedical Engineering in Vietnam*, Springer. Available at: https://link.springer.com/chapter/10.1007/978-3-642-32183-2_36

Bala, M. and Bashar, J.B. (2017) Analysis of nutritive contents of some Nigerian fruits. *Bayero Journal of Pure and Applied Sciences* 10(2), 204. DOI: 10.4314/bajopas.v10i2.34.

Barber, T.M., Kabisch, S., Pfeiffer, A.F.H. and Weickert, M.O. (2020) The health benefits of dietary fibre. *Nutrients* 12(10), E3209. DOI: 10.3390/nu12103209.

Basri, D.F. and Bakar, N.F.A. (2014) *Canarium odontophyllum* fruit as a rich source of nutritionally important minerals and beneficial trace elements. *International Journal of Medicinal Plants* 107, 590–595.

Basri, D.F., Fudholi, A. and Ruslan, M.H. (2012) Drying characteristics of the Borneo canarium odontophyllum (dabai) fruit. *American Journal of Agricultural and Biological Sciences* 7(3), 347–356. DOI: 10.3844/ajabssp.2012.347.356.

Basri, D.F., Rahman, N.S.A., Ali, S.S. and Zainalabidin, S. (2018) The vasorelaxant effect of Canarium odontophyllum Miq. (Dabai) extract in rat thoracic aorta. *Egyptian Journal of Basic and Applied Sciences* 5(1), 75–79. DOI: 10.1016/j.ejbas.2017.11.004.

Basri, D.F., Fairuzlshak, S. and Mohamad Zin, N. (2014a) Shell extract of seed from *Canarium odontophyllum* Miq. (dabai) fruit as potential source of antibacterial agent. *International Journal of Pharmaceutical Sciences Review and Research* 28, 257–262.

Basri, D.F., Saidi, N., Mahari, H., Saari, S. and Santhanam, J. (2014b) Preliminary screening for antimicrobial activity of the pulp of Canarium odontophyllum Miq. (dabai) fruit. *Global Journal of Pharmacology* 8, 213–220.

Basri, D.F., Mohd Shabry, A.S. and Chan, K.M. (2015) Leaves extract from *Canarium odontophyllum* Miq. (Dabai) exhibits cytotoxic activity against human colorectal cancer cell HCT 116. *Natural Products Chemistry & Research* 3(1), 166. DOI: 10.4172/2329-6836.1000166.

Cangao, C.A. (2014) The dabai story: Experience in commercializing an underutilized fruit. Available at: https://www.itfnet.org/v1/2014/03/the-dabai-story-experience-in-commercializing-an-underutilized-fruit/ (accessed 1 November 2020).

Chew, L.Y., Prasad, K.N., Amin, I., Azrina, A. and Lau, C.Y. (2011) Nutritional composition and antioxidant properties of *Canarium odontophyllum* Miq. (dabai) fruits. *Journal of Food Composition and Analysis* 24(4–5), 670–677. DOI: 10.1016/j.jfca.2011.01.006.

Chew, L.Y., Khoo, H.E., Amin, I., Azrina, A. and Lau, C.Y. (2012) Analysis of phenolic compounds of dabai (*Canarium odontophyllum* Miq.) fruits by high-performance liquid chromatography. *Food Analytical Methods* 5(1), 126–137. DOI: 10.1007/s12161-011-9217-1.

Chua, H.P., Nicholas, D. and Adros Yahya, M.N. (2015) Physical properties and nutritional values of dabai fruit (*Canarium odontophyllum*) of different genotypes. *Journal of Tropical Agriculture and Food Science* 43, 1–10.

Ding, P. (2011) Dabai (*Canarium odontophyllum* Miq). In: Yahia, E.M. (ed.) *Postharvest Biology and Technology of Tropical and Subtropical Fruits Cocona to Mango*. Woodhead Publishing Limited, Cambridge, United Kingdom, pp. 34–40.

Ding, P. and Diana, J. (2013) Physico-chemical changes in dabai (*Canarium odontophyllum* Miq.) fruit during modified atmosphere storage. *International Food Research Journal* 20(6), 3033–3040.

Ding, P. and Tee, Y.K. (2010) Application of exogenous ethylene on postharvest quality of dabai (*Canarium odontophyllum* Miq.) fruit. *African Journal of Agricultural Research* 5(24), 3483–3492.

Ding, P. and Tee, Y.K. (2011) Physicochemical characteristics of dabai (*Canarium odontophyllum* Miq.) fruit. *Fruits* 66(1), 47–52. DOI: 10.1051/fruits/2010040.

Dominic, J.J. (2006) Study on the effect of packaging and temperature on improving the shelf life of dabai (*Canarium odontophyllum*). In: *Proceedings of the Technical Session, Research Officers' Conference 2006, 7-9 September 2006*, pp. 181–185.

FAO (2019) The Biodiversity that is Crucial for our Food and Agriculture is Disappearing by the Day. Available at: http://www.fao.org/news/story/en/item/1180463/icode/ (accessed 2 November 2020).

Faridah Hanim, S., Azrina, A., Khoo, H.E. and Amin, I. (2015) Protective effects of pulp and kernel oils from *Canarium odontophyllum* fruit in normal and hypercholesterolemic rabbits. *International Food Research Journal* 22, 1318–1326.

Hassan, N.A., Jun, C.H., Razak, A.F.A., Afandi, M.S.M., Sani, M. *et al.* (2019) Physicochemical, microbiological and sensory properties of dabai cake. *Borneo Journal of Science & Technology* 1(2), 69–73.

Hassan, N.A., Afandi, M.S.M., Razak, A.F.A. and Jamalti, J.S. (2020) Characterization of dabai (*Canarium odontophyllum* Miq.) cookies. *Borneo Journal of Science & Technology* 2(2), 47–53.

Jelani, N.A.A., Azlan, A., Ismail, A., Khoo, H.E. and Alinafiah, S.M. (2017) Fatty acid profiles and antioxidant properties of dabai oil. *ScienceAsia* 43(6), 347. DOI: 10.2306/scienceasia1513-1874.2017.43.347.

Khoo, H.E., Azlan, A., Ismail, A. and Abas, F. (2012) Influence of different extraction media on phenolic contents and antioxidant capacity of defatted dabai (*Canarium odontophyllum*) fruit. *Food Analytical Methods* 5(3), 339–350. DOI: 10.1007/s12161-011-9250-0.

Khoo, H.E., Azlan, A., Nurulhuda, M.H., Ismail, A., Abas, F. *et al.* (2013) Antioxidative and cardioprotective properties of anthocyanins from defatted dabai extracts. *Evidence-Based Complementary and Alternative Medicine* 2013, 434057. DOI: 10.1155/2013/434057.

Khoo, H.E., Azlan, A., Kong, K.W. and Ismail, A. (2016) Phytochemicals and medicinal properties of indigenous tropical fruits with potential for commercial development. *Evidence-Based Complementary and Alternative Medicine* 2016, 7591951. DOI: 10.1155/2016/7591951.

Lau, C.Y. and Fatimah, O. (2007) Cold storage of dabai fruit (*Canarium odontophyllum* Miq.): effect of packaging methods and storage duration on quality. Technical paper no. S4-2. In: *Proceedings of the Sarawak Fruit Seminar*, Kingwood Hotel, Sibu, Malaysia, August 8–9, 2007.

Liew, C.Y. and Lau, C.Y. (2012) Processing of dabai nuts (*Canarium odontophyllum* Miq.) as potential snack products: Nutritional, physical and sensory analysis. In: *7th International Postharvest Symposium 2012 (IPS 2012)*, Putra World Trade Centre, Kuala Lumpur, Malaysia. 25–29 June 2012.

Malaysia Food Regulation Act (1985). Available at: https://extranet.who.int/nutrition/gina/sites/default/filesstore/MYS%201985%20Food%20Regulations_0.pdf (accessed 15 April 2021).

Maldonado-Celis, M.E., Yahia, E.M., Bedoya, R., Landázuri, P., Loango, N. *et al.* (2019) Chemical composition of mango (*Mangifera indica* L.) fruit: nutritional and phytochemical compounds. *Frontiers in Plant Science* 10, 1073. DOI: 10.3389/fpls.2019.01073.

Mokiran, N.N., Ismail, A., Azlan, A., Hamid, M. and Hassan, F.A. (2014) Effect of dabai (*Canarium odontophyllum*) fruit extract on biochemical parameters of induced obese–diabetic rats. *Journal of Functional Foods* 8, 139–149. DOI: 10.1016/j.jff.2014.03.007.

Morais, D.R., Rotta, E.M., Sargi, S.C., Bonafe, E.G., Suzuki, R.M. *et al.* (2017) Proximate composition, mineral contents and fatty acid composition of the different parts and dried peels of tropical fruits cultivated in Brazil. *Journal of the Brazilian Chemical Society* 28(2), 308–318. DOI: 10.5935/0103-5053.20160178.

Nagendra Prasad, K., Chew, L.Y., Khoo, H.E., Kong, K.W., Azlan, A. *et al.* (2010) Antioxidant capacities of peel, pulp, and seed fractions of *Canarium odontophyllum* Miq. fruit. *Journal of Biomedicine & Biotechnology* 2010, 871379. DOI: 10.1155/2010/871379.

Pin, C.H. and Daniel, N. (2017) Dabai, specialty fruit of Sarawak. MARDI Official blog, 13 February. Available at: https://blogmardi.wordpress.com/2017/02/13/dabai-specialty-fruit-of-sarawak/ (accessed 29 October 2020).

Pinnamaneni, R. (2017) Nutritional and medicinal value of papaya (*Carica papaya* linn.). *World Journal of Pharmacy and Pharmaceutical Sciences* 6(8), 2559–2578. DOI: 10.20959/wjpps20178-9947.

Rahman, M.N.A. and Thajudin, N.A.M. (2015) Product development of kembayau (*Canarium Odontophyllum*) exotic fruit sauce. *Journal of Tropical Resources and Sustainable Science (JTRSS)* 3(1), 19–28. DOI: 10.47253/jtrss.v3i1.681.

Richardson, D.P., Ansell, J. and Drummond, L.N. (2018) The nutritional and health attributes of kiwifruit: A review. *European Journal of Nutrition* 57(8), 2659–2676. DOI: 10.1007/s00394-018-1627-z.

Rowland, I., Gibson, G., Heinken, A., Scott, K., Swann, J. *et al.* (2018) Gut microbiota functions: metabolism of nutrients and other food components. *European Journal of Nutrition* 57(1), 1–24. DOI: 10.1007/s00394-017-1445-8.

Saari, S.M., Basri, D.F., Budin, S.B. and Warif, N.M.A. (2017) Effects of *Canarium odontophyllum* leaves on plasma glucose and T lymphocyte population in streptozotocin-induced diabetic rats. *Saudi Journal of Biological Sciences* 24(2), 320–323. DOI: 10.1016/j.sjbs.2015.09.032.

Shakirin, F.H., Prasad, K.N., Ismail, A., Yuon, L.C. and Azlan, A. (2010) Antioxidant capacity of underutilized Malaysian *Canarium odontophyllum* (dabai) Miq. fruit. *Journal of Food Composition and Analysis* 23(8), 777–781. DOI: 10.1016/j.jfca.2010.04.008.

Shakirin, F.H., Azlan, A., Ismail, A., Amom, Z. and Yuon, L.C. (2012) Antiatherosclerotic effect of *Canarium odontophyllum* Miq. fruit parts in rabbits fed high cholesterol diet. *Evidence-Based Complementary and Alternative Medicine* 2012, 838604. DOI: 10.1155/2012/838604.

Tan, C.X. and Azlan, A. (2016) Nutritional, phytochemical and pharmacological properties of *Canarium odontophyllum* Miq. (Dabai) fruit. *Pertanika Journal of Scholarly Research Reviews* 2(1), 80–94.

Ting, J.-C., Ho, W.-S. and Tawan, C. (2017) Flora morphology and sex specific marker development of a Sarawak's specialty fruit species *Canarium odontophyllum*. *International Journal of Development and Sustainability* 6(11), 1621–1637.

USA NIH (National Institute of Health) (2021) Nutrient recommendations: Dietary reference intakes (DRI). Available at: https://ods.od.nih.gov/HealthInformation/Dietary_Reference_Intakes.aspx (accessed 12 April 2022).

Wee, M.S.M. and Henry, C.J. (2020) Reducing the glycemic impact of carbohydrates on foods and meals: Strategies for the food industry and consumers with special focus on Asia. *Comprehensive Reviews in Food Science and Food Safety* 19(2), 670–702. DOI: 10.1111/1541-4337.12525.

Yuon, L.C. and Brooke, P. (2013) Changes to dabai fruit during maturation process. Available at: https://doa.sarawak.gov.my/page-0-270-283-ARC-ARTICLES-Archives.html (accessed 2 November 2020).

31 Mamey Sapote (*Pouteria sapota* Jacq.)

Víctor M. Jiménez[1,2]*, Tania Chacón-Ordóñez[1] and Patricia Esquivel[3]
[1]CIGRAS, Universidad de Costa Rica, San Pedro, Costa Rica; [2]IIA, Universidad de Costa Rica, San Pedro, Costa Rica; [3]Escuela de Tecnología de Alimentos, Universidad de Costa Rica, 11501-2060 San Pedro, Costa Rica

Abstract

Pouteria sapota, commonly known in some countries as mamey sapote, is a member of the Sapotaceae family native to the Mesoamerican region. Awareness about this tree and consumption of its fruit have been mostly restricted to producing countries (Mexico and Central America), with only a few reports outside the American continent. In this chapter, we will discuss features related to botany and cultivation of the plant, as well as characteristics of the fruit, considering chemical composition, nutritional properties and health potential. Special emphasis is given to the main bioactive compounds present in the fruit, which include volatiles, phenolics and carotenoids, as well as the bioaccessibility and bioavailability of the latter. Traditional uses, development of processed products and innovative processing alternatives are also among the issues reviewed. Usually, the fruit is harvested when the pulp begins to change colour to its characteristic orange/salmon hue due to the carotenoid profile. Once harvested, the fruit has a postharvest shelf life of three to seven days, depending on the degree of maturity and storage temperature. Postharvest fruit shelf life has only been slightly extended by reducing storage temperature due to damage caused by chilling injury; therefore, other alternatives to improve shelf life are described in this chapter. The fruit is mainly consumed fresh or processed in ice cream, smoothies and jam, among others, due to its sweet flavour. Mamey sapote fruit contains nutritionally important constituents, such as carotenoids, vitamin C, thiamine, riboflavin, niacin, calcium, potassium and iron. The presence of nutritionally valuable carotenoids in mamey sapote fruit has recently drawn attention due to rarely occurring keto-carotenoids, namely sapotexanthin and cryptocapsin. Potassium is the main mineral, in terms of quantity, present in the fruit pulp. Mamey sapote is also a fruit rich in polyphenols, gallic acid being one of the main compounds found. Volatiles are also relevant because of the sensory properties they confer to the fruit. Although the fruit are traded and consumed mostly fresh, their commercialization in the form of frozen pulp is becoming popular, especially for export. Because of fruit deterioration due to enzymatic browning reactions, alternatives for processing have been evaluated as well. Since traditional thermal blanching has shown negative effects on texture, and is detrimental for the nutritional and sensory properties, non-thermal processing technologies, such as the use of microwave technologies, to ensure the safety of the processed pulp, have been studied. Other innovative tools, such as high hydrostatic pressure processing, are also being considered as promising and innovative alternatives for the preservation of the mamey sapote pulp without affecting its sensory and nutritional quality. *Pouteria sapota* has the potential to expand out of the current traditional and ethnic markets to a wider and international context, especially if the pulp can be included as an ingredient in fruit mixtures and juice formulations.

*Corresponding author: victor.jimenez@ucr.ac.cr

DOI: 10.1079/9781789248067.0031

Introduction

Pouteria sapota (syn. *Achras zapota*, *Achras mammosa*, *Calocarpum mammosum*, *Calocarpum sapota*, *Calospermun mammosum*, *Pouteria mammosa* and *Sideroxylum sapota*), a member of the Sapotaceae, a family with worldwide distribution and about 1250 species, 450 of them in the neotropics, is native to the Mesoamerican region, with a range extending from Veracruz and Tabasco, in Mexico, down to Panama (Swenson and Anderberg, 2005; Azurdia, 2006; Lim, 2013; Martínez-Castillo *et al.*, 2019). Today, its cultivation area has expanded to several South American and Caribbean countries, as well as to South Florida. It is commonly found under commercial cultivation, associated with agroforestry systems, or in home gardens (Galluzzi *et al.*, 2010; Avendaño-Arrazate *et al.*, 2019). The tree and fruit are known by different names according to the country/region of cultivation. The most common are mamey sapote, sapote mamey, zapote colorado, mamey and zapote (Lim, 2013).

Being a tropical species, the tree is sensitive to cold. Freezing temperatures can kill young trees, while those below 15°C can seriously diminish production and fruit development. Altitudinal range goes from sea level to up to ca 1400 m, without a marked preference for a particular soil type, although well-drained soils might have an advantage because the tree is susceptible to flooded conditions (Lim, 2013).

This climacteric fruit is usually left to ripen attached to the tree. Since the right maturity index is difficult to assess visually, because the external appearance does not change much, fruit are usually picked when a change in the pulp colour, from green to orange/pink, is noticed by a scratch in the outer rough skin (Díaz-Pérez *et al.*, 2003), which usually occurs close to 400 days after blooming (Alia-Tejacal *et al.*, 2007). Due to the generally large size of unpruned trees, harvesting is very laborious and time-consuming. There are reports of workers having to climb the trees with ropes and of them using long sticks, with a basket and knife attached, to manually collect the fruit one by one (Alia-Tejacal *et al.*, 2007; Balerdi *et al.*, 2013).

The fruit usually need to continue their ripening process once detached, to achieve best eating quality (pulp colour fully reddish or deep orange, and soft). This is frequently achieved by storing the fruit at room temperature (25–27°C) for three to seven days. Higher temperatures of 35°C, not rarely found in the producing regions, however, can inhibit maturation. Shelf life can be extended up to two weeks at 10–13°C. Storage for longer periods is usually not possible because of cold damage, as will be detailed below (Alia-Tejacal *et al.*, 2007). Treatments with the ethylene inhibitor 1-methylcyclopropene alone or in combination with a carnauba emulsion extended the shelf life to 14 days at 20°C (Ergun *et al.*, 2005).

Ripening can be accelerated by applying gaseous ethylene, but ethrel, its liquid formulation, did not cause any effect, probably due to the suberized skin that hampers product penetration to release gaseous ethylene inside (Alia-Tejacal *et al.*, 2007). Loss of firmness clearly occurred during ripening, especially at the climacteric phase, while curiously very low correlation were noticed between softening and activity of the pectinmethylesterase, polygalacturonase and β-galactosidase enzymes, as commonly observed in other fruit (Arenas-Ocampo *et al.*, 2003).

Chilling injury, caused by storage below 10°C, needs to be considered during postharvest handling. Damaged fruit display pulp browning, lignification and adherence to the seed, asymmetrical softening and off odours and flavours, especially when kept at 5°C. The use of controlled atmospheres has been evaluated and showed a reduction in weight loss and delayed ripening due to reduced ethylene production, without noticeable physiological disorders when transferred to room temperature (Alia-Tejacal *et al.*, 2007; Saucedo-Reyes *et al.*, 2012).

Although few pests have been described for *P. sapota*, this fruit has been reported as a major host of the fruit flies *Anastrepha serpentina* and *A. ludens* (Díaz-Pérez *et al.*, 2001; Alia-Tejacal *et al.*, 2007). To avoid further dissemination of these and other potential damaging pests, fresh fruit must not be transported to areas/regions/countries where they have not been reported (yet). Hot air (43°C) and hot water (60°C) treatments have been described as options to manage these pests without serious damage to

Fig. 31.1. Mamey sapote fruit (*Pouteria sapota*).

the fruit quality (Díaz-Pérez *et al.*, 2001; Yahia and Ariza, 2003).

Botany

The mamey sapote tree grows with an erect, thick central trunk, sometimes with a diameter of over 1 m, usually reaching 12 to 18 m height (but with reports of up to 50 m), with few branches in an open crown. The large leaves, 10–30 cm long by 4–10 cm wide, usually clustered at the distal extremes of thin branches, are simple, obovate to oblanceolate in shape, and with a conspicuous depressed midrib and lateral veins. Their green colour is lighter in the abaxial surface, which becomes glabrous with age. The flowers are small, whitish, bisexual, almost sessile and pentamerous. They are abundantly produced in clusters of five to six, usually towards the tips of thin branches. The fruit (Fig. 31.1), commonly weighing between 0.3 to 2.7 kg, is a berry, subglobose, ovoid to ellipsoid in shape, with a length ranging from 7.5 to 25 cm and 8 to 12 cm wide. The woody skin is thick (1–2 mm) and rough. The typical colour of the ripe pulp is salmon pink, although it can vary from orange to reddish-brown. Flesh is commonly soft and smooth, sometimes finely granular, and low in fibre. Seeds are usually 5–8 cm long, glossy-dark brown, ellipsoid and every fruit contains between one and four (Azurdia, 2006; Balerdi *et al.*, 2013; Lim, 2013).

Traditional Use

The edible fruit has a pleasant and sweet flavour, resembling a mixture of chocolate, pumpkin and sweet potato tastes, among others. The pulp can be eaten fresh or prepared into shakes, smoothies, ice cream or other desserts (Quesada, 1996; Alia-Tejacal *et al.*, 2007). In the export market, the fruit needs to be commercialized as frozen pulp due to the high perishability.

Furthermore, other parts of the fruit can be used. The seeds can be boiled, roasted and ground for confectionary or beverage production (Lim, 2013). The seed oil has potential uses, including as a substitute for cocoa butter in the food industry (Hernández-Santos *et al.*, 2017), and it has also been used in the cosmetic industry to prepare several products, soaps, shampoos and hair dyes among them (Alia-Tejacal *et al.*, 2007). As for the tree, due to the nice reddish colour and the hardness of its wood, it is used to make expensive furniture (Morera, 1992; Azurdia, 2006).

Nutritional Composition

The pulp of the mamey sapote fruit contains nutritionally important components, such as vitamin C, thiamine, riboflavin, niacin, calcium, potassium and iron (Alia-Tejacal *et al.*, 2007) (Table 31.1). Additionally, it is rich in simple carbohydrates, hence a great source of energy. The

Table 31.1. Nutrient composition of mamey sapote fruit as reported by the USDA (USA Department of Agriculture) database (USDA, 2018).

Component	Per 100 g
Water	64.87 g
Protein	1.45 g
Total lipids	0.46 g
Carbohydrates	32.10 g
Total dietary fibre	5.4 g
Sugars	20.14 g
Energy	124 kcal
Minerals	
Potassium	454 mg
Phosphorus	26 mg
Calcium	18 mg
Magnesium	11 mg
Sodium	7 mg
Iron	0.78 mg
Zinc	0.19 mg
Vitamins	
Vitamin C	23 mg
Vitamin E	2.11 mg
Niacin	1.43 mg
Vitamin B-6	0.72 mg
Riboflavin	0.116 mg
Thiamin	0.013 mg

total soluble solids found in mamey sapote are remarkably high in comparison to other fruit, changing during maturation from 10 to 12 °Brix up to 30–34 °Brix when ripe (Díaz-Pérez *et al.*, 2000; Villanueva-Arce *et al.*, 2000). Prevalent sugars are sucrose (up to 70%), followed by glucose and fructose. The high percentage of total soluble solids explains the low moisture present in the fruit (Alia-Tejacal *et al.*, 2007).

Phytochemicals

Regarding polyphenol compounds present in mamey sapote fruit, Ma *et al.* (2004) compared the polyphenolic antioxidants present in three *Pouteria* species (*P. sapota*, *P. viridis* and *P. campechiana*). They determined that *P. sapota* fruit have the highest amount of these compounds and, therefore, a higher antioxidant activity than the other two species evaluated. Gallic acid, dihydromyricetin, (+)-gallocatechin and (-)-epicatechin were among the most predominant polyphenols found. A further study also identified the presence of high concentrations of p-hydroxybenzoic acid (484 mg/100 g dry weight – DW), along with the previously described polyphenols (Yahia *et al.*, 2011). The main phenolics are listed in Table 31.2.

Early analyses of carotenoid content attributed the orange flesh colour to the presence of α- and β-carotene (Alia-Tejacal *et al.*, 2007). Lutein and violaxanthin are carotenoids additionally reported that do not contribute to this colour (Yahia *et al.*, 2011). More recent studies have revealed the presence of keto-carotenoids, like sapotexanthin and cryptocapsin, as the main pigments responsible for the red colouration (Murillo *et al.*, 2011, 2013). Due to chemical structure, they are suspected to possess potential provitamin A activity.

As a result of continuous propagation by seeds, mamey sapote trees present considerable genotypic diversity, giving rise to differences in fruit shape, size and flesh colour (Bayuelo-Jiménez and Ochoa, 2006; Pinto *et al.*, 2016). These differences have also been translated into diverse carotenoid profiles according to pulp colouration. Fruit with a more yellowish pulp are rich in neoxanthin and violaxanthin esters, while those with a more deep-orange or red colouration owe their hue to the presence of the above-mentioned rare occurring keto-carotenoids sapotexanthin and cryptocapsin esters (Murillo *et al.*, 2013; Chacón-Ordóñez *et al.*, 2017). However, a large number of minor keto-carotenoid derivatives have also been identified (Murillo *et al.*, 2012; Turcsi *et al.*, 2015; Cacciola *et al.*, 2016; Chacón-Ordóñez *et al.*, 2017; Murillo *et al.*, 2020). The main carotenoids are listed in Table 31.3.

Besides the potential that the fruit has to become a relevant source of vitamin A, vitamin E activity, in form of δ(delta)-tocopherol, has been reported as well, and its contribution to the antioxidant capacity of the pulp and derived products has been postulated (Yahia *et al.*, 2011).

Another group of compounds relevant for the sensory properties of the fruit is that of

Table 31.2. Phenolic compounds reported in *P. sapota* pulp.

Compound	Reference(s)
(+)-catechin	Ma *et al.* (2004); Torres-Rodríguez *et al.* (2011)
(+)-catechin-3-O-gallate	Ma *et al.* (2004)
p-coumaric acid	Yahia *et al.* (2011)
Dihydromyricetin	Ma *et al.* (2004)
(-)-epicatechin	Ma *et al.* (2004); Torres-Rodríguez *et al.* (2011)
Epicatechin dimer	Yahia *et al.* (2011)
Gallic acid	Ma *et al.* (2004); Torres-Rodríguez *et al.* (2011); Yahia *et al.* (2011)
(+)-gallocatechin	Ma *et al.* (2004)
Gallocatechin-3-gallate	Torres-Rodríguez *et al.* (2011)
p-hydroxybenzoic acid	Yahia *et al.* (2011)
p-hydroxybenzoic acid dimer	Yahia *et al.* (2011)
Hydroxybenzoic acid derivative	Yahia *et al.* (2011)
Myricitrin	Ma *et al.* (2004)
Protocatechuic hexose-malonate	Yahia *et al.* (2011)
Syringic acid	Yahia *et al.* (2011)

the volatiles. Less research has been conducted on the volatile composition of ripe *P. sapota* fruit. Volatile compounds can be categorized into four main groups: terpenoids, compounds with aromatic rings, fatty acids and amino acid derivatives. Different studies have identified 56 volatile compounds in mamey sapote pulp (Pino *et al.*, 2006; Rodríguez *et al.*, 2018; Martín and Osorio, 2019). Among them, there are compounds belonging to the terpenoids group such as cedrol and α-pinene, aldehydes such as benzaldehyde, hexanals and 3-methylbutanal, and fatty acid esters including methyl oleate, methyl octanoate and methyl myristate. The methods used for the extraction and analysis of this kind of compound are known to have a great impact on the aromatic profile – therefore, some differences in the reported data are to be expected. A complete list of the reported volatile compounds is presented in Table 31.4.

Implications of Food Processing on Nutrients and Phytochemicals

Pouteria sapota fruit deterioration is characterized by changes in colour due to enzymatic darkening, softening of the tissues, reduction of nutritional value and alteration of sensory properties. Enzymatic browning is one of the main aspects limiting the shelf life of processed products. For commercial use, fruit quality decreases rapidly during storage, handling and processing (Palma-Orozco *et al.*, 2012). Darkening is mainly due to the activity of two polyphenol oxidases already purified and characterized (PPO1 and PPO2). PPO1 activity can be reduced with inhibitors such as ascorbic acid and sodium metabisulfite (Palma-Orozco *et al.*, 2011).

Non-thermal technologies for processing delicate fruit products are novel alternatives to preserve nutritional and sensory characteristics. Palma-Orozco *et al.* (2012) reported suitable enzyme inactivation in mamey sapote pulp by applying a non-thermal microwave treatment, which also reduced the tissue damage usually produced by conventional thermal blanching. The better preservation of colour, flavour and aroma observed with the microwave treatment reflects a greater stability of bioactive compounds (vitamins, volatiles, carbohydrates, pigments, etc.) and nutrients. While cell wall integrity was only lightly affected by this new technology, some changes at the chloroplast level generated non-homogeneous colour

Table 31.3. Carotenoids reported in *P. sapota* pulp.

Compound	Reference(s)
Capsanthin	Gulyás-Fekete *et al.* (2013)
Capsanthin-5,6-epoxide	Cacciola *et al.* (2016); Murillo *et al.* (2016); Chacón-Ordóñez *et al.* (2017)
Capsoneoxanthin	Cacciola *et al.* (2016); Murillo *et al.* (2016); Chacón-Ordóñez *et al.* (2017); Agócs *et al.* (2018)
Cis-capsoneoxanthin	Cacciola *et al.* (2016)
Capsoneoxanthin laurate-myristate	Chacón-Ordóñez *et al.* (2017)
Capsorubin	Murillo *et al.* (2012); Gulyás-Fekete *et al.* (2013); Murillo *et al.* (2016)
α-carotene	Murillo *et al.* (2013)
β-carotene	Murillo *et al.* (2013); Cacciola *et al.* (2016); Murillo *et al.* (2016); Chacón-Ordóñez *et al.* (2017)
β-carotene epoxide (isomer)	Yahia *et al.* (2011)
β-carotene-5,6, 5',6'-diepoxide	Chacón-Ordóñez *et al.* (2017)
β-carotene-5,6, 5',8'-diepoxide	Cacciola *et al.* (2016); Murillo *et al.* (2016); Chacón-Ordóñez *et al.* (2017)
β-carotene-5,6-epoxide	Cacciola *et al.* (2016); Murillo *et al.* (2016); Chacón-Ordóñez *et al.* (2017)
β-carotene-5,8-epoxide	Murillo *et al.* (2013); Cacciola *et al.* (2016); Murillo *et al.* (2016); Chacón-Ordóñez *et al.* (2017)
β-carotene-5',8'-epoxide	Cacciola *et al.* (2016)
Cis-β-κ-carotene-6'-one	Cacciola *et al.* (2016)
Cryptocapsin	Gulyás-Fekete *et al.* (2013); Turcsi *et al.* (2015); Cacciola *et al.* (2016); Murillo *et al.* (2016); Chacón-Ordóñez *et al.* (2017)
13 *Z*-cryptocapsin	Cacciola *et al.* (2016); Murillo *et al.* (2016)
13' *Z*-cryptocapsin	Cacciola *et al.* (2016); Murillo *et al.* (2016)
Cryptocapsin-5,6-epoxide	Gulyás-Fekete *et al.* (2013); Cacciola *et al.* (2016); Murillo *et al.* (2016); Chacón-Ordóñez *et al.* (2017)
Cryptocapsin-5,6-epoxide laurate	Chacón-Ordóñez *et al.* (2017)
Cryptocapsin-5,6-epoxide myristate	Chacón-Ordóñez *et al.* (2017)
Cryptocapsin-5,8-epoxide	Gulyás-Fekete *et al.* (2013)
Cryptocapsin laurate	Chacón-Ordóñez *et al.* (2017)
Cryptocapsin myristate	Chacón-Ordóñez *et al.* (2017)
Cryptocapsin palmitate	Chacón-Ordóñez *et al.* (2017)
Cryptocapsin stearate	Chacón-Ordóñez *et al.* (2017)
β-cryptoxanthin	Murillo *et al.* (2013)

Continued

Table 31.3. Continued

Compound	Reference(s)
β-cryptoxanthin-5,6,5',6'-diepoxide	Murillo *et al.* (2016); Chacón-Ordóñez *et al.* (2017)
β–cryptoxanthin-5,6, 5',8'-diepoxide	Cacciola *et al.* (2016); Murillo *et al.* (2016); Chacón-Ordóñez *et al.* (2017)
β–cryptoxanthin-5,6,5',6'-diepoxide laurate	Chacón-Ordóñez *et al.* (2017)
β–cryptoxanthin-5,6,5',8'-diepoxide laurate	Chacón-Ordóñez *et al.* (2017)
β–cryptoxanthin-5,6,5',8'-diepoxide myristate	Chacón-Ordóñez *et al.* (2017)
β–cryptoxanthin-5,6-epoxide	Turcsi *et al.* (2015); Murillo *et al.* (2016); Chacón-Ordóñez *et al.* (2017)
β–cryptoxanthin-5',6'-epoxide	Turcsi *et al.* (2015)
β–cryptoxanthin-5,6-epoxide laurate	Chacón-Ordóñez *et al.* (2017)
β–cryptoxanthin-5,8-epoxide	Murillo *et al.* (2016); Chacón-Ordóñez *et al.* (2017)
β–cryptoxanthin-5,8-epoxide laurate	Chacón-Ordóñez *et al.* (2017)
β–cryptoxanthin-5',8-epoxide	Murillo *et al.* (2016)
β-cryptoxanthin-5,8,5',6'-diepoxide	Murillo *et al.* (2016)
β–cryptoxanthin-5,8, 5',8'-diepoxide	Murillo *et al.* (2016)
3'-deoxycapsanthin	Turcsi *et al.* (2015); Cacciola *et al.* (2016)
iso-3'-deoxycapsanthin	Cacciola *et al.* (2016)
3'-deoxycapsanthin-5,6-epoxide	Gulyás-Fekete *et al.* (2013); Murillo *et al.* (2016)
3'-deoxycapsorubin	Murillo *et al.* (2012); Murillo *et al.* (2016)
3,3'-dideoxycapsorubin	Murillo *et al.* (2012); Murillo *et al.* (2016)
5,6-diepilatoxanthin	Murillo *et al.* (2016)
(9*Z*/9'*Z*)-5,6-diepilatoxanthin	Murillo *et al.* (2016)
6-epilatoxanthin	Murillo *et al.* (2016)
Lutein (isomer)	Yahia *et al.* (2011)
Lutein epoxide	Yahia *et al.* (2011)
Lutein-myristate-laurate	Yahia *et al.* (2011)
Luteoxanthin	Cacciola *et al.* (2016); Murillo *et al.* (2016); Chacón-Ordóñez *et al.* (2017)
Luteoxanthin laurate-stearate	Chacón-Ordóñez *et al.* (2017)
Neochrome	Murillo *et al.* (2016)
Neoxanthin	Cacciola *et al.* (2016); Murillo *et al.* (2016); Chacón-Ordóñez *et al.* (2017); Agócs *et al.* (2018)
(9'*Z*)-neoxanthin	Agócs *et al.* (2018)
Sapotexanthin	Murillo *et al.* (2011); Gulyás-Fekete *et al.* (2013); Cacciola *et al.* (2016); Murillo *et al.* (2016); Chacón-Ordóñez *et al.* (2017)
(9*Z*/9'*Z*)-sapotexanthin	Murillo *et al.* (2016)
(13*Z*/13'*Z*)-Sapotexanthin	Murillo *et al.* (2016)

Continued

Table 31.3. Continued

Compound	Reference(s)
Sapotexanthin 5,6-epoxide	Murillo *et al.* (2020)
Violaxanthin	Yahia *et al.* (2011); Murillo *et al.* (2016); Chacón-Ordóñez *et al.* (2017)
(9*Z*)-violaxanthin	Murillo *et al.* (2016)
Zeaxanthin	Murillo *et al.* (2013)
Zeaxanthin 5',6'-5,8-diepoxide	Cacciola *et al.* (2016)
Zeaxanthin esters with unsaturated fatty acids	Murillo *et al.* (2013)
β-zeacarotene	Cacciola *et al.* (2016)

mixtures in the pulp (Palma-Orozco *et al.*, 2012). Appropriate application of microwaves as a non-thermal treatment to preserve the quality of a processed product requires defining the dielectric properties of the material to be processed. These properties vary according to the state of maturity of the samples, which is to be considered for fine-tuning the implementation of this type of technology for the processing of mamey sapote fruit pulp (Kataria *et al.*, 2017).

Bioaccessibility and Bioavailability

Studies assessing the absorption of nutrients from mamey sapote fruit have been limited to carotenoid bioaccessibility and bioavailability. As mentioned above, this is a fruit with a particularly interesting carotenoid profile including the rare keto-carotenoids sapotexanthin and cryptocapsin (esterified forms). Both compounds are of the utmost interest because of their potential provitamin A activity (Chacón-Ordóñez *et al.*, 2016, 2017, 2019).

Results from *in vitro* digestion models have shown that sapotexanthin and cryptocapsin esters were not only able to overcome simulated gastric conditions but also were incorporated into mixed micelles (Chacón-Ordóñez *et al.*, 2016). These are two prerequisites, at least theoretically, to be considered as being potentially available for absorption by human enterocytes, i.e., bioavailable. Further bioavailability assays in humans were able to prove the absorption of sapotexanthin and cryptocapsin (Chacón-Ordóñez *et al.*, 2017). Additionally, an increase of retinyl palmitate in the triglyceride blood fractions after consumption of mamey sapote fruit was observed (Chacón-Ordóñez *et al.*, 2019). This confirmed their conversion into vitamin A and, therefore, the provitamin A activity of carotenoids present in mamey sapote fruit.

Health Benefits

A widely accepted strategy to mitigate widespread global micronutrient deficiencies in humans is to maintain a diversified diet rich in fruit and vegetables (FAO/WHO, 2002). Several bioactive phytochemicals present in these sources have been associated with health-promoting effects and prevention of numerous chronic diseases that currently affect a wide population range worldwide (WHO/FAO, 2003). Mamey sapote could easily be included in the list of relevant plants, since it has been identified as a good source of carotenoids, vitamin C, polyphenols and minerals.

In past years, the presence of the nutritionally valuable carotenoids of mamey sapote fruit mentioned above has drawn attention, especially since sapotexanthin and cryptocapsin have been found to be responsible for the mamey sapote characteristic deep-orange pulp colouration (Murillo *et al.*, 2011; Chacón-Ordóñez *et al.*, 2017). Carotenoids are of particular interest because of their potential health benefits when consumed, the most prominent being the metabolic conversion of some carotenoids to vitamin A. Deficiencies of this vitamin can lead to a series of adverse health conditions, such as an impaired immune system, xerophthalmia,

Table 31.4. Volatile compounds reported in *P. sapota* pulp.

Compound	Reference(s)
Acetic acid	Martín and Osorio (2019)
Azulene	Rodríguez *et al*. (2018)
Benzaldehyde	Pino *et al*. (2006); Rodríguez *et al*. (2018); Martín and Osorio (2019)
Benzophenone	Rodríguez *et al*. (2018)
Benzyl nitrile	Pino *et al*. (2006)
Cadalene	Rodríguez *et al*. (2018)
β-cadinene	Rodríguez *et al*. (2018)
δ-cadinene	Rodríguez *et al*. (2018)
Camphene	Pino *et al*. (2006)
α-cedrene	Rodríguez *et al*. (2018)
Cedrol	Rodríguez *et al*. (2018); Martín and Osorio (2019)
α-copaene	Martín and Osorio (2019)
p-cymene	Pino *et al*. (2006)
(*E*)-β-damascenone	Martín and Osorio (2019)
Dihydroactinidiolide	Rodríguez *et al*. (2018)
Ethanol	Rodríguez *et al*. (2018)
Ethyl palmitate	Rodríguez *et al*. (2018)
Hexadecanoic acid	Pino *et al*. (2006)
Hexadecanol	Pino *et al*. (2006)
Hexanal	Pino *et al*. (2006)
Hexanol	Pino *et al*. (2006)
(*E*)-2-hexenal	Pino *et al*. (2006); Martín and Osorio (2019)
(*Z*)-3-hexenal	Martín and Osorio (2019)
(*Z*)-3-hexenol	Pino *et al*. (2006)
4-hydroxy-2,5-dimethyl-3(2 H)-furanone	Martín and Osorio (2019)
4-hydroxy-4-methyl-2-pentanone	Martín and Osorio (2019)
β-ionone	Rodríguez *et al*. (2018)
Isopropyl palmitate	Rodríguez *et al*. (2018)
Limonene	Pino *et al*. (2006)
Linalool	Martín and Osorio (2019)
trans-linalool oxide furanoid	Martín and Osorio (2019)
trans-linalool oxide pyranoid	Martín and Osorio (2019)
cis-2-p-menthen-1-ol	Pino *et al*. (2006)
Methyl butanal	Martín and Osorio (2019)
3-methyl butanoic acid	Martín and Osorio (2019)
Methyl hexadecanoate	Pino *et al*. (2006)
Methyl isomyristate	Rodríguez *et al*. (2018)

Continued

Table 31.4. Continued

Compound	Reference(s)
Methyl myristate	Rodríguez *et al.* (2018)
Methyl octanoate	Pino *et al.* (2006)
Methyl oleate	Rodríguez *et al.* (2018)
Methyl palmitate	Rodríguez *et al.* (2018)
2-methyl-1-propanol	Martín and Osorio (2019)
β-myrcene	Martín and Osorio (2019)
Naphthalene	Rodríguez *et al.* (2018)
(*Z*)-β-ocimene	Martín and Osorio (2019)
2-phenylethanol	Martín and Osorio (2019)
α-pinene	Pino *et al.* (2006); Rodríguez *et al.* (2018)
α-pinene oxide	Pino *et al.* (2006)
β-pinene	Pino *et al.* (2006)
Trans-sabinene hydrate	Pino *et al.* (2006)
γ-terpinene	Pino *et al.* (2006)
Terpinen-4-ol	Pino *et al.* (2006); Rodríguez *et al.* (2018)
Tetradecanoic acid	Pino *et al.* (2006)
Thujopsene	Rodríguez *et al.* (2018)
Tridecane	Pino *et al.* (2006)
Undecane	Pino *et al.* (2006)
Cis-verbenol	Pino *et al.*, 2006

night blindness and anaemia (FAO/WHO, 2002). Unfortunately, the human body is not able to synthetize this important vitamin on its own and requires it from the diet. Some fruit, such as mamey sapote, can be a good source of provitamin A carotenoids that are converted in the human body to vitamin A. As previously mentioned, a human study found an increase in retinyl palmitate, a form of vitamin A, in human blood triacylglycerol-rich lipoprotein fractions after the consumption of mamey sapote fruit (Chacón-Ordóñez *et al.*, 2019). Therefore, the consumption of mamey sapote can be suggested to attain the recommended daily allowance (RDA) of provitamin A carotenoids.

Since mamey sapote contains a diverse range of rare and unique carotenoids, further investigation could help to elucidate whether these compounds may influence human health and general wellbeing, as well as their potential therapeutic use. For example, recent studies have shown that cryptocapsin and cryptocapsin-5,6-epoxide, both found in mamey sapote, appear to have neuroprotective properties, specifically in regard to prevention and/or treatment of Alzheimer's disease (Lakey-Beitia *et al.*, 2017, 2019).

Mamey sapote is also a fruit rich in polyphenols. Phenolic compounds are known for the high antioxidant capacity they possess and, therefore, a series of health benefits have been associated with the consumption of fruit and vegetables rich in these compounds. Some of these, including phenolic acids found in mamey sapote fruit, can act against free radicals and protect DNA (Ozcan *et al.*, 2014). Gallic acid, one of the main polyphenols found in mamey sapote fruit, has been found to have not only a high antioxidant capacity but numerous therapeutic activities in metabolic, gastrointestinal, neuropsychological and cardiovascular disorders (Zanwar *et al.*, 2014; Mirshekar *et al.*, 2018; Yang *et al.*, 2020). Others, like catechin and epicatechin and their derivates, are considered as neuroprotective agents against oxidative stress-caused degenerative diseases and brain-ageing processes (Isemura, 2019). Some phenolic compounds also have antidiabetic properties since they have been shown to reduce the levels of glucose in blood by inhibiting enzymes of the carbohydrate metabolism, such as α-amylase and α-glucosidase (Hanhineva *et al.*, 2010). An *in vitro* study conducted by Uuh-Narváez *et al.* (2021) found that a phenolic-rich ethanolic extract from mamey sapote fruit was able to inhibit α-amylase activity up to 98.9%, thus proving its potential for dietary therapeutic strategies to prevent type 2 diabetes mellitus.

Regarding the minerals present in mamey sapote, potassium seems to be the main one. According to the daily recommended dose by the WHO, 200 g of mamey sapote pulp could contribute to fulfil one third of the daily requirement for potassium (Table 31.1). Deficiencies of this essential nutrient have been associated with hypertension and cardiovascular disease (Young, 2001). Besides its preventive role against heart disease, the consumption of an appropriate dosage of potassium is required in order to maintain the normal cell function by ensuring a good balance of total body fluid volume, acid and electrolytes (WHO/FAO, 2003). Finally, mamey sapote can also be considered a good source of other vitamins such as vitamin C, vitamin E and niacin, thereby contributing to the maintenance of general good health.

Novel Products/Formulations

Mamey sapote fruit, as previously mentioned, are highly perishable. Due to their susceptibility to postharvest constraints, an alternative for their wider commercialization is to minimally process them. Options include preparation of pulp slices to be used in salads or creams, in addition to processing overripe fruit into jam (Cavalcante *et al.*, 2012).

Frozen pulp, mainly for the export market, is in some cases thermally pasteurized (Alia-Tejacal *et al.*, 2007). It is subsequently used to manufacture ice cream, yogurt, milkshake and smoothies, among others. Freezing is useful for marketing products that are sensitive to heat treatment, but it is always necessary to ensure the safety of the frozen product. An alternative for sensitive products such as mamey sapote is the use of high hydrostatic pressure in order to reduce the deterioration caused by heat treatments. It was observed that the use of high hydrostatic pressure allowed the inactivation of microorganisms in the mamey sapote pulp even at temperatures of 25°C without affecting product quality (Saucedo-Reyes *et al.*, 2017).

Further use of the by-products of processing, mainly the seeds, has been the subject of some research. A content of 40% crude oil (DW with 10% humidity) with prevalence of oleic (53–60%), stearic (22–26%), palmitic (7–10%) and linoleic (5.6–10.1%) acids has been reported (Moo-Huchin *et al.*, 2013; Solís-Fuentes *et al.*, 2015; Hernández-Santos *et al.*, 2017).

Conclusions and Future Trends

Mamey sapote is a fruit with great nutritional value due to its numerous bioactive compounds, including carotenoids, polyphenols, minerals and vitamins. Research has shown that mamey sapote carotenoids are not only absorbed in the human body but exert provitamin A activity. Therefore, consumption of the fruit can be recommended to help reduce vitamin A deficiency in tropical and subtropical countries. Additionally, polyphenols found in mamey sapote fruit appear to have potential for therapeutic use to prevent some oxidative stress-related diseases, such as metabolic, gastrointestinal, neuropsychological and cardiovascular disorders, and type 2 diabetes mellitus. Nevertheless, further research is necessary to confirm such biological activities and to determine the potential health benefits of mamey sapote consumption or the extraction of bioactive compounds for therapeutic use.

In addition to its nutritional value, mamey sapote is highly appreciated by consumers due to its unique and characteristic sweet flavour. Unfortunately, its high perishability greatly limits its marketing opportunities. Until now, the only method of preservation, which has allowed some sort of commercialization without losing product quality, is as frozen pulp. Some studies have evaluated technological alternatives to process the fruit and extend its shelf life. However, rapid deterioration of the fruit, which causes changes in pulp colour, flavour and nutritional value, is still the main limitation to the commercialization of processed products made from mamey sapote fruit. The use of innovative non-thermal technologies could be further evaluated as an alternative for its processing.

Diverse studies have addressed the great genetic variations that can be found between *P. sapota* trees. Given predominant sexual reproduction, the low level of domestication and the presence of many wild types in their native regions, a broad genetic diversity is to be expected. The characterization of different genotypes of mamey sapote trees can be greatly advantageous for selecting those genotypes with higher concentrations of promising bioactive compounds and including them in breeding programmes. Additionally, this could assist in planning and establishing germplasm banks to preserve valuable specimens. Some efforts have been made to genetically characterize trees from different regions and countries; however, studies addressing the physical–chemical characterization of fruit are still scarce. Since genetic variation might influence phytochemical composition in fruit and given the link between the compounds present in the mamey sapote fruit and possible health benefits, it is of great importance that we explore these genetic differences and establish further potential uses for this fruit.

Pouteria sapota has the potential to expand out of the current more traditional and ethnic markets and achieve wider relevance, especially if the pulp can be included as an ingredient in fruit mixtures and juice formulations.

Acknowledgements

Support of the University of Costa Rica under the research projects VI-735-C0-142 and VI-735-C0-750 is gratefully acknowledged.

References

Agócs, A., Murillo, E., Turcsi, E., Béni, S., Darcsi, A. *et al.* (2018) Isolation of allene carotenoids from mamey. *Journal of Food Composition and Analysis* 65, 1–5. DOI: 10.1016/j.jfca.2017.04.004.

Alia-Tejacal, I., Villanueva-Arce, R., Pelayo-Zaldívar, C., Colinas-León, M.T., López-Martínez, V. *et al.* (2007) Postharvest physiology and technology of sapote mamey fruit (*Pouteria sapota* (Jacq.) H.E. Moore & Stearn). *Postharvest Biology and Technology* 45(3), 285–297. DOI: 10.1016/j.postharvbio.2006.12.024.

Arenas-Ocampo, M.L., Evangelista-lozano, S., Arana-Errasquin, R., Jiménez-Aparicio, A. and Dávila-Ortíz, G. (2003) Softening and biochemical changes of sapote mamey fruit (*Pouteria sapota*) at different development and ripening stages. *Journal of Food Biochemistry* 27(2), 91–107. DOI: 10.1111/j.1745-4514.2003.tb00269.x.

Avendaño-Arrazate, C.H., Campos-Rojas, E., Andres Agustin, J., Espinosa-Zaragoza, S. and Sandoval Cancino, G. (2019) Crop wild relatives of Avocado, Mamey Sapote, and Cherimoya. In: Greene, S.L., Williams, K.A., Khoury, C.K., Kantar, M.B., and Marek, L.F. (eds) *North American Crop Wild Relatives, Volume 2: Important Species*. Springer International Publishing, Cham, Switzerland, pp. 387–415.

Azurdia, C. (2006) *Tres Especies de Zapote en América Tropical (Pouteria campechiana, P. sapota y P. viridis)*. University of Southampton, International Centre for Underutilised Crops, Southampton, UK.

Balerdi, C.F., Crane, J.H. and Maguire, I. (2013) Mamey Sapote Growing in the Florida Home Landscape. Horticultural Sciences Department, UF/IFAS Extension, University of Florida. Available at: https://edis.ifas.ufl.edu/publication/MG331 (accessed 12 April 2022).

Bayuelo-Jiménez, J.S. and Ochoa, I. (2006) Caracterización morfológica de sapote mamey [pouteria sapota (jacquin) H.E moore & stearn] del centro occidente de Michoacán, México. *Revista Fitotecnia Mexicana* 29(1), 9–17. DOI: 10.35196/rfm.2006.1.9.

Cacciola, F., Giuffrida, D., Utczas, M., Mangraviti, D., Dugo, P. *et al.* (2016) Application of comprehensive two-dimensional liquid chromatography for carotenoid analysis in red mamey (*Pouteria sapote*) fruit. *Food Analytical Methods* 9(8), 2335–2341. DOI: 10.1007/s12161-016-0416-7.

Cavalcante, Í.H.L., Martins, A.B.G., de Sousa Miranda, J.M. and Cavalcante, L.F. (2012) *Physical and Chemical Characteristics of Tropical and Non-Conventional Fruits*. IntechOpen, London.

Chacón-Ordóñez, T., Esquivel, P., Jiménez, V.M., Carle, R. and Schweiggert, R.M. (2016) Deposition form and bioaccessibility of keto-carotenoids from mamey sapote (*Pouteria sapota*), red bell pepper (*Capsicum annuum*), and sockeye salmon (*Oncorhynchus nerka*) filet. *Journal of Agricultural and Food Chemistry* 64(9), 1989–1998. DOI: 10.1021/acs.jafc.5b06039.

Chacón-Ordóñez, T., Schweiggert, R.M., Bosy-Westphal, A., Jiménez, V.M., Carle, R. *et al.* (2017) Carotenoids and carotenoid esters of orange- and yellow-fleshed mamey sapote (*Pouteria sapota* (Jacq.) H.E. Moore & Stearn) fruit and their post-prandial absorption in humans. *Food Chemistry* 221, 673–682. DOI: 10.1016/j.foodchem.2016.11.120.

Chacón-Ordóñez, T., Esquivel, P., Quesada, S., Jiménez, R.R., Cordero, A. *et al.* (2019) Mamey sapote fruit and carotenoid formulations derived thereof are dietary sources of vitamin A - A comparative randomized cross-over study. *Food Research International* 122, 340–347. DOI: 10.1016/j.foodres.2019.04.009.

Díaz-Pérez, J.C., Bautista, S. and Villanueva, R. (2000) Quality changes in sapote mamey fruit during ripening and storage. *Postharvest Biology and Technology* 18(1), 67–73. DOI: 10.1016/S0925-5214(99)00062-9.

Díaz-Pérez, J.C., Bautista, S., Villanueva, R. and López-Gómez, R. (2003) Modeling the ripening of sapote mamey [*Pouteria sapota* (Jacq.) H.E. Moore and Stearn] fruit at various temperatures. *Postharvest Biology and Technology* 28(1), 199–202. DOI: 10.1016/S0925-5214(02)00123-0.

Díaz-Pérez, J.C., Mejía, A., Bautista, S., Zavaleta, R., Villanueva, R. *et al.* (2001) Response of sapote mamey [Pouteria sapota (Jacq.) H.E. Moore&Stearn] fruit to hot water treatments fruit. *Postharvest Biology and Technology* 22(2), 159–167. DOI: 10.1016/S0925-5214(00)00191-5.

Ergun, M., Sargent, S.A., Fox, A.J., Crane, J.H. and Huber, D.J. (2005) Ripening and quality responses of mamey sapote fruit to postharvest wax and 1-methylcyclopropene treatments. *Postharvest Biology and Technology* 36(2), 127–134. DOI: 10.1016/j.postharvbio.2004.12.002.

FAO/WHO (2002) Vitamin A. FAO/WHO, Human vitamin and mineral requirements. Report of a joint FAO/WHO expert consultation, Bangkok, Thailand.

Galluzzi, G., Eyzaguirre, P. and Negri, V. (2010) Home gardens: neglected hotspots of agro-biodiversity and cultural diversity. *Biodiversity and Conservation* 19(13), 3635–3654. DOI: 10.1007/s10531-010-9919-5.

Gulyás-Fekete, G., Murillo, E., Kurtán, T., Papp, T., Illyés, T.-Z. *et al.* (2013) Cryptocapsinepoxide-type carotenoids from red mamey, *Pouteria sapota*. *Journal of Natural Products* 76(4), 607–614. DOI: 10.1021/np3007827.

Hanhineva, K., Törrönen, R., Bondia-Pons, I., Pekkinen, J., Kolehmainen, M. *et al.* (2010) Impact of dietary polyphenols on carbohydrate metabolism. *International Journal of Molecular Sciences* 11(4), 1365–1402. DOI: 10.3390/ijms11041365.

Hernández-Santos, B., Sánchez-Ortega, E., Herman-Lara, E., Rodríguez-Miranda, J., Gómez-Aldapa, C.A. *et al.* (2017) Physicochemical and thermal characterization of seed oil from Mexican mamey sapote (*Pouteria sapota*). *Journal of the American Oil Chemists' Society* 94(10), 1269–1277. DOI: 10.1007/s11746-017-3038-3.

Isemura, M. (2019) Catechin in human health and disease. *Molecules (Basel, Switzerland)* 24(3), 528. DOI: 10.3390/molecules24030528.

Kataria, T.K., Olvera-Cervantes, J.L., Corona-Chávez, A., Rojas-Laguna, R. and Sosa-Morales, M.E. (2017) Dielectric properties of guava, mamey sapote, prickly pears, and nopal in the microwave range. *International Journal of Food Properties* 20(12), 2944–2953. DOI: 10.1080/10942912.2016.1261154.

Lakey-Beitia, J., Doens, D., Jagadeesh Kumar, D., Murillo, E., Fernandez, P.L. *et al.* (2017) Anti-amyloid aggregation activity of novel carotenoids: implications for Alzheimer's drug discovery. *Clinical Interventions in Aging* 12, 815–822. DOI: 10.2147/CIA.S134605.

Lakey-Beitia, J., Kumar D, J., Hegde, M.L. and Rao, K.S. (2019) Carotenoids as novel therapeutic molecules against neurodegenerative disorders: chemistry and molecular docking analysis. *International Journal of Molecular Sciences* 20(22), 5553. DOI: 10.3390/ijms20225553.

Lim, T.K. (2013) *Pouteria sapota*. In: Lim, T.K. (ed.) *Edible Medicinal and Non-Medicinal Plants*, Vols. 1–6. Springer, Dordrecht, Netherlands, pp. 138–142.

Ma, J., Yang, H., Basile, M.J. and Kennelly, E.J. (2004) Analysis of polyphenolic antioxidants from the fruits of three *Pouteria* species by selected ion monitoring liquid chromatography-mass spectrometry. *Journal of Agricultural and Food Chemistry* 52(19), 5873–5878. DOI: 10.1021/jf049950k.

Martín, D.A. and Osorio, C. (2019) Identification of aroma-active volatile compounds in *Pouteria sapota* fruit by aroma extraction dilution analyses (AEDA). *Química Nova* 42, 607–610. DOI: 10.21577/0100-4042.20170369.

Martínez-Castillo, J., Blancarte-Jasso, N.H., Chepe-Cruz, G., Nah-Chan, N.G., Ortiz-García, M.M. *et al.* (2019) Structure and genetic diversity in wild and cultivated populations of Zapote mamey (*Pouteria sapota*, Sapotaceae) from southeastern Mexico: its putative domestication center. *Tree Genetics & Genomes* 15(4), 61. DOI: 10.1007/s11295-019-1368-z.

Mirshekar, M.A., Sarkaki, A., Farbood, Y., Gharib Naseri, M.K., Badavi, M. *et al.* (2018) Neuroprotective effects of gallic acid in a rat model of traumatic brain injury: behavioral, electrophysiological, and molecular studies. *Iranian Journal of Basic Medical Sciences* 21(10), 1056–1063. DOI: 10.22038/IJBMS.2018.29639.7165.

Moo-Huchin, V., Estrada-Mota, I., Estrada-León, R., Cuevas-Glory, L.F. and Sauri-Duch, E. (2013) Chemical composition of crude oil from the seeds of pumpkin (*Cucurbita* spp.) and mamey sapota (*Pouteria sapota* Jacq.) grown in Yucatan, Mexico. *CyTA - Journal of Food* 11(4), 324–327. DOI: 10.1080/19476337.2012.761652.

Morera, J.A. (1992) *El zapote*. Centro Agronómico Tropical de Investigación y Enseñanza, CATIE, Turrialba, Costa Rica.

Murillo, E., McLean, R., Britton, G., Agócs, A., Nagy, V. *et al.* (2011) Sapotexanthin, an A-provitamin carotenoid from red mamey (*Pouteria sapota*). *Journal of Natural Products* 74(2), 283–285. DOI: 10.1021/np1006982.

Murillo, E., Mosquera, Y., Kurtán, T., Gulyás-Fekete, G., Nagy, V. *et al.* (2012) Isolation and characterization of novel capsorubin-like carotenoids from the red mamey (*Pouteria sapota*). *Helvetica Chimica Acta* 95(6), 983–988. DOI: 10.1002/hlca.201100493.

Murillo, E., Giuffrida, D., Menchaca, D., Dugo, P., Torre, G. *et al.* (2013) Native carotenoids composition of some tropical fruits. *Food Chemistry* 140(4), 825–836. DOI: 10.1016/j.foodchem.2012.11.014.

Murillo, E., Turcsi, E., Szabó, I., Mosquera, Y., Agócs, A. *et al.* (2016) Carotenoid composition of the fruit of red mamey (*Pouteria sapota*). *Journal of Agricultural and Food Chemistry* 64(38), 7148–7155. DOI: 10.1021/acs.jafc.6b03146.

Murillo, E., Agócs, A., Nagy, V., Király, S.B., Kurtán, T. *et al.* (2020) Isolation and identification of sapotexanthin 5,6-epoxide and 5,8-epoxide from red mamey (*Pouteria sapota*). *Chirality* 32(5), 579–587. DOI: 10.1002/chir.23206.

Ozcan, T., Akpinar-Bayizit, A., Yilmaz-Ersan, L. and Delikanli, B. (2014) Phenolics in human health. *International Journal of Chemical Engineering and Applications* 5(5), 393–396. DOI: 10.7763/IJCEA.2014.V5.416.

Palma-Orozco, G., Ortiz-Moreno, A., Dorantes-Alvarez, L., Sampedro, J.G. and Nájera, H. (2011) Purification and partial biochemical characterization of polyphenol oxidase from mamey (*Pouteria sapota*). *Phytochemistry* 72(1), 82–88. DOI: 10.1016/j.phytochem.2010.10.011.

Palma-Orozco, G., Sampedro, J.G., Ortiz-Moreno, A. and Nájera, H. (2012) In situ inactivation of polyphenol oxidase in mamey fruit (*Pouteria sapota*) by microwave treatment. *Journal of Food Science* 77(4), C359-65. DOI: 10.1111/j.1750-3841.2012.02632.x.

Pino, J.A., Marbot, R., Sauri, E. and Zumárraga, C. (2006) Volatile components of sapote [*Pouteria sapota* (Jacq.) H. E. Moore et Stern] fruit. *Journal of Essential Oil Research* 18(1), 22–23. DOI: 10.1080/10412905.2006.9699375.

Pinto, R.M., Nakamura, L.R., Ramires, T.G., López Bautista, E.A. and Dos Santos Dias, C.T. (2016) Genotype selection of *Pouteria sapota* (Jacq.) H.E. Moore & Stearn, under a multivariate framework. *Acta Agronómica* 65(3), 312–317. DOI: 10.15446/acag.v65n3.49877.

Quesada, P. (1996) Inventario de la variabilidad genética del zapote (P*outeria sapota*) en Costa Rica. *Boletín Técnico Estación Experimental Fabio Baudrit M* 29, 48–67.

Rodríguez, C., Durant-Archibold, A.A., Santana, A., Murillo, E. and Abuín, C.M.F. (2018) Analysis of the volatile components of *Pouteria sapota* (sapote mamey) fruit by HS-SPME-GC-MS. *Natural Product Communications* 13(8), 1027–1030. DOI: 10.1177/1934578X1801300826.

Saucedo-Reyes, D., Carrillo-Salazar, J.A., Reyes-Santamaría, M.I. and Saucedo-Veloz, C. (2012) Effect of pH and storage conditions on *Listeria monocytogenes* growth inoculated into sapote mamey

(*Pouteria sapota* (Jacq) H.E. Moore & Stearn) pulp. *Food Control* 28(1), 110–117. DOI: 10.1016/j.foodcont.2012.05.003.

Saucedo-Reyes, D., Carrillo-Salazar, J.A., Román-Padilla, L., Saucedo-Veloz, C., Reyes-Santamaría, M.I. *et al.* (2017) Modeling the pressure inactivation of *Escherichia coli* and *Salmonella typhimurium* in sapote mamey (*Pouteria sapota* (Jacq.) H.E. Moore & Stearn) pulp. *Food Science and Technology International* 24(2), 117–131. DOI: 10.1177/1082013217735472.

Solís-Fuentes, J.A., Ayala-Tirado, R.C., Fernández-Suárez, A.D. and Durán-de-Bazúa, M.C. (2015) Mamey sapote seed oil (*Pouteria sapota*). Potential, composition, fractionation and thermal behavior. *Grasas y Aceites* 66(1), e056. DOI: 10.3989/gya.0691141.

Swenson, U. and Anderberg, A.A. (2005) Phylogeny, character evolution, and classification of Sapotaceae (Ericales). *Cladistics: The International Journal of the Willi Hennig Society* 21(2), 101–130. DOI: 10.1111/j.1096-0031.2005.00056.x.

Torres-Rodríguez, A., Salinas-Moreno, Y., Valle-Guadarrama, S. and Alia-Tejacal, I. (2011) Soluble phenols and antioxidant activity in mamey sapote (*Pouteria sapota*) fruits in postharvest. *Food Research International* 44(7), 1956–1961. DOI: 10.1016/j.foodres.2011.04.045.

Turcsi, E., Murillo, E., Kurtán, T., Szappanos, Á., Illyés, T.-Z. *et al.* (2015) Isolation of β-cryptoxanthin-epoxides, precursors of cryptocapsin and 3'-deoxycapsanthin, from red mamey (*Pouteria sapota*). *Journal of Agricultural and Food Chemistry* 63(26), 6059–6065. DOI: 10.1021/acs.jafc.5b01936.

USDA (2018) Nutrient Database for Standard Reference. Database Release of April 1, 2018. Available at: https://data.nal.usda.gov/dataset/usda-national-nutrient-database-standard-reference-legacy-release (accessed 12 April 2022).

Uuh-Narváez, J.J., González-Tamayo, M.A. and Segura-Campos, M.R. (2021) A study on nutritional and functional study properties of Mayan plant foods as a new proposal for type 2 diabetes prevention. *Food Chemistry* 341(Pt 1), 128247. DOI: 10.1016/j.foodchem.2020.128247.

Villanueva-Arce, R., Evangelista-Lozano, S., Arenas-Ocampo, M.L., Díaz-Pérez, J.C. and Bautista-Baños, S. (2000) Cambios bioquímicos y físicos durante el desarrollo y postcosecha del mamey (*Pouteria sapota* (Jacq.) H.E. Moore & Stearn). *Revista Chapingo Serie Horticultura* 6(1), 63–72. DOI: 10.5154/r.rchsh.1998.10.070.

WHO/FAO (2003) *Diet, Nutrition, and the Prevention of Chronic Diseases*. World Health Organization, Geneva.

Yahia, E.M. and Ariza, R. (2003) Postharvest hot air treatments effect on insect mortality and quality of sapote mamey fruit (*Pouteria sapota*). *Acta Horticulturae* 604(604), 691–695. DOI: 10.17660/ActaHortic.2003.604.86.

Yahia, E.M., Gutiérrez-Orozco, F. and Arvizu-de Leon, C. (2011) Phytochemical and antioxidant characterization of mamey (*Pouteria sapota* Jacq. H.E. Moore & Stearn) fruit. *Food Research International* 44(7), 2175–2181. DOI: 10.1016/j.foodres.2010.11.029.

Yang, K., Zhang, L., Liao, P., Xiao, Z., Zhang, F. *et al.* (2020) Impact of gallic acid on gut health: focus on the gut microbiome, immune response, and mechanisms of action. *Frontiers in Immunology* 11, 580208. DOI: 10.3389/fimmu.2020.580208.

Young, D.B. (2001) Potassium depletion and hypokalemia symptoms. In: Young, D.B. (ed.) *Role of Potassium in Preventive Cardiovascular Medicine*. Springer US, Boston, Massachusetts, pp. 51–67.

Zanwar, A.A., Badole, S.L., Shende, P.S., Hegde, M.V. and Bodhankar, S.L. (2014) Role of gallic acid in cardiovascular disorders. In: Watson, R.R., Preedy, V.R., and Zibadi, S. (eds) *Polyphenols in Human Health and Disease*. Academic Press, San Diego, pp. 1045–1047.

32 Araçá Fruit (*Psidium cattleianum* Sabine)

Juliana Vinholes*, Elisa dos Santos Pereira and Márcia Vizzotto
Embrapa Clima Temperado, Pelotas, Rio Grande do Sul, Brazil

Abstract

Araçá is a Brazilian native fruit described as having a juicy, sweet to a sub-acid pulp with a spicy flavour. The fruit are normally consumed fresh but, due to the short shelf life, are mostly processed into liquors, juices, nectars, jellies, ice cream, mousses or cakes. The species is generally resistant to disease and pests due to the environmental growth conditions. The plant's defence system produces high amounts of secondary metabolites (phenolic compounds, volatile compounds and carotenoids) with high bioactivity. However, some of these compounds are antinutritive, such as flavonoids; therefore, a balance between their beneficial and harmful effects must be defined. Araçá processing, such as enzymatic juice production, fruit frozen for 30 days, frozen pulp+L-ascorbic acid, jellies and chewable candies have showed good results in maintaining their nutrients and phytochemicals. Araçá is described as having antioxidant, antiageing, antiproliferative, antidiabetic, anti-inflammatory, antithrombotic, antimicrobial and cerebral properties. Although red araçá extract showed low bioaccessibility, with 28% recovery of total phenolic content (TPC) and 5% of individual compounds remaining after *in vitro* digestion, its potential biological activities (antioxidant and α-glucosidase inhibition) was high. In this sense, the microencapsulation of araçá pulp/extract can improve the stability of phytochemicals and antioxidants. Araçá has also been proposed as an antioxidant to biodiesel protection. In general, araçá is a fruit with a high antioxidant potential that can be applied in different areas.

Introduction

Psidium cattleianum is a Brazilian native species which produces an edible fruit known as araçá, and can be found in different regions in the country (from Rio Grande do Sul to Bahia) but also in other countries (Uruguay, Hawaii and the Caribbean islands) (Galho *et al.*, 2007; Patel, 2012). Araçá plantations are generally spontaneous but can also be cultivated in home gardens or by small producers. The plant has great prospects for short-term cultivation because its production is precocious and permanently resistant to diseases and pests (Raseira and Raseira, 1990). However, araçá fruit are very perishable, with a shelf life of one to two days at 20°C, and are hence difficult to commercialize (Medina *et al.*, 2011). To overcome this problem, fruit can be harvested during the pre-climacteric stage or be stored under refrigeration right after harvest to decrease respiratory rates, or processed to obtain subproducts (Drehmer and Amarante, 2008; Pereira *et al.*, 2018).

Araçá fruit have high levels of secondary metabolites such as phenolic compounds, which can be produced in response to the adverse environmental conditions where the plant

*Corresponding author: julianarochavinholes@gmail.com

 Handbook of Phytonutrients in Indigenous Fruits and Vegetables (eds D. Sivakumar *et al.*)
DOI: 10.1079/9781789248067.0032

grows (drought, floods, sunlight and heat) (Reynertson *et al.*, 2008). Phenolic compounds are well known antioxidants involved in biological protection against free radicals and reactive oxygen species (ROS) (Verma *et al.*, 2013). Fruit also contain other secondary metabolites such as volatile compounds and carotenoids that can protect against different pathologies. Araçá is a low-caloric fruit containing high levels of vitamin C and some important minerals and amino acids. Due to these characteristics, araçá fruit have been exploited for applications in different areas.

Botany

Araçá (*Psidium cattleianum* Sabine) belongs to the class Magnoliopsida, order Myrtales, family Myrtaceae and genus *Psidium* (Mattos, 1989). The araçá plant is an evergreen small fructiferous tree/shrub, 1–4 m high. Plants grow in environments with high moisture and luminosity. Araçá leaves are opposite, obovate to elliptic in outline, coriaceous, glabrous and have a deep-green colour. Flowers are white with 1.5 to 2.5 cm diameter, pentamerous, hermaphrodite and zygomorphic. The period of flowering in the south of Brazil occurs at two main times, September to October and December. A third flowering may also occur in March (Raseira and Raseira, 1996). Araçá fruit are ovoid or oblong with a diameter of 2.2–5 cm, pyriformis, with a flat or oval shape, crowned by a chalice weighing approximately 20 g (Castro *et al.*, 2004; Biegelmeyer *et al.*, 2011). Fruit can have yellow or red epicarp (Fig. 32.1). The endocarp can be cream, white or, in rare cases, red (Castro *et al.*, 2004; Fernandes-Santos *et al.*, 2010). The fruit pulp is translucent and characterized as juicy, sweet to sub-acid, aromatic with a spicy touch, filled with seeds. An araçá orchard can produce 10 ton/ha, reaching 2 kg of fruit per plant, when produced under specific conditions (0.5 m between plants and 4.0 m between rows of plants) (Franzon *et al.*, 2009). In Brazil, the fruit are commonly known as araçá, araçá-amarelo, araçá-vermelho, araçá-rosa, araçá-de-comer, araçá-da-praia, araçá-de-coroa or araçá-do-campo, while in other countries araçá is called cattley guava, Chinese guava, purple guava, yellow strawberry guava, red strawberry guava, guayaba, cherry guava and lemon guava (Fig. 32.1). (Bezerra, 2010; Lisbôa *et al.*, 2011; Mitra *et al.*, 2012). Important synonymies for *P. cattleianum* are *Psidium littorale* Raddi, *Guajava obovata* (Mart. ex DC.) Kuntze, *P. variabile* O. Berg, *P. ferrugineum* C. Presl, *P. indicum* Bojer, *Eugenia ferruginea* Sieber ex C. Presl, *P. obovatum* Mart. ex DC. and *G. cattleiana* (Afzel. ex Sabine) Kuntze.

Traditional Use

Araçá is a juicy fruit with a pleasant flavour and sweet pulp, and is traditionally consumed *in natura* or as part of jellies, sweets, ice creams, beverages and jams (Santos *et al.*, 2007; Biegelmeyer *et al.*, 2011; Reissig *et al.*, 2016). Different properties have been reported for araçá in recent years, and different food formulations

Fig. 32.1. Araçá fruit (*Psidium cattleianum* Sabine).

with additional health benefits have been developed.

Nutritional Composition

Araçá is a low caloric fruit (27 to 37 kcal/fruit) and can afford 5.9–12.1% of the recommended daily allowance (RDA) for carbohydrates and 0.8–1.4% of RDA for proteins (Table 32.1).

Araçá contains the amino acids leucine, lysine, threonine, isoleucine, valine and phenylalanine. These nutrients are important to cell signalling, regulation of gene expression and protein phosphorylation cascade. They are precursors for synthesizing other important compounds, such as hormones and low molecular weight nitrogen substances (Wu, 2009). Araçá can afford 382 mg of total amino acids and 154 mg of essential amino acids per 100 g of fruit (Hall *et al.*, 1980), leucine being the main essential amino acid, and glutamine the main non-essential (Table 32.1). Linoleic acid is the major fatty acid present in araçá. Fatty acids are essential for normal physiological function. A diet high in mono- and polyunsaturated fatty acids is associated with lower risk of developing cardiovascular disease and atherogenesis (Biegelmeyer *et al.*, 2011). Araçá contains a high amount of dietary fibre; the consumption of one fruit (20 g) contributes to 6.28% of the RDA. Potassium is the main mineral found, and this element is essential to the human diet, being the element with the highest intracellular osmotic activity. This mineral is responsible for the intra- and extracellular distribution of water, acts on acid-base balance regulation and establishes membrane potential, which is essential to nerve and muscle fibre electrical activity. It is also involved in cellular energy metabolism, hormonal secretion, protein and glycogen synthesis (Pohl *et al.*, 2013). Vitamin C (ascorbic acid) has different physiological functions in the human body. This vitamin is responsible for the increase of iron absorption by the organism and the protection of cell membranes (lipoproteins) against lipid peroxidation due to its high antioxidant activity (Figueroa-Méndez and Rivas-Arancibia, 2015). The consumption of one fruit of 20 g might provide more than four times the RDA of vitamin C for adults.

Phytochemicals

Carotenoids

Carotenoids are natural pigments reported as having antioxidant activity, ingestion of which has been associated with decreased risk of macular degeneration and cataracts, some types of cancers and cardiovascular events (Olson, 1992; Perera and Yen, 2007; Fiedor and Burda, 2014). These effects are probably associated with the ability of carotenoids to increase the immune system defence, inhibit mutagenesis and premalignant lesions, and act as a pigment in the eye's primate macule (Perera and Yen, 2007). The total carotenoid content (TCC) in araçá is in the range of 364.4–1134.0 μg and 389.0–1084.0 μg of β-carotene equivalents/100 g fresh weight (FW), for red and yellow araçá, respectively (Medina *et al.*, 2011; Denardin *et al.*, 2015; Vinholes *et al.*, 2017). More recently, Meregalli *et al.* (2020) reported that conventional and ultrasound-assisted araçá peel extracts contain 4.45–4.47 μg of β-carotene/ml of extract, respectively. In general, yellow and red araçá fruit contain high concentrations of all-*trans*-β-cryptoxanthin, all-*trans*-lutein, β-carotene and all-*trans*-zeaxanthin (Table 32.2).

The ingestion of araçá fruit may provide considerable amounts of dietary carotenoids with provitamin A activity such as β-carotene and β-cryptoxanthin (Table 32.2). These compounds help prevent vitamin A deficiency, which can impact vision, cellular differentiation and embryo development (Olson, 1992).

Phenolic compounds

Phenolic compounds are metabolites of plants that play an essential role in their reproduction, growth and protection against infections, wounds and UV radiation, among other things (Naczk and Shahidi, 2004). The consumption of fruit and vegetables rich in phenolic compounds has been reported to reduce the risk of several diseases (Shahidi and Ambigaipalan, 2015).

Table 32.1. Nutritional composition of yellow and red araçá fruit.

Nutritional composition	Values[a]	RDA[b]
Water (g/100 g)	82.36–84.2	3.7 l
Calories (kcal)	27–37	2000
Carbohydrates (g/100 g)	7.67–15.08	130 g
Total sugar	4.32–22.74	
Sucrose	6.69–9.52	
Glucose	2.09–3.61	
Fructose	3.39–6.27	
Protein (g/100 g)	0.5–0.8	56 g
Total amino acids (mg/100 g)	382	
Total essential amino acids (mg/100 g)	154	
Leucine (g/100 g of total recovered amino acids)	10.2	42 mg/kg
Lysine (g/100 g of total recovered amino acids)	5.8	38 mg/kg
Threonine (g/100 g of total recovered amino acids)	5.7	20 mg/kg
Isoleucine (g/100 g of total recovered amino acids)	5.5	19 mg/kg
Valine (g/100 g of total recovered amino acids)	5.3	24 mg/kg
Phenylalanine (g/100 g of total recovered amino acids)	3.7	33 mg/kg
Tyrosine (g/100 g of total recovered amino acids)	1.8	
Histidine (g/100 g of total recovered amino acids)	1.2	14 mg/kg
Methionine (g/100 g of total recovered amino acids)	0.8	19 mg/kg
Cysteine (g/100 g of total recovered amino acids)	0.2	
Glutamine (g/100 g of total recovered amino acids)	19.8	
Asparagine (g/100 g of total recovered amino acids)	9.7	
Alanine (g/100 g of total recovered amino acids)	7.5	
Glycine (g/100 g of total recovered amino acids)	7.5	
Proline (g/100 g of total recovered amino acids)	4.6	
Serine (g/100 g of total recovered amino acids)	4.5	
Arginine (g/100 g of total recovered amino acids)	3.9	
Hydroxyproline (g/100 g of total recovered amino acids)	2.3	
Lipids (g/100 g)	0.42–0.9	
Linoleic acid (g/100 g)	61.01–75.42	17 g
Oleic acid (g/100 g)	10.83–14.99	
Palmitic acid (g/100 g)	9.12–19.92	
Stearic acid (g/100 g)	3.57–4.63	
Crude fibre (g/100 g)	3.87	
Dietary fibres (g/100 g)	8.65–11.95	38 g
Insoluble fibre	11.55	
Minerals (g/100 g)	0.6–1.5	
Ca (g/100 g)	0.14–0.21	1000 mg

Continued

Table 32.1. Continued

Nutritional composition	Values[a]	RDA[b]
Mg (g/100 g)	0.08–0.11	420 mg
N (g/100 g)	0.85-.91	
P (g/100 g)	0.11–0.12	700 mg
K (g/100 g)	1.30–1.52	4.7 g
S (g/100 g)	0.06–0.07	
Na (g/100 g)	0.05–0.21	1.5 g
B (g/100 g)	0.0011–0.0012	
Cu (g/100 g)	0.0006–0.0011	900 µg
Fe (g/100 g)	0.0016–0.0043	8 mg
Mn (g/100 g)	0.0011–0.0018	2.3 mg
Ni (g/100 g)	0.0001	
Vitamin C (mg/g)	200–242	90 mg

[a] References: Hall *et al.*, 1980; Mukai *et al.*, 1994; Luximon-Ramma *et al.*, 2003; Kinupp and Inchausti, 2008; Silva *et al.*, 2008; Biegelmeyer *et al.*, 2011; Lim, 2012; Pereira *et al.*, 2012; Adrian *et al.*, 2015; [b] RDA recommended for a group of male individuals aged between 31–50 years (Institute of Medicine, 2005; Padovani *et al.*, 2006).

The total phenolic content (TPC) in studies with the whole fruit varied considerably, and results are not consistent (Table 32.3). Biegelmeyer *et al.* (2011) and Chaves *et al.* (2018), for example, found that TPC for red genotype was 1.7- to 1.9-fold higher than TPC for yellow araçá, while in other studies no differences were found between yellow and red varieties (Table 32.3) (Luximon-Ramma *et al.*, 2003; Vinholes *et al.*, 2017). Variations in TPC were also found among red araçá samples dried under different conditions (Bombardelli *et al.*, 2019). The total flavonoid content (TFC) was found to be 2.3- to 2.9-fold higher in red araçá than yellow araçá, probably due to a higher concentration of cyanidins in red araçá varieties. Total proanthocyanidin and anthocyanin content (TPAC, TAC) were also higher in red araçá varieties than in yellow.

The TPC and TAC of different parts of the fruit were also studied. TPC for yellow araçá was in the range of 581–632 mg of GAE/100 g fresh pulp (fp), while for red araçá the variation was 403–768 mg of GAE/100 g fp (Medina *et al.*, 2011). Values of TAC were higher for red (4.82–6.29 mg CGE/100 g fp) compared to yellow (0.21–0.55 mg CGE/100 g fp) varieties (Medina *et al.*, 2011). Similar results between TPC of yellow and red pulp/peel extracts (1.9–2.1 g CAE/100 g dry basis (db)) and their seed extracts (1.4–1.5 g CAE/100 g db) were found (Pereira *et al.*, 2020). Moreover, ultrasound-assisted extraction resulted in higher TPC (589.5 mg GAE/100g of peel (p)), TFC (374.1 mg of catechin equivalents (CE)/100g p) and TAC (121.8 mg of CGE/100g p) when compared to conventional extraction (TPC = 477.5 mg GAE/100g p, TFC = 351.8 mg CE/100g p and TAC = 116.8 mg CGE/100g p) (Meregalli *et al.*, 2020).

The phenolic composition of the whole araçá fruit is shown in Table 32.4. The total number of extractable and non-extractable phenolic compounds in both samples was similar, but the profile was different (Mallmann *et al.*, 2020). The most representative group of extractable compounds present in the yellow variety was the flavonoids (54%), followed by tannins (40%), phenolic acids (6%) and lignins (1%). In comparison, for the red variety, the most representative group was tannins (52%), followed by flavonoids (38%), phenolic acids (10%) and lignins (1%). The most representative extractable compounds for the yellow variety were catechin, proanthocyanidin dimer B I, taxifolin-hexoside, galloyl-bis-HHDP-glucoside II, HHDP-hexoside, eriodictyol-hexoside and (epi)catechin derivative. For the red variety, the most representative were catechin, gallocatechin, cyanidin-O-glucoside, castalagin, taxifolin hexoside and vescalagin (Mallmann *et al.*, 2020).

Table 32.2. Carotenoids in yellow and red araçá fruit.

	Concentration[a]			
	Yellow genotype			Red genotype
	Skin	Pulp	Whole fruit	Skin + pulp
All-*trans*-antheraxanthin	9.0[a]	1.6[a]		
Apocarotenoid		1.9[b]		
All-*trans*-lutein	10.0[a]	1.9[a], 15.7[b]	26.4[c]	557.8[d]
5,6-epoxy-β-cryptoxanthin	7.0[a]	2.4[a]		
5,8-epoxy-β-cryptoxanthin		3.5[b]		
All-*trans*-β-cryptoxanthin	6.0[a]	3.4[a], 26.4[b]		1029.8[d]
All-*trans*-α-cryptoxanthin		3.6[b]		
13′-*cis*-β-cryptoxanthin phytoene		0.6[b]		
Cryptoxanthin			0.9[c]	
All-*trans*-β-carotene	5.9[a]	3.7[a], 20.0[b]		
9-*cis*-β-carotene		3.0[b]	1.3[c]	
13-*cis*-β-carotene			1.2[c]	
5,6-epoxy-β-carotene			1.2[c]	
α-carotene			4.0[c]	60.8[d]
β-carotene			2.9[c]	512.6[d]
All-*trans*-zeaxanthin		1.7[a]	137.5[c]	137.5[d]

[a] References: Pereira *et al*., 2012; Dalla Nora *et al*., 2014; Ribeiro *et al*., 2014; Silva *et al*., 2014.
[b] Expressed as μg/g of extract.
[c] Expressed as μg/100g fw.
[d] Expressed as g/g dry matter (dm).
[e] Expressed as μg/g of extract.

The flavonoids quercetin-glucuronide, taxifolin-hexoside and eriodictyol-hexoside and the tannins galloyl-bis-HHDP-glucoside I, galloyl-bis-HHDP-glucoside II and HHDP-hexoside were common to both studies (Table 32.4) (Vinholes *et al.*, 2018; Mallmann *et al.*, 2020). The phenolic composition of different parts of yellow and red araçá fruit has also been studied. The phenolic composition of yellow araçá skin and pulp showed that compounds were more concentrated in pulp than in skin (Ribeiro *et al.*, 2014). The most representative compounds in skin and pulp were ellagic acid (2213–3818 μg/g extract), ellagic acid-deoxyhexoside (1475–2070 μg/g extract), gallic acid (464–1510 μg/g extract), an ellagitannin-like compound (77–1022 μg/g extract), cinnamoyl-galloyl-hexoside (573–880 μg/g extract) and methyl-ellagic acid-deoxyhexoside (475–646 μg/g extract) (Ribeiro *et al.*, 2014). Other compounds were also found but in lower concentrations. Yellow araçá pulp showed a high concentration of epicatechin (720–2560 μg/g dry fruit (df)) and chlorogenic acid-hexoside (298–727 μg/g df), and smaller amounts of coumaric and ferulic acid, myricetin and quercetin (Medina *et al.*, 2011). Another study reported that gallic acid (12.2 mg/100 g fp) and taxifolin-hexoside (11.7 mg/100 g fp) were the most representative compounds in yellow araçá pulp (Silva *et al.*, 2014). Another 13 compounds were identified, including galloyl and digalloyl-hexoside, di-HHDP-hexoside and its isomer, HHDP-digalloyl-hexoside and its isomer, HHDP-hexoside, quercetin-hexoside, eriodictyol-hexoside, methyl-ellagic acid-hexoside and methyl-ellagic acid-pentoside in concentrations ranging from 2.3 to 8.1 mg/100 g fp (Silva *et al.*, 2014). Yellow araçá juice contains most of the compounds found in other studies, and the compounds in higher concentration were gallic acid, epicatechin, quercitrin, isoquercitrin

Table 32.3. TPC, TFC, TPAC and TAC in whole fruit of yellow and red araçá genotypes.

	Concentration	
	Yellow	Red
	Whole fruit	
TPC	292[a], 5372[b], 603[c], 2671[d], 3820[e]	501.3[a], 5638[b], 606[c], 2698[d], 7190[e], 347-556[f]
TFC	35.1[g], 308[h]	100.2[g], 712[h]
TPAC	2473[i]	2561[i]
TAC	1.3[j], 0.14[k]	29.3[j], 1.7[k]

[a] mg GAE/100 g fresh fruit (ff) (Biegelmeyer *et al.*, 2011).
[b] µg GAE/g of fw (Luximon-Ramma *et al.*, 2003).
[c] mg chlorogenic acid equivalents(CAE)/100 g fw (Vinholes *et al.*, 2017).
[d] µg CAE/ml of extract (Vinholes *et al.*, 2018).
[e] mg GAE/g ff (Chaves *et al.*, 2018).
[f] mg GAE/100 g dw (Bombardelli *et al.*, 2019).
[g] mg quercetin equivalents (QE)/100g fw (Biegelmeyer *et al.*, 2011).
[h] µg QE/g fw (Luximon-Ramma *et al.*, 2003).
[i] µg cyanidin chloride equivalents/g fw (Luximon-Ramma *et al.*, 2003).
[j] mg cyanidin-3-*O*-glucoside equivalents (CGE)/100g fw (Vinholes *et al.*, 2017).
[k] mg CGE/100 g of ff (Chaves *et al.*, 2018).

and quercetin (Hister *et al.*, 2017). More recently, 23 phenolic compounds were quantified in yellow araçá skin +pulp of samples in different maturity stages (Schulz *et al.*, 2020). After sample acid hydrolysis, catechin, isoquercitrin, quercetin, gallic and syringic acids were found in all stages, with values ranging from 479.59 to 12,795.50µg/100 g dm. Phenolic acids were in higher concentration in more mature fruit, while flavonoids were in higher concentrations in less mature fruit (Schulz *et al.*, 2020). Pulp/peel and seeds differ quantitatively mainly due to the higher concentration of catechin in the pulp/peel for both yellow and red araçá (Pereira *et al.*, 2020). The yellow and red araçá differ due to the higher concentration of cyanidin-3-*O*-glucoside in the pulp/peel of red (Pereira *et al.*, 2020). The red araçá pulp has high concentrations of epicatechin (264–2130µg/g fp) and gallic acid (193–801µg/g fp) and lower concentrations of coumaric and ferulic acids, myricetin and quercetin (0.2–32µg/g fp) (Medina *et al.*, 2011). The presence of coloured pigments cyanidin-3-*O*-glucoside (355µg/g df), malvidin-3-*O*-glucoside (244µg/g df) and cyanidin (88µg/g df) were reported in the pulp of red araçá (Dalla Nora *et al.*, 2014).

Volatile compounds

The volatile composition of fruit is responsible for their aroma characteristics, and some compounds also have biological properties (Vinholes *et al.*, 2014a, b; Ayseli and İpek Ayseli, 2016). The major volatile compound found in yellow and red araçá by different authors was (*E*)-β-caryophyllene (Pino *et al.*, 2001; Marin *et al.*, 2008; Biegelmeyer *et al.*, 2011; Egea *et al.*, 2014). More recently, 5-hydroxymethylfurfural was identified as the major compound of red araçá pulp dried under different conditions (Bombardelli *et al.*, 2019). This compound is formed in carbohydrate-rich food exposed to acid-catalysed dehydration or heat (Maillard reaction from reducing sugars) (Husøy *et al.*, 2008). Nevertheless, authors found (*E*)-β-caryophyllene as the second major compound followed by 2,3-Dihydro-3,5-dihydroxy-6-methyl-4H-pyran-4-one, another sugar degradation product (Bombardelli *et al.*, 2019). Other compounds such as hexadecanoic acid, (*Z*)-3-hexenol, α-pinene, α-selinene, β-selinene, neo-intermedeol, α-humulene and geraniol were the major compounds in yellow and red araçás (Pino *et al.*, 2001; Marin *et al.*, 2008; Biegelmeyer *et al.*, 2011; Egea *et al.*, 2014; Bombardelli *et al.*, 2019).

A common flavour pattern of citric, green and sweet was found for yellow and red araçá fruit (Pino *et al.*, 2001). Yellow and red araçá fruit were however distinct by tropical and tomato characteristics, respectively (Egea *et al.*, 2014). The authors could not identify a specific compound related to this distinction.

Table 32.4. Phenolic compounds in the whole fruit of yellow and red araçá.

	Yellow		Red		
	Extractable compounds	Non-extractable compounds	Extractable compounds	Non-extractable compounds	Ethanolic extract
Compounds	µg/g freeze-dry fruit		µg/g freeze-dry fruit		µg/ml of extract
Phenolic acids and derivatives					
Coumaric acid hexoside	6.0		57.9		
p-coumaroyl-feruloyl glycerol I		26.3		23.9	
p-coumaroyl-feruloyl glycerol II		51.7		9.4	
p-coumaroyl-feruloyl glycerol III		22.6		34.4	
Isofraxidin-hexoside I	28.2		16.0		
Isofraxidin-hexoside II	5.0		9.3		
Aconitic acid		9.3		14.3	
Ferulic acid isomer		6.8		9.9	
Diferulic acid I	11.0		11.0		
Diferulic acid II	6.5		14.5		
3-*O*-feruloylquinic acid		17.6		25.8	
5-*O*-feruloylquinic acid		30.7		7.2	
Methyl-5-*O*-feruloylquinate		11.9		6.9	
Hydroxyisophthalic acid		10.2		13.9	
Benzoyl-benzoic acid III		6.2		2.0	
2,4-dihydroxybenzoic		7.5		5.3	
3,4-dihydroxybenzoic acid		6.2		8.1	
3,5-dihydroxybenzoic acid		6.5		5.2	
Valeryloxybenzoic acid I		15.8		25.1	
Valeryloxybenzoic acid II		5.4		8.2	

Continued

Table 32.4. Continued

Compounds	Yellow		Red		
	Extractable compounds	Non-extractable compounds	Extractable compounds	Non-extractable compounds	Ethanolic extract
	µg/g freeze-dry fruit		µg/g freeze-dry fruit		µg/ml of extract
Valeryloxybenzoic acid III		38.0		19.2	
Benzoyl-benzoic acid I		8.3		6.5	
Benzoyl-benzoic acid II		8.1		4.8	
Benzoyl-benzoic acid III		6.2		2.0	
1,3-benzene-dicarboxylic acid		9.1		2.4	
1,4-benzene-dicarboxylic acid		12.3		14.8	
Abscisic acid	8.5		11.7		
Abscisic acid derivative I	9.0		7.5		
Abscisic acid derivative II	5.2		7.3		
Caffeic acid isomer		34.1		30.8	
3-caffeoyl-shikimic acid		8.5		87.1	
4-caffeoyl-shikimic acid		19.2		6.7	
5-*O*-caffeoyl-shikimic acid		11.3		15.5	
Gallic acid	7.4	11.1	9.9	34.1	
Sinapic acid I		4.9		5.5	
Sinapic acid II		6.2		10.4	
Syringic anhydride		6.8		4.2	
Flavonoids					
Catechin	229.3		168.7		
(Epi)catechin isomer		5.9		9.4	

Continued

Table 32.4. Continued

	Yellow		Red		
	Extractable compounds	Non-extractable compounds	Extractable compounds	Non-extractable compounds	Ethanolic extract
Compounds	μg/g freeze-dry fruit		μg/g freeze-dry fruit		μg/ml of extract
(Epi)catechin derivative	14.8		7.9		
(Epi)catechin derivative	75.3		3.5		
(Epi)catechin-(epi)catechin	32.6		21.8		
Epicatechin gallate	3.0		4.8		
Myricetin-pentoside I			5.4		
Myricetin-hexoside	9.5				
Myricetin-pentoside II			5.8		
Luteolin-7-*O*-glucuronide		18.4		12.8	
Quercetin-glucuronide	38.5		17.2		1.3
Quercetin-hexoside	23.6		16.3		
Quercetin-malonyl-glucoronide I	16.5		12.4		
Quercetin-malonyl-glucoronide II	11.4		6.4		
Taxifolin-hexoside	109.2		83.1		1.3
Eriodictyol-hexoside	76.8		41.6		0.7
Cyanidin-3-*O*-hexoside			92.9		
Cyanidin				32.5	
Delphinidin				21.0	
Proanthocyanidin dimer B I	132.3		47.6		
Proanthocyanidin B dimer II	29.8		14.8		
Proanthocyanidin B gallate	17.4		8.7		
Proanthocyanidin B dimer	20.5		9.9		

Continued

Table 32.4. Continued

	Yellow		Red		
	Extractable compounds	Non-extractable compounds	Extractable compounds	Non-extractable compounds	Ethanolic extract
Compounds	µg/g freeze-dry fruit		µg/g freeze-dry fruit		µg/ml of extract
Tannins					
Methyl-gallate		171.5		253	
2,3,4-trihydroxy-2-methylbutyl gallate I		10.9		5.7	
2,3,4-trihydroxy-2-methylbutyl gallate II		6.8		13.6	
Gallagyl-derivative	25.9		30.9		
Galloyl-di-HHDP-hexoside I	20.0		12.4		
Methyl-galloyl-glucopyranoside I		3.9		3.7	
Methyl-galloyl-glucopyranoside II		6.0		8.1	
Galloyl-bis-HHDP	13.2		18.7		
Galloyl-bis-HHDP-glucoside I	34.5		16.4		76
Galloyl-bis-HHDP-glucoside II	80.5		45.7		53
Galloyl-di-HHDP-hexoside II	15.2		25.7		
Galloyl-di-HHDP-hexoside III	9.5		16.5		
Galloyl-di-HHDP-hexoside IV	6.7		20.2		
Digalloyl-HHDP-hexoside I	16.0		44.0		
Digalloyl-HHDP-hexoside II	19.5		15.1		
Digalloyl-(epi)catechin	13.5		24.9		
Trisgalloyl-HHDP-glucoside isomer					72
Trisgalloyl-HHDP-glucoside isomer					42
Trisgalloyl-HHDP-glucoside isomer					32
Trisgalloyl-HHDP-glucoside isomer					42

Continued

Table 32.4. Continued

	Yellow		Red		
	Extractable compounds	Non-extractable compounds	Extractable compounds	Non-extractable compounds	Ethanolic extract
Compounds	µg/g freeze-dry fruit		µg/g freeze-dry fruit		µg/ml of extract
Gallocatechin	45.7		143.8		
Brevifolin		20.9		11.8	
Ellagic acid	12.9	6.4	27.2	22.0	
Ellagic acid-hexoside	10.0		12.7		
Ellagic acid-pentoside			60.5		
Methyl-ellagic acid-pentoside	3.5				
HHDP-hexoside	76.8		43.7		7
HHDP-Hexoside isomer					112
Di-HHDP-hexoside I	6.3		5.1		24
Di-HHDP-hexoside II	10.5		7.6		
Di-HHDP-hexoside III	60.5		51.2		
Di-HHDP-hexoside IV	54.0		42.8		
Castalagin	43.8		85.5		
Vescalagin	42.9		77.1		
Castalagin/vescalagin isomer					57
Castalagin/vescalagin isomer					24
Unknown compound					18
Unknown compound					32
Lignins					
Vanillic acid		53.8		39.5	
Vanillic acid hexoside	8.5		8.0		

They attributed the flavour of araçá as the result of the combination of different volatile compounds found in the fruit (Egea *et al.*, 2014).

Antinutritive Compounds

Antinutrients are commonly found in plant foods and have adverse effects on health. A common example is phytic acid, which forms insoluble complexes with calcium, zinc, iron and copper. Oxalates, lectins, saponins and flavonoids (tannins), along with enzyme inhibitors (amylase and protease) are also antinutrients (Reyden and Sel Vendran, 1993). Araçá contains some antinutrients (Table 32.4). These compounds can exhibit antinutritional properties by impairing the digestion of various nutrients. Tannins can link to proteins, forming tannin–protein complexes, which can inactivate digestive enzymes. However, the inactivation of enzymes can have a positive effect when enzymes such as α-amylase, responsible for carbohydrate metabolism, are inhibited with consequent interruption of carbohydrate hydrolysis and absorption (Reyden and Sel Vendran, 1993; Pereira *et al.*, 2020). This principle is used for the control and treatment of type-2 diabetes mellitus. Recently, the presence of cyanidin-3-*O*-glucoside and malvidin-3-*O*-glucoside in araçá was related to the inhibition of α-amylase (Pereira *et al.*, 2020). Antinutrients can be a powerful tool for the management of different diseases. However, the ingesting dosage must be adjusted to balance the beneficial and harmful effects of bioactive and plant antinutrients. The chemical structure, the time of exposure and the interaction with other diet components should also be considered. A high content of antinutrients in the body can cause symptoms such as nausea, bloating, headaches, rashes and nutritional deficiencies (Popova and Mihaylova, 2019). The reduction of antinutrients in vegetable food can be reached by heat, thus increasing the isolated protein's nutritional values (Seena *et al.*, 2006).

Implications of Food Processing

Red araçá juice, prepared using mechanical and enzymatic treatments, was monitored for four months (Santos *et al.*, 2007). The enzymatic treatment yielded 73% of juice, reduced the viscosity and increased the extraction of phenolic compounds, resulting in a final product with good sensorial aspects. The effect of different processing methods on red araçá fruit phytochemicals was studied (Nora *et al.*, 2014). Fruit freeze-dried for 30 days showed similar TCC compared to fresh fruit; however, the content of individual compounds was distinct. All-*trans* lutein increased 1.5–1.6-fold at 30 days for frozen fruit and freeze-dried fruit, respectively. β-cryptoxanthin decreased 1.3-fold for freeze-dried araçá, while β-carotene decreased 1.2-fold for fruit frozen for 30 days. All carotenoids were reduced for araçá fruit frozen for 90 days and dried by hot air (70°C). Red araçá anthocyanins were more susceptible to processing and storage as cyanidin-3-glucoside, malvidin-3-glucoside and cyanidin were reduced when comparing fresh fruit treatments. The effect of L- ascorbic (1% w/w) addition on red and yellow araçá pulp stored at −20 ± 2°C for 1, 60 and 120 days was also studied (Vergara *et al.*, 2018). Overall, the addition of L-ascorbic acid increased the TPC of araçá pulp compared to controls and maintained the levels along the storage time. A significant increase was observed for anthocyanins for pulp +L-ascorbic acid at day 60 of storage, but the results were not significantly different among storage methods. The antioxidant addition maintained the levels of carotenoids for both genotypes when compared to controls.

The production of araçá jellies without the addition of sugar (aspartame, saccharin and cyclamate, and acesulfame and sucralose) increased the values of lipids and protein and decreased the carbohydrate content and calories when compared to the araçá conventional sugar jelly (sucrose). Moreover, conventional jelly and no added sugars jelly showed similar total phenols, total anthocyanins and ascorbic acid. TCC were significantly higher in no added sugar jelly (Reissig *et al.*, 2016). The bioactive compounds of red and yellow araçá chewable candies (conventional and low-calorie) were evaluated during the storage time (0, 60, 120 and 180 days) (Vergara *et al.*, 2020). Low-calorie candies showed the highest amounts of carotenoids, but at the end of the storage, these values dropped for both red (46.83%) and yellow (12.5%) formulations. High phenolic content was observed for red araçá chewable candies, and the yellow chewable candies showed higher amounts of monomeric anthocyanins after 180 days of storage (Vergara *et al.*, 2020).

Bioactive Properties

Antioxidant

Nora *et al.* (2014) reported that rats fed with red araçá showed a decrease in parameters altered by oxidative stress induced by cisplatin, such as lowered levels of glucose, LDL cholesterol, oxidized LDL cholesterol and total cholesterol when compared with control animals. A decrease in fat deposition in the liver showed an improvement in the lipid profile of animals fed with araçá. In another study, insulin-resistant rats fed with araçá extract (200 mg/kg/day, for 21 days) showed a decrease in liver lipid peroxidation and the formation of ROS (de Souza Cardoso *et al.*, 2018). Similar results were observed when red araçá extract (200 mg/kg/day, for 150 days) was given to rats fed with a highly palatable diet (Oliveira *et al.*, 2018). Araçá extract prevented the alterations of metabolic indicators of oxidative stress caused by the diet. The extract decreased TBARS, nitrite and ROS levels and prevented the reduction of antioxidant enzymes superoxide dismutase and catalase in cerebral structures. Moreover, water and acetone araçá extracts of both genotypes protected *Saccharomyces cerevisiae* against hydrogen peroxide at a concentration of 25% with yeast survival rates above 80% (Medina *et al.*, 2011). These results are supported by the *in vitro* inhibition reported for araçá for different free radicals, reactive oxygen and nitrogen species such as DPPH$^{\bullet}$, $O_2^{\bullet-}$, HOCl, NO$^{\bullet}$ and hydroxyl radicals among others (Fetter *et al.*, 2010; Medina *et al.*, 2011; Ribeiro *et al.*, 2014; Vinholes *et al.*, 2017; Betta *et al.*, 2018; Chaves *et al.*, 2018; Bombardelli *et al.*, 2019; Mallmann *et al.*, 2020; Meregalli *et al.*, 2020; Pereira *et al.*, 2020; Schulz *et al.*, 2020)

Antiageing activity

Adult mice fed with araçá extract (1000 mg/kg/day, eight months) had a different hippocampus gene expression profile when compared to control animals (Ramirez *et al.*, 2012). Araçá extract administration altered the expression of genes involved in different physiological functions, (e.g. metabolism, transport, signal transduction, etc.). It was observed that cell cycle, proliferation, differentiation signal transduction and inflammatory genes (inhibiting inflammatory process) were regulated, and genes encoding for proteins involved in radical scavenging were up-regulated.

Antiproliferative activity

The antiproliferative effect of red and yellow araçá extracts was studied on mammary cancer cells (MCF-7) and colon cancer cells (Caco-2 cells) (Medina *et al.*, 2011). Two extracts (water and acetone) were tested at different concentrations (40, 60 and 80 µg/ml), which resulted in a reduction in proliferation of both cancerous cells without affecting the survival of control cells (rat embryonic fibroblast cells 3T3). The antiproliferative effect was dependent on the concentration but independent of the genotype.

Antidiabetic activity

Araçá extract administration (200 mg/kg/day, for 21 days) on rats with insulin resistance induced by dexamethasone prevented hyperglycaemia and hypertriglyceridemia (de Souza Cardoso *et al.*, 2018). The effect of red araçá extract administration (200 mg/kg/d for 150 days) on metabolic parameters in Wistar rats fed with a highly palatable diet also prevented the alterations caused by the diet such as glucose intolerance, increased weight gain and visceral fat, high serum levels of glucose, triacylglycerol, total cholesterol, LDL-cholesterol and interleukin-6 (Oliveira *et al.*, 2018). *In vitro* studies showed that araçá extracts are potential inhibitors of the digestive enzymes α-glucosidase and α-amylase, which may be the mechanism of action of araçá (Pacheco, 2015; Vinholes *et al.*, 2017, 2018; Pereira *et al.*, 2020).

Anti-inflammatory and antithrombotic activities

Oliveira *et al.* (2020) studied the effect of red araçá fruit extracts on inflammatory and thromboregulatory parameters using a male Wistar rats metabolic syndrome model induced by a highly palatable diet. Araçá extract (200 mg/kg/d for 150 days) prevented the increase of the inflammatory parameters interleukin-6 and butyrylcholinesterase in serum and acetylcholinesterase in lymphocytes and prevented the decrease of thromboregulatory parameters NTPDase activity in lymphocytes and platelets and 50-nucleotidase in platelets. The *in vitro* anti-inflammatory activity of araçá hexane and ethyl acetate extracts (250 μg/ml) inhibited 18.3% and 26.5% cyclooxygenase-2 enzyme, respectively (McCook-Russell *et al.*, 2012).

Antimicrobial activity

Medina *et al.* (2011) studied the antibacterial activity of yellow and red araçá extracts against *Salmonella enteritidis*, a foodborne pathogen responsible for human toxinfection. The minimum inhibitory concentration was achieved at 5%, and extracts showing higher metabolite concentrations were those with high antibacterial effect. An intermediate inhibitory activity was reported for araçá against *Bacillus subtilis* and *Staphylococcus aureus* (McCook-Russell *et al.*, 2012). More recently, de Lima *et al.* (2020) evaluated the antibacterial activity and mechanism of action of hydroalcoholic extract in araçá (mixture of both yellow and red araçá ripe fruit, 75% ethanol) against *S. aureus*. Extracts of minimum inhibitory concentration and minimum bactericidal concentration were 9 and 18 mg/ml, respectively. The *S. aureus* cell membrane was compromised, losing its integrity and releasing important cellular constituents. The extract was able to intercalate with the bacterial DNA, indicating that araçá antibacterial activity can occur on more than one target in the bacteria.

Brain activity

Oliveira *et al.* (2018) studied the effects of red araçá extract (200 mg/kg/d for 150 days) on behavioural and neurochemical parameters in Wistar rats fed with a highly palatable diet. An antidepressant-like effect was observed due to the prevention of immobility in the forced swimming test. Moreover, araçá treatment prevented the increase of acetylcholinesterase activity in the prefrontal cortex, restored the Ca2+-ATPase activity in the prefrontal cortex, hippocampus and striatum, and restored the Na+,K+-ATPase activity in the prefrontal cortex and hippocampus.

Bioaccessibility

Yellow and red araçá fruit extract showed a recovery of 28% of TPC after a simulated *in vitro* digestion (Vinholes *et al.*, 2018). Both digested extracts were able to inhibit the α-glucosidase enzyme with 85.16 and 89.74% for yellow and red extracts, respectively, values similar to the non-digested extract. DPPH inhibition followed the same trend, while a reduction in the inhibition of hydroxyl radicals was observed. Compound characterization after digestion showed that red araçá extract maintained only 5% of compounds, the tannins being the most affected. Citric acid showed the highest recovery (49%), followed by unknown compound (17%), eriodictyol-hexoside (16%), quercetin-glucuronide (13%), trisgalloyl-HHDP-glucoside isomer and Galloyl-bis-HHDP-glucoside (8%), taxifolin-hexoside (6%) and trisgalloyl-HHDP-glucoside isomer (6%). The authors suggested that ellagitannin hydrolysis may form other undetected compounds, with high α-glucosidase and antioxidant inhibitory activities.

Novel Products/Formulations

Araçá has been used to develop new products and formulations. Red araçá juice was produced using mechanical and enzymatic process, and both were monitored for four months. The enzymatic treatment of juice resulted in a higher yield (73%) and preserved its chemical composition with an increase in phenolic compound concentration, resulting in a juice with remarkable sensorial aspects (Santos *et al.*, 2007). Reissig *et al.* (2016) produced red and yellow araçá jellies without the addition of sugar to promote healthier options for consumers. Four formulations were prepared: conventional jelly (sucrose) and no added sugar (aspartame/saccharin and cyclamate/acesulfame and sucralose). The results showed that important bioactive compounds (phenolic compounds, anthocyanins and ascorbic acid) were not lost compared with the conventional jelly, and these new products could be a good option for the use of araçá pulp.

Red araçá pulp was microencapsulated to minimize the loss of bioactive compounds caused by pulp processing (Rosário *et al.*, 2020). Different coating materials in different blends and equal proportions were tested. High encapsulation efficiency was obtained for carotenoids, anthocyanins and antioxidant activity using hydrolysed pinhão starch (PS), tara gum (TG), Arabic gum (AG), TG and AG, and PSTG, respectively. The coating material improved the retention of bioactive compounds and the stability of antioxidant activity. Araçá pulp extract was also evaluated as an antioxidant in biodiesel (Chendynski *et al.*, 2020). The biodiesel +araçá pulp extract oxidation was induced in Rancimat equipment at 110°C, with 10 l/h airflow. Araçá alone showed an induction period of 7.18 h, a value above the minimum of six hours provided in Brazilian legislation. Araçá combined with other extracts was also an efficient antioxidant with induction periods varying from 7.24 to 7.69 h.

Conclusions and Future Trends

Psidium cattleianum is a species resistant to diseases and pests, which can adapt and be cultivated under different and severe environmental conditions. Fruit produced under these conditions show high amounts of secondary metabolites with high antioxidant activity and bioactive properties, along with a good nutritional profile. Nevertheless, this is a highly perishable and underutilized fruit, mainly produced wild or in small family orchards. Therefore, the cultivation process (techniques of vegetative propagation, varieties, cultural practices, mineral nutrition and fertilization) and postharvest handling must be investigated. To avoid waste and better use of the fruit and to maintain the fruit characteristics for food and pharmaceutical applications, different processing methods have been studied. Araçá frozen pulp and microencapsulation seem to be efficient methods for the retention of bioactive compounds. In this sense, further research is needed to apply microcapsules loaded with araçá pulp/extracts for developing new food/pharmaceutical formulations, mainly because digestion can significantly reduce the content of bioactive compounds.

Acknowledgement

This study was supported by the National Council for Scientific and Technological Development (CNPq), Brazil, (process number 400201/2014-3, grant number 313712/2014-0), and the Institute of International Education (CAPES), Brazil (process number 1591849, grant number 2016/2018).

References

Adrian, J.A.L., Arancon, N.Q., Mathews, B.W. and Carpenter, J.R. (2015) Mineral composition and soil-plant relationships for common guava (*Psidium guajava* L.) and yellow strawberry guava (*Psidium cattleianum* var. lucidum) tree parts and fruits. *Communications in Soil Science and Plant Analysis* 46(15), 1960–1979. DOI: 10.1080/00103624.2015.1069310.

Ayseli, M.T. and İpek Ayseli, Y. (2016) Flavors of the future: Health benefits of flavor precursors and volatile compounds in plant foods. *Trends in Food Science & Technology* 48, 69–77. DOI: 10.1016/j.tifs.2015.11.005.

Betta, F.D., Nehring, P., Seraglio, S.K.T., Schulz, M., Valese, A.C. *et al.* (2018) Phenolic compounds determined by LC-MS/MS and *in vitro* antioxidant capacity of Brazilian fruits in two edible ripening stages. *Plant Foods for Human Nutrition* 73(4), 302–307. DOI: 10.1007/s11130-018-0690-1.

Bezerra, J.E.F. (2010) Araça. In: Vieira, R.F. (ed.) *Frutas Nativas Da Região Centro-Oeste Do Brasil.* Embrapa Informação Tecnológica, Brasilia, Brazil, pp. 42–62.

Biegelmeyer, R., Andrade, J.M.M., Aboy, A.L., Apel, M.A., Dresch, R.R. *et al.* (2011) Comparative analysis of the chemical composition and antioxidant activity of red (*Psidium cattleianum*) and yellow (*Psidium cattleianum* var. lucidum) strawberry guava fruit. *Journal of Food Science* 76(7), C991-6. DOI: 10.1111/j.1750-3841.2011.02319.x.

Bombardelli, M.C.M., Johann, G., Machado, C.S., Torres, Y.R., Palú, F. *et al.* (2019) Antioxidant activity and volatile composition of red araçá pulp under different drying conditions. *International Journal of Chemical Engineering* 2019, 1–9. DOI: 10.1155/2019/9836929.

Castro, C., Raseira, M. do C.B. and Franzon, R.C. (2004) Descrição da Planta. In: Raseira, M. do C.B. (ed.) *Espécies Frutíferas Nativas Do Sul Do Brasil*, 1st edn. Embrapa Clima Temperado, Pelotas, pp. 13–28.

Chaves, V.C., Boff, L., Vizzotto, M., Calvete, E., Reginatto, F.H. *et al.* (2018) Berries grown in Brazil: anthocyanin profiles and biological properties. *Journal of the Science of Food and Agriculture* 98(11), 4331–4338. DOI: 10.1002/jsfa.8959.

Chendynski, L.T., Cordeiro, T., Messias, G.B., Mantovani, A.C.G., Spacino, K.R. *et al.* (2020) Evaluation and application of extracts of rosemary leaves, araçá pulp and peel of bacuri in the inhibition of the oxidation reaction of biodiesel. *Fuel* 261(June 2019), 116379. DOI: 10.1016/j.fuel.2019.116379.

Dalla Nora, C., Jablonski, A., Rios, A. de O., Hertz, P.F., de Jong, E.V. *et al.* (2014) The characterisation and profile of the bioactive compounds in red guava (*Psidium cattleyanum* Sabine) and guabiju (*Myrcianthes pungens* (O. Berg) D. Legrand). *International Journal of Food Science & Technology* 49(8), 1842–1849. DOI: 10.1111/ijfs.12493.

de Lima, A.S., Maia, D.V., Haubert, L., Oliveira, T.L., Fiorentini, Â.M. *et al.* (2020) Action mechanism of araçá (*Psidium cattleianum* Sabine) hydroalcoholic extract against Staphylococcus aureus. *LWT* 119, 108884. DOI: 10.1016/j.lwt.2019.108884.

de Souza Cardoso, J., Oliveira, P.S., Bona, N.P., Vasconcellos, F.A., Baldissarelli, J. *et al.* (2018) Antioxidant, antihyperglycemic, and antidyslipidemic effects of Brazilian-native fruit extracts in an animal model of insulin resistance. *Redox Report* 23(1), 41–46. DOI: 10.1080/13510002.2017.1375709.

Denardin, C.C., Hirsch, G.E., da Rocha, R.F., Vizzotto, M., Henriques, A.T. *et al.* (2015) Antioxidant capacity and bioactive compounds of four Brazilian native fruits. *Journal of Food and Drug Analysis* 23(3), 387–398. DOI: 10.1016/j.jfda.2015.01.006.

Drehmer, A.M.F. and Amarante, C.V.T. do (2008) Conservação pós-colheita de frutos de araçá-vermelho em função do estádio de maturação e temperatura de armazenamento. *Revista Brasileira de Fruticultura* 30(2), 322–326. DOI: 10.1590/S0100-29452008000200009.

Egea, M.B., Pereira-Netto, A.B., Cacho, J., Ferreira, V., Lopez, R. *et al.* (2014) Comparative analysis of aroma compounds and sensorial features of strawberry and lemon guavas (*Psidium cattleianum* Sabine). *Food Chemistry* 164, 272–277. DOI: 10.1016/j.foodchem.2014.05.028.

Fernandes-Santos, C.A., Cunha-Castro, J.M., de França-Souza, F., Alcântara-Vilarinho, A., do Ferreira, F.R. *et al.* (2010) Prospecting and morphological characterization of *Brazilian Psidium* germplasm. *Acta Horticulturae* 849(849), 63–68. DOI: 10.17660/ActaHortic.2010.849.6.

Fetter, M. da R., Vizzotto, M., Corbelini, D.D. and Gonzalez, T.N. (2010) Propriedades funcionais de araçá-amarelo, araçá-vermelho (*Psidium cattleyanum* Sabine) e araçá-pera (*P. acutangulum* D.C.) cultivados em pelotas/RS. *Brazilian Journal of Food Technology* 13(EE01), 92–95. DOI: 10.4260/BJFT20101304115.

Fiedor, J. and Burda, K. (2014) Potential role of carotenoids as antioxidants in human health and disease. *Nutrients* 6(2), 466–488. DOI: 10.3390/nu6020466.

Figueroa-Méndez, R. and Rivas-Arancibia, S. (2015) Vitamin C in health and disease: Its role in the metabolism of cells and redox state in the brain. *Frontiers in Physiology* 6(DEC), 1–11. DOI: 10.3389/fphys.2015.00397.

Franzon, R.C., Campos, L.Z. de O., Proenca, C.E.B. and Sousa-Silva, J.C. (2009) *Araçás Do Gênero Psidium: Principais Espécies, Ocorrência, Descrição e Usos - Documentos 266*. Embrapa Cerrados, Planaltina, DF, Brazil.

Galho, A.S., Lopes, N.F., Bacarin, M.A. and Lima, M. da G. de S. (2007) Composição química e respiração de crescimento em frutos de *Psidium cattleyanum*. Sabine durante o ciclo de desenvolvimento. *Revista Brasileira de Fruticultura* 29(1), 61–66. DOI: 10.1590/S0100-29452007000100014.

Hall, N.T., Smoot, J.M., Knight, R.J. and Nagy, S. (1980) Protein and amino acid compositions of ten tropical fruits by gas-liquid chromatography. *Journal of Agricultural and Food Chemistry* 28(6), 1217–1221. DOI: 10.1021/jf60232a066.

Hister, C.A.L., Boligon, A.A., Dail Laughinghouse, H. and Tedesco, S.B. (2017) Determination of phenolic compounds and assessment of the genotoxic and proliferative potential of *Psidium cattleianum* Sabine (Myrtaceae) fruits. *Caryologia* 70(4), 350–356. DOI: 10.1080/00087114.2017.1360571.

Husøy, T., Haugen, M., Murkovic, M., Jöbstl, D., Stølen, L.H. *et al*. (2008) Dietary exposure to 5-hydroxymethylfurfural from Norwegian food and correlations with urine metabolites of short-term exposure. *Food and Chemical Toxicology* 46(12), 3697–3702. DOI: 10.1016/j.fct.2008.09.048.

Institute of Medicine (2005) *Dietary Reference Intakes for Energy, Carbohydrate, Fiber, Fat, Fatty Acids, Cholesterol, Protein, and Amino Acids*. The National Academies Press, Washington, DC.

Kinupp, V.F. and Inchausti, I.B. (2008) Protein and mineral contents of native species, potential vegetables, and fruits. *Ciencia e Tecnologia de Alimentos* 28(4), 846–857.

Lim, T.K. (2012) *Edible Medicinal and Non Medicinal Plants Volume 3, Fruits*. pp. 1–159. DOI: 10.1007/978-94-007-2534-8.

Lisbôa, G.N., Kinupp, V.F. and Barros, I.B.I. de (2011) *Psidium cattleianum* Araçá. In: Coradin, L., Siminski, A. and Reis, A. (eds) *Espécies Nativas Da Flora Brasileira de Valor Econômico Atual Ou Potencial Plantas Para o Futuro - Região Sul*. Ministério do Meio Ambiente, Brasilia, Brazil, pp. 205–208.

Luximon-Ramma, A., Bahorun, T. and Crozier, A. (2003) Antioxidant actions and phenolic and vitamin C contents of common Mauritian exotic fruits. *Journal of the Science of Food and Agriculture* 83(5), 496–502. DOI: 10.1002/jsfa.1365.

Mallmann, L.P., Tischer, B., Vizzotto, M., Rodrigues, E., Manfroi, V. *et al*. (2020) Comprehensive identification and quantification of unexploited phenolic compounds from red and yellow araçá (*Psidium cattleianum* Sabine) by LC-DAD-ESI-MS/MS. *Food Research International (Ottawa, Ont.)* 131(January), 108978. DOI: 10.1016/j.foodres.2020.108978.

Marin, R., Apel, M.A., Limberger, R.P., Maria, C.B., Pereira, J.F.M. *et al*. (2008) Volatile components and antioxidant activity from some myrtaceous fruits cultivated in Southern Brazil. *Latin American Journal of Pharmacy* 27(2), 172–177.

Mattos, J.R. (1989) *Myrtaceae Do Rio Grande Do Sul*, 1st edn. Editora Gráfica CEUE, Porto Alegre.

McCook-Russell, K.P., Nair, M.G., Facey, P.C. and Bowen-Forbes, C.S. (2012) Nutritional and nutraceutical comparison of Jamaican *Psidium cattleianum* (strawberry guava) and *Psidium guajava* (common guava) fruits. *Food Chemistry* 134(2), 1069–1073. DOI: 10.1016/j.foodchem.2012.03.018.

Medina, A.L., Haas, L.I.R., Chaves, F.C., Salvador, M., Zambiazi, R.C. *et al*. (2011) Araçá (*Psidium cattleianum* Sabine) fruit extracts with antioxidant and antimicrobial activities and antiproliferative effect on human cancer cells. *Food Chemistry* 128(4), 916–922. DOI: 10.1016/j.foodchem.2011.03.119.

Meregalli, M.M., Puton, B.M.S., Camera, F.D., Amaral, A.U., Zeni, J. *et al*. (2020) Conventional and ultrasound-assisted methods for extraction of bioactive compounds from red araçá peel (*Psidium cattleianum* Sabine). *Arabian Journal of Chemistry* 13(6), 5800–5809. DOI: 10.1016/j.arabjc.2020.04.017.

Mitra, S.K., Irenaeus, T.K.S., Gurung, M.R. and Pathak, P.K. (2012) Taxonomy and importance of Myrtaceae. *Acta Horticulturae* 959(959), 23–34. DOI: 10.17660/ActaHortic.2012.959.2.

Mukai, H., Takagi, T., Nakamura, Y. and Suzuki, T. (1994) 722 PB 438 Fruit quality of strawberry guava. *HortScience: A Publication of the American Society for Horticultural Science* 29(5), 536e–5536. DOI: 10.21273/HORTSCI.29.5.536e.

Naczk, M. and Shahidi, F. (2004) Extraction and analysis of phenolics in food. *Journal of Chromatography. A* 1054(1–2), 95–111. DOI: 10.1016/S0021-9673(04)01409-8.

Nora, C.D., Danelli, D., Souza, L.F., Rios, A. de O., Jong, E.V. de *et al*. (2014) Protective effect of guabiju (*Myrcianthespungens* (O. Berg) D. Legrand) and red guava (*Psidium cattleyanum* Sabine) against cisplatin-induced hypercholesterolemia in rats. *Brazilian Journal of Pharmaceutical Sciences* 50(3), 483–491. DOI: 10.1590/S1984-82502014000300006.

Oliveira, P.S., Chaves, V.C., Soares, M.S.P., Bona, N.P., Mendonça, L.T. *et al.* (2018) Southern Brazilian native fruit shows neurochemical, metabolic and behavioral benefits in an animal model of metabolic syndrome. *Metabolic Brain Disease* 33(5), 1551–1562. DOI: 10.1007/s11011-018-0262-y.

Oliveira, P.S., Soares, M.S.P., Bona, N.P., da Silva, P.G., Mendonça, L.T. *et al.* (2020) Brazilian native fruit extracts act as preventive agents modulating the purinergic and cholinergic signalling in blood cells and serum in a rat model of metabolic syndrome. *Archives of Physiology and Biochemistry* 1–8. DOI: 10.1080/13813455.2020.1743723.

Olson, J.A. (1992) Carotenoids and Vitamin A: An overview. In: Ong, A.S.H. and Packer, L. (eds) *Lipid-Soluble Antioxidants: Biochemistry and Clinical Applications*. Birkhäuser Basel, Basel, Switzerland, pp. 178–192.

Pacheco, S.M. (2015) Frutos da família Myrtaceae: Caracterização físico-química e potencial inibitório daatividade das enzimas digestivas. Universidade Federal de Pelotas. Available at: http://guaiaca.ufpel.edu.br:8080/handle/prefix/3055

Padovani, R.M., Amaya-Farfán, J., Colugnati, F.A.B. and Domene, S.M.Á. (2006) Dietary reference intakes: aplicabilidade das tabelas em estudos nutricionais. *Revista de Nutrição* 19(6), 741–760. DOI: 10.1590/S1415-52732006000600010.

Patel, S. (2012) Exotic tropical plant *Psidium cattleianum*: a review on prospects and threats. *Reviews in Environmental Science and Bio/Technology* 11(3), 243–248. DOI: 10.1007/s11157-012-9269-8.

Pereira, M.C., Steffens, R.S., Jablonski, A., Hertz, P.F., Rios, A. de O. *et al.* (2012) Characterization and antioxidant potential of Brazilian fruits from the Myrtaceae family. *Journal of Agricultural and Food Chemistry* 60(12), 3061–3067. DOI: 10.1021/jf205263f.

Pereira, E. dos S., Vinholes, J.R., Camargo, T.M., Nora, F.R., Crizel, R.L. *et al.* (2020) Characterization of araçá fruits (*Psidium cattleianum* Sabine): Phenolic composition, antioxidant activity and inhibition of α-amylase and α-glucosidase. *Food Bioscience* 37, 100665. DOI: 10.1016/j.fbio.2020.100665.

Pereira, E.D.S., Vinholes, J., C Franzon, R., Dalmazo, G., Vizzotto, M. *et al.* (2018) *Psidium cattleianum* fruits: A review on its composition and bioactivity. *Food Chemistry* 258, 95–103. DOI: 10.1016/j.foodchem.2018.03.024.

Perera, C.O. and Yen, G.M. (2007) Functional properties of carotenoids in human health. *International Journal of Food Properties* 10(2), 201–230. DOI: 10.1080/10942910601045271.

Pino, J.A., Marbot, R. and Vázquez, C. (2001) Characterization of volatiles in strawberry guava (*Psidium cattleianum* Sabine) fruit. *Journal of Agricultural and Food Chemistry* 49(12), 5883–5887. DOI: 10.1021/jf010414r.

Pohl, H.R., Wheeler, J.S. and Murray, H.E. (2013) Sodium and potassium in health and disease. *Metal Ions Life Science* 13, 29–47. DOI: 10.1007/978-94-007-7500-8_2.

Popova, A. and Mihaylova, D. (2019) Antinutrients in Plant-based Foods: A Review. *Open Biotechnology Journal* 13, 68–76. DOI: 10.2174/1874070701913010068.

Ramirez, M.R., Zanchin, N.I.T., Henriques, A.T. and Zuanazzi, J.A.S. (2012) Study of the effects of *psidium cattleyanum* on gene expression from senescent mouse hippocampus. *Boletim Latinoamericano y Del Caribe de Plantas Medicinales y Aromaticas* 11(2), 127–137.

Raseira, A. and Raseira, M. do C.B. (1990) Fruteiras nativas de clima temperado. *Hortisul* 1(2), 47–51.

Raseira, M.C.B. and Raseira, A. (1996) *Contribuição Ao Estudo Do Araçazeiro, Psidium Cattleyanum*. EMBRAPA-CPACT, Pelotas. Available at: https://www.embrapa.br/busca-de-publicacoes/-/publicacao/742996/contribuicao-ao-estudo-do-aracazeiro-psidium-cattleyanum (accessed 12 April 2022).

Reissig, G.N., Vergara, L.P., Franzon, R.C., Rodrigues, R.D.S. and Chim, J.F. (2016) Bioactive compounds in conventional and no added sugars red strawberry guava (*Psidium cattleianum* sabine) jellies. *Revista Brasileira de Fruticultura* 38(3), 1–7. DOI: 10.1590/0100-29452016062.

Reyden, P. and Sel Vendran, R.R. Macrae, R., Robinson, R.K. and Sadler, M.J. (eds). (1993) *Phytic Acid: Properties and Determination*. Academic Press, London, pp. 3582–3587. DOI: 10.1016/B978-0-12-384947-2.00544-4.

Reynertson, K.A., Yang, H., Jiang, B., Basile, M.J., Kennelly, E.J. *et al.* (2008) Quantitative analysis of antiradical phenolic constituents from fourteen edible Myrtaceae fruits. *Food Chemistry* 109(4), 883–890. DOI: 10.1016/j.foodchem.2008.01.021.

Ribeiro, A.B., Chisté, R.C., Freitas, M., da Silva, A.F., Visentainer, J.V. *et al.* (2014) *Psidium cattleianum* fruit extracts are efficient in vitro scavengers of physiologically relevant reactive oxygen and nitrogen species. *Food Chemistry* 165, 140–148. DOI: 10.1016/j.foodchem.2014.05.079.

Rosário, F.M., Biduski, B., Santos, D.F.D., Hadlish, E.V., Tormen, L. *et al.* (2020) Red araçá pulp microencapsulation by hydrolyzed pinhão starch, and tara and arabic gums. *Journal of the Science of Food and Agriculture* 101(5), 2052–2062. DOI: 10.1002/jsfa.10825.

Santos, M.D.S., Petkowicz, C.L. de O., Wosiacki, G., Nogueira, A., Carneiro, E.B.B. *et al.* (2007) Caracterização do suco de araçá vermelho (*Psidium cattleianum* Sabine) extraído mecanicamente e tratado enzimaticamente. *Acta Scientiarum. Agronomy* 29(5), 617–621. DOI: 10.4025/actasciagron.v29i5.737.

Schulz, M., Seraglio, S.K.T., Della Betta, F., Nehring, P., Valese, A.C. *et al.* (2020) Determination of phenolic compounds in three edible ripening stages of yellow guava (*Psidium cattleianum* Sabine) after acidic hydrolysis by LC-MS/MS. *Plant Foods for Human Nutrition* 75(1), 110–115. DOI: 10.1007/s11130-019-00792-0.

Seena, S., Sridhar, K.R., Arunb, A.B. and Young, C.C. (2006) Effect of roasting and pressure-cooking on nutritional and protein quality of seeds of mangrove legume Canavalia cathartica from southwest coast of India. *Journal of Food Composition and Analysis* 19, 284–29. DOI: 10.1016/j.jfca.2005.05.004.

Shahidi, F. and Ambigaipalan, P. (2015) Phenolics and polyphenolics in foods, beverages and spices: Antioxidant activity and health effects – A review. *Journal of Functional Foods* 18, 820–897. DOI: 10.1016/j.jff.2015.06.018.

Silva, M.R., Lacerda, D.B.C.L., Santos, G.G. and Martins, D.M. de O. (2008) Caracterização química de frutos nativos do cerrado. *Ciência Rural* 38(6), 1790–1793. DOI: 10.1590/S0103-84782008000600051.

Silva, N.A.da, Rodrigues, E., Mercadante, A.Z. and de Rosso, V.V. (2014) Phenolic compounds and carotenoids from four fruits native from the Brazilian Atlantic Forest. *Journal of Agricultural and Food Chemistry* 62(22), 5072–5084. DOI: 10.1021/jf501211p.

Vergara, L.P., Reissig, G.N., Franzon, R.C., Carvalho, I.R., Zambiazi, R.C. *et al.* (2020) Stability of bioactive compounds in conventional and low-calorie sweet chewable candies prepared with red and yellow strawberry guava pulps. *International Food Research Journal* 27(4), 625–634.

Vergara, L.P., Reissig, G.N., Franzon, R.C., Rodrigues, R. da S. and Chim, J.F. (2018) Bioactive compound retention in frozen red and yellow Strawberry guava pulps added with L-Ascorbic acid. *Revista Brasileira de Fruticultura* 40(6). DOI: 10.1590/0100-29452018032.

Verma, A.K., Rajkumar, V., Banerjee, R., Biswas, S. and Das, A.K. (2013) Guava (*Psidium guajava* L.) powder as an antioxidant dietary fibre in sheep meat nuggets. *Asian-Australasian Journal of Animal Sciences* 26(6), 886–895. DOI: 10.5713/ajas.2012.12671.

Vinholes, J., Gonçalves, P., Martel, F., Coimbra, M.A., Rocha, S.M. *et al.* (2014a) Assessment of the antioxidant and antiproliferative effects of sesquiterpenic compounds in *in vitro* Caco-2 cell models. *Food Chemistry* 156, 204–211. DOI: 10.1016/j.foodchem.2014.01.106.

Vinholes, J., Rudnitskaya, A., Gonçalves, P., Martel, F., Coimbra, M.A. *et al.* (2014b) Hepatoprotection of sesquiterpenoids: A quantitative structure-activity relationship (QSAR) approach. *Food Chemistry* 146, 78–84. DOI: 10.1016/j.foodchem.2013.09.039.

Wu, G. (2009) Amino acids: Metabolism, functions, and nutrition. *Amino Acids* 37(1), 1–17. DOI: 10.1007/s00726-009-0269-0.

Vinholes, J., Lemos, G., Lia Barbieri, R., Franzon, R.C. and Vizzotto, M. (2017) *In vitro* assessment of the antihyperglycemic and antioxidant properties of araçá, butiá and pitanga. *Food Bioscience* 19, 92–100. DOI: 10.1016/j.fbio.2017.06.005.

Vinholes, J., Reis, S.F., Lemos, G., Barbieri, R.L. and de Freitas, V. (2018) Effect of *in vitro* digestion on the functional properties of *Psidium cattleianum* Sabine (araçá), *Butia odorata* (Barb. Rodr.) Noblick (butiá) and *Eugenia uniflora* L. (pitanga) fruit extracts. *Food & Function* 9(12), 6380–6390. DOI: 10.1039/c8fo01329b.

33 Indian Gooseberry (Amla) (*Phyllanthus emblica* L.)

Hosakatte Niranjana Murthy* and Guggalada Govardhana Yadav
Department of Botany, Karnatak University, Dharwad, India

Abstract

Phyllanthus emblica L. (family Phyllanthaceae) is commonly known as Indian gooseberry, which yields edible fruit. Indian gooseberry is eaten raw and used for culinary purposes. The fruit are rich in nutrients, vitamins, amino acids and minerals. Fresh or dried fruit, seeds, leaves, roots and bark are used in the preparation of herbal medicines in the Ayurveda, Siddha, Unani, Tibetan and Chinese systems of medicine. It is used as the primary ingredient in the preparation of Ayurvedic polyherbal formulations such as chyawanprash and triphala. It is also used in shampoos and hair oils. The fruit and other parts of the plant are rich in flavonoids, tannins, phenolic acids, sterols and terpenoids. The fruit, seeds, leaves, roots and bark of the plant are used against varied ailments like diabetes, hypertension, inflammation, cancer, respiratory, neurological and other infectious disorders. This review summarizes the traditional uses, nutritional components, phytochemicals and biological activities of Indian gooseberry. Bioaccessibility and bioavailability of bioactive compounds, novel products or formulations of Indian gooseberry are also discussed.

Introduction

Phyllanthus emblica L. (*Emblica officinalis* Gaertn.) is popularly known as Indian gooseberry or amla, and yields edible fruit. The fruit of the Indian gooseberry are rich in nutrients and valuable bioactive phytochemicals. Indian gooseberry is native to India and widely distributed in India, Nepal, Sri Lanka and southern China. It is also generally cultivated in India, Thailand, Malaysia, Mauritius, Japan, Honduras, Costa Rica, Singapore, southern China and the West Indies for its fruit. It grows well in varied types of soil including sandy loam to clay soils, and humid to arid conditions. Indian gooseberry is propagated through budding and grafting, and even by seeds. Plants are normally planted at a distance of 6–10 m in 1 m pits along with soil, farmyard manure and neem cake. Grafted plants establish quickly and start bearing fruit within three to four years of planting, whereas plants established by seedling start bearing fruit after five to eight years. Mature trees (usually after ten years) will yield 100–300 kg of fruit. The average weight of the fruit is 60–70 g. Several varieties have been developed in India and released for cultivation. Some of the popular varieties are Banarasi, Chakaiya, Francis, NA-4 (Krishna), NA-6 (Kanchan), NA-7, NA-19 and BSR-1. Treating Indian gooseberry fruit with calcium nitrate (1%) or borax (4%) will enhance shelf life up to 20 days (Nath *et al.*, 1992). Similarly, the shelf life of Indian gooseberry fruit was extended to 15–20 days with waxol (6%) and growth retardant 2-chloroethyl-trimethylammonium chloride (400 ppm) and packed in perforated

*Corresponding author: hnmurthy60@gmail.com

 Handbook of Phytonutrients in Indigenous Fruits and Vegetables (eds D. Sivakumar *et al.*)
DOI: 10.1079/9781789248067.0033

Fig. 33.1. (a) Indian gooseberry (*Phyllanthus emblica*) tree; (b) leaves; (c) bark; (d) fruit; (e) transverse section of fruit.

polyethylene bags (Singh *et al.*, 2014). Nylon or perforated polythene bags, ventilated corrugated fibre boxes or wooden crates have been used for packing and storage. The fruit are harvested at maturity and can be stored for 20 days in cold storage at 0–2°C and 80–90% relative humidity.

Botany

Indian gooseberry belongs to the family Phyllanthaceae, is a medium-sized, deciduous tree, grows up to 5–25 m tall (Fig. 33.1a) and the trunk is up to 30–50 cm in diameter. Branches are horizontally spread and trees are normally pruned to not grow very tall. Leaves are borne on the main branches which are deltoid-squamiform. The branchlets are glabrous with few hairs, 10–20 cm long, usually bearing biseriate, alternate, thin, simple, light green, subsessile, closely-oppressed leaves. Leaves are narrowly oblong, 5–25 mm × 1–5 mm lamina, rounded, with a more or less oblique base, acute or obtuse and mucronate apex and pinnatinerved (Fig. 33.1b). The bark is thin, smooth, grey, peeling in patches (Fig. 33.1c). Flowers are greenish-yellow, fascicled in axils of leaves, and unisexual. The female flowers borne on the top of the twig are sessile, with six perianth lobes, a cup-shaped disk and a superior three-celled ovary crowned by three styles. Stigma are deeply bifid. Male flowers are numerous, usually present at the base of the twig, with a short pedicel, bearing a pale green perianth. Male flowers have three entirely connate stamens. The fruit are globose (drupe), 2–4 cm in diameter, pale green in colour, changing to yellow when mature, and with six vertical striations (Fig. 33.1d). The fruit skin is thin, translucent and adherent to the very crisp, juicy, concolorous flesh. The hexagonal stone (Fig. 33.1e) is tightly embedded in the centre of the flesh, with three slightly dehiscent compartments, each usually containing two trigonous seeds (5 × 3 mm) (Lim, 2012).

Traditional Uses

Indian gooseberry fruit are edible when fresh, dried, cooked and pickled. The fruit are rich in vitamin C and hence have both a savoury and a bitter taste. In India, these berries are soaked in sugar syrup for a long time and then eaten daily. In certain parts of India, the fruit are cooked along with lentils and served. Fresh fruit are also baked into tarts. Ripe and half-ripe fruit are used in the preparation of sweetmeats, jams, jellies, candies, chutney, pickles and relishes. The fruit juice is used to flavour vinegar. The fruit are dried and made into chips as well as being used in the preparation of various health foods such as 'Chyavanprash' and 'Triphala', along with other natural ingredients. In Indonesia and Malaysia, the fruit are used to sour seasonings in sambal and as a savoury in place of tamarind pulp. In China, wine is prepared from the fruit extract.

Indian gooseberry is one of the major ingredients used in the preparation of health food and medicine in the Ayurveda, Siddha, Unani, Tibetan, Thai and Chinese systems of medicine (Yadav *et al.*, 2017). The fruit has high medicinal value and is commonly used as a tonic to restore energy and strength. In Ayurvedic literature, it is mentioned that the extract of dried fruit is useful for a range of ailments including anaemia, leprosy, cough, asthma, hemorrhages, liver and digestive problems. The fruit are commonly used in the treatment of fever, jaundice, hemorrhage, vomiting, leprosy, pox, diabetes, eye disease, diarrhoea, dysentery and helminthic infections (Yadav *et al.*, 2017).

Nutritional Composition

The fresh fruit of the Indian gooseberry consists of 87.02% moisture, 0.34% protein, 4.39% carbohydrates, 0.34% ash and 7.75% dietary fibre, with an energy value of 23.66 (Table 33.1; Longvah *et al.*, 2017). The fruit has a high amount of vitamin C (252 mg/100 g; Longvah *et al.*, 2017), however, the content of vitamin C varies from 164 to 1074 mg/100 g, depending upon the genotype/variety (Barthakur and Arnold, 1991; Judprasong *et al.*, 2013). The fruit of the Indian gooseberry also contain thiamine, riboflavin, niacin, pantothenic acid, biotin and carotenoids (Table 33.1; Longvah *et al.*, 2017). They

Table 33.1. Nutritional composition of Indian gooseberry fruit (Longvah *et al.*, 2017).

Proximate composition (%)	
Moisture	87.02 ± 0.78
Protein	0.34 ± 0.03
Fat	0.16 ± 0.04
Carbohydrate	4.39 ± 1.09
Ash	0.34 ± 0.05
Total dietary fibre	7.75 ± 0.64
Insoluble fibre	6.20 ± 0.43
Soluble fibre	1.55 ± 0.43
Energy (kcal/100 g)	23.66 ± 4
Water soluble vitamins (mg/100 g FW)	
Thiamine (B1)	0.01 ± 0.009
Riboflavin (B2)	0.03 ± 0.012
Niacin (B3)	0.12 ± 0.02
Pantothenic acid (B5)	0.35 ± 0.04
Total B6	0.27 ± 0.054
Biotin (B7) (in µg)	1.42 ± 0.35
Total folates (B9) (in µg)	7.86 ± 1.27
Vitamin C	252 ± 30.4
Fat soluble vitamins (mg/100 g FW)	
Ergocalciferol (D2) (in µg)	0.27 ± 0.02
Tocopherol alpha	0.11 ± 0.01
Tocopherol gamma	0.01 ± 0.00
Tocopherol delta	0.06 ± 0.01
Tocotrienol alpha	0.05 ± 0.01
α-tocopherol equivalent	0.12 ± 0.02
Phylloquinones (K1)	1.64 ± 0.31
Carotenoids (µg/100 g FW)	
Lutein	38.7 ± 9.78
Zeaxanthin	2.86 ± 1.61
β-carotene	1.58 ± 0.09
Total carotenoids	62.01 ± 6.45
Minerals and trace elements (mg/100 g FW)	
Aluminium	0.08 ± 0.02
Calcium	20.14 ± 2.51
Chromium	0.007 ± 0.003
Copper	0.12 ± 0.08
Iron	1.25 ± 0.08

Continued

Table 33.1. Continued

Lead	0.004 ± 0.002
Lithium	0.001 ± 0.001
Magnasium	6.50 ± 0.93
Manganese	0.11 ± 0.02
Molebdeum	0.001 ± 0.000
Nickel	0.01 ± 0.003
Phosphorus	21.85 ± 1.87
Potassium	223 ± 12.6
Sodium	1.37 ± 0.19
Zinc	0.05 ± 0.01
Starch and sugars (g/100 g FW)	
Total starch	0.56 ± 0.03
Fructose	2.10 ± 0.02
Glucose	1.29 ± 0.02
Fatty acid composition (mg/100 g FW)	
Palmitic (C16:0)	26.48 ± 3.16
Stearic (C18:0)	11.99 ± 0.26
Oleic (C18:1n9)	23.93 ± 1.89
Linoleic (C18:2n6)	47.41 ± 2.23
α-linolenic (C18:3n3)	27.83 ± 2.51
Total saturated fatty acids	38.48 ± 3.26
Total mono unsaturated fatty acids	23.93 ± 1.89
Total polyunsaturated fatty acids	75.24 ± 4.62
Amino acid composition (g/100 g protein)	
Alanine	6.47 ± 1.67
Arginine	3.87 ± 0.45
Aspartic acid	14.90 ± 2.63
Cystine	0.61 ± 0.16
Glutamic acid	23.90 ± 2.39
Glycine	5.01 ± 0.47
Histidine	2.40 ± 0.26
Isoleucine	1.94 ± 0.34
Leucine	4.25 ± 0.42
Lysine	4.06 ± 0.50
Methionine	0.73 ± 0.17
Phenylalanine	4.85 ± 2.04
Proline	4.48 ± 0.48
Serine	4.95 ± 0.23

Continued

Table 33.1. Continued

Threonine	3.30 ± 0.69
Tryptophan	0.75 ± 0.08
Tyrosine	2.60 ± 0.39
Valine	2.99 ± 0.39
Organic acids (mg/100 g FW)	
Citric acid	751 ± 8.5
Fumaric acid	1.65 ± 0.28
Mallic acid	7.58 ± 1.17
Succinic acid	44.33 ± 6.23
Tartaric acid	14.06 ± 1.11

contain varied minerals, the amounts being 223 mg/100 g of potassium, 21.85 mg/100 g of phosphorus, 20.14 mg/100 g of calcium, 6.50 mg/100 g magnesium, 1.37 mg/100 g sodium, 1.25 mg/100 g of iron, 0.11 mg/100 g of manganese, 0.05 mg/100 g of zinc and 0.12 mg/100 g of copper (Table 33.1; Longvah *et al.*, 2017). Fatty acids present in Indian gooseberry fruit are linoleic acid (47.41 mg/100 g), α-linolenic acid (27.83 mg/100 g), palmitic acid (26.48 mg/100 g) and oleic acid (23.93 mg/100 g) (Table 33.1; Longvah *et al.*, 2017). Varied amino acids are present: alanine (6.47 mg/100 g of protein), arginine (3.87 mg/100 g of protein), aspartic acid (14.90 mg/100 g of protein), cystine (0.61 mg/100 g of protein), glutamic acid (23.90 mg/100 g of protein), glycine (5.01 mg/100 g of protein), histidine (2.40 mg/100 g of protein), isoleucine (1.94 mg/100 g of protein), leucine (4.25 mg/100 g of protein), lysine (4.06 mg/100 g of protein), methionine (0.73 mg/100 g of protein), phenylalanine (4.85 mg/100 g of protein), proline (4.48 mg/100 g of protein), serine (4.95 mg/100 g of protein), threonine (3.30 mg/100 g of protein), tryptophan (0.75 mg/100 g of protein), tyrosine (2.60 mg/100 g of protein) and valine (2.99 mg/100 g of protein) (Table 33.1; Longvah *et al.*, 2017). Diverse organic acids such as citric acid (751 mg/100 g), fumaric acid (1.65 mg/100 g), mallic acid (7.58 mg/100 g), succinic acid (44.33 mg/100 g) and tartaric acid (14.06 mg/100 g) have also been found in the fruit (Table 33.1; Longvah *et al.*, 2017).

Table 33.2. Fatty acid composition of Indian gooseberry seed oil (Arora *et al.*, 2011).

Fatty acid	Composition (%)
Myristic (14:0)	3.6
Palmitic (16:0)	2.3
Stearic (18:0)	3.1
Oleic (18:1)	26.4
Linoleic (18:2)	51.0
Linolenic (18:3)	11.8
Total saturated fatty acids	9.0
Total unsaturated fatty acids	89.2

Minor fruit are a rich source of essential nutrients including carbohydrates, proteins and dietary fibre (Murthy and Bapat, 2020), for example, *Bertholletia excelsa* fruit are reported to possess 6.05% carbohydrate, 13.36% protein and 9.53% fibre (Fernandes *et al.*, 2015). The concentration of carbohydrate, protein, fibre and energy in Indian gooseberry was foud to be 4.39%, 0.34%, 7.75%, and 23.66 kcal, respectively, and these values are comparable with other minor fruit of the world (Murthy and Bapat, 2020). Indian gooseberry is rich in vitamins especially vitamin C (252 mg/100 g) (Longvah *et al.*, 2017) and some of the genotypes are reported to possess even higher levels of this vitamin (575 mg/100 g) (Judprasong *et al.*, 2013). The vitamin C content of Indian gooseberry is higher than other vitamin-rich fruit such as ber (39–166 mg/100 g) and jujube (188–544 mg/100 g) (Longvah *et al.*, 2017).

Indian gooseberry seeds were reported to contain 18% oil. Linoleic acid (51%), oleic acid (26.4%) and linolenic acid (11.8%) were the major fatty acids. The amount of total unsaturated fatty acids was 89.2% and saturated fatty acids were 9.0% (Table 33.2; Arora *et al.*, 2011). Linoleic acid along with oleic acid represent vital unsaturated fatty acids present in the majority of edible oils such as groundnut, cotton and maize oil, which have more than 75% of unsaturated fatty acids. It was reported that linoleic acid acts as an anti-inflammatory, acne reductive, skin-lightening and moisture retentive agent. Linoleic acid-rich oils are used in the manufacturing of beauty products since they show beneficial properties on the skin (Ando *et al.*, 1998). Therefore, the seed oil of Indian gooseberry is useful for both human consumption and industrial application.

Phytochemicals

Chemical constituents such as flavonoids, tannins, phenolic acids, sterols and terpenoids have been isolated from the fruit, leaves/branches and roots of Indian gooseberry: (+)-catechin; apigenin 7-O-(6″-butyryl-β-glucopyranoside); emblirol A; emblirol B; epicatechin; epicatechin 3-O-gallate; epigallocatechin; epigallocatechin 3-O-gallate; eriodictyol 7-O-(6″-O-galloyl)-glucoside; eriodictyol 7-O-(6″-O-trans-coumaroyl)-g lucoside; eriodictyol 7-O-glucoside; fisetin; gallocatechin; kaempferol; kaempferol-3-O-glucoside; kaempferol 3-O-rhamnoside; myricetin 3-O-rhamnoside; naringenin; naringenin 7-O-(6″-O-galloyl)-glucoside; naringenin 7-O-(6″-O-trans-coumaroyl)-glucoside; naringenin 7-O-glucoside; quercetin; quercetin 3-O-(6″-O-glucosyl) glucoside; quercetin 3-O-glucoside; quercetin 3-O-rhamnoside; rutin; and taxifolin 7-O-glucoside (Zhang *et al.*, 2004; El-Desouky *et al.*, 2008; Liu *et al.*, 2008, 2012; Luo *et al.*, 2009; Chugh and Bharti, 2014; Xu *et al.*, 2016; Yan *et al.*, 2017; Srinivasan *et al.*, 2018; Li *et al.*, 2020).

The tannins extracted from Indian gooseberry have been identified as: 1,2,3,4,6 -Penta-O-galloyl-β-D-glucose; 1,2,3,6-tetra-O-galloyl-β-D-glucose; 1,2,4,6-tetra-O-gallo yl-β-D-glucose; 1,6-di-O-galloyl-β-D-glucose; 1-O-digalloyl-β-D-glucose; 1-O-galloyl-β-D-glucose; carpinusnin; chebulagic acid; chebulanin; corilagin; elaeocarpusin; ellagic acid 4-O-glucoside; ellagic acid 4-O-rhamnoside; ellagic acid 4-O-xyloside; emblicanin A; emblicanin B; flavogallonic acid bislactone; furosin; geraniin; isocorilagin; isomallotusinin; L-malic acid-2-O-gallate; mallonin; mallotusinin; mucic acid 1,4-lactone methyl ester 5-O-gallate; mucic acid 1,4-lactone 3-O-gallate; mucic acid 1-ethyl 6-methyl ester 2-O-gallate; mucic acid 2-O-gallate; mucic acid 3-O-gallate; mucic acid dimethyl ester 2-O-gallate; neochebulagic acid; phyllanemblinin A; phyllanemblinin C; phyllanemblinin D; phyllanemblinin E; phyllanemblinin F;

phyllemtannin; prodelphinidin A-1; prodelphinidin B-1; prodelphinidin B-2; punicafolin; putranjivain A; putranjivain B; and tercatain (Zhang *et al.*, 2004, 2017; Kumaran and Karunakaran, 2006; Pozharitskaya *et al.*, 2007; Liu *et al.*, 2008, 2012; Xiang *et al.*, 2010, 2011; Luo *et al.*, 2011, 2012; Puppala *et al.*, 2012; Xu *et al.*, 2016; Rose *et al.*, 2018;). Some of the phenolic acids isolated from Indian gooseberry are chebulic acid, cinnamic acid, de-carboxy ellagic acid, ellagic acid, gallic acid and pyrogallol (Khan *et al.*, 2002; Kumaran and Karunakaran, 2006; Pozharitskaya *et al.*, 2007; Luo *et al.*, 2009, 2011; Chugh and Bharti, 2014; Xu *et al.*, 2016; Zhang *et al.*, 2016, 2017; ; Rose *et al.*, 2018; Li *et al.*, 2020). Gallates isolated from Indian gooseberry are digallic acid, ethyl gallate, gallic acid 3-O-(6'-O-galloyl)-glucoside, gallic acid 3-O-glucoside, methyl gallate and propyl gallate (Kumaran and Karunakaran, 2006; Chugh and Bharti, 2014; Xu *et al.*, 2016; Zhang *et al.*, 2016, 2017; Rose *et al.*, 2018). Multifidol 1-O -(6″-O-β-D-apiofuranosyl)-β-D-glucoside and multifidol glucoside are acylphloroglucinols which are sequestered from Indian gooseberry (Xu *et al.*, 2016). Some of the phytosterols extracted from Indian gooseberry are 5α,6β,7α-trihydroxysitosterol, 7-ketositosterol, 7-α-hydroxysitosterol, stigmast-4-en-3,6-dione, stigmast-4-en-6β-ol-3-one, stigmast-4-ene-3β,6α-diol and β-daucosterol (Luo *et al.*, 2009; Qi *et al.*, 2013). Some of the terpenoid compounds isolated from Indian gooseberry are glochicoccinoside-D, phyllaemblic acid, phyllaemblic acid methyl ester, phyllaemblicin B, phyllaemblicin C, phyllaemblicin F, phyllaemblicin G6, phyllaemblicin G7, phyllaemblicin G8, phyllaemblicin H1, phyllaemblicin H2, phyllaemblicin H6, phyllaemblicin H8 and phyllaemblicin H12 (Zhang *et al.*, 2004; Liu *et al.*, 2009; Lv *et al.*, 2014, 2015). The polysaccharide isolated from Indian gooseberry is α-pyran acidic heteropolysaccharide (Li *et al.*, 2018). The furans extracted from Indian gooseberry are 1,2-dihydroxyethyl-3 ,4-dihydroxyfuran-2(5 H)-one, 2,2-dimethy l-5-(2-oxopropyl)-2H-furo[3,4 H]chromen-7(9 H)-one, 5-(2-hydroxyethyl)-2,2-dimeth yl-2H-furo[3,4-h]chromen-7(9 H)-one and 5-hydroxymethylfurfural (Luo *et al.*, 2009; Chugh and Bharti, 2014; Gao *et al.*, 2019).

Table 33.3. Antinutritional factors of Indian gooseberry fruit (Longvah *et al.*, 2017).

Factor	Composition (mg/100 g FW)
Oxalate	7.96 ± 0.91
Phytate	49.34 ± 3.36
Saponins	400 ± 12.0

Antinutritive Compounds

Antinutritive compounds present in fruit, vegetables and food grains are reported to interfere with the absorption of nutrients and act to reduce nutrient digestion and utilization. Phytic acid, phenolic compounds, enzyme inhibitors, saponins, lectins and hemagglutinins are some of the compounds which are considered antinutrients (Muramoto, 2017). Phytic acid forms complexes with metal ions and inhibits their absorption and bioavailability. Phenolic compounds decrease the bioavailability of amino acids. Enzyme inhibitors reduce protein digestibility. The antinutrient factors of Indian gooseberry have been estimated by Longvah *et al.* (2017). The level of oxalate, phytate and saponin content was 7.96 mg/100 g, 49.34 mg/100 g and 400 mg/100 g, respectively (Table 33.3). These values are lower when compared to other minor fruit crops (Murthy and Bapat, 2020). Various traditional methods of food preparation such as fermentation, cooking, soaking and puffing are followed to improve the biological values of various food crops (Samtiya *et al.*, 2020). Indian gooseberry fruit are usually eaten after cooking, fermentation or processing in sugar syrup, and these processes will help in lowering the concentration of antinutrients.

Implications of Processing on Nutrients and Phytochemicals

Indian gooseberry fruit can be stored naturally for up to ten days, however the fruit are highly perishable and require processing for increasing shelf life and value addition. Prajapati *et al.* (2011) evaluated the effect of pretreatment and drying on the quality of Indian gooseberry shreds. They carried out blanching of shreds (i.e. the shreds were dipped in boiling water for three minutes or in normal water containing 0.1% potassium metabisulphite for three minutes). After blanching, the shreds were treated with either 3% common salt or 3% black salt, or 10% ginger juice for 12 hours. After that, the shreds were dried in trays in solar or hot-air oven dryers (55–60°C) until a constant weight was achieved. They tested the fresh and dried gooseberry shreds for ascorbic acid, acidity, tannin, reducing and non-reducing sugars. Their experimental results revealed that the best product was obtained with 0.1% potassium metabisulphite blanching and drying in a solar dryer with added common salt at 3%. The shreds had ascorbic acid content of 298.3 mg/100 g, tannin 2.4%, acidity 2.6%, reducing sugar 3.0%, non-reducing sugar 21.0% and total sugar 24.0%. In another study, Pareek and Kaushik (2012) carried out studies using various drying methods to establish the quality of Indian gooseberry powder (5 mm thickness) during storage. The fruit slices were dried using different methods (direct solar drying, oven drying at 65°C, microwave drying at 65°C and fluidized bed drying at 65°C, air velocity of 90 (m/min) until the moisture level was below 15%. Among the four drying methods tested, they showed that the fluidized bed drying method was the best. With fluidized bed treatment, the gooseberry powder had the highest amount of ascorbic acid (272.74 mg/100 g), total sugars (39.41 g/100 g), reducing sugar (33.69 g/100 g), tannin (8.8 g/100 g) and acidity (9.94 g/100 g). The above-mentioned processing methods are useful for enhancing the shelf life of Indian gooseberry shreds and powder. With such treatments, material could be stored for up to 90 days. The stored material contains 205.5 mg/100 g of ascorbic acid (Pareek and Kaushik, 2012).

Indian gooseberry is processed into value-added products such as juice, candy, jam, pickle and sauce, of which juice is the popular one. Preservation of juice is important and thermal pasteurization at 75–95°C for 1–20 minutes has been followed for Indian gooseberry juice. However, a high temperature has an adverse effect on nutrition, especially in terms of vitamin C degradation and browning. To overcome these problems, Raj *et al.* (2019) carried out a thermal assisted high-pressure process (200–500 MPa, 1 s–20 min and 30–60°C) and studied its effect on the physicochemical properties, bioactive compounds and antioxidant capacity of Indian gooseberry juice. Their results revealed that the process had not affected the total soluble solids (TSS) and titrable acidity of the juice and about 85% of ascorbic acid was retained. This study showed that high pressure treated Indian gooseberry juice was nutritionally superior compared to thermal beverage in terms of biochemical attributes.

Bioactive Properties

Indian gooseberry has remarkable traditional uses in Ayurveda, Siddha, Unani, Tibetan, Thai, Chinese and other traditional systems of medicine and this has led to researchers analysing its biological activities. Various parts of Indian gooseberry such as the fruit, leaves, stem branches and roots are reported to have varied biological activities including antimicrobial, antioxidant, antidiabetic, anticancer and immunomodulatory properties. The biological activities of extracts and phytochemicals isolated from Indian gooseberry are now highlighted (Tables 33.4–33.6).

Antioxidant, hepatoprotective and cardioprotective activity

Aqueous extract of Indian gooseberry fruit was reported to be a potent inhibitor of lipid peroxide production and a scavenger of hydroxyl and superoxide radicals. Jose and Kuttan (1995) demonstrated the inhibition of peroxide formation from *in vitro* experiments using the photo-reduction method, xanthine-xanthine oxidase method, Fe^{2+}/acerbate system, Fe^{3+}/ADP-ascorbate system and hydroxyl radical scavenging activity. Indian gooseberry fruit, leaves and roots possess a plethora of phytochemicals which have exhibited very good antioxidant properties (Tables 33.4–33.6). Flavonoid compounds – namely (+)-catechin, epicatechin 3-O-gallate, epigallocatechin 3-O-gallate, epigallocatechin, eriodictyol 7-O-(6″-O-galloyl)-glucoside, eriodictyol 7-O-(6″-O-trans-coumaroyl)-gluco side, eriodictyol 7-O-glucoside, gallocatechin, kaempferol, kaempferol 3-O-glucoside, kaempferol 3-O-rhamnoside, myricetin 3-O-rhamnoside, naringenin 7-O-(6′-O-galloyl)-glucoside, naringenin 7-O-(6″-O-trans-coumaroyl)-glucoside,

Table 33.4. Biological activities of compounds isolated from Indian gooseberry fruit.

Compound	Activity	References
Flavonoids		
Fisetin	Anti-inflammatory	Li *et al*. (2020)
Kaempferol	Antioxidant, cytotoxicity, antiproliferation	Liu *et al*. (2008, 2012)
Kaempferol 3-O-rhamnoside	Antioxidant	Xu *et al*. (2016)
Kaempferol 3-O-glucoside	Antioxidant, cytotoxicity and antiproliferation	Liu *et al*. (2008, 2012)
Myricetin 3-O-rhamnoside	Antioxidant	Xu *et al*. (2016)
Quercetin	Antioxidant, antiproliferative, cytotoxicity	Liu *et al*. (2008, 2012); Luo *et al*. (2009); Srinivasan *et al*. (2018)
Quercetin 3-O-glucoside	Antioxidant, antiproliferation, cytotoxicity	Liu *et al*. (2008, 2012); Xu *et al*. (2016)
Quercetin 3-O-rhamnoside	Antioxidant	Xu *et al*. (2016)
Rutin	Antiproliferation, cytotoxicity	Liu *et al*. (2012)
Taxifolin 7-O-glucoside	Antioxidant	Xu *et al*. (2016)
Tannins		
1,6-Di-O-galloyl-β-D-glucose	Antioxidant, melanogenesis-inhibition	Xu *et al*. (2016); Zhang *et al*. (2017); Rose *et al*. (2018)
1-O-digalloyl-β-D-glucose	Antioxidant	Xu *et al*. (2016)
1-O-galloyl-β-D-glucose	Aldose reductase inhibitor, antioxidant, antiproliferative	Puppala *et al*. (2012) ; Zhang *et al*. (2004); Xu *et al*. (2016); Rose *et al*. (2018)
Chebulagic acid	Antioxidant, antiproliferative	Zhang *et al*. (2004); Luo *et al*. (2011, 2012); Xu *et al*. (2016)
Chebulanin	Antioxidant, antiproliferative	Luo *et al*. (2011, 2012); Xu *et al*. (2016)
Corilagin	Antioxidant, antiproliferative	Zhang *et al*. (2004); Kumaran and Karunakaran (2006); Xu *et al*. (2016); Rose *et al*. (2018)
Elaeocarpusin	Antioxidant, antiproliferative	Zhang *et al*. (2004); Xu *et al*. (2016)
Ellagic acid 4-O-glucoside	Antioxidant, neuroprotective	Rose *et al*. (2018)
Ellagic acid 4-O-rhamnoside	Antioxidant	Rose *et al*. (2018)
Ellagic acid 4-O-xyloside	Antioxidant	Rose *et al*. (2018)
Emblicanin A	Antioxidant	Pozharitskaya *et al*. (2007)
Emblicanin B	Antioxidant	Pozharitskaya *et al*. (2007)

Continued

Table 33.4. Continued

Compound	Activity	References
Furosin	Antioxidant	Kumaran and Karunakaran (2006)
Geraniin	Antioxidant, cytotoxicity and antiproliferative	Kumaran and Karunakaran (2006); Liu *et al.* (2008, 2012)
Isocorilagin	Antioxidant, cytotoxicity, antiproliferative	Liu *et al.* (2008, 2012); Luo *et al.* (2011, 2012)
Isomallotusinin	Antioxidant	Luo *et al.* (2012)
L-malic acid 2-O-gallate	Antioxidant, antiproliferative	Zhang *et al.* (2004); Xu *et al.* (2016)
Mallonin	Antioxidant	Xu *et al.* (2016)
Mallotusinin	Antioxidant, antiproliferative	Luo *et al.* (2011)
Mucic acid 1,4-lactone methyl ester 5-O-gallate	Antioxidant, melanogenesis inhibition	Zhang *et al.* (2017)
Mucic acid 1,4-lactone 3-O-gallate	Antioxidant , antiproliferative	Luo *et al.* (2011, 2012)
Mucic acid 1-ethyl 6-methyl ester 2-O-gallate	Antioxidant, melanogenesis inhibition	Zhang *et al.* (2017)
Mucic acid 2-O-gallate	Antioxidant, antiproliferative	Zhang *et al.* (2004); Xu *et al.* (2016)
Mucic acid 3-O-gallate	Antioxidant	Rose *et al.* (2018)
Mucic acid dimethyl ester 2-O-gallate	Antioxidant, melanogenesis inhibition	Zhang *et al.* (2017)
Phyllanemblinin A	Antioxidant	Xu *et al.* (2016)
Punicafolin	Antioxidant	Xu *et al.* (2016)
Putranjivain A	Antioxidant, antiproliferative	Zhang *et al.* (2004); Xu *et al.* (2016)
Tercatain	Antioxidant	Xu *et al.* (2016)
Phenolic acids		
Cinnamic acid	Antioxidant	Luo *et al.* (2009)
Ellagic acid	Antioxidant, antiproliferative, melanogenesis inhibition, cytotoxic	Pozharitskaya *et al.* (2007); Luo *et al.* (2009, 2011); Zhang *et al.* (2017); Rose *et al.* (2018)
Gallic acid	Antioxidant, antiproliferative, anti-inflammatory, melanogenesis inhibition, cytotoxic	Kumaran and Karunakaran (2006); Pozharitskaya *et al.* (2007); Luo *et al.* (2009, 2011); Zhang *et al.* (2016, 2017); Rose *et al.* (2018); Li *et al.* (2020)
Pyrogallol	Antiproliferative	Khan *et al.* (2002)

Continued

Table 33.4. Continued

Compound	Activity	References
Gallates		
Ethyl gallate	Antioxidant, cytoprotective	Zhang *et al.* (2016)
Methyl gallate	Antioxidant, melanogenesis inhibition, cytotoxic	Kumaran and Karunakaran (2006); Zhang *et al.* (2016, 2017); Rose *et al.* (2018)
Sterols		
β-daucosterol	Antioxidant	Luo *et al.* (2009)
Polysaccharides		
α-pyran acidic heteropolysaccharide	Antioxidant	Li *et al.* (2018)
Furans		
2,2-dimethyl-5-(2-oxopropyl)-2H-furo[3,4 h]chromen-7(9 h)-one	Antioxidant	Gao *et al.* (2019)
5-(2-hydroxyethyl)-2,2-dimethyl-2H-furo[3,4 h]chromen-7(9 h)-one	Antioxidant	Gao *et al.* (2019)
5-Hydroxymethylfurfural	Antioxidant	Luo *et al.* (2009)

naringenin 7-O-glucoside, naringenin, quercetin 3-O-(6'-O-glucosyl)-glucoside, quercetin 3-O-glucoside, quercetin 3-O-rhamnoside, quercetin and taxifolin 7-O-glucoside – were isolated from fruit, leaves, branches and roots of Indian gooseberry and all demonstrated potent antioxidant activities (Tables 33.4–33.6; El-Desouky *et al.*, 2008; Liu *et al.*, 2008; Xu *et al.*, 2016). Several tannins which were extracted from fruit, leaves, branches and roots, such as 1,6-di-O-galloyl-β-D-glucose, 1-O-digalloyl-β-D-glucose, 1-O-galloyl-β-D-glucose, chebulagic acid, chebulanin, corilagin, elaeocarpusin, ellagic acid 4-O-glucoside, ellagic acid 4-O-rhamnoside, emblicanin A, emblicanin B, furosin, geraniin, isocarilagin, isomallostusinin, L-malic acid 2-O-gallate, mallonin, mallostusinin, mucic acid 1,4-lactone methyl ester 5-O-gallate, mucic acid 1-ethyl 6-methyl ester 2-O-gallate, mucic acid 2-O-gallate, mucic acid 3-O-gallate, mucic acid dimethyl ester 2-O-gallate, phyllanemblinin A, punicafolin, putranjivain A, teractain, 1,2,3,4,6-penta-O-galloyl-β-D-glucose, 1,2,3,6-tetra-O-galloyl-β-D-glucose, 1,2,4,6-tetra-O-galloyl-β-D-glucose, carpinusnin, neochebulagic acid, phyllanemblininC, phyllanemblinin D, phyllanemblininE, phyllanemblininF, prodelphinidin B-1, prodelphinidin B-2, putranjivain B, flavogallonic acid bislactone and prodelphinidin A-1 have showed antioxidant potential (Tables 33.4–33.6; Kumaran and Karunakaran, 2006; Pozharitskaya *et al.*, 2007; Liu *et al.*, 2008; Luo *et al.*, 2011, 2012; Xu *et al.*, 2016; Zhang *et al.*, 2017; Rose *et al.*, 2018). Cinnamic acid, ellagic acid, gallic acid, chebulic acid and decarboxyellagc acid are certain phenolic acids that have displayed antioxidant properties (Tables 33.4–33.6; Kumaran and Karunakaran, 2006; Pozharitskaya *et al.*, 2007; Luo *et al.*, 2009, 2011; Xu *et al.*, 2016; Zhang *et al.*, 2017; Rose *et al.*, 2018; Li *et al.*, 2020). Ethyl gallate, methyl gallate, digallic acid, gallic acid 3-O-(6'-O-galloyl)-glucoside and gallic acid 3-O-glucoside are certain gallate compounds that exhibited antioxidant activities (Tables 33.4–33.6; Kumaran and Karunakaran, 2006; Xu *et al.*, 2016; Zhang *et al.*, 2016, 2017; Rose *et al.*, 2018). β-aucisterol (sterol), α-pyran acidic heteropolysaccharide (polysaccharide), 2,2-dimethyl-5-(2-oxopropyl)-2H-furo[3,4 h]chromen-7(9 h)-one, 5-(2-hydroxyethyl)-2,2-dimethyl-2H-furo[3,4 h]chromen-7(9 h)-one, 5-hydroxymethylfurfural (furans) and multifidol 1-O-(6″-O-β-D-apiofuranosyl)-β-D-glucosideandmultifidol 1-O-(6″-O-β-D-apiofuranosyl)-β-D-glucoside (acylphloroglucinols) are certain other compounds which were isolated from Indian gooseberry which have exhibited antioxidant activities (Tables 33.4–33.6; Luo *et al.*, 2009; Xu *et al.*, 2016; Li *et al.*, 2018; Gao *et al.*, 2019).

Table 33.5. Biological activities of compounds isolated from leaves and branches of Indian gooseberry.

Compound	Activity	References
Flavonoids		
Apigenin 7-O-(6″-butyryl-β-glucopyranoside)	Antiproliferative	El-Desouky *et al*. (2008)
Epicatechin 3-O-gallate	Antioxidant	Xu *et al*. (2016)
Epigallocatechin 3-O-gallate	Antiproliferative	Zhang *et al*. (2004)
Eriodictyol 7-O-(6″-O-galloyl)-glucoside	Antioxidant	Xu *et al*. (2016)
Eriodictyol 7-O-(6″-O-trans-coumaroyl)-glucoside	Antioxidant, antiproliferative	Zhang *et al*. (2004); Xu *et al*. (2016)
Eriodictyol 7-O-glucoside	Antioxidant	Xu *et al*. (2016)
Naringenin	Antioxidant	Xu *et al*. (2016)
Naringenin 7-O-(6″-O-galloyl)-glucoside	Antioxidant	Xu *et al*. (2016)
Naringenin 7-O-(6″-O-trans-coumaroyl)-glucoside	Antioxidant	Xu *et al*. (2016)
Naringenin 7-O-glucoside	Antioxidant	Xu *et al*. (2016)
Quercetin	Antifungal, antioxidant	Chugh and Bharti (2014); Xu *et al*. (2016)
Quercetin 3-O-(6″-O-glucosyl)-glucoside	Antioxidant	Xu *et al*. (2016)
Tannins		
1,2,3,4,6-penta-O-galloyl-β-D-glucose	Antioxidant	Xu *et al*. (2016)
1,2,4,6-tetra-O-galloyl-β-D-glucose	Antihepatitis B virus, anti-herpes simplex virus	Xiang *et al*. (2010, 2011)
1-O-galloyl-βD-glucose	Antiproliferative	Zhang *et al*. (2004)
Carpinusnin	Antioxidant	Xu *et al*. (2016)
Chebulagic acid	Antiproliferative	Zhang *et al*. (2004)
Chebulanin	Antioxidant	Xu *et al*. (2016)
Corilagin	Antioxidant, antiproliferative	Zhang *et al*. (2004); Xu *et al*. (2016)
Furosin	Antioxidant	Xu *et al*. (2016)
Geraniin	Antioxidant, antiproliferative	Zhang *et al*. (2004); Xu *et al*. (2016)
Neochebulagic acid	Antioxidant	Xu *et al*. (2016)
Phyllanemblinin C	Antioxidant, antiproliferative	Zhang *et al*. (2004); Xu *et al*. (2016)
Phyllanemblinin D	Antioxidant	Xu *et al*. (2016)
Phyllanemblinin E	Antioxidant, Antiproliferative	Zhang *et al*. (2004); Xu *et al*. (2016)
Phyllanemblinin F	Antioxidant	Xu *et al*. (2016)
Prodelphinidin B-1	Antioxidant, antiproliferative	Zhang *et al*. (2004); Xu *et al*. (2016)
Prodelphinidin B-2	Antioxidant, antiproliferative	Zhang *et al*. (2004); Xu *et al*. (2016)

Continued

Table 33.5. Continued

Compound	Activity	References
Putranjivain A	Antiproliferative	Zhang *et al.* (2004)
Putranjivain B	Antioxidant	Xu *et al.* (2016)
Phenolic acids		
Ellagic acid	Antifungal	Chugh and Bharti (2014)
Gallic acid	Antifungal	Chugh and Bharti (2014)
Gallates		
Propyl gallate	Antifungal	Chugh and Bharti (2014)
Acylphloroglucinols		
Multifidol glucoside	Antioxidant	Xu *et al.* (2016)
Sterols		
5α,6β,7α-trihydroxysitosterol	Cytotoxic	Qi *et al.* (2013)
7-ketositosterol	Cytotoxic	Qi *et al.* (2013)
7α-hydroxysitosterol	Cytotoxic	Qi *et al.* (2013)
Stigmast-4-en-3,6-dione	Cytotoxic	Qi *et al.* (2013)
Stigmast-4-en-6β-ol-3-one	Cytotoxic	Qi *et al.* (2013)
Stigmast-4-ene-3β,6α-diol	Cytotoxic	Qi *et al.* (2013)
Furans		
1,2-dihydroxyethyl-3,4-dihydroxyfuran-2(5 h)-one	Antifungal	Chugh and Bharti (2014)

Various experimental evidence on rat models has revealed that Indian gooseberry is useful in hepatoprotective activity and this could be due to the radical scavenging and antioxidant potential of varied phytochemicals present in this plant. Feeding of aqueous fruit extract of Indian gooseberry improved the levels of antioxidant enzymes in kidney and liver against cyclophosphamide-induced oxidative stress (Haque *et al.*, 2001). Reddy *et al.* (2010) studied the effect of Indian gooseberry fruit extract against alcohol-induced hepatic damage in rats. They administered rats with alcohol (5 g/kg body weight/day) for 60 days which resulted in increased lipid peroxidation, accumulation of protein carbonyls, nitrate and nitrite levels. Alcohol treatment also significantly lowers the activities of superoxide dismutase, catalase, glutathione peroxidase and glutathione S-transferase. Administration of Indian gooseberry fruit extract (250 mg/kg body weight) to alcoholic rats significantly brought the plasma enzymes toward normal levels and reduced the levels of lipid peroxidation. From these experiments, Reddy *et al.* (2010) concluded that the tannins and flavonoids which were present in gooseberry fruit extract were responsible for the protection against hepatic injury. A similar mode of action was observed by Chaphalkar *et al.* (2017) and they reported that hydroalcoholic extract of Indian gooseberry at a dose of 500 mg/kg restores the deranged liver enzymes like alanine transaminase, aspartate transaminase, alkaline phosphatase and total protein in ethanol-induced hepatotoxicity in rats.

The hypolipidemic effect of Indian gooseberry fruit juice was evaluated by Mathur *et al.* (1996) in cholesterol-fed rabbits. They administered Indian gooseberry fresh juice at a dose of 5 mg/kg body weight per day for 60 days and observed a reduction in serum cholesterol, triglycerides, phospholipids and cholesterol by 82, 66, 77 and 90%, respectively, with treated animals. In another study, Variya *et al.* (2018) studied the antihyperlipidemic potential of Indian gooseberry fruit juice in rats. They experimentally induced hyperlipidemia in rats through the administration of poloxamer-407,

Table 33.6. Biological activities of compounds isolated from Indian gooseberry root.

Compound	Activity	References
Flavonoids		
(+)-catechin	Antioxidant	Xu *et al.* (2016)
Emblirol A	Anti-TMV	Yan *et al.* (2017)
Emblirol B	Anti-TMV	Yan *et al.* (2017)
Epicatechin	Antioxidant	Xu *et al.* (2016)
Epicatechin 3-O-gallate	Antioxidant	Xu *et al.* (2016)
Epigallocatechin	Antioxidant	Xu *et al.* (2016)
Epigallocatechin 3-O-gallate	Antioxidant, antiproliferative	Zhang *et al.* (2004); Xu *et al.* (2016)
Gallocatechin	Antioxidant	Xu *et al.* (2016)
Tannins		
1,2,3,6-tetra-O-galloyl-β-D-glucose	Antioxidant	Xu *et al.* (2016)
1,2,4,6-tetra-O-galloyl-β-D-glucose	Antioxidant	Xu *et al.* (2016)
Corilagin	Antiproliferative	Zhang *et al.* (2004)
Flavogallonic acid bislactone	Antioxidant	Xu *et al.* (2016)
Prodelphinidin A-1	Antioxidant	Xu *et al.* (2016)
Phyllemtannin	Antiproliferative	Zhang *et al.* (2004)
Phenolic acids		
Chebulic acid	Antioxidant	Xu *et al.* (2016)
Decarboxyellagic acid	Antioxidant	Xu *et al.* (2016)
Gallic acid	Antioxidant	Xu *et al.* (2016)
Gallates		
Digallic acid	Antioxidant	Xu *et al.* (2016)
Gallic acid 3-O-(6'-O-galloyl)-glucoside	Antioxidant	Xu *et al.* (2016)
Gallic acid 3-O-glucoside	Antioxidant	Xu *et al.* (2016)
Acylphloroglucinols		
Multifidol 1-O-(6″-O-β-D-apiofuranosyl)-β-D-glucoside	Antioxidant	Xu *et al.* (2016)
Terpenoids		
Glochicoccinoside D	Antiviral	Lv *et al.* (2015)
Phyllaemblic acid	Antiviral	Lv *et al.* (2014)
Phyllaemblic acid methyl ester	Antiviral	Liu *et al.* (2009)
Phyllaemblicin B	Antiviral, antiproliferative	Zhang *et al.* (2004); Liu *et al.* (2009); Lv *et al.* (2015)
Phyllaemblicin C	Antiviral, antiproliferative	Zhang *et al.* (2004); Liu *et al.* (2009); Lv *et al.* (2015)
Phyllaemblicin F	Antiviral	Lv *et al.* (2014)

Continued

Table 33.6. Continued

Compound	Activity	References
Phyllaemblicin G6	Antiviral	Lv *et al.* (2014)
Phyllaemblicin G7	Antiviral	Lv *et al.* (2014)
Phyllaemblicin G8	Antiviral	Lv *et al.* (2014)
Phyllaemblicin H1	Cytotoxic	Lv *et al.* (2015)
Phyllaemblicin H12	Antiviral	Lv *et al.* (2015)
Phyllaemblicin H2	Cytotoxic	Lv *et al.* (2015)
Phyllaemblicin H6	Antiviral	Lv *et al.* (2015)
Phyllaemblicin H8	Antiviral	Lv *et al.* (2015)

tyloxapol and high-fat-diet. Subsequently, they treated rats with gooseberry fruit juice (2 ml/kg/day, p.o) to one group of animals and gallic acid (100 mg/kg/day, p.o) to another group of animals. Treatment of fruit juice as well as gallic acid decreased plasma cholesterol and reduced oil infiltration in the liver and aorta. They also observed increased peroxisome-activated receptor-α expression and increased activity of lipid oxidation through carnitine palmitoyltransferase. From this experimental evidence, Variya *et al.* (2018) concluded that gallic acid, a chief phytoconstituent present in Indian gooseberry, is responsible for antihyperlipidemia potential. Similarly, Ojha *et al.* (2011) demonstrated the cardioprotective effect of Indian gooseberry against isoproterenol-induced cardiotoxicity in rats. Their results showed significant improvement in hemodynamic and contractile functions when rats were orally administered in the dose range of 100 to 500 mg/kg body weight.

Anti-inflammatory activity

Extracts of Indian gooseberry have been used as anti-inflammatory drugs including for inflammatory bowel disease, rheumatoid arthritis, osteoarthritis, atherosclerosis and psoriasis. Jaijoy *et al.* (2010) studied the anti-inflammatory activity of aqueous fruit extract of Indian gooseberry in rats using ethyl phenylpropionate (EPP) and arachidonic acid (AA)-induced ear edema, carrageenan-induced paw edema as well as cotton pellet-induced granuloma models. The extract at 1 mg/ear displayed an anti-inflammatory effect on EPP-induced ear edema. Oral administration of the aqueous extract at the doses of 150, 300 and 600 mg/kg body weight caused a dose-dependent inhibition of carrageenan-induced rat paw edema. A similar finding was reported by Muthuraman *et al.* (2011), who studied the anti-inflammatory activity of phenolic compounds of Indian gooseberry against acute and chronic inflammation at two oral doses, 20 and 40 mg/kg body weight. Indian gooseberry showed dose-dependent healing action against indomethacin (18 mg/kg) induced gastropathy in mice and restored the level of inflammatory cytokines in a biphasic manner at 60 mg/kg (Chatterjee *et al.*, 2011). Sripanidkulchai and Junlatat (2014) assessed the bioactivities of ethanol and methanol extracts of branches of Indian gooseberry using *in vitro* and *in vivo* models. They used lipopolysaccharide (LPS)-induced RAW 264.7 murine macrophages and carrageenan-induced paw edema in rats for *in vitro* and *in vivo* studies respectively. From the *in vitro* RT-PCR results, they showed the significant inhibitory effect of both methanol and ethanol extracts on several inflammatory gene expressions including cyclohexogenase-2 (COX-2), inducible nitric oxide synthase (iNOS), tumour necrosis factor-α (TNF-α), interlukemin-1-β (IL-1β) and interlukemin-6 (IL-6) in LPS-induced RAW 264.7 murine macrophage cells. Their study also reveals the anti-inflammatory effects of methanol and ethanol extract in carrageenan-induced paw edema in rats. Recent studies by Li *et al.* (2020) depicted the role of fisetin and gallic acid as anti-inflammatory compounds that were isolated from Indian gooseberry (Table 33.4).

Antidiabetic activity

Indian gooseberry was tested and used for the management of diabetes and related metabolic complications. Various studies have been performed on streptozotocin (STZ)-induced diabetic rats and these established the protective role of Indian gooseberry against diabetes. For example, Rao *et al.* (2005) administered polyphenol-rich fruit extract of Indian gooseberry (10–20 mg/kg of body weight/day) to the STZ-induced diabetic rats for 20 days. They noticed a reduction in levels of 5-hydroxymethylfurfural which is a glycosylated protein, an indicator of oxidative stress in the serum of experimental rats. Furthermore, increased albumin levels were recorded in treated rats. Similarly, many research studies have established the antidiabetic potential of several extracts of Indian gooseberry using STZ-induced diabetes in rats (Mehta *et al.*, 2009; Nain *et al.*, 2012; Patel *et al.*, 2013). Phytochemicals such as gallic acid, gallotannins, ellagic acid and corilagin were correlated for the plant's antidiabetic activity (Nampoothiri *et al.*, 2011; D'souza *et al.*, 2014). Clinical studies on diabetic patients showed that treatment with Indian gooseberry significantly reduced fasting and two-hour postprandial blood glucose levels. Furthermore, healthy and diabetic human volunteers showed significant improvement in high-density lipoprotein-cholesterol and reduction in low-density lipoprotein-cholesterol levels with the consumption of Indian gooseberry (2–3 g/per day) (Akhtar *et al.*, 2011).

Neuroprotective activity

Varied *in vivo* and *in vitro* studies have revealed the therapeutic activities of Indian gooseberry against neurological disorders (Husain *et al.*, 2019). Thenmozhi *et al.* (2016a, b) studied the memory-enhancing effect of Indian gooseberry extract against aluminum chloride-induced Alzheimer's disease in rats. They observed that intraperitoneal injection of aluminum chloride (100 mg/kg body weight) for 60 days significantly elevated acetylcholinesterase activity in the hippocampus and cortex. Furthermore, they demonstrated that treatment of rats with an extract containing the phytochemicals emblicanin A (37%), emblicanin B (33%), punigluconin (12%) and penduculagin (14%) with 50, 100 and 200 mg/kg bodyweight for 60 days significantly reverted acetylcholinesterase activity. Spatial learning, memory, and locomotor impairment were also attenuated in the treated animals. Improvement of cognitive function was reported by Golechha *et al.* (2010, 2011) by hydroalcoholic extract of Indian gooseberry in experimental rats. They recorded attenuation of kainic acid and elevated tumour necrosis factor-alpha (TNF alpha) in the brain, as a result of the antioxidant and anti-inflammatory potential (Golechha *et al.*, 2010, 2011). Ellagic acid 4-O-glucoside isolated from Indian gooseberry has been reported to reduce ß-amyloid-induced neurotoxicity in transgenic *Caenorhabditis elegans* (Table 33.1; Rose *et al.*, 2018). Similarly, Xie *et al.* (2012) established the antiepileptic potency of epigallocatechin-3-gallate against pentylenetetrazole-induced epilepsy.

Anti-mutagenic, cytotoxic, antiproliferative and anticancer activity

Several phytochemicals isolated from plants have antimutagenic properties (Słoczyńska *et al.*, 2014). Aqueous extract of Indian gooseberry was found to inhibit 2-acetamidofluorene, N-methyl-N-nitro-N-nitrosoguanidine and 4-nitro-O-phenylenediamine induced mutagenicity (Jose *et al.*, 1997). Similarly, Haque *et al.* (2001) demonstrated the protective role of Indian gooseberry against cyclophosphamide-induced mutagenicity in mice. In another study, Thakur and Rao (2014) established *in vitro* genoprotective activity of Indian gooseberry extract on human blood lymphocytes. They demonstrated the reduction in sister chromatic exchanges and regulation of cell cycle proliferation of human lymphocytes with the treatment of Indian gooseberry extract.

Ngamkitidechakul *et al.* (2010) tested the anticancer potential of aqueous extract of Indian gooseberry on human cancer cell lines, A549 (lung), HepG2 (liver), HeLa (cervical), MDA-MB-231 (breast) and SW620 (colorectal) *in vitro*. HeLa cell line showed DNA fragmentation, increased activity of caspase-3,7 and 8 and up-regulation of *Fas* protein with the treatment of gooseberry extract indicating activation of the death receptor pathway for apoptosis. Various other research studies have also revealed that the antiproliferative and cytotoxic effect of Indian gooseberry extract is mediated by the activation of apoptosis (Jose *et al.*, 2001; Piva *et al.*, 2009). Several studies have shown that bioactive compounds present in extracts of fruit/leaves/branches/roots such as apigenin 7-O-(6″-butyryl-β-glucopyranoside), epigallocatechin 3-O-gallate, eriodictyol 7-O-(6″-O-trans-coumaroyl)-glucoside, kampferol, kaempferol 3-O-glucoside, quercetin, quercetin 3-O-glucoside, rutin (flavonoids), 1-O-galloyl-β-D-glucose, chebulagic acid, chebulanin, corilagin, elaeocarpusin, geraniin, isocorilagin, L-malic acid 2-O-gallate, mallotusinin, mucic acid 1,4-lactone 3-O-gallate, mucic acid 2-O-gallate, phyllanemblinin C, phyllanemblinin E, phyllemtannin, prodelphinidin B-1, prodelphinidin B-2, putranjivain A (tannins), ellagic acid (phenolic acid), 5α,6β,7α-trihydroxysitosterol, 7-ketositosterol, 7α-hydroxysitosterol, stigmast-4-en-3,6-dione, stigmast-4-en-6β-ol-3-one, stigmast-4-ene-3β,6α-diol (phytosterols), phyllaemblicin B, phyllaemblicin C, phyllaemblicin H1 and phyllaemblicin H2 (terpenoids) are responsible for antiproliferative, cytotoxic and antitumour activities (Tables 33.4–33.6; Zhang *et al.*, 2004, 2016, 2017; El-Desouky *et al.*, 2008; Luo *et al.*, 2011; Liu *et al.*, 2012; Qi *et al.*, 2013; Lv *et al.*, 2015).

Dermoprotective and wound-healing activity

Several *in vitro* and *in vivo* studies have shown that Indian gooseberry possesses skincare properties including antiwrinkling, antimelanogenic and wound-healing effects. Fujii *et al.* (2008) investigated the effects of Indian gooseberry extract on human skin fibroblasts, especially for the production of procollagen and matrix metalloproteinases, *in vitro*. The results showed an improved proliferation of fibroblasts and a decrease in matrix metalloproteinases. This study established the mitigative, cosmetic and therapeutic uses of Indian gooseberry. In another study, Majeed *et al.* (2011) tested Indian gooseberry extract on human dermal fibroblasts against ultraviolet-B (UV-B) irradiation-induced oxygen species and showed protective effect against UV-B induced damage. Recent studies by Zhang *et al.* (2017) have revealed that several compounds isolated from Indian gooseberry including 1,6-di-O-galloyl-β-D-glucose, mucic acid 1,4-lactone methyl ester 5-O-gallate, mucic acid 1-ethyl 6-methyl ester 2-O-gallate, mucic acid dimethyl ester 2-O-gallate, ellagic acid and methyl gallate possess melanogenesis-inhibitory activities without toxicity (Table 33.4). Sumitra *et al.* (2009) studied the wound-healing action of Indian gooseberry through *in vivo* studies and showed the healing action through up-regulation of collagen and extracellular signal regulated kinases. Further, they demonstrated that ascorbic acid and tannins (embalicanin A and embalicanin B) present in Indian gooseberry were responsible for strong antioxidant action and proposed that these antioxidants supported the repair process of the wound.

Gastroprotective and antiulcerogenic activity

The antiulcerogenic activity of fruit juice and methanol extract of Indian gooseberry was established by Rajeshkumar *et al.* (2001) in ethanol, indomethacin and histamine-induced experimental ulcers in rats. With the administration of ethanol, indomethacin and histamine, the authors observed severe gastric damage, intraluminal bleeding and reduction of glutathione content of gastric mucosa. However, treatment with fruit juice and extract restored the gastric anomalies in experimental rats. Similarly, experimental evidence conducted by Al-Rehaily *et al.* (2002) showed gastroprotective effects of Indian gooseberry extract (250 and 500 mg/kg body weight) in experimental rats and fruit extract inhibited the development of gastric lesions, pyloric-ligation-induced basal gastric secretion, acidity and gastric mucosal injury. Several other studies also showed the antiulcerogenic effect of Indian gooseberry (Sairam *et al.*, 2002; Chatterjee *et al.*, 2011). Several experimental studies have revealed the curative effects of Indian gooseberry for acute pancreatitis (Thorat *et al.*, 1995; Sidhu *et al.*, 2011; Aruna *et al.*, 2014). For example, Aruna *et al.* (2014) proved the protective action of Indian gooseberry against cerulean-induced pancreatitis in rats and demonstrated reduced levels of collagen, caspase-1 and myeloperoxidase activities.

Antimicrobial, antiplasmodial and larvicidal activity

Plants are the source of antimicrobial agents, especially against microbes that cause infectious diseases. Several *in vitro* and *in vivo* studies have demonstrated the antibacterial, antifungal and antiviral activities of chemical compounds isolated from many plants (Khameneh *et al.*, 2019). Various reports suggest that extracts and compounds isolated from Indian gooseberry possess potent antimicrobial activities. For example, alkaloids extracted from Indian gooseberry exhibited significant antimicrobial activity against several bacteria including *Bacillus cereus*, *Bacillus subtilis*, *Escherichia coli*, *Pseudomonas aeruginosa*, *Salmonella typhi*, *Salmonella paratyphi*, *Sarcina lutea*, *Staphylococcus aureus*, *Shigella boydii* and the fungi *Aspergillus niger*, *Candida albicans* and *Saccharomyces cerevisiae* (Vijayalakshmi *et al.*, 2007). A previous study by Wang *et al.* (2009) showed coxsackievirus B3 activity of phyllaemblicin B isolated from the Indian gooseberry. Similarly, Lv *et al.* (2014, 2015) isolated several sesquiterpenoid glycosides from Indian gooseberry which have exhibited potent antiviral activities against hepatitis B, influenza-A and foot-and-mouth viruses. Various compounds isolated from the leaves and roots of Indian gooseberry such as quercetin, emblirol A, emblirol B (flavonoids), ellagic acid, gallic acid (phenolic acids), propyl gallate (gallate), 1,2,4,6-tetra-O-galloyl-β-D-glucose (tannin), 1,2-dihydroxyethyl-3,4-dihydroxyfuran-2(5 h)-one (furan), glochicoccinoside D, phyllaemblic acid, phyllaemblic acid methyl ester, phyllaemblicin B, phyllaemblicin C, phyllaemblicin F, phyllaemblicin G6, phyllaemblicin G7, phyllaemblicin G8, phyllaemblicin H6, phyllaemblicin H8 and phyllaemblicin H12 (terpenoids) have demonstrated antifungal and antiviral activities (Liu *et al.*, 2009; Xiang *et al.*, 2010, 2011; Chugh and Bharti, 2014; Lv *et al.*, 2014, 2015; Yan *et al.*, 2017).

Malaria is a major health problem in the world. The availability of the limited number of drugs and drug resistance by the causative agents has led to the search for plant based antimalarial compounds as antiplasmodial agents. Bagavan *et al.* (2011) evaluated the ethyl acetate and methanol extract of Indian gooseberry leaves for antimalarial activity against chloroquine-sensitive and chloroquine-resistant strains of *Plasmodium falciparum*, and the results showed promising antimalarial activity against both the tested strains. Similarly, the antiplasmodial activity of aqueous fruit extracts was demonstrated by Pinmai *et al.* (2010) using *in vitro* and *in vivo* experiments.

Jeyasankar *et al.* (2012) tested the larvicidal activity of Indian gooseberry leaf extract (50–250 ppm) against *Aedes aegypti* and *Culex quinquefasciatus* which are the vectors of *Plasmodium falciparum*. They demonstrated that extract showed significant results against *Culex quinquefasciatus* larvae (LC_{50} value of 80 ppm). A previous study by Bagavan *et al.* (2009) showed adulticidal activity of methanol extract of Indian gooseberry against cattle tick *Haemaphysalis bispinosa* and sheep fluke *Paramphistomum cervi*.

Bioaccessibility and Bioavailability

Bioaccessibility of bioactive compounds in food is defined as the quantity or fraction which is released from the food matrix in the gastrointestinal tract and becomes available for absorption. This includes the digestive transformation of food into material ready for assimilation into intestinal epithelium cells and its metabolism (Carbonell-Capella *et al.*, 2014). Bioavailability of bioactive compounds includes the fraction of ingested nutrient or compounds that reach the systematic circulation and is utilized (Carbonell-Capella *et al.*, 2014). Various approaches have been followed to study the bioaccessibility and bioavailability of bioactive compounds in food, including *in vitro* methods (simulated gastrointestinal digestion, Coco-2 cells cultures), *ex vivo* techniques (gastrointestinal organs in laboratory conditions), *in situ* assays (intestinal perfusion in animals) and *in vivo* models (animal and human studies). Several studies have been carried out regarding the bioaccessibility and bioavailability of bioactive compounds of Indian gooseberry, for example, Patel and Goyal (2013) studied biotransformation of gallotannins from fresh fruit juice *in vitro* and their study illustrates that gallotannins were transformed into gallic acid in the presence of gastrointestinal fluids. In another study, the bioavailability of vitamin C in Indian gooseberry and its combinations with piperine, ginger and synthetic vitamin C was also tested using New Zealand rabbits (Somasekhar *et al.*, 2016). They administered Indian gooseberry extract, synthetic vitamin C, Indian gooseberry extract with piperine and with ginger separately at 100 mg/kg to each group of New Zealand rabbits orally and analysed serum samples by high-performance liquid chromatography at 0, 1, 3, 4, 6, 8, 10, 12 and 24 h. Bioavailability studies revealed that Indian gooseberry extract with piperine combination has a higher concentration of vitamin C when compared to synthetic vitamin C and the authors concluded that the presence of piperine acted as a bioavailability enhancer. Chaudhary *et al.* (2020) encapsulated emblicanin-rich Indian gooseberry extract using the inner phase of water-in-oil-in-water type double emulsion and studied the controlled release of emblicanin enduring phagocytosis and *in vitro* digestion. They reported the release of bioactive compounds from the inner phase of the double emulsion during intestinal digestion followed by gastric and salivary digestion. Angiotensin-converting enzyme inhibitory activity and antioxidant activity was also recorded in the intestinal digests of Indian gooseberry extract. Phagocytic activity of digesta of Indian gooseberry extract was also recorded.

Health Benefits

Indian gooseberry is used in various forms. The fruit are used for culinary preparations, made into a paste, flakes, powder and pickles. The fresh juice is used as cooling and refrigerant sherbet, decoction and wine. The fruit, leaves, seeds and roots are used in the preparation of decoctions used as astringent medicines in diarrhoea and hemoptysis. Indian gooseberry fruit possess antioxidant, antibacterial, antifungal and antiviral properties, as shown by several investigations (Haque *et al.*, 2001; Vijayalakshmi *et al.*, 2007; Chugh and Bharti, 2014; Lv *et al.*, 2014, 2015). It is used as an aphrodisiac and diuretic, for the treatment of diabetes, fever, gonorrhea, nausea, indigestion, mouth ulcers, respiratory problems, scurvy, skin sores/wounds and sore eyes (Srivasuki, 2012). Indian gooseberry is used in the preparation of 'Rasayanas' which are non-toxic Ayurvedic complex herbal preparations or individual herbs used to rejuvenate or attain complete potential of an individual to prevent diseases and degenerative changes that lead to disease (Vayalil *et al.*, 2002). The emerging data suggest that the Rasayanas are good for immune stimulation, quenching free radicals, enhancing cellular detoxification mechanisms, repairing damaged non-proliferating cells, cell proliferation and self-renewal of damaged proliferating tissues (Vayalil *et al.*, 2002).

Novel Products/Formulations

Indian gooseberry is used as food and for several herbal preparations to maintain health and to treat disease. 'Triphala', a herbal formulation composed of three fruit, *Terminalia chebula*, *Terminalia bellirica* and *Phyllanthus emblica*, is an important drug in Ayurveda (the Indian system of traditional medicine). Triphala is an antioxidant-rich herbal formulation used in various forms such as 'churna' (fine powder), 'kwatha' (decoction), 'mashi' (ash), 'taila' (oil) and 'ghritha' (triphala cooked with clarified butter or ghee) (Baliga *et al.*, 2012). Triphala is prescribed for many ailments but most importantly to treat various gastrointestinal disorders. Recent scientific studies suggested that triphala possess free radical scavenging, antioxidant, anti-inflammatory, antipyretic, analgesic, antimicrobial, antistress, anticancer, adaptogenic and hypoglycemic effects (Baliga *et al.*, 2012).

'Rasayana' is a complex mixture of various herbs, forming a health supplement that improves memory, mental strength, cognition, immunity and general wellbeing (Vayalil *et al.*, 2002). 'Chyawanprash' is classified under the category of rasayana and is prepared by incorporating around 50 herbs including Indian gooseberry. Herbs used in the preparation of formulations are boiled in water, and then dried extract is combined with honey. The finished product, which has the consistency of fruit jam, is sour and spicy (Sharma *et al.*, 2019). Another herbal preparation called 'Brahma rasayana' contains more than 40 herbs, and the main ingredients are *Phyllanthus emblica* and *Terminalia chebula*, which constitute about 20 and 6.7%, respectively, of the total ingredients present (Vayalil *et al.*, 2012). Various animal and clinical studies have demonstrated that 'Brahma rasayana' has immunomodulatory, antioxidant and anticancer properties (Vayalil *et al.*, 2012).

Wine has been prepared from Indian gooseberry, and is similar in terms of its composition, taste and aroma (Soni *et al.*, 2009). Further, Indian gooseberry wine is reported to possess a high amount of total phenolic (11.02 μg gallic acid equivalent/ml) and flavonoid (59.46 μg quercetin equivalent/ml) content. Also, gallic acid, syringic acid, protocatechuic acid and caffeic acid were present in good amounts. Antioxidant assay, low-density lipoprotein oxidation prevention, foam cell prevention and nitrite scavenging activities were found to be highest in the wine (Nambiar *et al.*, 2016). Thus, the wine of Indian gooseberry could be used as one of the functional fermented foods.

A herbal formulation called Ophthacare eye drops is available for the management of various ophthalmic disorders, which is prepared by using *Phyllanthus emblica*, *Carum copticum*, *Terminalia bellirica*, *Curcuma longa*, *Ocimum sanctum*, *Cinnamomum camphora*, *Rosa damascena* and *Meldespum apum*. Biswas *et al.* (2001) conducted clinical trials of Ophthacare in patients suffering from various ophthalmic disorders (conjunctivitis, conjunctival xerosis, acute dacryocystitis, degenerative conditions – pterygium or pinguecula – and postoperative cataract patients) and reported an improvement with the herbal treatment with no side effects in most cases.

Indian gooseberry is rich in vitamin C and is employed in hair preparations as an antidandruff agent, hair growth promoter and to strengthen hair (Srivasuki, 2012). Therefore, fruit of this plant are used in the preparation of herbal shampoos and conditioners (Al Badi and Khan, 2014; Utane *et al.*, 2017). *In vitro* and *in vivo* studies have demonstrated that Indian gooseberry promotes hair growth by inhibiting 5α-reductase activity (Kumar *et al.*, 2012). In view of this, the proprietary DA-5512 formulation has been prepared using six herbal extracts: *Phyllanthus emblica*, *Pinus densiflora*, *Pueraria thunbergiana*, *Thea sinensis*, *Tribulus terrestris* and *Zingiber officinale* (Yu *et al.*, 2017). Preclinical and clinical studies demonstrate that the proprietary herbal extract DA-5512 effectively stimulated hair growth and promoted hair health (Yu *et al.*, 2017).

Conclusions and Future Trends

Indian gooseberry is an underexploited fruit which is rich in nutrients and phytochemicals. It has been used since time immemorial for the treatment of various ailments in the Ayurveda, Siddha, Unani, Tibetan and Chinese systems of medicine. Scientific data suggest that Indian gooseberry has antioxidant, hepatoprotective, cardioprotective, anti-inflammatory, antidiabetic, neuroprotective, antimutagenic, cytotoxic, antiproliferative, anticancer, dermoprotective, wound-healing, antimicrobial, antiplasmodial and larvicidal properties. These activities are due to valuable phytochemicals in Indian gooseberry, including flavonoids, tannins, phenolic acids, sterols and terpenoids which are present in the fruit, leaves, stem branches and roots. Several Indian gooseberry-based health foods such as triphala, chyawanprash, juice and wine are available on the market. Indian gooseberry is also used in varied formulations such as shampoo, ointment and skincare products. There is still scope for isolation and characterization of bioactive compounds from this plant and this will help in the treatment of specific diseases. Another area in which research is needed is bioaccessibility and bioavailability of bioactive compounds of Indian gooseberry after the consumption of fruit and its products. Clinical studies are essential to validate the effects of bioactive molecules, especially analysis of immunomodulation, antidiabetic, anticancer and neuroprotective properties.

References

Akhtar, M.S., Ramzan, A., Ali, A. and Ahmad, M. (2011) Effect of amla fruit (*Emblica officinalis* Gaertn.) on blood glucose and lipid profile of normal subjects and type 2 diabetic patients. *International Journal of Food Sciences and Nutrition* 62(6), 609–616. DOI: 10.3109/09637486.2011.560565.

Al Badi, K. and Khan, S.A. (2014) Formulation, evaluation and comparison of the herbal shampoo with the commercial shampoos. *Beni-Suef University Journal of Basic and Applied Sciences* 3(4), 301–305. DOI: 10.1016/j.bjbas.2014.11.005.

Al-Rehaily, A.J., Al-Howiriny, T.S., Al-Sohaibani, M.O. and Rafatullah, S. (2002) Gastroprotective effects of 'Amla' *Emblica officinalis* on *in vivo* test models in rats. *Phytomedicine: International Journal of Phytotherapy and Phytopharmacology* 9(6), 515–522. DOI: 10.1078/09447110260573146.

Ando, H., Ryu, A., Hashimoto, A., Oka, M. and Ichihashi, M. (1998) Linoleic acid and α-linolenic acid lightens ultraviolet-induced hyperpigmentation of the skin. *Archives of Dermatological Research* 290, 375–381. DOI: 10.1007/s004030050320.

Arora, A., Kumar, I., Sen, R. and Singh, J. (2011) *Emblica officinalis* (Amla): physico-chemical and fatty acid analysis from arid zone of Rajasthan. *International Journal of Basic and Applied Chemical Sciences* 1, 89–92.

Aruna, R., Geetha, A., Suguna, P. and Suganya, V. (2014) Rutin rich *Emblica officinalis* Geart. fruit extract ameliorates inflammation in the pancreas of rats subjected to alcohol and cerulein administration. *Journal of Complementary & Integrative Medicine* 11(1), 9–18. DOI: 10.1515/jcim-2013-0023.

Bagavan, A., Kamaraj, C., Elango, G., Abduz Zahir, A. and Abdul Rahuman, A. (2009) Adulticidal and larvicidal efficacy of some medicinal plant extracts against tick, fluke and mosquitoes. *Veterinary Parasitology* 166(3–4), 286–292. DOI: 10.1016/j.vetpar.2009.09.007.

Bagavan, A., Rahuman, A.A., Kaushik, N.K. and Sahal, D. (2011) *In vitro* antimalarial activity of medicinal plant extracts against *Plasmodium falciparum*. *Parasitology Research* 108(1), 15–22. DOI: 10.1007/s00436-010-2034-4.

Baliga, M.S., Meera, S., Mathai, B., Rai, M.P., Pawar, V. *et al*. (2012) Scientific validation of the ethnomedicinal properties of the Ayurvedic drug Triphala: A review. *Chinese Journal of Integrative Medicine* 18(12), 946–954. DOI: 10.1007/s11655-012-1299-x.

Barthakur, N.N. and Arnold, N.P. (1991) Chemical analysis of the emblic (*Phyllanthus emblica* L.) and its potential as a food source. *Scientia Horticulturae* 47(1–2), 99–105. DOI: 10.1016/0304-4238(91)90031-S.

Biswas, N.R., Gupta, S.K., Das, G.K., Kumar, N., Mongre, P.K. *et al*. (2001) Evaluation of Ophthacare® eye drops--a herbal formulation in the management of various ophthalmic disorders. *Phytotherapy Research* 15(7), 618–620. DOI: 10.1002/ptr.896.

Carbonell-Capella, J.M., Buniowska, M., Barba, F.J., Esteve, M.J. and Frígola, A. (2014) Analytical methods for determining bioavailability and bioaccessibility of bioactive compounds from fruits and vegetables: a review. *Comprehensive Reviews in Food Science and Food Safety* 13(2), 155–171. DOI: 10.1111/1541-4337.12049.

Chaphalkar, R., Apte, K.G., Talekar, Y., Ojha, S.K. and Nandave, M. (2017) Antioxidants of *Phyllanthus emblica* L. bark extract provide hepatoprotection against ethanol-induced hepatic damage: a comparison with silymarin. *Oxidative Medicine and Cellular Longevity* 2017, 3876040. DOI: 10.1155/2017/3876040.

Chatterjee, A., Chattopadhyay, S. and Bandyopadhyay, S.K. (2011) Biphasic effect of *Phyllanthus emblica* L. extract on NSAID-induced ulcer: An antioxidative trail weaved with immunomodulatory effect. *Evidence-Based Complementary and Alternative Medicine* 2011, 146808. DOI: 10.1155/2011/146808.

Chaudhary, N., Sabikhi, L. and Hussain, S.A. (2020) Emblicanin rich *Emblica officinalis* extract encapsulated double emulsion: controlled release of bioactive during phagocytosis and in vitro digestion. *Journal of Food Science and Technology* 57(4), 1371–1381. DOI: 10.1007/s13197-019-04171-0.

Chugh, C.A. and Bharti, D. (2014) Chemical characterization of antifungal constituents of *Emblica officinalis*. *Allelopathy Journal* 34, 155–178.

D'souza, J.J., D'souza, P.P., Fazal, F., Kumar, A., Bhat, H.P. *et al*. (2014) Anti-diabetic effects of the Indian indigenous fruit *Emblica officinalis* Gaertn: active constituents and modes of action. *Food & Function* 5(4), 635–644. DOI: 10.1039/c3fo60366k.

El-Desouky, S.K., Ryu, S.Y. and Kim, Y.K. (2008) A new cytotoxic acylated apigenin glucoside from *Phyllanthus emblica* L. *Natural Product Research* 22, 91–95.

Fernandes, D.C., Alves, A.M., Castro, G.S.F., Jordao Junior, A.A. and Naves, M.M.V. (2015) Effects of Baru almond and Brazil nut against hyperlipidemia and oxidative stress In vivo. *Journal of Food Research* 4(4), 38. DOI: 10.5539/jfr.v4n4p38.

Fujii, T., Wakaizumi, M., Ikami, T. and Saito, M. (2008) Amla (*Emblica officinalis* Gaertn.) extract promotes procollagen production and inhibits matrix metalloproteinase-1 in human skin fibroblasts. *Journal of Ethnopharmacology* 119(1), 53–57. DOI: 10.1016/j.jep.2008.05.039.

Gao, Q., Xiang, H.-Y., Chen, W., Ling, J., Chen, X. *et al.* (2019) Two new isobenzofuranone derivatives from *Phyllanthus emblica* and their bioactivity. *Chemistry of Natural Compounds* 55(5), 847–850. DOI: 10.1007/s10600-019-02829-x.

Golechha, M., Bhatia, J. and Arya, D.S. (2010) Hydroalcoholic extract of *Emblica officinalis* Gaertn. affords protection against PTZ-induced seizures, oxidative stress and cognitive impairment in rats. *Indian Journal of Experimental Biology* 48(5), 474–478.

Golechha, M., Bhatia, J., Ojha, S. and Arya, D.S. (2011) Hydroalcoholic extract of *Emblica officinalis* protects against kainic acid-induced status epilepticus in rats: evidence for an antioxidant, anti-inflammatory, and neuroprotective intervention. *Pharmaceutical Biology* 49(11), 1128–1136. DOI: 10.3109/13880209.2011.571264.

Haque, R., Bin-Hafeez, B., Ahmad, I., Parvez, S., Pandey, S. *et al.* (2001) Protective effects of *Emblica officinalis* Gaertn. in cyclophosphamide-treated mice. *Human & Experimental Toxicology* 20(12), 643–650. DOI: 10.1191/096032701718890568.

Husain, I., Zameer, S., Madaan, T., Minhaj, A., Ahmad, W. *et al.* (2019) Exploring the multifaceted neuroprotective actions of *Emblica officinalis* (Amla): a review. *Metabolic Brain Disease* 34(4), 957–965. DOI: 10.1007/s11011-019-00400-9.

Jaijoy, K., Soonthornchareonnon, N., Panthong, A. and Sireeratawong, S. (2010) Anti-inflammatory and analgesic activities of the water extract from the fruit of *Phyllanthus emblica* Linn. *International Journal of Applied Research in Natural Products* 3, 28–35.

Jeyasankar, A., Premalatha and Elumalai, K. (2012) Larvicidal activity of *Phyllanthus emblica* Linn. (Euphorbiaceae) leaf extracts against important human vector mosquitoes (Diptera: Culicidae). *Asian Pacific Journal of Tropical Disease* 2(Supplement 1), S399–S403. DOI: 10.1016/S2222-1808(12)60190-X.

Jose, J.K. and Kuttan, R. (1995) Antioxidant activity of *Emblica officinalis*. *Journal of Clinical Biochemistry and Nutrition* 19(2), 63–70. DOI: 10.3164/jcbn.19.63.

Jose, J.K., Kuttan, G., George, J. and Kuttan, R. (1997) Antimutagenic and anticarcinogenic activity of *Emblica officinalis* Gaertn. *Journal of Clinical Biochemistry and Nutrition* 22(3), 171–176. DOI: 10.3164/jcbn.22.171.

Jose, J.K., Kuttan, G. and Kuttan, R. (2001) Antitumour activity of *Emblica officinalis*. *Journal of Ethnopharmacology* 75(2–3), 65–69. DOI: 10.1016/S0378-8741(00)00378-0.

Judprasong, K., Charoenkiatkul, S., Thiyajai, P. and Sukprasansap, M. (2013) Nutrients and bioactive compounds of Thai indigenous fruits. *Food Chemistry* 140(3), 507–512. DOI: 10.1016/j.foodchem.2013.01.057.

Khameneh, B., Iranshahy, M., Soheili, V. and Fazly Bazzaz, B.S. (2019) Review on plant antimicrobials: a mechanistic viewpoint. *Antimicrobial Resistance and Infection Control* 8, 118. DOI: 10.1186/s13756-019-0559-6.

Khan, M., Lampronti, I., Martello, D., Bianchi, N., Jabbar, S. *et al.* (2002) Identification of pyrogallol as an antiproliferative compound present in extracts from the medicinal plant *Emblica officina*lis: Effects on *in vitro* cell growth of human tumor cell lines. *International Journal of Oncology* 20, 187–192. DOI: 10.3892/ijo.21.1.187.

Kumar, N., Rungseevijitprapa, W., Narkkhong, N.A., Suttajit, M. and Chaiyasut, C. (2012) 5α-reductase inhibition and hair growth promotion of some Thai plants traditionally used for hair treatment. *Journal of Ethnopharmacology* 139(3), 765–771. DOI: 10.1016/j.jep.2011.12.010.

Kumaran, A. and Karunakaran, R.J. (2006) Nitric oxide radical scavenging active components from *Phyllanthus emblica* L. *Plant Foods for Human Nutrition* 61(1), 1–5. DOI: 10.1007/s11130-006-0001-0.

Li, W., Zhu, H.-W., Chen, Y.-J., Xiao, H., Ge, Y.-Z. *et al.* (2020) Bioactivity-guided isolation of anti-inflammatory components from Phyllanthus emblica. *Food Science & Nutrition* 8(6), 2670–2679. DOI: 10.1002/fsn3.1553.

Li, Y., Chen, J., Cao, L., Li, L., Wang, F. *et al.* (2018) Characterization of a novel polysaccharide isolated from *Phyllanthus emblica* L. and analysis of its antioxidant activities. *Journal of Food Science and Technology* 55(7), 2758–2764. DOI: 10.1007/s13197-018-3199-6.

Lim, T.K. (2012) *Phyllanthus emblica*. In: Lim, T.K. (ed.) *Edible Medicinal and Non-Medicinal Plants*, Vol. 4. Springer, Dordrecht, Netherlands, pp. 258–296.

Liu, Q., Wang, Y.-F., Chen, R.-J., Zhang, M.-Y., Wang, Y.-F. *et al.* (2009) Anti-Coxsackie virus B3 norsesquiterpenoids from the roots of *Phyllanthus emblica*. *Journal of Natural Products* 72(5), 969–972. DOI: 10.1021/np800792d.

Liu, X., Cui, C., Zhao, M., Wang, J., Luo, W. *et al.* (2008) Identification of phenolics in the fruit of emblica (*Phyllanthus emblica* L.) and their antioxidant activities. *Food Chemistry* 109(4), 909–915. DOI: 10.1016/j.foodchem.2008.01.071.

Liu, X., Zhao, M., Wu, K., Chai, X., Yu, H. *et al.* (2012) Immunomodulatory and anticancer activities of phenolics from emblica fruit (*Phyllanthus emblica* L.). *Food Chemistry* 131(2), 685–690. DOI: 10.1016/j.foodchem.2011.09.063.

Longvah, T., Ananthan, R., Bhaskaracharya, K. and Venkaiah, K. (2017) *Indian Food Composition Tables 2017*. National Institute of Nutrition, Hyderabad, India.

Luo, W., Zhao, M., Yang, B., Shen, G. and Rao, G. (2009) Identification of bioactive compounds in *Phyllenthus emblica* L. fruit and their free radical scavenging activities. *Food Chemistry* 114(2), 499–504. DOI: 10.1016/j.foodchem.2008.09.077.

Luo, W., Zhao, M., Yang, B., Ren, J., Shen, G. *et al.* (2011) Antioxidant and antiproliferative capacities of phenolics purified from *Phyllanthus emblica* L. fruit. *Food Chemistry* 126(1), 277–282. DOI: 10.1016/j.foodchem.2010.11.018.

Luo, W., Wen, L., Zhao, M., Yang, B., Ren, J. *et al.* (2012) Structural identification of isomallotusinin and other phenolics in *Phyllanthus emblica* L. fruit hull. *Food Chemistry* 132(3), 1527–1533. DOI: 10.1016/j.foodchem.2011.11.146.

Lv, J.-J., Wang, Y.-F., Zhang, J.-M., Yu, S., Wang, D. *et al.* (2014) Anti-hepatitis B virus activities and absolute configurations of sesquiterpenoid glycosides from *Phyllanthus emblica*. *Organic & Biomolecular Chemistry* 12(43), 8764–8774. DOI: 10.1039/c4ob01196a.

Lv, J.J., Yu, S., Xin, Y., Cheng, R.R., Zhu, H.T. *et al.* (2015) Anti-viral and cytotoxic norbisabolane sesquiterpenoid glycosides from *Phyllanthus emblica* and their absolute configurations. *Phytochemistry* 117, 123–134. DOI: 10.1016/j.phytochem.2015.06.001.

Majeed, M., Bhat, B., Anand, S., Sivakumar, A., Paliwal, P. *et al.* (2011) Inhibition of UV-induced ROS and collagen damage by *Phyllanthus emblica* extract in normal human dermal fibroblasts. *Journal of Cosmetic Science* 62(1), 49–56.

Mathur, R., Sharma, A., Dixit, V.P. and Varma, M. (1996) Hypolipidaemic effect of fruit juice of *Emblica officinalis* in cholesterol-fed rabbits. *Journal of Ethnopharmacology* 50(2), 61–68. DOI: 10.1016/0378-8741(95)01308-3.

Mehta, S., Singh, R.K., Jaiswal, D., Rai, P.K. and Watal, G. (2009) Anti-diabetic activity of Emblica officinalis in animal models. *Pharmaceutical Biology* 47(11), 1050–1055. DOI: 10.3109/13880200902991532.

Muramoto, K. (2017) Lectins as bioactive proteins in foods and feeds. *Food Science and Technology Research* 23(4), 487–494. DOI: 10.3136/fstr.23.487.

Murthy, H.N. and Bapat, V.A. (2020) Importance of underutilized fruits and nuts. In: Murthy, H.N. and Bapat, V.A. (eds) *Bioactive Compounds in Underutilized Fruits and Nuts, Reference Series in Phytochemistry*. Springer, Cham, Switzerland, pp. 3–19.

Muthuraman, A., Sood, S. and Singla, S.K. (2011) The antiinflammatory potential of phenolic compounds from *Emblica officinalis* L. in rat. *Inflammopharmacology* 19(6), 327–334. DOI: 10.1007/s10787-010-0041-9.

Nain, P., Saini, V., Sharma, S. and Nain, J. (2012) Antidiabetic and antioxidant potential of *Emblica officinalis* Gaertn. Leaves extract in streptozotocin-induced type-2 diabetes mellitus (T2DM) rats. *Journal of Ethnopharmacology* 142(1), 65–71. DOI: 10.1016/j.jep.2012.04.014.

Nambiar, S.S., Venugopal, K.S., Shetty, N.P. and Appaiah, K.A.A. (2016) Fermentation induced changes in bioactive properties of wine from *Phyllanthus* with respect to atherosclerosis. *Journal of Food Science and Technology* 53(5), 2361–2371. DOI: 10.1007/s13197-016-2208-x.

Nampoothiri, S.V., Prathapan, A., Cherian, O.L., Raghu, K.G., Venugopalan, V.V. *et al.* (2011) In vitro antioxidant and inhibitory potential of *Terminalia bellerica* and *Emblica officinalis* fruits against LDL oxidation and key enzymes linked to type 2 diabetes. *Food and Chemical Toxicology* 49(1), 125–131. DOI: 10.1016/j.fct.2010.10.006.

Nath, V., Singh, I.S. and Kumar, S. (1992) Evaluation of aonla cultivars for their shelf-life at ambient temperature. *Narendra Deva Journal of Agriculture Research* 7(1), 117.

Ngamkitidechakul, C., Jaijoy, K., Hansakul, P., Soonthornchareonnon, N. and Sireeratawong, S. (2010) Antitumour effects of *Phyllanthus emblica* L.: induction of cancer cell apoptosis and inhibition of *in vivo* tumour promotion and *in vitro* invasion of human cancer cells. *Phytotherapy Research* 24(9), 1405–1413. DOI: 10.1002/ptr.3127.

Ojha, S., Golechha, M., Kumari, S. and Arya, D.S. (2011) Protective effect of *Emblica officinalis* (amla) on isoproterenol-induced cardiotoxicity in rats. *Toxicology and Industrial Health* 28(5), 399–411. DOI: 10.1177/0748233711413798.

Pareek, S. and Kaushik, R.A. (2012) Effect of drying methods on quality of indian gooseberry (emblica officinalis garrtn.) powder during storage. *Journal of Scientific & Industrial Research* 71, 727–732.

Patel, S.S. and Goyal, R.K. (2013) Biotransformation of gallotannins from fresh fruit juice of Emblica officinalis in in-vitro System. *Research Journal of Phytochemistry* 7(1), 18–23. DOI: 10.3923/rjphyto.2013.18.23.

Patel, S.S., Goyal, R.K., Shah, R.S., Tirgar, P.R. and Jadav, P.D. (2013) Experimental study on effect of hydroalcoholic extract of *Emblica officinalis* fruits on glucose homeostasis and metabolic parameters. *Ayu* 34(4), 440–444. DOI: 10.4103/0974-8520.127731.

Pinmai, K., Hiriote, W., Soonthornchareonnon, N., Jongsakul, K., Sireeratawong, S. *et al.* (2010) In vitro and in vivo antiplasmodial activity and cytotoxicity of water extracts of phyllanthus emblica, terminalia chebula, and terminalia bellerica. *Journal of Medical Association of Thailand* 93(Suppl 7), S120–S126.

Piva, R., Penolazzi, L., Borgatti, M., Lampronti, I., Lambertini, E. *et al.* (2009) Apoptosis of human primary osteoclasts treated with molecules targeting nuclear factor-kappaB. *Annals of the New York Academy of Sciences* 1171, 448–456. DOI: 10.1111/j.1749-6632.2009.04906.x.

Pozharitskaya, O.N., Ivanova, S.A., Shikov, A.N. and Makarov, V.G. (2007) Separation and evaluation of free radical-scavenging activity of phenol components of *Emblica officinalis* extract by using an HPTLC-DPPH method. *Journal of Separation Science* 30(9), 1250–1254. DOI: 10.1002/jssc.200600532.

Prajapati, V.K., Nema, P.K. and Rathore, S.S. (2011) Effect of pretreatment and drying methods on quality of value-added dried aonla (*Emblica officinalis* Gaertn) shreds. *Journal of Food Science and Technology* 48(1), 45–52. DOI: 10.1007/s13197-010-0124-z.

Puppala, M., Ponder, J., Suryanarayana, P., Reddy, G.B., Petrash, J.M. *et al.* (2012) The isolation and characterization of β-glucogallin as a novel aldose reductase inhibitor from *Emblica officinalis*. *PloS One* 7(4), e31399. DOI: 10.1371/journal.pone.0031399.

Qi, W.Y., Li, Y., Hua, L., Wang, K. and Gao, K. (2013) Cytotoxicity and structure activity relationships of phytosterol from *Phyllanthus emblica*. *Fitoterapia* 84, 252–256. DOI: 10.1016/j.fitote.2012.12.023.

Raj, A.S., Chakraborty, S. and Rao, P.S. (2019) Thermal assisted high-pressure processing of Indian gooseberry (*Embilica officinalis* L.) juice – Impact on colour and nutritional attributes. *LWT* 99, 119–127. DOI: 10.1016/j.lwt.2018.09.051.

Rajeshkumar, N.V., Therese, M. and Kuttan, R. (2001) *Emblica officinalis* fruits afford protection against experimental gastric ulcers in rats. *Pharmaceutical Biology* 39(5), 375–380. DOI: 10.1076/phbi.39.5.375.5902.

Rao, T.P., Sakaguchi, N., Juneja, L.R., Wada, E. and Yokozawa, T. (2005) Amla (*Emblica officinalis* Gaertn.) extracts reduce oxidative stress in streptozotocin-induced diabetic rats. *Journal of Medicinal Food* 8(3), 362–368. DOI: 10.1089/jmf.2005.8.362.

Reddy, V.D., Padmavathi, P., Gopi, S., Paramahamsa, M. and Varadacharyulu, N.C. (2010) Protective effect of *Emblica officinalis* against alcohol-induced hepatic injury by ameliorating oxidative stress in rats. *Indian Journal of Clinical Biochemistry* 25(4), 419–424. DOI: 10.1007/s12291-010-0058-2.

Rose, K., Wan, C., Thomas, A., Seeram, N.P. and Ma, H. (2018) Phenolic compounds isolated and identified from amla (*Phyllanthus emblica*) juice powder and their antioxidant and neuroprotective activities. *Natural Product Communications* 13(10), 1934578X1801301. DOI: 10.1177/1934578X1801301019.

Sairam, K., Rao, C.V., Babu, M.D., Kumar, K.V., Agrawal, V.K. *et al.* (2002) Antiulcerogenic effect of methanolic extract of *Emblica officinalis*: an experimental study. *Journal of Ethnopharmacology* 82(1), 1–9. DOI: 10.1016/s0378-8741(02)00041-7.

Samtiya, M., Aluko, R.E. and Dhewa, T. (2020) Plant food anti-nutritional factors and their reduction strategies: an overview. *Food Production, Processing and Nutrition* 2(1). DOI: 10.1186/s43014-020-0020-5.

Sharma, R., Martins, N., Kuca, K., Chaudhary, A., Kabra, A. *et al.* (2019) Chyawanprash: A traditional Indian bioactive health supplement. *Biomolecules* 9(5), E161. DOI: 10.3390/biom9050161.

Sidhu, S., Pandhi, P., Malhotra, S., Vaiphei, K. and Khanduja, K.L. (2011) Beneficial effects of *Emblica officinalis* in L-arginine-induced acute pancreatitis in rats. *Journal of Medicinal Food* 14(1–2), 147–155. DOI: 10.1089/jmf.2010.1108.

Singh, B.K., Singh, S. and Yadav, S.M. (2014) Packaging, post harvest treatments, storage and protection of Aonla and their role in health care: An overview. *Research Journal of Microbiology* 9(5), 216–220. DOI: 10.3923/jm.2014.216.220.

Słoczyńska, K., Powroźnik, B., Pękala, E. and Waszkielewicz, A.M. (2014) Antimutagenic compounds and their possible mechanisms of action. *Journal of Applied Genetics* 55(2), 273–285. DOI: 10.1007/s13353-014-0198-9.

Somasekhar, V., Ashok, P., Kameswari, S.A.R., Rajendran, R. and Singh, R. (2016) Comparative antioxidant and bioavailability studies of Vitamin C in *Phyllanthus emblica* Linn. and its combinations with *Piper nigrum* Linn. and *Zingiber officinale* Roscoe. *Brazilian Journal of Pharmaceutical Sciences* 52(1), 35–43. DOI: 10.1590/S1984-82502016000100005.

Soni, S.K., Bansal, N. and Soni, R. (2009) Standardization of conditions for fermentation and maturation of wine from amla (emblica officinalis gaertn). *Natural Product Radiance* 8, 436–444.

Srinivasan, P., Vijayakumar, S., Kothandaraman, S. and Palani, M. (2018) Anti-diabetic activity of quercetin extracted from *Phyllanthus emblica* L. fruit: *In silico* and *in vivo* approaches. *Journal of Pharmaceutical Analysis* 8(2), 109–118. DOI: 10.1016/j.jpha.2017.10.005.

Sripanidkulchai, B. and Junlatat, J. (2014) Bioactivities of alcohol based extracts of *Phyllanthus emblica* branches: antioxidation, antimelanogenesis and anti-inflammation. *Journal of Natural Medicines* 68(3), 615–622. DOI: 10.1007/s11418-014-0824-1.

Srivasuki, K.P. (2012) Nutritional and health care benefits of amla. *Journal of Pharmacognosy* 3, 147–151.

Sumitra, M., Manikandan, P., Gayathri, V.S., Mahendran, P. and Suguna, L. (2009) *Emblica officinalis* exerts wound healing action through up-regulation of collagen and extracellular signal-regulated kinases (ERK1/2). *Wound Repair and Regeneration* 17(1), 99–107. DOI: 10.1111/j.1524-475X.2008.00446.x.

Thakur, S.B. and Rao, M.V. (2014) *In vitro* fluoride induced genotoxic effect on human blood lymphocyte cells and its amelioration by *Emblica officinalis* extract. *Molecular Cytogenetics* 7(Suppl 1), 48. DOI: 10.1186/1755-8166-7-S1-P48.

Thenmozhi, A.J., Dhivyabharathi, M., Manivasagam, T. and Essa, M.M. (2016a) Tannoid principles of *Emblica officinalis* attenuated aluminum chloride induced apoptosis by suppressing oxidative stress and tau pathology via Akt/GSK-3β signaling pathway. *Journal of Ethnopharmacology* 194, 20–29. DOI: 10.1016/j.jep.2016.08.047.

Thenmozhi, A.J., Dhivyabharathi, M., William Raja, T.R., Manivasagam, T. and Essa, M.M. (2016b) Tannoid principles of *Emblica officinalis* renovate cognitive deficits and attenuate amyloid pathologies against aluminum chloride induced rat model of Alzheimer's disease. *Nutritional Neuroscience* 19(6), 269–278. DOI: 10.1179/1476830515Y.0000000016.

Thorat, S.P., Rege, N.N., Naik, A.S., Thatte, U.M., Joshi, A. *et al.* (1995) *Emblica Officinalis*: A novel therapy for acute pancreatitis an experimental study. *HPB Surgery* 9(1), 25–30. DOI: 10.1155/1995/51310.

Utane, R., Deo, S. and Itankar, P. (2017) Preparation of herbal shampoo (HS) by green method and their characterization. *International Journal of Researches in Social Sciences and Information Studies* 5, 254–258.

Variya, B.C., Bakrania, A.K., Chen, Y., Han, J. and Patel, S.S. (2018) Suppression of abdominal fat and anti-hyperlipidemic potential of *Emblica officinalis*: Upregulation of PPARs and identification of active moiety. *Biomedicine & Pharmacotherapy* 108, 1274–1281. DOI: 10.1016/j.biopha.2018.09.158.

Vayalil, P.K., Kuttan, G. and Kuttan, R. (2002) Rasayanas: evidence for the concept of prevention of diseases. *The American Journal of Chinese Medicine* 30(1), 155–171. DOI: 10.1142/S0192415X02000168.

Vayalil, P.K., Kuttan, R. and Kuttan, G. (2012) Immunomodulatory activity of brahma rasayana, an herbal preparation containing Phyllanthus emblica as the main ingredient. In: Kuttan, R. and Harikumar, K.B. (eds) *Phyllanthus Species- Scientific Evaluation and Medicinal Applications*. CRC Press, Boca Raton, Florida, pp. 315–323.

Vijayalakshmi, S., Arunkumar, V., Anju, D., Gunasundari, P., Moorthy, P. *et al.* (2007) Comparative antimicrobial activities of emblica afficinalis and ocimum sanctum. *Ancient Science of Life* 27(2), 1–6.

Wang, Y.F., Wang, X.Y., Ren, Z., Qian, C.W., Li, Y.C. *et al.* (2009) Phyllaemblicin B inhibits Coxsackie virus B3 induced apoptosis and myocarditis. *Antiviral Research* 84(2), 150–158. DOI: 10.1016/j.antiviral.2009.08.004.

Xiang, Y.-F., Ju, H.-Q., Li, S., Zhang, Y.-J., Yang, C.-R. *et al.* (2010) Effects of 1,2,4,6-tetra-O-galloyl-β-D-g lucose from *P. emblica* on HBsAg and HBeAg secretion in HepG2.2.15 cell culture. *Virologica Sinica* 25(5), 375–380. DOI: 10.1007/s12250-010-3144-y.

Xiang, Y., Pei, Y., Qu, C., Lai, Z., Ren, Z. *et al.* (2011) *In vitro* anti-herpes simplex virus activity of 1,2,4,6-tetra-O-galloyl-β-D-glucose from *Phyllanthus emblica* L. (Euphorbiaceae). *Phytotherapy Research* 25(7), 975–982. DOI: 10.1002/ptr.3368.

Xie, T., Wang, W., Mao, Z., Qu, Z., Luan, S. *et al.* (2012) Effects of epigallocatechin-3-gallate on pentylenetetrazole-induced kindling, cognitive impairment and oxidative stress in rats. *Neuroscience Letters* 516(2), 237–241. DOI: 10.1016/j.neulet.2012.04.001.

Xu, M., Zhu, H.T., Cheng, R.R., Wang, D., Yang, C.R. *et al.* (2016) Antioxidant and hyaluronidase inhibitory activities of diverse phenolics in Phyllanthus emblica. *Natural Product Research* 30(23), 2726–2729. DOI: 10.1080/14786419.2015.1137573.

Yadav, S.S., Singh, M.K., Singh, P.K. and Kumar, V. (2017) Traditional knowledge to clinical trials: A review on therapeutic actions of *Emblica officinalis*. *Biomedicine & Pharmacotherapy = Biomedecine & Pharmacotherapie* 93, 1292–1302. DOI: 10.1016/j.biopha.2017.07.065.

Yan, H., Han, L.R., Zhang, X. and Feng, J.T. (2017) Two new Anti-TMV active chalconoid analogues from the root of *Phyllanthus emblica*. *Natural Product Research* 31(18), 2143–2148. DOI: 10.1080/14786419.2017.1280487.

Yu, J.Y., Gupta, B., Park, H.G., Son, M., Jun, J.-H. *et al.* (2017) Preclinical and clinical studies demonstrate that the proprietary herbal extract DA-5512 effectively stimulates hair growth and promotes hair health. *Evidence-Based Complementary and Alternative Medicine* 2017, 4395638. DOI: 10.1155/2017/4395638.

Zhang, J., Miao, D., Zhu, W.-F., Xu, J., Liu, W.-Y. *et al.* (2017) Biological activities of phenolics from the fruits of *Phyllanthus emblica* L. (Euphorbiaceae). *Chemistry & Biodiversity* 14(12), e1700404. DOI: 10.1002/cbdv.201700404.

Zhang, Y.-J., Nagao, T., Tanaka, T., Yang, C.-R., Okabe, H. *et al.* (2004) Antiproliferative activity of the main constituents from *Phyllanthus emblica*. *Biological & Pharmaceutical Bulletin* 27(2), 251–255. DOI: 10.1248/bpb.27.251.

Zhang, Y., Zhao, L., Guo, X., Li, C., Li, H. *et al.* (2016) Chemical constituents from *Phyllanthus emblica* and the cytoprotective effects on H2O2-induced PC12 cell injuries. *Archives of Pharmacal Research* 39(9), 1202–1211. DOI: 10.1007/s12272-014-0433-2.

Index

Note: Page numbers in *italics* denote tables and figures